Roelof A. A. Oldeman

Forests: Elements of Silvology

With 215 Figures

Springer-Verlag Berlin Heidelberg GmbH

Professor Dr. ROELOF A. A. OLDEMAN
Gen. Foulkesweg 76
6703 BX Wageningen
The Netherlands

ISBN 978-3-642-75213-1 ISBN 978-3-642-75211-7 (eBook)
DOI 10.1007/978-3-642-75211-7

Library of Congress Cataloging-in-Publication Data. Oldeman, Roelof A. A., 1937– Elements of silvology / Roelof A. A. Oldeman. p. cm. Includes bibliographical references (p.). ISBN-13: 978-3-642-75213-1(U.S. : alk. paper) 1. Forest ecology. 2. Forests and forestry. I. Title. QK938.F6037 1990 581.5' 2642-dc20 90-9680

2131/3145-543210 – Printed on acid-free paper

I dedicate
this book to my wife
Wil Helder
with feelings that would
fill a book too

La forêt, c'est la collectivité immense où la vie besogne sans bruit, où un travail gigantesque s'accomplit et se poursuit sans trépidations, sans fracas et sans poussière; où de magnifiques choses visibles sortent de menues choses invisibles.

H. Biolley 1930, dans: Forêts de mon Pays

No hay que ponerse ante el Universo suspicaz como un aldeano. El positivismo fue una filosofía aldeana.

J. Ortega y Gasset 1929, ¿Qué es filosofía?

Fear no alder black! Heed no hoary willow!
Fear neither root nor bough! Tom goes on before you.

Tolkien 1954, The Lord of the Rings

"Ik geloof het niet", zei heer Ollie zorgelijk. "Het is jullie eigen schuld, bedoel ik. Ik geloof dat jullie er alleen uit kunnen komen als je het bos met rust wilt laten. Dat geloof ik."

Marten Toonder 1971, De Kwinkslagen

Preface

In matters of forest ecology, it has already become impossible to produce a comprehensive state of the art without omitting many times more than one text can possibly contain. The sheer historical richness of sources alone would account for this, but it is compounded by a quarter of a century of recent research having yielded an amazing richness of facts, viewpoints and models. I have actively participated in the research effort of these five-and-twenty years, and it has therefore been my intention to trace back the approaches that now exist to those admirable pioneers of a still somewhat simpler period, roughly between the years 1950 and 1970, who laid the basis for the actual and seemingly confusing deluge of publications in this field.

It is my great concern that scientists, perhaps in sheer desperation, will choose among methods instead of returning to the forests, where each method makes sense in its own limited way. Are we ever going to be able to arrive at one science of forests, in which different approaches appear as means to cast light on special aspects instead of competing as exclusive roads towards the truth? Some of the most valuable books that appeared in the 1980's increase my concern by too exclusively emphasizing such theories as population biology, architectural analysis or computer-simulated production ecology (cf. Table 2.1). In our preface to "Tropical Trees and Forests: an Architectural Analysis", which appeared in 1978, Hallé, Tomlinson and myself already gave an overview of such works one decade earlier.

The choice of one limited approach has been frequent in the history of science as a means to dig deeper in a special field of research. The rationale for concentrating upon one method is, that it may allow us to finally reach the universal core of the matter if we dig deep enough. Present-day conditions of research emphasise this tendency, because financing bodies insist on "research priorities" that are not too difficult to grasp. As a mirrored image of the scientific arguments for narrowing down research to such priority lines, we find the so-called "problem-oriented research". This leads to the tackling of practical problems showing a highly visible profile, often manifested today by intensive media coverage. Acid precipitations, deforestation of the humid tropics, rural forestry in dry tropical regions or wildlife conservation are examples of such issues.

Political lobbies defining problem statements and scientific lobbies translating these statements into scientific questions, according to the past successes of a certain school of thought, then combine their powers. These mighty forces exert high pressure upon the minds of scientists to make choices including methods easy to explain to a large public and research projects that carry little risk of generating even the impression of waste of public moneys. The same forces largely occur in privately financed research. Scientific gurus, political bankers, the media and a research field covered with a rich but very confusing array of theories and models practically drive the young researcher towards narrow options and specialisations.

This reasoning and this situation may appear quite healthy to those who hope that in this way - finally! - science may become fit to be managed. However, if this means that scientists and their methodological tools are to be organised by managers, so as to solve all successive problems that humanity may meet on its way, this is certainly an illusion. The success story of European forestry research is an illustration of this point.

For two centuries European forestry research has been "problem-oriented". It started by solving problems in the fields of reclamation of degraded lands and wood production, both by the planning and design of man-made forests and tree plantations. Later, it incorporated the use of older forests as biotopes for wild plants and animals and the principles of farmers' forestry. In each European country, the emphasis was placed on the most urgent national problems. Fashions in forestry succeeded each other after widely publicised local success stories: the Central European Plenterwald, the British and Dutch reafforestion ventures, the French conversion of coppice with standards into high forest or the Scandinavian large-scale coniferous forest management systems are some examples.

The whole terminology and the array of concepts in forestry research bear witness to this blending of practical management problems and aims with gradually added ecological notions. The word forest stand, meaning a management unit like an agricultural field, came to indicate any small to intermediate-sized surface covered by trees or forest. When a stand grows up, it arrives first at a pole phase, then at a ripe phase. Both denote ecological states of forest development, but the first one is also apt to yield poles to farmers, whereas the last phase is ripe for the harvesting of sawnwood. Average stand properties from the outset were calculated from tree properties directly linked to wood production, such as tree height and diameter. Only later were the ecological implications of these variables perceived.

European forestry research, including the work done by European researchers in other parts of the world, thus developed into an onion-like complex of knowledge, in which layer after layer of special, problem-oriented concepts became superposed upon their only communication point at the base, whence these layers were fed from the roots of practical problem statements and natural sciences. At the present day, both roots have proliferated. There are numerous unprecedented forest problems in the world and there is an unprecedented advance and diversification in basic sciences. None of the layers around 1970 contained the answers to problems like the impact of pollution, semi-natural tropical rainforest management, the use of strict forest reserves, genotype bank forest management, tropical rural energy forests or woodlots, and tens of others. The inputs from natural sciences were, at the same time, insufficient to rapidly add other layers of problem-solving concepts and so prolific that they started to identify new problems, even before these became manifest in day-to-day forest management.

Authors like Brünig (e.g. in Wiersum 1984) like to point out that the concepts united in this way during two centuries of European forestry research have become completely insufficient to solve today's forestry problems. Is this whole priceless body of knowledge, then, to be considered as a brake to progress rather than as an asset? It is my conviction that this entirely depends on its use. This, again, depends on the choice that researchers will make. If, on the one hand, the seemingly easiest way is chosen, research and management priorities will be defined in the traditional way, and this will lead to a lack of solutions for today's and tomorrow's forestry problems, and to sterilisation of progress in forestry science. If, on the other hand, this complex of knowledge is used as a treasure house, from which selected concepts and facts are used as judged necessary by courageous young researchers who do not fear the uncertainty of leaving well-trodden paths, progress in both forestry and forest science will result.

World-wide, this state of affairs is confirmed. The development of forest science cannot be conceived any further without the new notions, concepts and facts that saw the light in many countries, such as Mexico, Japan, the USA, Malaysia, Australia, Suriname or the Côte d'Ivoire, to name only a few. National or European forest research still yields valuable and necessary sources of knowledge, but any serious basis for advanced forest science will have to draw upon all that has been thought out in this field, the world over. This means an important shift in focus as well as a shift in gear, because a wider viewpoint also requires a lot of high-quality work.

This short tale on research in forestry could be paralleled with analogous case histories in various fields of botany, zoology, soil science and ecophysiology, all with long scientific traditions linked to human problems. Botany was an adventitious root of pharmacy, zoology of hunting and cattle sciences, and soil science or ecophysiology can be neatly linked to agriculture. None of the managers in these fields, the pharmacists, the hunters or the farmers, has ever been able to foretell the development in these sciences, nor to plan and regulate the workings of the scientific brains that made it possible. But all have benefited by these unpredictable and highly profitable scientific developments, most often generated only because some scientist had dropped the narrow viewpoints of immediate usefulness.

In other words, it proved to be necessary to develop the intelligent processing of information and knowledge in itself, before intelligent problem-solving action could be improved. Science in this way has become the contemporary intellectual tool for ranging information according to orderly lines of knowledge and insight. Scientific activity is an investment, both in the development of theory-building brains and in the accumulation of ordered and falsifiable theories or models. In this, it is an autonomous venture with its own rules. Scientific problem-solving is quite different from practical problem-solving. But adequate scientific models, on condition that they are well translated in everyday language, are one of the roots of high-quality solutions of practical or other human problems.

My diagnosis of the state of affairs in the sciences of forests has led me to undertake a long and hazardous voyage in this realm of the scientific mind. My ambitious aim was to establish a sort of map of this field of research, with signposts allowing the user to find his or her way towards more details, into greater depth, or from the scientific field to the practical aspects. Because all forest researchers and all foresters have one thing in common, i.e. some notion about the nature of forests, these notions became the focus of attention.

The complexity of forests is such that forest ecology as a natural science is on a par with such sciences as brain biology or the science of coral reefs. In a few hectares of forest, wherever in the world, the number of organisms lies between millions and billions, interacting with each other and within many abiotic gradients in a never-ceasing, swarming, highly dynamic and nevertheless organised play of feedbacks, life and death. There is only one computer able to process all the data within such a system: that computer is the forest itself. It was therefore necessary to order and select data according to the notions invented long ago, but made operational in new and efficient ways more recently, of hierarchical systems analysis. They allowed sketching a general framework, at the same time defining the entries into special approaches or more detailed aspects.

Because scientists have absolute freedom to choose the limits of systems, subsystems or supersystems in systems hierarchies, a decision had to be taken as to this choice. This decision was consciously made in favour of directly visible limits, liable to be mapped graphically with simple means. Many of the calculation models presented nowadays do the inverse by superposing a geometrical grid, in two or more dimensions, over an ecosystem and localising limits through calculated values for properties of grid squares, cubes or abstract spaces. This is being justified for two reasons. The first one is the necessity of "quantitative" models, a term monopolised formerly by statisticians and nowadays by computer-model builders. The second reason is the reduction of data needed to keep models manageable.

I concur with the gist of both arguments, but not with the way to solve them. For such "superposed" models, a doubt always remains in my mind whether some essential feature of the forest has been overlooked, so that the models may work but miss the point in explaining the working of the forest itself. A spectacular example of this having happened is the century-old controversy on the existence of "layers" or "strata" in the forest, which can be solved simply by looking at scale-drawings with a mind unbiased by pre-existing models on paper, as is shown in Chapter 6 of the present book. An understanding of forests in simple terms, which can be immediately validated in the field with simple means, according to my conviction has to precede and not to follow the establishment of computed simulation models. This is often admitted, but too seldom done in a thorough way.

Therefore, in this book the whole matter of forest ecosystems analysis has been taken up again from the base, i.e. the direct observation of forests and their component organisms. The original meaning of scientific terms has been explained wherever possible, and a large number of examples from over the whole world has been reproduced and surveyed. The vocabulary for distinguishing systems and their states had to be completed, and this has led to the creation of some neologisms and new terms, which I hope the reader will be able to assimilate easily and without grumbling, once their usefulness is conceded.

Once the map of forest ecosystems analysis had been established, two series of "signposts" have been added during the writing of the text. The first set points to the links with forestry and silviculture, by giving references to textbooks and articles or in the form of tables and illustrations, comparing terms and their meaning. The second set of signposts includes the sources where information can be found on the different schools of ecological and forest ecosystem research. If textbooks exist, detailed explanations were superfluous. In this way, compilation was avoided as far as possible and desirable.

Among the data presented, many are unpublished or have in many cases not yet appeared in overviews or textbooks. I obtained them from the numerous reprints that I had the privilege to receive these last years, from additional reading and from personal communications in some cases. My sources also include numerous unpublished doctoral and master's theses, and mimeographed documents from institutions in those parts of the world where the scientific culture does not include the printing of scientific results according to the rules prevailing in Europe or the USA. Graduate students and people doing doctoral research have impressed me, during the twelve years that I have been teaching, as having particularly free, open and keen minds. I have come to fully appreciate the continental European custom of publishing Doctor's theses.

The extensive use of what sometimes is called the "grey literature", outside the more prestigious scientific circuits, allowed me to find many scientific eye-openers and to use and make public many results of studies which would otherwise have ended too often in the grave of the seldom consulted, long rows of theses in a University library. Some of the theses cited were made intentionally in order to find facts needed for this book. Such studies then often concerned forests close at hand, i.e. in the Netherlands. Finally, each thesis contains a bibliography which may be ordered from the library where it is stocked, as a photocopy. This strengthens the signpost function of such works. I only regret that I did not have ready access to more theses, particularly outside the Universities of Montpellier and Wageningen, and some German and tropical ones.

During the adventurous voyage, which was the writing of this book, and even before, I met some of the most brillant, old and young minds of our century. This voyage fully confirmed my earlier conviction, that many more unconventional ideas are spawned in our scientific field than is commonly realised, but that the occurrence of crazy or cheating scientists is so extremely rare that it can in no way justify the myths on this subject that are common currency in popular belief. I do hope that the text, such as it has become, will convey this sense of excitement and richness to its readers and, most of all, that they will discover or rediscover the forest through the models presented.

ROELOF A. A. OLDEMAN

Acknowledgements

This book has been long in the cooking, and many people have helped me reach the stage where it could grow from a set of ideas into a set of lecture notes and finally into the present text. Now is the time to send the manuscript to the publisher, before I finish reading Prigogine and Stenger, Hofstadter or Egolfs Bakuzis' incomparable lecture notes and, who knows, might have to start all over again. Those who contributed to the ideas in the present book cannot all be mentioned on this page, but even if more are omitted than acknowledged explicitly, please let all accept my sincere thanks.

I learned a lot about what decision-makers, managers, bankers and forest-lovers want to know about forests from people met during my TROPENBOS-time. My thanks for this go to people like Dr. Van Spiegel, Mr. Heering, Dr. Ross, Mr. Van Dam, Dr. Clément, and UNEP's Dr. "Chuck" Lankaster. It is the latter who, in a discussion at the Hague, doubted whether the concepts of silvology and silvonomy would ever be printed. Here is one of them. In the same line of thought, my thanks go to WNF's Mr. Van der Giessen, FERN's Prof. Bourdeau, IIED's Prof. Poore and Mr. Rietbergen, AUW's Prof. van Maaren, GFF's Mr. Meester, Mr. Van der Meijden of SBH, and IUCN's Dr. Sayer.

Where are the scientific roots? I still think that the seed formed in my mind during a discussion with Dr. Jean-Louis Guillaumet at the Paris Museum, just before my departure to Ecuador in 1974. I thank him truly, and also those with whom I could discuss these matters since then, for instance Ing. Cañadas Cruz, Prof. Bourlière, *les* Hladik, Dr. Hadley, Prof. Mutch, Prof. De Wit, Dr. Whitmore, Dr. Koop, Prof. Barkman, Prof. Baudière, Dr. Ashton, Prof. Kojima, Prof. Ohkubo, Dr. Alvim, Mr. Cao, Prof. Hari, Prof. Tomlinson, Prof. Leikola, Dr. Saleh, Prof. Roux, Dr. Bancilhon, Prof. Soemarwoto and his wife, Dr. Ng, Prof. Mlinšek, Dr. Rossignol, Prof. Schütz, Ms. Teller, Prof. Mayer, M. Dubost, Prof. Szujecki, Dr. Charles-Dominique, Prof. Oliver, Dr. Gasc, Dr. De Granville, Prof. Fanta, Dr. Maître, Dr. Sombroek, Prof. Kio, Dr. Leeuwenberg, Dr. Huxley, both Dr. Lescure and Dr. Lescure, Prof. Kira, Prof. Sarukhán, Prof. Gomez-Pompa, Dra. Monasterio, Prof. Balslev, and so many others.

Without the encouragement of ESF's FERN, formulating the need for a state of the art in forest ecosystems research, this book might not have been written. It has not become a state of the art, however, in the usual sense. My special thanks go to Dr. Rolfe A. Leary (IUFRO 06.09) without whom the subject of methodology and philosophy of forest research would not have been made explicit, to Dr. Peter Raven (Missouri Botanical Gardens) for his contribution on yellow tropical flowers, to Dr. Pieter Schmidt, without whom the paragraphs on tree production would have looked quite different, to Prof. Francis Hallé for an ever-continuing collaboration although he doesn't agree with the diverse nature of secondary forest as postulated here, and to Ir. Paul Romeijn for his welcome and sharp comments to the speciation and selection paragraph. Prof. Eberhardt Brünig I thank for having been always and frequently stimulating in his own thorough, assertive and far-seeing way, whereas I am sincerely grateful to Ir. Jaap Kuper, Her Majesty's *Opperhoutvester*, for having unearthed the Backman publications.

The former Department of Silviculture and Forest Ecology at Wageningen University has suffered much by, but also contributed significantly to, the growth of this book. I thank all researchers and particularly the promovendi and students who took the trouble to produce a whole stack of useful comments. And I can never thank enough Ms. Boudewijn and Mr. De Bruijn, without whose whole-hearted cooperation this book would not have obtained its final shape.

Wageningen, May 1990 Roelof A. A. Oldeman

Contents

First Part: Introductory Matter

Chapters 1 and 2 introduce the reader to the field of silviculture as a part of forestry, to the work expected from a professional silviculturist and to some historical backgrounds of concepts used in this book. The necessity of a solid theoretical basis in forest ecology is stressed. General aspects of theory and model building are outlined.

1 The Forester as a Green Designer

1.1 Forestry is Not Self-Evident

Forestry is a performance of service. The profession of "forester" hence was born only in societies which needed this service, because they started lacking either forest products or forests, in this order. The production of wood and other forest products in many places and at many times was the business of farmers. The Roman Empire and its human society were great forest destroyers. Wood served to yield energy for energy-devouring installations, such as the public baths (thermae), central heating in villas, or bakeries. The construction of fleets and the war industry in general (bows and arrows; lances; clubs; fortifications and camps) required much wood of a high quality. The solution for growing wood, however, was not found in the planting or conserving of forests, but mainly in planting hedges and coppice on farms (Piussi pers. commun. 1983). Problems in regreening the much-polluted Roman cities were solved by planting plane trees *(Platanus orientalis),* which still are known to be very resistant, also to 20th-century chemical pollution, but which now are dying back from an introduced pest.

The Mayan civilisation, over 1000 years ago in Central America, can be said to have considered wood as a minor product from fruit trees or shadow trees (Gomez-Pompa 1985 b). The production of tree fruits was a life insurance against maize misharvests. When the trees produce less fruits because of advanced age or for other reasons, they yield wood of the right sizes and shapes for construction or, in smaller trees, as firewood. For this last purpose the Mayans, like the Romans, knew hedges (living fences). The breadfruit tree (*Artocarpus incisus,* Moraceae) fulfills this insurance function in the traditional village forest gardens of South East Asia (cf. Oldeman in Jacobs 1988), and on both borders of the Sahara such systems have their dry-region variant in the oasis, where date palms (*Phoenix dactylifera,* Palmae) assume this role. In projects for international cooperation in rural development these kinds of systems today are receiving a growing amount of attention (cf. Huxley 1983). The concept is revived under the name "agroforestry systems", in which "multiple-purpose trees" are used. Although these systems cannot be considered as "true" forests in a strict sense, their success heavily depends on an adequate contribution from silviculturists, if they are to be renewed and improved.

Until the beginning of the 19th century one may say that forests only marginally and locally served for sustained wood production, let alone for silviculture. If large quantities of wood were harvested from forests, this rarely was in the context of sustained yield management. The main functions of forests were cattle grazing, hunting, procurement of organic fertilizer (litter and humus) and a kind of piggy-bank function for evil times, such as wars or famines. Only when urbanisation and industrialisation led to a vastly increased demand for huge quantities of wood in standard sizes and grades, not even two centuries ago, did it become necessary to grow wood on such large surfaces that the plantations started to behave like forests. This is the tradition behind the vast majority of approaches in forestry education and practice today (cf. Groenendijk et al. 1985).

However, wood plantations were also multifunctional. Reclamation of degraded lands, such as wind-blown sand dunes (Netherlands, North Africa, SW France), eroded mountain sides (Central Europe, parts of the Andes) could only be financed because the wood could be sold as mine-props, telegraph poles, sawnwood or as firewood for local populations or for steam engines (USA, Turkey). The management of the remaining vestigial forests and of the large plantations, starting to behave as forests as they grew older, led to a demand for basic theory. The first handbooks came from Freiburg, around 1800, whereas the French are still busy with the "conversion de taillis-sous-futaie" (conversion of coppice-with-standards into high forest) since the adoption of their forest laws of 1827. One should know this context to understand the basic French concepts in silvicultural theory. Influences from other European countries, together with forestry activity in the colonial period which brought Europeans into contact with natural tropical rainforests, at the end of the 19th century led to a strong awareness of the necessity to respect the nature of forests. This was completely in line with the Romantic way of thinking that dominated this period in Europe. A.J. van Schermbeek, who had been a colonial conservator of forests and later became one of the first professors of silviculture in Wageningen wrote in 1898, in the preface of his translation of a German forestry treatise (p.10):

> "In Holland, since my return from Java, I have been struck by the sad monotony of our managed forests, which I have not been reluctant to expose in periodicals and newspapers. The more I became familiar with the harmful consequences of this tendency, and continued research kept confirming the fears expressed by Dr. K. Gayer, I held it my duty to attract attention to this subject among a wider public. Everybody who is working with the Dutch forest should be permeated by the necessity to leave as much as possible to the natural forces when managing a forest. Then alone the sustention of forest management is conceivable, whereas every wanton rape of nature will cause a punishing throwback on management itself."

During the 20th century, our century, there have been alternating schools of more "natural" and more "artificial" ideas in forestry.

The accent was on fast wood production, for evident reasons, during and after the two World Wars, whereas pre-war periods and the years just behind us (the 1970's) showed an increased attention for the nature of the forest. One fundamental reason for this recent shift, becoming apparent these days with the problems raised by acid precipitation, is the relation between a society which has been industrialised with **abiotic** means and a forestry sector that is completely dependent on **biotic** means: the lack of living forest also would mean a lack of forestry and forestry-based industry. In the tropics, and to a lesser degree in northern regions which have not yet lost all their forests (e.g. North America, Siberia), the situation again is different. Although there is increased awareness of the enormous value of those forests, warranting careful management for conservation and sustained yield, in practice they still are treated too often as wood-mines. The mistaken idea is that these forests are self-repairing natural resources. At best, they are spared as a savings-account for difficult times. But liquidation of such savings is already becoming apparent in countries with high international debts.

This historical perspective has been much simplified and condensed: it rapidly sketches the origin of a number of ideas to be found on the following pages. But it is enough to show three principles that probably will not change over the next centuries:

1. Without an adequate technical-economical basis, the conservation and sustained use of no forest area can be guaranteed, be it a natural forest reserve or a wood production estate.
2. Without an adequate socio-economic basis the continued existence of forests will not be accepted, and forest will be cleared as soon as people feel the need for products or for land.
3. Without an adequate biological basis of survival, the durability of no forest area, be it a natural forest reserve or a timber estate, can be secured.

The first two principles will not be expanded in this text, because they belong in the realm of forestry techniques, management, economics and policy. Of course, these two requirements do represent the conditions that determine the feasibility of any silvicultural system in a certain society.

This book concerns the basic aspects of the third principle. Once the forest technologist has designed the necessary system of tools and labour and defined the requirements for products, once the forest manager and the decision-makers have drafted their laws, rules and management plans, still nothing is possible unless a **green designer** conceives a living forest system that remains healthy and stable under all this pressure. An optimal harmonisation of these disciplines is indispensable. Their dosage and proportions determine the professional specialisation of each forester. This harmonisation moreover is a very personal affair. Objectively viewed, no one discipline is more or less important than the others, because the weakest link determines the strength of the chain. However, forestry is far too vast a field for one person to master it entirely. Every forester hence cannot but focus on some special points. The true specialist now can be recognized by his incessant claims that the worst and most difficult problems are in his or her special field. **One should not believe such specialists,** and only seek their strength in the capability to do a well-circumscribed job very competently, be it parks management, forest law, pine silviculture or paper manufacturing.

The silviculturist, being a specialist in the "green" aspects of forestry, concentrates on forests and trees as living systems, as well as water, soil and climate that determine their living conditions, these being very important aspects of forestry. A civil engineer designs plans for building artefacts with abiotic materials (glass, concrete, metals) that take the shapes defined on man-made blueprints. On the contrary, the green designer has to consider living materials that take the shapes defined by genetical or ecological instructions. **Silviculture is the design of steering**

processes in forest ecosystems, in such a way that developmental sequences are reached which maximise the chances of survival for that forest. Speaking with Bellamy (1978) one can refer to **greenprints.**

1.2 Silviculture

Silviculture has the following main aspects.

- Silviculture always requires thinking between a day and a millenium. People change their demands upon forests day by day, whereas forests need periods of 1000 years to develop completely and be self-sustaining.
- Silvicultural steering of forests is not free. It leads to a "target forest" in the service of a human "target group". In general one can say that the target group of virgin forest is Humanity as a whole, rural forests have farmers as their target groups, timber estates find their target groups in urban and industrial populations. More refined "target forest types", often multifunctional ones, can for instance be found in the national Dutch Long-Term Forestry Plan presented to parliament in 1984. However, no target groups have been defined in this plan.
- Both conditions mentioned above entail the necessity for fast work. A fast diagnosis of a forest most often is indispensable in order to produce a timely design that allows the survival of that forest in an acceptable way. This book has been written also with this in mind.
- Design is not the same thing as application. Designing is the gathering, selecting and weighing of information in order to process it into a step-by-step development plan. Application is the use of already available, selected and weighted information according to known directives. The designer starts with an open problem, the applied worker begins with the elements for a solution.
- Existing information always has been obtained in the past. But human problems are ever-changing. For the solution of predictable future problems, the silvicultural designer has a choice. Either very open and flexible systems can be conceived, or new research can be started. Usually, both approaches are combined.
- As far as they are known, the general principles and rules of forest biology and ecology structure an information-base that will still be valid after a century or more. If the silvicultural designer has a voice in the investment of means and brains in research, he should realize that so-called fundamental research yields this kind of information.
- Because knowledge and insight remain incomplete, every silvicultural design is partly a gamble. That is why every silvicultural design should contain openings in which cheap and easy monitoring research can be fitted as an accompanying activity. One should be aware of this aspect when considering forest and wood mensuration, forest inventory and other management sciences.
- Forest policy and requirements concerning forest products are limiting conditions in silvicultural design. But forestation (which is short for either afforestation, reforestation or tree planting - see Wiersum 1984) and forest conservation are targets for different kinds of design. The older a planted forest becomes, the

more it will behave as though it were a spontaneously grown forest. The optimisation of the initial forestation phase, so that the ecosystem balances out as soon as possible, is a property of a good forestation design. Optimisation of silvicultural treatments, in such a way that existing forest remains healthy and stable, taking into account the limiting conditions, is the main aspect of forest conservation design. Designs for the transformation or conversion of one forest type into another are in between these two.

- Finally, a design has to be comprehensible. A first requirement to be met is that field personnel can understand, read and discuss it. It may be that the designer needs all kinds of tools to enhance the quality of the design: computers and whole stacks of books and articles written by researchers for researchers in scientific jargon. But all trace of these tools should have disappeared in the final design. It should be a well-illustrated, simple, clear and concise document.

However, designing itself requires a very solid theoretical basis. That is what silvology is about.

1.3 Silvology

The theoretical basis for both understanding forests and designing silvicultural systems according to the lines sketched above is a special branch of ecology. It is proposed here to label it with its own name, silvology. There are several reasons for doing so.

First, forests are unlike any other terrestrial ecosystem because of their size, longevity and degree of complexity. Many of the notions in current forest ecology were borrowed from studies on other ecosystems, the rationale being that all ecosystems are variants on one theme, i.e., primary production by green leaves and its distribution among green plants, other organisms and dead biomass compartments. In its general sense, this statement is undoubtedly true. But the sheer volume of any forest patch creates structured room for many more organisms than a volume of other ecosystems covering an equal surface. Its long lifespan allows chains of life to be built such as cannot originate within shorter-lived vegetations. It permits processes of abiotic resource exploitation to proceed slowly but inexorably over time. Many minerals that are fixed and unavailable for annual vegetations, like many crops, become mobilised and available to a forest on the long run. Finally, both size and longevity allow for a degree of complexity in which patterns and processes occur which are vital for forests, and rather accessory in other ecosystems. Symbiosis is a key-word in this sense.

Second, forests are often considered to be the acme of Nature, as distinct from artificial, man-made ecosystems. This is the consistent follow-up of a line that started with the 19th-century myth of Frankenstein and continued with the 20th-century literature on robots and androids, suggesting that human bodies or beings cannot be imitated artificially without most horrible consequences. It was only the vocabulary that slowly changed, replacing the concept of God's creation by the impersonal notion of Nature as a super-power that differentiates the early from the later literature in this sense. The basic idea remained, that living systems cannot be made by humans. The un-making of so many living systems by human action over the last half-century, manifested by the rate of deforestation in the tropics (cf. Wiersum 1984) or the extinction rate of species populations (e.g. Myers 1979; but also see Simon 1986), coupled to the myths that originated with Frankenstein, led to the conclusion that forests, as opposed to all ecosystems-with-trees planted by Man, should be considered as sanctuaries of Nature, not to be touched (e.g. Jacobs 1988).

Earlier, Oldeman (1980) accepted this reasoning and coined the word "silvoid systems" for man-made ecosystems with a dominating tree compartment. But the question is not as simple as that. This may be illustrated by a man-made animal, the cow, which does not cease to be an animal, whatever may be the strings attached to its potential for survival, that is almost certainly limited to man-made environments. Another footnote is the fact that biological change never is completely reversible, be it spontaneous like speciation or species migration, or man-made like crop selection and the spread of exotic species. At a more fundamental level, that of the behaviour of living systems, comparable processes follow the same biological rules for survival or extinction of organisms or ecosystems, whether the factor inducing such processes be anthropogeneous or not.

The question therefore should be asked: what are we after? Is it the impact of human activities, and their dosage that should be analysed? Or are we interested in the fundamental rules which explain the behaviour of forests and their components under all kinds of influences? The order seems clear. If the fundamental aspects are tackled first, the human aspects can be solved afterwards as a variation on the theme of fundamental forest patterns and processes. This is the second reason to coin the word silvology.

A third reason is that forest sciences up to now have been largely isolated, or marginalised, or both. They have been isolated as far as they tackled specific problems relevant to a quite small and socially hidden group of professionals in forestry. They have been marginalised to some extent in biological and ecological sciences, as special and difficult cases, to be studied in the future as variations upon ecosystems patterns brought to light in the study of seemingly simpler cases, like lakes or meadows. And they have been isolated and marginalised at the same time because there is an ancestral fear of forests in the human mind, as proven by thousands of folk tales, myths and proverbs. Tolkien's "Lord of the Rings" contains many examples of this basic attitude, which leads human beings still now to consider the clearing or at least the taming of forests to be major victories.

Without further thorough study of this frequent attitude, which would require a thick volume on the symbolic meaning of forests in human psychology, one conclusion certainly is that we need to heighten the profile of the forest, in science as well as elsewhere. It should not be admitted that forests are isolated or marginalised because they have to be rendered "harmless", whatever that may be. The first step to do so is to give a specific name to the field of science that is involved.

For reasons of objectivity, the forest being a distinct ecological object, of philosophy, the fundamental aspects of forests having priority over derived practical aspects, and of psychology, the forest having to be taken out of hiding, this book is on silvology.

2 Build-Up and Use of Theories

2.1 Some Concepts

A **theory** should not be confused with a **hypothesis**. A hypothesis is a supposition that looks well but still has to be proven. Originally, "theory" (ϑεωρια) was the Greek term indicating the act of concentratedly and religiously looking at a theatre, its actors giving an ordered message. It has come to mean, in a figurative sense, looking at ordered facts so as to read the information which they convey. The order in a theory often is very abstract, and theories often claim universal validity. A theory concerning a limited system will here be called a model. This distinction is not made in this way by all authors.

A model is made after a system. This is a piece of reality that we delimit and distinguish within the whole of reality, and which is explained in terms of the subsystems it contains. A forest (system) often is explained by its trees (subsystems). This example also teaches us that the model is abstract, i.e. that many things are unconsidered, left out, abstracted: a forest contains many more subsystems than trees alone, e.g. animals, herbs, shrubs, fungi, mosses and ferns. Moreover, this example shows that the limits of the system may be freely chosen: "a forest" has the limits that one wishes to consider. Many forest models express their results per hectare. That is an example of an arbitrary delimitation, because neither forests nor their components know or follow the human metric system.

A common distinction is made between descriptive and explanatory models. Descriptive models only give the properties of a system seen from the outside. Explanatory models represent a system in terms of its subsystems. Descriptive: "this forest dies, as can be observed by its brown colour". Explanatory: "this forest dies **because** drought killed mycorrhizal fungi, which weakened the trees, and therefore these have dead, brown branches". A black box is a system that we do not know inside, by choice (we do not want to know) or by necessity (we cannot know it as yet). In that case we only know what goes into it (inputs) and what comes out of it (outputs). Unfortunately, for many decision-makers in forestry the forest has remained a black box: one sends people and machines in (inputs) and wood comes out of it (outputs). This should not be confused with forestry.

For models and theories it is commonly assumed that their validity can be scientifically proven. What is scientific proof? In all cases this can be reached following the "see-for-yourself"-principle. First, anybody who wishes to do so should be able to repeat the counting, measuring and reasoning that led to a statement, and reach the same conclusions. And second, reasoning according to the model or theory should allow predicting either future behaviour or unknown properties of the real system. If the system behaves according to the prediction, this is an indication of the scientific correctness of the model or theory. Mathematics are used as a kind of universal language to express all counts and measures. The last word has not been spoken as yet on scientific proof.

Three concepts that certainly emerge in that context are objectivity, subjectivity and exactitude. Most often, these words are quite sloppily used. To start with the last one: exactitude most often is constricted to statistical precision including measuring errors, standard deviations, correlation coefficients, etc. But, more generally, exactitude has to do with the object or phenomenon being studied, with the aim of the study, and with the instructions allowing the repetition of the same observations. Example: a forest transect as measured and drawn by a first reconnaissance team does not look very precise. For instance, the exact details of branching are lacking in the trees that are drawn to scale, and a mean measuring error of about 5% has to be accepted. But, for the aim of the study, which is the understanding of the forest mosaic with its eco-units, this is sufficient. Moreover, the handbook or manual used contains enough instructions for another group of researchers to repeat, with the same exactitude, the same observations a week or so later. So this research can also be repeated. Reasoning on the basis of these data, the second group can arrive at precisely the same conclusions as the first one. Hence this study can be checked. The influence of the observer (the subject) on the results has been eliminated: that is what is called objectivity. The object, not the subject determines the results. Whether this concerns a scale drawing or a statistical computation is unimportant. As long as a method allows checks, with a precision defined beforehand, objectivity is guaranteed.

Many misunderstandings have their origin in the sloppy use of "objective" in the sense of "good" and "subjective" in the sense of "bad", using at the same time the degree of precision or exactitude as a measure for objectivity without relating it to the aims of research. A correct statement, for instance, would be: "the conclusions are too far-reaching when considering the exactitude of the data", which indeed is a frequent error. But, "the research is subjective" is the incorrect statement often heard. Moreover, current scientific beliefs seem to imply that counting is "more objective" than measuring, and measuring is "more objective" than drawing to scale. This leads to the criteria used in establishing scientific data.

2.2 Criteria for Modelling

Since times immemorial, people have used three criteria to create understandable order in the world around them. It seems that no other criteria could be invented:

- numbers: how many eucalypts are there in this forest stand?
- structures: this eucalypt has a short contorted trunk and a deep, large crown;
- dosages: this eucalypt contains three tons of biomass.

To establish numbers, one has to count; to establish structures, one has to measure; and to establish dosages, one has to weigh.

Moreover, it was discovered a long time ago that statements using one criterion can be translated into statements using the other. For instance: how many tons of biomass are there in a stand with a hundred eucalypt trees? (numbers to dosage), or: eucalypts with deep crowns contain more biomass than with shallow crowns (structure to dosage). The translation into numbers seems to be universal: number of tons, number of times that a crown is higher than large, number of numbers of different trees in a stand, etc.

In forest theory, these three criteria are met in different fields of science. The other way round, these can be distinguished as to the criterion mainly used:

- **Numbers**: population biology, plant and animal sociology. Here, individual plants or animals, or species, are counted and calculations are made with the resulting numbers. Results are often expressed as formulae.
- **Structures** (forms, sizes, proportions): morphology, architectural analysis. Here, organs, organisms or ecosystems are measured or photographed. There is nowadays an active line of research expressing the results in numbers (digitalisation) in order to simulate the drawings on computerised plotters. Results most often are expressed as scale drawings, including transects and maps.
- **Dosages** (weight, sometimes volumes): physiology, ecophysiology and production ecology. Here, the quantities of matter (chemical substances), energy, biomass or displaced matter and energy (flows) are established. Because of these being expressed in numbers, computer processing and simulation of dynamics do not represent real difficulties.

Counting, measuring or weighing are means to gather information in an objective way. Photographing also is a method of gathering information: in this, it differs from drawing. Drawing and calculating are means to order information, which is a second step in theory building. An important aspect of forest ecosystems is that they contain so much information. They are more complex than most other ecosystems, and might be compared to the biological communities in coral reefs (cf. Dauget 1985).

It is of enormous intellectual and practical importance to plan the attack on so much information well. The intellectual confusion, the sense of chaos that assail people in the forest are expressed well in that nickname for the tropical rainforest: "the green hell". An ecophysiological model, capable of coping with flows and quantities of light, water, nitrogen and carbon in a forest stand is already quite complex. Adding one more element, phosphorus for instance, requires much study and ingenuity if the model is to remain manageable. Transects, such as those made by first-year students, already contain so much information that only part of it can be used in numerical models, e.g. ten spatial coordinates per tree (Koop 1985) or a few other variables (Faber 1985). And plant sociology has not succeeded in representing tropical rainforests in a fully satisfying way by counting the populations of the myriads of species.

In every forest analysis therefore, whether this is clearly stated or not, a **choice** has been made between the little amount of information to be gathered and the large amount of information left aside. This choice is called abstraction, literally the abstraction of the excess information from the total information. Furthermore, the nature of information to be gathered has been chosen by the researcher (counting, measuring, weighing). And finally, the method of data processing and reasoning also is a matter of choice, sometimes using translations from one set of data, according to one criterion, into another set that relates to a second criterion. Dosages expressed in numbers, numbers derived from structures, and so on, form the links between different approaches.

No single one of the scientific fields mentioned earlier can beget a complete, universal theory of forests. Their insights are complementary. But at this moment it is not even possible to make one integrated forest model, as shown by the discussions between 1982 and 1986 that have resulted in the establishment of FERN (the European, ESF-sponsored Forest Ecosystem Research Network). The present text is going as far as possible in this direction, but cannot succeed where a continent fails.

2.3 The Choice of Systems: A Game Without Frontiers?

2.3.1 Limits

It has been stated above that one is free to trace the limits of systems to be studied. This is a general rule from the physical sciences. But in the sciences of life there are several traditions. On one side of the spectrum, the approach dates from the beginning of biology. Here, the search is for **"natural limits"**, i.e. limits that are apparent and visible, and are thought to be formed by the living systems themselves. On the other side we find a more recent form of mathematical analysis working with grids built up by squares, each square being a subsystem of the grid. The grid then is explained by the properties of its squares. This approach depends on the establishment of mathematical coordinates in relation to mutually perpendicular axes.

2.3.2 Grids, Pixels and Power of Resolution

Someone who does not know grids should look at a newspaper photograph through a lens (Fig. 2.1a). Such a photograph consists of tiny image spots, some darker and others lighter. Together and seen from a distance, these spots make a recognisable image. There is a special branch of mathematics for the analysis of such phenomena: pattern recognition and image analysis. This approach becomes more and more important because remote sensing images (cf. Fig. 2.1b) are all built by such spots, in fact tiny surfaces. These are called pixels (**pic**ture **el**ements). Each pixel represents an observed piece of the surface of the earth; LANDSAT, for instance shows 90 x 90 m per pixel and newer satellites, like SPOT, cover 20 x 20 or even 10 x 10 m per pixel. Here we have our subsystems. From the combination of different properties, measured in each pixel: colour, clarity, sometimes infrared (temperature/humidity), the pattern of the whole image is built. In large pixels (e.g. 100 x 100 m) the hectare covered may include all kinds of phenomena. These can-

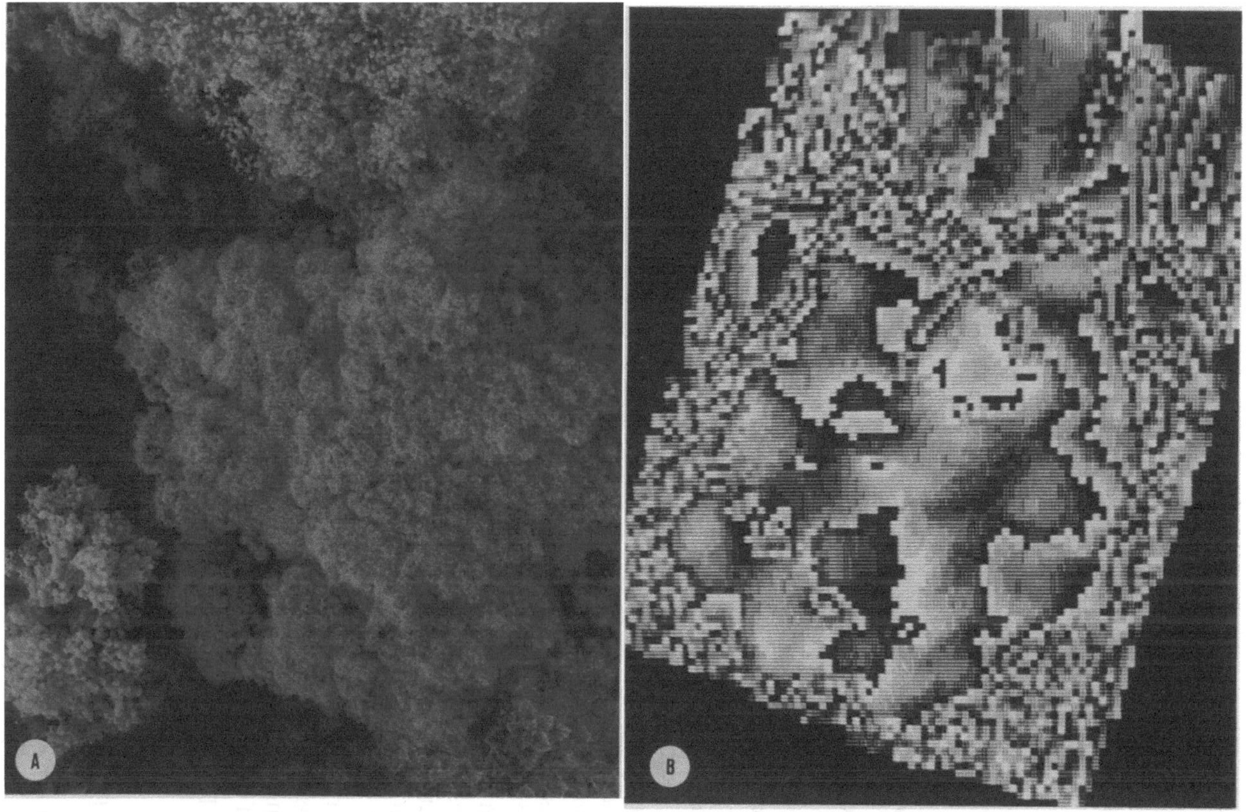

Fig. 2.1. Crown of *Piptadeniastrum africanum* (Mimosaceae, Troll's model) in the National Park of Taï, Côte d'Ivoire, 14 May 1984. **A** Black and white on infrared emulsion, photographed from an ultralight plane (Vooren and Offermans 1985), showing "natural", visible limits as perceived by eye. **B** Digital transposition of same picture, representing heights in pixels of 50 x 50 cm forming a grid (analysis in Offermans 1986). Digital document courtesy Dept. of Geodetics and Land Survey, Agricultural University Wageningen

not be seen, but may only be guessed from the data representing optical or other radiation properties. With pixels of 10 x 10 m, one can determine much better what is going on inside one hectare: this pixel covers the size of a larger tree-crown. Hence it may be said that the smaller are the pixels, the larger is the power of resolution. With a certitude of plus/minus 5 m, the limits can be established on such images between systems with different optical properties. Moreover, there is temporal resolution next to spatial resolution. Resolution obtained in the time dimension is greater in weekly than in monthly or yearly observations. The limits are established here between events, which are symptoms of various dynamic processes.

2.3.3 Ground Truth

One has to go in the field to understand the significance of such remote sensing images. The easiest way to organize this field research is to trace a grid in the forest, with the same square-size that appears on the image as one pixel. Properties of each square can then be established, for instance: "the distribution and clustering of dark blue pixels correspond in the field to the distribution and clustering of fully grown tree crowns". This working method shocks few people, because during countless centuries such squares have been used under names such as acres, hectares, morgens or bunders. These units have been used in forest management and therefore also in research. In branches of ecology that concern populations, species counts commonly are done with a low resolution power (averages per hectare) but sometimes with higher resolutions (numbers per quarter or half hectare - see Van Roosmalen 1980 - Fig. 2.2). Because forest estates express wood yield and increment in m^3/ha/y, the data in growth and yield studies, forest mensuration, forest inventories and forest statistics are expressed per square (or circle) that can easily be calculated back to mean values per hectare.

For many research aims, these methods are excellent. But for the designer of silvicultural systems or natural forest reserves there is one great disadvantage: squares or circles most often are irrelevant for the functioning of subsystems in forest ecosystems. Grids are pessimal means for the analysis of function in forest ecosystems. Grid-based analysis most often does not allow to establish explanatory models of forest ecosystems. The results are limited to statistical correlations between the properties of systems and images, and a correlation is no causal explanation.

2.3.4 The Eye and the Natural Limits

It is unknown how this works precisely, but the human eye seems to be an instrument with a very high power of spatial resolution. Moreover, this power has been enhanced by all sorts of instruments, such as microscopes or telescopes. Since

Fig. 2.2. Pixel-like mapping of food tree distribution over the home range of black spider monkeys in Suriname. Scale in central circle. (After Van Roosmalen 1980)

Home range showing distribution and density of food plants

a 110

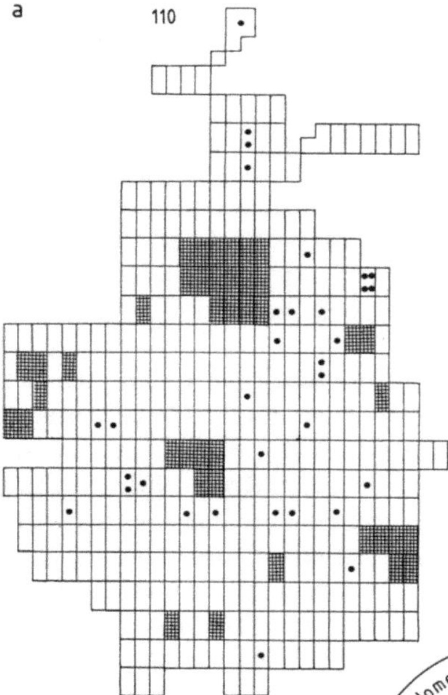

Pourouma mollis
'granbousi-papaya'
(33)

b 109

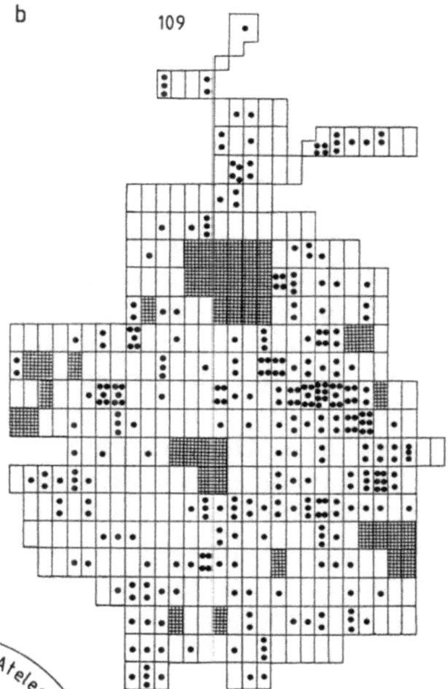

Virola melinonii
'hoogland baboen'
(313)

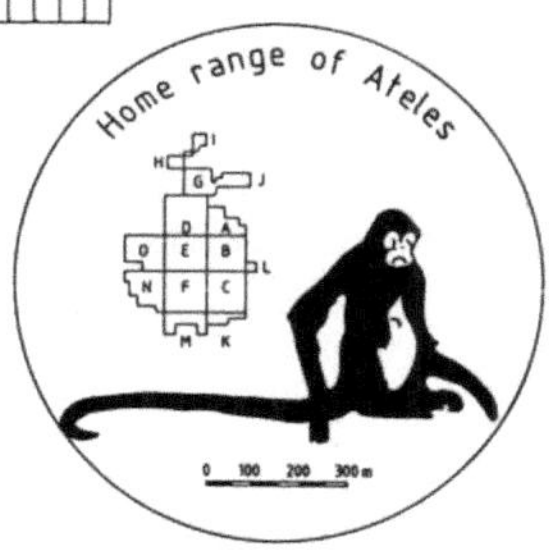

c 111

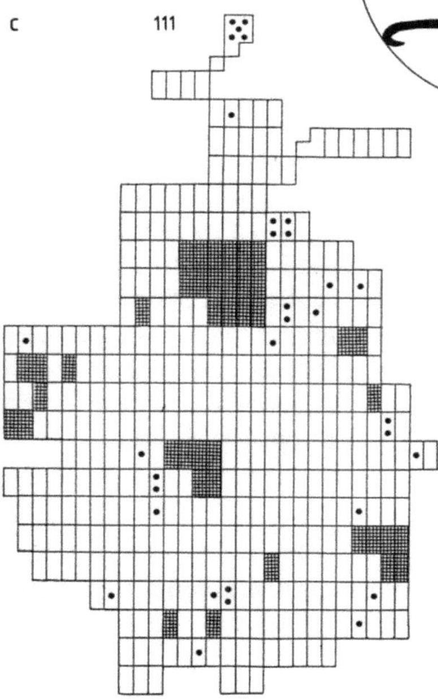

Virola surinamensis
'laagland baboen'
(32)

112 d

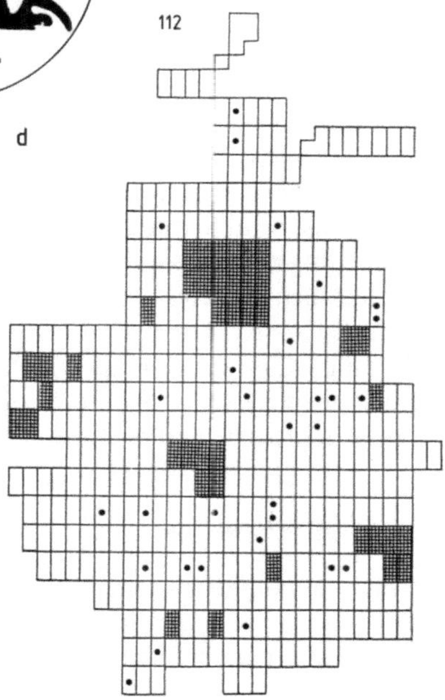

Minquartia guianensis
'alata-oedoe'
(30)

movie pictures were invented, it is known that the temporal resolution of human eyes lies somewhere between 0.10 and 0.05 seconds. The eye, aided by instruments, has observed very fine limits that seem to be given by Nature itself. First, there are the limits which are manifest in membrane, cuticula, epidermis, bark, skin and such. These limits define a rather clear "natural" system, which will be discussed later in more detail. Such a system is built in this way: cell nuclei are defined by membranes, cells by cell walls, organisms by some sort of skin. Often, organs also are well limited by some sort of envelope. Tissues are less well defined and one can in several cases have legitimate doubts about their exact limitations. However, in general few people have difficulties in delimiting subsystems and systems with "natural" limits, within organisms.

Although there have been precise drawings of forests for more than two centuries (Fig. 2.3), the impression is that these have not been made with very precise limits in mind. Only recently, since analytical scale drawings were made, (e.g. see Hallé and Oldeman 1970; Oldeman 1974a; Hallé et al. 1978; Koop 1981a; Torquebiau 1981; Mulder and De Waard 1984, and many others), there has been a growing awareness that there are individual, natural portions of forest, having been "born" at one moment, growing up as one system and dying at the end. Such portions of forests have spatial and temporal limits (cf. Oldeman 1986), in the form of bordering surfaces with functions comparable to but not identical with those of a skin. These limits often are not as easy to distinguish as those of organs, organisms or cells. Even if they can only be approximately established, however, a more logical hierarchy of living systems can be conceived.

2.4 Hierarchies

As said before, a system is composed of subsystems, which in their turn are composed of smaller subsystems, and so on. For instance, a biological forest community as a system is composed of its component plants and animals, which in their turn are composed of organs. These organs again are composed of tissues, the tissues of cells... and so one can go further. Chapter 3 contains such series: a tree consists of subsystems, the reiterated models, being composed in their turn of subsystems, leaf-bearing axes, which then again are built by a collection of organs. Or, at a higher level (e.g. Chap. 6), silvatic mosaics are composed of eco-units (regeneration units), their subsystems, which contain coexisting plants, animals and abiotic factors as subsystems. We may conceive a continuous series in this way, every "higher" system becoming more complex. This is a hierarchy (Table 2.1).

2.4.1 Added Information

Why is every higher system more complex? The reason is that it contains not only its subsystems, but also the information serving to keep these subsystems together in an orderly manner. Hence:

System = Collection of Subsystems + Cohesion

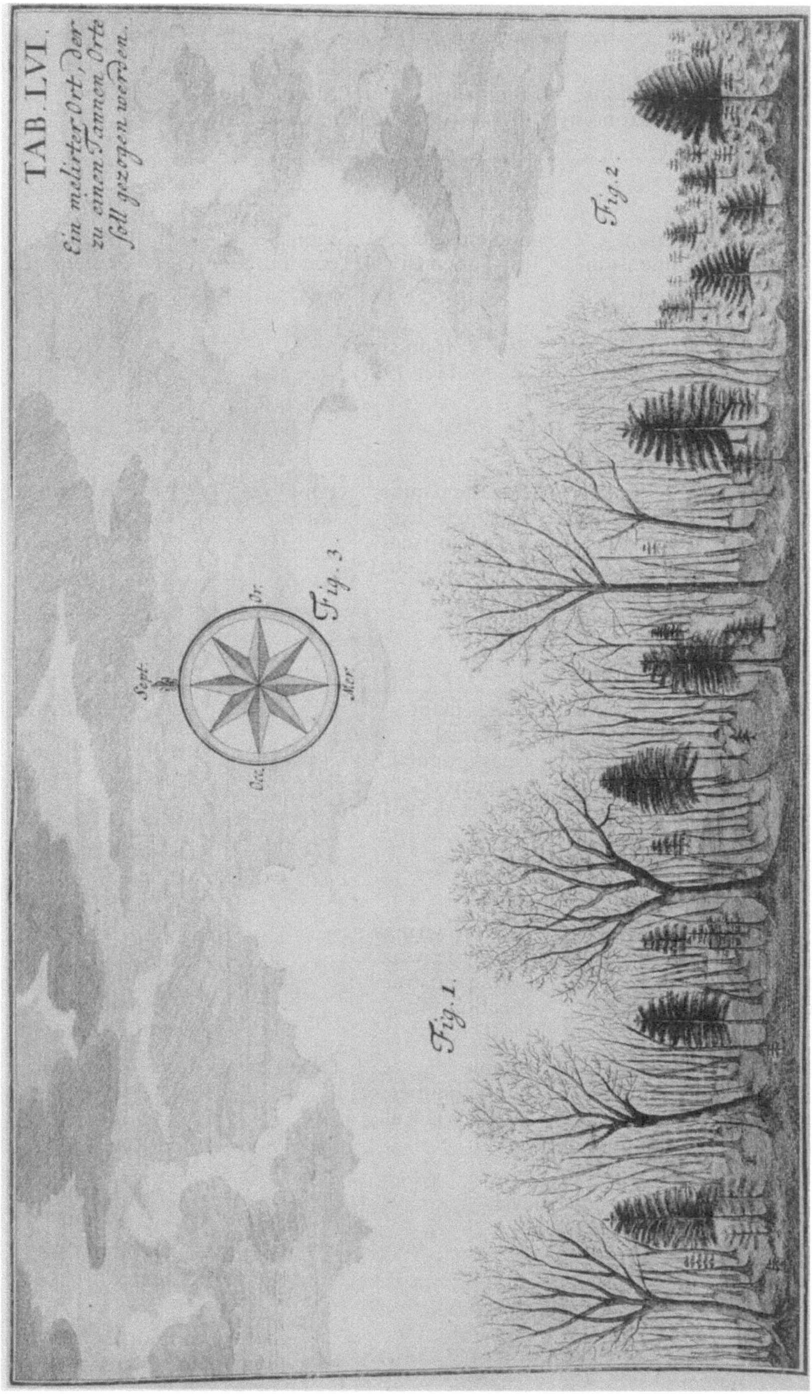

Fig. 2.3. One of the oldest forest transect drawings from Europe, showing the conversion of a broadleaved German forest into coniferous forest. (Cramers 1766 reproduction courtesy Central Library Agricultural University Wageningen)

Table 2.1.

A Architecture (Morphology, etc.)	B Populations (Syntaxonomy, etc.)	C Biomass (Production, etc.)	D After Odum, 1975, Fig. 1. modified		
Site mosaics (maps)		Potential production (resources)		Biosphere	
Silvatic mosaic (transect, map; cf. Koop 1981)	Coenon (Dimensionless; cf. Barkman 1970)	gap mosaic (cf. Shugart 1984)	Biotic) ⇋ community	energy	⇌ Ecosystem
ECO-UNIT (transect, map; cf. Oldeman 1983 b, c)	Microcoenon (dimensionless; cf. Barkman 1970)	Stand production (cf. Mohren 1987)			
Compartments = auxiliary levels					
(e.g. Structural sets, infrastructural sets, potential sets etc.)	(e.g. Populations, species, synusiae, guilds, etc.; cf. Freysen and Woldendorp 1978)	(e.g. stem mass, root mass, leaf mass, crop mass, weed mass, etc.)	Population ⇋		⇌ Population system
Limit between organisms / communities - - - - - - - - -					
ORGANISM (Component, occupying spatio-temporal ecotope; cf. Oldeman 1974 b)	ORGANISM (Linnean individual, i.e. population of organs; cf. floras and faunas)	Single plant or animal (mass, inputs, outputs; cf. Hari et al. 1981)	Organism ⇋		⇌ Organismic system
Compartments = auxiliary levels					
(e.g. Reiterated complex, model-conform complex, organ complex etc.)	(e.g. Inflorescences, phyllotaxis, life form, seasonality, etc.)	(e.g. crown mass, root system mass, sap stream translocation, etc.)			
ORGAN (Subsystem in organism; cf. Hallé et al. 1978)	ORGAN (Member of organism; cf. Handbooks botany, zoology)	Organ (Function-bearer; cf. De Wit et al. 1978)	Organ ⇌	matter	⇌ Organ system
lower levels - (tissue, cell, subcellular) - Omitted			Cell ⇌		⇌ Cellular system
			Gene ⇌		⇌ Genetic system
			Living components plus abiotic components equals biosystems		

This statement rests upon a number of biological necessities that will not be analysed in these pages. But generally it holds true, for instance:

Population = Members of that Population + Their Ecological Relations;
Tree = Collection of Leaf-Bearing Axes + Their Physiological Relations;
Silvatic Mosaic = Set of Eco-Units + Their Mutual Border Interactions.

The black box concept, as mentioned earlier, is a very useful scientific trick for avoiding too much work in reaching certain research aims, by using a particular style of abstraction. If the scientific problem is situated at one particular level of the hierarchy (see Table 2.1), information gathering can be limited to that level by treating its subsystems or the supersystems above as black boxes. This means that all details concerning the functioning of subsystems or supersystems may be left out: they do not have to be explained. For the explanation of one forest stand, for instance, one can treat the trees and other components as black boxes: their organs are of no interest and neither is their physiology. The only thing that we want to know is their input/output relationships within the stand. In the same way, the whole forest mosaic can be considered as a black box: the other stands are of no interest, and we consider the input/output relationship between the mosaic and our chosen stand alone.

It might be said that our model here is hemmed in with black boxes, so that no frays of unexplained aspects remain in the model. On the other hand, there are black boxes resulting from impotence: when scientists are powerless to explain the insides of a system, the model can only remain a black box. In that case it is at least useful to know that fact and work around it.

2.4.2 Components and Compartments

If, within a system, one distinguishes all individual subsystems that can be found in order to explain it, these subsystems are its components, e.g. the individual plants and animals in a forest stand. But for analytical purposes it may be necessary or easier to lump together several of these components in one compartment. Examples are the subterranean compartment of a forest stand including soil, root systems, soil flora and fauna, water, or the compartment of all endocrine glands in an animal body. Here again, there is a possibility for well-balanced abstraction, because the number of compartments in a system is smaller than the number of components.

Table 2.1. Hierarchies. Every system is composed by subsystems (below) **and** is part of a supersystem (above). From above to below, direct abiotic regulation weakens and transmitted regulation increases. Columns A to C are pure hierarchies according to specific criteria, i. e. form (A), numbers (B) and dosages (C). Capitals: systems limits are "natural" and are or can be made visible on a map. Note compartments as auxiliary levels, which can be defined and used with complete freedom by a researcher, who should only describe them well to be understood. Column D is an example of current combinations of the preceding columns. Criteria (form, numbers, dosages) are mixed in such hybrid hierarchies,here dosages (right-hand side), forms (lower left-hand side), numbers (upper left)

This is the case in very simple statements, e.g. "this forest stand is less vital because of insect feeding", in which the insects are the compartment referred to.

2.4.3 Disturbance and the Role of Hazard

A strong cyclone develops inside the system "world climate" and constitutes one of its subsystems. This cyclone has its impact upon the organisation within the system "forest stand" (eco-unit), because it influences the tree compartment. These kinds of interactions between far distant systems in the hierarchy can best be treated as chance factors, often "stress factors", within the lower system in the hierarchy. Most often, healthy living systems have repair mechanisms or "scar-forming" mechanisms in reaction to such stresses. Such interactions as storm, drought, fire, inundation which originate from high up in the hierarchy, or diseases and pests that come from lower levels, are very common in forest ecosystems.

2.4.4 Examples of Hierarchies

Table 2.1 presents the hierarchy of "natural" systems as used in this book, compared with other hierarchies from literature. Column A rests on structural criteria, column B on counted criteria and column C on criteria of mass and energy (cf. Sect. 2.2). Column D shows an often-used hierarchical approach, exemplified by a figure from E.P. Odum (1975). This author, and many others, unite the criteria of columns B and C at the community levels, but also use the criteria of column A implicitly at the levels of the organism and below, because their organisms and organs essentially Linnean and therefore morphologically determined.

Neither of the hierarchies shown in Table 2.1 is "wrong". It is emphasised once again that it is the privilege of each researcher to choose the limits and characteristics of the system or systems studied, and that this choice is closely linked to the scientific question asked and the convenience of the method used to provide an answer.

Table 2.1 presents the hierarchical systems chosen to structure the following chapters (columns A to C). This choice has been made for the following reasons.

2.4.4.1 Continuity of Criteria

In current hierarchies, such as Odum's, there is a discontinuity at the meeting place between organisms (below) and biological communities (above). From macromolecular biology to the level of the organism, the prime criterion is structural: build-up of organic molecules, of genes, of chromosomes, of cells, and so on. From the population level on, however, the main criterion is either numerical, or concerns biomass: population counts, frequency and rarity of species, root mass versus aboveground biomass and so further.

Our basic hierarchy (Table 2.1, column A) is defined by structure from one end to the other. This is useful primarily because in this way all systems concerned **have**

to be defined in space and time directly. This is a precondition for linking multidimensional systems and models (columns B to D) to the forest as we perceive it directly.

2.4.4.2 *Use of Numerous Levels*

One obvious difference, particularly between columns A and D, is the number of hierarchical levels used. This is accentuated by a high number of auxiliary levels, which are intensively used in the following chapters. The number of levels has profound methodological implications. On the one hand, numerous levels complicate the hierarchy, because data sets at more levels are needed in order to link together phenomena that are far apart, like photosynthesis at the organ level (leaves) and wood production at the stand level (eco-unit). But on the other hand, the hierarchy is simplified, relations between neighbouring levels being more simply established because they concern interactions between fewer subsystems, like the reiterated crownlets (not leaves or branches) in a tree (organism).

The number of interactions at a detailed level required to explain the behaviour of a system at a much more encompassing level is so huge that statistical processing of data is indispensable. This is exemplified by the use of the leaf area index (LAI, organ level) to directly explain ecosystem biomass production (stand level), as shown in column C. On the contrary, a "sandwich model" of an eco-unit explained in terms of architectural sets (compartments, auxiliary level) which are explained in their turn by their trees (organisms) can establish falsifiable explanations by gathering a few hundreds of data items at each subsystem level (column A). Such explanations are not statistical, i.e. they do not assess the probability of chance events, but they are structural, i.e. they establish the patterns of events not due to chance. Both methods would meet in a concept like "crown area index" (CAI; see Hallé et al. 1978), which refers to column C.

The aim of this book is not to provide the answer to one particular question concerning forest ecosystems, but to provide a silvological framework with the elements needed to prepare the answers to many different questions. Hence a dense hierarchy of closely placed system levels is more apt to serve our purposes than a wide one with few levels. Wide-spaced hierarchies can be mapped on dense ones, but the inverse is impossible.

2.4.4.3 *Architecture as the Prime Criterion*

As mentioned above, architectural criteria are defined in time and space and do not involve more dimensions directly. **Architecture is here used as spatio-temporal structure linked to a well-defined hierarchical level.** Structure is a passe-partout term, that has been used to denote many different phenomena including derived structures, like those of graphs representing processed numerical data in an arbitrary number of dimensions. Population numbers and biomass are dimensionless, but data according to these criteria are processed in models that do have structure. In order to keep the present text clear, it will be supposed that such models have structure, but no architecture, even if time and one or more spatial dimensions are used. In practice, this means that architecture excludes all data in time and space that are derived from models, not directly from living systems. However, the time dimension in architectural models, as presented in the following pages, too often remains hypothetical, and so it always will as far as measurements concern systems that live longer than a scientist. In architectural models the three spatial dimensions therefore are the most solid ones.

Architectural criteria, even if time is uncertain, allow delimiting living systems in space. This has not been done systematically and analytically before at the levels concerning biological communities. The excellent earlier maps and profile diagrams that were published are good physiognomical models, but could be used imperfectly in systems analysis. This has become possible by linking architectural properties of forest organisms, trees particularly, to forest ecosystem architecture, so that now column A can be used as a basic reference.

Moreover, architectural models are represented graphically. This makes them easy to validate in the field. The great lines can be seen by the trained eye, and a few selected measurements go a long way in establishing or denying the truth of such models. For the same reason they also provide an easy way of communication with practical users, such as silviculturists or forest officers, a communication that depends on the blending and reciprocal translation of empirical and theoretical knowledge. The abstract lines on a scale drawing, and the images caught and retained by the experienced eye are complementary and allow fruitful discussion.

Finally, because of the dimensional properties of the architectural models mentioned above, it is easier to hook up data from population or production models (columns B and C) onto architectural models (column A) than to do the inverse. In the next chapters this will be demonstrated by several examples, in which it always appears that "forest structure" as a derived property amounts to some artificial geometrical pattern superposed upon the forest with more or less success. The century-long discussion on forest strata, as derived from tree height frequency curves provides a beautiful case (cf. Oldeman 1974a).

2.4.4.4 Intentional Omission of Levels

The choice of levels as represented in columns A, B and C (Table 2.1) will rarely render necessary the addition of an ad-hoc level between two others, in order to solve some special problem. This would be a certain way for continuing the general tendency already too strong in ecological research, which consists in every researcher constructing a special hierarchy for every scientific question asked. Fortunately, this point is not reached as yet, but a recent analysis of a research programme (Oldeman in press) clearly confirmed that methodological heterogeneity in this sense is quite strong.

On the contrary, such levels in Table 2.1 as are empty of real systems can and should be omitted from systems analysis. One example is a site which is composed of so many very different and very small parts that each can host one forest eco-unit at best. In this case, the silvatic mosaic level is non-existent (cf. Chaps. 5 and 6). Another instance is the architecture of herbs, in which most often the auxiliary level of reiterated complexes can be left out, because herbs seldom reiterate (Chap. 4).

2.4.4.5 Precision of Limits

One disadvantage of column A in Table 2.1 is that systems' limits cannot be traced with enough precision in all cases (cf. Chap. 5.1.2). The borders between tissues are sometimes unclear and the limits between different ecosystems, at the eco-unit or the silvatic mosaic levels, are not clearcut planes but transition zones with a width that can be considerable.

The system levels at which such problems are least difficult have been printed in capitals in Table 2.1, whereas system levels at which they are liable to occur are in small print. These latter include all auxiliary levels, because they most often are either used or left out according to the clarity of the limits of the systems concerned.

2.4.4.6 Flexibility and Completeness

Table 2.1 has been carefully set up as a frame of reference for silvology. With this set of hierarchies (columns A,B,C) all current research lines can be covered and linked together, because all levels used are present. This warrants a systematic coverage of the field of silvology, as complete as it could be made for the moment.

On the other hand, this table can be used to pinpoint the most convenient method to tackle specific issues in research. For this purpose, it constitutes a flexible tool, because it allows an almost complete freedom of choice as to the levels needed in order to solve current questions with the least possible amount of work, with the degree of precision required by the question, and with the potential to link the results to those obtained in other research lines.

In the next chapters, the levels of forest components (organisms), eco-units and silvatic mosaics will be outlined according to these principles.

Second Part: Forest Components

In Chapters 3 and 4, forest components are treated as systems, to be explained by their organs, considered as black boxes. The forest is also treated as a black box, from which only the inputs into the components are important, and which is a sink for their outputs. Successively, trees, other plant life forms such as lianes, herbs and epiphytes, animals, and microflora and -fauna are treated. As far as possible, backgrounds are given in architecture, ecophysiology and reproductive biology. Forest components are not considered in a taxonomical way, and species or species lists are only used as examples.

3 Trees

3.1 What Is a Tree?

Trees are the components that define any forest ecosystem, although a forest is much more than a collection of trees. Therefore, trees deserve a chapter on their own.

It is much more difficult to define a tree than a forest. If no definition of forest can be conceived without implicitly or explicitly mentioning trees, the same is not true for the definition of a tree. There will always be exceptions when one tries to pinpoint characters common to all trees. For instance:

- **wood**: many trees have wood, but an anatomist will frown and doubt whether palms and tree ferns have wood in the true sense;
- **trunk**: many trees have one trunk, so much so that in Europe plants with more than one "trunk" are considered shrubs. But what is to be done with coppice? With many-stemmed palms? With the mathematical definition of a tree, stating that a tree is any set of points, between each pair of which there is only one pathway?
- **size**: quite a few definitions of trees use size as an element, for instance the height or stem diameter of a woody plant. But in the tropical rainforest, for instance, one may find 25-cm-high mini-trees, which in all senses but size do correspond to what is felt to be a tree, with one trunk and forming wood.

Like Hallé et al. (1978, pp. 1,2), who gave a number of tree definitions, we will use the word tree here in a large and rather vague sense, circumscribed as follows, although a circumscription is not a definition:

- we consider to be trees: all plant organisms that live their life cycle according to a strategy including maximisation of lifespan, investing matter and energy in the formation of self-supporting transport tissues in amounts that are important in relation to their total biomass, and the overtopping of surrounding vegetation or vegetation layers by an adapted growth programme.

In this way, the tree is conceived as a "permanent" forest component (see Third Part).

3.2 The Life Cycle of Trees: Architecture

Figure 3.1 shows the life cycles of different categories of trees. Here again, it is not possible to describe the cycles in total independence of the ecosystem in which they have evolved. Hence the names mentioned: "pioneer trees", "late pioneer trees" and "successor trees" (also see Sects. 5.2.4 and 6.2.4.2).

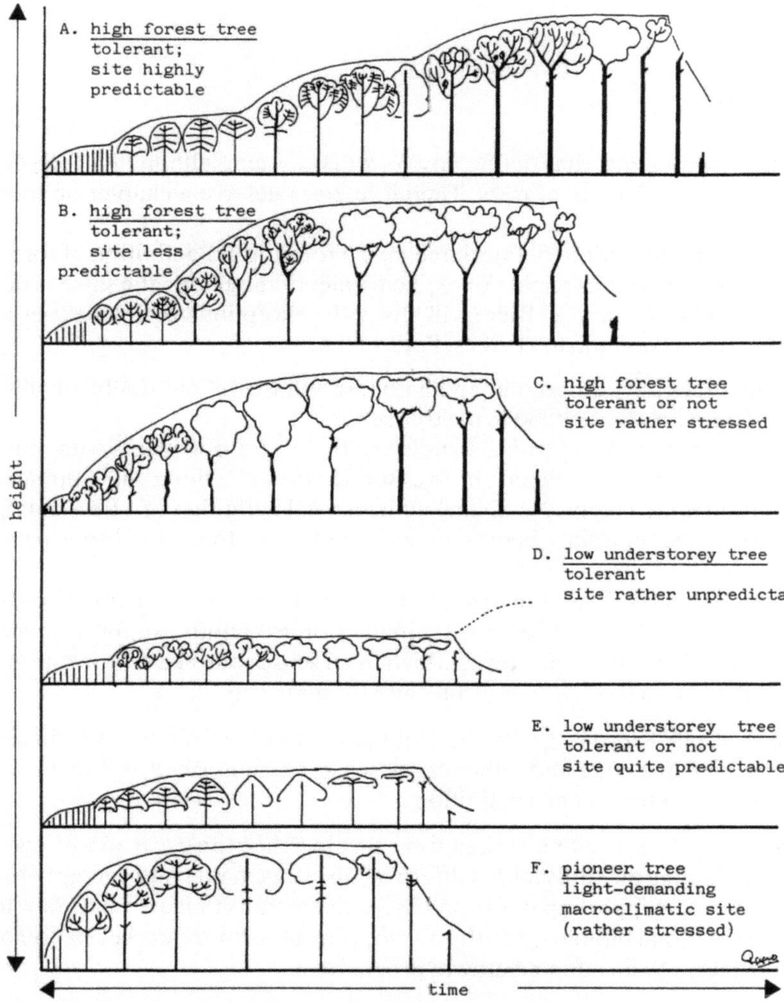

Fig. 3.1. Life cycle of different kinds of trees. Note that certain steps in growth may be emphasised or lacking, depending upon the environment (site). **A** High forest tree (successor), note long seedling phase and metamorphosis. **B** High forest tree (late pioneer), note absence of metamorphosis. **C** High forest tree (long-lived pioneer), note absence of both model-conforming phase and metamorphosis. **D** Low understorey tree in highly dynamic forest, note absence of model-conforming phase. **E** Low understorey tree in stable forest, or gaps, note long survival of model. **F** Pioneer tree, note rapid juvenile growth and predominance of the model. For examples see Kahn (1983), Suselo (1983). A more refined classification is described by Oldeman and Van Dijk (in Hadley 1988)

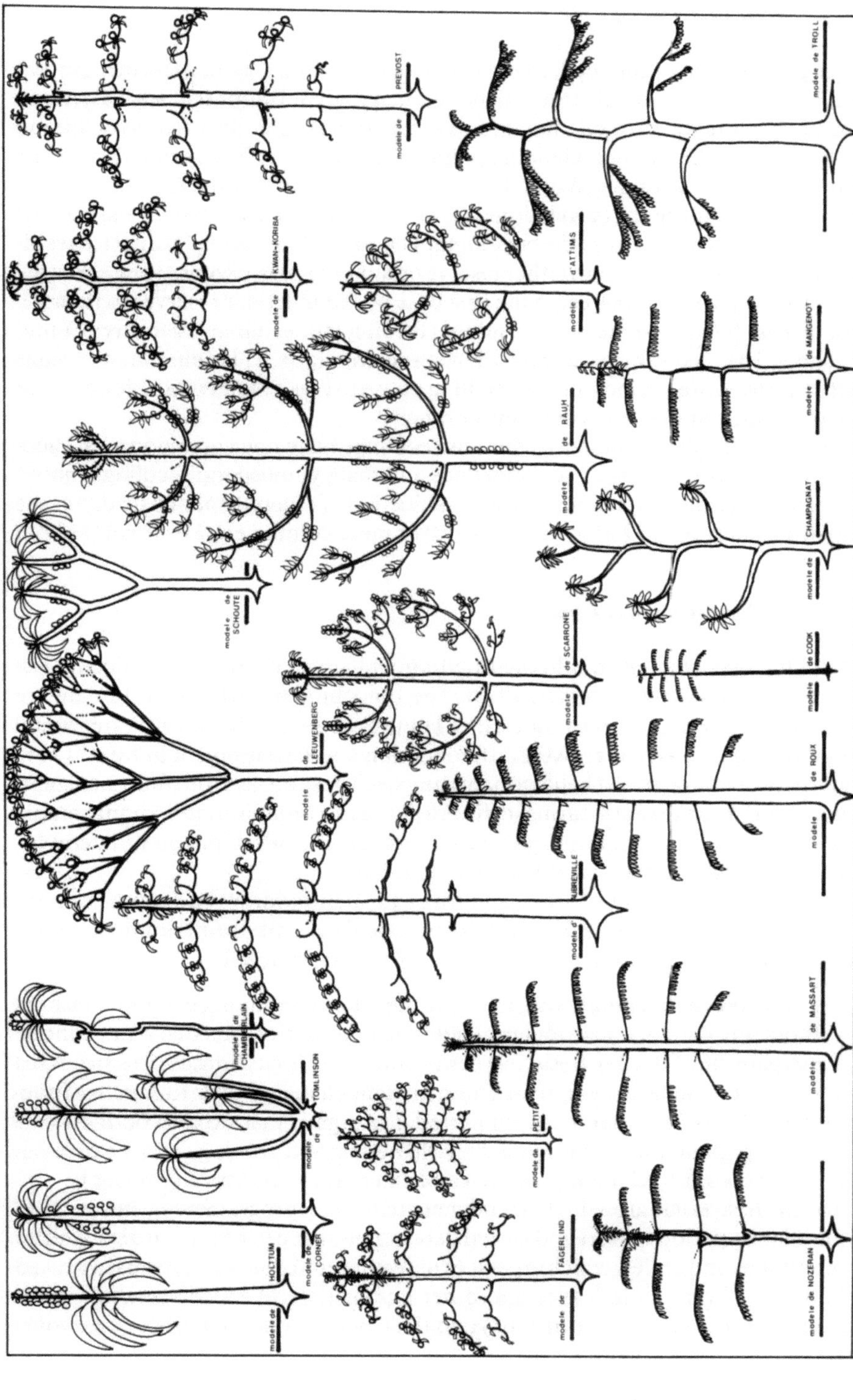

Fig. 3.2. Inherited tree models, according to Hallé and Oldeman (1970). Hierarchical level (see Table 2.1): branched complexes, except Holttum's and Corner's models. (Oldeman 1974a)

3.2.1 The Seedling Period

After germination of the seed, trees consist of one subterranean axis (the taproot) and one aboveground axis (the epicotyledonary axis). Both axes bear lateral organs. Under the soil surface, lateral roots most often are spirally inserted on the tap root, unless (as in monocotyledons: palms, bamboos etc.) the taproot disappears almost at once to be replaced by side roots. Above the ground, the epicotyledonary axis bears green organs (cotyledons, leaves, needles), most often in a spiral but sometimes in two rows from the beginning. The cotyledons, being the first leaves already present in the seed, are either reserve organs, or green organs. They allow the first growth phase of the tree, in the first case by feeding their reserves to it, in the second case by immediate photosynthesis feeding the seedling with energy bound in sugars. The seedling phase ends when branching starts. This delimitation means that some trees are prolonging the seedling phase to cover their whole life cycle, for instance those palm species that never branch.

It is recalled that monocotyledonous seedlings have one cotyledon, that dicotyledonous seedlings have two cotyledons and that gymnosperm seedlings show a whorl of numerous cotyledons. For a classification of dicotyledonous seedlings see Ng (in Tomlinson and Zimmermann 1978) or more completely, De Vogel (1979).

3.2.2 The Model-Conform Period

Branching occurs after an inherited pattern which has been given in detail elsewhere (e.g. Hallé and Oldeman 1970). The branching models found to date are summarised in Fig. 3.2. Tree models (see Hallé et al. 1978; Edelin 1977) are strictly programmed patterns of growth, differentiation and development in trees. These patterns are defined by the behaviour of the axes. These are leaf-bearing, meristem-bearing and often flower-bearing, cylindrical organs, built by their terminal meristem. This meristem both builds the axis and steers the development of lateral organs and meristems. It therefore has a double function.

The self-construction of trees according to such models is adaptive, but not strictly adaptive in the sense that there is one model per type of environment. Adaptive characters of the differentiated axes are, for instance (see Fig. 3.3):

- axes with **indeterminate** growth, their end meristem continuing to build such axes without any endogenously determined time limit, are organs that can continue the expansion of the tree body in a certain direction for as long as possible. Axes with a **determinate** growth, their end meristem changing over to another function, e.g. flowering after a certain period, are organs that expand both the tree and the population of the species by flowering and fruiting, in a flexible sequence reacting to changing internal and external conditions (Fig. 3.3a,b).
- axes with **rhythmic** growth, their end meristem being programmed genetically to alternating periods of rest and growth, are organs that can be recognized by short internodes and scales, which are traces of resting, and that allow the harmonised building of very large leaves and other organs (lateral inflorescences, for instance: cf. Hallé and Martin 1968) and the reaction to favourable or unfavour-

able seasons. Axes with **continuous** growth, their end meristems being programmed genetically to function roughly in the same way as long as they live, are organs better suited to climates without unfavourable seasons and to continuous expansion of the tree body (Fig. 3.3b).

- axes with **rhythmic branching,** their end meristem steering the behaviour of lateral meristems in such a way that periods of branch formation alternate with periods without branch formation, are organs that can precisely time their reaction to conditions favourable to height growth and to spreading. Axes with **continuous or diffuse branching,** their end meristem steering the behaviour of lateral meristems so that they always or irregularly start forming branches, are organs that are better adapted to more constant conditions inside and outside the tree (Fig. 3.3b).
- **orthotropic** axes, being vertical, most often with leaves in a spiral, are exploring organs, ensuring height growth, penetration of higher "empty" vegetation layers and overtopping. **Plagiotropic** axes, being horizontal, with leaves arranged in a dorsiventral pattern, are exploitation axes that are extended in spaces where they may intercept light, and with which a tree may outshadow other plants more effectively (Fig 3.3c).
- **sylleptic** axes, built by a meristem without a preceding resting period, are quick-profit organs, allowing the vegetative tree body to expand in one continuing sweep when the internal and external conditions are right, particularly in non-seasonal climates. **Proleptic** axes, built by a meristem after a resting period, often dormant inside a bud which is a meristem with its surrounding scales, are organs lying in wait for signals from inside the tree or outside that the conditions are right to resume the expansion of the tree body, particularly in climates with an unfavourable season, e.g. winter or dry season (Fig. 3.3d).

The model-conform period may be very short or may extend over the whole life cycle of the tree species considered. It is very short when environmental conditions are so quick-changing and unreliable that the programme of the model cannot cope with them. A good example is the genus *Quercus* (Fagaceae). In the Javanese mountains (Indonesia), with a constant climate, and in the North-Eastern United States, with a regular climate, oaks conform to Rauh's model (Fig. 3.2) for quite some years, until they reach a height of several meters. But in Western Europe and in the Mediterranean (Fig. 3.4), i.e. in regions with variable winters and drought periods, oaks show only a very short model-conform period if at all, before so many meristems die that the programme cannot be followed any more in a strict way. In other words, the importance and role of the model during the life cycle of the tree are in themselves adaptive characters of a species.

The tree models described by Hallé and Oldeman (1970, Engl. transl. 1975) and Hallé et al. (1978) are abstractions (see Chap. 2). In fact, each species builds its own specific variant of such a tree model, and this variant can be represented by a document concerning that species alone, which was called architectural diagram by Edelin (1977). In an architectural diagram, less data are abstracted from the whole organic pattern observed than in the architectural model of the same tree.

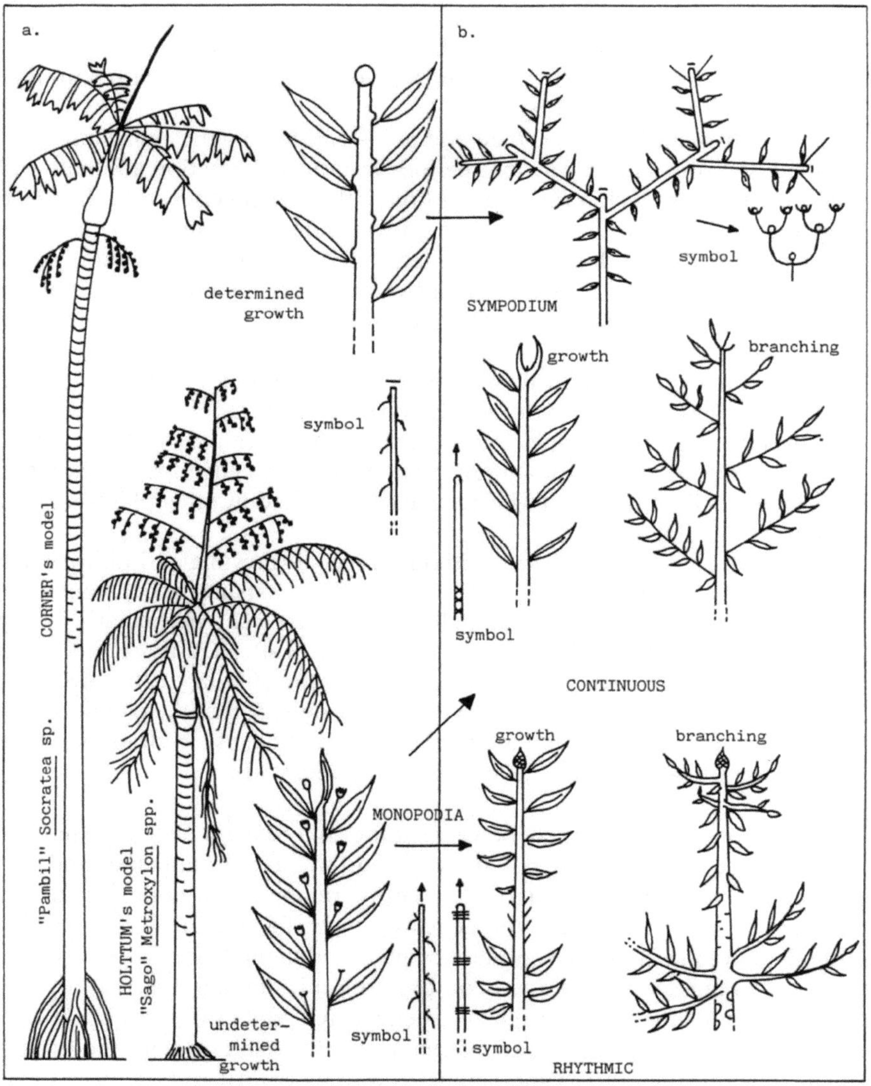

Fig. 3.3. Organ complexes and branching in trees. **a** Axes with determined versus undetermined growth build Holttum's and Corner's models (Fig 3.2) and **b** lead to sympodia or monopodia, the last ones growing and branching either continuously or rhythmically. **c** Orthotropic axes, leaves spiraled, vertical, versus plagiotropic axes, horizontal with either the leaves formed in two rows by the meristem (primary) or arranged horizontally by torsions (secondarily). **d** Meristems may function directly, without resting (syllepsis), or after resting (prolepsis); in the latter case the axis often shows scale leaves at the base. **e** Mixed axes change their orientation, the base having another function (trunk-like) than the top (branch-like) and so allow a very flexible tree architecture. Symbols: used in simplified drawings, e.g. at another hierarchical level (Table 2.1)

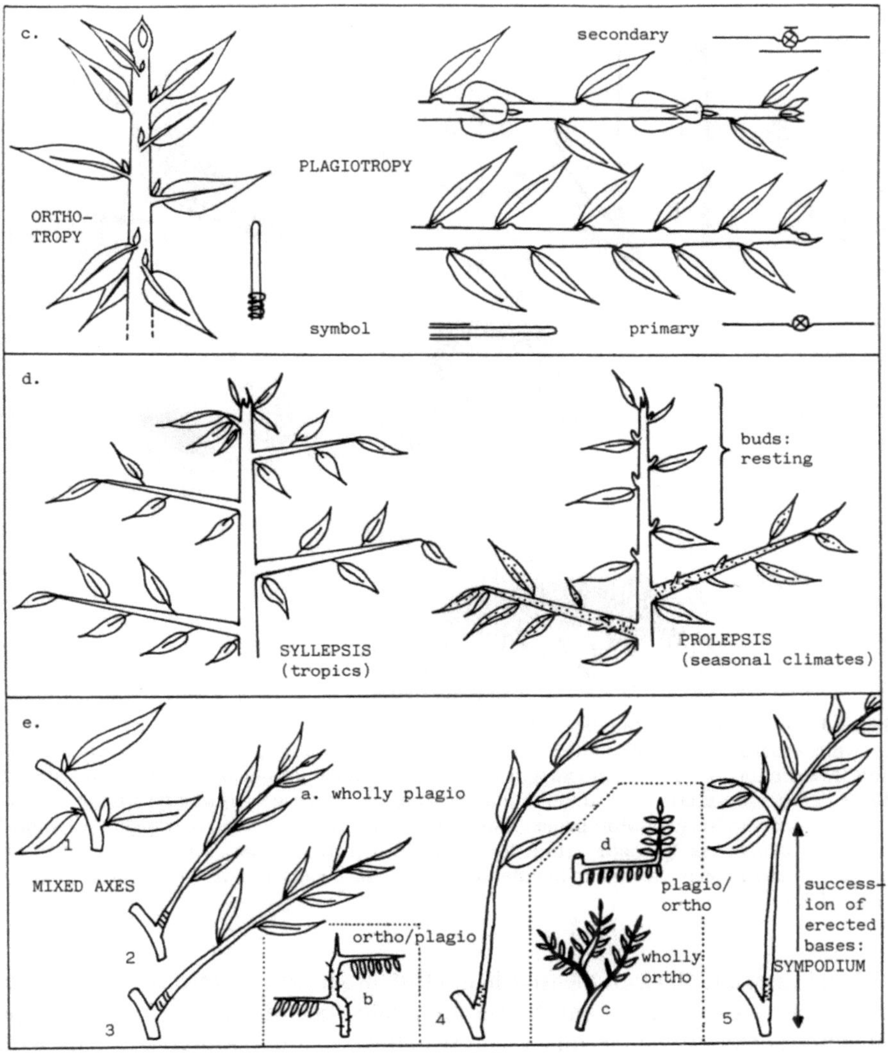

To explain this, the case of *Cupressus sempervirens* (Cupressaceae) is examined. This tree grows according to Attims' model defined by "axes with continuous growth, differentiated into a monopodial trunk and equivalent branches; branching takes place either continuously or diffusely. Flowering is always lateral and does not affect shoot construction." (Hallé et al. 1978 p. 228; Fig. 3.2). The architectural diagram showing the version of Attims' model that is proper to *C. sempervirens*, according to Edelin (1977, p. 199), first states that there are only five observed branching orders, the trunk being the first-order axis. Only the first two branching orders, determining the structural skeleton of the tree, are built by axes that correspond exactly to those defined for the model. The axes of the third order are a tran-

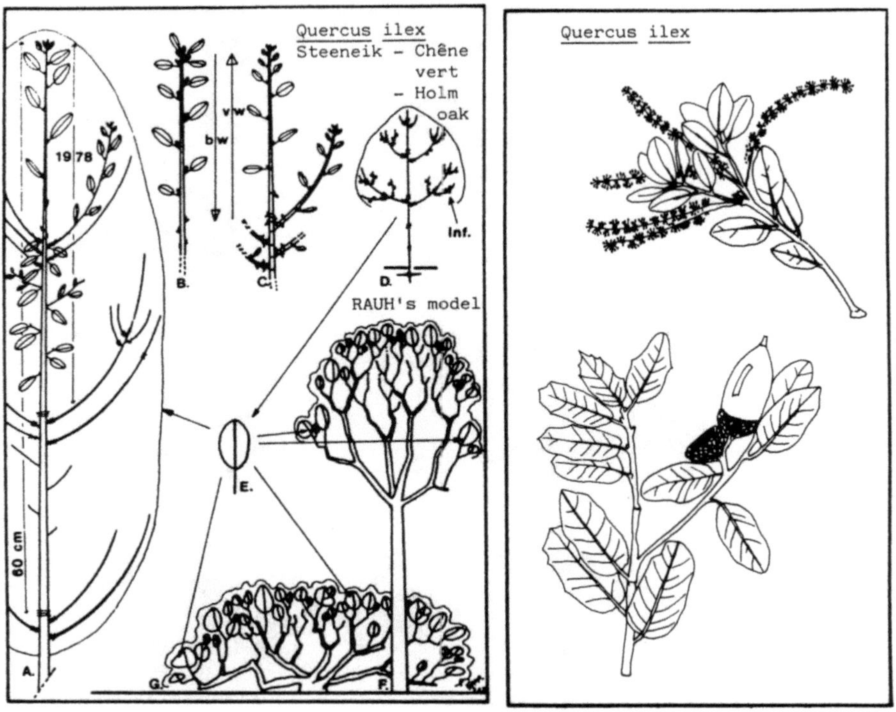

Fig. 3.4. *Quercus ilex*, an example of an oak (holm oak, Mediterranean). To the *left: A* Reiterated crownlet according to Rauh's model, two shoots in 1978; *B* terminal shoot, *BW* gradient of probable flowering; *C* diagram of branching, *VW* gradient of probable branching; *D* Rauh's model (cf. Fig. 3.2), *Inf.* inflorescence; *E* symbol of model, trunk + crown are model-conforming; *F* holm oak in high forest North of Montpellier, France: note numerous reiterated crownlets in expanded crown; *G* holm oak in "garrigue" (shrub forest, matorral) in Tunisia: note reduced primary trunk, abundant reiteration in crown and low stature. To the *right* details of lateral, subapical flowering. (Buis 1979)

sitional case, and those of the fourth and fifth orders are completely different because they are deciduous and flower terminally, bearing female flowers on the fourth-order axes and male ones on those of the fifth order. Moreover, on the axes of the first two orders, the leaves are scaly, whereas they are acicular on the axes of the last three orders.

If the above definition of the model states that flowering is lateral, it clearly concerns the skeleton shoots only, and both the details of higher-order branches and those of leaf morphology are needed to refine this notion of "lateral flowering" by way of short shoots being placed laterally on the skeleton shoots, which make an architectural cypress diagram out of the abstract Attims' model.

In architectural diagrams, one can also find, for exàmple, the difference in rhythmic branching between many trees from the humid tropics and from the temperate regions. In the rainforest, as in some gymnosperms (e.g. the genus *Abies*, Abietaceae), rhythmic branching leads to very compacted branch stories which

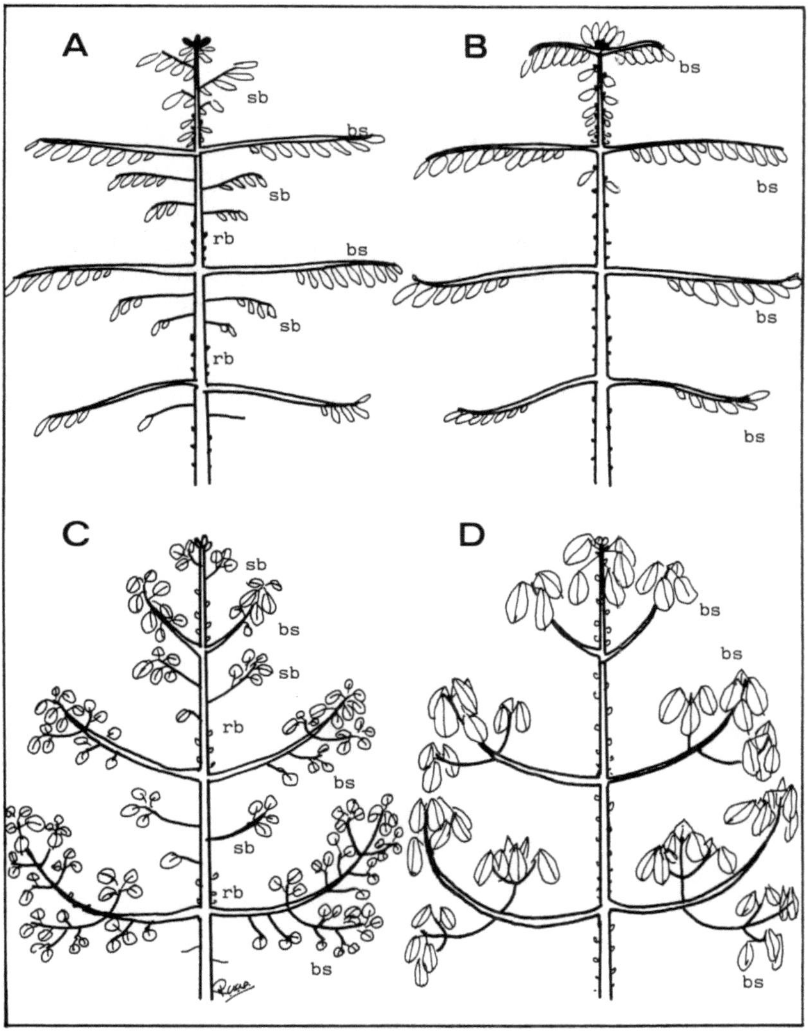

Fig. 3.5. Different tendencies in rhythmic branching in the temperate regions (**A, C**) and the humid tropics (**B, D**). **A** Massart's model, temperate version as in *Abies alba*, with branch storey (*bs*), small branches between the storeys (*sb*) and resting buds (*rb*). **B** Massart's model, humid tropics as in *Diospyros* spp. (Ebenaceae, ebony wood), no small branches. **C** Rauh's or Scarrone's model in the temperate regions, as in *Juglans regia* (Juglandaceae, walnut) or *Aesculus hippocastanum* (Sapindaceae, horse chestnut): branch storeys (*bs*), small branches (*sb*) and resting buds (*rb*). **D** Rauh's model in the humid tropics, without small branches, e.g. in *Cecropia* spp. (Moraceae; Bois Canon, Bousipapaya, Guarumo, Yarumo). Note that this is a rule-of-thumb only. Small branches between storeys are more frequent in temperate regions

may be considered as whorls or pseudo-verticils. On the contrary, trees such as poplars (*Populus* spp., Salicaceae), show branch tiers that start with small branches, the following ones increasing in size until a set of large structural branches ends the tier, followed by a branchless portion of the trunk (Fig.3.5).

The duration of the model-conform period and the character of the axes building the model are not the only adaptive features of a tree in this developmental phase. Figure 3.6 shows some examples of trees that build an architecture being intermediate between architectural models. This also occurs more frequently in environments with unfavourable seasons, weather or life spaces than in constant environments. The climate in the lower reaches of the tropical rainforests is remarkably constant, and in this environment the architectural tree models can be found to be most numerous and most purely expressed. This is the environment where Hallé and Oldeman (1970) studied tree architecture in the first place. Intermediate architectures were found there in some cases, but appeared to be much more frequent in dry and temperate regions.

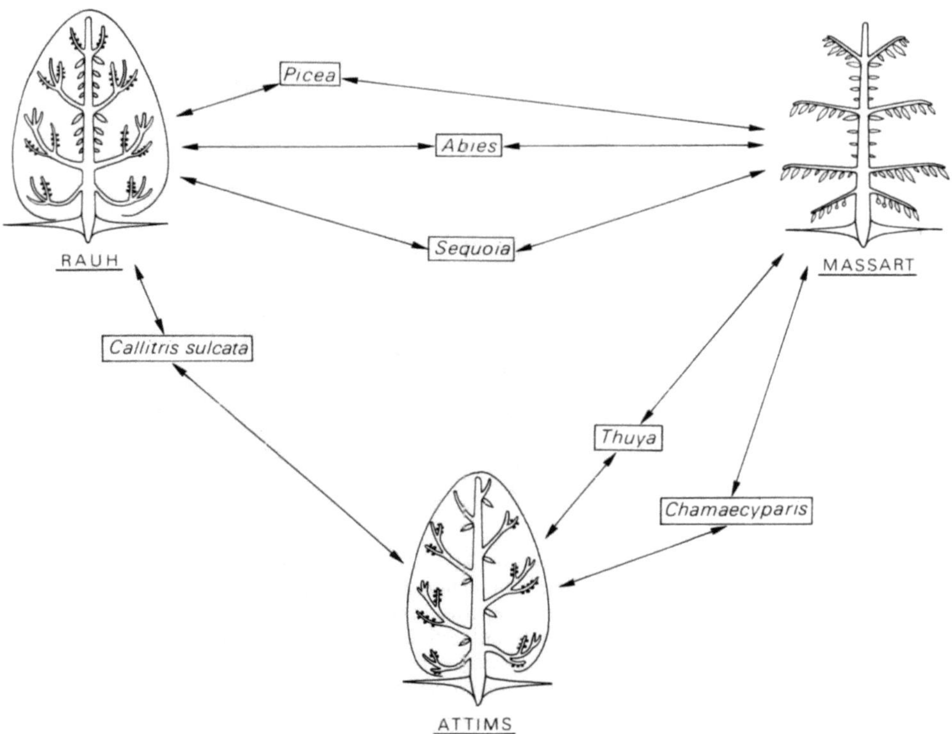

Fig. 3.6. Example of tree architecture intermediate between models (cf. Fig 3.2) in conifers. *Picea, Abies* and *Sequoia* are genera that may show plagiotropy in their branches to a varying degree, also within the same species according to provenance (cf. Chap. 6.3.8). *Callitris sulcata* shows a transition between rhythmic and continuous branching. *Thuya* and *Chamaecyparis* may be transitional both in branch orientation (ortho/plagio) and rhythm of growth and branching (rhythmic/continuous). The architectural models decidedly allow us to understand such growth patterns better. (Edelin 1977)

An example is a "weed species" introduced from the USA in Europe, and observed by us in the Netherlands, *Prunus serotina* (Rosaceae, "American black cherry"). In the open, this species conforms to a variant of Massart's model, but when growing in the lower parts of a forest it may lose much of its branching rhythm, while branches tend to shift from plagiotropy to orthotropic behaviour. The architecture of *P. serotina* hence is intermediate between Massart's, Roux's and Attims' models. The shift towards orthotropy may be linked to architectural metamorphosis (see Sect. 3.2.3).

Knowledge of the architectural model of a tree species is useful in order to understand the whole development of trees. Without it, it is impossible to clearly analyse the following periods of its life cycle because the inherited potential of each meristem or bud according to the model-conform differentiation programme always determines the possibilities of a tree. This is also important for vegetative multiplication: cuttings with plagiotropic meristems, for instance, yield rooted branches and not trees (Fig. 3.21A).

It may be noted that, according to Table 2.1, the seedling is situated at the auxiliary level of "organ complexes", whereas the architectural model is a "branched complex".

3.2.3 Metamorphosis and Programmed Adaptation

The metamorphic period has been discovered and described by Edelin (1984). In order to understand it clearly, one has to understand that the outer differentiated axes on the branches are not growing according to the latest step in the differentiation programme. This is shown in Fig. 3.7A,B. The first step in this programming in reality is, for instance, branching with plagiotropic branches (Fig. 3.7Bb). The second step then is the appearance of orthotropic axes in the same spiral on which the plagiotropic ones appeared first (Fig. 3.7Bc). The third step is manifest by the appearance of plagiotropic branches on the orthotropic axes (Fig. 3.7Bc; Fig. 3.8). Such an orthotropic branch hence is completely comparable to the first branched step (Fig. 3.7Bb) after the unbranched seedling (Fig. 3.7Ba). The same pattern can emerge later: if three or four kinds of differentiated axes play a role, the following differentiation has its place within the volume that is occupied by axes showing the first branch specialisation.

The result is that the outside of the tree crown is always covered with an envelope consisting of the same kind of axes, being specialised in the interception of radiation (photosynthesis) and often also in the bearing of flowers and fruits, a multiplication and dispersal function. This envelope of special branches (Fig. 3.7) may well be considered as the skin of the tree crown, a skin that remains constant in properties from the moment that the tree starts branching. In Section 2.3.1 (see also Table 2.1), the limits or skins of living systems were discussed at some length. Here, such a "natural" boundary can be precisely defined.

Metamorphosis therefore is a process that entails the inner reprogramming of an expanding tree crown, while the outer layer obtains its final properties in an early period of tree development. This is not always the case with flowering and fruiting, which may occur earlier or later according to the species, and which also

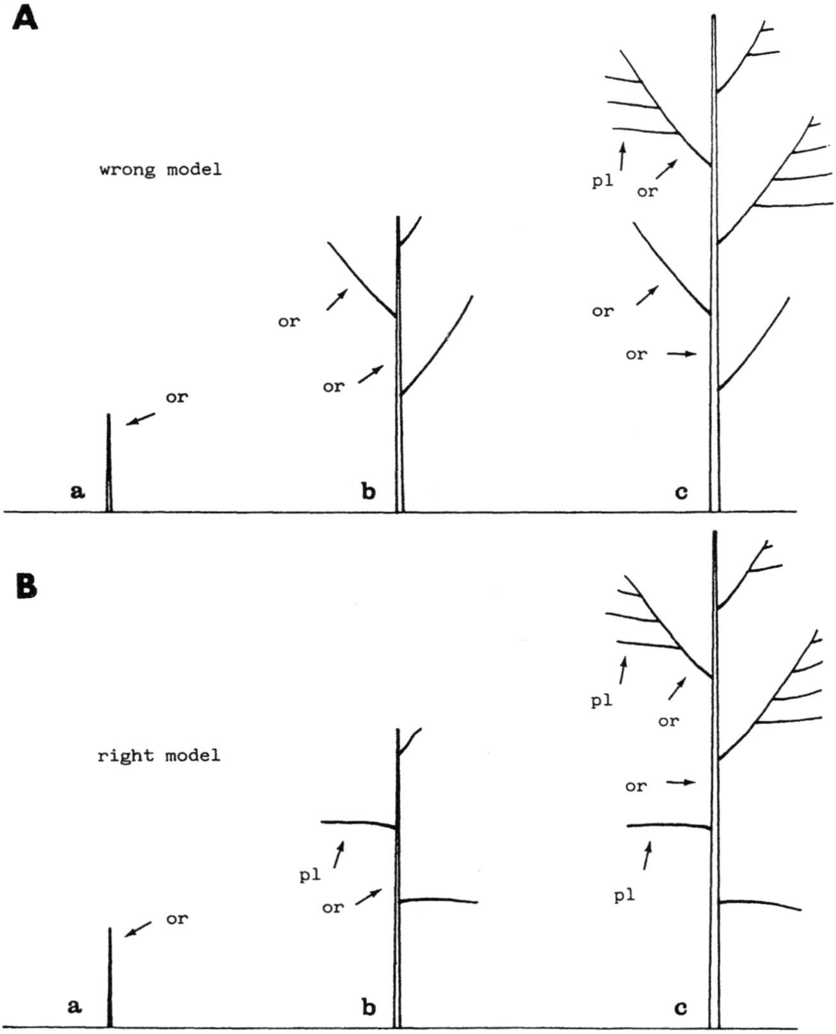

Fig. 3.7. Differentiation sequence in a branched tree model: wrong model (**A**) and right model (**B**). The "ultimate" differentiation of twigs, here plagiotropic (*pl*), is the ultimate *at each step* in growth. Less differentiated axes, here orthotropic (*or*) are *not* present in first branching (as in **Ab** wrong), but are interposed later (as in **Bb** right). By reasoning out the rest of the branching sequences, one can indeed see that the wrong model (**A**) has no future potential for crown expansion, because all branches of the order (3 + n) would be plagiotropic. (Edelin 1984)

may be localised elsewhere in the tree body. Cauliflory, the flowering on the old wood in trees such as cocoa (*Theobroma cacao*, Sterculiaceae, Nozeran's model) is a classical example of the last case (also see Figs. 6.22 and 6.23).

It is to be noted, therefore, that sexual maturity in trees is not a criterion for adulthood as is often seen in animals. Maturity of the boundary layer, maturity of the trunk and roots, maturity of the whole crown also inside, and sexual maturity are stages of adulthood that are not synchronised and there is no valid reason to call a tree mature only because it flowers and fruits. This would fit in with theories on plant evolution that consider an evolutionary stage of separated vegetative and sexual individuals in one species, still found in such groups as mosses and ferns, i.e. the occurrence of gametophytes and sporophytes. Trees may be considered in this context as the symbiosis of a sporophyte (the vegetative body) and gametophytes (the inflorescences), both still conserving traces of different cycles of maturation.

Metamorphosis is a programmed period of tree development, in which the repetition of the patterns displayed in early branching can be found back in each of the subcrowns that are building the whole crown of the tree. Not all trees show metamorphosis, although it seems to be frequent enough when observing large numbers of trees of different species, as has been done since Edelin published his findings in 1984. Figure 3.8, the analysis of the development of *Shorea leprosula* ac-

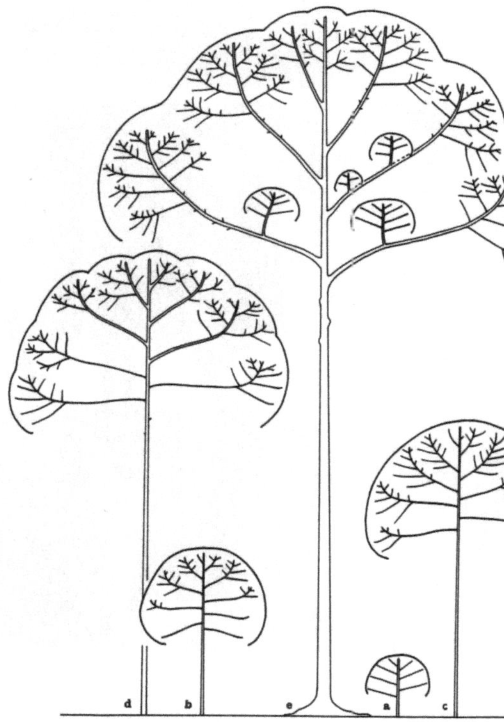

Fig. 3.8. Example of a tree species with metamorphosis, *Shorea leprosula* (Dipterocarpaceae, Indonesia, "meranti"). *a* Model-conforming phase, Roux's model (see Fig. 3.2). *b* At the top, first formation of orthotropic branches (see Fig. 3.7B), bearing plagiotropic branches: Roux's model again, but small. *c* No plagiotropic branches left on the trunk, all are in small Roux's models at the end of the orthotropic branches (first step in metamorphosis) *d* Process repeated on second-order orthotropic brances (second step in metamorphosis) *e* Adult tree with third step in metamorphosis, and reiteration (unprogrammed, opportunistic), according to Roux's model, on the old wood. (Edelin 1984)

Fig. 3.9. *Betula papyrifera* (Betulaceae, transition Roux's/Attims' models, paper birch; Orange, Massachusetts, USA, July 1985). General architecture (coppice) and scars of typically orthotropic axes that have played a role in metamorphosis: note narrow angle with trunk. (Photographs by R.A.A. Oldeman)

cording to this author, shows a concrete example of the general process represented in Fig. 3.1c. It is extremely frequent in this family, the Dipterocarpaceae, which contains the dominant trees from the Western part of the tropical Asian rainforests (Whitmore 1975). In Malaysia and parts of Indonesia, dipterocarps may make up 20 to 60% of all the trees per hectare.

Metamorphosis also occurs in trees of the temperate region. The example of *Prunus serotina*, an introduced species that may be observed all over North-Western Europe, has been mentioned above. Birches also show good examples of metamorphosis: Fig. 3.9 represents a paper birch *(Betula papyrifera*, Betulaceae) in Orange, Mass., USA, with clear traces of metamorphosis.

3.2.4 Reiteration and Opportunistic Adjustment

Once the tree encounters environmental circumstances which are either rapidly changing or very different from the conditions in the seedling or model-conforming period or after metamorphosis, meristems outside the programmed sequence become active. It may be that outside influences have killed part of the tree or only some of its meristems: top-boring insects, e.g. *Hipsypila* sp. in mahogany trees (Meliaceae) or *Phoracantha* spp. in *Eucalyptus* (Myrtaceae, Fig. 3.10) in the tropics, or frost in ash trees *(Fraxinus excelsior* - Oleaceae, Fig. 3.11) in Western Europe, or falling dead wood from the canopy in closed forests all over the world. It may also

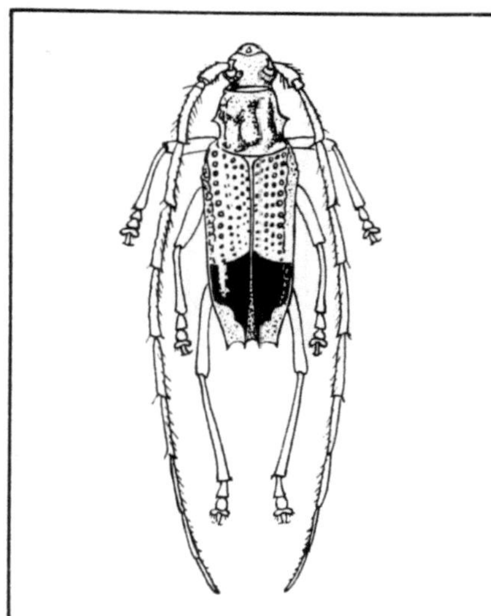

Fig. 3.10. *Phoracantha* sp., example of a top-borer, causing regenerative reiteration in trees. *Phoracantha* attacks end shoots of meristems in all *Eucalyptus* species (Myrtaceae, Australia; species that we know have Roux's model), and activates resting meristems that start regeneration of lost parts. This insect, after immigration from Australia to the Mediterranean region, has become a serious pest. (Buis 1979)

Fig. 3.11. *Fraxinus excelsior* (Oleaceae, Rauh's model, ash; near Lelystad, Flevoland-polder, Netherlands, April 1979), example of reiteration after damage.
A Model-conforming crown in undamaged trees. **B** Two reiterated trunklets from the upper bud pair under the end meristem, the latter frozen to death in the winter of 1977/1978. (Photographs by R.A.A. Oldeman)

Fig. 3.12. *Sambucus nigra* (Caprifoliaceae, Champagnat's model, elder; near Ede, Netherlands, Spring 1978). Note that *S. canadensis* grows in the same way in the Guyanas. Reiterated trunks can be recognised by their juvenile bark, as in many other trees. (Photograph by R. A. A. Oldeman)

be that a sapling of a large tree species grows into an illuminated opening, or encounters soil enrichment by falling organic matter or fertilisation. In these latter cases, the existing active meristems cease to have enough production capacity to absorb the available resources.

The answer to such unpredictable events is the activation of resting meristems. This has been called opportunistic behaviour of trees by Hallé et al. (1978). Such meristems are behaving in a younger way than the surrounding ones, which are or were active in the model-conform or metamorphic sequences. Younger behaviour of a meristem means that the axes being built resemble a younger axis in the same tree. The youngest of those was the seedling axis. When a meristem behaves in such a way, the result is not a branch but a structure that is equivalent to a whole young tree. In fact, the same pathway of the sequence from the seed is covered again, but this time starting from a vegetative meristem. This process of repetition of whole young structures in the same tree is called reiteration, because *iter*, in Latin, means pathway. A reiterated complex is a tree that has been built successively by a model-

Fig. 3.13. *Artocarpus incisus* (Moraceae, Rauh's model, breadfruit; Cayenne, French Guyana, cultivated; 1972). Reiteration in the form of root suckers and on the young trunk. **A** The root is thicker in the background, where it transports nutrients to sucker plus mother-tree, than in the foreground, where it only feeds the mother-tree. **B** On the root sucker, which is a reiterated complex, there are reiterated axes that can be recognised by their branching angle. Scale: pocket knife (8 cm, **A**) and French centime (13 mm, **B**). (Photograph by R. A. A. Oldeman)

conform sequence, sometimes a metamorphic sequence and several reiteration sequences. Such a tree can be analysed as a population of coherent small trees (Torquebiau 1979). The resemblance is heightened sometimes by the presence of aerial roots at their base (cf. Fig. 3.23D). How can one recognize a reiterated part on a tree in the field? There are a number of criteria, summarized by Torquebiau and some of which are most often clearly visible:

- the reiterated trunk has a more vertical position than surrounding branches;
- the reiterated trunk lives longer than a branch;
- the reiterated trunk has a larger diameter and a faster growth than a branch;
- sometimes, the reiterated trunk has juvenile bark, which can be distinguished from the "adult" bark structure of surrounding trunk and branches (e.g. Fig. 3.12, *Sambucus nigra*, Caprifoliaceae);
- the presence of juvenile leaves, also in the reiterated crownlet (e.g. *Juniperus virginiana*); their absence or rarity in eucalypt coppice as observed near Quito (Ecuador) shows that reiteration has not reached the youngest stage;
- a smaller branching angle than the branches, if reiteration took place on the trunk (Fig. 3.13) or a larger one on the branches;
- the reiterated axis is longer than a branch that one would expect at the same place in the tree.

3.3 The Life Cycle of Trees: Root Architecture

The subterranean part of a tree is difficult to study. However, there are several contraptions making it possible to observe roots or to follow root growth (Schuurman and Goedewagen 1965; Böhm 1979; Bosch 1984; Van Sonderen 1986). The great challenge has been and still is the observation of roots without destroying them or their environment. Washing out whole root systems (Lyford and Wilson 1964; Kahn 1982, 1983; Schmidt in prep.) allows an overview, but stops all processes by killing the trees and destroying the whole rhizosphere at the place where they grew. Figure 3.14 shows how the same effect has been reached by strong erosion in a loose sandy soil in Kalimantan (Indonesia).

3.3.1 Architectural Criteria in Root Growth

"A root is a root" is one of the erroneous concepts that might arise from such rough methods. In order to analyse whole tree root systems, the concept of root differentiation will be examined here first. It will allow us to improve the traditional notions, such as that of "taproot systems", "superficial root systems" or "heartroot systems". Figure 3.15 illustrates a more precise approach.

In root systems, one can distinguish root "organs" using several criteria:

- **positropic roots** (short for positively geotropic roots, "orthotropic" according to Kahn 1977, 1978) are growing downwards, in the direction of gravity;
- **diatropic roots**, short for diageotropic roots (plagiotropic according to Kahn 1978) are horizontal and grow perpendicularly to the direction of gravity;

Fig. 3.14. Root system, washed out by erosion of pure white sands (tree species unknown). Note the heavy sinkers, positropic macrorhizae that were the result of root reiteration, and the broken-off stumps of mighty, superficial, diatropic roots. Central Kalimantan, road from Palankaraya to Kasongan, Indonesia, 1982. (Photograph by R. A. A. Oldeman)

- **negatropic roots**, short for negatively geotropic roots (a variant of the plagiotropic, according to Kahn 1977, 1978) grow vertically upwards, against gravity.
- **mixed roots** start growing one way and change their orientation later, as in the knee-roots in Fig. 3.15D, where they are first negatropic, later positropic.

In contrast to Kahn's concept, the terms plagiotropy and orthotropy are here reserved for aboveground vegetative shoots. First, this is done because Kahn defines these properties not only structurally but also functionally, whereas only structures are considered here. And second, the general terms "horizontal" and "vertical" do not indicate that this is meant especially for roots. Hence terms were adopted that are derived from classical German root physiology.

Another set of criteria lumps together the size and longevity of roots, more or less as in plagiotropy for branches:

- **macrorhizae** or skeleton roots are massive, long-living and most often ensure support, transport and food-stocking functions, like the axes above the ground;
- **brachyrhizae** or feeding roots are slender, shorter-living and most often ensure absorbing and metabolic functions (rhizosynthesis), like the leaves aboveground;
- **root-hairs** are the shortest living root organs, which may correspond most closely to leaves in their absorbing functions: the feeding root category may be just an

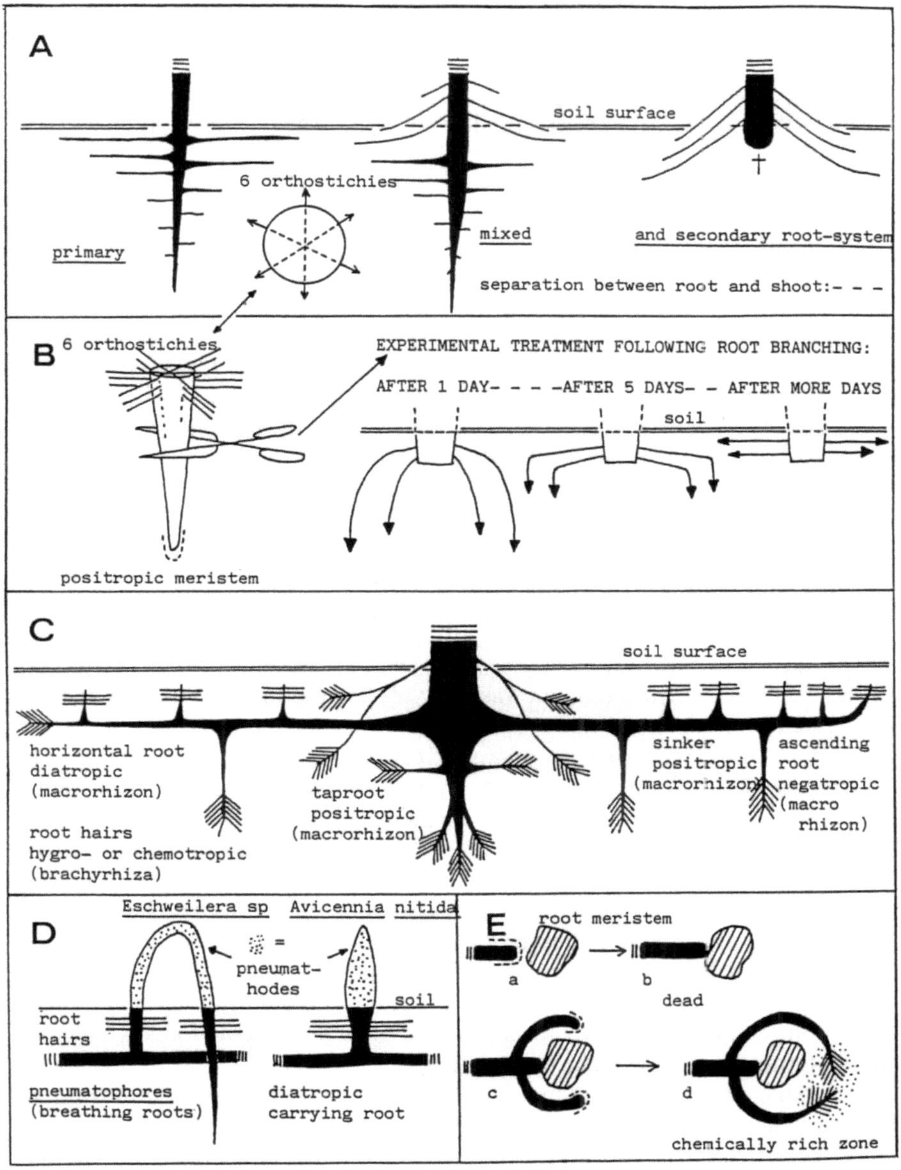

Fig. 3.15. Root systems in trees. **A** Three main types in root systems. The primary one is probably basic, with growth dynamics as in **B** according to Dyanat-Nejad's experiment (1971), showing a positropic meristem inducing diatropy in side roots and fixing it after 5 days. **C** Different root organs in a mixed root system. **D** Pneumatophores, after Oldeman (1971). **E** Growth determined by heterogeneous environment ("gravel-behaviour" and chemotropy)

intermediate that we need because we cannot distinguish a neat limit between skeleton and feeding roots.

A little excursion into ancient Greek is relevant here, because most root terms are derived from this language. Makros (μακρος) means big, large; brachus (βραχυς) means short; tropein (τροπειν) means to turn around or turn towards, as in the tropics where the sun turns around each year; and finally rhiza (ριζα) is root, plural rhizae (ριζαι), roots.

Figure 3.15 shows how the root system of an adult tree may be constructed from the root organs described above. The occurrence of negatropic, ascending roots has been known for a long time in "breathing roots" or pneumatophores (Fig. 3.15D). But since research (e.g. Lyford and Wilson 1964; Oldeman 1971) has shown that many trees have short negatropic roots that penetrate the litter layer at the soil surface from below, and that bear diatropic feeding roots exploiting this litter for nutrients, air, or both, root systems are considered to variously express the theme shown in Fig. 3.15C. This has technical consequences for soil compaction by machines or tourists.

Rhythmic growth in roots has not been proven to exist as yet. One should be prudent in assuming it, first because a root meristem does not function in the same way as an aboveground meristem: roots possess neither buds nor lateral meristems and no clear periods or mechanisms for endogenous latency or dormancy have ever been described in roots. Second, it is easy to provoke imposed pseudo-rhythmic behaviour, such as that of *Castanea sativa* roots (Fig. 3.17Ab) that grew in perforated soil, in which each perforation that was too wide corresponded to an unbranched root segment.

Root branching has one clearly organised aspect (Fig. 3.15A: orthostichies). No lateral meristems are formed on roots in any way comparable to axillary meristems on branches. Root branching occurs by the activity of adventitious lateral meristems. But this adventitious process occurs on lines that run parallel to the length of the root, the orthostichies. The number of these may be different per species. Cocoa shows a frequent case with six orthostichies per root, but palms may have tens or even hundreds. The position of each side root on the orthostichy is indeterminate, so that these lines can only be observed directly by looking along a carefully cleaned branched root (or an adventitious root aboveground) in the sense of the length of the cylinder. The exact localisation of these lines on the roots is linked to their anatomy, which will not be treated here.

Sympodial growth, in which the growth of an axis is relayed by another one, as in determinate axes aboveground (Fig. 3.3a,b) also may occur in roots. An example is the tropical mangrove tree, *Rhizophora mangle* (Rhizophoraceae, Attims' model) which can form extended tidal forests bordering the coasts of seas and lagoons or the banks of the lower parts of rivers (Fig. 3.16A). This may be contrasted to the monopodial growth pattern of the roots in another mangrove tree, *Avicennia nitida* (Verbenaceae, probably Roux's model, Lescure 1978b) as represented in Fig. 3.16B.

After having considered the differentiation in roots, defining "root organs", the architecture of the root system as a whole will be examined in all developmental

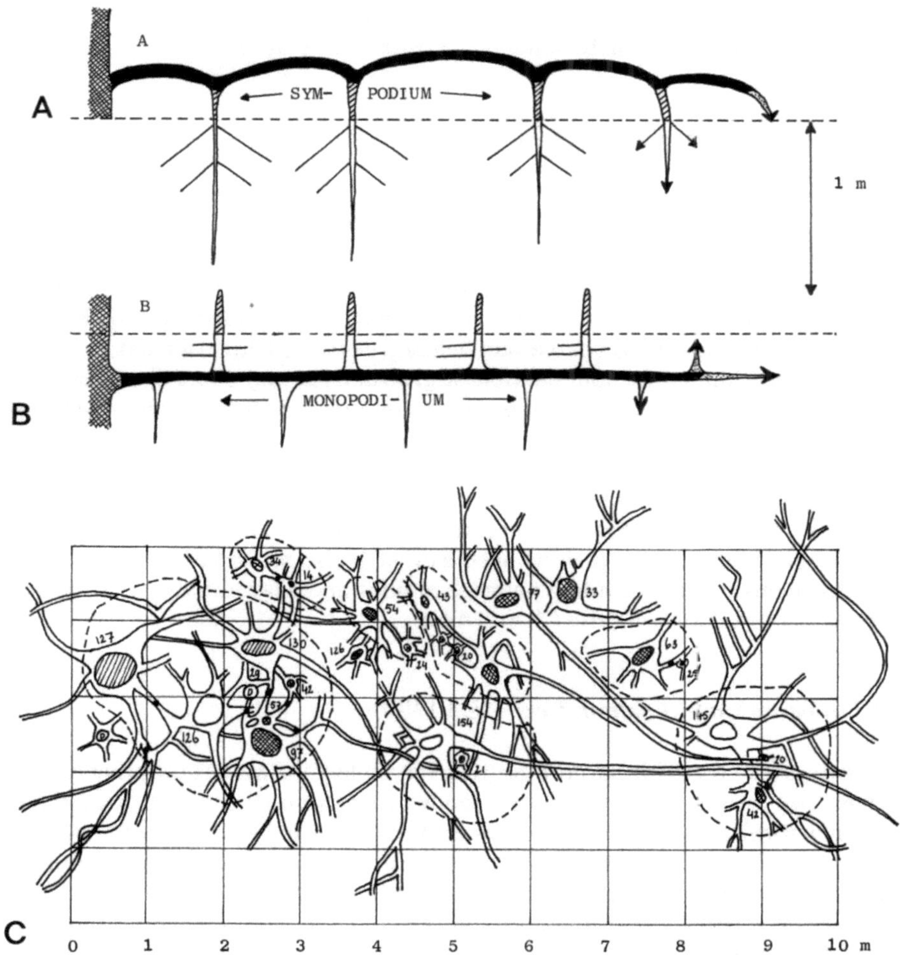

Fig. 3.16. Root system differentiation (**A, B**) and anastomosis (**C**). **A** *Rhizophora mangle* (Rhizophoraceae, Attim's model, tropical mangroves), aboveground transporting root system, sympodial; gas exchange in columnar positropic roots; **B** *Avicennia nitida* (Verbenaceae, model trans. Roux/Attims, tropical Atlantic mangroves), underground transporting root, monopodial: gas exchange in columnar negatropic pneumatophores. trunk; horizontal transport system; gas exchange system; anchoring and supporting roots, nutrient uptake; growing. (**A, B**, after Gill and Tomlinson 1977). **C** *Aucoumea klaineana* (Burseraceae, Rauh's model, Okoumé, pioneer rainforest, Gabon). Root anastomosis (see Chap. 5.3) in a 40 m^2 plot, in a natural stand of 20 to 25 years old. Anastomosing only among Okoumé trees. dominant Okoumé; suppressed Okoumé; stump; anastomosis (roots grown together); *126* girth Okoumé; other tree species; "ecophysiological cell" consisting of several linked Okoumé trees. (After Leroy-Deval 1973)

phases that were described earlier for the aboveground parts (Fig. 3.17B). For European tree species, this can be studied in Köstler et al. (1968).

3.3.2 ***Seedling Roots*** (Figs. 3.17A, Ba)

Most generally, seedlings form a positropic taproot, bearing slender absorbing diatropic roots that are formed in a spiral. Figure 3.15B shows that the growth model is regulated by the terminal meristem of the taproot, just as in model-conform branched crowns. One has to consider that possibly the root model is already present very early in the seedling and persists only during a rather short time-span. This would fit in with adaptation to the very heterogeneous soil environment, in which obstacles may hinder the continuation of growth over very short distances, and where roots are attracted to moist or fertile soil pockets (chemotropy - see Fig. 3.15E).

Kahn (1978, 1983), after whose data Fig. 3.17 has been drawn, and who studied these patterns in many tropical rainforest trees, claims that the architecture of root systems in tree seedlings (Fig. 3.17Ba) corresponds to that of unbranched herbs and small woody plants during their whole life cycle. It also has to be noted that monocotyledonous plants, including palms, bamboos and agavas, have a fast-disappearing taproot, if any (see also Chap. 4.4.6).

3.3.3 ***Roots in Model-Conform Trees*** (Fig. 3.17Bb)

When trees start branching according to their architectural model, the **primary root system**, that derives directly from the seedling root, comes into being. It contains more macrorhizae than the taproot alone, the others being side roots often with a diatropic character. It is not easy to distinguish a clearcut growth model in this stage. If the absorbing brachyrhizae of the seedling are considered as mere short-lived organs comparable to the leaves in the crown, it is in this model-conform phase that a clearly and sturdily branched root model appears. But within the volume that it already occupies, it has always met numerous barriers in the soil already, so that endogenous growth patterns are difficult to analyse. It is too early as yet to draw conclusions from ongoing research using perforons (earlier called "root-boxes": Fig. 3.18), as to the differentiation and development of whole tree root systems in homogenised soil blocks. Their development can be followed on a day-by-day basis with this non-destructive method, that has been reviewed by Van Sonderen (1986).

The images provided by Kahn have been shown here for a tree conforming to Rauh's model, but this is only an example: there are all kinds of variations belowground, just as there are different models in crown formation. Shrubs that do not develop their crown beyond this stage often do not develop their roots beyond the primary system either.

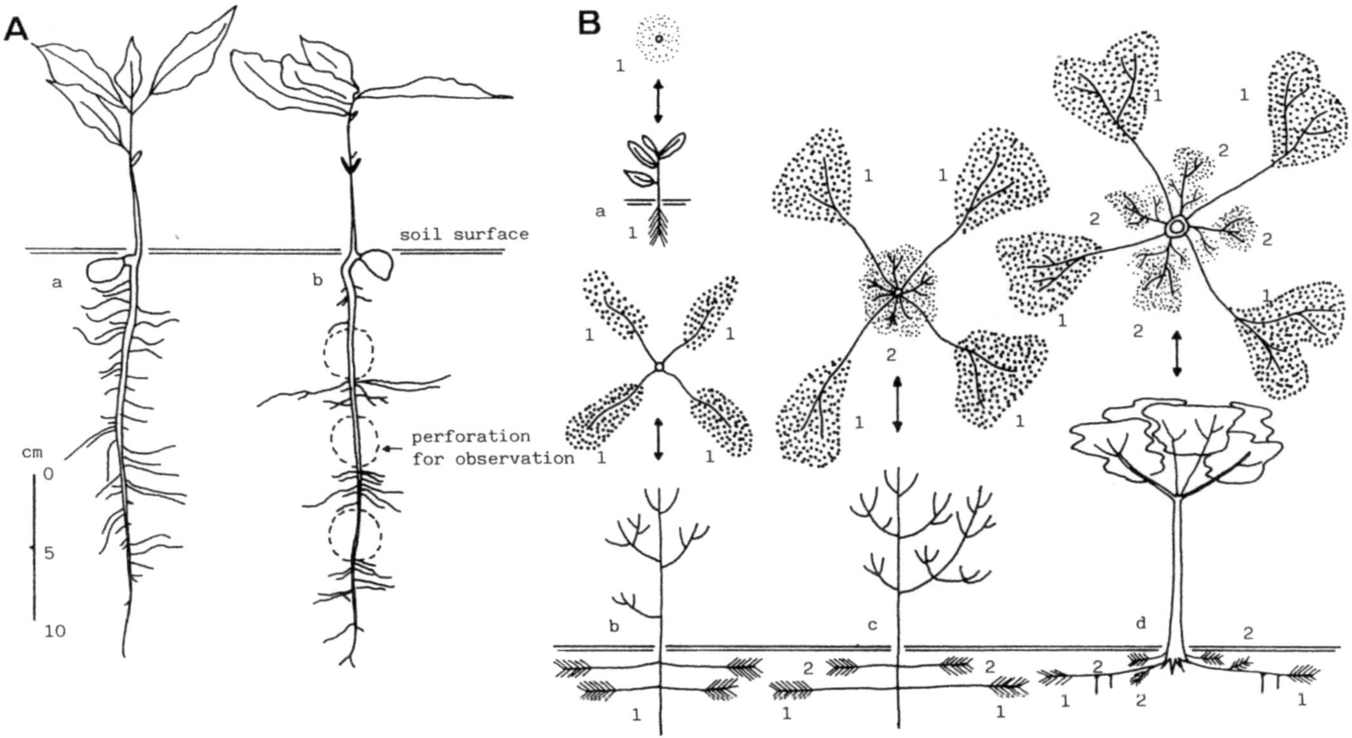

Fig. 3.17. Tree root systems. **A** *Castanea sativa* (Fagaceae, Europe), seedling root observed in perforon (Fig. 3.18). *a* Younger than 3 months, primary root system excavated, no perforations; *b* same, but with pseudo-rhythmic branching induced by too large perforations. (After Van den Tweel 1979). **B** Four steps in growth of root systems in high rainforest trees from Ivory Coast, which are the equivalent of four strategies from woody herb to big tree. *a* Seedling or unbranched treelet from undergrowth; *b* model-conforming phase or model-conform treelet from undergrowth; *c* tree in metamorphosis or starting reiteration; *d* abundantly reiterating, high forest tree. *1* primary root system; *2* secondary root system. (Free after Kahn 1982)

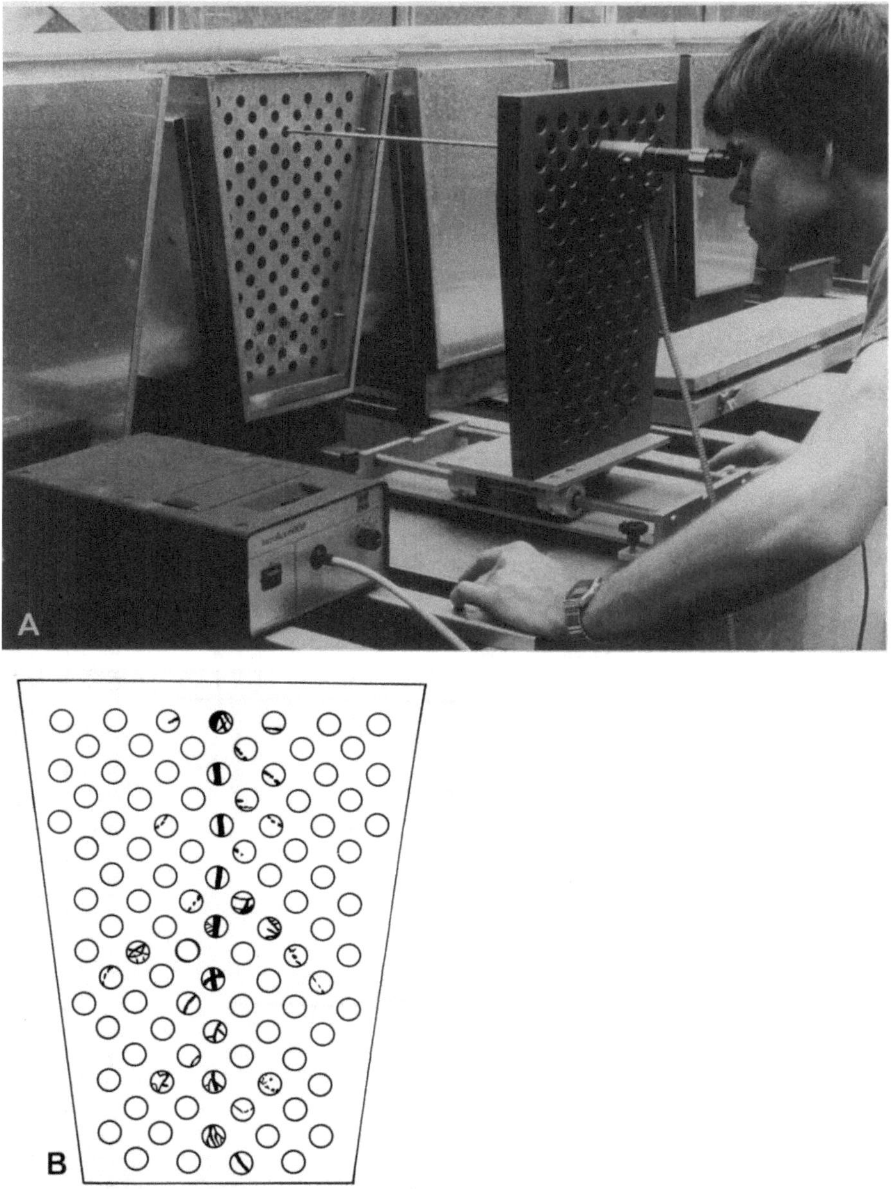

Fig. 3.18. Development of a whole tree root system in a homogenised perforated soil block (perforon, formerly called "root box"). **A** Perforon in the greenhouse at Wageningen, observations with intrascope via optical fibres (cf. Bosch 1984). **B** Results of monitoring root growth of 4-month-old oak seedlings (*Quercus petraea*, Fagaceae, Rauh's model). Root map is updated every week: *full lines* roots on 29 March 1986; *interrupted lines* roots formed between 29 March and 4 April 1986. Research perspectives include tridimensional computer modelling of growth patterns

3.3.4 Roots in Metamorphosing Trees (Fig. 3.17Bc)

The primary, superficial, diatropic part of the root system here has grown out from the tree at a considerable distance, so that an empty zone develops around the trunk base. All absorbing roots and root hairs are at the extremities of the primary root system, while at the base of the trunk there generally is an accumulation of organic matter resulting from the decomposition of litter and dead roots. This empty part of the root zone is now filled by the **secondary root system**, originating from adventitious root meristems at the lowest part of the trunk itself.

The link between architectural metamorphosis and the initiation of the secondary root system has not been formally established by research and hence is given here as a hypothesis that fits in with the general development patterns.

3.3.5 Roots in Reiterating Trees (Fig. 3.17Bd)

Reiteration in the crown has been documented in several species as being accompanied by reiteration of root systems, the "sinkers" that are branched like a primary root system. During this time, the development of the secondary root system continues. A sinker may be more than 1 m long, and sinkers may be very numerous (Fig. 3.19A). In other species they may be lacking altogether, as in northern gymnosperms such as Douglas fir (*Pseudotsuga menziesii*, Pinaceae, Massart's model) or Norway spruce (*Picea abies*, Pinaceae, transition model Massart/Rauh), as shown in Fig. 3.19B.

A very general image of root systems, resembling the classical forestry classification is given by Kahn (1977, p. 351; some terms ours):

- **primary root systems** are built by a positropic taproot and a diatropic system of lateral roots; this corresponds to the forester's "taproot-system";
- **mixed root systems** show the replacement of an insufficient primary system by an adventitious secondary system, built from the orthotropic trunk base or sometimes from aerial roots (mangroves, for instance); this would seem to correspond to the "heartroot system" in forestry;
- **secondary root systems** precociously replace a primary root system suppressed very early – this system is merely distinguished from the mixed one by losing the primary system so very early; this probably includes the "superficial root systems" in forestry.

3.3.6 Specialised Roots

Some special differentiated roots have to be mentioned for the sake of completeness.

- **Aerial roots** are cable-like roots that hang down from the canopy, where they were built by adventitious root meristems on high branches (e.g. in *Ficus*

Fig. 3.19. Tree root systems. **A** Unidentified species, Saül, French Guyana, 1971: note numerous reiterated sinkers, positropic taproots that are equivalent to seedling taproot. **B** *Pseudotsuga menziesii* (Pinaceae, Massart's model, Douglas fir, Western United States, introduced to Europe). Uprooted, wind-thrown Douglas fir: note absence of sinkers, the whole root system having become diatropic with age. Royal Domain, Netherlands, 1979. (Photographs by R. A. A. Oldeman)

spp., Moraceae) or in epiphytes (see Chapter 4). When they reach the soil, they branch and form an absorbing root mat in the organic horizon.

- **Pneumatophores** or "breathing roots" (Fig. 3.15D) are negatropic roots, or mixed roots that start as negatropic roots and later bend back to a positropic position, but that always rise above the ground and are provided with breathing organs (pneumathodes, lenticels) so as to admit air to the underlying diatropic roots, often in marshy or tidal forests with oxygen-poor soils.
- **Stilt roots** are stout positropic roots that arise from adventitious root meristems higher up the trunk or from lower branches, and bear dense clusters of superficial diatropic side-roots and root hairs once they penetrate in the upper soil or litter horizons; they may also function as breathing roots.
- **Buttresses** or **plankroots** are thickened flanges on the trunk that result from higher cambial activity at the junction of major side roots (Fig. 3.20). They are fairly common in tropical rainforest trees, when these reach the reiteration phase, but have also been described in other forests, like the magnificent *Ulmus laevis* (Ulmaceae, Troll's model) in the forests bordering the Rhine in France (Walter 1974).

Fig. 3.20A
(Fig. 3.20B see page 56)

Fig. 3.20. Tree root systems. Buttresses. **A** Buttresses in *Peltogyne pubescens* ("bois violet", Papilionoidae, ident. local guide). Note light-coloured crest of each buttress, showing increased cambial activity there, so that younger bark pierces older, outer bark. Saül, French Guyana, 1971. (Photograph by R. A. A. Oldeman). **B** Buttresses in *Ulmus laevis* (Ulmaceae, Troll's model), France, semi-natural riverside forests along the Rhine near Strasbourg. (Photograph by J.-M. Walter 1972/1974).

3.4 Application of Architectural Tree Properties

The architectural properties of trees have been known and used in many societies and since the dawn of the agricultural period. Figure 3.21 shows some current applications of different natural processes in trees, based upon these properties. They can best be understood when realising that reiteration is the common denominator of many phenomena that have different names: root suckers, stem suckers ("epicormic shoots"), cuttings, trunk and branch regeneration, layering, grafting or sprouting of tree stumps. Reiteration here fulfills three biological functions:

- **replacement** of lost parts;
- **vegetative multiplication**;
- **adjustment** of undamaged trees (see Chap. 5).

3.4.1 Provoking Reiteration for Shoot Production

The following applications rest on the replacement of lost parts.

3.4.1.1 Coppicing (Fig. 3.21B) is used whenever wood of small diameters is to be produced. This is a very important category of wood in rural communities, because it is easy to process as firewood and it can be used as stakes, stalks and other farm

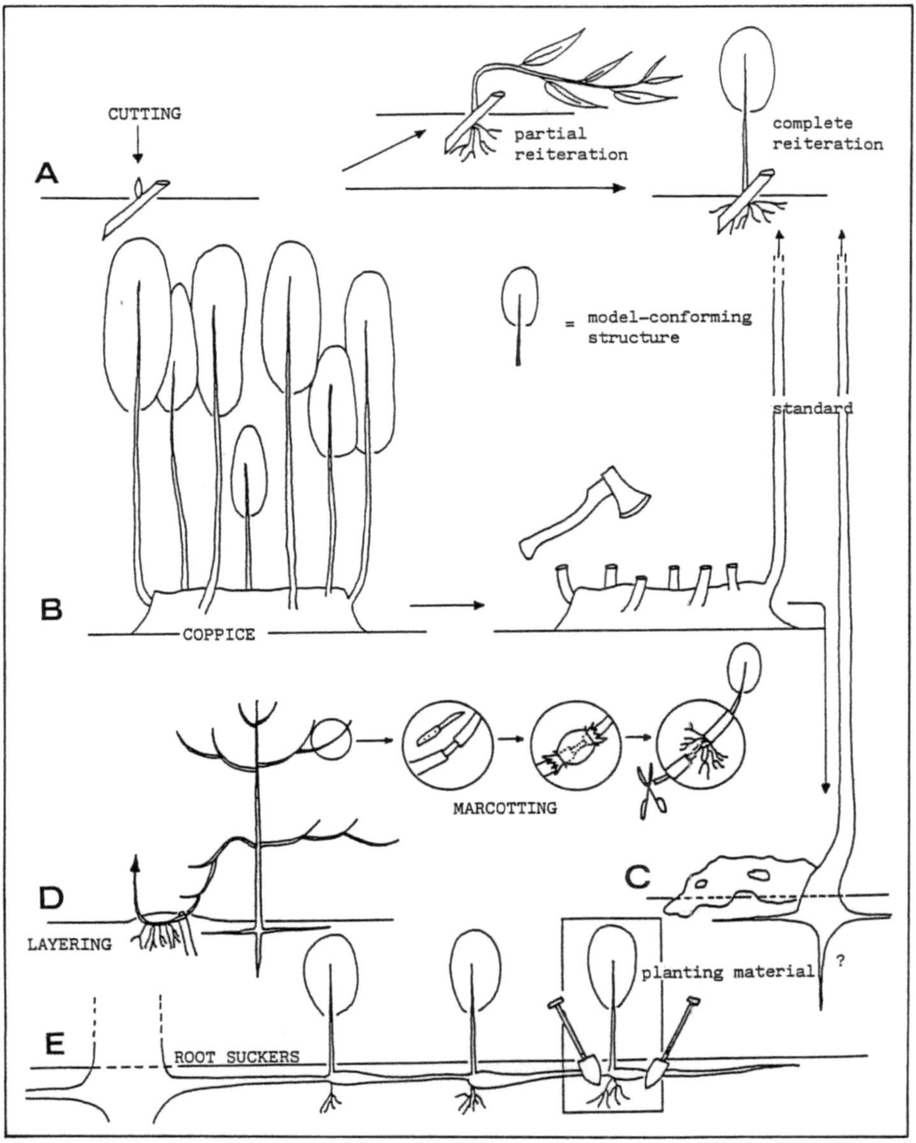

Fig. 3.21. Silvicultural application of reiteration. **A** Cuttings may yield rooted branches (partial reiteration) or young trees (complete reiteration) depending on original bud differentiation. **B** Coppice with stump suckers (shoots); if only one is left (standard) it can become a high tree again - root reiteration possible, but unproven (?). **C, D** Layering or marcotting is successful in little differentiated branches, otherwise neither roots, nor complete crowns are formed. **E** Root suckers are sometimes used to obtain planting material (valuable rare trees): note that supporting rootare thickest at the outside (cf. Figs. 3.13, 3.26)

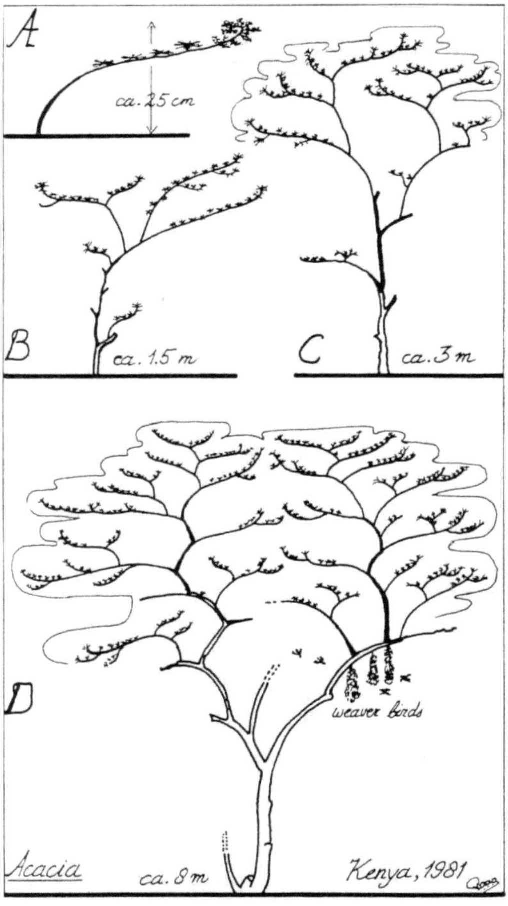

Fig. 3.22. The genus *Acacia* (Mimosoideae) includes several semi-arid region species with an architecture that is apt to form natural coppice. It conforms to a version of Champagnat's model (see Fig. 3.2) with short shoots, i.e. short, very specialised axes with very short internodes, bearing many leaves and ensuring flower formation. **A** Original model, from seed: one mixed ortho/ortho axis (Fig. 3.3e) with short shoots. **B** Champagnat's model, but note traces of reiteration at base. **C** Start of reiteration: three model-conforming crownlets, of which one is small, in the crown. **D** Adult tree, abundantly reiterated. This development is used in coppicing and lopping (firewood, fodder) in semi-arid African regions and in other continents. (Oldeman 1989a in press)

commodities. Only such species can be coppiced that are able to reiterate from old or adventitious meristems at the trunk base, and many trees can do so: Fagaceae from the Javanese mountains and from temperate regions (*Quercus* spp., Rauh's model, cf. Fig. 3.4), *Olea europea*, the olive tree (Oleaceae, model unknown), the Date Palm (*Phoenix dactylifera*, Palmae, Tomlinson's model) from the borders of the Sahara, and a great number of *Eucalyptus* species (Myrtaceae, often Attims' model) or *Acacia* species (Mimosaceae) from Australia are examples.

However coppicing is also useful to obtain bark in large quantities. Oak bark from Europe or wattle bark from Australia (*Acacia*) contains tannins used in tanning leather, and many small stems contain a much higher proportion of bark than one big trunk with the same volume. Throughout Europe, and from the Roman times on, oak coppice was hence one of the success stories of the rural regions, and yielded supplies to urban markets (cf. Houtzagers 1954; Oosterhuis 1979). Less than a century ago, oak coppice, for instance, still represented the largest Dutch forest area. Coppicing took place with rotations between 10 and 15 years, the bark went to special mills where it was crushed and sent to the tanneries, whereas the stripped thin wood was much appreciated as fuel by bakeries. Willow species (*Salix* spp., Salicaceae) were coppiced to produce long and supple shoots, used for weaving baskets and, in somewhat stouter diameters, to weave the foundation pieces for the Dutch dikes. Willows occur in marshy riverbank forests, so the material was grown close to the users. Coppice shoots from several tree species were used in Europe to weave the walls of farmer's huts ("wattle"), which were covered with mud afterwards to make them wind- and watertight ("wattle and daub"). The Australian *Acacia* species obtained their name from this use during the first colonial settlements in the 18th century. Ash (*Fraxinus excelsior*, Oleaceae, i.e. the olive family, see Fig. 3.11) is still coppiced on small areas in order to provide special wooden stirring rods for the aluminium smelters. The ash tree burns into ash that does not react with the molten metal.

In the tropics, large quina plantations (*Cinchona* sp., Rubiaceae, South. America) continue the success story of coppice. The anti-malaria drug quinine is extracted from the bark. In rural forestry and agroforestry, coppice plays an increasing role also. Many *Acacia* species (Mimosaceae, Fig. 3.22, see also Oldeman 1989 in press) are coppiced in order to yield firewood and fodder, for instance in North Africa and the Sahel region. Other leguminous trees (e.g. *Calliandra calothyrsus* in Indonesia, *Leucaena leucocephala* in somewhat drier tropical regions) are coppiced in the same way and for the same reasons.

Selective coppicing is sometimes used to transform coppice woods into high forest again. One shoot is left as a standard (Fig 3.21C), which then develops into a tree. Such trees can be recognised by their irregular stem base, the former tree stump (Fig. 3.23B).

3.4.1.2 Pollarding (Fig. 3.23C) is a technique that differs from coppicing and by the fact that the trunk is cut at a higher level, most often between 1 and 2 m high. The products are the same as those from coppice, but the ecological role as a forest component is different. Moreover, from a practical viewpoint, one can walk under a pollard and not under a coppiced tree. This is a reason of convenience to do the first and not the second. Willows (*Salix* spp.) are among the most common pollarded trees.

3.4.1.3 Lopping (Fig. 3.24) is a common farmer's technique the world over to obtain firewood. The branches of a tree are lopped off, but the top is left intact, which gives the tree the aspect of a paint-brush or pinsel. Later, the lopped branches are replaced by reiteration from the buds left on the branch stumps, and a secondary crown is formed under the rest of the original one. Such trees can be recognised, if regularly lopped, by their crowns in the form of an 8. In Europe this may be ob-

Fig. 3.23. Coppicing and pollarding. **A** *Tilia vulgaris* (linden, Tiliaceae, Troll's model), natural reiteration at stem base (photograph by J.W. Rottier, Rhenen, 1980). **B** *Quercus* cf. *robur* (Summer oak, Fagaceae, Rauh's model), old coppice of which standards are left, near Rhenen, 1981 (photograph by L.V.M. Meyssen). **C** *Tilia cordata* (linden, Tiliaceae, Troll's model), pollards along a path near Nijmegen, 1980 (photograph by W. Snieders). **D** *Populus* sp. (poplar, Salicaceae, Rauh's model), detail of pollarded tree, note root formation at the base of reiterated stems. Bennekom, 1982 (Photograph by H. Muys)

Fig. 3.24. Lopping for firewood in *Altingia excelsa* (Hamamelidaceae, model unknown) by local populations in Java (Indonesia). Note the adventitious reiterated crowns one below the other, in a double-8 shape. (Oldeman 1983a)

served in rural regions in oaks, and in the tropics many species are so treated: eucalypts (observed in Kenya), or *Altingia excelsa* (Hamamelidaceae, Fig. 3.24) in Indonesia are examples.

The combination of these three techniques, in which one may dominate, can result in quite sophisticated systems, like the "energy hedges" designed by Van de Winckel (1986) after experiments in the Dutch province of Brabant. These designs may well be adapted to other regions, e.g. in view of the "poor man's energy crisis" in the semi-arid tropics.

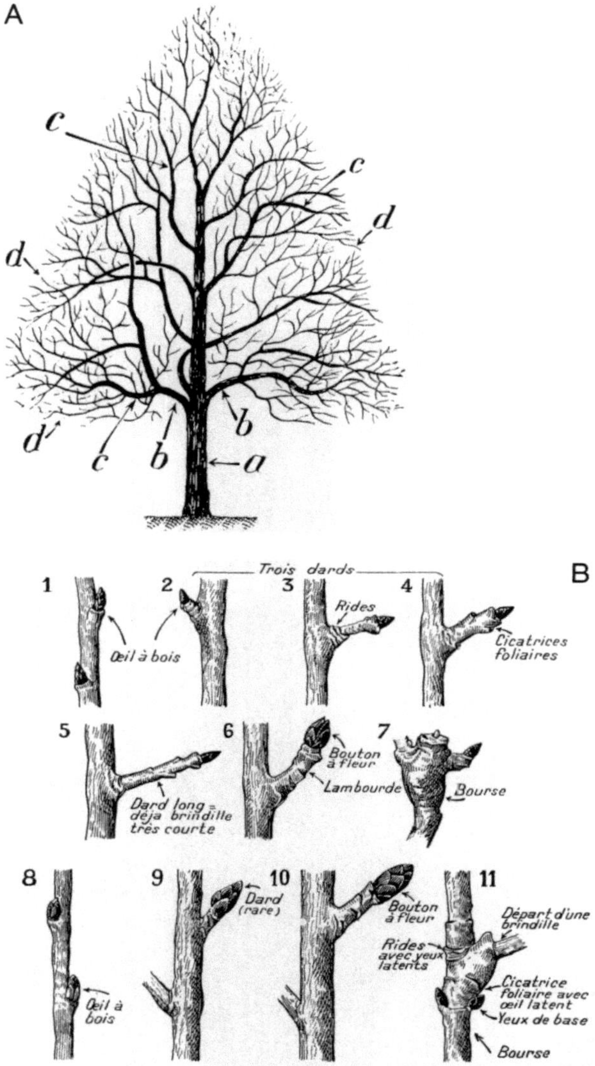

Fig. 3.25. Principles of pruning, as applied in the French tradition. Example of the pear tree, *Pyrus communis*, Rosaceae, model trans. Rauh/Massart, with short shoots that bear the fruits. **A** Tree as perceived by a farmer, with trunk (*a*), main branches, in which we recognise trunks by reiteration or metamorphosis (*b*), smaller main branches, also partially or completely reiterated (*c*) and twigs of the 3rd, 4th and 5th order (*d*). **B** The different buds (meristems plus scales), vegetative or flowering, and short shoots among which farmers recognise very precisely the ones to stimulate or to remove by fine pruning. French terms left untranslated. (Champagnat 1949)

3.4.1.4 Pruning, clipping, trimming (Fig. 3.25) are more precise techniques to provoke reiteration of very well-defined shoots. These methods are traditionally used in fruit trees. They are mentioned here, because in many rural forestry or agroforestry systems wood is a minor product that is harvested from old trees that have ceased to produce fruits. Both the Japanese (e.g. miniature "bonsai" trees) and the French have very old and sophisticated traditions of pruning.

Figure 3.25 represents a fruit tree as perceived by traditional French farmers. One may recognize the axes that have resulted from metamorphosis (B,C) or reiteration, whereas the architectural model is unclear (in this species probably Rauh's model, see Fig. 3.2). Each of the buds occurring on the outer branches has a special name indicating its future and well-known function, left untranslated. Pruning is done carefully, so as to activate a series of fruit-producing buds during a series of years. All these techniques have been described by Champagnat (1949).

3.4.2 Provoking Reiteration for Vegetative Multiplication

The application of reiteration in vegetative multiplication is most often done as follows.

3.4.2.1 Cuttings (Fig. 3.21A) are used to form new individuals from reiterating buds on parts of older trees. In this way, if rooting takes place, a population of genetically identical trees can be obtained. As shown by Fig. 3.23D rooting of reiterated shoots is a naturally occurring process. In cuttings, it is sometimes stimulated by the use of phytohormones. Figure 3.26B shows a population obtained in this way; such populations are called clones. Being genetically identical to the tree that yielded the cuttings, growth differences within a clone can be attributed solely to environmental conditions (site factors).

The reiterating meristem behaves according to its programmed differentiation. If this meristem is orthotropically disposed, the cutting yields an orthotropic axis that behaves like a trunk, and a complete tree is the result: complete reiteration. Such meristems or buds can most often be found as laterals on orthotropic axes. But if one takes cuttings from a strongly plagiotropic branch, chances are high that the meristem activated on the cutting also is plagiotropic. The result is then a rooted, horizontal branch. This has been experimentally demonstrated in several Asian dipterocarps (Roux's model when young, cf Fig. 3.8a and Smits in press) and many gymnosperms (*Abies*, but also in some larches, *Larix* spp.). Much later, adventitious meristems are formed near the basis of the rooted branch. They behave orthotropically and start forming a new trunk. But in that case, most often years of production have been lost and such badly formed trees are useless for forestry plantations. In physiological literature (cf. Evers 1984, who cultivated Douglas fir from buds in vitro) this programmed behaviour of meristems, depending on their place in the tree's architecture, is called topophysis.

3.4.2.2 Grafting (Fig. 3.27A) is a natural phenomenon occurring from time to time when branches sweep against each other, remove each other's bark and then are

Fig. 3.26. Reiteration and vegetative multiplication in *Populus nigra* (Salicaceae, Rauh's model, Europe). **A** *P. nigra* var. *italica*, root suckers, note abundance and root diameter (cf. Fig. 3.21). Goidschalxoord, Netherlands, 14 Feb. 1982 (photograph by R.A. Kleijweg). **B** Variety unknown, clone, i.e. all trees descending from one tree (cf. Sect. 6.3.8) by large cuttings (= barbatellas), note remarkably similar branching and reiteration pattern (cf. Fig. 3.42). Growth differences (*stars*) caused by site quality. Wageningen, Netherlands, 7 May 1983. (Photograph by B. R. Werkman)

Fig. 3.27. Grafting and layering as natural processes, forms of reiteration. **A** *Fagus sylvatica* (Fagaceae, Troll's model, Western Europe), two neighbouring branches, each ca. 1 cm in diameter, being grafted upon each other. If the base of the one is cut, it may start behaving as a side branch of the other, equivalent to partial reiteration. Nuenen, Netherlands, 30 April 1983 (photograph by S.M. Smeulders). **B, C, D** *Populus canadensis* (Salicaceae, Rauh's model, introduced in Europe); natural layering in the past by a broken tree (**B**) has led to the extremity now being a new tree, still connected to the old one (**C**); in other parts several reiterated trunks have been added (**D**). Meyendel, The Hague, Netherlands, 15 Aug. 1982. (Photograph by B. Koeze)

pressed close together. One can then cut off one of these branches at the base, and its extremity behaves as if it were a lateral branch upon the other. Grafting as a technical process is done by opening the bark of a branch or young tree trunk and inserting a living twig or a bud (oculation) by means of a special scalpel. The grafted organ then is fixed with raphia string or resinous products that also may desinfect the wounds. The stem of a young tree also may be cut off and the top of another young tree may be grafted on the root-stock.

The grafted organs, when developing, do so outside the differentiation sequence of the receiving tree. That is to say that they show a form of reiteration, transplanted from one tree upon another. Grafting is used very often in horticulture. In forestry, it most often serves to obtain a population of good seed trees by grafting branches of one particularly good seed tree upon numerous root-stock of the same species, which are planted out later to form seed orchards. Grafted plus-trees may be planted out in hedges, which are cut back frequently, yielding cuttings and reiterating profusely after each pruning, thus yielding ever more cuttings in an optimal, physiologically young state. This method has been recently introduced in Indonesian dipterocarp silviculture (Smits et al. 1987). Grafting may also be used to obtain populations with some desirable property. Many red beeches (*Fagus sylvatica* var. *atropunicea*, Fagaceae, Troll's model) have originated from red beech branches grafted on normal beech rootstock when there was a large market for red beech, owing to fashion.

3.4.2.3 Layering or marcotting (Fig. 3.21D) is also derived from natural phenomena, i.e. root formation by reiterated crownlets. These techniques are sometimes used to obtain clones for special uses that do not require mass production. Mass production can be done much more cheaply by cuttings and, more recently, by in vitro multiplication from meristems (cf. Evers 1984).

However, in natural forests layering can be a mechanism for vegetative multiplication also. Hallé and Oldeman (1970, their Fig. 44) show this to be the case for mangrove trees of the genus *Rhizophora*, mentioned earlier, for certain *Pandanus* species (Pandanaceae, humid Old World tropics) and in a more specialised way in tree ferns of the genus *Cyathea* (Africa, South America). Edelin (1977) described it for several gymnosperm species (*Cupressus*), but in the *Pinus sylvestris* trees on wind-blown sands in the Netherlands, the dense mats of branches spread out on the ground beneath the trees have never been demonstrated to take root.

3.4.2.4 Root suckers (Fig. 3.21E) are quite common in many trees, e.g. *Artocarpus incisus* (Moraceae, Rauh's model - breadfruit; Fig. 3.13A), pantropical, or *Liquidambar styraciflua* (Hamamelidaceae, model unknown, sweetgum), from the South-Eastern USA and Mexico (see Zimmermann and Brown, 1971, their Figs. I-13/14), or several species of poplars (*Populus* sp., Salicaceae, Rauh's model, Fig. 3.26A) from Europe and North America.

Root suckers have to be treated in the same context as root anastomosis (Fig. 3.16C), because here the limit between individual trees and populations becomes vague. Root suckers reiterate the architectural model entirely. Hence they can be isolated and transplanted after having taken root (Fig. 3.21E). This is done sometimes if there is no nursery stock available, but this technique is expensive and therefore infrequently used. It might become more important in agroforestry, when farmers want to exchange a few useful trees among them, and labour costs are not taken into account.

Figure 3.21E shows the phenomenon described by Zimmermann and Brown (1971), who indicated that the diameter of the diatropic root on which root suckers originate is larger on the side away from the original tree than on the side towards it. This can be explained by the sap stream (see Sect. 3.5.4), which is feeding the sucker plus the tree at the distal side, and only the tree on the proximal side.

3.5 The Life Cycle of Trees: Energy and Nutrients

During all phases of development defined on the former pages, trees have an intensive exchange with their environment. This environment is treated here as a black box (see Chap. 2), the inputs of which towards the trees will be taken for granted. The environment will also be taken as a black box when considered as a sink for the output of the trees. Of course the inputs, e.g. sunlight, or the outputs, e.g. transpiration, will be mentioned but they will not be analysed further. Much of the matter in Sections 3.5 and 3.6 has been united by Schmidt (1980).

3.5.1 Concepts

Production in any living system is the incorporation of matter from nutrients and energy into that system after their absorption from the environment. Growth is the situation when the greater part of these inputs remains in the tree, whereas the smaller part disappears as outputs: there is a net gain. Growth is always followed by a phase of equality of inputs and outputs, when production of the whole tree becomes zero even if it continues to be positive in certain organs, such as the trunk. This balancing phase is here called biostasis, because the often-used terms of homeostasis or steady state originate from theory concerning closed systems in physics. In a third and last phase of the life cycle of a tree, the outputs exceed the inputs. Net production is negative, although once again it can still remain positive in some organs. This phase is here indicated as decay. This development can be found in all living systems, over the whole hierarchy (Table 2.1). Here, it will be treated for trees.

The basic processes playing a role in production are:

- **Photosynthesis, assimilation, water absorption, ion absorption, rhizosynthesis** (nitrogen reduction and synthesis of vitamins, amino acids and proteins in roots);
- **Transport** in xylem and phloem (translocation, e.g. see Zimmermann 1983);
- **Photorespiration, other forms of respiration, transpiration;**
- **Cell formation** in primary and secondary meristems, **formation of tissues** and **organogenesis, hormonal regulation.**

These basic processes will not be treated here. They are the subjects of botany, plant physiology and plant anatomy textbooks, e.g. Bidwell (1974), Mitscherlich (1970,1971,1975) or Givnish (1986).

The feedback between these processes and the architectural development that has been reviewed in the first parts of this chapter will be kept in mind. Production is an orderly process and it takes place according to the organisational hierarchy of Table 2.1. It has to be remembered that changes at one level often have consequences at other hierarchical levels (also see Sect 6.3.8). Except for time, the following spatial dimensions are most often used for the study of production by trees.

- **Height or length (one-dimensional)** are used to assess the result of the activity of root or shoot meristems, i.e. the axes that are produced. The length of an axis gives an indication for the exploration of the environment of the tree in the direction of that axis.
- **Surfaces (two-dimensional)** are used to assess exchange, production or transport processes. Leaf or root surface indicate exchange capacity for gases, water or nutrients. The surface of the cross-section of active transport tissues, xylem and phloem, gives an indication for sapstream flow capacity (cf. Braun 1963, 1982; Zimmermann 1983). The surface of the cambium indicates its capacity for the production of wood and bark.
- **Volumes (three-dimensional)** serve to assess the global result of living matter production as derived from length and surface. For axial volumes, the organs determining volume at a moment in time are the meristems (length) and the cambium (surface), or in other trees, like palms, the top meristem alone, because the final cross-section is reached in the seedling by meristem expansion.
- **Cumulated volume (four-dimensional)** serves to assess the production process in time, i.e. during parts of the life cycle or over the whole life cycle. Volume is three-dimensional, the added dimension is time. In seasonal regions, a cumulated volume graph often can be reconstructed by growth-ring analysis. Each growth ring may not be a year ring, however. Cumulated volume can be determined for wood, because inactive wood is conserved within the tree. Bark is shed (e.g. *Pinus sylvestris*, Pinaceae, *Platanus* spp., Platanaceae or eucalypts are spectacular examples). Twigs, leaves or brachyrhizae are short-lived, too. For these tissues and organs, cumulated volume hence cannot be determined precisely, except in closed environments.
- **Mass, or biomass (one-dimensional)** is often used instead of volume, because it can be determined more easily and precisely. Mass is a fundamental physical property of matter, that can be approximately determined by weighing because the force of gravity is approximately equal in the whole world. In the term biomass, the prefix indicates only that the determined mass is part of a living system. It does not mean that the biological properties of that mass are considered. Hence the term most often is qualified: root biomass, wood biomass, leaf biomass, etc.. In biomass studies, exact methodological instructions (cf. Sect. 2.1) are very important: fresh weight or dry weight, estimated weight as in approximations of leaf biomass in litter traps or root biomass in washed root systems are examples. Direct weighing is often accompanied or replaced by calculations

based on statistical correlations between developments and their results. This is called allometry: an easily determined variable is used as an indicator (parameter) for parallel change in a less easily measured variable. Current parameters are diameter at breast height (dbh) for biomass or volume of stem wood, diameter under the first branch or side-root for biomass of leaves, branches, short roots.
- **Cumulated biomass (two-dimensional)** is used in the same way as cumulated volume.

The accent on long-term growth processes is characteristic for forest ecology and forestry. Forests cannot be understood if their long life span is not taken into account, and this can only be done through cumulated volumes and biomass. And wood is usually not harvested year by year, ring by ring, like a hollow cylinder! As a popular Dutch saying has it: only thick wood can be sawn, and a forester will add: wood grows on wood. The economic calculations to prevent harvesting more than the mean yearly increment and their consequences for timber estate management are the basis for classical forest management plans. Long-term growth programmes with a yearly yield have only recently received a renewed interest, for energy-woodlots or -hedges and in agroforestry. Earlier, they most often were limited to fruit trees, in which another part of biomass is harvested, i.e. not the wood but the fruits.

Using the criteria mentioned above, several production functions should be considered, in which the black-box approach for environment and organs has been maintained.

3.5.2 Production Systems and Functions

- **Input-output relationships.** Every producing, living system depends upon its inputs from the environment: the higher the inputs, the higher is the production. This relation between inputs and production is not linear. With increasing inputs, the performance of the system approaches an asymptotic value. This value represents the maximum performance, which is different for each kind of production.

Photosynthesis receives inputs of light and carbon dioxide from the atmosphere and water plus organic nitrogen compounds from the roots by the sapstream, while its outputs are oxygen and sugars. Increasing light intensity leads first to fast increase, then to slowing increase and finally to near-constancy of photosynthetic activity (e.g. Douglas fir needles, Fig. 3.28A). The example of beech (*Fagus sylvatica*, Fig. 3.28B, C, D) also shows this phenomenon, while indicating the influence of the state of the photosynthetic system. Production is different in shadow- leaves and sun-leaves, and appears different when related to different leaf properties, such as leaf surface, chlorophyll contents, or dry matter content.

In rhizosynthesis, inputs are water and nitrate ions from the environment, and sugars from the crown. Outputs contain vitamins and substances rich in amino acids, that are distributed throughout the tree by the sapstream. Direct effects of an increased nitrogen offer from the environment include the increase of amino acid

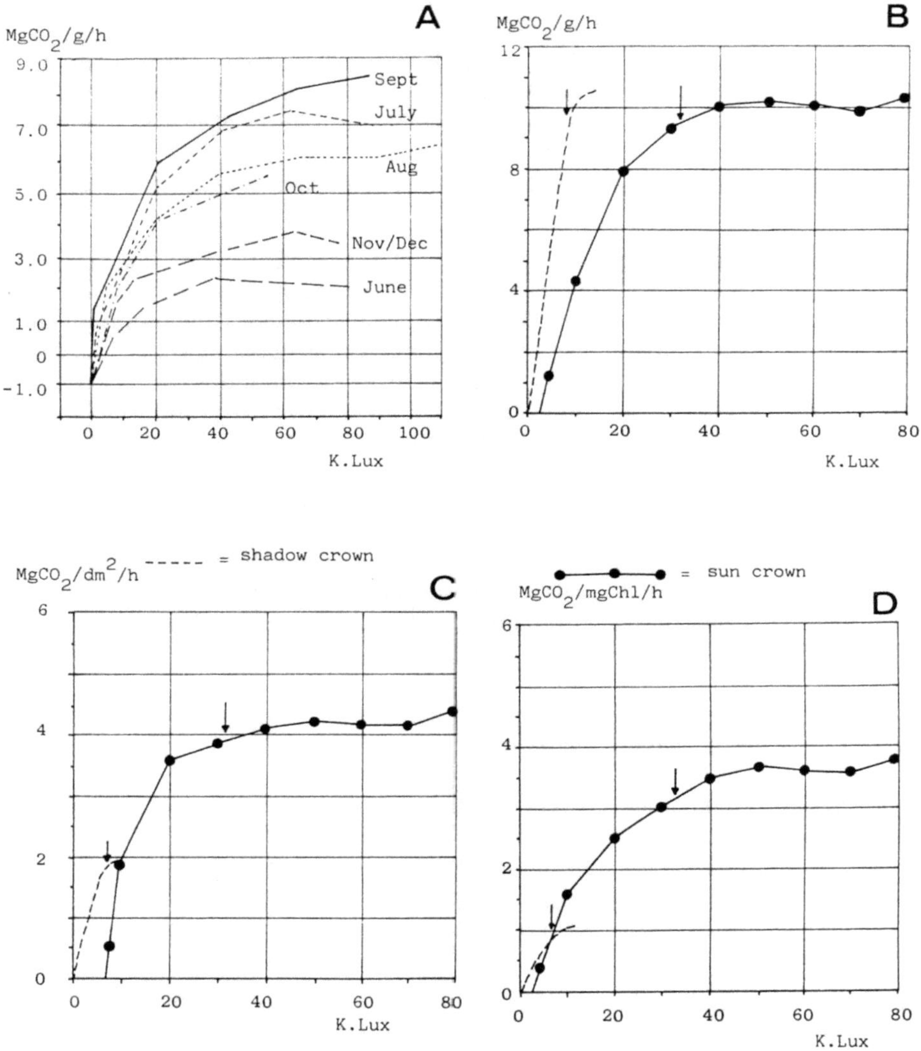

Fig. 3.28. Input-output relationships, examples of photosynthesis in temperate region trees. **A** *Pseudotsuga menziesii*, Douglas fir (Pinaceae, Massart's model), needles observed in Germany in different development stages during the growing season. Temperature 15 to 20°C; rel. humidity of the air 50 to 60%. (After Künstle 1971). **B** *Fagus sylvatica*, beech (Fagaceae, Troll's model, sun and shadow leaves). Outputs expressed as dry matter content. **C** Same species, outputs expressed as leaf surface units. **D** Same species, outputs expressed as chlorophyll content. (After Schulze 1970)

concentration in the xylem sap (Fig. 3.29A). Rhizosynthesis is not active throughout the year in the temperate region, where it peaks in spring.

In cambial production, the inputs mainly arrive with the sapstream in the form of water, sugars, nitrogen compounds and so on, whereas outputs are xylem and phloem cells. There probably is no direct exchange with the environment (cf. Philipson and al. 1971) and no direct environmental influence (cf. Larson; Richardson, both in Zimmermann 1964). Still, the cambial system is not altogether immune to the environment. Direct, strong sun radiation on the bark may kill the cambium in *Fagus sylvatica* (Fagaceae, beech), for instance. It remains unclear in how far the phloem and bark respiration include or mask cambial respiration. The cambium probably also has a maximal performance (cf. Hallé and al. 1978), but further research is still needed.

- **Production systems and their products.** A tree contains two primary production systems: the photosynthetic system and the rhizosynthetic system. They prepare absorbed nutrients and energy for use by the secondary production systems, the meristems and cambium, that ensure the incorporation of matter in the tree. Photosynthesis provides a tree with sugars that contain all energy and most of the structural substances needed for growth, functioning and maintenance. Rhizosynthesis plus water and mineral uptake furnish water, minerals and amino acids. Figure 3.30 shows a qualitative flow chart for the distribution of these substances. An example of the quantitative distribution of photosynthetic products is given in Fig. 3.29B. Odum (in Odum and Pigeon 1970, see Fig. 5.77) gives the distribution of calories in tropical trees from Puerto Rico.

3.5.3 Production and Function of Organs and Tissues

Matter and energy, once absorbed by a tree, are used for building, maintenance and functioning (respiration) by the secondary production systems. They make:

- **Leaves**, which are built either with reserve carbohydrates after a leafless period, or with sugars produced by the leaves below the new ones in evergreen trees. Even young leaves are photosynthesising, but initially this is insufficient to cover their own needs. The leaf starts exporting sugars and ceases to import them some time later. Exporting may be first towards the end meristem above it and the younger leaves, but finally it only sends its products towards the roots (Fig. 3.30; Oldeman in Tomlinson and Zimmermann 1978). This investment of matter and energy in leaves, that are emptied and shed later, (Givnish in Tomlinson and Zimmermann 1978, refers to "throwaway organs") is needed to build the photosynthetic system that energises the whole growth process.

- **Xylem** (wood tissue) is produced by the cambium towards the inside of the axes over their whole length. In groups of trees without a cambium, e.g. palms or tree ferns (see Chap. 4), this tissue is formed in longitudinal strands after the meristematic zone has reached its final diameter in the seedling stage. But in most trees, the cambium forms cylindrical sheaths of wood that are not equally wide in all

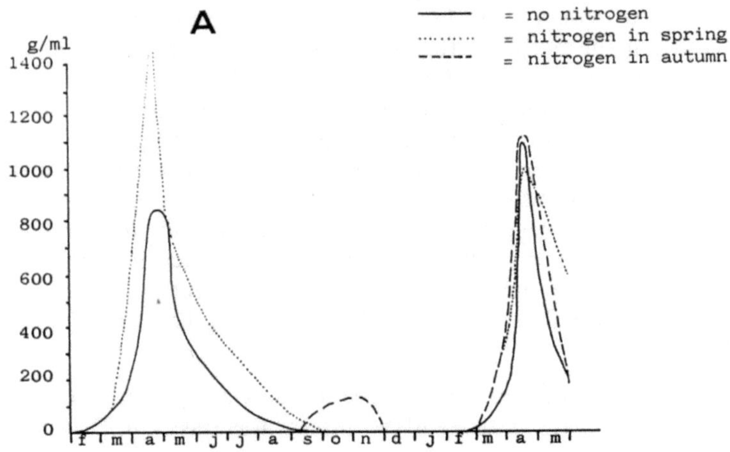

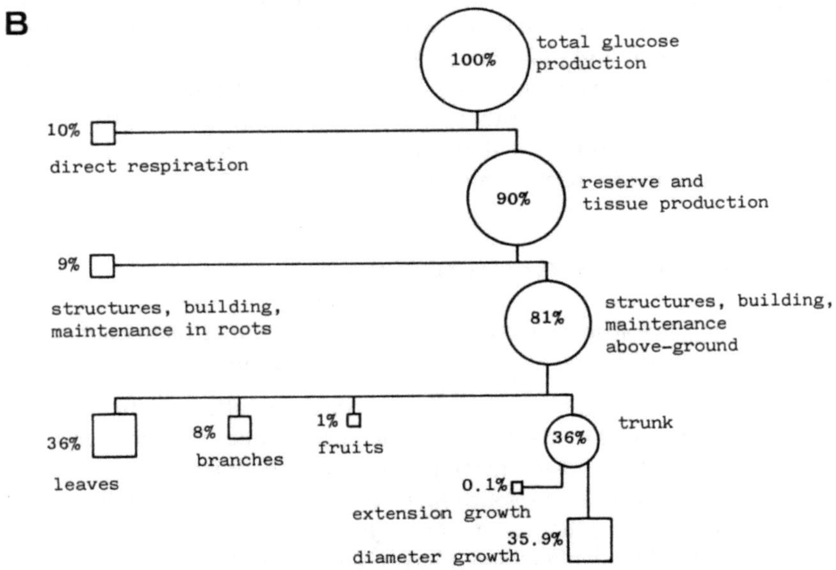

Fig. 3.29. Production and distribution of substances from rhizosynthesis (nitrogen) and photosynthesis (carbon: sugars). **A** Concentration of asparagin acid in xylem sap in *Malus domestica* cv. M II (apple cultivar, Rosaceae, model trans. Massart/Rauh) with and without nitrogen fertilisation of 200 kg/ha. The rhizosynthetic aminoacids glutamin, asparagin and arganin show comparable effects. (After Tromp and Ovaa 1969). **B** Distribution of photosynthetic sugar among different organs and functions. (After Fritts 1976)

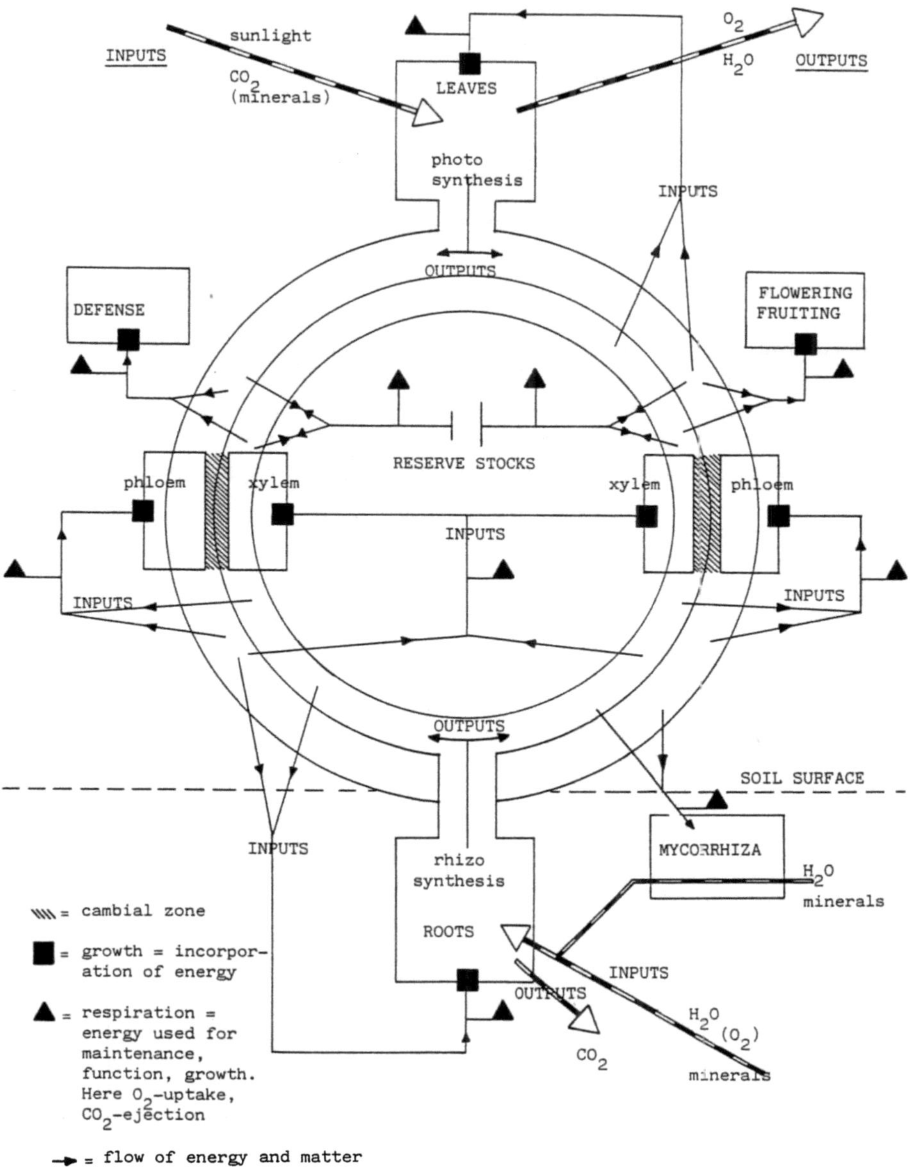

Fig. 3.30. Flow chart of the distribution of inputs (energy and matter) among the primary and secondary production systems in the tree, and outputs ejected into the environment. For the mycorrhizal compartment see Section 5.3.4.1: this is a closely connected and vital ecological interaction system (symbiosis) which acts as an input multiplier into the roots. (Illustration courtesy Dr. P. Schmidt)

parts of an axis (Fig. 3.31). Xylem production is fed from the sapstream that contains photosynthetic and rhizosynthetic substances. The xylem may remain active from one year (e.g. *Robinia pseudoacacia*, Leguminosae, Attim's model with rapid metamorphosis, see Maas and Gubbels 1986b, a) to several years (e.g. *Picea abies*, Pinaceae, transition model Massart/Rauh, see Edelin 1977) in transporting a watery solution upwards and in stocking reserve assimilates (cf. Braun 1963, 1970). After heartwood formation, the centre of the xylem cylinder has only a support function (Fig. 3.32B). The structure and functioning of xylem have been treated completely by Zimmermann (1983). For xylem in fossil trees see Creber and Challoner (1984a, b).

- **Phloem** (rind and bark) is produced by the cambium towards the outside of the axis, at the same time as xylem formation towards the inside. In cambiumless trees, the phloem is in the same longitudinal strands as the xylem. Tree species of the present time make more xylem than phloem; in fossil trees this seems to have been the inverse. Braun (1979) shows that trees of the order of Lepidodendrales

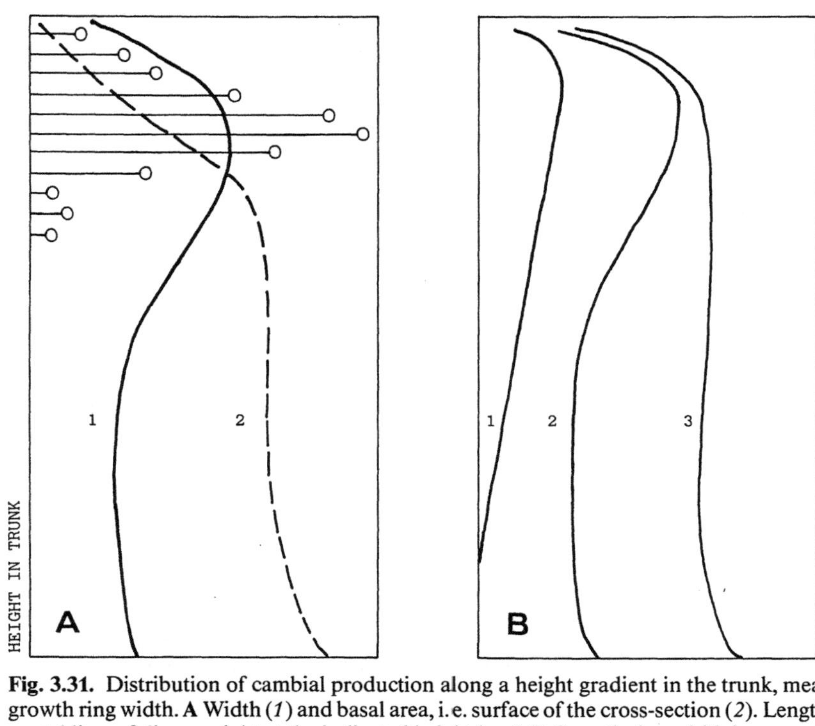

Fig. 3.31. Distribution of cambial production along a height gradient in the trunk, measured by growth ring width. **A** Width (*1*) and basal area, i.e. surface of the cross-section (*2*). Length of horizontal lines: foliage weight at the indicated height level. **B** Growth ring width in severely lopped trees or trees in dense stand (*1*), in dominant trees in a young stand (*2*) and in a free-standing tree with living branches down to ground level (*3*). (After Farrar 1961)

A

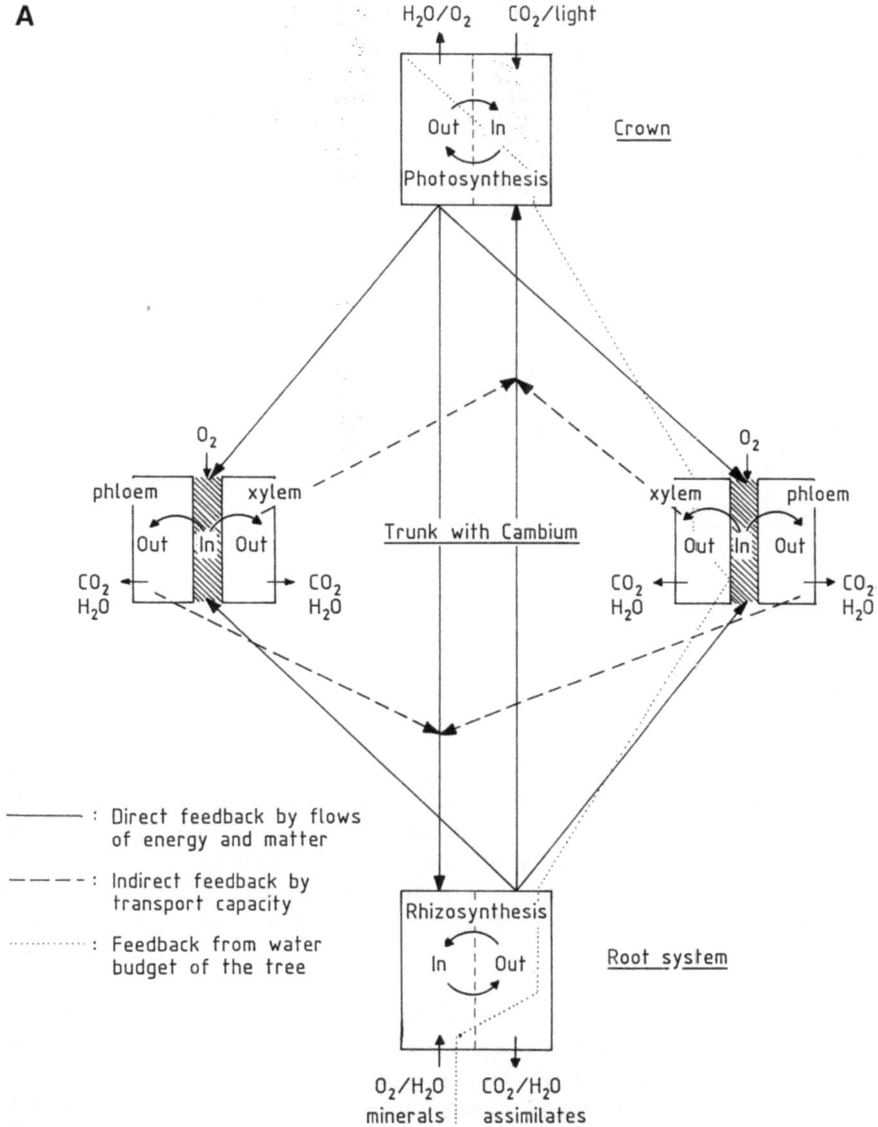

Fig. 3.32. The sapstream and its feedback relationships in production as a whole. **A** Feedback relations between the three main primary and secondary production systems in a tree, by inputs and outputs: "source-sink relationships". Direct relation between cambial system and environment hypothetical. (Illustration courtesy Dr. P. Schmidt, derived from Fig. 3.30). **B** Simplified version of **A.** Schematic visualisation of the sap stream (*3*), structure of a trunk or branch (*2*) and rule-of-thumb for feedback relationships in the tree (*4*). With this rule-of-thumb the reaction of trees to silvicultural treatments or other influences can be roughly predicted. (Oldeman and Westra 1980) (Fig. 3.32B see page 76)

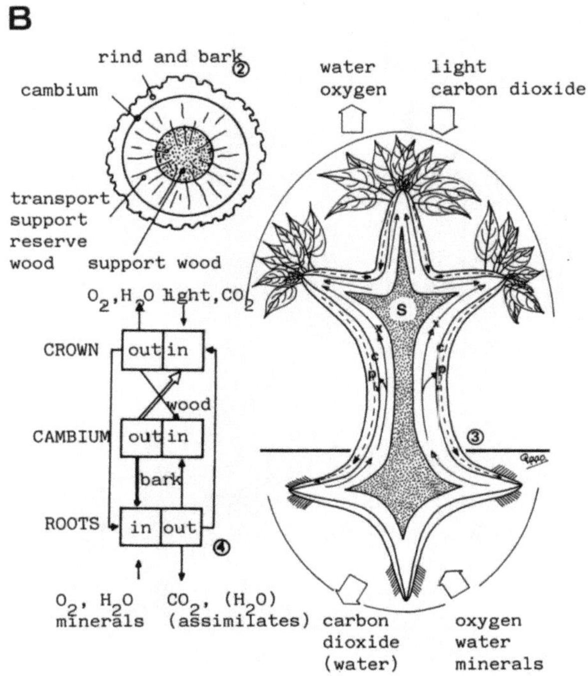

Fig. 3.32B

(Schoute's model, Hallé et al. 1978), occurring in the Carboniferous period, had axes that were built for 99% by phloem tissues. Phloem production is fed by the sapstream, just like xylem production. Phloem, in temperate climates, remains active in transport (translocation) for about 1 year and in reserve stocking for some years. Afterwards, it is transformed into a protecting layer (periderm, bark), often with complicated processes of cork formation. Finally, it is shed.

- **Roots**, including macrorhizae and brachyrhizae, are built by meristems and cambia. Macrorhizae ensure anchorage, transport and stocking of food reserves. Brachyrhizae and root hairs, the latter certainly being "throwaway organs", ensure absorption, symbiosis and rhizosynthesis. For symbiosis see Section 5.3.4. Functioning and production of roots are fed by the crown through the descending sapstream, or from reserves, and obtain their energy by root respiration. It is important to note that roots consume oxygen and produce carbon dioxide, just like any respiring organ. Moreover, roots may "sweat out", i.e. exudate other metabolic products that are instrumental in creating a special biochemical environment in the rhizosphere.

- **Defense mechanisms** use all sorts of biochemical substances, tissues or organs that are built in order to protect a tree against many outside attacks. In fact these are the tools of special forms of ecological interaction (see Sect. 5.6.8). The formation of periderms (see above: phloem), resins, tannins and special chemical substances like taxine (in *Taxus baccata*, Taxaceae, common yew, Massart's model) or cumarin (in the South American *Dipteryx odorata*, Caesalpiniaceae, "Gaïac de Cayenne", Attims' model) or thorns (genus *Acacia*, Mimosaceae) cost energy and matter. So do the production of tissues for scar formation on wounds (callus) or regenerative reiteration after damage. Hladik (1978) provided lists of biochemical defense substances in the rainforest of Gabon, Central Africa. But it is very difficult to define the part of the energy-and-matter budget of the tree that is spent on defense, the more so because defense substances may be just inseparable by-products of other processes (Hallé et al. 1978).

- **Reproduction** also requires energy and matter. Part of the photosynthetic and rhizosynthetic production is invested in favour of this biological function. That this may be a quite important shift in the use of energy, from one secondary production system to another, is clear in trees that show "mast years" (see Sect. 6.2.4.8). Such species flower and fruit abundantly once every 2 to 10 years. In such mast years, wood production is reduced in order to "pay for" fruit production, e.g. in European beech (*Fagus sylvatica*). This effect may be expected to occur in mast-flowering tropical tree families, such as Dipterocarpaceae (tropical Asia) or Vochysiaceae (tropical America). In fruit trees like apples and pears this also happens. Sometimes, all reserves are used in flowering and fruiting, as is the case in monocarpic (hapaxanthic) trees (Holttum's model, Fig. 3.2) that end their life with a "big bang" fruiting (Hallé et al. 1978). Fruits and seeds often require extra investments to build assimilates not only to support later germination, but also to feed dispersing animals (edible fruits). Functions here again may be mixed: the non toxic arillus in the otherwise completely poisonous berries of *Taxus baccata* remains behind in the stomach of the bird that disperses them (cf. Sect. 5.6.8).

- **Reserve stocks** are part of the photosynthetic and rhizosynthetic assimilates that are not directly used, but stocked in the form of starch or fats. Special tissues (e.g. rays) in wood and bark, and special parenchymatous cells in woody axes and roots serve to stock these reserves. In temperate regions this happens after the growing season, at the end of summer and the beginning of autumn (Braun 1970). These reserve stocks may be considerable. In beech, *Fagus sylvatica*, Gäumann (1935) found them to be large enough to initiate three periods of new shoot formation. Reserve stocks may remain chemically active. For instance, in cold winters, temperate region trees may temporarily transform the osmotically inactive starch into sugars and so increase their resistance to frost. But reserve stocks are necessary in all climatic regions, because during their long lives trees always encounter unfavourable periods, such as the dry "freak" years in the rain forest zones of S.E. Asia and West Africa in 1982/1983, or the yearly inundations of valley forests in the humid tropics during the rainy season, when root respiration is blocked for some weeks or months. Biotic hasards also have to be met by using reserves (see "Defense, Fig. 3.30; Sect. 5.6.8").

- **Water flows** in the xylem are energetically independent of production systems.This water flow system is ascending (Zimmermann 1983), contrarily to the part of the sapstream that descends through the phloem and that is chemically propulsed (Zimmermann & Brown, 1971). The ascending water flow proceeds by the following steps:

  A. Water absorption by the roots;
  B. Water transport through the xylem in root, axis and leaf;
  C. Water transport through cell walls and cytoplasm through the leaf;
  D. Water transpiration through the leaf cuticula and stomata.

The tree does not need the energy bound by its primary production systems to keep the water flow moving. Root pressure, if any, plays a negligible role. The energy used for the water flow is the difference in water potential between the soil and the air (cf. Meidner and Sheriff 1976; Zimmermann 1983). The driving, or rather the pulling force is the very low water potential of the unsaturated air around the leaves. The largest difference in potential lies between the leaf and the atmosphere. This is the largest resistance to be overcome. It is codetermined by the transition of water from the liquid to the gaseous phase, which costs much energy. This energy, used in evaporation, comes directly from the sun and its amount is much higher than that which is bound by photosynthesis. According to Baumgartner (1967), the ratio between evaporation energy (E_e), photosynthetic energy (E_p) and respiration energy (E_r) is:

$$E_e : E_p : E_r = 120 : 3 : 1.$$

There is an amount of energy that is still lower, that would figure in this equation as 0.002, and which can be considered to be lost or spent during the upward transportation of water.

A small part of the absorbed water is conserved in the plant for physiological functions. It serves as imbibition water, as a component in the osmotic system, as a solution medium and as donor or receptor of electrons in several chemical reactions. The water also procures the aqueous environment needed by these reactions and for translocation. In the ascending stream it is the transport medium for dissolved nutrients and amino acids. The largest part of the water is directly transpired, which has the indispensible side-effect of cooling the leaves (also see Meidner and Sheriff 1976).

3.5.4 The Sapstream

All preceding paragraphs concerned subsystems inside a whole tree. In the following part of the text, the whole organism will be considered and explained in terms of its physiological subsystems or, rather, subprocesses. The central feature that integrates the whole tree is the sapstream. The sapstream is a continuous, flowing network of watery solutions that permeates and links together all parts of a tree (Fig.

3.32). Moreover, because trees are open systems that are in communication with their environment, the sapstream has its entries and exits, too.

Through the sapstream, the primary production processes, photosynthesis and rhizosynthesis are dependent on each other by mutual regulation. Their feedback relationship, in which the outputs of the one are the inputs of the other, has been discussed by Borchert (in Tomlinson and Zimmermann 1978). This feedback relationship explains most of the growth of herbaceous crops (cf. De Wit et al. 1978), as well as that of cambiumless trees such as palms.

However in most of the forest trees that are the concern of this chapter, there is a third system involved in the global regulation of growth by mutual feedback. This is a secondary production system, the cambial system. It quantitatively regulates the sapstream itself, because it continuously or rhythmically renews the conducting tissues, the phloem and the xylem; but it is fed by the roots and the crown, as explained earlier, so it functions according to the "diet" received.

Figure 3.32 is a simplified version of Fig. 3.30, stripped down to show the main feedback mechanisms. In Fig. 3.32B (2,3) this is mapped upon a schematic image of a tree, and Fig. 3.32B4 reduces the diagram to the level of a rule-of-thumb, like a compass showing the direction taken by the reaction of trees to all inputs that one may care to consider, including fertilisation, thinning effects and chemical pollution. Of course, flows may temporarily change direction in some part of the tree (e.g. while flushing or fruiting, as said above), but the overall picture is given here.

In a wide sense, the thesis may be correct that (ecological) "primary production in green plants always takes place in exactly the same way regardless of the vegetation type (forestry, agriculture, meadows)", as claimed by Rabbinge and Mohren (1984). In a certain sense, it has been made true in their theoretical production model, inspired upon those of De Wit, and in which the whole plant does **not** represent an independent integration level. Referring to Table 2.1, the primary production of an eco-unit (e.g. forest stand, agricultural field) is explained by the organs (e.g. leaves, roots) that feed it. The individual tree, or other plant, is interpolated between these two levels by calculations. Hence the very particular physiological feedback mechanisms inside such plants are left aside. For many silvological purposes this should not be done, for instance when it is necessary to clearly understand wood formation or the vitality of trees. Production ecology models are clear examples of impeccable scientific models, tailor-made to answer very precise questions in a reliable way, but leaving other questions aside. The same methodological issue will be raised for the JABOWA forest regeneration model (see Sect. 5.6.6).

The sap stream in a tree has been partly outlined in the earlier paragraph on water flows. Here, it is completed and regulation processes are included.

A. Water enters in the roots, as a solution containing nutrients and biochemical substances; this input is regulated by osmotic and other absorption properties of roots: there is **selective** absorption.
B. Rhizosynthesis regulates the chemical contents of the very diluted watery solution that is admitted into the xylem; this regulation determines the effects of the solution everywhere during its ascent.

C. The ascending sap, on its way up through the xylem, feeds the cambial system and so regulates, by the "diet" offered to the cambium, the formation of new xylem and phloem that determine the transport capacity later.
D. The ascending sap, on its way up, is drained of certain substances to be used in the building of reserve stock tissues and cells and so regulates, by the "diet" offered, the later buffer capacity of the tree against stress.
E. The ascending sap, with what is left of its chemical load, feeds the photosynthetic tissues in the leaves and the building of new photosynthetic organs, and regulates, both by the "diet" offered and by the available quantities of water to be transpired that determine the opening of stomata and amounts of CO_2, the photosynthetic process itself.
F. Photosynthesis regulates the chemical contents of the water that is left in the very concentrated solution admitted into the phloem – this regulation determines the effects of the solution everywhere during its descent.
G. The descending sap, on its way through the phloem, where it is biochemically propulsed (see Zimmermann and Brown 1971), feeds the cambial system and so regulates, by the "diet" offered, the formation of new phloem and xylem that determine the transport capacity later.
H. The cambial system regulates the harmonisation of the proportions of xylem and phloem that are built, and most of the reserve stock tissues too (e.g. rays), but for this harmonisation the chemical contents of both the ascending and descending parts of the sapstream also have to be harmonised.
I. The descending sap, on its way down, is drained of certain substances to be stocked as reserves and hence also regulates, by the "diet" offered, the buffer capacity of the tree against future stress.
J. The descending sap, with what is left of its chemical load, feeds rhizosynthesis in the roots and the building of new rhizosynthetic organs and regulates, both by the "diet" offered and by its osmotic properties, rhizosynthesis and water uptake by the roots.
K. Back to A, the cycle is closed.

3.5.5 ***Stimulation, Stress and Sapstream Feedbacks***

A balance between the three main production processes, rhizosynthesis, photosynthesis and cambial activity, in which the outputs of one process balance the inputs of the others, is the normal situation as long as the environment does not change. Environmental change is change in inputs: light, minerals, water or stress factors such as storms that break part of the tree. The tree reacts to increased inputs into one of its production systems by increasing the outputs of that production system, i.e. crown, root system or cambium. These increased outputs of any one system in their turn provide increased inputs into the other two. The consequence of this is either an increased intensity of all processes, or an increase in volume of the tree, or both, until a new balance is reached.

If one or more environmental inputs decrease, or become denaturated (pollution), the consequences are the same, but in a negative sense. The processes then can become less intensive, or biomass may decrease by the death and shedding of axes, organs or both. Irregular, chaotic and unstable reiteration patterns accompany such stressed behaviour of trees. Figure 3.33 shows this in three European tree species, having been exposed to drought and pollution. The same behaviour as in Fig. 3.33 (B2, B3) could be observed in 1985 in broadleaved street trees in San Juan de Puerto Rico that had been exposed to abnormal drought and high concentrations of exhaust fumes. The bottom line (Fig. 3.33A2b, B3, C3) shows that the death of the cambium is the ultimate reaction to durable stress. This precedes the death of the tree. In most cases of stress, e.g. by acid precipitations, it is difficult to pinpoint the system that receives the first impact: crown? roots? The closed flow in the sapstream can translate impacts on one system very fast into aberrant behaviour of another production system.

The cambium, enclosed between solid layers of wood and bark, is the less flexible production system. It is well possible that buttresses (cf. Fig. 3.20A) and fluted trunks (cf. *Minquartia guianenis*, Olacaceae, "alata-oudou" in Creole, Nozeran's model, Guyanas and Amazonia - see Hallé and al., 1978, their Fig. 80) are cambial reactions to changes in inputs. Even if the cambium becomes quite underfed (low inputs), growth rings are formed over part of the trees. De Kort and Baas (1986) showed this to be the case in Douglas fir of low vitality, having grown in heavily polluted stands. This phenomenon is also illustrated in Fig. 3.31B. Even then, the sapstream keeps transporting inputs towards the roots.

In the preceding paragraphs, regulation of growth processes by phytohormones has not been mentioned. It fits into this global image as follows. The sapstream has been treated as a transport system that may be compared to the roads and traffic flows in a town. Its analysis shows where the roads (tissues) are, where they lead (production systems), and the nature of the loads transported (metabolic products). Phytohormones fit into this picture as if they were the (chemical) traffic lights and traffic policemen guiding, stopping or accelerating the flow of traffic. In the "chemical economy" of a tree, their "cost" in energy and nutrients may be compared to that of defense substances, mentioned earlier. The role of biological and ecological regulators of a biochemical nature is considered further in Section 5.6.8.

3.5.6 Cambial Activity: A Parameter for Vitality and Production

In this paragraph, the inputs from the ecological compartment "soil" and the water supply will be considered as being non-limiting, which of course is an oversimplification. It may be corrected by more thorough site studies than are included in this text, e.g. in Kozlowski (1981), or Armson (1977, Chap. 15). Climatic factors determining the weather and the seasons are much more variable, also in the short term of a year or less. The influence of climatic inputs in trees, for instance during a year that is very dry or very wet, is very complex if one considers all subsystems, as has been done by Fritts (1976) in his standard text on tree rings and climate.

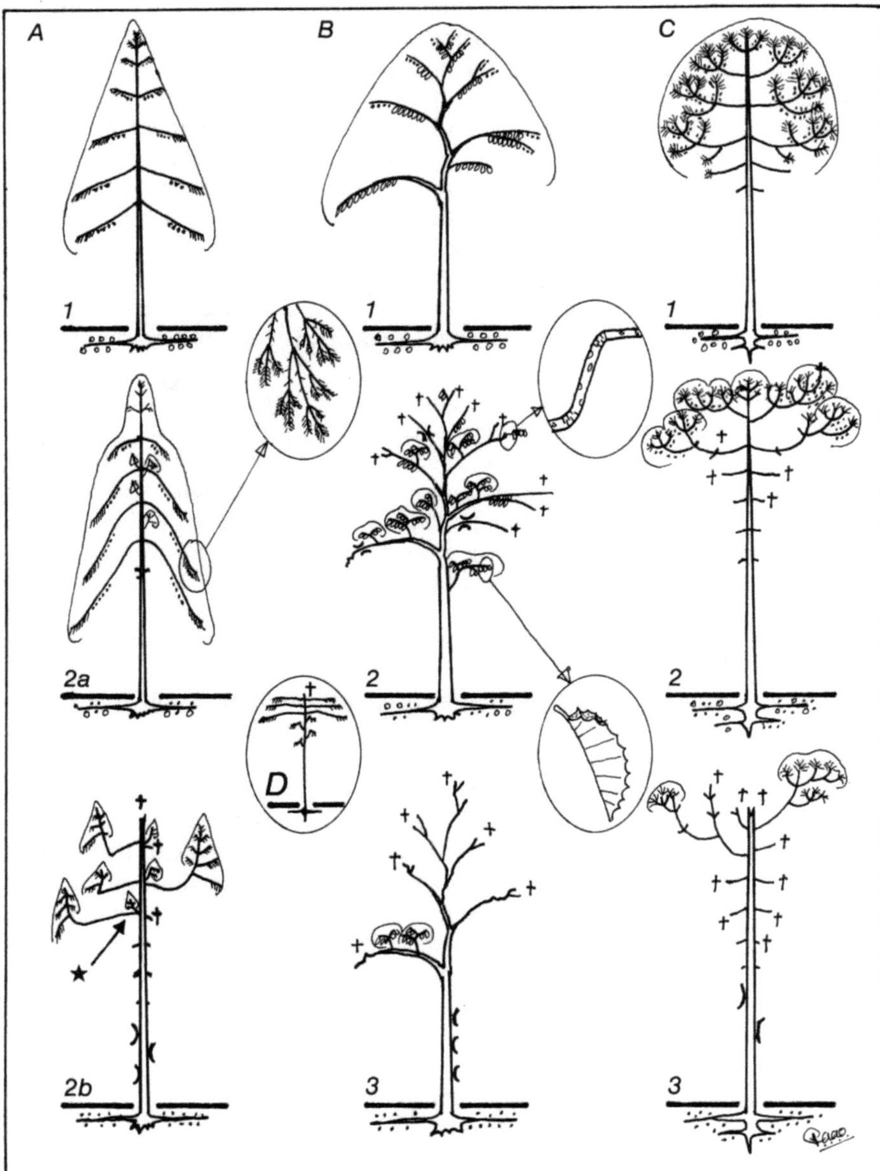

Fig. 3.33. Stress symptoms in trees, e.g. exposed to drought or acid rain, that can be explained by sapstream diagrams (cf. Fig. 3.32). **A** *Picea abies* (Norway spruce, Pinaceae, Massart's model) starts sacrificing old needles to form more new ones (*1*), then loses its top by competition for sap between trunk and branch extremities; branches become too long and droop, also small reiteration on trunk (*2a*) and finally reiterate while bark dies off (cambial dysfunction), as is also common in Douglas fir, *Pseudotsuga menziesii* (*2b*). **B** In beech (*Fagus sylvatica*, Fagaceae, Troll's model, Goderie 1984; Roloff 1985), as in other broadleaved trees, sapstream competition in the beginning of the growing season is between young leaf formation and cambial activity, so that insufficient

The ultimate effect of all these processes, however, is expressed by the diameter growth of a tree. This is also clear from Fig. 3.32, which shows that all inputs into the cambial systems are transmitted from the environment through the other production systems. Therefore, diameter growth is for tree vitality and activity what a thermometer is for temperature: a reliable indicator, if one knows how to read it.

This reading is complicated, because inputs in year t not only determine growth processes in that year, but also in the next years (Fig. 3.34). Except for the cambium, these inputs are used also in meristem or bud formation, establishing reserve stocks, and phytohormone production in preparation of next year's growth. When these are functioning in year t+1, they in their turn prepare growth in year t+k, which may include fruiting. In Fig. 3.34, all these interactions between the tree's subsystems have been indicated by the term autocorrelation between the years t-1 and t. The ring made in year t can be traced back and analysed. Trees that produce growth rings that can be distinguished anatomically, which even in the humid tropics seem to be frequent, write their own history that can be read back from ring width. This dendrochronology among other things has led to the description of earlier climates (cf. Fritts 1976; Matthews 1976; Creber and Challoner 1984b). In this way, wooden objects from past centuries can be dated. The developmental history of a tree or a stand can be reconstructed by reading the rings. For instance, these can clearly indicate the alternation of open and closed canopies, of dominance and suppression (Fig. 3.35).

Figures 3.36 and 3.37 show some examples of analyses of growth rings or diameter growth rates (Fig. 3.37B) in tree trunks observed in different climatic zones, and at different time intervals. The influence of pollution (Fig. 3.36) on yearly ring width series, the influence of monthly rainfall on rainforest trees in Suriname (Fig. 3.37B) are examples of different ways to use growth ring analysis in assessing the cambial activity as an indicator for whole tree behaviour.

3.6 The Life Cycle of Trees: Cycles and Variants

Each individual tree has its individual development. This is the result of a different genetic potential and a differing sequence of inputs, linked to its site. On the preceding pages, the architectural and physiological inheritance of trees has been

wood is formed and twigs die early (*1*), then reiteration takes over production of green mass, which is confirmed by juvenile leaves (*2* and *insert*); finally dying bark announces death (*3*). **C** *Pinus* spp. (Pinaceae, Rauh's model), same explanation as in *Picea abies*, but the model with all axes equivalent (orthotropic, Fig. 3.3) makes diagnosis of reiteration slightly more difficult. **D** *Abies alba* (silver fir, Pinaceae, Massart's model) expresses competition between top and branches, accompanied by lower branch shedding in the remarkable "bird's nest" shape, cited from Germany. Tree architecture defines the pathways of the sapstream as a spatial pattern, the degradation of which indicates diminishing feedbacks and the location of handicapped organ complexes

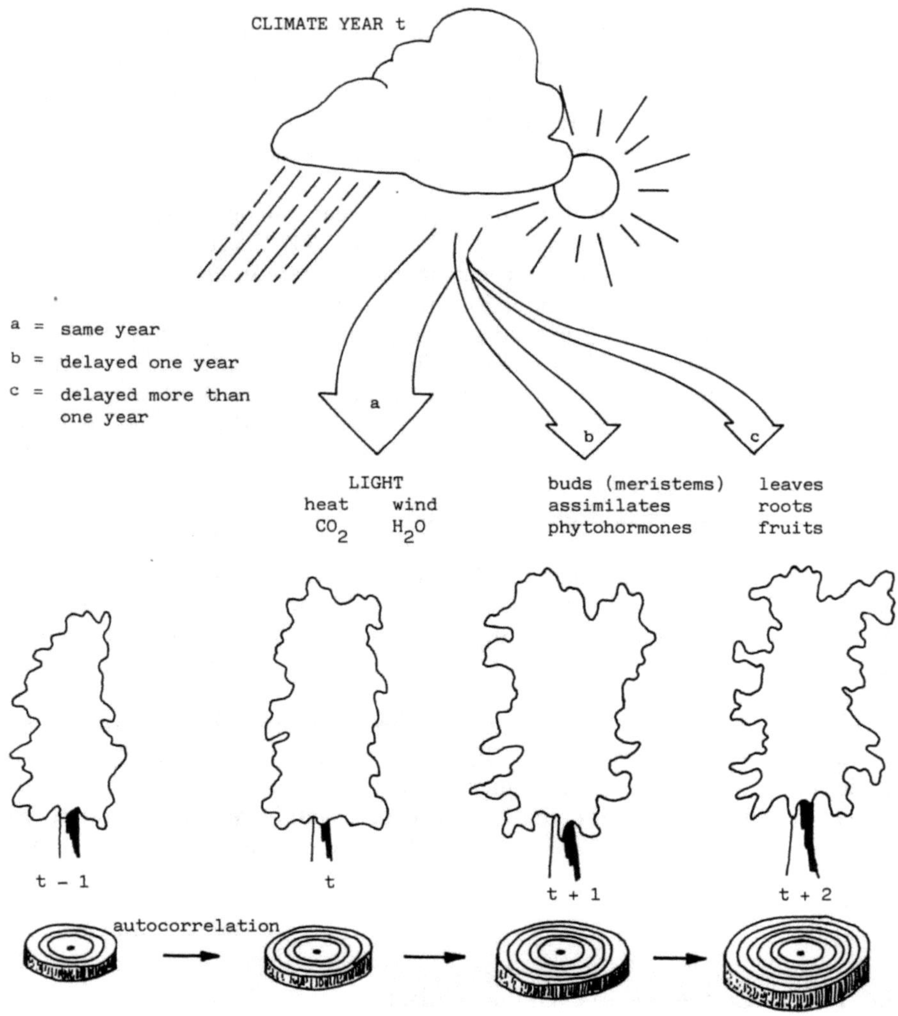

Fig. 3.34. Cambial activity as an indicator of tree activity and vitality. Because there is a delay that may cover several years in the transmission of inputs towards the cambium, and because other secondary production systems may compete, growth ring width cannot be used straightforwardly without taking these delays into account. (After Fritts 1976)

explained: all tree behaviour is a variant upon these structural and physiological patterns and processes.

3.6.1 The Backman Theory

What precisely is the life cycle of a tree? This question has occupied scientists since the times of Leonardo da Vinci (1452-1519, see Gibbs-Smith 1978, pp. 87-103; and Zimmermann 1983, his Chap. 4.1), including Goethe (1749-1832), who coined the

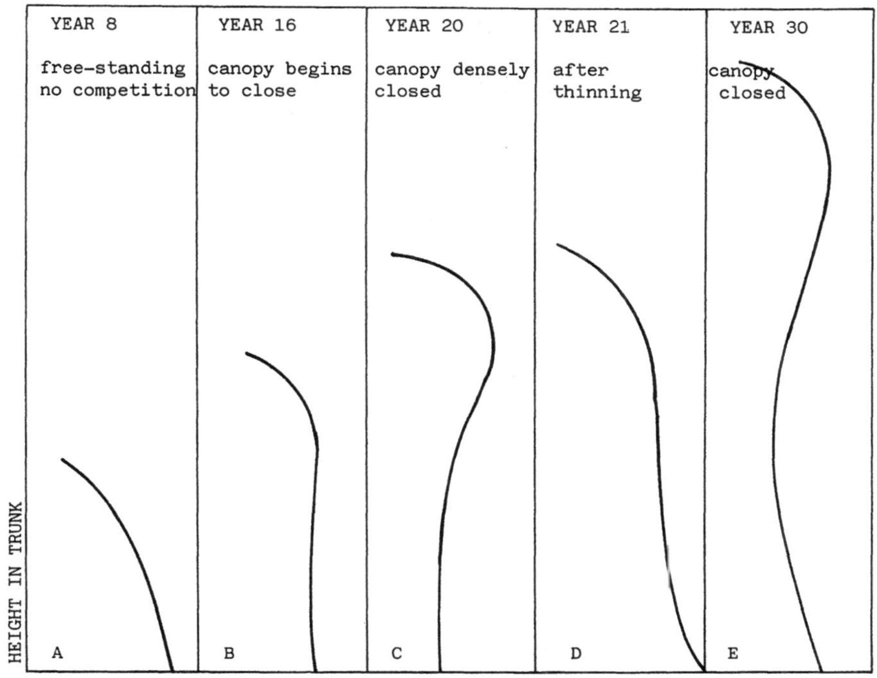

Fig. 3.35. Influence of outside inputs upon the cambial activity at different height levels (cf. also Fig. 3.31). Example of *Pinus resinosa* (Pinaceae, Rauh's model, red pine, Great Lakes, North America), selected years. **A** Living branches with needles over the whole height. **B** After the 15th year the lower branches are in shade. **C** Between the 16th and the 20th year the lower branches die. **D** Thinning gave the whole tree light again, which is apparent from the graph at the 21st year, after thinning. **E** In the 30th year shade effects beneath a closed canopy are very apparent again. The horizontal scale (width) is about 1000 times the vertical scale (height). (After Farrar 1961)

concept of plant metamorphosis (Metamorphose der Pflanzen) and Darwin (1809-1882, see Moorehead 1971 or Leakey 1979), who compared the architecture of trees with that of corals. Around the turn of the century, D'Arcy Thompson (1917) made his famous study "on growth and form". In the first half of our century, hundreds of studies have been done on this subject, but since the Second World War no really new ideas seem to have emerged. Therefore the following pages will be used to summarize the theories on growth and organic time by Gaston Backman (1942, 1943), who seems to have had the last word, at a time when it was difficult to let it be heard internationally because of the war.

In 1943, Backman wrote about the conditions that should be fulfilled in order to construct biologically acceptable growth curves. These requirements were (see Backman, 1943, p.5):

- a *cumulated growth curve* should start at the origin of the graph, where size or mass and time all equal zero;

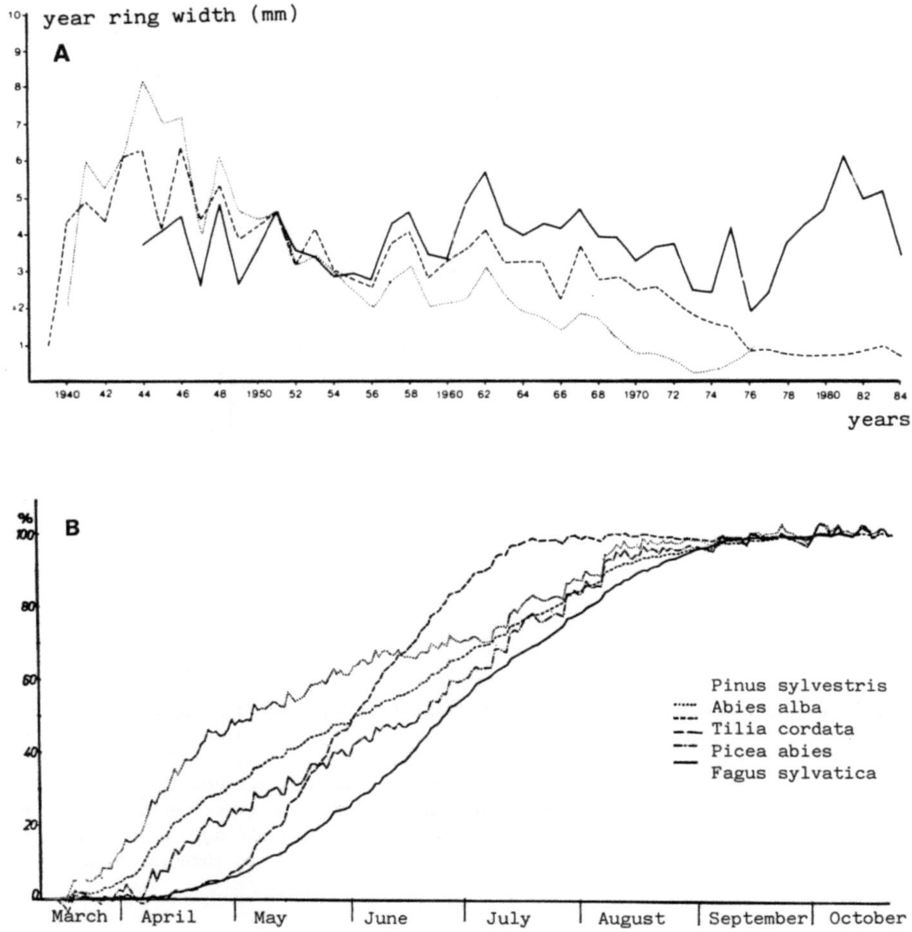

Fig. 3.36. Growth ring width as a parameter for tree growth. **A** *Pseudotsuga menziesii* (Douglas fir, Pinaceae, Massart's model) in an environment with increasing NH_4 pollution (Veluwe, the Netherlands). Average ring widths from 1940 to 1984 in the classes (1) healthy, (—— 3 trees), (2) diseased (- - - ,4 trees) and (5) dying (. . . . ,4 trees). (De Kort and Baas 1986). **B** Daily diameter increment in five tree species near Freiburg (FRG). *P. sylvestris* starts growing early, *F. sylvatica* is late. Both close their growth late in the season. *T. cordata* starts rather early and finishes early. Note daily modulation of the growth curves. (After Mitscherlich 1970)

- it should tend asymptotically towards a final size or mass;
- it should inflect, so that the curve has an S-shape and generally the inflection should be situated asymmetrically towards the left;
- *growth rate curves* should have two inflections at both sides of a maximum, the inflection to the right generally being further away from the maximum than that to the left;
- growth rate curves should also start at the origin;

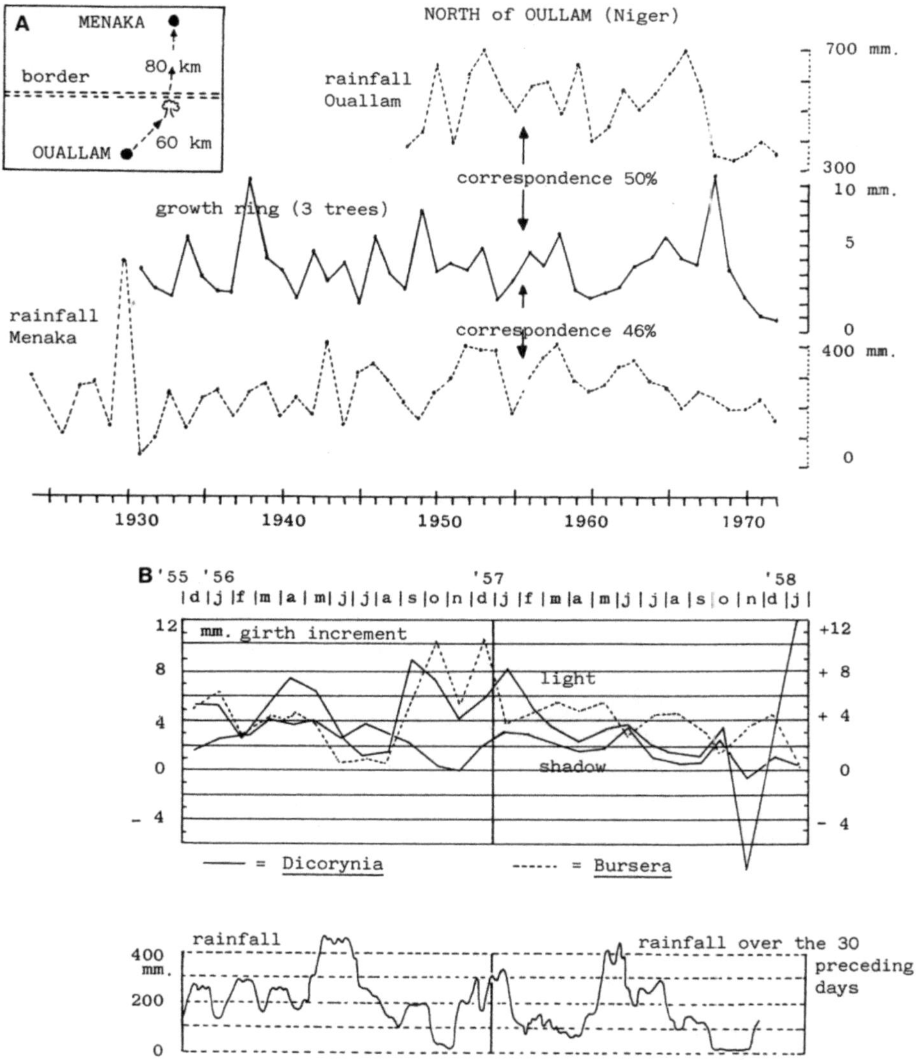

Fig. 3.37. Growth ring width as a parameter for tree growth. **A** *Acacia raddiana* (Mimosoideae, Champagnat's model probably, cf. Fig. 3.22). Growth ring width depends upon rainfall in this semi-arid region in Niger. (After Mariaux 1975). **B** *Dicorynia* sp. (Papilionoideae, probably Troll's model) and *Bursera amara* (Burseraceae, model unknown), growth ring width as correlated with the rainfall over the preceding 30 days in Suriname rainforest. (After Schulz 1960)

- growth rate curves should leave the x-coordinate under a stronger angle than returning to it after the maximum.

All generally accepted growth curves from more recent years seem to fulfill these requirements (cf. Fig. 3.38A, averages). Now these requirements rested upon the assumption (Backman 1942, p.455) that: "the logarithm of the growth rate is negatively proportional to the square of the logarithm of time, if the growth rate is measured by the maximum rate and the time by the time of the same." Mathematically expressed:

$$\log H = k_2 \cdot \log^2 T, \tag{3.6.1}$$

in which H is the maximum growth rate, T the time at the maximum, and k_2 a constant.

This formula can be transformed into physical units measuring time and space, then expressed as a differential function dy/dt with elimination of the logarithms, and solved, which yields three constants, i.e. c_0, c_1 and c_2. The math-

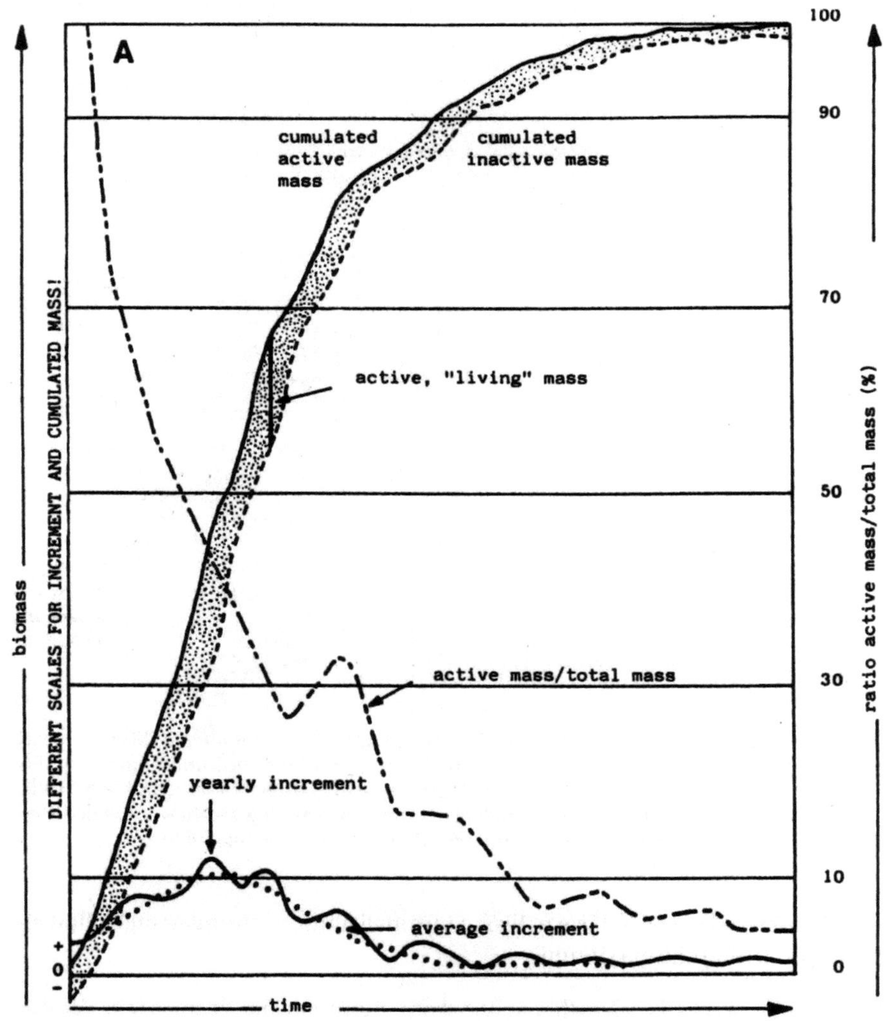

Fig. 3.38A

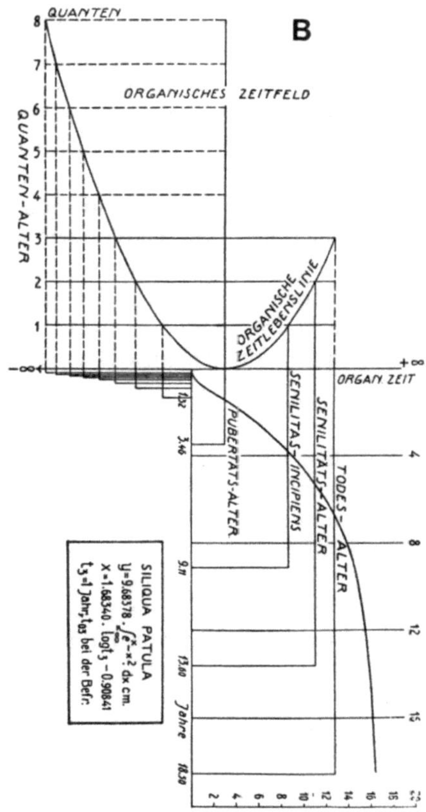

Fig. 3.38. Tree development over the whole life cycle. **A** Cumulated growth, real and averaged increment, and metabolically active biomass, assumed to survive for 2 years. The tree is assumed to go through nine unfavourable periods during its life. Schematic curve, of which real curves are variants. (Oldeman 1974a). **B** Transformation of symmetrical curves expressed in a theoretical "organic" time-space continuum into a real growth curve for one species (*Siliqua patula*) by Backman (1943). German words: Pubertätsalter = age of puberty; Senilitäts-Incipiens = senescence; Senilitätsalter = age of senility; Todesalter = age at death; Zeitfeld = time field; Zeitlebenslinie = temporal life-line. Backman's theory should be studied precisely with the data available now

ematical treatment is not reproduced here. The significance of c_0 is that of a scale of physical space, indicating the proportion between the theoretical space and cm, mm, g, kg and so on. Derived from function (3.6.1) after reducing the solved differential equation by an integration, in which c_0 appears, the measuring stick for the organic time dimension then becomes x:

$$x = c_1 \cdot \log t + c_2, \qquad (3.6.2)$$

in which c_1 is a scale of organic time, and c_2 measures the physical time used relatively to c_1, logarithmically. It may be concluded that organic time is a logarithmic function of physical time.

The integral function mentioned, but not reproduced here, is symmetrical and expressed in organic time. Hence, the half-time of this function is important, because it is a reference point to be used when transforming this curve into a real, asymmetrical growth curve that fulfills the requirements enumerated above. This transformation has been visualised for a plant species in 1943 (Backman, his Fig. 4), reproduced here as Fig. 3.38B. It shows in the "organisches Zeitfeld" (organic time field) that indeed the assumption about maximum growth rate H and time at that moment T has been used. The concept of "biological quanta" remains undiscussed here. Backman and his students worked 20 years to check the transformation of the theoretical "organic time curve" towards real-time growth curves by establishing the c-values for different organisms, among which are trees. They found that biological key moments in a growth curve were reached at values for x when in

$$x = \pm \sqrt{n/2} \qquad (3.6.3)$$

n is a whole positive number or zero. These key moments are:
$x = 0/2$: sexual maturity;
$x = 3/2$: natural death.

The significance of $n = 1$ and $n = 2$ is probably linked to stages of ageing. The whole numbers above 3 do not seem to have biological significance.

These principles once being established, Backman and his collaborators found that the growth of organisms happens in cycles, each of which follows a growth curve that occupies increasing amounts of real time. Considering an enormous amount of literature and observations, Backman (1942) concludes that trees grow in three cycles: tricyclic growth. The first one should correspond to the "seedling period" mentioned above (Sect. 3.2.1). The second one is called by this author "primordial growth", in analogy with his observations on animals; after his observations on trees from the temperate regions we can find no apparent and neat correspondence with an architectural growth period. The third cycle takes the bulk of the real time of the whole life of the tree. Table 3.1 gives an example studied by Backman (1942), in which he compares calculated and measured values of height growth in a tree species.

The hypothesis is advanced here that there may be closer correspondance between Backman cycles and architectural growth phases than is now apparent (also compare Fig. 3.1). The values $n = 1$ and $n = 2$ in particular might be relevant. The testing of this hypothesis would be extremely interesting, especially in the humid tropics where architectural sequences can be more easily distinguished, where more tree species occur and where seasonal effects are minimal. It would also be interesting because this theory seems to have power of prediction, as empirically confirmed by the foresters of the German school of "Nature-conforming Forestry" (Naturgemässe Forstwirtschaft). These predictions by Backman follow the rules enumerated hereafter:

- The total life span of a tree is longer if: maximal growth rate is reduced; maximal growth rate is delayed; sexual maturity is delayed; the end size is larger;

Table 3.1. Growth of *Fraxinus excelsior* (Oleaceae, Rauh's model, ash) as calculated by Backman and measured by other authors. (Backman 1942, p. 468). Details and calculations: see original publication. This is the example with the largest deviations between calculated and measured height among all species considered by Backman (but cf. Sect. 3.6.2)

Years	Cycles			Growth	
	I m	II m	III m	Calculated m	Measured m
0	0	0	0	–	–
1	0.07	0.00	0.00	0.07	0.07
2	–	0.08	0.01	0.16	0.18
3	–	0.26	0.07	0.40	0.41
4	–	0.37	0.23	0.67	0.57
5	–	0.41	0.53	1.01	1.02
6	–	0.42	0.97	1.46	1.21
7	–	0.43	1.55	2.05	–
8	–	–	2.23	2.73	–
9	–	–	3.01	3.51	–
10	–	–	3.84	4.3	4.9
15	–	–	9.55	10.1	8.7
20	–	–	12.31	12.8	12.5
25	–	–	15.45	16.4	15.7
30	–	–	17.8	18.3	17.9
35	–	–	19.52	20.0	19.7
40	–	–	20.78	21.3	21.3
45	–	–	21.70	22.2	22.2
50	–	–	22.39	22.9	22.7
55	–	–	22.9	23.4	23.0

maybe if cycles I and II have a lesser share.

- Maturity occurs later if: maximal growth rate is reduced; maximal growth rate is delayed; the end size is larger.
- The end size is larger if: the life span is longer; maximal growth rate is reduced; maximal growth rate is delayed; maturity is delayed; maybe cycles I and II have a more important share.

"Naturgemässe" foresters hence speak of a "Backman-Fichte" (*Picea abies*, Pinaceae, model transition Massart/Rauh, Norway spruce) if the tree has been waiting a long time in the shadow under closed forest, for example with branch tiers 1 cm apart and a height of about 1.5 meters at an age of 40 years (Fig. 3.39). They expect, because the maximal growth rate and sexual maturity are delayed, that such a tree will live longer and yield a larger mass of wood of higher quality, if liberated by opening up the canopy above it. Poplars or eucalypts are grown in completely op-

Fig. 3.39

posite conditions on large clearcut surfaces in the full sun, pushing them to reach their maximum growth rate as soon as possible. The prognosis according to Backman's rules then is quite different.

3.6.2 Modulated Life Cycle Curves

In reality all growth curves are modulated by environmental influences (Figs. 3.36; 3.37). The aberrations in Table 3.1 have been ascribed by Backman to growth deviations under environmental influences, as is also implicit in his rules. Growth modulations therefore are important. They can be considered over different time spans: the whole life cycle, a year, a month or even a day. Growth differences due to different environmental inputs may be truly enormous (Fig. 3.39), so that tree size may never be taken as a parameter for chronological age (also see Oliver 1978). The criteria used to represent modulated growth curves are those outlined in Section 3.5.1.

Backman created a theory of growth, because his "organic time" claims to have universal validity for all living, that is growing organisms. In the following pages, we will review various models of modulated growth curves, because they refer to a limited part of reality only: the tree or one of its subsystems (see Table 2.1).

Modulations over the whole life cycle (Fig. 3.38) are linked to the inherited properties (architectural dynamics, physiology) and the inputs over a whole life span. Anticipating Chapter 5, and referring to Fig. 3.1, it may be stated that the sequence of inputs is not altogether due to chance factors. These have been considered in Section 3.5.5, e.g. the direct influence of unpredictable climatic factors. The "temperament" of a tree, as represented in Fig. 3.1, indicates the environmental margins within which that tree can function: for instance, a pioneer is able to grow only if 60% to 100% of gross macroclimatic irradiation is available and on soils that are mineralised to a certain degree, also between limiting values. These conditions are often regulated by the ecosystem.

Growth performance of a tree over its whole life span hence oscillates between the maximum performance at a certain age, at optimal ecosystem conditions, and the minimum performance still allowing survival, at a certain age at pessimal ecosystem conditions but not lethal ones. This affects the balance between metabolically active biomass and over-aged (or shed) metabolically inactive biomass. Now all metabolically inactive mass, considered "dead", is shed sooner or later, except for the wood that remains captive, after becoming inactive, within the "sapwood"

Fig. 3.39. "Backman-Fichte", "Backman spruce" (*Picea abies*, Pinaceae, model trans. Massart/Rauh). These spruces germinated on a fallen log in the shadow, some 40 years ago. Branch storeys are a few centimeters apart, and age may correspond to that of the large spruces in the background. The wood of such Backman spruces is supple and hard as steel (very narrow growth rings) and is sold at a high price for special purposes. According to Backman's theory, the first two growth phases have been much prolonged and the maximum growth rate much delayed, so that these spruces may become very old and big. Hyytiälä, Finland, natural forest reserve Susimaki, August 1988. (Photographs by R. A. A. Oldeman)

that still is active in sapstream transport. For this reason, and because of the indicator function of the cambium, the balance between activity and inactivity can be most precisely measured and interpreted in wood increment studies or growth ring research.

For certain species, data exist on the period of activity of each yearly wood layer (cf. Braun 1963, 1970). Wood remains active for a period of 1 to 10 years. In Fig. 3.38, it is assumed that the wood remains active for 2 years, and that the tree concerned passes through nine unfavourable periods in its life.

The curve of yearly increment (Fig. 3.38, below) hence oscillates nine times around the mean curve, and this oscillation can be found back in a corresponding modulation in the cumulated growth curve. Both may be expressed, for instance, in dm^3 of wood. Now in order to obtain the quantity of active wood at any one moment, the sum of the increment of the 2 years before is substracted from the cumulated curve. The lower line then indicates the cumulation of "dead" wood in the tree over time, and the shaded zone between the two cumulated curves shows the part of metabolically active wood at each phase of the life cycle. Dieback of wood gradually overtakes growth during the life cycle, the ratio becoming lower as life goes on, but also in a modulated way. In the senescent stage of the tree (Fig. 3.1), the amount of active wood becomes insufficient to keep the tree alive. This may take a long time. *Pinus aristata* (Pinaceae, "Bristlecone pine", California, Rauh's model, see Leathart 1977, p. 88 for illustration; also Matthews 1976) may live up to 5000 years and *Sequoiadendron giganteum* (Taxodiaceae, "giant sequoia", model unknown, see Engbeck 1976) can reach 3300 years of age. Even if the present graph represents wood, the same type of diagram might be established for other tissues (bark) or organs (leaves, brachyrhizae) if one could accurately determine the quantities of dead and shed biomass over the whole life cycle.

Parallel to the gradual shift in the ratio between "living" and "dead" biomass, a general change in growth processes occurs. In the first part of life, growth increases, i.e. on the average more matter and energy are incorporated into the tree in year t + 1 than in year t. This can be found back in an increased average increment, i.e. the value of the cumulated mass divided by age. In a second phase, growth will decrease. These oscillations in growth are translated into higher and lower cumulated mass increments over time.

Both height growth and thickening show this development, which conforms to Backman's requirements [see Eq. (3.6.1)]. Height growth cumulates earlier and steeper than diameter increment (Fig. 3.40). Both curves are characteristic in their expression per tree species. Height growth depends more on the species, whereas diameter or girth increment can be influenced to a greater degree by the environment, in particular by the "social position" of the tree in the forest (cf. Leibundgut 1970; Mitscherlich 1970, also Sect. 5.2.4.1). Therefore, height is used as a criterion for silvicultural classes of productivity (yield classes). Maybe this criterion should be replaced by diameter in older trees (Delvaux pers. commun.), but this has not been done in a systematic way.

In Fig. 3.38A, a shorter modulation, covering periods of several years, corresponding to nine hypothetical unfavourable periods, is superposed upon the whole

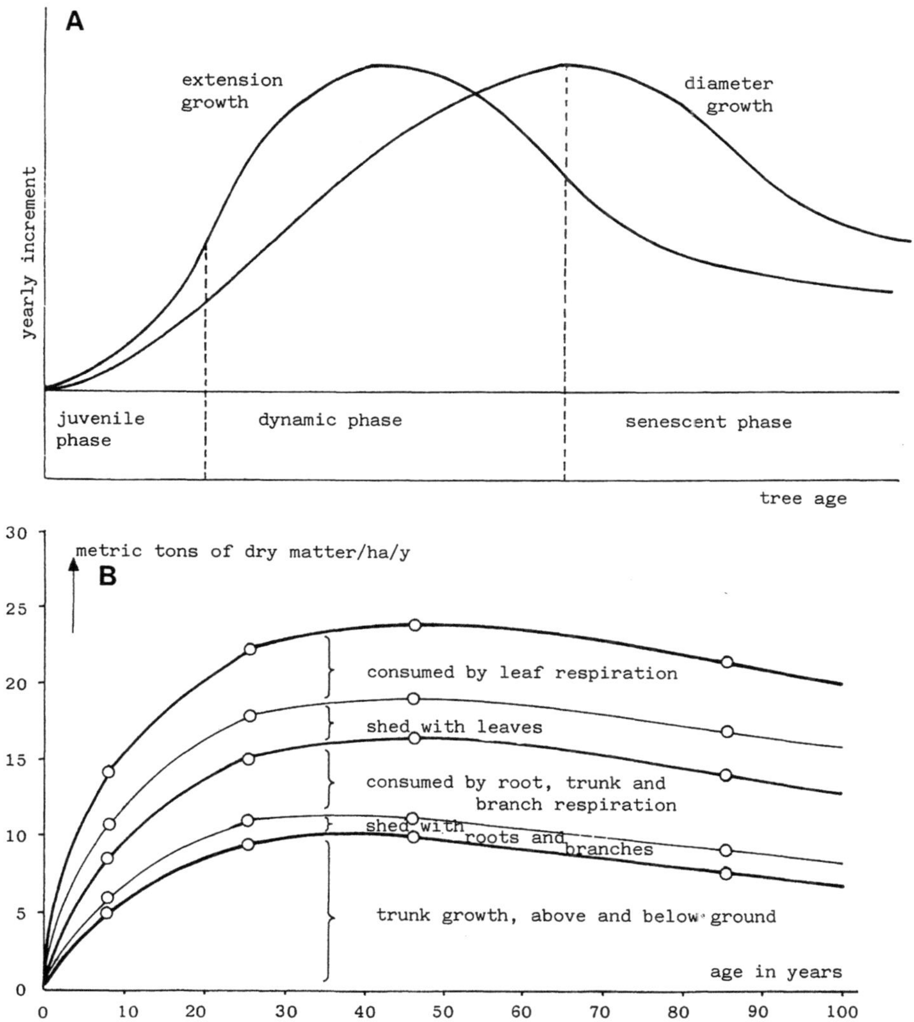

Fig. 3.40. Tree development over a whole life cycle. **A** Growth phases, according to production patterns (also compare architecture). Extension growth reaches a maximum before secondary thickening. These graphs shed a doubt on the use of one single dimension, e.g. height, in theories like Backman's. (Illustration courtesy Dr. P. Schmidt). **B** Dry matter production per year per hectare, and its distribution over different organs in even-aged stands of *Fagus sylvatica* (beech, Fagaceae, Troll's model) of second yield class in Denmark. The *upper, heavy line* indicates total production, the *lowest one* shows the part that is incorporated into the tree as cumulated mass. (After Möller et al. 1954; also cf. Fig. 5.79)

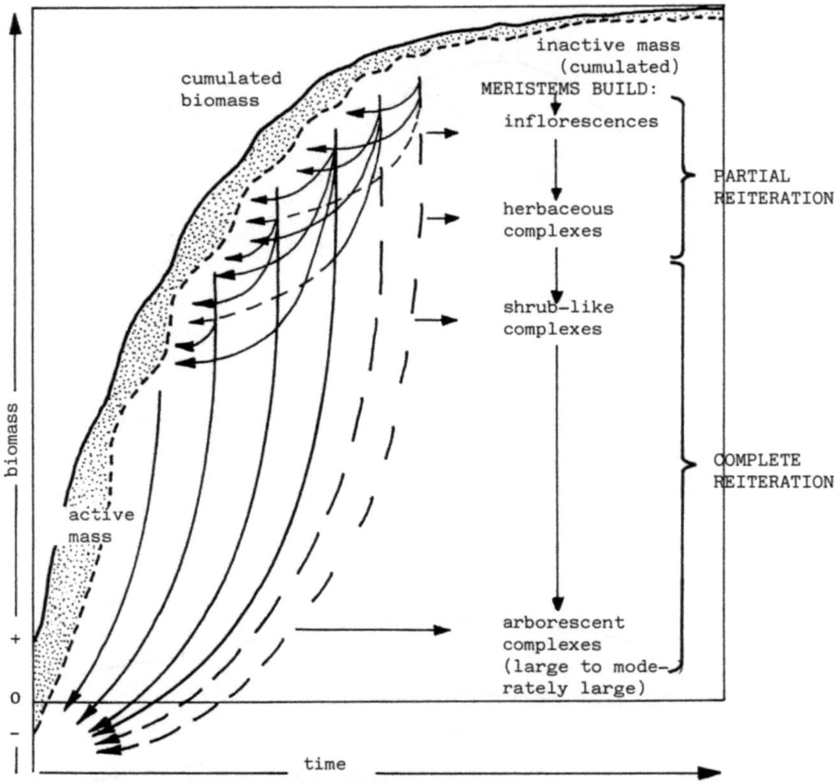

Fig. 3.41. Production patterns and tree architecture (cf. Fig. 3.1), over the whole life cycle. The amount of active mass being a parameter for vitality (vigour), each time that favourable circumstances lead to an increase of that amount, the tree becomes reiteration-prone. The older the tree, the more meristems become involved that each build smaller reiterated complexes. These function biologically like herbs, shrubs or whole trees, but without roots, depending on their size and physiology. (Oldeman 1974a)

S-shaped cumulated curve and its asymmetric bell-shaped increment counterpart. This modulation is most current in successor trees (Fig. 3.1), that grow through alternating periods of suppression and liberation, also in natural forests. But in pioneer trees, such modulations may also be observed during series of dry or stressed years. The general trend of Fig. 3.36A is an example of this kind of modulation, also expressed as an oscillation of the current yearly increment around the average increment.

During the unfavourable periods, the replacement of inactive tissues by active ones may be so reduced that the production systems in the tree are irreversibly damaged. Trees will then die, and this is the fate of many young trees that stand too long in the shadow or in dry conditions. But if they survive, liberation may cause very fast growth. A following unhappy period may then reduce growth again. This alternation of fat and meagre periods ends when the tree becomes dominant, but then it

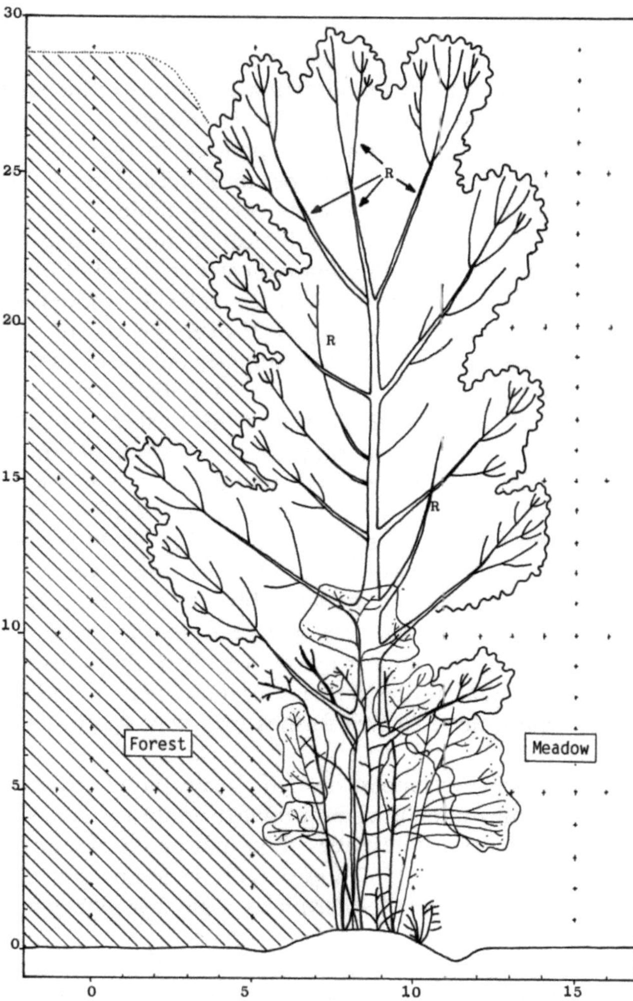

Fig. 3.42. *Populus x euramericana* (Salicaceae, Rauh's model) in a living hedge separating forest from meadow. Note the reiteration pattern (*R*) conforming to the principles shown in Fig. 3.41. (Scale drawing made on the Verwolde Estate, Laren, the Netherlands, courtesy I. van Vliet 1979)

is exposed to macroclimatic impacts. These production patterns can be found back in tree architecture (Fig. 3.41): when energy and matter are plentiful, as can be seen from the relative increase in active biomass, extra meristems are mobilised and reiteration takes place. The size and biological behaviour of the reiterated parts make them either comparable to mere inflorescences, the architectural model being reduced to almost nothing except its flowers, or to herbs, shrubs or small trees, depending on the quantity of extra inputs available. One could compare, for instance, Figs. 3.41 and 3.42.

Modulations with a yearly period are illustrated by *Acacia raddiana* in Sahelian Africa (Fig. 3.37A). They are usually found in nearly all trees, because successive years are different. This modulation cannot be influenced by silvicultural treatments, contrary to those that cover series of years.

Modulations per season are apparent in the monthly girth increments of two species from the Suriname rainforest (Fig. 3.37B). They translate an influence that is essential for the correct choice of tree species in silviculture: the length and favourable ecological factors of the growing season. Height growth is the first criterion that indicates seasonal effects. For instance, *Populus x euramericana* cv. Robusta (Salicaceae, Rauh's model, Fig. 3.42) under normal circumstances shows height growth over the whole growing season. *Quercus robur* (Fagaceae, Rauh's model, summer oak) often shows a second flush later in the growing season. This is the "Lammas shoot", around St. John's day at the end of June. Another result of this flush may be the induction of another growth ring, a "false year ring". This is one of the indications that temperate Fagaceae have a tropical background, as is also apparent from the geographical distribution of this family. In the tropics, there often is more than one growth ring per year, accompanying each flush, or rings are lacking altogether when growth is continuous. An example is *Hevea brasiliensis* (Euphorbiaceae, Rauh's model, rubber tree; Hallé and Martin 1968) with six flushes and six growth rings per year. In the greenhouse, without a winter, oaks behave in the same way. In the open, they flush once, sometimes twice and exceptionally thrice per season, indicating an endogenous growth rhythm, only skipping three to five flushes in a way synchronised with the temperate seasons.

Extreme circumstances (drought, heavy cloud covers reducing light, inundations) may sometimes bring the growth to an abrupt halt in the middle of the season. In some tropical rainforest trees, there may be shrinking of the diameter (*Dicorynia guianensis*, Oct./Dec. 1957, Fig. 3.37B.). In temperate trees this also may lead to false year rings. In the Mediterranean region, growth usually stops during the dry summer.

Daily modulations (Fig. 3.43) have more physiological than ecological significance, unless such a day shows a peak-load stress that influences the whole behaviour of the tree in the future ("poisoning"). But normal daily modulations are most often linked to the sapstream. When transpiring strongly, as in daylight, the trunk and branches contain less water. They shrink in diameter. At night, transpiration diminishes and the axes expand again. This oscillation is strong when the water budget is abundant (Fig. 3.43: 15 April or 22 October 1966), and weak when this budget is scarce (4 June, 1966).

3.7 The Life Cycle of Trees: Proportional Diagnosis

A tree body shows proportions that most often change during its life. These variable ratios between dimensions such as height, partial height, diameter at normalised levels (above the roots, breast height) indicate phases of the develop-

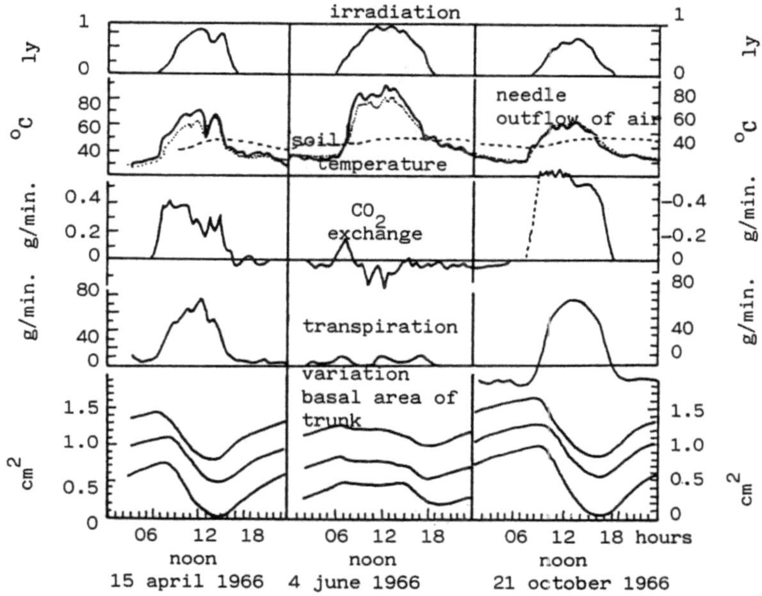

Fig. 3.43. Daily modulation of growth curves, as monitored by the change in basal area in 110 years old *Pinus resinosa* (cf. Fig. 3.35) and compared to transpiration, photosynthesis, temperature of needles and soil and irradiation, before (15 April), during (4 June) and after (21 October) the summer drought. (After Fritts 1976)

ment of a tree, or vitality classes or other properties of a tree at one moment of its life. Another aspect is that these proportions can be experimentally studied. They are parameters, or symptoms, that allow us to arrive at a diagnosis of the state of a tree.

Some of these ratios were mentioned earlier, e.g. the ratio between active and inactive biomass. Another, that has not yet been treated, is the shoot-root ratio, indicating the proportions between aboveground and subterranean parts of the tree. The root system having been relatively little studied in trees, the shoot-root ratio is better known in herbaceous plants. In trees, the aboveground mass is partly inactive, however, and therefore the shoot-root ratio is variable. The often advanced "constant shoot-root ratio" seems to be a myth, at least for trees. Even in the seedling phase, this ratio varies and is used in the vitality diagnosis of seedlings in nurseries (cf. De Fraiture 1985). Maybe it would be better to abandon the shoot-root ratio altogether, except for specific empirical use in certain circumstances, and rather consider the ratio between physiologically active organs, for instance the leaves and root-hairs. One indication for the usefulness of such an approach is the close relation between foliage weight and the cross-sectional surface area of sapwood found in *Pinus sylvestris* (Pinaceae, Rauh's model) by Whitehead (1978).

However, it is much easier to measure and process the ratios between linear measures. There are many.

The h/d ratio. This ratio is related to the architecture of the tree. Oldeman (1974a) found that nearly all model-conforming trees in French Guyana showed the ratio of

$$h = 100 \cdot d \qquad (3.7.1)$$

in which h is expressed in meters, d in centimeters. This has been confirmed since then in the gallery forests of Côte d'Ivoire (Devineau 1975; Valkeman 1985), in several forest types in India (Pascal 1983), in submontane forest in Papua New Guinea (Louman, 1986) and in the Netherlands for beech (*Fagus sylvatica*, Fagaceae, Troll's model; Van Alfen 1981).

If a young tree breaks off, the ensuing regenerative reiteration pushes this ratio upwards:

$$h > 100 \cdot d \qquad (3.7.2)$$

And when a tree grows out by abundant reiteration within the crown (cf. Fig. 3.1 and Chap. 5.2.4.: trees of the present), the ratio becomes:

$$h < 100 \cdot d \qquad (3.7.3)$$

Faber (1975) considered this ratio in Douglas fir stands in the Netherlands (*Pseudotsuga menziesii*, Pinaceae, Massart's model) and found that during storms no windthrow occurred in stands where $h < 50 \cdot d$.

Figure 3.44 shows how this information can be used graphically. The line $h = 100 \cdot d$ here is a reference line and **not** a regression curve. The method is not statistical, although the reference line has been established according to statistical principles by considering populations of model-conforming individuals per tree species (cf. Oldeman 1974a). Considering the points on the graph, and indicating by symbols the architecture of each tree as "model-conforming", "regenerated after damage" and "expanding by reiteration", their position in relation to the reference line is diagnostic.

Just as an analogous cloud of graphical points for individuals of one tree species gives an indication for the development of that tree species, the presence of trees in different developmental stages in a forest stand (eco-unit, cf. Sect. 5.1) is symptomatic for the development of that stand. In other words, the h/d ratio (Fig. 3.44) is an element of explanation of a forest ecosystem in terms of its component trees (cf. Sect. 2.1; Table 2.1). The other way round, human impacts on such forest units can be explained by this method (Louman 1986), which rests on the understanding of tree behaviour.

The h/d ratio is certainly linked to the different growth curves for both measures, as represented in Fig. 3.40A. It is interesting to note that Backman (1942) used only height growth as a parameter. The use of ratios, linked to tree architecture and production, may be a means to improve the Backman theory.

The linear relation h/d is a symptom that rests upon "deeper" ratios which can be linked to biological necessities in the functioning of a tree. These more complex functions have been studied for a long time. They may be shortly summarised as follows.

The h/d^2 ratio. The square of the diameter is proportional to the cross-section of the trunk. And the cross-section of the trunk, or rather its increment, is a parameter for the transport capacity of the sapstream (cf. Fig. 3.38A). Transport capacity is

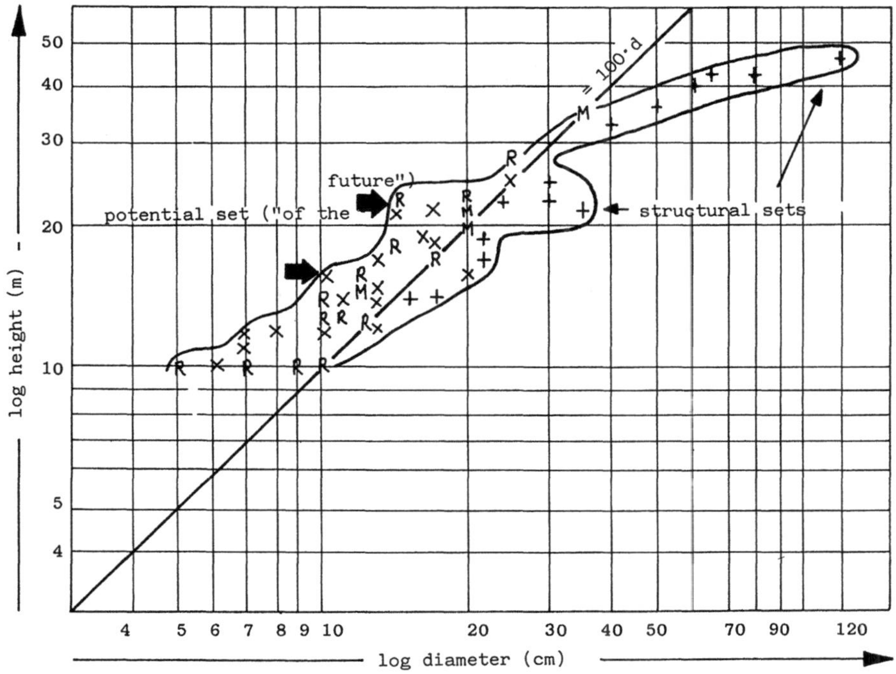

Fig. 3.44. Changing proportions of the tree during its life cycle. Model-conform trees (M) show a height-diameter ratio of h ≈ 100d. The trees marked (x) cannot be classified exactly with architectural criteria because they are damaged and misshapen; most show h = 100d. In the graph presented here, the trees marked (+), showing h > 100d and that are sustaining the biostatic forest (see Sect. 5.2.4) on a 20 x 40 rainforest plot in French Guyana can be diagnosed as to their ecological role by their h/d ratio, indicating their ecological function and giving a suggestion as to their vitality. Note that the line *h = 100d is not a regression curve*, but a line of reference. (Oldeman 1974a)

one of the key functions determining tree vitality. Variations of this ratio hence are symptoms of the state of the tree because they refer to both cambial activity and overall extension growth. Given the different ways of height growth (Figs. 3.1 and 3.2) and the great differences in hydrological properties of wood (Braun 1963, 1970), the interpretation of this ratio as a symptom for tree vitality will still need much study. The importance of the subject is clear from the use of "basal area" figures, which represent the sum of the cross-sectional areas of all tree trunks per hectare, as one of the oldest and most frequently used parameters in silviculture.

The h/d³ ratio. The third power of the diameter is a mechanical function, that indicates the support function of the trunk, or any other vertical axis, for lateral branches, according to the formula

$$d^3 = k \cdot (F \cdot x), \tag{3.7.4}$$

in which d is the diameter, k is a constant, F the horizontal force and x the distance between the diameter considered (e.g. at breast height, 1.30 m) and the point of application of force F, exerted by the weight of the branch. This relation, leading to more complex exponents of the type n/3 can be studied in more detail in McMahon (e.g. 1975), or McMahon & Kronauer (1976).

The h/d^4 ratio. The fourth power of the diameter, as expressed in more complex exponents, is probably also a mechanical function. It would be linked to the elasticity of axes in a tree. Elasticity is very important for a tree to withstand varying wind effects, from gentle breezes to gale force and being applied over a continuous period or in gusts that may cause a rhythmic oscillation of the whole tree. These aspects have been studied, for instance, by Bosman (1979) or Kuiper and Van Schooten (1985).

General aspects of h/d relations. According to Stevens (1974), trees build their body with a minimum of matter to fulfill their biological functions. The seeming miracle of tree growth is that, in one way or another, trees have adopted growth patterns that accommodate all relations between height and diameter. As yet, there are no numerical models that allow us to calculate how this is done, but it is certain that the assimilating body has to be structured (h/d), water has to be provided to all its parts (h/d^2), it has to keep upright (h/d^3) and it should dampen swinging movements of the crown (h/d^4). Given the fact that the resulting growth patterns have evolved along different evolutionary lines and in different environments, from the boreal regions (e.g. Larsen 1980) to the tropical rainforest (e.g. Hallé et al. 1978), it is the more amazing that architectural tree analysis shows them to be variants upon a very limited number of geometrical structures. The same architectural models can be found in fossil trees.

The h_f/h_t ratio. This ratio compares the height of the free trunk (h_f) to the total height (h_t) of a tree. Japanese research (e.g. Ogawa et al. 1965a, b) considered this ratio by means of **crown depth diagrams**. Their definition of free trunk height is the distance between the soil surface and the first living branch. Hence their diagrams compare all developmental periods of a tree (cf. Fig. 3.1).

Oldeman (1974a) limited his study of this ratio to trees that had started their final expansion by strong, tree-sized crown reiteration. His definition of free height therefore was the distance between the soil surface and the first living large structural fork in the tree. He did not consider small branches or suckers below this fork, even if they were still alive.

This difference in definitions of course is linked again to the scientific questions asked (Sect. 2.1). It may have been linked also to the tree species considered. In Japan, this research was started with coniferous trees, that begin reiterating rather late and which are commonly used in silvicultural systems that most often cut short the reiteration period by harvesting the trees. In French Guyana, where Oldeman worked, large Dicotyledonous rainforest trees were considered, in virgin forest. However, there is no contradiction as to the principle, and Fig. 3.1 shows a compromise: free height is considered from the model-conforming phase to the end of the tree's life, but adventitious, reiterated shoots are not taken into account and neither are "rest branches" from an earlier developmental period.

Crown depth of course is a direct symptom of the photosynthetic activity in a tree, if its general architecture is known, particularly as to its tendencies to spread the crown in a lateral direction. The point of insertion upon the central trunk of the lowest living branch or reiterated trunk is called the morphological inversion point.

It owes this name to the fact that a tendency is inverted: below, the trunk grows bigger and bigger (diameter growth; free height increase by branch shedding), whereas above the inversion point the axes become smaller and smaller, ending with the slender twigs at the periphery of the crown. This point is also important in forest stand diagnosis (Sect. 5.2.4.1).

If the ratio between free height and total height is written as

$$h_f / h_t = k, \qquad (3.7.5)$$

in which k is a variable during tree development, it would be interesting to know the value of k during the developmental process, i.e. the crown depth at different ages. This would provide a symptom (parameter) for photosynthesis. Such a parameter is easier to determine than the ratio between active and inactive tissues (Fig. 3.38A). One could then numerically express the vitality of a tree as a function of crown depth, and make calculations.

Unfortunately, there is as yet no certain way to determine the normal value of k during undisturbed tree development. A deep crown indicates a vital tree and a shallow crown is a sign of low vigour. As long as no normal values had been established, Oldeman (1974b) took an empirical value of $k = 0.5$ for "normally vital" trees in French Guyana. Figure 3.45 shows the application of this concept in the analysis of data from Thailand (Ogawa et al. 1965a, b). The conclusions appear not to be nonsensical, but the use of this value, claiming that "a tree with average vigour (vitality) has a crown depth of half its total height" should be taken for nothing else than the empirical guesstimate that it is. The reference line

$$h_f = 0.5\, h_t \qquad (3.7.6)$$

is in no way as solidly established as the reference line $h = 100 \cdot d$.

Other ratios. Many other allometric relations have been established, but generally they either do not fit in with monitoring results over longer periods of time (diameter, height etc. in forestry), or they belong to other integration levels (Table 2.1). Those relations that concern organs, e.g. length versus width of leaves, will not be treated here, and such as refer to the interaction of trees in a stand or forest ecosystem (e.g. Faber 1983, 1985) will be treated later, in Section 5.6.7.

3.8 The Life Cycle of Trees: Populations

Population biology is the analysis of organisms as grouped in species, referring in particular to number and frequency of individuals of any one species, their genetic relations, and their distribution among different ecosystems as well as inside ecosystems where they occur. As such, this chapter is not concerned with populations, because it focuses on individual trees.

However, the life cycle of any tree contains features that directly depend on its belonging to a population. This is particularly true of its reproductive organs, the flowers, fruits, cones; or in tree ferns sporangia. In Section 3.5.3, some remarks were made as to the amounts of matter and energy spent on reproduction (cf. Fig. 3.30), and to the periodicity of flowering and fruiting, for instance in mast years.

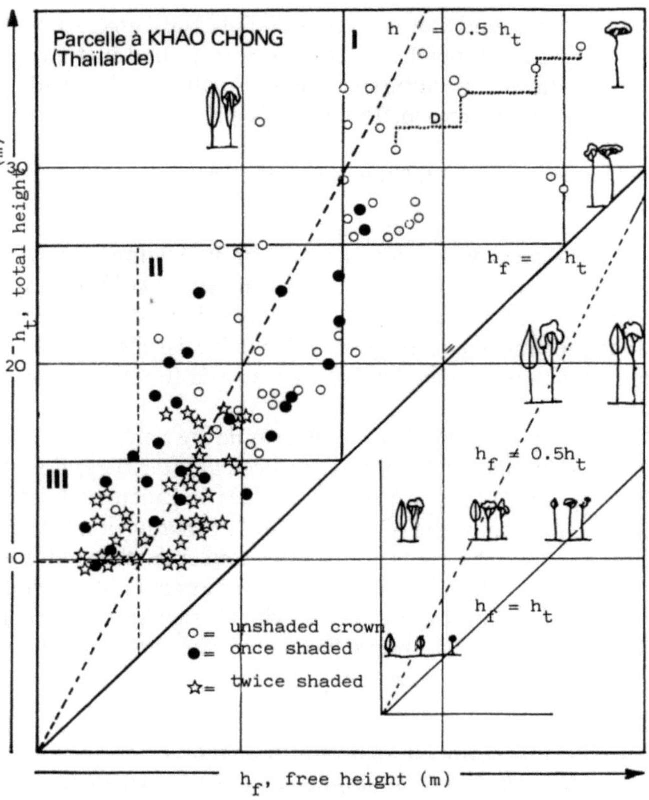

Fig. 3.45. Changing proportions of a tree during its life cycle. Data gathered by Ogawa et al. (1965a) in Thailand over a forest mosaic were compared to the empirical reference line $h_t = 0.5\ h_f$, indicating the crown depth of a tree with average vitality in French Guyana according to Oldeman (1974a). These crown depths are visualised in the *lower right-hand corner* of the figure. The interrupted line D (*upper right*) indicates the development of a crown when shedding lower reiterated crownlets and losing vitality. Ogawa et al. (1965a) placed the statistical points in three groups (*I*, *II* and *III*), which they assume to represent forest canopy layers (see Fig. 6.9). The tree vitality diagnosis from this diagram suggests that many eco-units of this mosaic (Table 2.1) are old and less vigorous. The empirical character of the line $h_t = 0.5\ h_f$ should be remembered, because the constant employed still has to be studied more closely, and explained. (Oldeman 1974b)

3.8.1 Trees as Populations of Organs or Crownlets

In a more general sense, an organism may be conceived as a population of organs. In his study on growth rhythms of the mango tree (*Mangifera indica*, Anacardiaceae, Scarrone's model, pantropical), Scarrone (1969) conceived the tree as a population of shoots, behaving either in a coherent way or erratically, i.e. independently of each other.

Torquebiau (1979) did a thorough study on trees as populations of reiterated models. Except for techniques that use trees as if they were exactly that, such as coppicing and cloning (see Sect. 3.4.1), there are a number of arguments that justify this viewpoint.

- The phenology, i.e. the chronology of biological phenomena such as flushing or flowering, often is not the same for every reiterated crownlet. One may have shed its leaves, the other may still have its old leaves, as may be observed in individuals of *Paulownia tomentosa* (Scrophulariaceae, Koriba's model, China). This species, introduced in N.W. Europe, shows every autumn that reiterated crowns shed their leaves later than the branches of the initial model and at different moments. Many tropical trees (e.g. *Tabebuia* spp., Bignoniaceae, Trop. America, Koriba's model) show the same phenomenon of desynchronised behaviour of reiterated crownlets.
- Root formation at the base of reiterated trunks (Fig. 3.23D) is quite common, according to Torquebiau's observations. Physiologically, a reiterated complex with roots is the equivalent of a rooted tree that originated from a seed. In tropical trees with buttresses, a closer analysis showed that buttressing in many species starts at the same time as reiteration. In a certain sense, this corresponds to root formation, because buttresses represent a privileged part of the sapstream that feeds one branch complex (cf. Zimmermann and Brown, 1971).
- Torquebiau (1979) argues that reiterated complexes can also be conceived as a population of smaller trees, because their mortality is linked to natural selection. His Fig. 14 and 15 indeed show that some reiterated complexes are eliminated, whereas others survive. But this does not prove that mortality depends upon natural selection. If one considers reiteration as an auxiliary level of organisation (Table 2.1), this organisation can well be programmed to favour reiteration in certain places and not in others. This would be regulation, not natural selection. Experimental research alone can solve this problem.
- Crown shyness, when crowns do not touch each other but keep at a distance one from the other, occurs not only between individual tree crowns, but also among reiterated crownlets (Offermans 1986; Fig. 2.1). Here also, reiterated crowns might be conceived as individuals which are part of a population that is a tree. The article by Offermans reviews the hypotheses on the origin of crown shyness that have been advanced in literature. None seems to offer a complete explanation for the fact that in many tree species the crowns will not touch each other (e.g. whole crowns of *Pinus sylvestris*, and both whole crowns and crownlets of Dipterocarpaceae and Vochysiaceae; see Oldeman and Fundter 1986).

3.8.2 Some Definitions

This small review of the concept of trees as populations of "subtrees" is also inserted to make the following point. Population biology is usually linked to the species, in the taxonomical and genetical sense. It is possible, and probably useful, to enlarge this concept in order to cover all integration levels in Table 2.1, or whatever other hierarchy may be chosen. The definition would then become:

Fig. 3.46A

Population biology includes every model that explains living systems in terms of numbers, frequencies and reproduction rates of biologically equivalent component subsystems.

This calls for a second definition:

Systems are biologically equivalent if they show both identical architecture (i.e. structure at one hierarchical level) and identical ecophysiological or genetical regulation processes.

These definitions apply to classical population biology, but also include other approaches.

3.8.3 Transfer of Functions

In this paragraph, the transfer of functions within a tree will be introduced. This transfer of a function from one component subsystem to another occurs at all hierarchical levels. It is particularly important for silviculturists in the analysis of forest stands. The generality of the concept is not sufficiently recognised as it is, however. Therefore some examples of transfer of functions in trees will be given here.

Mangenot (1973) or Corner (1964) in their fascinating treatises on plant life mainly present examples of the transfer, or transference, of sexual functions from one tissue or organ to another. The classical example is the

Fig. 3.46B

Fig. 3.46. Transfer of functions. In *Euphorbia* cf. *ingens* (tree euphorbia, candelabra tree, Euphorbiaceae, model unknown, all axes orthotropic), the photosynthetic function is transferred from the small green leaves to the young branches. Later, a thick smooth bark covers the once photosynthetic surfaces. The phenomenon is also known, and persists until a higher age, in the branches of Cactaceae from the drier regions in the neotropics. Kenya, road Nairobi-Nakuru, 1981. (Photographs R. A. A. Oldeman)

evolution of the seed from the spore, in which, for instance, the protective function has been transferred successively from the epidermis (around a naked embryo, such as in *Ginkgo biloba*) via the seedcoat (around an embryo plus reserve food) towards the exocarp (around a fruit, including seeds and sometimes pulp).

This example concerns evolutionary time-scales, as does the following one. In African tree euphorbs (Fig. 3.46, *Euphorbia* cf. *ingens*, Euphorbiaceae, probably Rauh's model, East African savanas), the photosynthetic function has been largely transferred from the leaves to the stems, the green colour of which testifies to the presence of chlorophyll. The same kind of transfer is well known in Cactaceae. As in the evolution of the seed, the function here is transferred from a smaller organ (leaf) to a larger one (axis).

This also occurs in Compositae, which are considered as an advanced family, as judged by their sophisticated inflorescences. Here, miniaturisation of the flowers was the process by virtue of which transfer of several functions took place towards the inflorescence as a whole. The most apparent functions that have been taken over by the inflorescence are the formation of a complex with forms and colours that attract pollinators, mainly insects, and the construction of a dispersal platform for the miniature fruits.

The transfer of functions in leguminous trees may be mentioned because it takes place on a much shorter time-scale (Oldeman 1989a, in press). By transferring height growth and diameter growth throughout the tree's body towards such axes or reiterated crownlets where it most enhances the chances of survival, the flexibility of these trees becomes truly amazing. They may even take the form of shrubs or lianas if and when necessary. Here, knowledge of the transfer of functions is of direct interest to the silviculturist. This professional has to take it into account in his species choice for plantations, in order to obtain the biological type that is wanted, e.g. coppiced shrubs for firewood, high trees for sawnwood production, or various forms for forests that should counter erosion. On the other hand, silvicultural treatments may be conceived that lead to the desired form of the woody legume species.

3.9 Between Trees and Forests

This chapter gives only a rough outline of the life history of trees, in all its complexity and wonderful ingeniousness. They are the giants among land plants, and if miniaturisation as in orchids (see Sects. 4.5 and 4.6) is quite a complex and interesting matter in itself, the building and maintenance of the enormous woody bodies of trees is a biological "engineering" feat that includes both processes of miniaturisation and gigantisation (cf. Hallé 1986). The silvologist has to understand at least roughly how this works in order to understand the forests that are structured by trees.

Only a thorough understanding of this matter will allow better forests to cover a large enough part of the earth in the future. And still, too little is known about trees. Research should still be intensified if we want to reach the stage where isolated blocks of particular, often empirical knowledge are replaced by more general concepts by virtue of which particular problems can be approached.

4 Forest Components Other Than Trees

4.1 Introduction: The Trees and the Others

One can imagine one isolated tree in the middle of a seemingly endless, barren landscape. This picture has yielded some of the more spectacular press photos (e. g. Adams et al. 1978). Although it is doubtful whether such a tree lives entirely without organic companions, in the form of micro-organisms at least, it is relatively easy to describe its architecture and functioning as if it were all alone.

This image of trees is one of the weak points in many silvicultural monoculture systems. However, it is an important image for the present chapter, because only shrubs and herbaceous plants may also be described as if they were isolated, whereas all other forest organisms, particularly animals, are so dependent on ecosystem interactions that they can barely be considered as isolated elements. Therefore, this chapter will be shorter than the former one and mostly take the form of brief notations on the properties of forest component groups. The reader is referred to handbooks on botany and zoology for a more thorough treatment of the taxa, and to Chapters 5 and 6 for the ecological interactions.

In order to understand why most components are so dependent upon ecosystems, Table 2.1 is evoked once more. It shows how abiotic influences, in forests, are filtered by more living systems the further one descends from one hierarchical level to another. For instance, sunlight is processed by the forest mosaic before arriving at the stand, by the stand before arriving at the grass growing on the soil, by the grass before arriving at its tissues, and by these tissues before arriving at its cells in the form of energy bound in sugars. It is not to be forgotten that only forests are considered here, not other other ecosystems, in which organisation at some levels can be lacking.

Most often also, the size and longevity of living systems are smaller and shorter when one descends in the hierarchy. Trees are big, cells are very small. This certainly holds true when all levels are represented in one organism: a tree contains reiterated crowns, that contain organ complexes such as leaved branches and inflorescences, that contain organs, that contain tissues, that contain cells and so on, always smaller and shorter-lived. To a lesser degree, this is also true for monocellular organisms which are smaller and live shorter lives than monotissular organisms such as certain fungi and algae, that are smaller and shorter-lived in their turn than mosses or herbs, and so on.

If miniaturisation occurs, organisms that are at the organisation level of the branched model, or at least the organ complex (e. g. insects), may show sizes and life spans that correspond to those on lower levels, e. g. a cell or complex of cells (tissue). A very small insect may be barely bigger than a quite large monocellular protozoan.

In that case, miniature organisms undergo the effects of the filtering properties of all larger forest components, at whatever level of organisation they may be. A small mosquito is organised as an organ complex, but its short life is largely determined by the organs of animals (veins, for instance) or plants (a leaf, for instance) that regulate their facilities for food and shelter.

These ecological interactions will be treated in more detail later (Sect. 6.3.8), but it is important to note already that they often are so overbearing that no property of small-sized forest components is autonomous. Their feedback action inside a forest ecosystem, then, can only be explained by their huge numbers. Chauvin (1967) cites the example of *Formica polyctena,* a red forest ant, of which more than 1 million nests were found in the Italian Alps, containing more than 300 billion ants that during the 200 days of yearly activity dispose of 15.000 metric tons of prey insects, often noxious to mankind.

The present chapter accommodates the subjects treated in such a way that literature can be most easily consulted. Shrubs, herbs, lianes, epiphytes, monocotyledons, animals, cryptogams and some animal groups are reviewed in this order. For each of these groups, some brief notes on architecture, functioning and life cycle will be given, and some examples will serve as illustration.

4.2 Shrubs

Architecture of shrubs. Shrubs are most often considered as woody plants that remain low and have more than one trunk, all trunks coming from the stem base. Figure 4.1 (A,B) shows the image of shrub architecture as given in the classical German and French growth form analysis. As compared with Fig. 3.2, the tree models of Troll (Fig. 4.1A), Champagnat (Fig. 4.1B) and Roux (Fig. 4.1C) are easily recognised. The classical European schools term the branching pattern at the base basitony and branching at or near the top of the trunk, as in trees, acrotony. These patterns are usually explained by bud inhibition patterns, or priority patterns (Champagnat, in Tomlinson and Zimmermann 1978) caused by phytohormonal control, which accounts for phenomena such as apical dominance or its absence. These processes have been studied in detail (see also Zimmermann and Brown 1971).

In fact, Fig. 4.1C shows that the concept of basitony versus acrotony cannot explain the whole life of a shrub. In *Casearia* (Flacourtiaceae), *Myrciaria* (Myrtaceae), several Annonaceae and other tropical shrubs, there is both acrotony in the model-conforming crown of each shoot, and basitony in the plant as a whole. It is as though the branched growth period were model-conforming, whereas the reiteration period were shifted to the base of the first trunk (Fig. 4.1D). In fact, the whole architecture of a shrub has several points in common with coppice (Fig. 3.21).

Coppicing, however, is provoked. It is a reiteration pattern in reaction to damage of the tree, i. e. the cutting of the trunk. Can shrubs be explained as trees without trunks, that start the processes of crown expansion at ground level? This brings

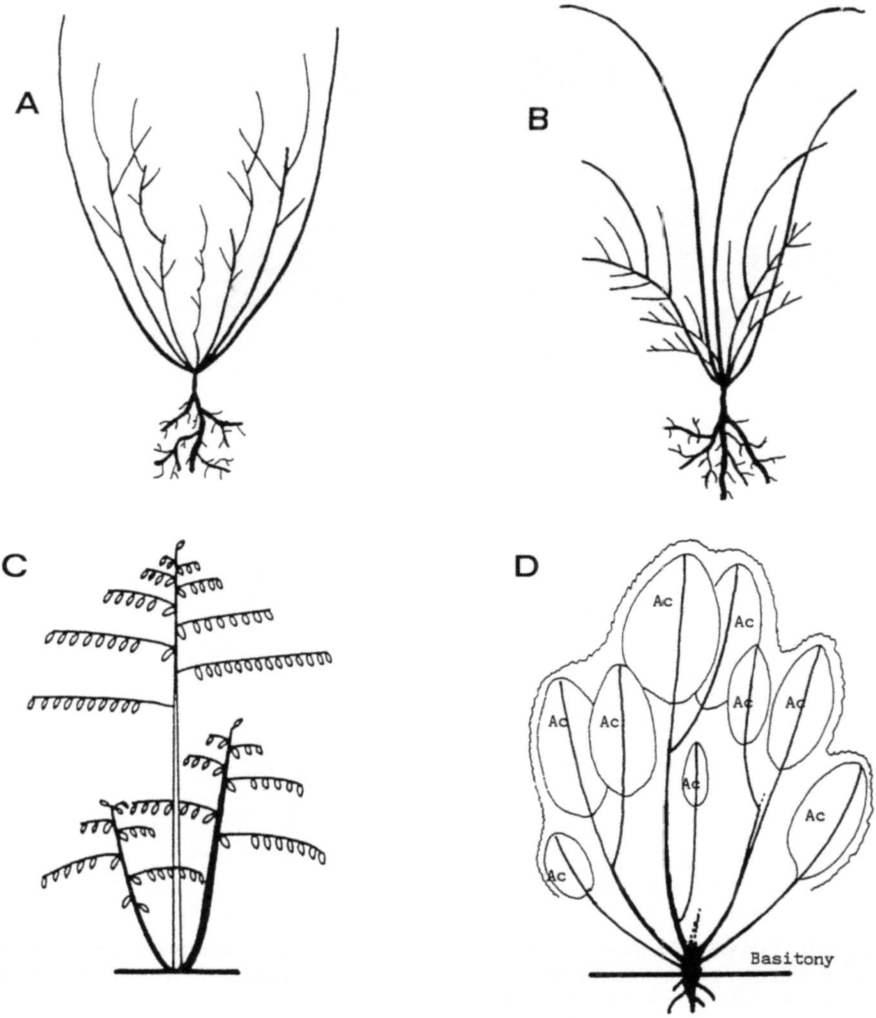

Fig. 4.1. Shrub architecture. **A** Basitonic branching pattern, each axis developing a version of Troll's model (hazel, *Corylus avellana*, Europe, Betulaceae). **B** Basitonic branching pattern, each axis developing a version of Champagnat's model (elder, *Sambucus nigra*, Europe, Caprifoliaceae, cf. Fig. 3.12). (**A** and **B** after Rauh in Braun 1980). **C** Basitonic branching pattern, each axis developing a version of Roux's model (*Casearia* sp. pl., Flacourtiaceae, pioneer forest in French Guyana after Oldeman 1974a). **D** Principle of shrub architecture, hypothetical as long as nó world-wide study is available. The whole plant presents a basitonic growth pattern, whereas each of the basitonic axes is the trunk of some architectural model (Fig. 3.2), that may often be acrotonic. Cutting all except one of these axes, in a true shrub, would provoke new formation of basitonic axes, not height growth in the only model left

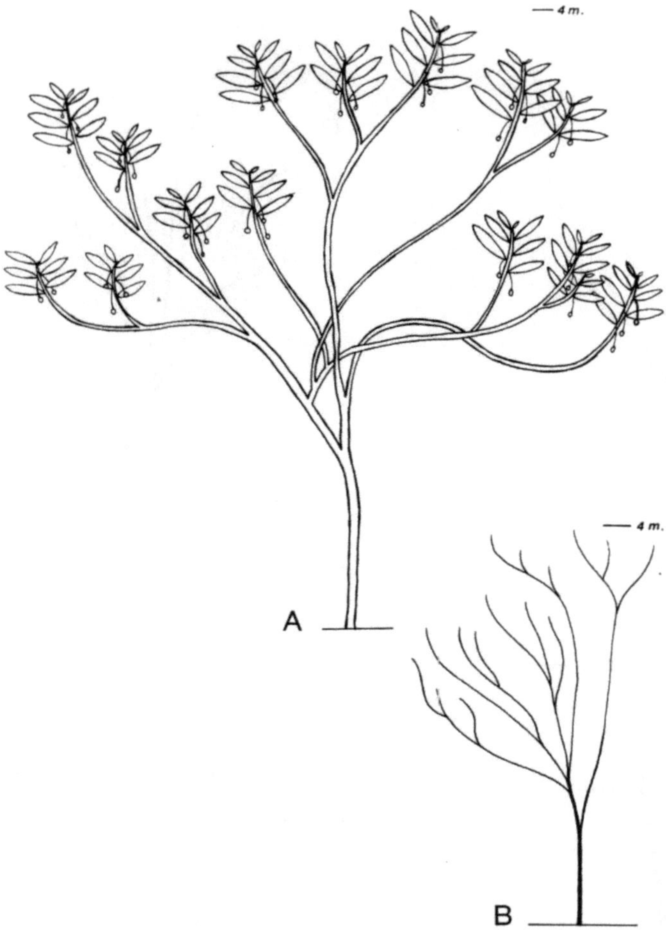

Fig. 4.2. Two Mexican shrubs, that may not be true shrubs because they may not have a strict overall basitonic growth pattern and may prolong their short unique trunk whenever it is ecologically opportune. **A** *Myriocarpa longipes* (Urticaceae, Champagnat's model). **B** *Urera caracassana* (Urticaceae, Champagnat's model). (Torquebiau 1981)

shrubs close to the small trees with a life cycle like that in Fig. 3.1 (D,E). If the reiteration pattern were programmed so that it always preferentially takes place at the base, the situation would be close to that in Tomlinson's model (see Fig. 3.2 and Sect. 4.6). In the former chapter we have seen, however, that one property of reiteration is its unprogrammed, opportunistic character.

It looks as though there are all kinds of transitions between trees that reiterate at their base (base suckers) and shrubs that reproduce branched models from their base in a programmed way. Such trees can be reduced to a shrub form by natural or man-made impacts, for instance in *Quercus ilex* (Fig. 3.4). On the contrary, "real" shrubs cannot be brought to limit themselves to one high trunk, but always build numerous small trunks from their base. This is true for shrubs that can grow in the open as pioneer shrubs, e. g. elder (*Sambucus* spp., Champagnat's model, Fig. 3.12), gliricidia (*Gliricidia sepium*, Papilionaceae, pantropical, Champagnat's model) or some Ericaceae from the Andes. But it also is the case in tolerant shrubs, such as hazel (*Corylus avellana*, Betulaceae, Troll's model?), found up to 9 m high under dense forest in the Alsace (France), or the more light-demanding *Myriocarpa longipes* and *Urera caracassana*, both Urticaceae with Champagnat's model from the undergrowth of Mexican secondary forest, described by Torquebiau (1981) and represented here in Fig. 4.2. It may well be representative for tropical rainforests that these last species show a short trunk, because no clearcut shrub species like those that occur in temperate regions have been cited from the lower reaches of those forests.

Production in shrubs. Not many thorough studies on the development of biomass, diameter, height, volume or on the sapstream in shrubs seem to have been done, if any. This is somewhat astonishing, as a comparison between such models for natural shrubs and coppice seems to be useful. But neither for shrubs, nor for coppice production, tables or chronological series of the development of measured parameters seem to exist for the tropics. For temperate regions see Rollinson and Evans (1987). Given the increasing importance of both shrubs and coppiced or pollarded trees in agroforestry, such research should urgently be undertaken. It is also a missing link in the assessment of the role of shrubs in semi-natural or natural forests, although sometimes the "shrub compartment" has been considered for its biomass (e. g. in Belgium by Duvigneaud and Denaeyer-Desmet in Reichle 1970).

The difficulties in determining meaningful ratios and measures to be used as parameters in such studies should not be underrated. What is "the" diameter in a shrub? Is it the summation of the diameters of all small trunks? Is it the diameter of the irregular base? The same questions can be raised for variables such as height, volume or biomass.

Still, shrubs play a very important role as components in certain growth phases of eco-units (or regeneration units; Chap. 5), and they should always be considered as a potential asset in the design of silvicultural or agroforestry systems.

4.3 Herbs

Just like shrubs, herbs may build vegetations on their own, if the site or human influence are limiting to tree growth. In forest ecosystems, they are important components because they may use a considerable part of the ecological production potential and so be indicators for the state of the whole forest. It goes almost without saying that herbaceous components are of the utmost importance in man-made or utilised forests in which cattle or secondary, annual crops play a role. As will become apparent in Chapter 6, the herbaceous vegetation gains in importance on more stressed or less fertile and drier sites.

Architecture of herbs. The studies by Jeannoda-Robinson (1977) and Ndiaye (1977) concern the architecture and growth dynamics of herbaceous plants in general. The monocotyledons, among which are the grasses, will be treated in Section 4.6 and mainly dicotyledonous herbs are examined here.

In order to understand the architecture of herbs, some aspects should be clear.

- The life cycle is adapted to abundant flowering rather than to building a persistent vegetative body. Herbs live for the next generation, forming much seed, rather than for long individual survival. Of course this is a generalisation which has numerous exceptions, but it may be taken as a general trend.

- If individual survival in perennial herbs is part of the life cycle, the subterranean parts often play a very important part in it. They tend to be specialised, and in many species one can also find subterranean stems which are not roots but adapted axes: the stolons.

Many features of herbs can be derived from these two rules, e. g.:

- Reiteration is rare among herbs, except maybe in the form of root suckers, because reiteration in trees and shrubs serves the expansion of the vegetative body.
- The architecture of the inflorescence obtains an increased importance in herbs, because their life is oriented towards offspring.
- Wood formation is either absent or little pronounced. If it exists, it most often occurs at the base of the stem. There is some doubt whether "woody herbs" should be included among the herbs at all, or be considered as mini-shrubs. This is very apparent in the humid tropics.
- Miniaturisation of the vegetative body is a common process.
- The differentiation sequence of the branched model, if any, may be modified. Miniaturisation finally leads to unbranched organ complexes, except for the roots (e. g. Figs 4.3A, 4.4A). If branching occurs, as in tree models, it may be either miniaturised (Fig. 4.3C), or both miniaturised and adapted (Fig. 4.3E). Or finally, it may show a wholly changed version of a tree model (Fig. 4.4B) or behave in original and purely herbaceous ways (see Jeannoda-Robinson 1977).
- A phenomenon called neoteny, which is a concept with a zoological origin and means the appearance of adult characters in juvenile organisms, is frequent in small herbs. These may flower when still in a seedling-like state (Fig. 4.3A). This is a combination of miniaturisation and change in the differentiation sequence.

Hallé et al.(1978) gave a general outline of herbaceous architecture as compared with trees and climbers.

Figures 4.3 and 4.4 show some few examples of herbs, their architecture and their ecological strategy. These figures illustrate the immense flexibility displayed by this group of plants, that may range from autonomous colonisers of empty biotopes (pioneers) to infiltrators of already full environments, such as the undergrowth of the tropical rainforest.

In fact one may - and should - expect herbs to play a role in forest ecosystems of all sizes and ages (Chaps. 5 and 6). Colonizers of empty places, e. g. clearcuts, landslides, or eroded parts of the landscape, have many and various strategies. Some come in numbers: they are small and produce many seeds (Fig. 4.3A), often after a rosette stage that allows them to wait until circumstances are optimal for

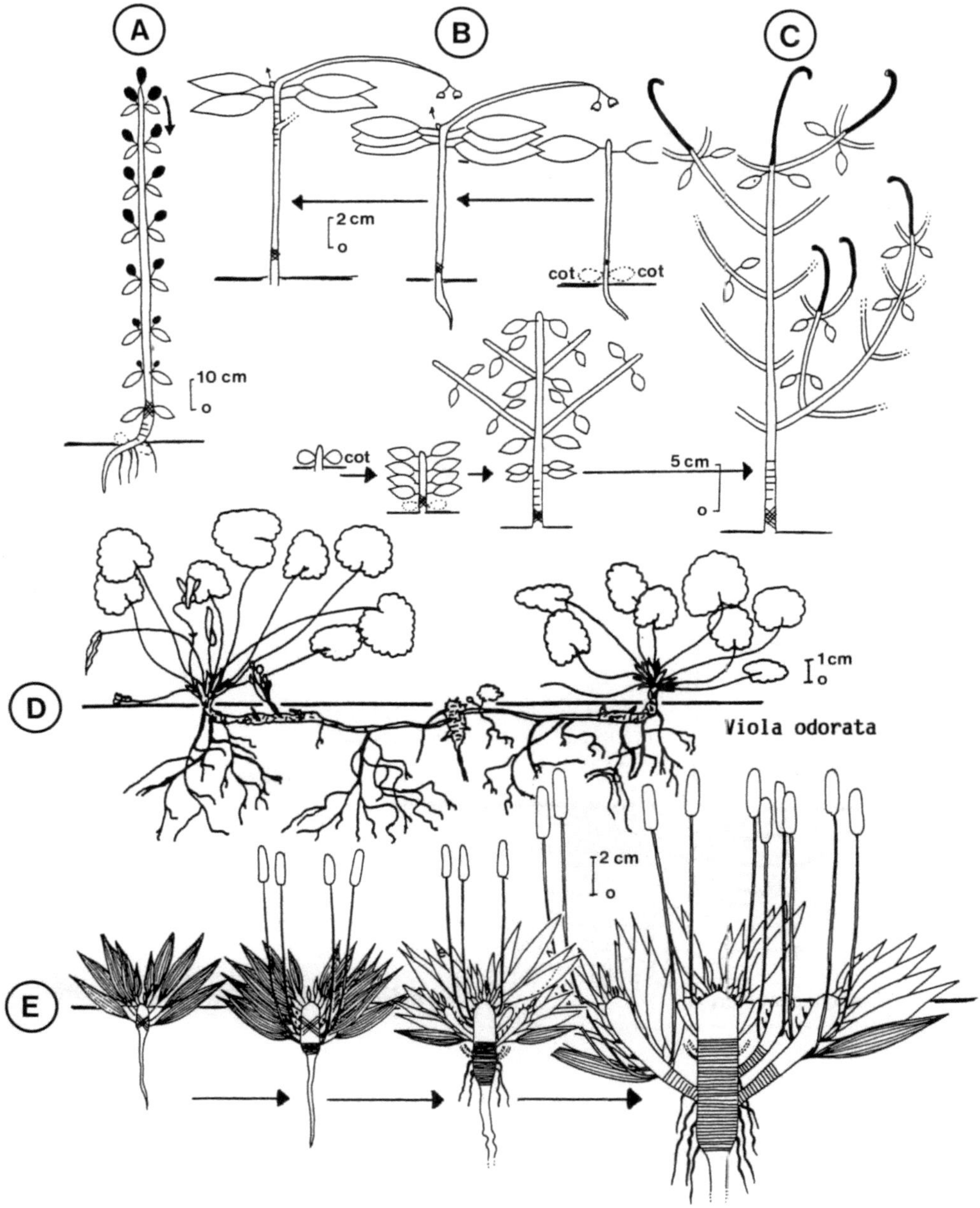

Fig. 4.3. Examples of herbaceous architecture in different environments. **A** *Achyrospermum* sp. (Labiatae), version of Holttum's model with first terminal, then descending lateral inflorescences. Plant flowers and dies. **B** *Pavonia* aff. *flavispina* (Malvaceae) from the deep shade under the Guyanese rain forest, a herb or a mini-tree? Chamberlain's model. **C** *Stachytarpheta jamaicensis* (Verbenaceae), wide-spread tropical weed of pioneer vegetations, a version of Scarrone's model. **D** *Viola odorata* (Violaceae), architecture with rhizomes, not directly comparable with trees, a herbaceous component of half-shaded open places in Europe. **E** Development of *Plantago lanceolata* (Plantaginaceae) in open vegetations, Southern Europe, an underground version of Rauh's model. Note tractile roots that draw the plant down. (**A**, **B**, **C**, **E** after Jeannoda-Robinson 1977, **D** after Ndiaye 1977)

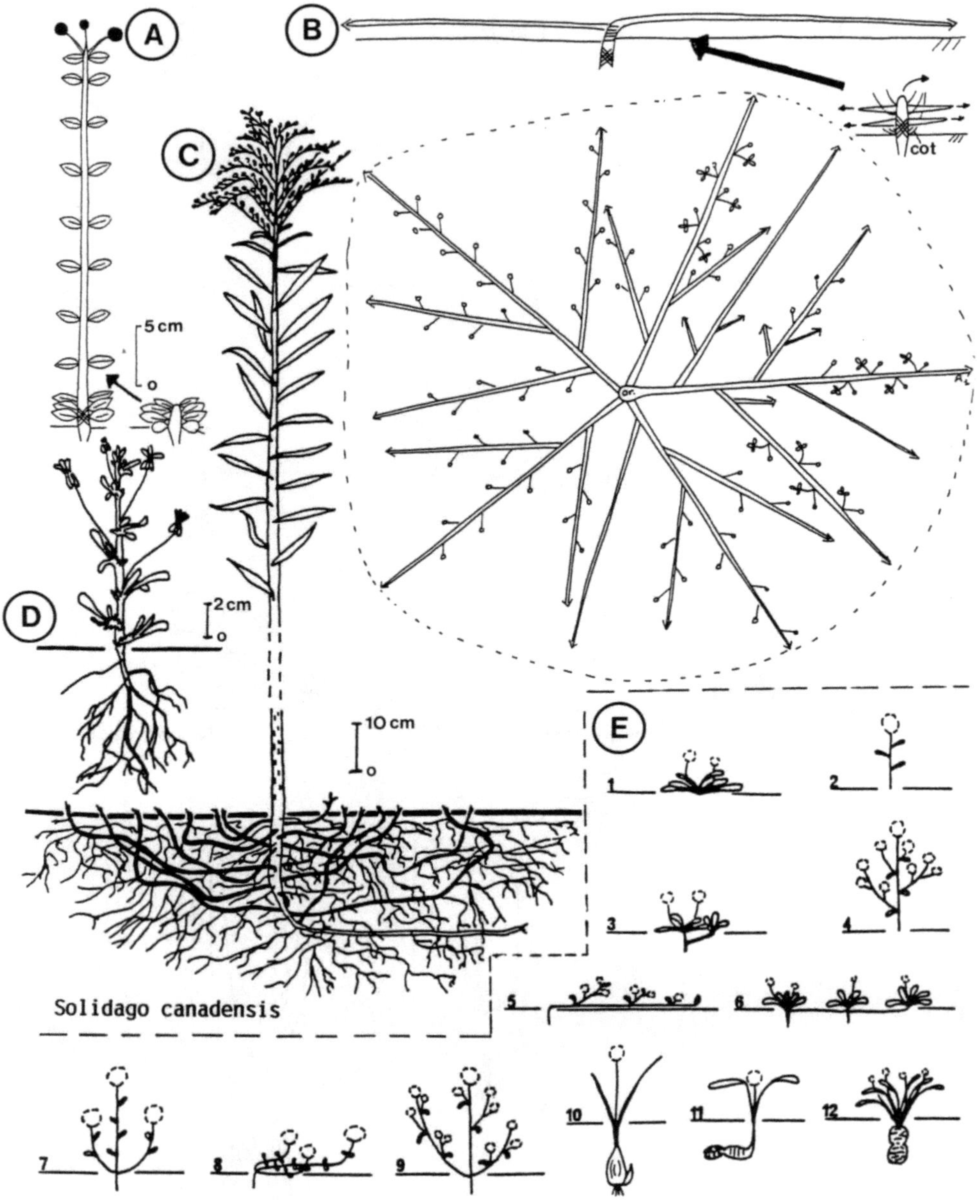

Fig. 4.4. Examples of herbaceous architecture in different environments. **A** Two stages in development of *Erythrea centaurium* (Gentianaceae) from Mediterranean shrub forests, note rosette stage. **B** *Medicago lupulina* (Papilionoidae), a prostrated version of Troll's model. Many models show prostrated forms in herbs and so cover large stretches of open terrain. **C** *Solidago canadensis* (Compositae), underground strategy with dense rhizome mat, version of Tomlinson's model in open places in Canadian forests. **D** *Viola tricolor* (Violaceae), along brooks in Europe, note lack of rhizomes (cf. Fig. 4.3D). **E** Morphological series of herb architectures according to Ndiaye, for dicotyledonous herbs mainly (also see Fig. 5.31/32). (**A**, **B** Jeannoda-Robinson 1977, **C**, **D**, **E** Ndiaye 1977)

seed dispersal (Fig. 4.4A). Others form a creeping carpet of plagiotropic stems (Fig. 4.4B), sometimes also sympodia (Jeannoda-Robinson 1977), and cover quite large areas before dispersing their seeds. In less inhospitable places, they may grow into a recognisable model, e. g. the ruderal pantropical herb *Stachytarpheta jamaicensis* (Fig. 4.3C).

In sites with inhospitable seasons, herbs may hide underground. The remarkable *Plantago lanceolata*, a common weed the world over, that has spread from Europe to regions such as the Andes, conforms to Rauh's model (Fig. 3.2), but draws its body down into the earth by "tractile roots" that grow out and contract afterwards in a longitudinal direction, once they are well anchored (Fig. 4.3E). Another example of herbaceous species being active underground is that of *Solidago canadensis* (Fig. 4.4C), a common and showy garden plant forming innumerable stolons that emerge to form new orthotropic shoots with yellow terminal inflorescences. Underground specialisation of this kind is frequent in dry regions (Sahel, e. g. Oldeman 1983b, his Fig. 6). But it may also occur in certain niches of tropical forests (e. g. gingers, Zingiberaceae - monocotyledons), or in temperate forests (Fig. 4.3D).

Among the undergrowth species from various forests, there are once again numerous architectures, principles of growth dynamics and ecological strategies. Here one finds forest components so finely attuned to their place in the ecosystem, that they may barely survive outside it. An example is *Pavonia* aff. *flavispina* (Malvaceae, Chamberlain's model, Fig. 4.3B) from the lowest reaches of high, adult, closed Guyanese rainforests where it grows in sun spots, very slowly and forming wood at its base. It might be conceived as a miniature tree. The violets from the European forests, *Viola odorata* (Fig. 4.3D) and *V. tricolor* (Fig. 4.4D) are certainly adapted to different spaces in forests, the first one more open, the second one more shaded and humid. Still, the normal Floras of the regions give no clearcut answers to questions about biotope differences for these two species, that nevertheless so clearly appear to be differently adapted. In the ecological flora by Weeda et al. (1985, p. 234) there is a suggestion that there may be a seasonal factor implied, *V. odorata* belonging to the species that flower early in spring.

Figure 4.4E shows a series of herb architectures as distinguished by Ndiaye (1977) in his study on weed vegetations in Montpellier (Mediterranean France). These are shown here, because Figs. 5.31 and 5.32 reproduce the analysis of a vegetation of pioneer herbs, in which they play a role. In Fig. 4.4E, 1 and 2 are unbranched herbs with short and long epicotyledonary axes. Branching patterns 3 and 4 are erected herbs with short and long epicotyledonary axes, whereas 5 is also erected but forms stolons with reduced aerial axes. Number 6 corresponds to the same pattern as 5, but the aerial axes take root and finally form a clone, a behaviour also common in lianes (see Sect. 4.4). Pattern 7 is an erected variant with unbranched aerial axes and basitonic branching (see Sect. 4.2), 8 represents a basitonic prostrated pattern, the term prostrated having been coined by Jeannoda-Robinson (1977), and 9 is also basitonic but with branched erected axes. Finally, the series from 10 to 12 represents erected herbs with very specialised underground strategies, i. e. bulbs (10), prostrated rhizomes (11) and erected rhizomes (12).

Production patterns in herbs. These will not be treated here. The interested reader is referred to the extensive agricultural and botanical literature on the subject. Some aspects, however, have to be remembered. In the first place, herbs have always been favourite subjects in plant physiology, because their short life allows experiments covering whole life cycles. Hence they have been studied in great detail. In the second place, herbs generally remain small, so that production studies have been done more often on herb populations than on individual herbs. And third, most staple food plants are herbaceous. They are often cultivated as very homogeneous crop populations. This favours the establishment of production models for a whole population, in which an average part may be allocated to each indi-

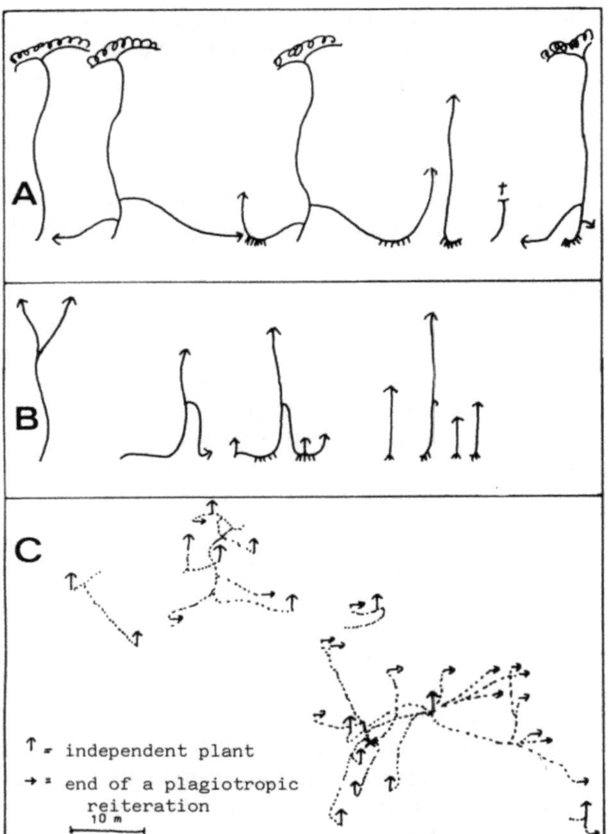

Fig. 4.5. Lianes, growth models (**A, B, C**), and an example (**D**). **A** Vegetative multiplication following rooting of plagiotropic shoots, that later separate and form a clone, like *Rubus* spp. in temperate regions. **B** Vegetative multiplication following orthotropic reiteration on fallen stem. After rooting and separation a clone is born. **C** Clonal group of Bignoniaceae sp. in dynamic forest. (**A, B, C** after Beekman 1981a, French Guyana). **D** *Tetracera alnifolia* (Dilleniaceae), the water liane from Gabon, synoptic view of different growth phases. (After Caballé 1980)

vidual herb because these individuals have been selected and bred to be almost equal. This agricultural approach is known in silviculture when wood production in monocultures is emphasised. It can be traced back to the principles upon which forest yield tables are established.

4.4 Lianes

Lianes are often thought to be an exclusive feature of tropical forests, but this idea is a mistake. Even if the many herbaceous lianes from the temperate regions are excluded (many Papilionaceae, for example), one may find some woody species that follow exactly the same strategies as tropical lianes, such as ivy (*Hedera helix*, from the tropical tree family of Araliaceae), *Lonicera* spp. (honeysuckle or woodbine, Caprifoliaceae) or that best of all European lianes, *Vitis vinifera* (vine, Vitaceae) and its cousin *Parthenocissus quinquefolia*, one of the last wild specimens of which is carefully preserved in a German forest reserve near the Rhine at Freiburg. The lianes in these forests along the Rhine ("Auerwald") have been studied by

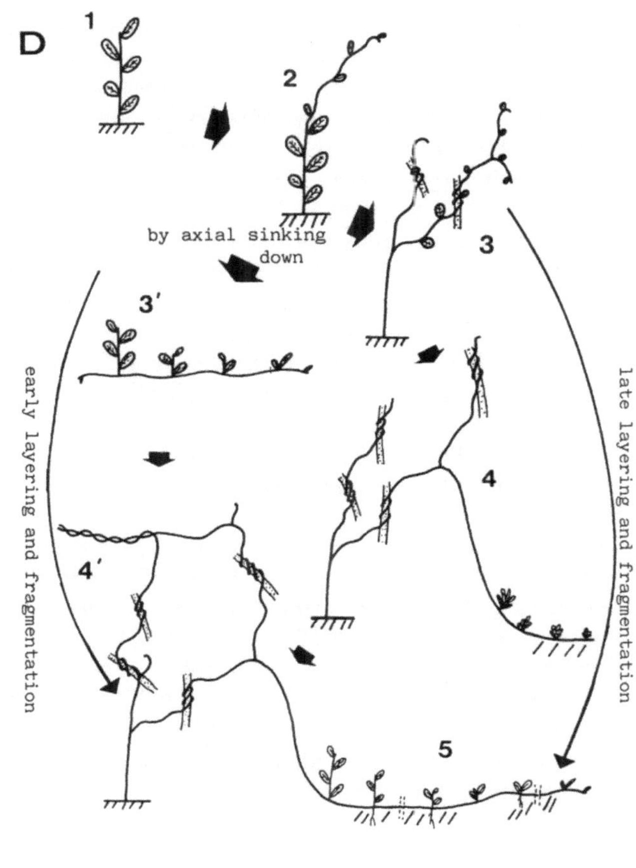

Fig. 4.5D

Beekman (1981b). Plants like the brambles (*Rubus* spp.) or climbing roses (*Rosa* spp.), all Rosaceae, should be included in the lianescent plants from the temperate regions. Even this incomplete list should convince a temperate-region silvologist that lianes are much more important in "his" forests than is generally admitted. In the following paragraphs, examples from the tropics will be treated, but for nearly all tropical liane strategies there are parallels in species of other bioclimates (Beekman 1981a,b).

Architecture of lianes. In lianes, as in herbs, one may find architectural models that correspond to those in trees (Fig. 3.2) and architectures that are properly lianescent (see Hallé et al. 1978). All lianes have in common that their axes have reduced amounts of supporting tissues so that, by various mechanisms, they have to climb up trees, houses or rocks to reach the light, or anchor themselves in a growing tree crown that carries them upwards. Beurel and Doumenge (1982) give a survey of the astonishing array of anatomical features that allow these supple stems to function, and of the hooking, clutching and grappling organs that they form.

Except for the book on lianes by Caballé (1986), the sources cited here are Beekman (1981a, b, bibliography), Jacobs (1976, bibliography) and some articles by Caballé. The rattans will be mentioned in the paragraphs on monocotyledons.

By far the most interesting feature of lianes as forest components is not so much their architectural model (auxiliary level: branched complex, see Table 2.1) as their reiteration behaviour. Lianes are to coppice (Fig. 3.21) what a supersonic aircraft is to a propellor plane. Once established, woody lianes are almost totally resistant to eradication because they have a capacity for reiteration that seems to be inexhaustible. They form stump suckers, they layer and they reiterate on all portions of their body whence a new stem with its package of leaves can profitably ascend.

In forestry, attention has been focussed strongly upon the twining stems themselves. In the tropics, they are considered to be a nuisance in tree harvesting, because they link together the crowns of neighbouring trees, so that a chosen tree will not fall down, but remains suspended after being felled (see literature review by Jacobs 1976). In the temperate region, lower, bushy, creeping lianes such as *Rubus* species are considered a nuisance in forest regeneration, because they outshade the young trees. Because these effects have their direct cause in the climbing stems, these stems have been cut back, poisoned, ringed and so on. Results generally have been poor, whereas such operations are expensive.

This is so because the whole liane strategy and architecture have not been considered. Lianes all have the organs they need to ascend, whether they do so by twining the stems themselves around supporting trees or other lianes, or by forming hooks, grapnels or tendrils (Fig. 4.6). But, biologically speaking, these ascending stems are vulnerable. The supporting tree may collapse (Fig. 4.6B). The stems may break off, notwithstanding the in-built flexibility and elasticity due to an adapted anatomy (cf. Jacobs 1988; Beurel and Doumenge 1982). Occasions to start climbing, or to do so once again after having fallen to the ground, arise in quite unpredictable and hazardous ways.

One group of lianes therefore has seedlings showing many of the characteristics of very tolerant tree seedlings. Barely growing, they keep waiting in the undergrowth of the forest, among tree seedlings that are very similar-looking: i. e. a self-supporting vertical axis of some decimeters high, with a generalised type of juvenile leaves. Once the occasion presents itself, the top of such a liane seedling will differentiate and form a very fast-growing, specialised stem, often with scale-leaves only, and with long internodia (Fig. 4.5D). In other cases, this stem bears the first clutching organs. Branching of this stem occurs whenever necessary, most often in two cases: either ascension continues on more than one support by different climbing axes of the same liane, or the top of the liane penetrates a zone where a liane crown can expand because of abundant light. Lianes that show one version or another of an architectural tree model (Fig. 3.2) form the extra stems by reiteration. The model is then manifested most clearly in the reiterated parts of the liane crown and not, as in trees, in the saplings. In some models, the model and its reiteration already may be difficult to separate, e. g. in Troll's and Champagnat's models in lianescent Leguminosae (Oldeman 1989a, in press).

Fig. 4.6A
(Fig. 4.6B see page 122)

Fig. 4.6. Images of unidentified liane species in Makokou (Gabon). **A** One of the common climbing devices (thorns), described by Beurel and Doumenge (1982); note other lianes climbing along the thorns. **B** Contorted, thick stems, the support tree of which has probably died, note reiterated stem climbing up; the dense cluster of thin lianes in the background can be remarked. (Photographs 1981 by R. A. A. Oldeman)

If and when the sequence, going from germination to arrival in the expansion zone, becomes specialised and shows strict programmation, the position of the crown model becomes subservient to the whole life cycle. Its clear, programmed differentiation is more or less superfluous and may get lost. The result then is one of the specialised lianescent growth models described by Cremers (1974; in Hallé et al. 1978). In whatever way this may happen, the ascending stems always have their main physiological function in supplying the expanding or established crown with sap, or plain water. They are mainly water pipes. This is spectacularly evident in the lianes of the pantropical genus *Tetracera* (Dilleniaceae, Fig. 4.5D) that yield several decilitres of clear drinking water from a cut fragment of a stem, 5 to 10 cm in diameter and 1 m or so long. These are the famous "water-vines" of the tropical rainforest, with their reddish, flaky bark that is so easy to recognise. Figure 4.6B should be looked at with this in mind. It represents an enormous ecological mechanism of vertical water transport, containing numerous wide and small pipes, some straight and some tortuous.

The combination of the watering function, the vulnerability, and the unpredictable environments constitutes a strong selection pressure for vegetative replacement mechanisms. Figure 4.5 shows how this happens most often. Woody forest lianes have a strong tendency towards basitonic branching (see Sect. 4.2), and do so in a reiteration pattern that yields numerous new climbing stems. To cut back lianes as a silvicultural measure therefore has the opposite effect to what is desired. Most often it results in the multiplication of climbers instead of their reduction, especially if a thinning of the trees has created a microclimate with more light. An example of the plant sizes involved is given by Caballé (1977) in his development analysis of *Entada scelerata* (Mimosaceae) in the Gabonese rainforest, where one individual covers ca. 0.02 ha. With 50 of such lianes, one hectare is completely covered, which corresponds to tree densities for the largest forest trees. Coppicing such climbers is clearly counterproductive.

Most of these liane species also show pronounced aptitude to spread horizontally by layering. This seems to be programmed in their differentiation sequence. Bramble (*Rubus* spp.) and ivy (*Hedera helix*) can be observed everywhere in Western Europe to do just this. In *Rubus*, the tips of sinking stems that touch the soil will root nearly immediately, as in the Guyanese lianes represented in Fig. 4.5A,B. And in ivy, a climber thoroughly studied by Beekman (1981b), the principle of Fig. 4.5A is programmed. It always forms adventitious and adhesive roots on its plagiotropic creeping stems, that form large carpets on the forest floor until a tree is met that lets through enough light. An ascending stem, still plagiotropic, is then formed. Once it has arrived in the canopy, orthotropic, fertile axes ensure fructification and seed dispersal.

The example of a clonal group of one of the numerous woody liane species from the Bignoniaceae family in the Guyanas (Fig. 4.5C) shows that vegetative multiplication is a companion of the above-mentioned reiteration and layering patterns.

Climbing plants include a continuum from herbaceous to woody forms. For the herbaceous lianes, Section 4.3 should be consulted. Often, a herbaceous climber can be understood as a herb in which a prostrated stem is replaced by a climbing one. These are important in low vegetations without trees, or in which only tree seedlings are present. On that scale, they interact with seedlings in a way analogous to woody lianes interacting with saplings and adult trees. The lifespan of herbs being limited, however, interactions here will stretch over some years at most, if the tree seedlings are not killed but grow on.

No precise studies on liane root systems are known to the present author, although individual roots have been studied botanically and anatomically. This is a lacuna that should be urgently filled, because the strength of lianes may well lie below the ground, or in the formation of adventitious root systems, rather than in their climbing stems. These latter seem to play a role of throw-away organs, at least to a certain extent.

Production in lianes. No complete production models of lianes seem to exist. If analysed by using a sapstream model (cf. Fig. 3.30), several aspects should be taken into account that are very different from sapstream models in trees. It will have to be assessed whether or not the ascending sapstream, in the xylem, is more or less important than the descending stream, as compared with trees. The anatomical analysis of woody liane stems (Beurel and Doumenge 1982) suggests that these axes, which are often non-circular in cross-section and include phloem bundles, may work in another hydrological way than tree trunks and branches. The presence of adventitious root systems, also in many aboveground parts of lianes, suggests that the input-output feedback diagram (Fig. 3.32) needs to be quite strongly adapted.

4.5 Epiphytes

Επι is Greek for "upon", φυτον for "plant". Epiphytes are plants that grow on other plants. Epiphytes should not be confused with symbionts, such as **parasites** that feed on their host plants, or **mutualists** that exchange nutrients and energy with other organisms (cf. Sect. 4.7). Epiphytes just grow upon the exterior of other plant species, wherever they find a convenient support. The outside of other plants often being quite dry, most epiphytes are found in climates with a humid atmosphere and abundant rainfall, such as the upper flanks of the Andes (Fig. 4.8), or the tropical rainforests (Oldeman, 1974a, his Fig. 83). But in every nook and cranny where some moisture is conserved and some nutrients are available, one may expect epiphytes to occur in every bioclimate (cf. Oldeman et al. 1983, in the Netherlands).

Epiphytes have no other definition than that they grow upon other plants. Hence they include the complete gamut of life forms, from epiphytic trees to epiphytic shrubs, lianes, herbs and spore plants, such as mosses, lichens, ferns and fungi. The closer their biotope is situated to the bark of the supporting tree, the smaller they are. In the temperate regions, this layer may be 1 or 2 mm thick: here, small cryptogams predominate among the epiphytes. On plant **leaves** in all bioclimates, the biotope layer is extremely thin. Epiphytes living there, called epiphylls because they live on (επι) leaves (φυλλα), generally consist of fungi and very flattened lichens.

In the tree crowns of the tropical rainforest, on the contrary, the volume that can be occupied by epiphytes is quite large. Here, one may find epiphytic trees, e. g. in the genera *Ficus* and *Coussapoa* (Moraceae) and *Clusia* (Clusiaceae) in tropical America. Epiphytic trees cannot live and produce on the scarce nutrient and water budget on the branches of the host tree. Therefore they most often form adventitious roots that descend to the forest soil and form mats of brachyrhizae and root hairs in the superficial horizons. Several *Ficus* species occurring all over the tropics and known as strangling figs go one step further. They germinate on a branch. Their adventitious roots form anastomoses. First, a dense interlinked root network is formed, that surrounds the host tree. Finally, this network closes itself completely and becomes a kind of secondary, hollow trunk that strangles the host tree. The *Ficus* then has become a forest tree, that can reach enormous sizes (Figs. 6.22 and 4.8C).

It is impossible to describe and analyse the richness of epiphytes in a few pages. The most spectacular ones seem to belong to the monocotyledons (e. g. Bromeliaceae, Fig. 4.8B, Orchidaceae, the orchids, Fig. 4.7C, or Araceae, Fig. 4.7A), if one excepts epiphytic trees. The dense mat of epiphytes on a tree stump (Fig. 4.7A) or on a thick branch (Fig. 4.8B) may be seen in a less developed stage in Fig. 4.7B. The more mosses, ferns and small seed plants (e. g. Gesneriaceae, Fig. 4.7B) become established, the more humidity and organic nutrients are accumulated. A mat of epiphytes like that in Fig. 4.8B, frequent in certain forests (Fig. 4.8A) is like a sponge that conserves all water and accumulates nutrients. It so becomes a complete miniature ecosystem, the components of which may be very numerous and very diverse. They always include animals. Such miniature but very complex systems can best be treated as compartments of the whole forest (see Chap. 2).

These compartments are so dependent upon the workings of the forest ecosystem as a whole that their analysis can be used as a means to discover symptoms for the state of a forest. This includes the use of epiphytes as parameters for regulation mechanisms. It is important to note that many species of epiphytes, because of their very dependence upon specific environmental factors, are used as environmental quality indicators.

In Western Europe, epiphytic algae, lichens and mosses are often precise indicators for the acidity of the environment. Experts can tell the pH of a micro-environment, with a precision of one tenth of a unit, by the presence or absence of certain species (Barkman pers. commun.). In the Netherlands, a large array of such species has disappeared with the general pollution of the environment that occurred during the years 1950/1960. On the contrary, forest dieback in the Black Forest (Schwarzwald, Germany) is accompanied by the proliferation of certain lichens that hang like beards from the yellowing boughs of trees that lose their vitality, perhaps because of increased pollution. During the last quarter of a century, the epiphytic flora on the trees bordering the Kebun Raya (Botanical Garden) in Bogor, Indonesia, is said to have changed dramatically as a presumed reaction to the changed chemical environment, due to the vastly increased motorised traffic on the roads around that garden (pers. commun., different sources). In most of the examples cited, the formal proof is lacking as to the precise environmental factors causing changes among the epiphytes. However, as soon as such changes occur this may be taken as a sign that monitoring of the environmental factors has become necessary, also with regard to the conditions prevailing in the human ecosystem.

Another aspect of epiphytes has to be considered here. These plants have been named "plants-that-grow-upon-plants" very probably because early botanists marveled at the discovery of this kind of biotope, particularly in the humid tropics. In reality, epiphytic substrates include other supports, such as wet, moss-covered rocks in half-shadow (cf. Fig. 4.7C), dead tree stumps (cf. Fig. 4.7A) or human artefacts such as houses, telephone poles, electricity masts and others. Epiphytes here imperceptibly blend into a category of plants that may colonise inhospitable environments. In ecological literature such environments are often linked to primary succession, that is a succession that starts in places where there has been no preceding vegetation at all. Naked rocks, recent alluvions or other sediments, and sterile sands are examples.

Fig. 4.7A–C

There probably is no better place in this chapter to evoke the Ericaceae. These include the whole gamut of dwarf shrubs colonising, or at least participating in later stages of colonising such environments as skeletal soils in the boreal regions (*Vaccinium* spp., cf. Larsen 1980), sterile, degraded sandy soils throughout temperate Europe (*Vaccinium* spp.), Alpine and Andean high mountain environments such as the páramos, but also the crowns of Amazonian and sub-Himalayan rainforest trees as epiphytes (Temple 1975). Ericaceae are also mentioned as an important epiphytic family in the Malesian rain forests (Whitmore 1975). Moreover, they contain shrubs (*Azalea, Rhododendron* for example) that form understoreys in temperate and submontane forests (North America, Himalayas). Because of the shadow, this also is a stressed environment.

In the neotropics, Cactaceae fulfill an analogous role, but they range from the very dry regions, such as the Mexican and Peruvian deserts, along physiologically dry environments on the South-American coasts, to epiphytic habitats in the rain forest. The Amazonian epiphytic genus *Rhipsalis*, that also has been cited from West Africa, is an example. The seeds of the neotropical *R. cassutha* might have made an intercontinental voyage with the help of birds, says a hypothesis explaining the presence of this plant as the only Cactaceae in the Old World.

Production of epiphytes. Given the biological heterogeneity of this group, the limits of which are unclear with regard to other ecological plant groups, no comments will be given as to their production patterns.

Epiphytes as populations of organs. By virtue of their specialised and flexible functions in forest ecosystems, epiphytes might well yield numerous examples of the transfer of functions from one organ complex to another, such as the transfer of a trunk function (support and sapstream transport) to the adventitious roots in the strangling figs mentioned earlier.

4.6 Monocotyledons

Monocotyledons include some 20% of all the species of angiosperms in the world, and contain herbs, shrubs, climbers, epiphytes, trees. To cover them in an elementary way only, a book would be needed that exceeds the present text. If this immense group of 40,000 to 50,000 species is mentioned here in a few pages and in a few notes, this is because monocotyledons are so often neglected by silvologists

◄

Fig. 4.7. Examples of epiphytes, which contrary to parasitic plants do not extract nutrients from their host tree, which is a mere support. **A** Large broken trunk covered with epiphytes, note *Monstera* sp. (Araceae, a common house plant in Europe) on top and ferns below, Costa Rica, Atlantic rainforest. **B** Within or next to clusters as shown on **A**, miniature epiphytes like the two Gesneriaceae and the fern are living on a carpet of small mosses and lichens, among the adventitious roots and creeping stems of lianescent epiphytes. French Guyana. **C** Orchids [*Zygosepalum labiosum* (L.C.Rich.) Garay; ident. R.Ek - also see Ek 1989] are among the most spectacular and commercially sought epiphytes, this one growing on a shadowed riverside rock as a support. French Guyana. (Photographs 1969-1979 by R. A. A. Oldeman)

and foresters alike. Their special properties as forest components are rarely acknowledged in ecological studies (but see De Granville 1978). In forestry, they are currently considered as trouble-makers, because they often occur as very effective and tenacious weeds. In silviculture there also are exceptions, notably in palms (e. g. rattans) and bamboos (Gramineae), but most often the productive species are treated as agricultural crops, e. g. the African oil palm (*Elaeis guineensis*). Some can also be traced back to designs for agroforestry systems (e. g. Alvim in Huxley 1983).

For general information on monocots, Corner (1964, 1966), Castro-Dos Santos (1981), Descoings for tropical savannas (1976) and McClure (1966) for bamboos are useful sources. Information can also be found in Ndiaye (1977) and Jeannoda-Robinson (1977), who included herbaceous monocotyledons in their texts.

Monocotyledons as a particular group of plants were distinguished by the Scottish school-teacher and amateur botanist John Ray (1628-1705). They not only differ from dicotyledonous plants in that the seedling bears one seedling-leaf (cotyledon), but a whole set of characters is correlated with this. The single cotyledon is nearly always sheath-like. The primary root system (taproot) generally aborts early and the secondary root system predominates. Most often, the roots are not capable of secondary thickening. The leaves show parallel nerves, and usually have sheath-like bases. The flowers are organised in whorls of three: tepals (two whorls), stamens (one or more whorls) and carpels. The pollen has a special development. And, maybe most important for silvicultural design and silvology, only rare monocots have a cambium. This also is the case in most monocotyledonous trees, e. g.

Fig. 4.8A

Fig. 4.8. Examples of epiphytes and epiphyte compartments. **A** Montane tropical forest in the Andes, 1900 m above sea level, Finca Jaramillo, road Quito-San Miguel de los Bancos, Ecuador. Note trunks and branches covered by dense mats of epiphytes. **B** Detail of **A**. The spongy epiphytic mat is formed by spore plants and small seed plants, among which "Spanish moss" (Bromeliaceae, *Tillandsia*) and miniature Piperaceae. Larger epiphytes rising up, a Bromeliaceae half hiding a fern (to the *left*) and an Orchidaceae (to the *right*). (Photographs 1975 by R. A. A. Oldeman). **C** *Ficus* sp. (Moraceae, strangling fig) in French Guyana, rainforest near Saül. Note the lady, about 1.60 m high. (Photograph 1976 courtesy De Granville)

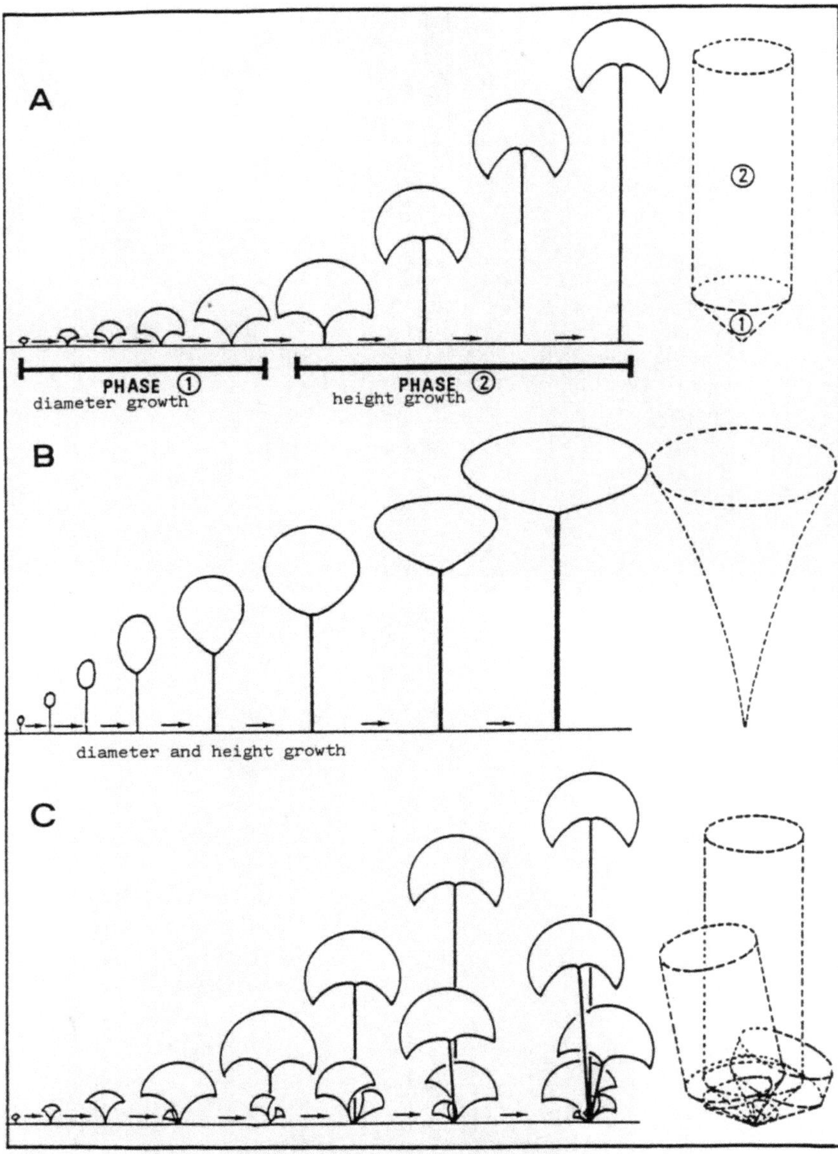

Fig. 4.9. Monocotyledons. Growth patterns of common monocotyledonous trees compared with the growth of dicotyledons. **A** The meristem first builds the final diameter (phase 1) then starts height growth (phase 2). In phase 1 there are often juvenile leaves. **B** Dicotyledonous tree. Diameter growth and height growth occur together. **C** Same principle as **A**, but in monocotyledonous trees of Tomlinson's model (see Fig 3.2) that branch at the base. Note the volume occupied in each case (*dotted*). (De Granville 1978)

palms, the whole vegetative body of which in fact is built by the apical meristems of the axes (Fig. 4.9).

4.6.1 *Architecture of Monocotyledons*

The following main points have been taken from Castro-Dos Santos (1981). Her study gives the best and most thorough summary of this matter to the present time. Figure 4.10 shows her synoptical view of the architectural characteristics in monocots. It is to be compared with Fig. 3.2. These structures already show that monocotyledonous growth has some preponderant characteristics.

- Meristems are organised in a special way, so as to ensure both production functions of dicotyledonous meristems-cum-cambium. Buds are never formed, the meristem being protected by numerous young organs surrounding it. In monocot trees, the meristem is flattened out, so that the meristematic area covers the whole top of the axis, the periphery being occupied by young leaves or other appendages.
- Lateral branching, in the absence of a cambium, cannot be accommodated by secondary thickening of a xylem cylinder, so as to transport more sap to the new and growing organs. Therefore, branching is strictly programmed, just as the formation of other lateral organs (cf. palm leaves). In general, if a new branch or leaf develops, an old leaf or branch is dropped (see Corner 1966).
- Reiteration therefore is very rare in monocots, except when adventitious roots are formed. This indeed is the principle that is apparent in all growth models in Fig. 4.10, except for the numbers 1 to 7. Moreover, the number 4 includes species that are exceptional in monocots because they have a cambium, like *Dracaena* spp. (Agavaceae, see Fig. 4.11D). The formation of adventitious roots is very strongly developed in monocots. It is one of the keys to understanding their agressive behaviour in forests (Fig. 4.11B).
- Plagiotropy is rather rare in monocots. This can be explained because plagiotropy implies an element of indeterminate growth, that may require more resources than can be provided by the constant sap stream. The plagiotropic organs hence are most often the leaves themselves, strictly programmed. In bamboos, each plagiotropic shoot arises underground, is fed by its adventitious roots, and shows a strict development programme (McClure's model, Fig. 4.10-9). Creeping monocotyledonous plants are always either sympodial, each axis in the sympodium forming adventitious roots, (Fig. 4.10-7a,8) or if they are monopodial, each internode may be sustained by its own adventitious roots.
- Vegetative multiplication is frequent in monocots. One cannot consider their role as forest components without taking this aspect into account. Each part of a monocotyledonous plant that has adventitious roots can in principle be separated and become a new individual. Clones are therefore common. It is well known that this is one of the reasons why bamboos and grasses (Gramineae) can barely be eradicated once they are established as components in an ecosystem.
- The vascular system in monocots is built in quite a different way than in dicots. Instead of one xylem cylinder in the centre of an axis, surrounded by a hollow cy-

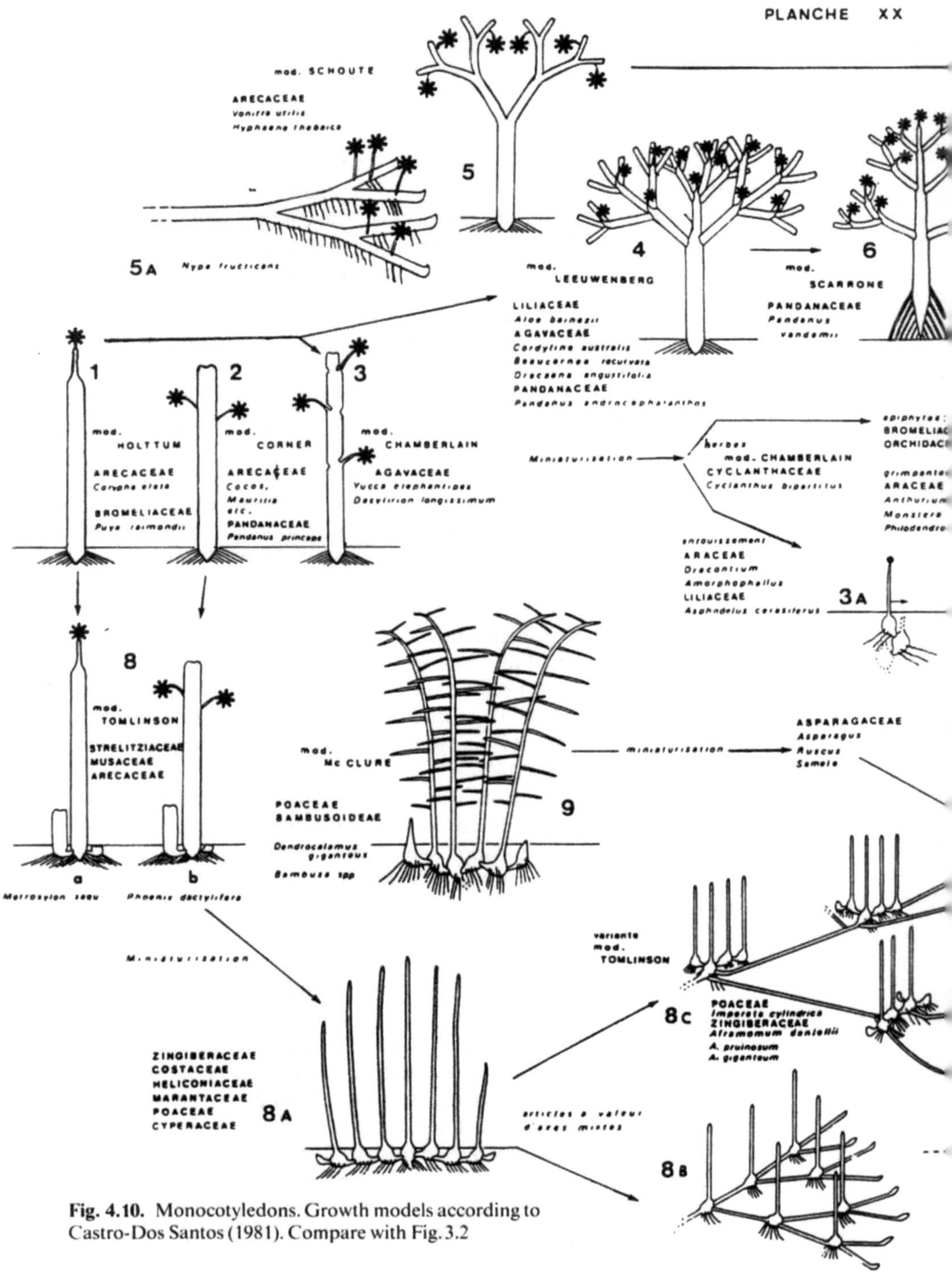

Fig. 4.10. Monocotyledons. Growth models according to Castro-Dos Santos (1981). Compare with Fig. 3.2

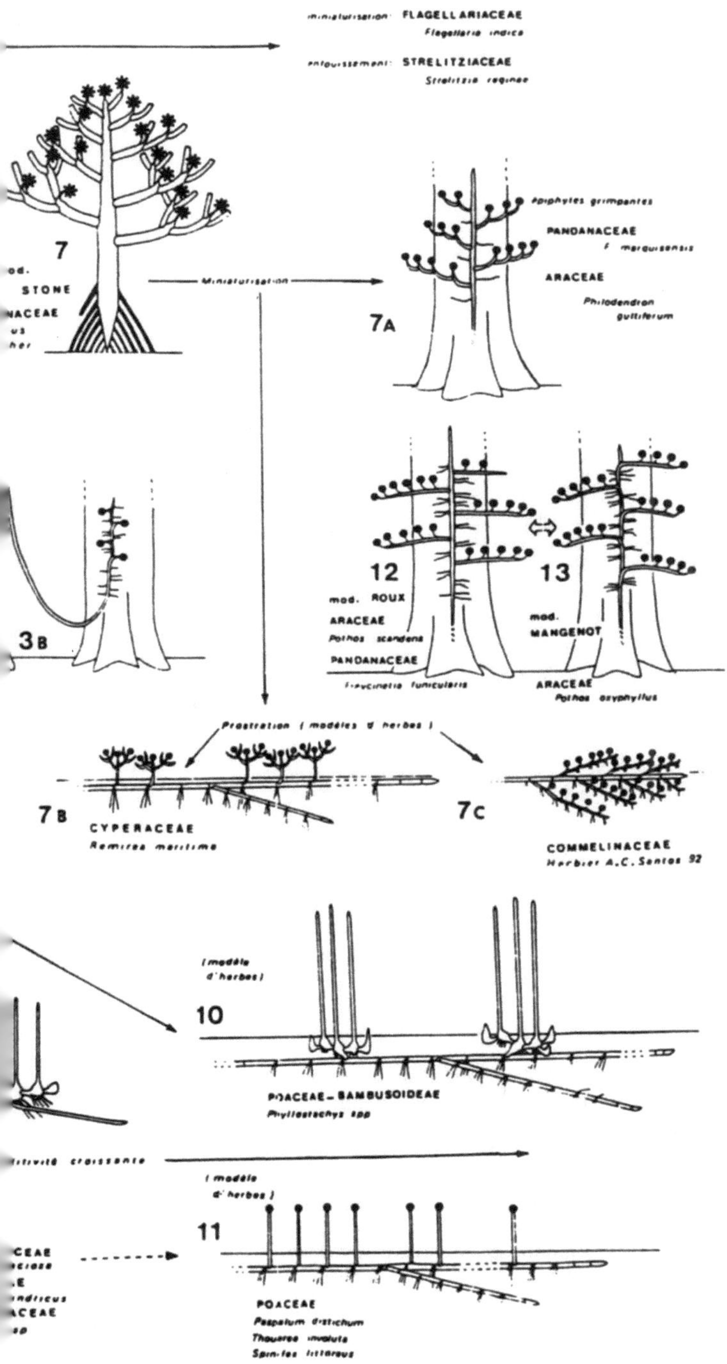

FLAGELLARIACEAE
Flagellaria indica
STRELITZIACEAE
Strelitzia reginae
7
STONE
miniaturisation
7A
PANDANACEAE
ARACEAE
Philodendron
guttiferum
3B
12
13
mod. ROUX
ARACEAE
Pothos scandens
PANDANACEAE
mod.
MANGENOT
ARACEAE
Pothos oxyphyllus
Prostration (modèles d'herbes)
7B
7C
CYPERACEAE
Remirea maritima
COMMELINACEAE
Herbier A.C. Santos 92
(modèle
d'herbes)
10
POACEAE – BAMBUSOIDEAE
Phyllostachys spp
croissante
(modèle
d'herbes)
11
POACEAE
Paspalum distichum
Thouarea involuta
Spinifex littoreus

Fig. 4.11. Monocotyledons. Some examples. **A** *Ancistrophyllum* sp. (Palmae), an African rattan-like climbing palm that shows the typical rattan-image of its crown silhouetted above the forest canopy. Makokou, Gabon, 1981. **B** Invasion by grasses of the undergrowth in a Dutch forest in openings with much light. Veluwezoom, sandy soil, 1979. **C** *Euterpe oleracea* (Palmae, Tomlinson's model) forming the skeleton of a marshy forest in French Guyana. The species is harvested for its edible palm cabbage (palm heart). Crique Arataye, 1967. **D** *Dracaena draco* (Agavaceae), introduced in the botanical garden of Cibodas (Indonesia, 1979). The tree, showing Leeuwenberg's model (see Fig. 3.2) belongs to one of the few monocotyledonous genera having a cambium, also in the roots. (Photographs by R. A. A. Oldeman)

linder of phloem, monocots show axes with a hard epidermis, within which xylem and phloem occur side by side in longitudinal strands that are surrounded by cellular tissues and that go from a precise source to a precise sink in the sap stream. Monocots do not have the kind of wood that is usual in forestry. In human societies that use woody monocotyledons as their main resource, e. g. the South American Indians, no sawn wood is included in buildings. Palm trunks either are used as poles, or are split. The same is true for bamboo-based societies in Asia.

- Flowering quite often occurs late in the life cycle. There are bamboo species that multiply vegetatively during a century, after which the whole clone starts flowering at the same time. In large palms, such as *Corypha elata* (Holttum's model) flowering occurs at the end of several decennia of vegetative life: the end meristem starts building a gigantic inflorescence, which will bear millions of fruits. The tree dies afterwards. This phenomenon is called hapaxanthy (see Hallé et al. 1978). But when the life cycle is short, as in herbaceous monocots such as grasses, each component axis may flower profusely before dying, leaving at the same time numerous seeds and numerous vegetative offspring. If, at the same time, the roots produce chemical products that impede the germination of other seeds (allelopathy see Chap. 5.6.8, e. g. in alang-alang from SE Asia, *Imperata cylindrica*, Gramineae), the species may become the absolute ruler over an area for long years. The situation only changes if the biotope becomes unsuitable. Alang-alang may be eliminated by planting shadow trees of species that withstand the allelopathic influences, like *Pinus merkusii* in Indonesia.

4.6.2 Production in Monocotyledons

As far as is known by the present author, no complete studies on production in arborescent monocots have been done. There are yield studies for the more common crop palms, such as cocos, oil palm or sugar palm, but these do not rest as yet on complete ecophysiological models.

All sapstream models for monocots should take into account that there is no cambium. In comparison with Fig. 3.32, there are only two primary production systems to be considered: the crown and the root system. Moreover, the root system is no one-sided production system, the inputs of which arrive through the base of the main stem. Given the wide-spread tendency in monocots for "unlimited adventitious root formation" (Castro-Dos Santos 1981), the input of the root system is diffuse and occurs all over the plant body, wherever there are adventitious roots.

The following hypothesis is stated here as a precursor to the development of ecophysiological production models for monocotyledons. For production purposes, these plants should each be considered as a set of production units, every one of those including its adventitious root system and its green crownlet. In Fig. 4.10-9, for instance, four such units are fully represented. The whole production of the individual then is the sum of the unit productions, plus the energy and matter spent to have them function together as one individual. Such models can be developed very well on the base of the concepts put forward by Harper (1977), because one monocotyledonous individual is here treated as a population of production units. Harper counts his populations either in "ramets", modules of potential clonal growth comparable to modules of a sympodium in dicots and to the production units mentioned here in monocots, or in "genets", individuals that are the result of the development of a zygote, and which therefore are genetically different.

It might be said that in dicots there is a tendency to unite larger and larger populations of axes or models (reiteration) as ramets inside one physiologically individual plant, whereas monocots show the inversed tendency to divide individuals into populations of ramets, as soon as such individuals become too large for the strictly regulated production system. Many food crops, e. g. all grain crops, are monocots: maize, rye, millet, rice or wheat. Current ecophysiological production models have been built largely by studying such crops. The emphasis on populations, and the subservient role of individual plants in such models may be explained in part because monocotyledons were among the first plants to be modelled.

On the other hand, grasses such as those shown in Fig. 4.10-8C, 9, 10, 11 may develop large and long-lived horizontal individuals, the connecting parts of which are buried in the ground. Pasture-lands therefore have more in common with forests than one might think at first sight, with development covering tens of years implying relatively long-lived plant species.

4.6.3 Ratios and Proportions in Monocotyledons

The ratios between sizes, volumes and biomass weight, as enumerated in Chapter 3, were all related to dicots. Castro-Dos Santos cites three quite different ratios that are relevant for monocotyledonous growth and development, and which have been thoroughly studied. These are the following:

- *stem diameter versus leaf dimensions;*
- *leaf dimensions versus leaf shape* (length/ width);
- *stem diameter versus diameter of adventitious roots;*

However, according to Castro-Dos Santos there is no explanation linking these ratios to production patterns. They change with the development in diameter in the juvenile stage, and reach certain values once the plant starts growing in height (Fig. 4.9A). In other development series (Castro-Dos Santos, p. 58) these ratios change in a different way.

When much-needed further work on monocotyledonous production models starts it can be expected to emphasise the secondary production system of flowers and fruits, both for the biological reasons mentioned earlier, and for economic reasons because many monocotyledonous crops are grown for their seeds and fruits (coconuts, bananas, dates, grains etc.). One may therefore predict that other ratios will be studied, each of them including the numbers, weights or sizes of the inflorescences or fruits.

4.6.4 Some Monocotyledonous Groups Important for Silviculture

A few paragraphs will be added in order to pinpoint important monocots in forest ecosystems.

- Grasses (Gramineae) are important both in the temperate regions and in the tropics because they may be the most tenacious weeds in silviculture. Excess grass in forests usually is a sign that something is amiss either with badly dosed treatments like wrong planting distances or excessive thinning, or with outside influences. Acid precipitations are reputed to stimulate "grassification" ("Vergrasung", Fig. 4.11B) of forest stands. As for lianes, such grass problems should be tackled through the roots, rhizomes or stolons, i. e. belowground, and not by weeding or poisoning the culms.

  But grasses also contain useful forest plants. In the Sahel regions, where a combination between rural forestry and grazing has to be established in many places, modes of coexistence between trees and fodder grasses should be designed (e. g. Bruggink 1986); and bamboos, which are woody grasses, are valuable forest components in the Far East. They have been described and analysed, together with their use, by McClure (1966).

- Palms are never to be forgotten in tropical forestry or in agroforestry. Species like the date palm (*Phoenix dactylifera,*) have for thousands of years been the backbone of an agroforestry system in very dry regions, known as the "oasis". In marshy regions in South America, the ouasseye or pinot palm (*Euterpe oleracea,* Fig. 4.11C), which is harvested for the edible young leaf initials known as "palm cabbage", might play an analogous role. Corner (1966) cites a description of all palm products used by an Amerindian village in Brazil: this village can live on palms alone for food, bedding, construction, weapons, musical instruments and clothing. One might speak of a "palm civilisation". In an analogous way, several peoples in South-East Asia can cover all their needs by using the climbing palms known as rattans (Fig. 4.11A): "rattan civilisations". Rattans include the longest land plants on earth. One stem was measured being over 125 m long. Rattans certainly are not so easy to cultivate, but systems are now being designed in Malaysia and Indonesia, making use of secondary forests (e. g. Weinstock 1983).
- Many monocotyledonous families yield products that can very well be produced in tropical silvicultural systems, making such systems both ecologically and economically richer and better balanced. Food crops are provided by several giant herbaceous species of Araceae that form thick, edible roots, such as taro or dacheen (*Colocasia* and *Xanthosoma* spp.) or the lianescent yam (*Dioscorea* spp., Dioscoreaceae). Supple adventitious roots descending from monocotyledonous epiphytes (e. g. *Carludovica* spp., Cyclanthaceae) are used for their fibres in the Guyanas. Orchids are one of the most interesting families from a commercial viewpoint (cf. Fig. 4.7C), and so are some Araceae (*Monstera, Philodendron,* Fig. 4.7A) and Bromeliaceae (cf. *Vriesea splendens,* not taking into account the pineapple or ananas). Bananas (*Musa sapientium,* Musaceae) come from a mon-

ocotyledonous family that contains commercially interesting decorative plants, such as the South American *Heliconia* or the malagasy *Strelitzia nicolai.*

Monocotyledons should never be forgotten, either in ecosystem analysis or in silvicultural design. They are components that will remind one of their existence if they are not considered in time.

4.7 Cryptogamous Plants

These plants may be very small, such as yeasts, small algae or fungi. They may also reach sizes of some 10 m high, as in tree ferns, but the small-sized species are a vast majority. Cryptogams differ from other plants in their multiplication, which involves no seeds, as in phanerogams, but spores. A spore is a reproductive body, often consisting of one cell but sometimes of more. It separates itself from the parent organism and develops, directly or indirectly, into a new individual. Spores may take many shapes and forms. They may be vegetative, and the resulting individual then is genetically identical to the parent. They may also be differentiated in microspores that produce male gametes, and macrospores that produce female gametes, the distinction not always being so clear, however. Gametes fuse and yield a new individual that is genetically different from both parents, like in all sexual reproduction processes. In order to understand the properties of spore plants as forest components, it is useful to know that there is often an alternation of generations: sporophytes yield asexual spores, out of which come gametophytes that yield gametes, that fuse and give rise to sporophytes again. The number of variations on this theme is amazing, and allows spore plants to survive in a very large array of biotopes.

Spore plants are thought to be among the oldest plant species on Earth, real and tough survivors. Their small spores can be borne everywhere by the wind. Many such spores are very resistant to drought, high temperatures, freezing and other influences that often kill seeds. The smaller these plants are, the larger is their geographical area, from the cosmopolitan fern *Pteridium aquilinum*, that can be found both in temperate and humid tropical regions, to many fungi. *Armillaria* species (honey fungus, one of the "root rots") have been found both in Europe and the Philippines.

Spore plants are very important forest components. They may be small, but they assume important regulation mechanisms in the forest, and may be very precise indicators for environmental factors such as the pH (see Sect. 4.5).

The science of spore plants again is a field in itself that will not be thoroughly treated here. (e. g. see PUDOC 1982). Some few notes are presented to convey only an impression of these very important groups of forest components.

- **Ferns** are green plants, ranging from herbs to tree-like forms (Fig. 4.12A) or lianes (Fig. 4.12D). They can generally be recognised because their leaves ("fronds") are rolled up when young (Fig. 4.12A, below). Their spores are often borne on the lower side of the leaves in sporangia, the very small brownish or greenish box-like structures, that identify a fern leaf. Some tree ferns, e. g. in the genera

Fig. 4.12. Examples of spore plants. **A** *Cyathea* sp., a tree fern several meters high in the undergrowth of Mont Tonkouy, 1000 m above sea level (Ivory Coast, 1965, photograph courtesy F. Hallé). **B** Tree stump covered on one side by lichens, beside which (to the *left*) mosses are growing. Verwolde, Netherlands, 1979. **C** Polyporaceous fungi on dead birch stem. They reduce dead wood, sporocarps shown here. Hyytiälä, Finland, 1982. **D** *Gleichenia* sp., a fern that occurs all over the tropics and forms carpets on steep eroded soils, e.g., along roads. French Guyana, 1971. (Photographs **B**, **C**, **D** by R. A. A. Oldeman)

Cyathea and *Alsophila*, may occur either as pioneer trees, as in Puerto Rico, or as undergrowth species. According to De Granville (1978) they replace palms in certain stressed environments. However, most ferns are herbaceous and grow on the forest soil, on dead wood or on naked rocks. There also are many epiphytic ferns, in all parts of the world (cf. Koop 1981a, for NW Germany), the most spectacular one probably being the African *Platycerium alcicorne*, that is a much appreciated decorative plant.

Except for some local uses, e. g. in flowering pots made from the root-cuff at the base of tree ferns, building poles from tree ferns, or decoration, ferns do not yield many commercial products. But generally, their presence in a forest is both a sign that the ecosystem is rather complete and a symptom of the health of that forest. Both absence and proliferation of ferns are warning signs for foresters.

- **Mosses** are smaller than ferns, and show gametophytes built very differently from the sporophytes. They often form carpets, sung by poets. Mosses (e. g. Fig. 4.12B, to the left) also contain many indicator species for micro- or macro-environments. They may contribute to form an optimal germination environment for tree seeds, particularly because they accumulate much water, and dry out very slowly. Mosses have no roots, which is the reason for this water accumulation in the green parts. Any moss carpet therefore plays a role in the hydrological cycling in forests. Mosses are among the first to colonise fallen trunks and branches, but also living branches (see Sect. 4.5 and Fig. 4.8). There is always a microfauna associated with such carpets.
- **Fungi** are the most common, and ecologically among the most important forest components belonging to the spore plants. Their sporocarps ("fruiting bodies") include the well-known toadstools or mushrooms. Figure 4.12C shows the sporocarps of fungi that live in dead or dying wood. These fungi often belong to the family of Polyporaceae, that is also particularly frequent in the humid tropics.

Most fungi have the same kind of basic building unit, a multicellular very thin thread that contains no chlorophyll. Such a filament is called hypha (plural hyphae). Usually, the fungi that are important in forests build a network of hyphae, that is called mycelium.

The chlorophyll-less mycelium may be considered as the basic structure in fungi, even if its appearance may be either reduced (e. g. one cell) or extended, either simple or specialised (e. g. rhizomorphs). The direct environment of hyphae may be mainly inorganic, e. g. mineral soil, but it has to be always partly organic. Interactions between fungi and dead or living organic matter, in terms of systems hierarchy (Table 2.1), occur from cell to cell, or from tissue to hypha. These interactions are of several types.

a) **Saprophytic life cycle.** Because it is not green, a fungus disposes of no autonomous carbon synthesis (photosynthesis). Saprophytic fungi obtain their carbon synthates from *dead* plant and animal matter, of which there is enough in a forest. In Chapter 5.3.4, saprophytic fungi will be ecologically considered.
b) **Symbiotic life cycle.** The same lack of photosynthetic capabilities makes hyphae able to weave through the tissues of living plants or animals and find their energy

there, in the form of energy-rich photosynthates or other substances. Inorganic nutrients can be directly absorbed by hyphae from the abiotic environment, and may be passed on to the green plants or to the animals that serve as symbiotic partners. There are several kinds of symbiosis, depending on the balance of payments between the partners:

- parasitism: the fungus takes more than it provides;
- mutualism: the fungus exchanges energy, nutrients and water with its host on an equal basis, each partner providing what the other lacks;
- resistance: the green plant, or animal, is protected against exchange with a fungus (chemical protection, anatomical protection), or eats it to obtain what it has got. Nematode grazing upon mycorrhizal fungi (see Chap. 5) is an example, people eating mushrooms another.

The mycelium, or its cells, can become specialised for such fuctions. Rhizomorphs are strands of many hyphae that are surrounded by a resistant layer, often with remarkable colours such as glistening black, or orange, and that serve to penetrate difficult environments, or to cross them. Haustoria are mycelium cells or branches specialised to penetrate plant and animal cells for the uptake of products.

This is the place to note that some seed plants have adopted life styles that are remarkably close to that of fungi. In the monocotyledons the Burmanniaceae, and in the dicots the Gentianaceae are examples of families containing genera that are without chlorophyll, are very small and live as saprophytes, e.g. in the South American rainforests. Parasitic phanerogams (seed plants), such as the Asian Rafflesiaceae, should be mentioned here, too.

Fungi are cryptogams that may form both types of spores, the asexual ones and the gametes, on mycelium structures that are very different. Spores may just be modified mycelium cells, or may be the end product of some very elaborate differentiation sequence that leads to structures such as sporocarps. This in its turn leads to intricated life cycles, weaving through the forest ecosystem from one host or partner in one state of the mycelium to another, when another type of mycelium is formed. The transition from one host or partner to another may be achieved by migrating spores or by travelling mycelium.

The term "cryptogams" ("those that marry in hidden places") may be taxonomically old-fashioned, but in silvology it is particularly relevant. Patient studies in poorly accessible places of these very small plants, that therefore can barely be observed without damaging them, have nevertheless led to an impressive array of knowledge concerning fungi in forests.

Without forgetting to mention that parasitism, mutualism and resistance may be three states of the same relationship between identical organisms, depending on the "balance of payments" between them, i. e. depending upon the state of the ecosystem, it is clear that fungi are organisms that play a key role in tree pathology. In the branch of silviculture that is indicated as "forest pathology", fungi are indicated as diseases when they cause tree populations to die back (see Fig. 5.74).

Because they are so closely linked to other organisms, ecophysiological production models of fungi are most often treated as subsystems of the production system of their partner. They are then considered as regulatory mechanisms (cf. Odum and Pigeon 1970).

Other cryptogamous groups exist, but will not be treated in this rapid survey of non-arborescent forest components.

4.8 Animals

4.8.1 General Characteristics and Ecological Functioning

It is, of course, impossible to cover the animal kingdom in a few pages. But animals are such important components of forest ecosystems that they have at least to be rapidly surveyed, just as the plant groups in the preceding pages.

All common forest animals have three things in common: they have finite sizes, a dorsiventral architecture, and they are mobile. This holds true as a rule-of-thumb, with enough exceptions to prove that animals are living systems: only abiotic, man-made systems seem to keep strictly to the rules.

The mobility of animals is well illustrated by Kaufmann (1972) in his book "Insect Travellers", in which he documents transcontinental and transoceanic migrations of insects. Birds yield other spectacular examples. The range of animals does not depend entirely on size. Elephants, with territories of roughly 60 km in diameter, travel lesser distances than tiny insects voyaging with the atmospheric jet streams at some 13 km altitude. Other, very small animals may on the contrary be limited in their movements to the microhabitats of one small forest valley. But move they all do.

Their role in forest ecosystems therefore is very different from that of plants. Animals first accumulate biomass, but once they have attained their final, adult size, they transfer organic matter, water and nutrients from one place in their biotope to another. These biotopes often cover several ecosystems as defined by their plant components, in other words, several vegetation types. If ecosystems have such a thing as organised transport of matter and energy between their components or compartments, animals are the vectors. Up till now, this aspect of animal life in forest ecosystems seems hardly to have been taken into account in any forest production model. In Chapters 5 and 6 (e. g. Fig. 6.52), it will be seen that this consideration might lead to very different ecosystem definitions from those based on plants.

Finally, Hallé (pers. commun.) suggested the notion of "plagiotropy" in animals, if such a concept has any sense outside the plant realm. Seen from the outside, the vast majority of the land animals have a back (dorsum, in Latin) and a front or belly (venter, in Latin). They are dorsi-ventrally built. Even animals such as snakes (Fig. 4.14D), earthworms or nematodes show some outward dorsiventrality by the way in which their eyes, mouth parts or genitals are placed. As an unproven hypothesis, dorsiventrality could be linked to movement, because this most often takes place perpendicularly to the direction of gravity. Animals that fly are among the most dorsiventral of all, with a horizontally expressed structure that supports them in the air. Fish also are dorsiventral but mostly in the other, vertical sense to cleave the waters (Fig. 4.14B). Worms frequently move in diagonal directions and have a more radial symmetry, whereas snakes have a very horizontal pattern of movement: their backs and fronts are very different if they do not live in trees.

In any case, a logically coherent image of animals in general emerges when linking the three properties they have in common: structurally, mobility is favoured by dorsiventrality, and functionally, their steady adult size allows them to move without accumulating an important amount of the energy and matter consumed.

As a curiosity, reiteration should be mentioned. In zoology, it is called regeneration, a term that should not be confused with "regeneration" in forestry. The lizard that has lost its tail sometimes reforms two tails. In plants, one would speak of reiteration. Earthworms also can form new fore and hind parts if cut in two. Still, such regeneration phenomena are both rarely perfect and perfectly rare. Compared with plants, where reiteration is frequent because it is also an expansion mechanism, its rarity in animals might be linked to their steady adult size that does not require reiteration. Theoretically interesting though they may be, animals like corals or hydras are not considered here since they are not forest components.

The role of animals in the food chain will be considered in Chapters 5 and 6. It explains many of their specific properties, that are left unmentioned here. The following pages of this chapter concern a few examples of important animal groups. Many data were obtained from Storer and Usinger (1955) and Ayensu (1980).

4.8.2 Mammals

Mammals, which are viviparous, warm-blooded, four-limbed and more or less covered with hair or fur, are important forest components. Their metabolism is rapid and food requirement is high. They are commonly considered, together with birds, some reptiles and amphibians that all are striking to the naked eye, as "wildlife". In the newspaper sense, this seems to be a loose group of animals that is rather liked by humans for a variety of reasons, from aesthetic to culinary ones.

Like all animals, mammals are linked to specific forest compartments or regeneration units, in which they fulfil many functions. But whether or not a compartment is used by a particular functional group of animals depends upon the forest type. An example are the monkeys (cf. Fig. 2.2), that are present in the crown canopy of tropical forests, but not in that of temperate region forests.

On the contrary, other categories of mammals seem to be present in nearly all forests of the Earth. Except for humans and rats this certainly is the case of pigs (Suoidea), species of which are present on all continents except Australia. Figure 4.13C shows the example of the European species, *Sus scrofa*.

Very specialised mammals are exemplified by the South-American sloth (*Bradypus* sp., Edentata) that is shown in Fig. 4.14A, in Costa Rica. This three-toed sloth is a monophagous animal that eats only the leaves of *Cecropia* pioneer trees (Moraceae). Variety exists in its dense, matted pelt which is an ecosystem in itself with many algae, fungi and insects living in it.

It was mentioned in the beginning of this chapter that the life of many non-arborescent forest components cannot be explained independently of the forests in which they live. A spectacular series of such animals was discovered by Dubost (1968), who described rodents and artiodactyles from Africa and South America. The animals belong to totally different species, but show remarkably converging build and habits (Fig. 4.16). This was discussed also by Bourlière, (in Meggers et al. 1973), a book that yields much information on tropical animals. Such convergences have been known for a long time, of course, between the other mammals and the marsupials, e. g. in the case of opossums (Fig. 6.8B; Australia, the Americas) and rats (Eurasia, Africa, the Americas).

4.8.3 Birds

Birds are animals that lay eggs, are warm-blooded, most often can fly because two of their four limbs are differentiated into wings, are covered with feathers and have a highly developed voice, hearing and sight. Body temperature is relatively high and consequently food requirements are high also. They are among the components of all forests, in which they fulfil many functions that are referred to later (cf. Sect. 6.2.4.6). For birds also, many characteristics cannot be understood well without considering their biotope.

Their degree of mobility and their power, more than their size, can yield some insight in the distinction of functional groups in birds. Powerful birds, but not necessarily the largest ones, can cover larger areas than others. The ecosystem of which they are components then most often is larger than a forest mosaic, especially in more densely populated countries with large agricultural areas. This is the case for birds of prey, such as the European *Falco peregrinus* (falcon, Fig. 4.13B).

The highest mobility is found in migratory birds, that pass one part of the year in an entirely different region of the world than the other part. In some species, that pass the cold winter of the northern regions in the warm south, it is known that they return the following year to the territory they occupied the previous summer. Such birds are components of geographically widely separated ecosystems. It is probable that they fulfil a precise ecological function in both. Van Dorp (in Gomez-Pompa and Del Amo 1985) showed that three out of the most important seed dispersal agents in the Veracruz rain forests (México) were migratory birds from North America.

On the other hand, very small and weak non-migratory birds may have restricted territories. This is the case of hummingbirds (Trochilidae) that pollinate flowers and that are capable of stocking so little energy-rich honey that they pass part of the cool night in a cataleptic, death-like state, being incapable of producing enough body heat. The sun literally brings them back to life.

With very mobile organisms like birds it has to be emphasised that the size of their territories should be carefully compared to vegetational units, so as to assess their forest component functions in a correct way.

Both mammals and birds, and some groups of amphibians and insects, belong to the forest animals that produce sounds. Except for the transport of matter and energy, they also transmit messages, sometimes in a kind of international language, such as the alarm cry of birds of song that is the same the world over. These sounds should be considered as mechanisms in ecosystem functioning. Forest dwellers know this, and can judge the state of their forest territory by carefully listening.

4.8.4 Reptiles, Amphibians and Fish

These three groups of cold-blooded, egg-laying vertrebrates should be put in the opposite order if the evolutionary context were considered. Fish are seen as the most primitive, amphibians as the animals that started to live on the land, and reptiles as those that achieved terrestrial life. The order given in the title is that of their estimated importance as forest components. These groups, like the preceding ones, contain many species that ensure a large array of functions in forest ecosystems.

Having a variable body temperature, these animals have energy and food requirements lower than those of mammals and birds. Their feeding dynamics vary between the fast and mobile patterns displayed by fish and the slow tempo of the reptiles, interrupted by short and violent episodes of prey hunting.

Relatively few species of these groups are vegetarians. One such case is the fruit-eating South American *Iguana iguana*, a large lizard, but most are predators. This makes them regulators of other animal groups. Frogs and toads (amphibians) regulate insect populations, and are among the more noisy forest animals, as everyone knows who has spent a night in the humid tropics (but also elsewhere, Fig. 4.13A). Snakes most often are regulators of populations of larger prey animals, such as rodents or birds, but also of amphibians. It is to be emphasised that the regulation function by snakes proceeds with biochemical means as far as these animals are poisonous (cf. Sect. 5.6.8).

Saurians (Loricata) regulate fish populations, among others. Sound fish populations may play a very important role in forest ecosystems, and crocodiles, caimans and alligators eliminate the sick and weak among the fish.

Fish are not usually considered as forest components, but it has to be strongly emphasised that they play an often underestimated role in seed dispersal of forest trees. For instance, the South American Pakou (Pacú) shown in Fig. 4.14B is described by Goulding (1980) as one of the most active seed-eaters of Amazonia and the Guyanas, dispersing at least 15 tree species.

4.8.5 Insects and Other Arthropodes

Storer and Usinger (1955) very concisely stated (p. 356): "The Phylum *Arthropodia* (Gr. αρθροσ, jointed + ποδ -, of foot) contains fully 740,000 of the more than 900,000 known species of animals. It includes the crabs, shrimps, barnacles

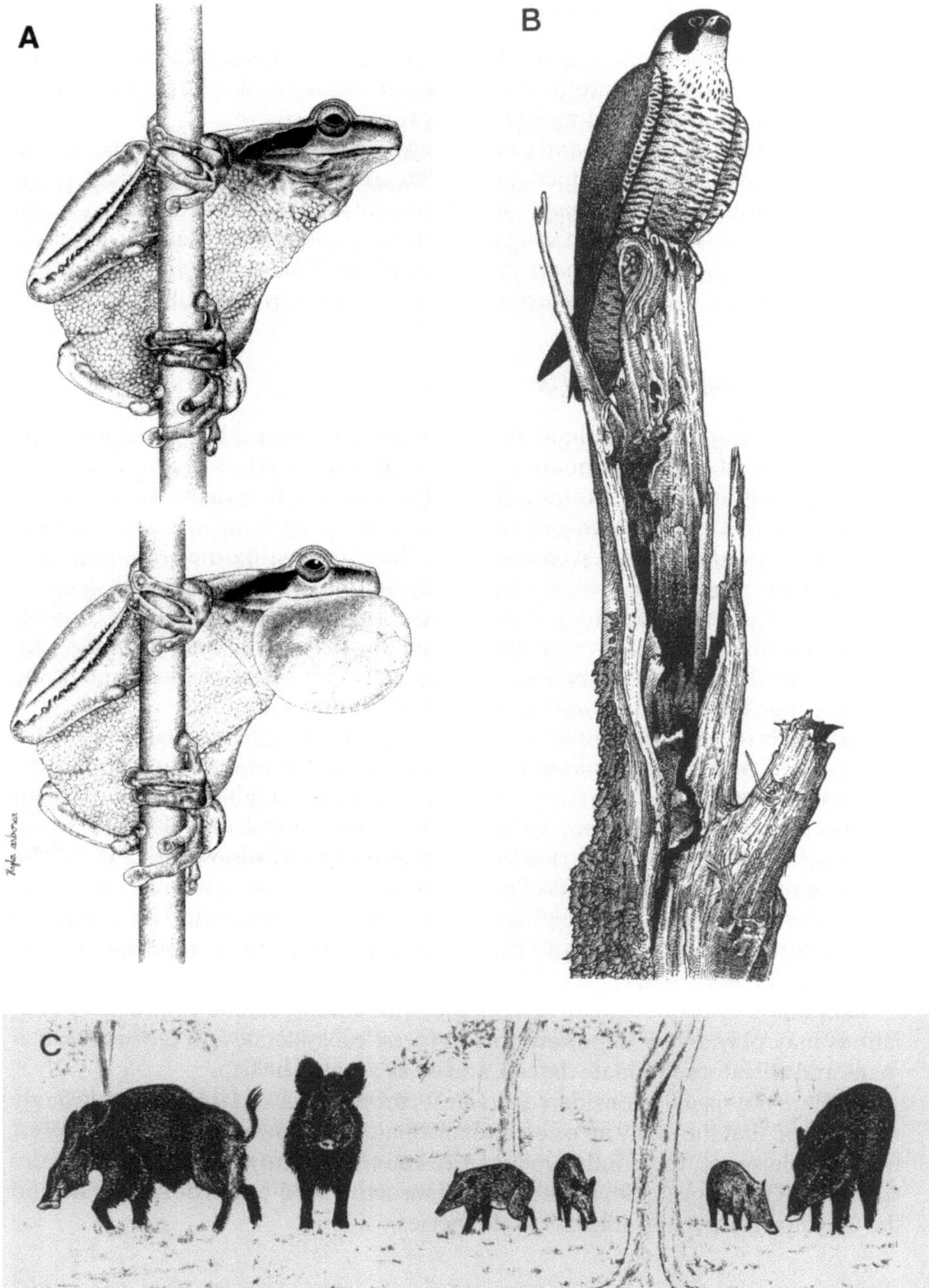

Fig. 4.13. Examples of European forest animals. **A** *Hyla arborea*, the green tree frog, singing below. **B** *Falco peregrinus*, one of the mighty birds of prey covering large areas with more than one silvatic mosaic (cf. Chap. 6); **C** *Sus scrofa*, the wild boar roaming the forest and moving the soil. This animal prepares the biotope for tree seed germination. ("La Hulotte", numbers 23, 42/43 and 53)

Fig. 4.14. Examples of South American forest animals. **A** *Bradypus* sp., the three-toed sloth, crossing an asphalt road (Costa Rica, 1979). **B** Pacou, a fish which eats and disperses tree seeds (River Approuague, French Guyana, 1968). **C** Bee, of the superfamily Apidoidea, ready to harvest honey and pollinate the rosaceous *Hirtella* flowers (French Guyana, 1968). **D** Tree snake of the Colubridae (cf. Boiginae), Saül, French Guyana, 1971. (Photographs by R. A. A. Oldeman)

and other crustaceans (Class *Crustacea*), the insects (Class *Insecta*), the spiders, scorpions, ticks, etc. (Class *Arachnoidea*), the centipedes (Class *Chilopodia*), the millipedes (class *Diplopoda*), and some less familiar and fossil forms". According to more recent estimates (Ayensu 1980), the insects alone count one million described species, and may well contain twice as many that are still unknown.

Arthropods have an exoskeleton made up of chitine (insects, spiders) or calcium (crabs, lobsters). Their body is segmented, as well as their legs which gave them their name. All segments and joints are specialised for some biological purpose, which has resulted in an amazing array of forms adapted to all thinkable biotopes on earth.

Arthropods reproduce by eggs. After hatching ("nymphs"), the larvae ("instars") may be very different from the adult ("imago"). This is one of the effects of the rigid exoskeleton that does not allow growth, so that it has to be shed each time when a new growth period starts, after which a new exoskeleton is formed. This profound reorganisation at every step in the life cycle has been adapted by strategies in which every larval stage and the imago may survive in completely different parts of their ecosystem, and look completely different. The best-known example is the caterpillar which is the larva of a butterfly. But such radical change in body architecture ("metamorphosis") is not universal. A cockroach or grasshopper larva (Orthoptera in general), for instance, can easily be recognised as an "incomplete" version of the imago.

Such richness of form and such flexibility in development make it impossible to draw anything but a few main ecological lines concerning arthropods in the present paragraph. For a more systematic treatment of forest insects the reader is referred to the relevant literature (e. g. Brauns 1976).

What should be said is that the development through larval stages to the adult imago yields enormous potentialities to use different forest biotopes. In this regard, insects may be compared with the fungi that can travel from one niche to another in forms that are adapted to each (cf. Sect. 4.7). Insect eggs are tough. Each larval stage may be adapted to a new mini-environment, e. g. another host plant, another ecological function or another season. In each specific case, the life cycle should be known entirely if one is to assess the exact ecological function of a species (e. g. PUDOC 1982).

Insects are the only non-vertebrates that fly, and so their range of migration is large. But flight is limited to one stage in their life, the adult one. However, this stage allows them to travel extensively (Kaufmann 1972). If given the occasion, these tough organisms will proliferate and may even become the main producers of biomass in a given ecosystem (cf. Sect. 5.6.9). When this happens in a managed forest, they are referred to as pests or biotic catastrophes (see Fig. 5.74).

If present in normal proportions, insects fulfil many regulatory functions. They pollinate flowers, thus facilitating the fruiting of plants (Fig. 4.14C). On the other hand, other insect species act as seed predators and so limit the demographic expansion of plant species (Janzen 1970a, 1978). Insects regulate each other, and they regulate other animal populations in the role of parasites,

Fig. 4.15. Examples of African and South American insects. **A** Termite colony, note rainwater-evacuating properties of roof. Gabon, Makokou, 1981. **B** Wasps (Vespoidea) that pollinate flowers, here ready for battle around nest. French Guyana, 1972. **C** Army ants (fourmis maniants) that eat all animals around. Gabon, Makokou, 1981. Note large warriors. (Photographs by R. A. A. Oldeman)

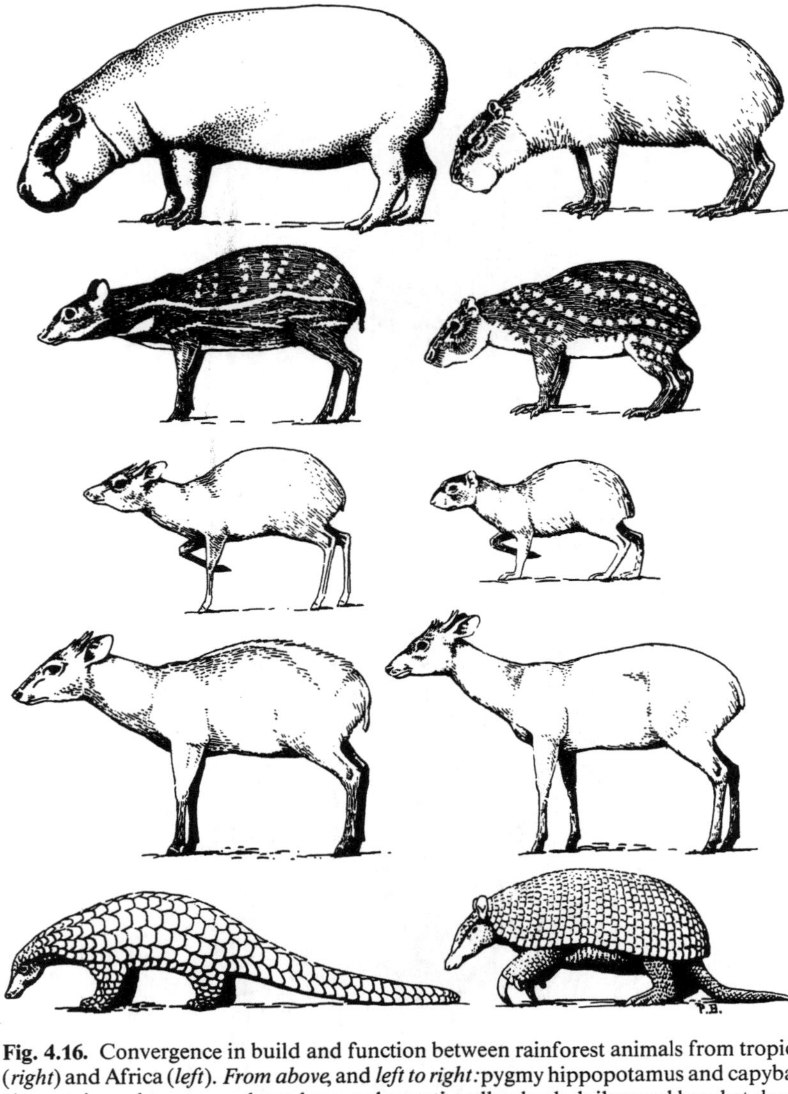

Fig. 4.16. Convergence in build and function between rainforest animals from tropical America (*right*) and Africa (*left*). *From above*, and *left to right*: pygmy hippopotamus and capybara; African chevrotain and paca; royal antelope and agouti; yellowback duiker and brocket deer; terrestrial pangolin and giant armadillo. Each *pair* of animals drawn to the same scale, scale may be different between pairs. (Dubost 1968)

flies or mosquitos for instance, that also are vectors of microbial diseases. Insects are active in the breakdown of dead or living wood, e. g. termites (Fig. 4.15A). They may be both pollinators and attack other animals that come too near, e. g. bees and wasps (Fig. 4.15B). Ants are a fascinating group of insects, that may determine the functioning of a forest to a great extent because they operate in large numbers (see Sect. 6.2.3.3). The army ants (fourmis maniants) of

Africa, that every once in a while spread out and eat all animals in the neighbourhood, including humans (Fig. 4.15C) are powerful components of the African rainforests.

Spiders regulate insect populations, and may sometimes be large enough to eat other animals as well. Their function also implies biochemical regulation, because spiders are poisonous (cf. Sect. 5.6.8). Crabs may play an unsuspectedly important role in forests, e. g. in the Malaysian mangroves, where they eat tree seedlings.

These few examples have to suffice in this context, but they should serve to remind foresters and silvologists alike that a forest ecosystem can neither be explained nor used in a balanced way without paying great attention to the arthropods. If the dangers of insect and fungal proliferation for tree crops are appreciated realistically, an American statement might well be true that silviculture is for 90% forest protection against such potential pests and epidemics.

4.8.6 Worms

The term "worm" in popular speech lumps together all animals with long, round bodies but without conspicuous appendages, sometimes even including certain insect instars. As a matter of fact, true worms are dorsiventral and have a front end which meets the environment, and a back end. Two groups of such worms are mentioned here as particular important forest components (cf. Sect. 5.3.4.4. and Fig. 5.28).

The most primitive of the two is the class of Nematoda (roundworms). They have slender, cylindrical or tapered bodies, a resistant cuticle, a complete digestive tract and no segmentation. Few roundworms can be observed with the naked eye, and most of the very numerous species are microscopical. They are perceived as minute threads, in Greek νημα (nema), which gave them their name. Their role in forests has only recently been perceived as the important ecological factor that it is, knitting together the whole underground food pyramid. There are vegetarian nematodes, consuming roots or grazing on fungi, predatory nematodes and parasitic nematodes (e. g. Bongers 1985).

The second group of worms is much further advanced in the evolutionary sense. This is the phylum Annelida, containing the class of Oligochaeta, the earthworms. Their bodies are segmented, the segments or somites are essentially similar and numerous. They structure the worms both externally and internally. Their skin is much more developed than in nematodes and bears small, chitinous hair-like bristles. In some aspects of their biological organisation earthworms are comparable with insects, which also are segmented.

Forest earthworms are very important in the processing of dead organic biomass into humus. Their role is the fragmenting of larger particles by eating them and secreting the fragments that are left after digestion. These fragments then are further decomposed by micro-organisms. If forest soils are completely without earthworms, something is seriously amiss.

4.9 Between Forests and Non-Arborescent Components

This short survey of non-arborescent forest components has been written as a necessary step in between the trees, which define forests, and the smallest forest ecosystems that are the subject matter of the next chapter. However small and incomplete it may be, the list of groups and examples of different organisms united in the preceding pages constitutes a much-needed reminder. Too many forest studies do not render justice to the enormous biological richness of these ecosystems, by just considering the trees or in any case by selecting a minority of the component species on pragmatic grounds. Pragmatism not only concerns the use of forests by humans, leading to emphasis on organisms that are directly useful (crop plants, game) or noxious (crop pests and diseases). It also is related to the reduction of forest models to scientifically manageable proportions by reducing the number of subsystems to be considered (cf. Sect. 2.4.4.).

Both reasons are perfectly valid. However, a serious silvology cannot be allowed to continue ignoring the real complexity of forest ecosystems. The struggle to avoid undue simplification while still keeping models manageable pervades the next chapters. This venture cannot be perceived in its true dimensions, including myriads of species and interactions, without an overview of at least some of the important groups of organisms involved.

Third Part: Forest Ecosystems

The following chapters concern the living systems that are called forest ecosystems. They are no longer treated as black boxes, as in the preceding chapters, but are explained. The lower-level forest ecosystem is called *eco-unit*. This abstraction includes all forest regeneration units: gaps, clearcut surfaces, stands, burnt surfaces, etc. It is explained in terms of its component organisms or its compartments as subsystems, and according to criteria of architecture, production ecology and species composition. The higher-level forest ecosystem, a forest mosaic, is called *silvatic mosaic*. It is explained in terms of its subsystems, the eco-units that build it and in terms of certain biotic compartments linking the eco-units together. The site largely remains an unexplained black box.

5 Forest Eco-Units

5.1 What Are Forest Eco-Units?

5.1.1 Elements of Definition

The formal definition of different kinds of eco-units is not so easy as it appears in a forest scale-drawing (cf. Sect. 5.4). Oldeman (1983b, p. 176) defined an eco-unit as follows:

> ".. the unit of vegetation which started its development at the same moment and on the same surface."

He gave as examples a patch of forest growing in a gap, a *Pinus caribaea* stand, or a corn field, and mentioned the possibility of using analysis of populations of eco-units of the same kind, i.e. within one size range, at different ages but each with the same structural and functional sequential build-up.

Table 2.1 shows the place of an eco-unit in the systems hierarchy used in this book. Its fundamental subsystems or components are organisms. These will be most often lumped together into compartments, because even the most simple forest eco-unit has too many component organisms for a component model to be manageable. The supersystem that is built by eco-units is called a silvatic mosaic. The term forest mosaic is also used in other models, e.g. landscape models, in which it does not have the same meaning (e.g. Shugart 1984, and his bibliography). In particular, most forest mosaics as seen by other authors either include or neglect the abiotic site factors. The particular kind of mosaic meant here, that is built by eco-units, is hence indicated in Chapter 6 as a *silvatic mosaic* (Oldeman 1983c). A silvatic mosaic is assumed to occur on one homogeneous site of sufficient surface area: if this area is not sufficiently large, corrections are to be made in the model.

The same assumption has a more general validity in the case of the eco-unit, because an eco-unit most often is small enough to fit into one unit of even a detailed map of soil, climate or a combination of both. Therefore the present chapter mostly neglects the cases in which this is not so.

This is emphasised, since a real problem in analysing eco-units is their delimitation, in the field as much as in the model.

5.1.2 Spatial Limits and Gradients

Walter (1974) mentions scales of a "decametric, hectometric or kilometric" order (tens, hundreds or thousands of meters respectively) when analysing patterns and processes in a forest. Shugart (1984, his Fig. 7.4) uses a scale between 0.01 and 0.1 ha, approximately, as the surface area for the application of "gap models". These examples point to a grid-like approach (see Chap. 2). Such grids are necessary if one is to make calculations of the type that is usual in growth and yield models, in phytosociological studies (cf. Godron 1971) or in remote sensing inventories. These approaches have in common that they aim at the establishment of the distribution pattern of biomass, species populations or other categories of organisms (cf. Raunkiaer's biological types, or Ellenberg-analysis; Sect. 5.5) by the use of statistics.

In Chapter 2, grid analysis has been mentioned as an alternative to the analysis based on "natural limits" that can be perceived by the human eye. One example of many such "natural limit" analyses is given in Fig. 5.1, representing the mosaic of eco-units in semi-natural forests in Germany, as mapped by Koop in 1978 (published 1981a). This method of forest mapping is quite old, as exemplified by Fig. 2.3, a transect published in 1766. The bibliographies given by Koop (1981), Whitmore (1982), or Shugart (1984) and the examples and considerations united during the Rinteln Vegetational Symposium of 1968 (Sommer and Tuexen 1974) give some idea of the long and intensive intellectual struggle with the concepts of vegetational limits.

As a prerequisite to understanding the concept of limits in eco-unit analysis, chosen among the many possible or existing notions, the concepts of gradients and limits are considered here, somewhat like the comparison between gradients and sequences in the analysis of tree morphogenesis by Oldeman (1974a, his Fig. 19). This is necessary in order to solve the apparent contradictions between such schools as treat vegetations as continuums *without* clearcut limits and others that treat them as mosaics in a large sense *including* interior limits.

Figure 5.2A shows how an ecological factor may change gradually over a distance. An example would be the mean temperature, as a running average, from pole to equator. Such a gradient may hide a biological limit, for example a mean temperature value above which it is possible for palms to grow. Such limits are seldom due to one factor alone, although one factor may be a limiting factor in the whole complex. In our example, palms may occur from very dry to very humid climates, from desert vegetations to rainforests, but they seem to be limited mainly by a critical temperature. This is so evident that the tropics are sometimes defined as those regions where palms can live (Hallé pers. commun. 1976). The functional limit here of course is in the living system "palm", and not in the factor temperature. "Placing" living systems between limiting values on such gradients by computer-aided ordination is a method used in the USA under the name "gradient modelling" (Kessell 1979).

Figure 5.2B shows a limit brought about by an abrupt change in the value of a factor. Many examples can be cited: the abundance of grass on one side of a wall

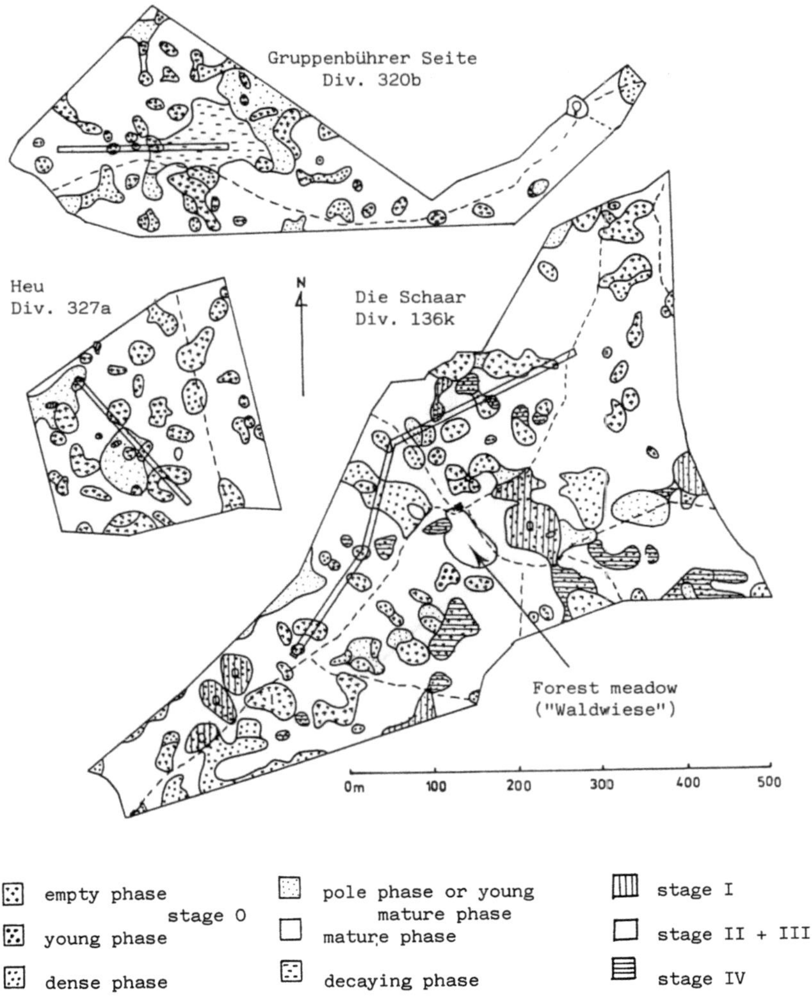

Fig. 5.1. Example of mosaic built by eco-units in the semi-natural forests "Hasbrucher Urwald" and "Neuenburger Urwald" (Germany). Note the difference in sizes of the eco-units, which have been distinguished according to their development phase and successional stage. The "mature phase" predominates, and smaller eco-units of other development status are included in it, which is not a general feature of all forests. (Koop 1981a)

and its rarity on the other side, the abundance of air above the surface of a stream and its rarity below, the frequency of roots above and below soil level in most ecosystems, or the change in temperature outside and inside the skin of a warm-blooded animal. This kind of change over small distances exists, but the distance approaches zero only in physical systems such as the water surface mentioned above. In living systems, abrupt changes most often point to the existence of some filtering layer.

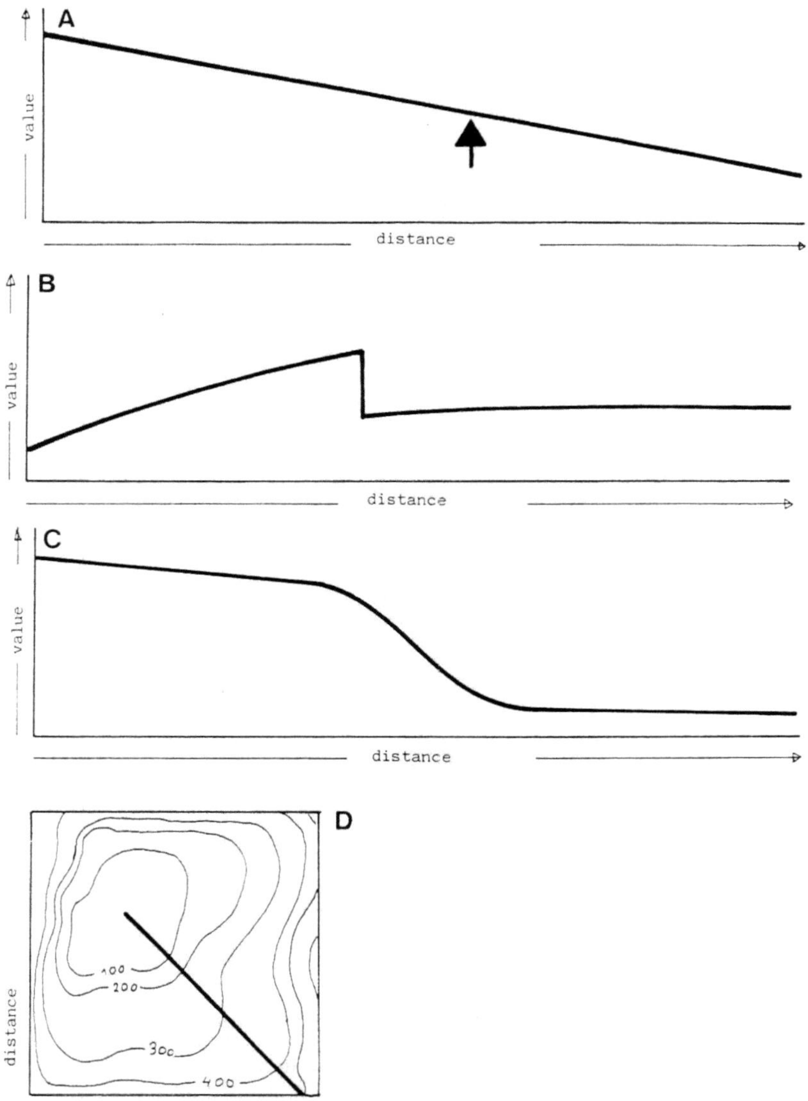

Fig. 5.2. Examples of spatial gradients and limits. **A** Linear gradient (e.g. mean annual temperature from equator to pole) and limit (e.g. critical temperature for palms). **B** Abrupt change in gradient value as a symptom of some structural limit. **C** Slower change in gradient value as a symptom of a limiting or boundary zone. **D** Gradient as a cross-section of isocurves (e.g. temperature, light intensity etc.), represented as a *thick line*

Figure 5.2C concerns a much more gradual change in the value of a factor over distance. One might think of a species becoming slowly rarer, a transition between desert and forest climates, or at another scale the intensity of sunlight reaching the forest floor over the distance of transition between open field and forest, e.g. in what the Germans call a "Mantelgesellschaft", a community cloaking a forest (cf. Westhoff in Sommer and Tuexen 1974). The same kind of modulated gradients can be found vertically in eco-units (cf. Oldeman 1974a,b, 1979).

Figure 5.2D indicates how such gradients look when viewed on a map. They then cut through lines, indicating a value of the factor considered. For temperature, such lines are called isotherms, for precipitation they are termed isohyetes, for atmospheric pressure isobars, and so on. Isoluxes, for light intensities, play an important part in defining eco-units (e.g. Florence 1981). It is useful to remember that lines linking points with the same value for some factor are always indicated with terms beginning with "iso-", derived from the ancient Greek word for "identical".

The shapes of iso-curves and gradients most often change with time, even when linked to structures, because their causes, such as flows of matter and energy and their directions, are not constant with time. If the temperature gradient in our first example (see Fig. 5.2A) seems to be constant in time, then this is a statistical artefact due to the use of averages. In reality, temperatures change with the seasons (winter!). In fact, this time dimension is the major difficulty when using palms or other groups of organisms to define the tropics.

Gradients and iso-curves can be measured, but this is a time-consuming and expensive activity. It should be done only if and when absolutely indispensable to check a model or for monitoring purposes, such as in weather forecasting. The diagnosis of the gradients separating forest eco-units one from the other therefore rests upon the indirect method of analysing the features of forest architecture that are known to cause fluctuations in gradients.

The whole of Fig. 5.2 shows that limits on a lower hierarchical level can be visualised as modulations upon gradients established for a higher level. This should be borne in mind when examining regression curves. These curves represent averages that characterise a higher level, whereas the scattering of the measured values, i.e. the statistical points, indicates both the influence of a lower level of organisation and the errors made by the observer. One should never try to understand regression curves without trying to understand the meaning of the scatter of the points that support them. The measuring method here of course is an important element in evaluating the role of errors.

5.1.3 Temporal Limits and the Zero-Event

In the preceding pages, the spatial limits of eco-units were reconnoitered. The original definition of an eco-unit (Oldeman 1983b) also includes the statement that, on the surface which it covers, its development starts at one moment in time. Many factors can induce the events that create such surfaces, on which new forest development may start. Examples are tree falls, fire, storms, inundations, biotically induced tree mortality, heavy rains, wood harvests, bulldozers or arboricides (see Sect. 5.2.5.3).

The effects of each of these events are quite different, and they cause quite different forest developments. They not only affect different surfaces, between an are (10 x 10 m) and a square kilometer, for instance, but they also determine the "heritage" of that surface from its previous state. The seeds stocked in the soil may be lost or not (fire!) and the same is true for tree stumps or root systems that may or may not be capable of reiteration (see Sect. 3.2.4) after the event. Animal habitat is changed to a greater or lesser degree. The general influence of forest fires on insects has been considered, for instance, by Leffef (1984). Organic matter may either accumulate, e.g. after storms, or disappear, e.g. after flooding or fire. If the zero-event is cessation of agricultural land use, the consequences can be extremely radical and permanently prevent the return to the former forested state of the land. Some quite other forest, if any, will prevail after such an event.

Traditional forest dwellers have always realised the importance of these events. According to Cañadas (pers. commun. 1975) and Sarukhán (pers. commun. 1975) tropical American Indian languages have whole vocabularies with terms indicating such happenings, going from a tree fall to more complicated events and to forms of clearing for shifting cultivation. Traces of such vocabularies in the European languages are rare. In French, the terms "chablis" and "volis" have survived. There also is a Finnish word for "chablis", i.e. "rytä" (Leikola pers. commun. 1982). In Dutch, the only words still existing, "traa" and "loo" (pronounce "low"), indicate human-made forest clearings. One of the reasons for the existence of such vocabularies certainly is that useful herbs and lianes are to be found in small-sized young forest patches shortly after the zero-event. In France, one of these may have been the variety of *Vitis vinifera* that yields the special wine called "Chablis" up to this day. Most of the useful plants (medicinal, paint, fibres, for instance) that the present author has found with Amerindian or Creole guides in tropical America came from such young forest patches.

There is a deep difference in meaning between these terms and the currently employed words in present-day scientific publications, such as "gap-phase", "disturbance" or "catastrophe". The traditional words indicate the whole event, with its immediate consequences and causes, for instance:

- *chablis:* the uprooting of a tree, the uprooted tree, the inaccessible heap of broken or surviving vegetation and branches, the opening (gap) in the forest canopy;
- *volis:* the breaking of a tree-trunk (most often by storm), the broken and fallen upper part of the tree, the mass of vegetation and branches and the gap.

The newer terms, on the contrary, show a result ("gap") or some disagreeable thing ("disturbance", "catastrophe"). Shugart (1984, p.167) analyses "the severity and frequency of disturbances" in terms of likelihood and magnitude, and states that magnitude is difficult to define. One cause of this difficulty certainly is that the whole event has been lost from sight: the very act of abstraction that allows the building of the valuable mathematical models discussed by Shugart has obscured from view several aspects of forest development that cannot be missed both for further silvological insight and for silvicultural design (also see Fig. 5.16).

Therefore, the kind of eco-unit that is meant should be indicated by the event that caused the start of its development and an indication of its size: a one-tree chablis eco-unit, a large-sized fire eco-unit, a medium-sized top-borer eco-unit, a medium-sized clearcut eco-unit, and so on. Several cases will be shown, in which different authors have used different abstractions to analyse different kinds of eco-units under different names. The present approach has been chosen also in order to avoid the creation of more neologisms than is strictly necessary.

5.1.4 Eco-Units Are Not Necessarily Even-Aged

In forestry, the development of stands is divided in even-aged and uneven-aged. These terms indicate that all trees in a stand either have the same age or do not. No other forest components than trees are considered in this distinction. The definition of an eco-unit as a surface on which forest development starts at the same moment seems to have caused some confusion with even-aged development. Therefore it is stated explicitly here that the two notions do not concern the same organisation level.

Even-aged versus uneven-aged development concerns selected forest components, the trees. Eco-unit development concerns the eco-unit, which is situated one hierarchical level higher (Table 2.1). The development of an eco-unit rests on the interactions between all of its components, not only the trees. For the general eco-unit definition, the distinction between even-aged and uneven-aged forest is thus scarcely relevant. A large-sized, clearcut, planted eco-unit will start its development with an even-aged tree compartment, like pine plantations the world over. A large-sized, clearcut, non-planted eco-unit, in which young trees have been spared for the future, and where seedlings and reiterated stump-shoots also play a role, will start its development with an uneven- aged tree compartment, like hardwood clearcuts in North Carolina (cf. McGee and Hooper 1970). Of course, the two developments are different, but this concerns the kind of eco-unit and not the eco-unit concept in itself. As a matter of fact, the comparison of many different forests suggests that there is a continuum between even-aged and uneven-aged development, the intermediates containing larger or smaller even-aged tree groups ("cohorts", cf. Sect. 5.5.9), that may or may not coincide with eco-units.

The term stand is here used in its original continental European meaning of a management unit often delimited by roads. A stand therefore may or may not coincide with an eco-unit. In the case of clearcutting it does, but in the case of selective or group cutting it may contain many or a few eco-units. One may say that a stand always has man-made limits, whereas an eco-unit may have either natural or man-made borders.

5.1.5 Eco-Units Are Living Systems but Not Organisms

In the past, there have been several attempts to explain ecosystems as a kind of super-organisms, so much so that phytosociologists (e.g. Westhoff in Sommer and Tuexen 1974) do not even want to refer to "individual" ecosystems, because an individual might be confused with an organism.

It has to be explicitly stated here, in order to prevent confusion, that eco-units are not to be conceived as organisms or super-organisms. If Oldeman (1986) may seem to suggest so, this is only in order to convey the idea that both organisms and forest ecosystems (eco-units) are living systems, just like all units mentioned in Table 2.1., i.e. cells, tissues, organs, etc. All living systems have some features in common, but in a very abstract sense.

a) All living systems are delimited by a filtering layer, which defines what is inside and outside. In cells this is the cell wall, in organisms it is a skin, epidermis or bark, and in forest eco-units it is a layer of plant biomass, like leaves aboveground and roots belowground.
b) All living systems show an internal communication system. In cells this is mainly electro-chemical. In plant organisms it is partly physical (sapstream), partly chemical (phytohormones), whereas in higher animals there is an electric system added (nerves). In forest eco-units, signals pass through chains of organisms as feedback-responses.
c) All living systems have mass translocation systems. In cells, molecules migrate in an ordered way. Sapstream and bloodstream are mechanisms ensuring translocation in organisms. In forest ecosystems, mass translocation takes place, for instance, by litter-deplacement, through the non-rooted components, mainly animals, and in symbiotic processes.
d) All living systems survive by adjusting their behaviour to change in exterior circumstances. Cells do so in response to signals transmitted (mainly) by neighbouring cells. Organisms do so by adjusting their growth pattern (plants), movement pattern (animals), or both. Ecosystems do so by adjusting the allocation of resources to their components, which may lead to the local extinction or proliferation of a component, but ensures survival of the eco-unit as a whole.
e) All living systems reproduce themselves. From cell to organism, and even on lower levels (e.g. viruses) this process has been intensively studied, to say the least. It is astonishing, therefore, that so few authors are willing to consider that ecosystems, which are organised with the same degree of complexity, do reproduce themselves. Empirically, the fact cannot be denied that the same kind of forest occurs again and again, from young to old, on a particular site. But instead of finding out how the information needed to build such a forest is transmitted and implemented by growth, many models of forest regeneration have been made that try to explain the recurrent patterns by random processes or component mortality, or both. Maybe this is linked to the fact, remarked upon by Van Miegroet in a public discussion in Wageningen in 1979, that the (European) forests had no apparent trouble regenerating during two hundred thou-

sand years, only during the last two centuries, when foresters have been regenerating them.

In organisms, reproduction shows some few important features. First, the information needed to build the organism has to be present in one circumscribed place and at the right time: a zygote, or the nucleus of a cell capable of vegetative multiplication, or a macromolecule. Second, in the process of uniting this information the vast majority of the information-bearers dies: gametes, most cells of an organism (leaving only a few cells in vitro) or molecules. Third, there are always "key-and-lock" mechanisms involved, from the meeting of molecular structures that fit one into the other, e.g. fertilisation, to the meeting between biological structure and environmental structure, e.g. the egg and the uterus or a megaspore and its structured environment of germination, or a meristem and its reiteration environment in vegetative multiplication.

For ecosystems in general, all three conditions are met, but the information-bearers and key-and-lock mechanisms are completely different from those in organisms. First, the information needed to build an eco-unit is present in the form of a propagule bank, containing for instance spores, seeds, eggs and vegetative propagation material. Out of this propagule-bank, the seed bank has been studied most intensively (e.g. Vazquez-Yanes in Gomez-Pompa and Vazquez-Yanes 1976; see Fig. 5.3; Bormann and Likens 1979; literature given by Shugart 1984). The set of available meristems and seeds together has been termed "infrastructural complement" by Oldeman (1974a). In terms of floristic composition, this information pattern is more or less akin to the "potential natural vegetation", a notion used by phytosociologists (e.g. Westhoff and Den Held 1969; Werger and Westhoff 1985). Second, in the process of uniting this information the vast majority of plant and animal propagules dies (e.g. Wheelwright and Orians 1982). Third, the key-event in which circumstances and information meet, so that eco-unit development or ecosystem development in general may start is the "zero-event" mentioned in Section 5.1.3.

For the eco-unit, both the quality of the information present, i.e. the composition and spatial distribution of the propagule-bank, and the character of the key-event are regulated to a certain degree by the surrounding silvatic mosaic and for another part by abiotic influences originating on still higher levels of the hierarchy of systems (see Sect. 6.2.2).

5.1.6 Eco-Units and Silvicultural Design

The former pages may seem purely theoretical. Therefore, the fields of application that may be served by working with these general concepts are outlined here

- *Forest regeneration* completely rests on the use and modification of the propagule bank. All regeneration problems are caused by the neglect of certain components, such as potential diseases and parasites, or by the choice of incompatible additions, such as the wrong exotic tree species that are eliminated rapidly or may proliferate too strongly.

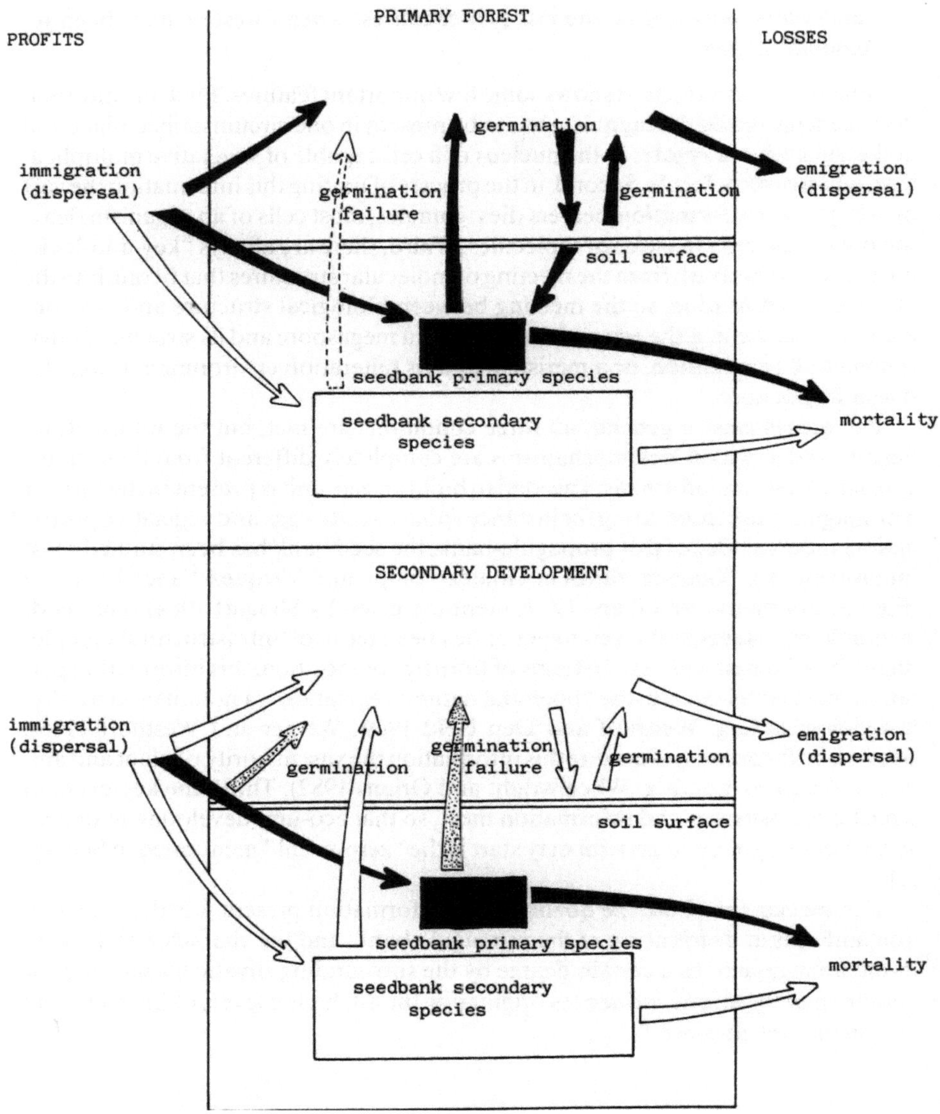

Fig. 5.3. The seed-bank in a Mexican forest, and its functioning. Note that the "zero-event" is small-sized in "primary forest", and large-sized in "secondary development". *Dark arrows* seeds "primary forest species"; *open arrows* seeds "secondary forest species". (After Vasquez-Yanes in Gómez-Pompa and Vazquez-Yanes 1976)

- *Wood harvesting methods* are too rarely designed to function in harmony with the commonest zero-events in the forest region concerned, which often leads to unexpected reactions from the propagule-bank, such as weed problems or secondarisation by soil compaction.
- *Thinning operations* influence the border layers (canopy, rhizosphere) of eco-units and so cause translocations of biomass, unpredicted functional signals which mobilise certain organisms and allocate resources to their development, which often leads to unexpected wood production losses and increased production of unwanted organisms, such as fungi, weeds or insects (also see Sect. 5.6.8).
- *Pollution*, too dense recreation or the modification of hydrological regimes (groundwater pumping, dams etc.) cause eco-units to adjust their behaviour much in the same way as thinnings or harvests, by changing the systems of signals and resource allocations and not necessarily favouring those forest components that the manager wants, like trees or wildlife.
- *Afforestation or reafforestation* design generally operates with a species choice that is only a tree species choice, and its rate of success would be astonishing in view of the neglect of the total propagule bank, if it were not for the fact that reforested terrains nearly exclusively are so denuded and destroyed that only very poor propagule banks are left, if any.

These examples are given in order to show that manageable eco-unit models can bring a synoptic view that indeed often rearranges existing knowledge, but does so in a way that allows it to be used in a large array of situations.

5.1.7 The Behaviour of Forest Eco-Units

There is trouble defining a forest (cf. Groenendijk et al. 1985). For instance, the Dutch Forest Law presents a formula that seems to be understood well enough by all interested people: "A forest is a community of plants and animals, the aspect of which is determined by tree-forming species". As has been mentioned in Chapter 3, no definition of forest can be conceived without mentioning the trees. An exception may be a definition that is extended to include all "forest lands". Definitions seem to be made not only to define the object, but also to please its users.

For silvicultural design, it would be useful to define or at least abstractly describe the forest unit that is most often manipulated, and because the start of every silvicultural system is forest regeneration or forest establishment, we will use the following circumscription of a forest eco-unit.

We consider to be a forest eco-unit (regeneration unit): every surface on which at one moment in time a vegetation development has begun, of which the architecture, ecophysiological functioning and species composition are ordained by one set of trees until the end.

This circumscription implies that the replacement of the trees automatically entails the death of the forest eco-unit in question, and its replacement by one or more other eco-units.

In this way, eco-unit development is treated as a subprocess in "succession". This distinction is not only practical for silviculture. It is also meant to express clearly that in a "forest patch" or a "stand", i.e. a forest eco-unit, there is considered to be no succession as long as it does not die. This avoids many problems that may be found in contemporary literature, where a sequence of fungi or insect species over some years and some square meters usually is described as a "succession", just as the replacement of trees over hundreds of years and many hectares. Succession is a concept that will be more explicitly treated in Chapter 6, and on which there are enough publications for several years' reading (e.g. bibliographies in Kessell 1979; Bormann and Likens 1979; Fanta 1982; Shugart 1984; Werger and Westhoff 1985). Giving eco-unit development its own place and its own hierarchical level of course does not replace the existing models and ideas, but it is a complement that can be used in their further development.

5.2 Development of Forest Eco-Units: Architecture

5.2.1 General Remarks

A simplified and general image of the development of one eco-unit is given in Fig. 5.4. Simplification concerns the limits, which have been assumed to be linear and to be surrounded by an unchanging forest. Neither of these assumptions corresponds to reality, and on this point the image will be corrected later (Sect. 5.4). Moreover, the number and diversity of forest components have been reduced in order to yield a clear image of the architectural phases in forest eco-unit development. This schematic way of introducing a subject has many scientific precedents, e.g. the "ideal gas" that was at the base of 19th century physical gas analysis as a first outline of the situation.

The development of an eco-unit may be divided in four phases. These will be compared in Table 5.1 with development phases distinguished by other authors (e.g. Mayer et al. 1979; Bormann and Likens 1979) and with the recently published antique Mayan concept of forest development (Chen 1987).

5.2.2 The Innovation Phase

All eco-unit development starts with an innovation phase (Fig. 5.4A). The term is derived from the Italian word to indicate regeneration, rinnovazione (cf. Amorini and Gambi 1976). Just after the zero-event, which has conditioned both the surface of the young eco-unit and its propagule bank, the information of the propagule bank is mobilised. At the same time, it may be enriched by propagules coming in from outside: seeds, spores and animals starting to reproduce in the new eco-unit. The name "seedling phase", sometimes used for the innovation phase, is limited to the tree compartment and in its strict sense covers only such cases as do not contain remnant trees of the former eco-unit (Fig. 5.4A, centre) or reiterating tree stumps. In

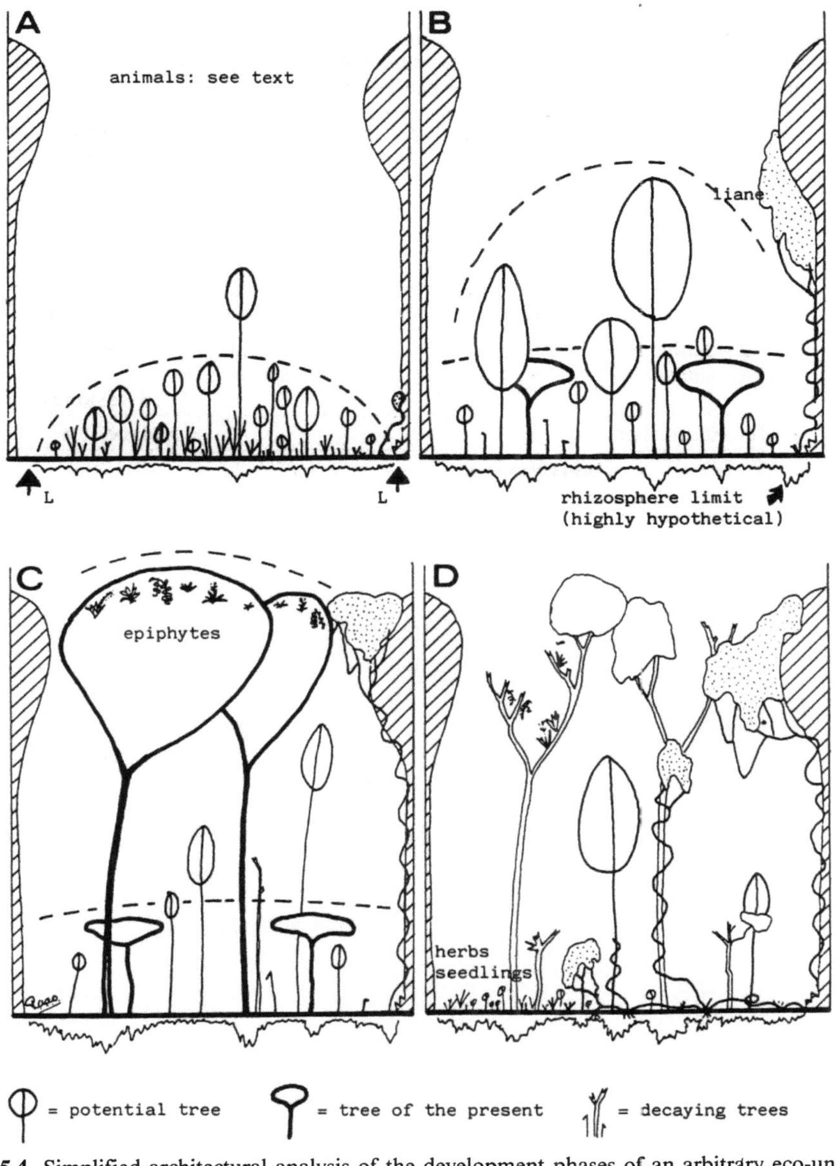

Fig. 5.4. Simplified architectural analysis of the development phases of an arbitrary eco-unit. Simplifications are linear limits (*L*), an unchanging surrounding forest and an irrealistically low number of components. Rhizosphere limits have never been studied as ecosystem limits and hence remain hypothetical. **A** Innovation phase, lens-shaped, in which the information of the propagule bank is made operational. **B** Aggradation phase, some years after formation of a lower structural ensemble (*interrupted line*) that ensured canopy closure; low trees of the present survive. **C** Biostatic phase, in which trees of the present have built structural ensembles (*interrupted lines*) that organise forest architecture and all other component behaviour. **D** Degradation phase, during which (in this case slowly) trees of the present decay, become trees of the past, and forest architecture breaks down to liberate many other forest components. Further explanation: see text

Table 5.1. Comparison of development phases according to some authors. No time scale implied: phases may stretch or shrink in time according to the forest. Urwald = virgin forest; Bestand = stand; Zerfall = decay; Verjüngung = rejuvenation (regeneration); Plenter (wald) = tree-wise mixed uneven-aged forest; ╪ = transition between phases

	This book	Bormann & Likens (1979)	Chen (1987)	Central European (Leibundgut 1978; Mayer 1980)	Traditional European (Leibundgut 1966; Cochet 1977; Mayer 1980)	
		After clearcut	Mayan concepts	(Virgin forest) Urwald	Stand management	
	Degradation phase		-	Zerfalls-phase	eliminated from silviculture	
	Zero-event				Harvest	
	innovation phase	reorganisation phase	sak'aab	Verjün-gungsphase	seedling phase (semis) (Jungwuchs)	
		(biomass devt.)				
	Canopy closure		sak'aab-kool		Canopy closure	
	(early) aggradation phase (late)		kambal-hubche'	Initialphase	dense phase (fourré) (Dickung)	
			kanal-hubche'		(gaulis)	
			kelenche'		pole phase . . . (Stangenort) (perchis)	
self-regeneration	(early) biostatic phase (late)	aggradation phase		Optimalphase	(Baumbestand, Baumphase)	stand regeneration
			ka'anal'k'aax	Terminal-phase	mature phase (futaie) . . . (Altbestand)	
	Shifting events (Chap. 6)		leading to fragmentation of eco-unit into mosaic			
	(eco-unit mosaic; see Sect. 6.1)	(steady state; see Sect. 5.6.6)		(Plenter-phase; see Sect. 6.3.3)	(other silvicultured systems, e.g. selection systems)	

reality, many forest components are mobilised: the vegetation is soon dominated by herbs and weedy climbers surrounding the seedlings of trees, by root and stump suckers and potential trees that survived the zero-event.

This vegetation attracts many and diverse animals: rodents, other herbivores and their predators as well as many insects that live on vegetative organs (leaves, stems, meristems) and which are predated in their turn by insectivores such as other insects, hedge-hogs or moles. For European forest mammals, this has been analysed by Van Vuure (1985). In the tropics, innovation phases attract snakes, which find both heat from the sun and many prey animals to feed on. Only a reduced number of fungi can develop in the beginning, because for most species the soil temperature and humidity are not optimal as yet. The development of these conditions depends to a high degree on the nature of the zero-event and the composition of the propagule bank. Smits (1982) describes the state of the propagule-bank in a dipterocarp forest in Kalimantan (Indonesia), after mechanical exploitation: most fungal species are eliminated by soil heating, so that no dipterocarp forest can immediately regenerate there, because these trees depend on symbiotic root fungi (mycorrhizae).

The eco-unit in this phase has no durable architecture. Its border layer has not yet been formed. Above the ground, it consists of a shifting mass of mainly herbaceous plants, the only permanent elements of which are the tree seedlings and other tree components. However, the whole eco-unit has a "lens-shaped" physiognomy (Fig. 5.4A) because whatever the surrounding forest may be like, it always casts shadow and removes water and nutrients from the root compartment. The centre of the innovation phase of the eco-unit is most exposed to the light and least to root competition. Therefore, the centre shows the highest biomass and height increment. However, this rule only applies up to a certain size limit: on very large eco-units the zone of maximum bioproduction reaches neither the centre, nor the borders (cf. Fig. 5.16).

The innovation phase is barely structured and very dynamic. The dynamics have most often been ascribed to forms of interaction between organisms termed "competition", "parasitism", "diseases", or "predation". On the organisation level of the eco-unit, these processes may be considered as the balancing out of the interactions that are going to determine the further development. There is quite a lot of "overshooting" in the cybernetical sense. Populations become too dense and fall back later on; too much biomass is produced in some compartments, e.g. larvae that feed on roots (cf. Brauns 1976, his Fig. 886), and this biomass decreases dramatically later; or the architecture shows precocious layers that collapse shortly after.

Part of the process of balancing interactions is not programmed. Much of its unpredictable dynamics are linked to the direct influence of climatic factors, e.g. drought or frost, that are occurring at random. Innovation phases contain quite some hardy, short-lived species of plants and animals that can resist such outside influences. These we call innovation species. The term "pioneer species" will be reserved for a successional stage, which has its own innovation phase in eco-unit development. Innovation species occur in each innovation phase of all types of eco-units: plants or animals that may be visible to the naked eye, such as innovation trees (cf. Fig. 3.1) or shrubs, or may not be so easy to perceive, e.g. innovation fungi or insects. Many of these innovation species are eliminated during the following developmental phases.

On the other hand there are predictable factors involved. The nature of the lateral border zones, being the limits between the eco-unit and the surrounding forest, has a regulating microclimatic influence because of their intercepting architecture. The smaller the eco-unit, the stronger is this regulation. As a source of mobile species, seeds and spores, the surrounding forest mosaic also regulates the processes in the young eco- unit. And finally, the genotypes of the species present in the propagule bank have evolved under selection pressure for mutual interactions. The propagule bank not only contains the sum of these genotypes, but also the information that allows a system of interactions to become operational.

The architecture of an eco-unit is represented as a transect in Fig. 5.4. If a map were added, this would show asymmetries in the disposition of the different zones near the borders or in the centre, because climatic influences are oriented. The sun goes from East through South to West, so that the shadow patterns of the surrounding forest are not equal. In most regions there is a dominant direction of the wind that causes an asymmetrical precipitation pattern. And if the terrain shows a slope, these and other factors also are oriented in a predictable way. With regard to the climatic influences, forest borders therefore are generally distinguished as to their exposition, which favours some components and handicaps others. Considering tree components, the importance of exposition for silviculture is clear. Given the exposition, one can or cannot plant or sow the tree species that is desired.

The fluidity of patterns and processes in the innovation phase of an eco- unit makes this development phase easily influenced. There is a rapid sequence of plant and animal species and interactions between them, before this sequence slows down towards more leisurely dynamics and a more permanent architecture. But during the few years of the innovation phase, these durable features are prepared. The sequence of niche-formation for ephemeral species prepares the way for more permanent niches. Given the influence of unpredictable factors, such as the climate, there remains an uncertainty concerning the final character of the eco-unit. In silviculture, this phase is both the most promising one for light steering interventions, and the most risky one in case of error.

5.2.3 The Aggradation Phase

After the innovation phase we distinguish an aggradation phase. Its beginning is well delimited in time, because it starts as soon as the canopy is closed (Fig. 5.4B). The term "aggradation phase", as opposed to degradation, is borrowed from Bormann and Likens (1979), whereas the time-limit criterion used by the present author replaces the biomass criteria used by Bormann and Likens. Architecturally, aggradation starts with the first step in structuring the eco-unit, which henceforth shows the first of its more persistent architectural characteristics.

In the development sequence distinguished by the Mayans (Chen 1987), there are clear terms for the innovation phase (sak'aab) and the early aggradation phase (kambal-hubche'). Canopy closure seems to have been considered so important by them that it was also distinguished as a phase (sak'aab-kool). This may be compared to the concept of reorganisation (Sect. 5.2.6) but it occurs at a different moment.

Canopy closure means that a closed upper filtering layer now covers the eco-unit. It blocks part of the outside influences (light, precipitation, wind) and selects what passes. The components that ensure this eco-unit function are the woody plants. In Fig. 5.4B, canopy closure is implemented by the crowns of both small trees or shrubs reaching their maximal expansion, and potential trees of larger species that fill up the spaces between them.

This image is not the classical one, which shows a layer of crowns of young individuals (potential trees, Fig. 5.8) of larger species at the moment of canopy closure. This classical image is derived from forest plantations, where most often it is correct. But in most natural forests inspected by the present author, canopy closure involves small woody species. In tropical rainforests these belong to families such as Euphorbiaceae, Moraceae, Burseraceae, Melastomaceae or Rubiaceae. In Europe, species like elderberry (*Sambucus nigra*), cherries (*Prunus* sp. pl.), rowan trees (*Sorbus aucuparia*), hazel (*Corylus avellana*) or common yew (*Taxus baccata*) may play a role in canopy closure in different eco-unit types. Figure 3.1 (D,E) shows the architecture during the life cycle of such small trees or shrubs. In man-made forests of the type "coppice with standards", coppiced trees (Fig. 3.21) play an analogous part, and the same occurs after clear-cutting in the Appalachian region (USA), when stump sprouts of red maple (*Acer rubrum*) and other species spontaneously cause a large proportion of the canopy to be closed. Several techniques are being tried out at present to prevent this process, that is less desirable in silviculture (Loftis 1985).

The role of lianes is indicated on Fig. 5.4B. It may be more important than the figure suggests. If liane crown development is strong, a liane blanket may take the lion's part in canopy closure (cf. Kahn 1982, for Côte d'Ivoire, Fig. 5.41) and retard the further development by years, because it leaves no space for tree crowns to develop further. *Rubus* species and some roses form such blankets in the temperate regions. Climbing bamboos, e.g. *Guadua* sp.pl. in the Andean countries, both form blankets and render the soil unfit for tree seed germination by their leaf litter, so that the aggrading phase is brought into stasis as long as the bamboos remain alive. Exotic liane species may be able to strongly increase the proportion of retarded aggradation phases, e.g. the kudzu (*Pueraria lobata*, Papilionaceae) from Japan, that has become a major weed in the southeastern United States (National Academy of Sciences 1979).

Shrubs or shrubby trees may play the same part when they proliferate. They block the development of higher trees, if they are so numerous as to occupy the whole closing canopy. This happened in Europe after introduction of a shrubby form of *Prunus serotina* (American black cherry) from the United States. The species had been introduced in order to improve litter quality in forest plantations, to create an understorey with more niches for wildlife and, hopefully, to survive as isolated trees with highly appreciated wood. The genotype that was introduced, however, had a shrubby habit with a high tendency for basal reiteration. After harvest of the old forest, the eco-units soon became low, closed vegetations of *P. serotina*, under the heavy shadow of which nothing else developed well. This occurred in several forest districts in Germany and the Netherlands, and earned the species the nickname "forest pest".

For all practical planning purposes, aggrading phases with a retarded or blocked development are a risk. This should be borne in mind in the design of new types of village forests or firewood plantations in the tropics. In its volumes on firewood species, the National Academy of Sciences (1980, 1983) refers to this problem when warning that fast-growing firewood trees are often very aggressive and have tendencies to behave like weeds. We add here that many of the "miracle species" for wood production, advocated in forestry, have such properties.

A certain proportion of eco-units may become blocked in the aggradation phase. But this is by no means the general case. Most often, enough larger tree species are involved in canopy closure to ensure that a higher forest eco-unit is built (cf. Fig. 5.4C). An example of a usual eco-unit in its aggrading phase is given in Fig. 5.5. It represents a spontaneous maple eco-unit (*Acer pseudoplatanus*) amidst a *Pinus nigra* plantation on very poor dune sands near the sea coast in the province of North Holland. The lens-shaped form of the eco-unit is still evident. So is the high number of potential trees that form the new canopy. Indeed, their stem number usually is much higher than indicated in the schematic Fig. 5.4B, a simplification that was mentioned already.

The aggradation phase persists much longer than the innovation phase, most often 10 years or more. Once closed, the canopy is very dense and the tree number is high. During the aggradation phase, many trees either die or remain behind in de-

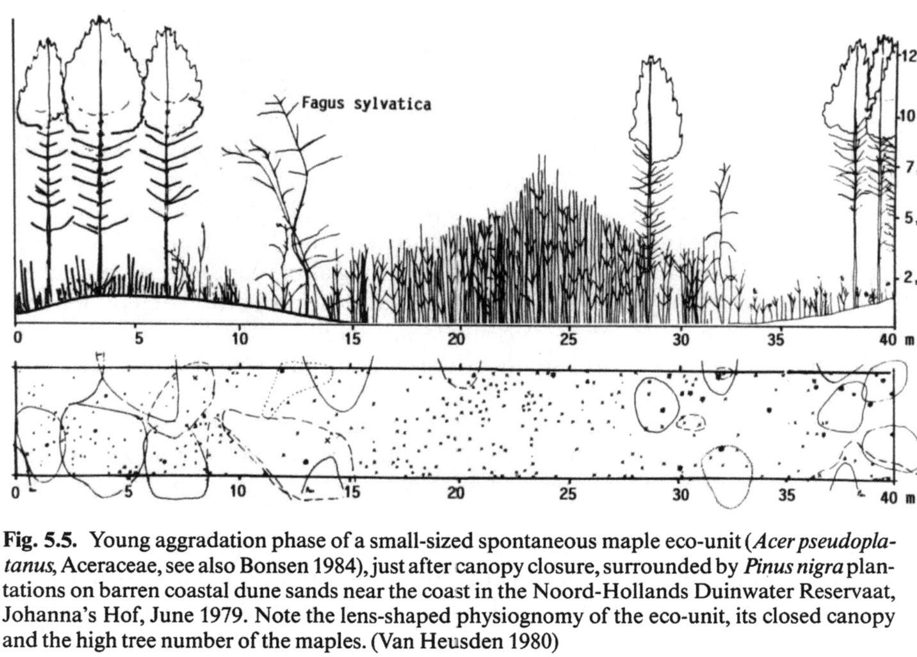

Fig. 5.5. Young aggradation phase of a small-sized spontaneous maple eco-unit (*Acer pseudoplatanus*, Aceraceae, see also Bonsen 1984), just after canopy closure, surrounded by *Pinus nigra* plantations on barren coastal dune sands near the coast in the Noord-Hollands Duinwater Reservaat, Johanna's Hof, June 1979. Note the lens-shaped physiognomy of the eco-unit, its closed canopy and the high tree number of the maples. (Van Heusden 1980)

velopment. At its beginning, the aggradation phase therefore has quite another aspect than it has later on. In literature (Table 5.1) it is often divided again, at least in two parts: a "dense phase" and a "pole phase". Figures 5.5 and 5.6A clearly show the dense phase. The term "pole phase" is an example of mixed ecological and economical concepts in forestry terminology: it indicates both an ecological state of a stand and the marketable product (poles). The poles are taken out of such stands by thinning, which is the man-made imitation of tree mortality during the aggradation phase.

In the Mayan terminology the aggradation phase contained three phases (Chen 1987). It was perceived as a sequence of a short, ca. 5-year, early phase (kambal-hubche'), a longer middle phase of about 20 years (kanal-hubche') and a long late phase of transition towards maturity of some 70 years (kelenche').

Many forest components of the innovation phase are eliminated from the aggrading phase, because their niches disappear once the eco-unit becomes self-regulating by virtue of its compartment of woody plants. At ground level, herbs disappear, to be slowly replaced by mosses in some forests. The mammals are also expulsed. Van Vuure (1985) indicates the disappearance of most small rodents from the dense phase, and the absence of mammals in the pole phase in European

Fig. 5.6A
(Fig. 5.6B see page 174)

Fig. 5.6. Aggradation phase and degradation phase. **A** Same maple eco-unit as Fig. 5.5. **B** Twelve hundred-year-old oak, senescent in 1978, that collapsed in 1981 (*Quercus robur*). It illustrates the decay of former trees of the present into trees of the past. Note hollow trunk and reiterated trunks, small reiterated crownlets on top. Semi-natural forest after a century of non-management, Urwald Hasbruch (North-West Germany), "Amaliëneiche", Spring 1978. See also Fig. 5.1. (Photographs by R. A. A. Oldeman)

forests. According to Brauns (1976), other insect species gradually take over the eco-unit, with a minimum in species number in the dense phase, although he takes together the pole phase and later development phases ("Stangen- und Altholz", i.e. poles and old wood). Szujecki (1983) analysed the soil insect fauna over a century following reafforestation of Polish farmland with pines, and Stocki (1981) mapped the June/July part of the spatial distribution pattern of insects in the dense phase of a Dutch man-made Douglas fir forest (*Pseudotsuga menziesii*), both authors contributing to the all-too-scarce knowledge of insect distribution shifts during forest eco-unit development. The aggradation phases might be less attractive to many bird species, according to the matrix of stand properties versus bird diversity given by Komdeur and Vestjens (1982), for N.W. European forests and avifauna.

Summarizing, the aggradation phase is the first to show an architectural pattern. The first closed canopy may persist as a low tree layer, or a shrub layer, if such species were involved in canopy closure and if they can survive overtopping and shadow, which many can. A persisting layer of plants having reached their maximum expansion (trees of the present Oldeman 1974a; Fig. 5.4) is called a structural ensemble, after Hallé et al. (1978). Not all eco-units build a low structural ensemble, as in Fig. 5.4B, but if they do so it happens before the formation of higher structural ensembles. This process of layering from below has been described by Kahn (1982) for the Ivory Coast, and by Van de Winckel (1980) for hardwood forests bordering the Rhine in France (also see Walter 1974). If no lower structural ensemble is built, the canopy of the aggrading phase is unstable and remains quite dynamic until a higher structural ensemble is established by the surviving higher trees. The layering pattern, as defined by trees of the present, will be discussed in Sections 5.2.4. and 6.2.3.1.

For eco-unit development, the important aspect of the formation of a canopy is that it acts as a filtering layer and, if persisting as a lower structural ensemble, that it will keep this function to a certain extent. Within the forest eco-unit in a later phase, the lower structural ensemble then spatially delimits a well-defined compartment. Most often, however, lower structural ensembles are not very dense and their filtering capacities are rather slight: their role should be assessed more precisely but seems to be subordinate.

5.2.4 The Biostatic Phase

5.2.4.1 Overview

The result of the aggradation phase finally is a biostatic phase (Fig. 5.4C). The term "biostasis" has been chosen because homeostasis concerns physical, closed systems (also see Sect. 3.5.1), whereas the term "homeorrhesis", coined by Waddington and meaning a "state of equal flows" does not indicate that it concerns living systems. This development phase is often referred to as the "mature phase" or "tree phase" (Baumphase). This last term again mixes ecology and economy: it is the mature forest where one can harvest "Baumholz" (sawnwood).

The concept of the "steady state phase" as used by Bormann and Likens (1979) has not been taken over here as such (Fig. 5.84). The reasons not to adopt this concept in the sense given earlier for biostasis (cf. Sect. 3.5.1) are related to the high probability that the forest ecosystem studied by these authors started its development as one huge eco-unit on a clearcut surface, but fragmented into smaller eco-units when biomass accumulation leveled off after aggradation. This is more thoroughly discussed in Section 5.6.6.

In the Mayan vocabulary, the "mature forest" is indicated by the term "ka'anal'kaax" (Chen 1987). This seems to be the end phase in the cycle of forest development as perceived by the Mayans, to be followed again by "sak'aab", the innovation phase. Reorganisation of the forest, from old to young (also see Sect.

5.2.5.3.) in the Mayan model is abrupt, pointing to a probable silvicultural origin of this model (cf. Gomez-Pompa 1985b). On the contrary, the South American Indian terms for zero-events (see Sect. 5.1.3) are derived from the natural forest as an integrated life resource (cf. Grenand 1980).

5.2.4.2 The Trees

In the biostatic phase, eco-units are most clearly organised. Their architecture is persistent, and barring accidents this phase may be very long-lived. The trees that determine this architecture are the trees of the present, because they have no potential left for crown expansion. These trees can be considered as a set of the present, and this set is subdivided in structural ensembles at different heights. In Fig. 5.4C, two structural ensembles are indicated. This vertical pattern ordains the life of all other forest components in this phase. The surviving potential trees, which do have a potential for future expansion, subsist as "suppressed" trees, which may not survive but may die before their potential can be expressed.

The pattern, established by considering the trees of the present as the only components relevant for the building and maintenance of eco-unit architecture, differs considerably from other approaches, in which layering is always defined for all plants, over surfaces that may or may not correspond to an eco-unit. This is the case for the A,B and C layers distinguished in the tropical rainforest in the classical work by Richards (1952). Godron (1971) distinguishes his vegetation layers according to normalised height limits and warns that these should not be considered as a straightjacket but should only be applied if there are indeed layers between them. He applies this concept inside "homogeneous" units of vegetation, homogeneity being defined by species composition analysed according to statistical methods. His limits do not coincide with eco-unit limits: phytosociological limits indeed are of another kind, established according to other methods (cf. Figs. 5.1 and 5.7). The current distinction of "tree layer", "shrub layer", "herb layer", as applied without taking eco-units into account, may lead to statements such as "there are mahoganies in the shrub layer". Now, mahoganies (the South American genus *Swietenia*, Meliaceae) are big forest trees and not shrubs. The statement from a forest development point of view would be: "there are potential mahogany trees in the spaces left open by the lower structural ensemble", or simpler:" the potential set contains mahogany".

This is not merely a question of terminology. The statements referring to sets and ensembles are more than descriptive, because they predict a future development according to an explanation by virtue of eco-unit system component behaviour. But this prediction only makes sense in its own context, i.e. that of eco-unit development. It is irrelevant at the level of the silvatic mosaic, or any kind of forest mosaic.

For concepts of layers or strata in large tracts of forest, the reader is referred to Sect. 6.2.3.1, where the difference between "Schichtung" (layering in the strict sense) and "Stufung" (stepwise architecture of mosaics) is discussed. It is essential to understand these different concepts well. For instance, the image of forest layers

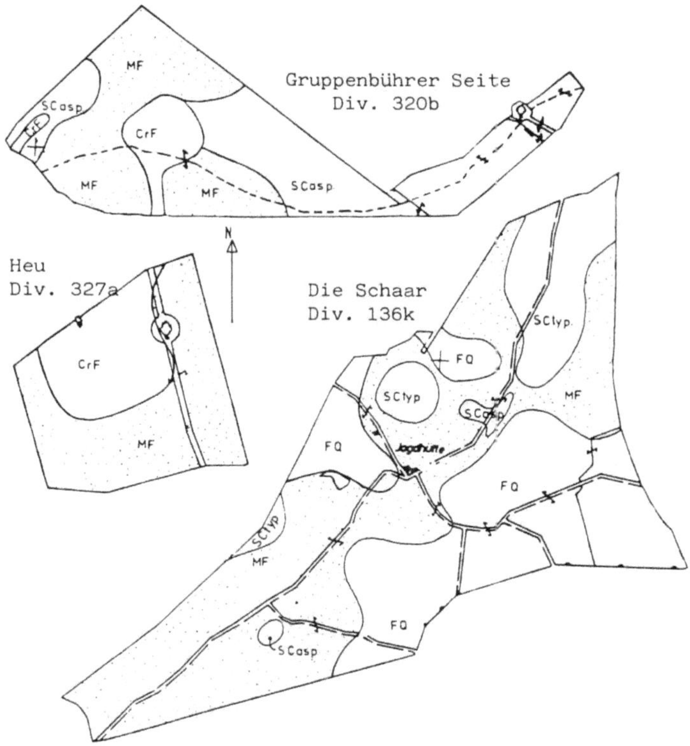

Fig. 5.7. Example of phytosociological limits between vegetation types in the semi-natural forests "Urwald Hasbruch" and "Urwald Neuenburg" (Germany). Compare with Fig. 5.1, representing eco-unit limits. According to Koop (1981a), from whom this figure is taken, eco-unit architecture is different in each vegetation type. *FQ* Fago-Quercetum; *MF* Milio-Fagetum; *SCtyp* Stellario-Carpinetum typicum; *SCasp* Stellario-Carpinetum asperuletosum; *CrF* Carici-remotae-Fraxinetum. *Jagdhütte* hunting lodge

used to localise animals in Ayensu (ed. 1980), as evidenced by the small reference images at the right upper corners of the pages, shows an eco-unit feature used in a forest mosaic.

The explanation of the development of the eco-unit rests on the definition, stating that its trees ordain its architecture, species distribution and ecophysiology. This statement now is made more precise by considering the fact that component trees have a changing function in forest eco-unit architecture (Fig. 5.8). When trees are in an early development period, they conserve a potential for further crown expansion. During this period, they may either express this potential in an aggrading phase, or leave it unused as long as they survive in a suppressed state in a biostatic phase. Potential trees (Fig. 5.8a–k) therefore play no persisting part as building blocks of forest architecture. They either help along in passing (aggrading phase) or they are subordinated to a forest architecture that is determined by other trees (biostatic phase). If and when a potential tree is liberated to resume its growth and

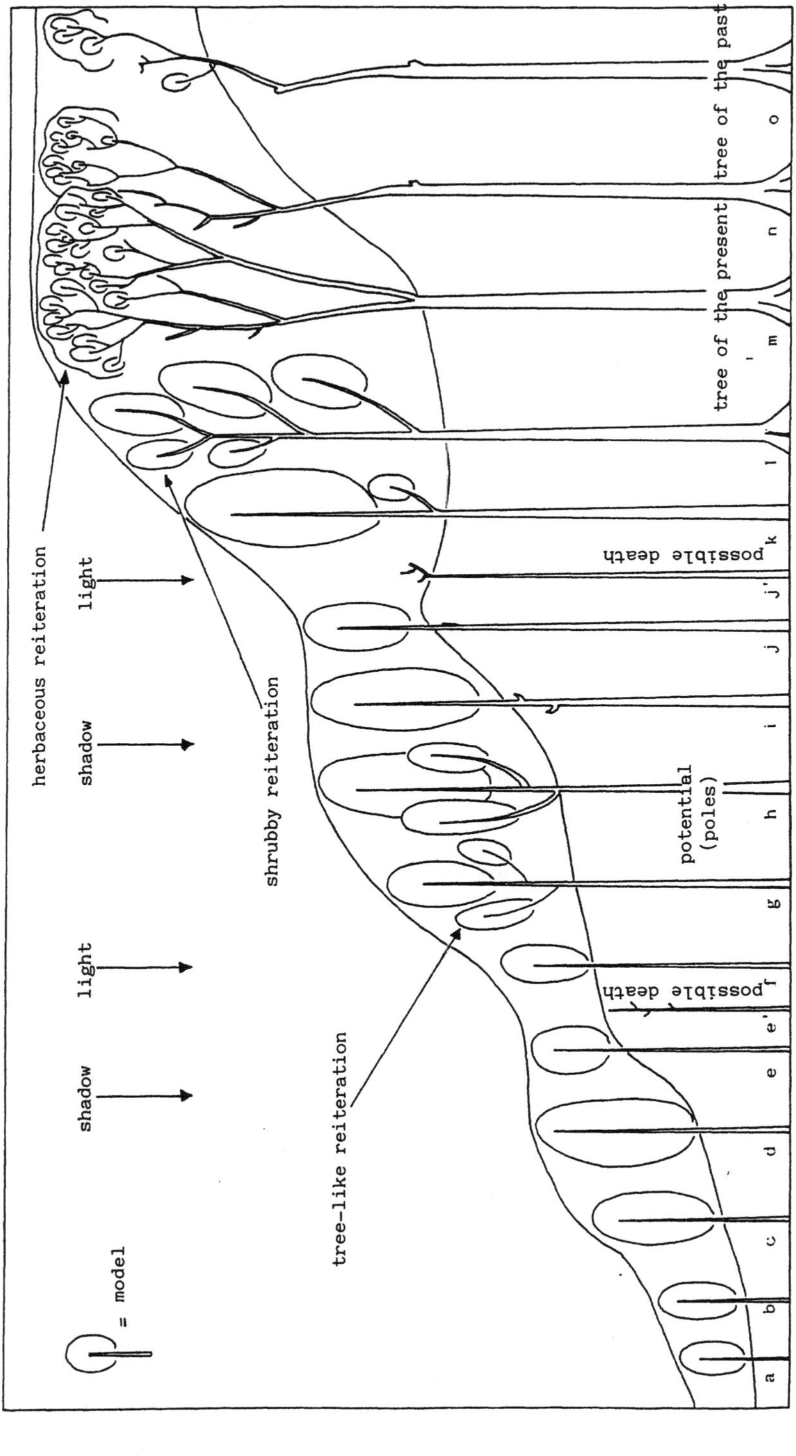

= model
shadow
light
shadow
light
herbaceous reiteration
shrubby reiteration
tree-like reiteration
potential
(poles)
possible death
possible death
tree of the present
tree of the past
a
b
c
d
e
e'
f
g
h
i
j
j'
k
l
m
n
o

Fig. 5.8

reaches its maximum expansion, it becomes a tree of the present (Fig. 5.8m), after having passed through a period of transition (Fig. 5.8l) when its growth generally is relatively fast. Trees of the present have no future except senescence and decay (trees of the past, Fig. 5.8o), but before decay sets in they can maintain themselves for a long time by replacing lost parts through some form of reiteration.

Figure 5.8 represents a hardwood tree from the tropical rainforest (cf. Oldeman 1974a; Hallé et al. 1978; Suselo 1983). This figure should be compared with Fig. 3.1, that shows the development of trees with a different temperament during their life cycle. The concept of "temperament" of a tree, which is an old European forestry notion (cf. Lanier 1986), will be discussed later (sees Sects. 5.5.11 and 6.2.4.2).

Figures 3.1 and 5.8 illustrate why it is easiest to establish the state of a tree ("potential", "of the present", "of the past") in the humid tropics, where model conformity and metamorphosis provide clear criteria. In lands with a pronounced unfavourable season, e.g. a winter or a dry period, the model-conforming period of a tree's life is often much shortened. Up till now, no clearcut architectural criteria exist to evaluate the state of temperate and dry tropical forest trees, although Pascal (1983) claims that the criteria for tropical rainforest trees can be used throughout a range of forest types in India, including drier forests. In the temperate region, the studies of tree architecture aiming at finding such criteria are still in the beginning (e.g. Goderie 1984; Bonsen 1984; Hilgen and van de Brenk 1984; Roloff 1985; Maas and Gubbels 1986b).

For the semi-arid regions, a beginning has been made by Van der Zandt and Sterck (1986) with their study on the growth dynamics of *Faidherbia (Acacia) albida*, to be compared with the work on *Acacia auriculiformis* in wetter regions by Edelin (1984) and Schreuel and Stegeman (1986). Two more categories of trees are difficult to judge as to their state. These are the gymnosperms (cf. Mulder and de Waart, 1985) and those pioneer trees that do not show reiteration or, on the contrary, start reiterating very early (cf. Oldeman et al. 1983; Lescure 1986, p. 56).

The diagnosis of the state of a tree in its eco-unit (or other forest stand) then falls back to a more empirical level, which makes use of proportional diagnosis (see Sect. 3.7). Total height, the ratio between free trunk height and total height (h_f/h_t) and the height/diameter ratio (h/d) are among the criteria that may be of help. The tree classification of IUFRO (International Union of Forest Research Organisations), accepted in conformity with a proposal by Leibundgut (1958) still is the con-

Fig. 5.8. The life cycle of a large hardwood species in the tropical rainforest (cf. Fig. 3.1) and the roles it plays in eco-unit architecture at different periods. The size of the reiterated crowns is here compared with other biological types, from which these crowns differ mainly by the absence of an autonomous root system. *a - k* potential trees that have not exhausted their expansion potential, but that may die early (*e'*, *j'*). In the eco-unit, these trees are either growing (a-d, g-h, k) or suppressed (*e-f*, *i-j*); *l* transition period; *m* tree of the present, fully expanded, long-persisting, strongly reiterated, building and maintaining the structural set that determines biostatic eco-unit architecture; *n* senescence; *o* tree of the past that is being eliminated during the degradation phase of the eco-unit. Time scale approximately logarithmic. Compare with Fig. 5.4, shade-tolerant. (Oldeman 1980)

ceptual framework on which there is most international agreement. It is reproduced here, with standardised numerical IUFRO-code and notes concerning the Assmann classification (1953). For examples of stand structure analysis according to such classifications, the reader is referred to Mayer (1980, p. 142–143), or Lanier (1986, p. 85 and following).

a) *Stand layer (height class).* A tree may be in the upper layer, height 2/3 to 1/1 of maximum height (code: 100), in the middle layer, height 1/3 to 2/3 of the maximum (code 200), or lower than 1/3 of the maximum height (code 300). Maximum height is defined as the average height of the 100 thickest trees per hectare in the upper layer. Assmann corrects these values to take into account the crown's exposure to light, and Mayer (1980) adds a suggestion that a finer differentiation may be needed locally. Our Fig. 5.4 is in agreement with this suggestion, particularly in the sense of eco-unit differentiation within the forest model.
b) *Tree vigour (vitality).* Trees may be strongly developed, i. e. more than the average (code 10), show an average development (code 20) or may be weakly developed or just lingering on (code 30). The criteria for the development are empirical, and concern such factors as "health" and "growth potential". Assmann (1953) used more precise criteria, such as crown volume, crown symmetry, crown density and leaf or needle coverage. These criteria are now used in vitality classifications for the assessment of pollution impacts on European forests (e. g. in Bauer 1985, p. 122 and following; or Van den Tweel and Den Boer 1986), adding criteria concerning the state of the trunk, the length of the yearly shoots, leaf or needle shape and colour, and reiteration pattern ("Ersatztriebbildung"; also compare Fig. 3.33).
c) *Developmental tendency (conversion tendency).* This is linked to forest conversion, because trees of one species may overtop those of another species, which is of utmost silvicultural importance. Trees may be *fore-runners* with a higher growth potential than the others (code 1), stationary with an average growth potential (code 2) or *laggers* when they grow slower than most others (code 3). This criterion is based on the comparison between a tree and its neighbours and aims at finding the "coming trees" in a stand, considering all relevant properties, such as height growth, crown depth and symmetry, and vitality. The code is then used among others to classify trees as to their "social position" in the stand, which since Kraft (1884) has been indicated as "dominant", "co-dominant" or "suppressed". Finer categories are sometimes used.

   Except for these ecological criteria, there are economic criteria: silvicultural position (in commercial value classes), trunk quality (quality of marketable wood) and a third one that we would prefer to include in the ecological criteria, although it is relevant also to the length of the commercial trunk. This is:
d) *Crown depth.* Trees may have deep crowns, covering 1/1 to 1/2 of total tree height (code 4), medium crowns, covering 1/4 to 1/2 of the total height (code 5) or shallow crowns, covering less than 1/4 of the total height (code 6). The biological significance of crown depth has been discussed in Sect. 3.7. Its ecological

significance is clear from Figs. 5.4 and 5.8. Crown depth changes with the state of a tree. A potential tree crown in the aggrading phase (Fig. 5.4B) is deep; when it becomes suppressed in the biostatic phase, such a tree has a shallow crown. The crown of a tree of the present is also deep in the beginning of the biostatic phase, or for small trees at the formation of a low structural ensemble. With age, its depth decreases until the crown degrades and the trees, of whatever structural ensemble, become trees of the past.

5.2.4.3 Components Other Than Trees

The biostatic phase contains trees in all three states (Fig. 5.4C), and therefore is the most apt to serve as an example in which to explain these states. Both the aggradation phase and the degradation phase (Fig. 5.4D) most often lack trees of the present. No trees of the past can usually be found in the aggrading phase, whereas they may linger in the innovation phase. If this architectural pattern in biostasis is well understood, it explains the spatial distribution pattern of the components other than trees. Gatsuk et al. (1980) have united data on plants from the Soviet Union, which allow to establish development states for non-arborescent forest components.

For this distribution pattern, the emphasis in biostatic eco-units is localised in the upper canopy, i.e. the upper structural ensemble, and the zone near the ground. In the large crowns of the highest trees of the present, the crown architecture yields a welter of horizontal axes, of forks, of holes in the main reiterated trunks where branch fall has left rotting knots, and of "suspended soils". This term seems to have been coined by Delamare Deboutteville (1951 in Trochain 1980), and indicates the accumulated mass of litter and humus on large axes and in forks or holes in tree crowns. The possibilities for accumulation are also heightened by a rough bark structure, smooth bark giving little hold to intercepted falling organic matter.

This environment is settled by epiphytes (see Sect. 4.5), most richly in humid climates. That the epiphytic distribution pattern in the canopy is an eco-unit property is clear from the studies by Johansson (1974; Fig. 5.9). His illustration typically represents a tree of the present with a much-reiterated crown, that is a member of the upper border layer of the eco-unit. The epiphytes in their turn influence the living space in such a way that there is a rich fauna of insects, amphibians and reptiles associated with them (e.g. Benzing in Sutton et al. 1983, for tropical rainforests). Small-sized epiphytes are to be found in places where small reiterated crownlets occur, and large epiphytes are associated with large, tree- or shrub-sized reiterated subcrowns. This is the case in the humid tropics. In the temperate regions there are scarcely any large epiphytes.

Moreover, big and profusely reiterated trees produce huge crops of fruits (Oldeman 1974a). These attract many frugivorous birds and mammals, from toucans and hornbills in the tropical rainforest to magpies in European oak forests and from squirrels all over the world to monkeys in the tropics. The tropical case is magnificently illustrated in the book "Jungles", edited by Ayensu (1980), which

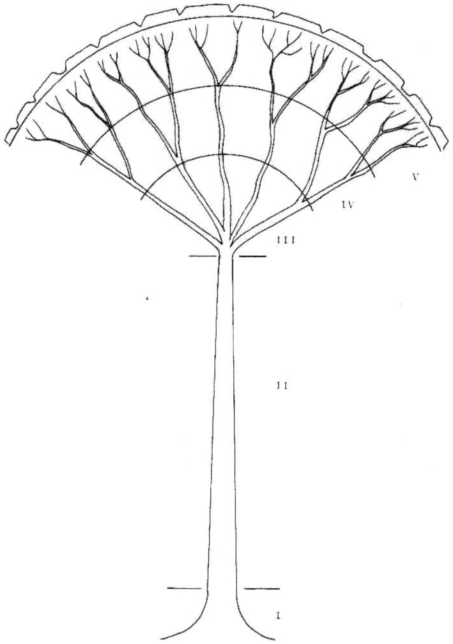

Fig. 5.9. Creation of niches for epiphytes by a West-African tree of the present, member of the upper structural set (forest canopy, cf. Fig. 5.4). A tree that is a "bearer of epiphytes" is termed a phorophyte. Each of the zones *I* to *V* has its own epiphytic flora, in *III* to *V* the life forms of epiphytes often correspond to the tree-like, shrub-like and herbaceous character of reiteration (Fig. 5.8). (After Johansson 1974; see also Ek 1989)

places much emphasis on animals, whereas Van Vuure (1985) included the biostatic phase in his analysis of the mammal fauna in European forests. In Europe, this phase is rather poor and for mammals the canopy fauna is limited to squirrels and one predator, the tree marten (e.g. Reuver and Van der Zee 1984). The European bird fauna of biostatic eco-units on the contrary is rather rich (Philippona et al. 1983). The liane crowns that may link together the trees of the present create micro-environments for many small organisms, and contribute with their fructification to that of the trees. Both these life forms also produce new leaves periodically or continuously and the diet of herbivores and frugivores may be very diverse (cf. Hladik 1978; Van Roosmalen 1980). Figure 5.10 shows the diet of the Suriname spider monkey during one month in the dry season. However, the food species that are depicted certainly come from different eco-units (cf. Fig. 2.2).

The life spaces of insects are also shaped by the eco-unit. Murphy (1973) mapped the different kinds of such spaces on a transect made in Singapore, representing two neighbouring eco-units (Fig. 5.11). The difference between both eco-units is clear. The difference in insect diversity between eco-unit development phases has been confirmed by the preliminary analysis of the insect fauna in rela-

Fig. 5.10. Diet of spider monkeys during one month, dry season. (Van Roosmalen 1980; cf. also Fig. 2.2)

Fig. 5.10. Diet of spider monkeys during one month, dry season. (Van Roosmalen 1980; cf. also Fig. 2.2)

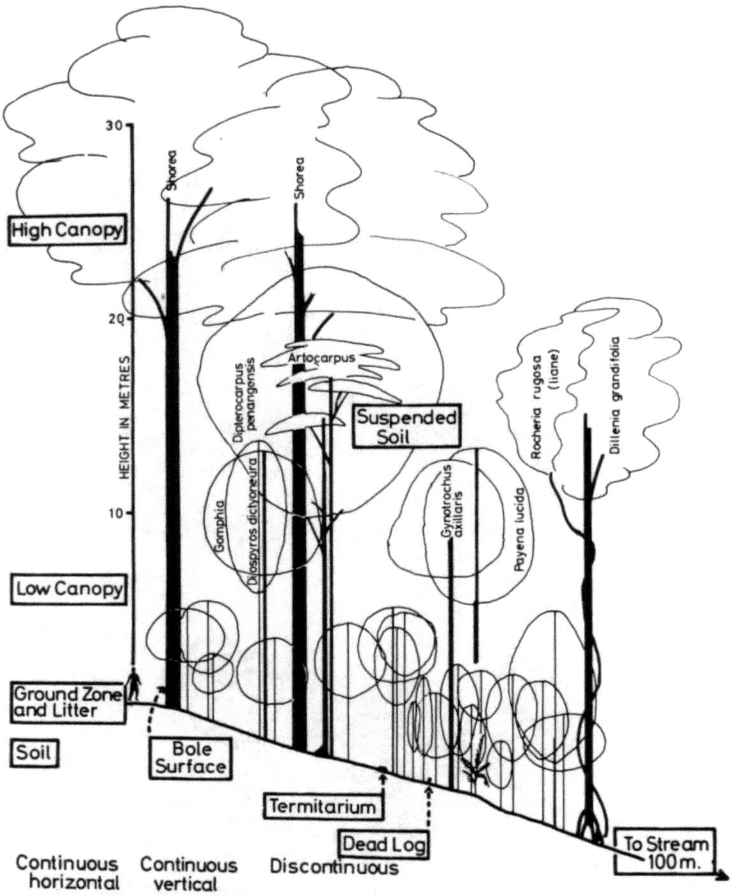

Fig. 5.11. Insect niches linked to the architecture of two neighbouring eco-units. The one to the *right* is in a biostatic phase, the one to the *left* in an aggradation phase. They are separated by a transitional zone where *Artocarpus, Dipterocarpus penangensis, Gomphia* and *Diospyros dictyoneura* build a vertical canopy. The dead log is a rest of an earlier eco-unit now being replaced by the aggradation phase of a new one (*left-hand side*). (Transect drawn to scale, Murphy 1973)

tion to Amazonian forest architecture by Akker and Groeneveld (1984). It will be extremely difficult, however, to include the thousands of insect species that occur in even a moderately complex forest mosaic, into one model that accounts for their life spaces. The maps drawn by Brauns (1976), in the form of forest sketches on which numbers indicate a typical habitat for a species, are among the most beautiful and manageable documents in this sense for North-Western European forests.

In the biostatic phase, a forest eco-unit accumulates the maximum of wood. The set of the present contains trees that start to approach the asymptote of accumulated woody biomass (cf. Fig. 3.41). Therefore the number of xylophagous insects increases, both on the outside of the trunks, e.g. arboricolous termites (Fig. 4.15A), or on the inside. Here often the cores of the trunks have already started to rot by fungi having penetrated through wounds caused, for instance, by imperfect scar formation after branch shedding, or by the wrenching off of reiterated crown parts.

The occurrence of heart-rot in virgin forests in the Philippines has been studied by Eusebio (1977). Because he considered trees of over 30 cm in diameter only, his figures would be more or less representative for the set of the present, as averaged over numerous biostatic eco-units of somewhat comparable sizes and structures. It is to be assumed that the set of the present shows a higher percentage still of trees affected by heart-rot, because the diameter limits certainly include potential trees. Indeed, Eusebio states that "mature" and "overmature" trees tend to be more susceptible. In these dipterocarp forests his averages, which have to be understood with the preceding notes in mind, are: five trees per ha in Eastern Quezon, two in Eastern Mindanao and five in Zambales. Such proportions might be normal for tropical rainforests. Vooren (pers. commun. 1986) confirms that many large trees in the Ivory Coast rainforest have hollow trunks, and De Graaf (1986) mentions a proportion of 10% to over 20% of the largest rainforest trees in Suriname. Data cited by Koop (1981a) suggest that between 4 and 12 hollow trees per ha occur in mixed northwestern European hardwood forests, but he links this to clearly diminished vitality (see Sect. 5.2.5).

The biostatic phase shows some important components at soil level. Herbs are very rare and if low plants exist in this phase they are either dwarf shrubs, such as heather (*Calluna vulgaris*), or blueberries (*Vaccinium* sp.) with an associated vole, or otherwise mosses with their associated insects. Larger animals are those that feed on roots, fallen fruits or fungal sporocarps, e.g. wild boar species. But this last case only applies in eco-units with dense structural ensembles, casting heavy shadow. The signals, i.e. the quantity and quality of the light that are let through by these canopies, determine the forest components at the forest floor level. The occurrence of many small creeping or erect herbs with variegated leaves on the tropical rain forest floor (e.g. Marantaceae, Rubiaceae; see Fig 5.88A and Lee in Givnish 1986) may be linked to the spectral properties of the transmitted light.

Patterns and pathways in biostatic forest architecture are clear. The organisation of the eco-unit is at its acme. Murphy (1973) remarks, for instance, that spider webs are tended across the pathways of flying insects within the forest. These pathways in their turn are linked to forest architecture (cf. Figs. 5.4C and 5.11). The biostatic phase of the eco-unit may persist for a long time if no accidents occur. Its internal organisation is clearly separated from its environment, both by the limits drawn by the upper canopy and the rhizosphere, and by the species composition that has resulted from the development going from the innovation phase and its many transitional species, through the competition of the aggradation phase with its high mortality, towards a biostatic state that rests on steady interactions between rather steady populations of species.

5.2.5 The Degradation Phase

5.2.5.1 Overview

When, for inside or outside reasons, biostasis breaks down, the eco-unit enters into its degradation phase (Fig. 5.4D). This phase has been termed "Zerfallsphase" (fall-apart phase) in German literature, "senescent phase" by Oldeman (1980) and, for large forest stands, "reorganisation phase" by Bormann and Likens (1979). This last term will be considered in all its implications, because the degradation of biostatic architecture causes ecological signals to descend to the forest floor and the potential set. These signals, e. g. light or organic matter, already activate part of the propagule bank to prepare a following innovation phase. As stated above (Sect. 5.2.4.), there is no Mayan term for this phase.

5.2.5.2 Dynamics Explained

Degradation prepares the zero-event, because it clears the space occupied by an eco-unit for re-occupation by one or more new and young eco-units. In Fig. 5.4D, degradation is a slow process that is switched on by the senescence of trees. This is exemplified by the senescent, 1200-year-old oak in Figure 5.6B. But degradation may be very rapid when caused by accidents, i. e. random events from the viewpoint of eco-unit organisation. Such accidents may be, among others, fire (cf. Kessell 1979, and Fig. 5.12A), waterlogging of the soil so that the roots no longer anchor the trees of the present adequately, strong winds or gusts, lightning (see the map of thunderstorm frequencies in the tropics published by Nieuwolt 1977), inundations in floodplain forests, forms of run-off or subsoil erosion (e. g. Hallé et al. 1978, p. 366 and following), avalanches (see Korpel 1980), or even the waterlogging of the epiphyte load on asymmetric crowns, causing trees to topple over. Degradation also may be biotically powered, when some forest component starts to monopolise or drain the whole production capacity of the eco-unit at the cost of the trees. Insects play this part rather often.

It seems, however, that the eco-unit has to be "ripe" to receive such biotic impacts. Sprugel (1976) thoroughly studied the degradation and regeneration in *Abies balsamea* forests on hillsides exposed to wind in the Northeastern United States (Fig. 5.12B). The degrading eco-unit has the shape of a long, crested wave along the hillsides, followed by eco-units in earlier development phases that have the same shape. At an age of 60 to 70 years, the eco-unit becomes exposed to factors such as wind, rime-ice, summer cooling and winter desiccation weakening it so that other biotic components may kill the trees. These components are several forms of rot (i. e. fungi, Sprugel 1976) or insects, such as the spruce budworm (*Chorisoneura fumiferana*), the major pest that destroys millions of cubic meters of valuable timber in the process (Schiltz et al. 1982). Once the trees have died, an innovation phase establishes itself on the vacated surface, and the process repeats itself through the consecutive development phases. It is to be emphasised that insects such as the

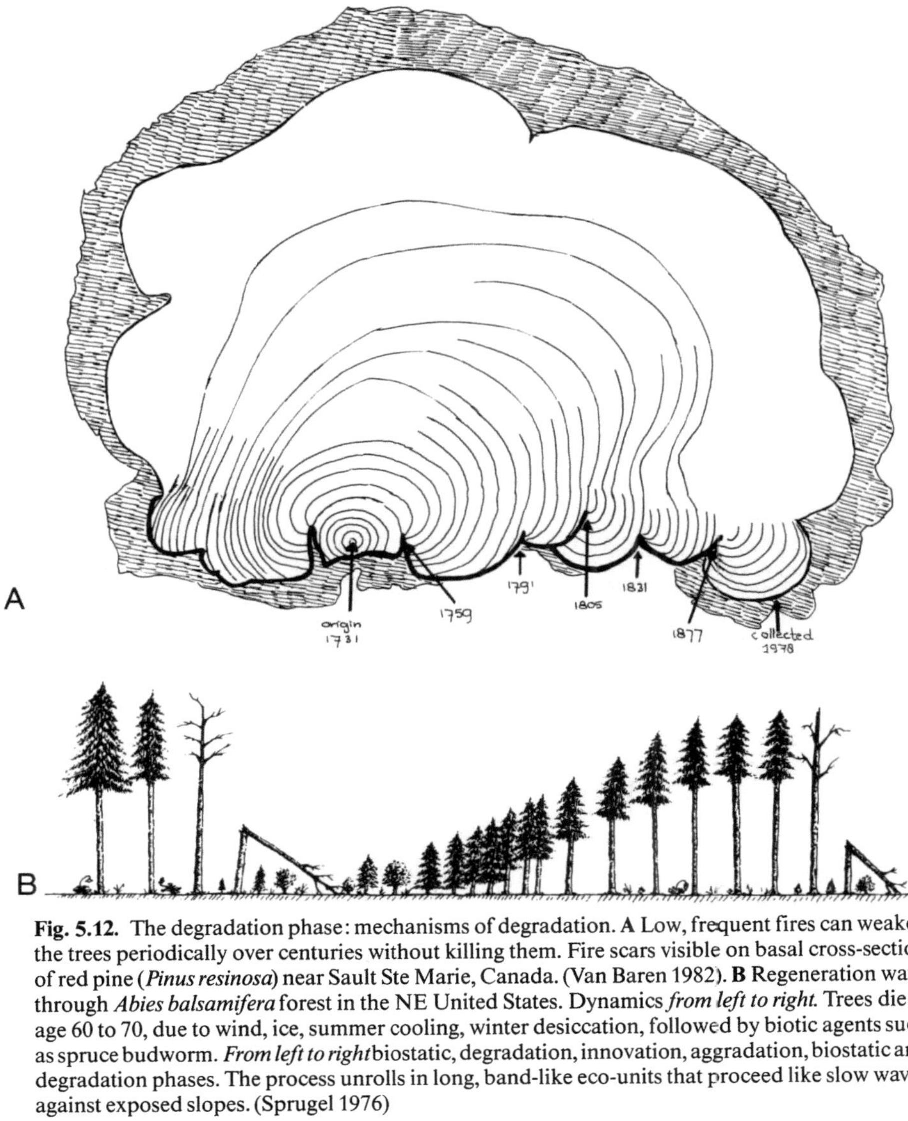

Fig. 5.12. The degradation phase: mechanisms of degradation. **A** Low, frequent fires can weaken the trees periodically over centuries without killing them. Fire scars visible on basal cross-section of red pine (*Pinus resinosa*) near Sault Ste Marie, Canada. (Van Baren 1982). **B** Regeneration wave through *Abies balsamifera* forest in the NE United States. Dynamics *from left to right*. Trees die at age 60 to 70, due to wind, ice, summer cooling, winter desiccation, followed by biotic agents such as spruce budworm. *From left to right* biostatic, degradation, innovation, aggradation, biostatic and degradation phases. The process unrolls in long, band-like eco-units that proceed like slow waves against exposed slopes. (Sprugel 1976)

spruce budworm should be considered as silvatic mosaic components rather than eco-unit components (see Sects. 4.8.5. and 6.2).

Forest degradation over large surfaces has an enormous emotional impact on humans, who are inclined to term them catastrophes, as exemplified by the dieback of *Abies fraseri* forest in the National Park on Mount Mitchell, North Carolina, USA (Fig. 5.13). Here, the impact of a parasite, the balsam woolly aphid, a sap-sucking organism that pierces the bark to attain the sapstream (Smathers 1978), occurs in 30- to 40- year-old trees. The effects of such massive attacks can be assessed with

Fig. 5.13. Degradation phase. Large-scale forest dieback in *Abies fraseri* above 1500 m altitude, Mount Mitchell, Blue Ridge Parkway National Park, N. Carolina (USA), June 1986. The immediate cause of this mortality is the woolly balsam aphid disturbing the sap stream in trees over 30 to 40 years old. Why these trees are so very vulnerable is unknown. (Photograph courtesy M. V. M. Maas)

Fig. 3.30. The reasons for the decline in vitality of the *A. fraseri* forests at about 1500 m altitude in the Great Smoky Mountains and on Mt Mitchell, a decline that is probably needed to bring eco-units into a state that is vulnerable to a large-scale aphid attack, are still unknown. Two factors of uncertainty are that the parasite is an exotic species, introduced from Japan or Europe around 1905, and that air pollution may or may not have played a part in diminishing the vigour of these forests.

Other examples of eco-unit degradation due to biotic factors are the death, due to the Dutch Elm disease, of elms (*Ulmus*) in the semi-natural hardwood forests bordering the Rhine in France and Germany (cf. Van de Winckel 1980) or the large-scale forest dieback in Japanese *Pinus thunbergii* stands due to nematodes (*Bursa-*

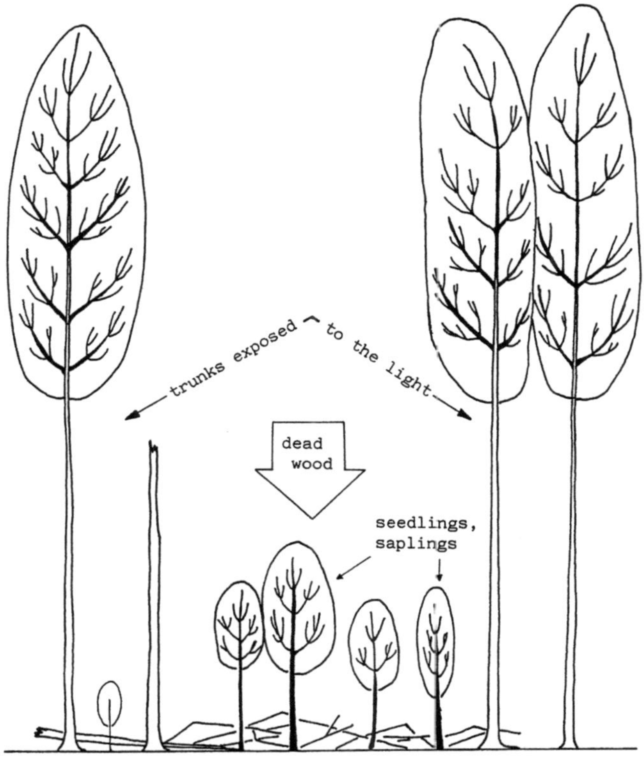

Fig. 5.14. Degradation phase. In European poplars (*Populus* sp.pl.), three guilds of xylophagous capricorn beetle larvae (guilds within the genus *Saperda*) are specialised on dead wood (many species), living wood in trunks exposed to light (some) and seedlings or saplings (a few). If all three guilds are present, this chablis is going to extend and no young poplars can survive: the large pioneer poplar eco-unit hence cannot be succeeded by the same kind of eco-unit (Oldeman 1980)

phelenchus lignicolus), transmitted by bark beetle larvae (*Monochamus* sp. pl.) that at the same time transmit the spores of a fungus (see Mamiya 1981; Shimazu and Karagiri 1981). In the case of the elms (see PUDOC 1982), the sheer agressiveness of the fungus (*Ophiostoma ulmi*) in combination with its vectors, bark beetles of the genus *Scolytus*, may explain why elms (*Ulmus* spp.) are attacked at very young ages, before the eco-units that they build reach their biostatic phase.

In large eco-units, accidents that cause a few trees to topple may set into movement processes that end in "rolling up" the whole stand. An example is that of European poplars (*Populus* sp. pl.; Fig. 5.14). Three groups of capricorn beetles (*Saperda* sp.pl.) have xylophagous larvae that are specialised either in dead wood consumption, or in the attack of trunks of large trees that are exposed to light, or finally to feed on the wood of seedlings and saplings. A chablis (see Sect. 5.1.3.) in a European poplar stand, which is often large-sized because these pioneer trees colonise empty places such as flood plains, activates first the dead wood beetles, then the large trunk beetles and finally the seedling-and-sapling animals. The result is that the old stand dies off and that no new poplar eco-unit can take its place: in nature, poplar eco-units are succeeded by eco-units of other kinds. It goes without saying that *Saperda* species are pests in populiculture.

As far as we know, no humid tropical examples of such processes have been documented. There exists a mysterious aerial photograph of circular mortality patches of quite large surface in the Amazonian rainforest, which has not been explained (ITC pers. commun.). The synchronous mortality of groups of the pioneer tree *Ateleia herbert-smithii* (Papilionaceae) in Central America, according to Janzen (1986b), is the central event in a web of insect-tree interactions, but it is also programmed by the tree itself, that flowers massively and then dies. The case studies on tropical rainforest insects included in the volumes edited by Meggers et al. (1973), Furtado (1980), Leigh et al. (1982), Fournier and Sasson (1983), Sutton et al. (1983) or Gomez-Pompa and Del Amo (1985) do not include any comparable phenomenon. In these publications it is stressed, on the contrary, that organisms such as Macrotermitinae in Malaysia (Collins in Furtado 1980) or leafcutting ants of the genus *Atta* in the neotropics (Cherrett in Sutton et al. 1983) show strategies of resource conservation by avoiding to kill trees. They become pests, i.e. tree killers, only in completely transformed ecosystems, such as over-exploited forests or timber and fruit tree plantations. Most data on insect "pests" given in Fournier and Sasson (1983) still represent the classical phytopathological view, describing an insect species that eats a crop, and neglecting the web of ecosystem interactions that leads to this event. The newer approach of biological or integrated pest control in phytopathology sees this in a different way, but is rather difficult to implement in the humid tropics with their innumerable insect species.

All these examples show that during the degradation phase the trees of the past, that once organised the biostatic forest architecture, fall apart and cease to build the environments needed by other forest components. On the contrary, spaces are liberated for many more components than there were in the biostatic phase. A whole array of decomposition organisms moves in or germinates, that live on the increased quantities of dead wood. Epiphytes show a marked increase in numbers and vitality, due to higher amounts of available sunlight, until they

Fig. 5.15. Dynamics of epiphytic and other niches, illustrated by monocotyledonous species in the Guyanas. (De Granville 1978). *I*, *II* Terrestrial heliophile monocots (*t*) dominating near the coast; *III Mauritia* palms (*T*) in galeries along creeks in coastal savanas; *IV* three categories of monocots in dry land forest: heliophilous epiphytes (*E*) in canopy, mobile semi-heliophilous epiphytes (*PM*) on trunks and moving up (*arrows*), and terrestrial species either tolerant or more heliophilous but in that case suppressed; (*p*) palms, traces of old chablis; *V* stemless palms and mobile creepers in the undergrowth below less dense canopy in lower structural ensembles; *VI* epiphytes fall down in chablis with the fall of large branches (*VI*) or whole trees (VI_2): *open arrows*. They keep alive as long as there is light on the forest floor. This light also activates suppressed monocots, particularly palms (*p'*); *VII* in marshy valleys, monocots occupy the whole vegetational space; *VIII* on naked rock faces, dominance of heliophilous saxicolous monocots (*S*), the substrate having much in common with an epiphyte substrate. Continuing exchange of epiphytes between forest canopy and rock surface. All these cases have been studied and documented in detail by De Granville (1978). Note that this is an explanatory architectural model, based upon eco-unit development in terms of Fig. 5.4

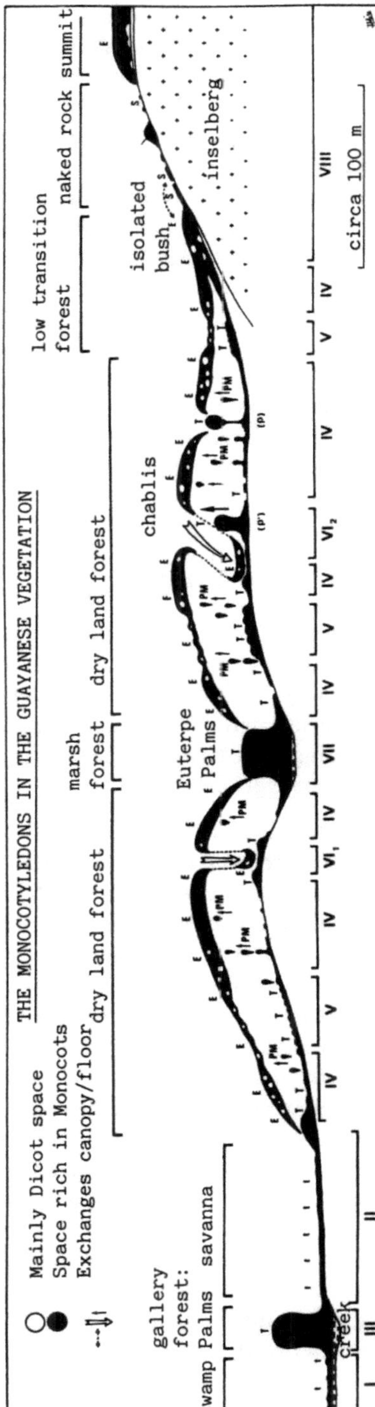
THE MONOCOTYLEDONS IN THE GUAYANESE VEGETATION
Mainly Dicot space
Space rich in Monocots
Exchanges canopy/floor
swamp
gallery forest: Palms
savanna
dry land forest
marsh forest
dry land forest
low transition forest
naked rock summit
Euterpe Palms
chablis
isolated bush
inselberg
creek
I
III
II
IV
V
IV
VI1
IV
VII
IV
V
IV
VI2
IV
V
IV
VIII
circa 100 m

Fig. 5.15

fall down from the rotting tree crowns. This has been exemplified by De Granville (1978) in his master diagram of the movement of monocotyledonous niches with forest development (Fig. 5.15). Potential trees that conserved sufficient vitality start new growth. Seeds and spores germinate, and a herbaceous carpet develops on the soil, patch by patch. All these organisms attract animals with which they enter into interaction. The accumulation of large-sized dead wood from large reiterated crowns and from trunks is at the centre of the spectacular increase in non-tree dynamics in the degrading phase. Therefore, Cosijn et al. (1983), who edited a special volume of the Netherlands Forestry Journal on dead wood in the forest, include many phenomena linked to degrading eco-units. Brauns (1976) provides insect species maps, on forest and tree sketches, for degrading European forest. J. Lescure (pers. commun.) found a specialised amphibian fauna in degrading eco-units in the Guyanese rainforest, including the spectacular arrow-poison frogs (*Dendrobates* sp.pl.). If and when these dynamics lead to the predominance of certain organisms, insects or rot fungi, the end of the degrading round is the gong, i.e. the rapid, biotically induced disappearance of trees.

5.2.5.3 The Zero-Events

If the degradation phase is induced by an accident, including wood harvesting, the immigration or germination of other forest components also takes place, but not in the same way. The character of the zero-event determines, together with the available information in the propagule bank, how the innovation phase of the next eco-unit is going to behave. It is important to know the character of this event better than by some general statement.

a) Fire may lead to very different results, according to its behaviour. Bibliographies are available in Kessell (1979), Kamminga (1983) or Cooper (1974). Figure 5.12A shows that periodic fires with little accumulated fuel may periodically weaken the trees that are nevertheless adapted to such burning environments and survive. But not only the trees are adapted. So are the eco-units which they build, including all other components (cf. Leffef 1984). Giant redwoods in California are very fire-resistant, so that they may reach ages of over 2000 years and sizes over a 100 m high, while nearer the soil the fires rage periodically and induce the formation of new generations of eco-units, including among other species the Douglas fir (*Pseudotsuga menziesii*); see Engbeck (1976) or Mulder and De Waart (1984). A special fire issue of Fir Report (see Helgerson 1988) summarises the periodicity of occurrence and the broad ecological and management consequences of forest fires in Oregon and California.

   Given the ecosystem adaptations, forest management with prescribed man-made burning is often the best silvicultural option in thinly populated regions, e.g. the Southern United States, Canada or Finland. Forest fire brigades, that prevent or extinguish such forest fires, are never 100% successful but decrease the frequency of fires.

The result is that more fuel accumulates than in natural circumstances. The rarer fires are also fiercer. No adaptive biological structures can withstand them, and whole forest regions that could have been maintained by frequent and moderate burning lose their forest cover completely. In the densely populated Mediterranean region, where forest fires account for over half a million ha of deforestation per year in the 1980's, the study of this zero-event is one of the greatest challenges of our time.

b) Water can also play many different parts. Waterlogging of soils after heavy rains is often localised and may lead to the uprooting of one or a few trees. If the whole forest eco-unit stands in a shallow soil, limited beneath by an impermeable root floor (see Sect. 5.3.3), underground runoff may destabilise its whole rhizosphere. Superficial runoff affects the roots of large standing trees less than the organic soil layer and the propagule bank, which are often eroded away even under high forest. This has been observed by the present author in Kalimantan (Indonesia), the Guyanas, Ecuador, Côte d'Ivoire and Gabon, on steeper slopes. Periodic inundation of an eco-unit in narrow tropical valleys during the rainy season asphyxiates the roots of large trees, to which they react by shedding their leaves (cf. Oldeman 1974a). Such periodic decreases in tree vigour may lead to destabilisation, either by rendering the trees more vulnerable to attacks by other forest components, or by making them more prone to uprooting. Inundations of flood-plains often cause the whole site to change, while at the same time eradicating whole eco-units of smaller or larger sizes (Vandeursen and Wisse 1985, along the Rhine, Alsace, France), and creating whole new abiotic sites by the processes of erosion and sedimentation.

c) Wind is an extremely variable factor. Local gusts, for instance during thunderstorms, topple trees over every year, from one to several per hectare. In regions without hurricanes or seasonal storms, this results in the presence of chablis (see Sect. 5.1.3.) in smaller or larger numbers. The strength and periodicity of the gusts codetermine whether a chablis (uprooting) or a volis (broken trunk) is the result. Most often, trees are not pushed over by a constant pressure of the wind, but successive gusts push them like a swing (e.g. Kuiper and Van Schooten 1985). This seesawing is accentuated if the gusts arrive with a certain periodicity, until finally the roots are torn out of the soil, or the trunk breaks off. The risk of both is increased if fungi live inside the axes or roots of the tree: heart-rot or root-rot. If the falling tree bowls over some of its neighbours, the resulting open surface may be very complex (cf. Mutoji-A-Kazadi 1977; Florence 1981). Hurricanes have quite different effects, and may almost completely clear large surfaces of forest periodically, as happens in New England every 30 to 40 years. In North-Western Europe this also occurs several times per century, e.g. the Dutch storms in 1972 and 1973. Whitmore (1974) analysed the role of cyclones in the Solomon Islands. In the Caribbean, hurricanes do not throw the trees over, but most often tear off their crowns partially or entirely. These trees most often regenerate new crowns by reiteration, but in the meantime the eco-unit architecture remains altered for several years. In Puerto Rico, for instance, we could observe no clearcut sets and ensembles of trees (see Sect. 5.3.4.1) in an otherwise mature-looking submontane rainforest of this kind in El Verde. Here, biostasis and degradation blend into each other in a more or less reversible way involving "natural pollarding", that deserves to be analysed more precisely.

d) The humidity of the air, as discovered by Vooren (in prep.) in Côte d'Ivoire, influences branch and trunk fall because the wood starts to work, like in a piece of wooden furniture that is transported into a centrally heated building. Every morning the dew saturates the wood and every afternoon at about 16.00 h it has been dried out again by the sun. These daily hours show relative maxima of tree and branch falls.

e) Earthquakes and landslides have been studied in the role of zero-events in Central America and Papua New Guinea by Garwood et al. (1979). The cases presented by these authors show how such large surfaces are cleared and settled again by pioneer forest eco-units, at such regular intervals that large eco-units occupy the largest proportion of such forest mosaics. However, small-scale landslides and earth tremors may also play their part in the creation of smaller eco-units. For instance, in Saül, French Guyana, we observed after one day with ca. 200 mm precipitation, that a patch of forest of about 1 ha had slid down, standing up, from a steep hill-side, leaving an empty surface of about half a hectare uphill. Andean people know that earth tremors always topple over some few trees. In this rough category of zero-events, volcanic eruptions should be mentioned also. The press has widely publicised the wholesale deforestation caused by the eruption of Mount St. Helens in 1980, when ca. 600 km^2 of forest burned (Findley 1981). On the other hand, we observed a marked decrease in vitality of trees on the eastern slopes of the Ecuadorian Andes, due to the aerial pollution caused by the gases liberated in the long series of eruptions of the Reventador in 1975.

f) Human impacts will not be treated as a category in itself. Every human-induced zero-event has its natural counterpart. Only the intensity and frequency are influenced by human intervention. For instance, clear-cutting in North-Western Europe and the British Isles is certainly more extended than would be the large-sized eco-units caused by locally prevailing wind, water and fire regimes. The intensity of clearcutting may rather be compared with a forest fire, burning away into gases a considerable part of the biomass, than with the effects of a heavy storm, after which the fallen biomass remains behind. Understanding the basic natural factors is enough to evaluate the consequences of silvicultural interventions or plain forest exploitation without silvicultural follow-up.

The tendency is for biotic degradation factors to be linked with abiotic ones, and to act slowly. Abiotic degradation factors would tend to act more abruptly and to be followed by biotic reactions. The event of separation between the degradation and innovation phases is the death of the former eco-unit, which is as difficult to pinpoint in time as the death of a hospital patient. The "clinical death" of a degrading eco-unit could somewhat arbitrarily be fixed at the moment of death of the trees of the present. This would be in agreement with our circumscription of an eco-unit (see Sect. 5.1.7). Moreover, this symptom is easy to perceive in the field for users of the forest model.

5.2.6 Clinical Death of Eco-Units, Size, Succession

The "clinical death" of an eco-unit is prepared earlier and has immediate consequences afterwards. This leads to better understanding of Bormann and Likens' term "reorganisation phase" for both the degradation and innovation phases, taken together. If the zero-event is accidental, the preparation phase is relatively short, and the innovation phase will be long because much of the reorganisation will be included in it. If the preparation phase is long, at the moment of clinical death of the eco-unit the new eco-unit may nearly be formed below already, so the innovation phase is short. Each bioclimate and soil, that is each abiotic site, shows its predominant asymmetry in these relations. Oldeman (in Cosijn et al. 1983) gave some ratios between dead tree numbers due respectively to chablis, volis or dying upright. These ratios are different from site to site, even if these sites are as close to each other as a hilltop, a slope and a valley bottom. This latter sequence of sites is called a toposequence. The ratio between different modes of tree mortality is a symptom that is relatively easy to read in the field, and that shows the predominant local zero-events (cf. Van Baren and Hilgen 1983). Such a diagnosis is important if one aims at creating silvicultural systems according to the principles of Brünig (e.g. in Wiersum 1984). Such systems conserve their dynamic stability (resilience) against the impact of local zero-events: "catastrophes", "pests" and other "disturbances".

Furthermore, it should be emphasised explicitly that rare, large-sized, high-energy and unpredictable zero-events, such as fire in the tropical rain forest during dry freak years (Indonesia, West Africa in 1982 and 1983), earthquakes in Holland (two tremors up to now in the 20th century), or series of relatively wet years with increased erosion in the Sahelian countries (such as the 1950's) are indispensable to understand the developmental stages of eco-units found in a certain region. In current practice, silvicultural design only takes into account a certain development model for stands and disregards all events that are rarer than once in 5 to 10 years, whereas rotations of several decades are aimed at. The same is true for many management plans in nature conservation, which have no answer to rare natural events that induce drastic change in the ecosystems they must conserve. In the case of forest fire, measures are taken to make the event so rare that it will not interfere within a rotation. We saw earlier that this may literally backfire. In forest protection in general, the risk of backfiring is present if the causes of proliferation of pathogens (pests, diseases) are not taken away and measures are directed only at specific killer organisms that are components of rare processes, such as epidemics.

In the third place, it should be clearly stated that no zero-event is simple. It is always composed of synergetic factors, e.g. waterlogging, wind, fungus proliferation in trees plus a tormented microrelief of the soil (cf. Brünig 1985). These factors combine their action on a certain surface and at a certain moment, within a certain forest mosaic. No single "gap phase model" can encompass the very different innovation phases that are spawned by the numerous possible combinations of so many factors. Figure 5.16 shows a tentative model of the arrangement of innovation phases that could result from these factors, if the results of their action complexes were put in an order of increasing size (after Riera in Gillon 1984). This model should be worked out more precisely, after which its numerical expression can yield the basic point of departure for a computer-aided simulation model.

As a fourth remark, the degrading phase of an eco-unit, and the character of the zero-event that terminates it, lead to the replacement of an old eco-unit by one or

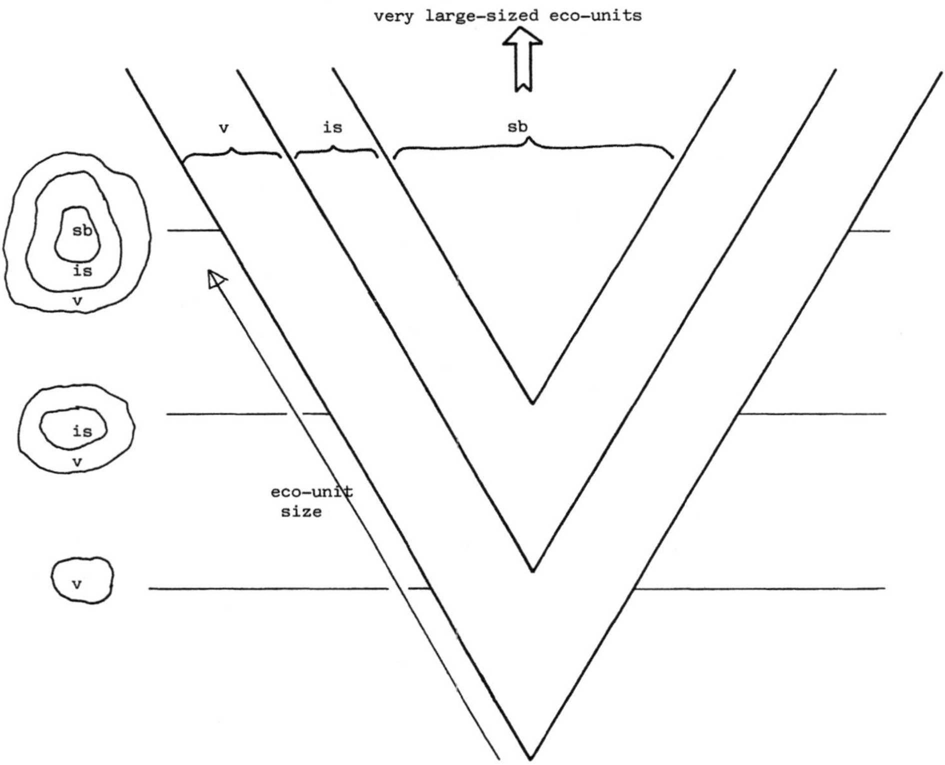

Fig. 5.16. Zero-event, resulting eco-unit size and reaction of propagule bank. On the outer periphery, the vegetative part of the propagule bank (*v*) is most active, on the inner periphery many immigrated seeds (*is*) may germinate and in the centre of the chablis the seed bank in the soil (*sb.*, cf. Fig. 5.3) dominates the event. The smaller the size of the new eco-unit, the more the periphery gains in importance. Under a certain size-limit first the seed bank remains inactive, below a second size-threshold both seed bank and immigrated seeds do not function. This diagram does not take into account the quality of the propagule bank as a whole. If a moderate fire destroys the seed bank, for instance, the immigrated-seed-zone covers the centre, and with a heavier fire it extends also to the border. In forest regions other than the tropical rainforest (Guyana), for which this diagram was made, the vegetative part of the propagule bank may be important even in large eco-units, e.g. in the Appalachians, where spontaneous coppicing occurs all over clearcuts (see text). If corrected for such factors, this diagram can be expressed numerically and serve as a starting point for a more general eco-unit regeneration model. (After Riéra in Gillon 1984)

more new ones. This is the only process that, in Chapter 6, will be indicated as succession. If the degrading eco-unit was large-sized, the zero-event may leave fragments intact, while other fragments are already dead. This process will be called fragmentation of an eco-unit. Indeed, because the eco-unit development processes become desynchronised, the larger eco-unit is henceforth divided into smaller ones that each correspond to our definition. If the old eco-unit was very small-sized, some strong zero-event may destroy it together with the neighbouring eco-units. There is, then, a large empty surface on which one larger eco-unit may develop. A mosaic of small-sized units has been replaced by one large eco-unit. This process will be called regression. This is what happens, for instance, if stretches of tropical rainforest are exploited, bulldozered and transformed into monocrop plantations (Oldeman 1983c, his Fig. 4); but it also may happen after fire or shifting cultivation.

A final remark concerns the terminology that has been compared on Table 5.1. From left to right, one may observe that the biostatic phase and its equivalents are seen as beginning earlier in stand development. Although this depends much on the forest type concerned, as a rule of thumb it holds true because the further one goes to the right side of the table, the more these concepts include economic aspects, and economic maturity occurs earlier than ecological maturity. One should realise that a later silvicultural maturity means a different end-use of the wood: trees with higher diameters yield high-quality sawnwood if well tended, but pulpwood plantations reach their economic maturity very early (also cf. Sect. 3.6.1, end).

5.3 The Life Cycle of Forest Eco-Units: Root Compartment

The development of the subterranean part of an eco-unit deserves a special section. This is one of the most difficult subjects to tackle, because up to now it has been technically impossible to draw rhizosphere transects. Of course, the roots of a whole stand have been washed out with the help of high-pressure fire hoses (Zimmermann and Brown 1971; Schmidt in prep.; Fig. 3.16C). This yields an image that has lost everything but the most general traits, because most of the fine roots and all other rhizosphere components are lost. Klinge (1973a, b) described the difficulties met in tearing out the roots in an Amazonian rainforest for biomass studies. And this only concerned their weight! The underground structure was disregarded. The only document known to us to date concerning the development of the root compartment in an eco-unit was produced by Kahn (1983). It is a diagrammatic and generalised reconstruction of the root-spaces in a humid tropical forest eco-unit of unspecified characteristics (Fig. 5.17). Compared with the structural detail of the forest above the ground, the paucity of information in the subterranean compartment is striking.

In order to gain some insight into the architecture of the rhizosphere in an eco-unit, we will proceed along two lines. Figure 5.17 has one outstanding feature: roots move in and towards all zones where they can penetrate and find air, moisture and

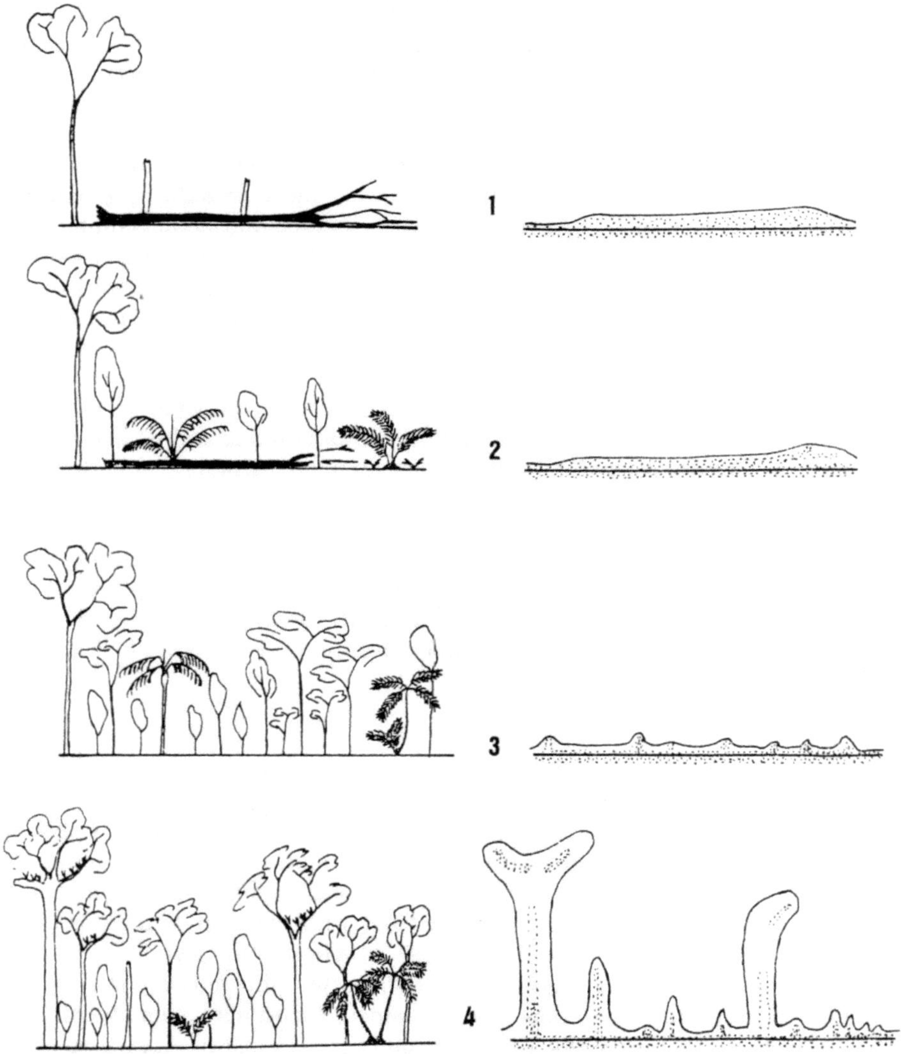

Fig. 5.17. Rhizosphere development during eco-unit development in the humid tropics. *1* Chablis, the roots (to the *right*) cover the fallen tree, soil, litter; *2* innovation phase, roots invading rotten wood and propagule bank mobilised; *3* aggradation phase, rotten wood in elimination, soil rhizospere somewhat thicker; *4* biostatic phase, no underground change, but roots invade aerial niches (rotten trunks, suspended soils etc.). Kahn (1983), from whom this figure is taken, does not use the same terminology

nutrients. Hence, in the following pages some general aspects of forest soil profiles will be combined with the individual root strategies that can be found back in the Chapters 3 and 4 for most forest plant components. This includes plants that are not only rooted below the forest floor, but also, for instance, in suspended soils (epiphytes) or in dead wood (Fig. 5.17.4).

5.3.1 Architecture and Horizons

Figure 5.18 represents a synoptic view by Kahn (1983), derived from data gathered by Huttel (in Lemée 1975) in the Ivory Coast. This concerns the biomass of macrorhizae (thick roots) and brachyrhizae (thin roots, see Fig. 3.15C) against depth in the soil. This distribution is variable according to the sites, from one forest to another and from one eco-unit to another, but all comparable curves in literature agree that more than half of total root biomass is in the upper 30 cm of the soil profile (cf. Hinckley et al. in Kozlowski 1981, p. 165). Sometimes there are additional root biomass maxima at greater depths. Figure 5.19 gives an example of three leguminous tree species from the Californian desert that would yield such curves, because they also expanded their roots deep down. This is often linked to groundwater; but in all forests, the superficial root zone remains very important.

Therefore this zone is the first one that will be presented here. Most often, one can consider it as a transition between the abiotic and the biotic part of the eco-unit. The soil here is rich in organic matter. For a century, this has been analysed according to the theories of P. Müller (1887 ex Babel in Ellenberg 1971; see also Armson 1977; or Houtzagers 1956). The following soil layers are recognised in this superficial part of the soil (Fig. 5.20):

- *L-layer*: **Litter** of freshly fallen dead parts of plants and animals, to which small animals (insects!) sometimes contribute more than large animals. This layer may be subdivided for finer studies. The subdivisions are not treated here.
- *F-layer*: **Fermentation** layer in which the origin of the plant residues is still easily recognised with the naked eye. This layer is also subdivided if a finer analysis is to be made.
- *H-layer*: **Humus** substance layer where residues cannot be identified any longer. Organic matter, finely fragmented, may be mixed with up to 50 vol.% of mineral matter. This layer again can be subdivided.

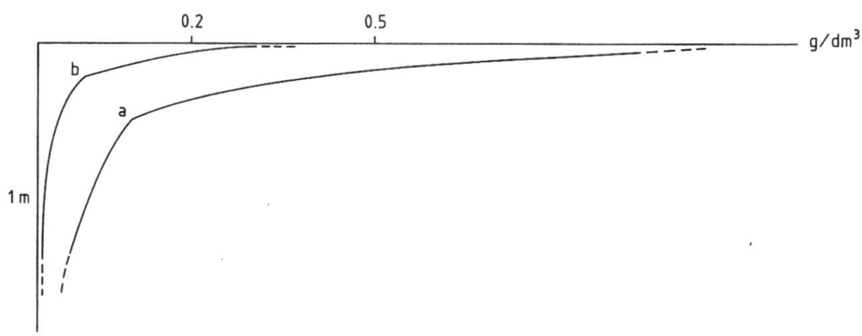

Fig. 5.18. Vertical root distribution by diameter, grams of dry matter per dm^3 soil. The shape of this curve varies from site to site, from eco-unit to eco-unit, and from development phase to development phase. Often there is a second maximum deeper down (see Fig. 5.19). *a* Root diameter over 10 mm; *b* root diameter under 2 mm. (Taken from Kahn 1983, who established it after earlier data from Huttel)

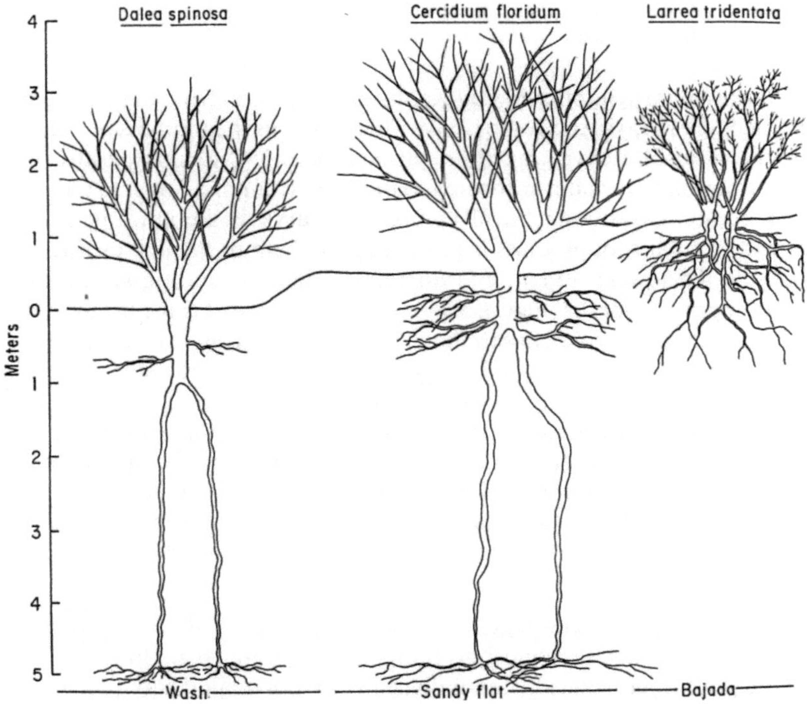

Fig. 5.19. Roots of three legume species in the Sonoran Desert (California). *Dalea spinosa* has root nodules and fixes nitrogen, *Cercidium floridum* does not. Note that roots grow deeper than tree height aboveground in these inhospitable sites. There is a second maximum of root biomass deep down (see Fig. 5.18). (Courtesy P.W. Rundel, from Sharifi and Rundel 1986)

- *Ah-layer*: part of the mineral soil horizon enriched in humus, i.e., clearly coloured by humus;
- *A-horizon*: **mineral** horizon without traces of humus (but not necessarily without organic matter, cf. De Bakker and Schelling 1966);
- *B-horizon*: **weathered mineral** horizon where substances are washed in and accumulate, e.g. organic matter (cf. podzols) and/or metals such as iron or aluminium, that are immobilised in an insoluble form (cf. red tropical soils).

The processes that lead to this pattern, in which layers may be lacking, are the fall of litter, its fragmentation by animals such as earthworms, termites, millipedes or other arthropodes, its humification by micro-organisms among which the fungi are important, and finally its mineralisation into H_2O, CO_2 and inorganic compounds that are often ionised. These processes do not always occur in this order. If humification occurs fast, for instance, there scarcely is any F-layer. Humification is accompanied by the formation of humic acids and mineralisation, which may be partly or wholly parallel to humification and result in ionised substances and protons. With Ulrich (e.g. in Ulrich and Pankrath 1983) one may therefore say that mineralisation is acidifying the soil. On the contrary, the production of phytomass causes such substances to become assimilated in plants. This desacidifies the soil. In a simple formula, Ulrich presents these interactions as follows:

$$CO_2 + H_2O + X^+Y \quad \underset{\text{mineralisation}}{\overset{\text{phytomass production}}{\underset{\longleftarrow}{\overset{\longrightarrow}{\text{energy}}}}} \quad CH_2OX^+Y^- + O_2.$$

Mull humus. Litter decomposition fast: no F- or H-horizons, in some seasons all litter gone (temperature regions). Ah with well-mixed organic matter and minerals, often in granular (korrel-) structure. Least acid, with burrowing animals (e.g. earthworms).

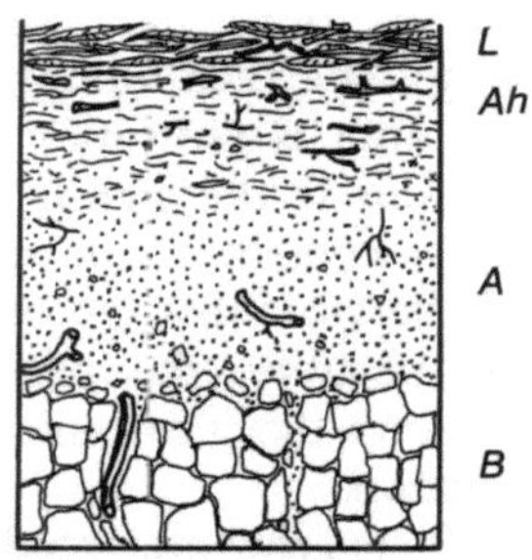

Moder humus. Litter decomposition slower. Stages in decomposition permanently present in thin F- and H-horizons. Shallow mixing in Ah-horizon, by mechanical action (mites, collemboles). Acid humus.

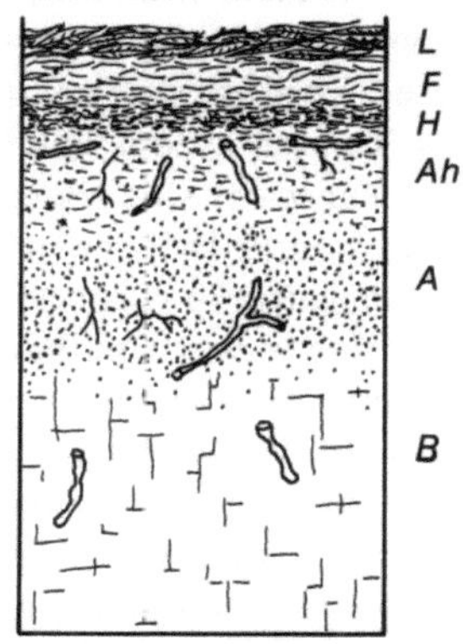

Mor humus. Fresh litter (L) over partly decomposed but still recognisable plant parts (F), often with clearly visible white strands of hyphae (fungi). H-horizon with finely fragmented organic matter that is sharply separated from the A-horizon. Animal action (mixing) is insignificant. This is the most acid humus.

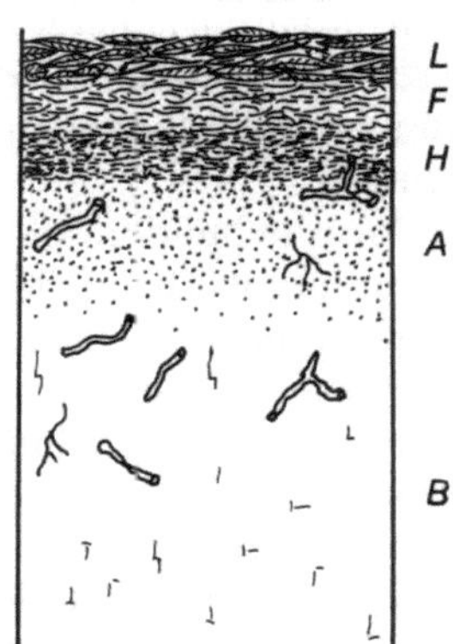

Fig. 5.20. The three main patterns due to humus formation. Meaning of the letters for soil horizons: see text. (Redrawn after different sources. With thanks to J.J. Westra for screening)

Destabilisation of these relations plays a large part in forest dieback due to "acid precipitation" or other acidifying influences from outside.

5.3.2 Root Occupation Patterns

Figure 5.20 shows the general patterns in temperate region humus profiles. They will not be discussed further here except to stipulate that usually mull humus is considered to be optimal under forest, because nutrients are recycled rapidly, biological soil activity is high, and site quality in general is enhanced. Mor humus is considered to offer the least possibilities for silviculture. For the tropics, most soil manuals (e.g. Young 1976; Doley in Kozlowski 1981) do not cite such humus profiles. However, Burnham (in Whitmore 1975) uses the same classification of Fig. 5.20 in tropical Asian rainforests and states that, due to the high level of biotic activity, the litter layer is thin and organic matter content of the soil is rather low, at least in well-drained situations. It will be clear that, according to different climates and differences in geological substrate, the basic patterns of Fig. 5.20 show numerous variations. However, they form an extremely useful reference framework to understand such variations.

Roots grow preferentially in those layers that are rich in organic matter. This explains the fact that most root biomass is found so superficially. Furthermore, these layers are well aerated, so that root respiration can proceed in an optimal way. This is the zone in which long and sturdy diatropic macrorhizae are formed, that bear many side roots with brachyrhizal absorption complexes (Fig. 5.21). If the L, F, and H layers are thick and increase in thickness, the horizontal roots form negatropic roots dorsally, that penetrate the higher organic layers to bring absorbing roots (with their mycorrhizae!) in this freshly formed and well-aerated environment. In tropical tree-savannas and semi-arid woodlands, Doley (in Kozlowski 1981) gives examples of lateral roots extending up to 18 m from the bases of trees that are relatively small when compared with those in tropical rainforests. In the white-sand forests of northern Suriname, De Graaf in the field showed us diatropic tree root systems just under the thin, loose litter layer and resting directly upon the barren mineral soil. Figure 3.14 shows the importance of lateral, horizontal root penetration of "Kerangas forest" trees in Kalimantan (cf. Whitmore 1975; Brünig 1974) into the thick organic peat layer covering a deep, pure white sand horizon.

Now the build-up of the horizon of organic matter in the topsoil is an eco-unit property. This is demonstrated, for instance, by the data on litter fall and composition of surface litter in plantations of *Pinus sylvestris* over 50 years, published by Ovington (1965). It should never be considered as the equivalent of the gross macroclimatic sunlight and precipitation that finally are the factors in response to which eco-units form upper structural ensembles, i.e. canopies. The formation of the superficial root-mat by diatropic roots and brachyrhizae is an **interior** feature of the eco-unit, that also depends on the "organic heritage" from previous eco-units, left on the surface where a new one starts developing. The nature of the zero-event determines this heritage, as stated earlier.

On the contrary, there are superficial horizons that are enriched in nutrients and aerated by abiotic events. The most common of these is sedimentation, as in mangroves on sea-coasts or in river deltas (Tomlinson 1986) and other periodically flooded forest. Most often, aeration is insufficient in these environments. Roots are formed essentially following the same diatropic strategy as in organic topsoil, but here one often finds pneumatophores (Figs. 3.15D, 3.16B) that offer both pneumathodes (air-ways) at the top and concentrations of brachyrhizae in the most recent and nutrient-rich soil layer. Trees with stilt-roots, like *Rhizophora* spp. in the tropics, send positropic stilts downwards into the mud, and form an absorption complex at the end. Pneumathodes are then localised on the aboveground part of the stilt-root (cf. Oldeman 1971). Many of the root system types

Fig. 5.21. Maximum root biomass in superficial, organic soil layer in tropical rainforest. Note numerous diatropic roots. Aboveground forest architecture unknown. Okoumé zone near Ndjolé, Gabon, 1980. (Photograph by R. A. A. Oldeman)

described by Jeník (1978 in Tomlinson and Zimmermann) express such root strategies.

In silvicultural design, both cases should be clearly distinguished, because they require different treatments. The first case can be modified by intervention in the eco-unit, the second case depends on maintenance of outside influences, such as the hydrological regimen, that have to be kept intact if the forest is to survive.

Stilt-root strategies may also occur when the top-soil structure is determined by the organic processes in the eco-unit. Examples are to be found in the tropical rainforest: the palms of the genera *Iriartea* and *Socratea* (tropical America) are such forest components, but there are also many hardwood trees forming stilt-roots. A special place is taken by trees that germinate above the ground on dead logs ("seed-logs"). When these trees grow up, their diatropic roots become positropic when they reach the sides of the log and then go down into the soil. When the log has rotted away, the tree keeps standing on stilts. Examples are birches (*Betula* sp.pl.) in North America (Fig. 5.22A) or *Cecropia* species (Mo-

Fig. 5.22A

Fig. 5.22. Roots and decaying trees. **A** *Betula alleghaniensis* (Betulaceae, yellow birch), has germinated in a natural forest on a fallen trunk, that now has nearly decomposed. The birch rests on stilt roots that once spanned that trunk. Joyce Kilmer Memorial Forest, North Carolina, USA 1986. (Photograph by R. A. A. Oldeman). **B** *Fagus grandifolia* (Fagaceae, American beech), that has consumed the dead parts of its own trunk with inside roots that continue into the soil (cf. Fig. 3.23D). Transversal cut through the trunk base. Mount Mitchell, North Carolina, 1986. (Photograph by M. V. M. Maas)

raceae) that form pioneer eco-units in the neotropics (De Graaf pers. commun. 1981).

The preference of roots for organic soil is very apparent in Fig. 5.17. Figure 5.22B gives the example of adventitious roots that were "cannibalising" the rotten core of the trunk of their own tree before penetrating into the soil at the centre of the hollow trunk (*Fagus grandifolia*, American beech, Fagaceae). An analogous observation was made by us in Dutch *Betula verrucosa* of about 80 years old, cannibalising its own dead trunk parts at the tree base by negatropic roots. In the tropics, the phenomenon has been observed, too.

Two more cases are to be briefly mentioned. Dead logs, dead fallen wood in general and "snags", are invaded by tree and other plant roots as soon as they are sufficiently rotten by fungi (cf. Oldeman in Cosijn et al. 1983). The roots first penetrate into the soft parts of such logs, e. g. between the bark and the wood. This invasion of rotting wood by roots (Fig. 5.23A) is in principle no different from penetration in dead wood inside a living tree, mentioned above.

In regions with litter and humus accumulation, that may lead to peat formation, these thick packages of organic matter are generally occupied by many roots. The root systems then are not concentrated in diatropic, superficial structures, but go down much deeper with positropic root formation. In such zones as the Andean páramos, which are covered by a vegetation that always contains woody plants but sometimes no trees (Cuatrecasas 1977; Cleef 1981), even the aerial axes of the plants may specialise to live under the ground, as exemplified by the "cryptofrutex" of Leeuwenberg's model (Fig. 3.2) in Fig. 5.23B.

Finally, according to Ulrich (in Ulrich and Pankrath 1983), the distribution of organic matter in the soil changes with destabilisation due to acidifying influences. With the input of acids the buffering soil substances, such as clay minerals, are destroyed in the process. The resulting soil structure in the A and B horizons (cf. Fig. 5.20) is one of soil aggregates in which toxic metals are mobilised in a soluble form that can be absorbed by plants. Between these aggregates, humus is washed in. Plant roots avoid the toxic aggregates, and become concentrated in the spaces between them, where the organic matter is. The destruction of the chemical buffer characterises the beginning of the first of Ulrich's destabilisation phases, whereas the resulting state of the soil allows a second, but unstable balanced state of a forest. Such processes may easily lead to weakening of an eco-unit, the trees of which then become vulnerable to abiotic and biotic decay, as described in the pages on the degradation phase. The symptoms have been explained in Fig. 3.33.

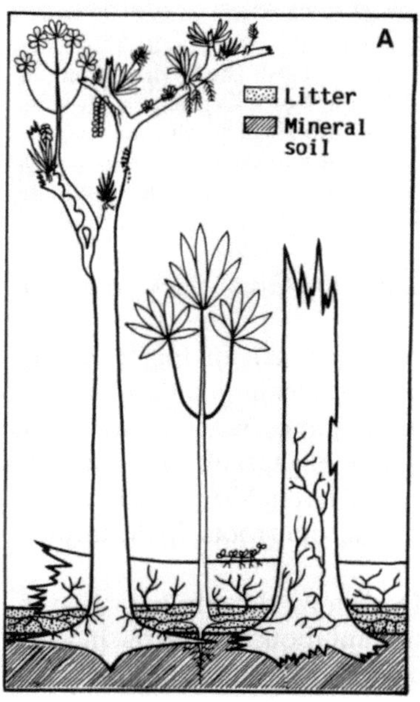

Fig. 5.23A

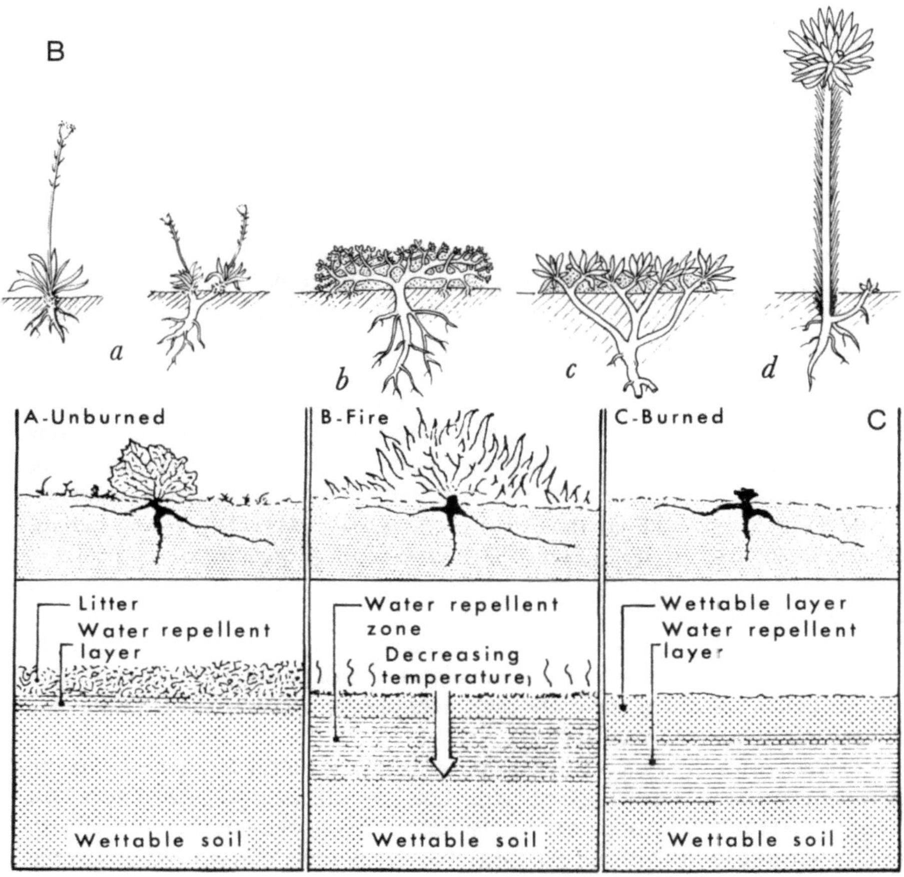

Fig. 5.23. Substrate, soil layers and root penetration patterns. **A** Tree and epiphyte roots in fallen or standing dead trunks and on a decaying tree. (Oldeman 1974a). **B** Root systems and biological types in deep organic peatlike soils of the páramos in the Andes: *a* "rosula"; *b* and *c* "cryptofrutex" = hidden shrub; *d* "caulirosula" = rosette plant with stem. (Cuatrecasas 1968). **C** By burning chaparral shrub vegetation in California, the cost is the burnt litter and the benefit is the lowering of a water-repellent layer so that a deeper horizon becomes available for roots: the root-floor is lowered. (DeBano et al. 1979)

The examples of the páramo and the polluted forests of Europe lead us to the second set of characteristics that determine the architecture of the rhizosphere in an eco-unit. They concern the vertical penetration of roots and most often depend on water conditions and the density of the soil horizons.

5.3.3 Root Floors

Root floors is the term coined by Oldeman (1980) to indicate all **horizontally structured subterranean features that prevent deeper development of root systems.**

A root floor does not have to be impenetrable to each and every individual root, but blocks the architectural deployment of the whole root system as such. Root floors can show fissures and holes, through which some individual roots can pass. If there is a better root environment below, a new part of the root system may be reiterated there, as is often the case in semi-arid regions (cf. Fig. 5.19). Our field observations on *Acacia* species in Kenya and Ecuador, when part of their root system had been washed clean by erosion, indicate that these trees have one developed part of the root system in the superficial, organic horizons, linked by strong positropic roots of many meters long to another, diatropic part that developed close to the groundwater. This agrees with the cases mentioned by Doley (in Kozlowski 1981), and with personal observations (1957) of denuded root systems of *Pinus sylvestris* that had grown on blown sands of the Veluwe (Netherlands) and formed a taproot of over 10 m long.

Two functions are often attributed to positropic roots in the strategy of subterranean occupation of space by trees. The first function is anchoring, the second is water retrieval from deeper layers. These notions are certainly too general, and have to be nuanced. In the first place, root architecture is as different as aboveground architecture, although there are not enough data as yet to make an analytical classification (see Sect. 3.3). Hence it depends on the plant species whether it has the power to form positropic roots or not. If not, the two functions mentioned above are ensured by other, diatropic roots. Second, water retrieval from deep soil layers depends on the aeration of the environment. Without respiring, roots cannot function. And third, the water uptake usually is at least partly linked to symbiosis between plants and fungi (mycorrhizae), whereas the necessary microflora does not necessarily exist at great depth.

In West Africa, loggers tell the story of a bulldozer, blade pushed into the ground at some 15 m from a big forest tree, that shoves the standing tree before it on its superficial root plate without taproots below. This story is maybe exaggerated, but it might well be true in its principle. Trees do not absolutely need positropic roots for anchoring. But the story indicates another frequent factor in the humid tropics. This is the presence of a hard, dense, impermeable layer at some decimeters (1 dm = 10 cm) depth that acts as a root floor. Experiments done by Dr. L.K. Wiersum (Westra pers. commun. 1978) prove that at least one of the reasons for such layers to act as root floors is the very small pore diameter. Wiersum grew plants in artificial substrates of small glass beads welded together. In each experiment, the beads had different diameters and hence the pores between them had different sizes. For every plant species studied, there appeared to be a minimum accessible diameter. If pores were smaller, root tips could not enter them.

One frequently occurring soil type with a shallow root floor in the temperate regions is called podzol. Many afforestation projects have been implemented on such soils, which are less useful for agriculture. The A-horizon (cf. Fig. 5.20) of a podzol generally shows very poor and lightly coloured sand, whereas the B-horizon is dense and contains accumulated metals (iron!) and organic matter. The organic matter does not necessarily come from the vegetation in the recent past, but may have accumulated from other sources in the geological past (De Bakker and Schelling 1966). But theories on podzolisation have changed over the years. Tropical soils that resemble this profile are sometimes also indicated as podzols.

Because forest researchers and foresters will often encounter podzolic soils, it is useful to understand that tree roots seldom penetrate the B-horizon. The species mentioned above as having a very deep taproot in blown sands, *Pinus sylvestris*, has a horizontally expanded root system with few, stunted positropic roots on Dutch podzols. At the level of the eco-unit, this influences the number of trees that can grow well per hectare. If depth is limiting, each tree needs more surface. A spectacular example of the effect of a root floor is given by Blancaneaux (1973) for French Guyana (Fig. 5.24). When chablis are formed, the whole root clod slips away over the root floor, here a subhorizon of the B-horizon, called B_{22}. The resulting hole in the ground is very persistent, so much so as to deserve a Creole name, "jugung-paytay" (djougoung-pété). In this hole, water accumulates and only a sparse monocot vegetation (Rapateaceae) is established, without tree seedlings. In the long term, this influences eco-unit development over extended surfaces on such sites. For European examples on another soil type with less pronounced root floors, and with a gradual reconstitution of the upper horizons in the root-holes see Koop (1981a). In general, **a soil horizon is more liable to function as a root floor if its structure contrasts more sharply with that of the horizon above (Westra's rule I,** unpublished).

The second determinant for root-floors is the groundwater-table. Under the groundwater-table the soil has no aeration and hence is inaccessible to the roots of most species. The water-table may be aboveground, at least part of the time, as in mangrove forests and some willow formations (*Saliceta*) near rivers in temperate regions. If it is permanently aboveground it forms an aquatic environment that is irrelevant for our subject matter. Most generally it is the groundwater-table that is considered in forests. In all cases when the water level is at or above the level of the ground, root systems have some breathing device (pneumathodes, pneumatophores, Fig. 3.15) or may conceivably breathe through symbiotically associated organisms, such as root bacteria in alder (*Alnus* sp. pl., Betulaceae) in temperate regions and tropical mountains.

With a descending water-table, the soil layers may remain saturated with water for a long time. In Dutch polders, this kind of drainage, leading to a permanent lowering of the groundwater, is called "ontwatering", a term that will be used here. Species like alders and willows (*Alnus* sp.pl, *Salix* sp. pl.) have been discovered to contribute with their root functioning to the ripening of these ontwatered soils. Ripening is the irreversible extraction of water and the admission of gases, leading to one of the types of soil structure found in non-inundated soils. This is by no means an exceptional and artificial situation. In many natural environments, "verlanding" (= land formation) takes place. A British example of this, in Shropshire, is given by Cousens (1974, his Fig. 3.1) under the name "hydrosere". An example of the traces of tropical verlanding is shown in Fig. 5.25, representing the cylinder of ripened soil around a hole that has contained the root of such a plant in naturally verlanding marshes near Kourou (French Guyana). The Dutch polder vocabulary is used here, because it contains specific terms for such situations. Studies on ontwatering, verlanding processes and forests are receiving increased interest in Europe (see Fanta 1986).

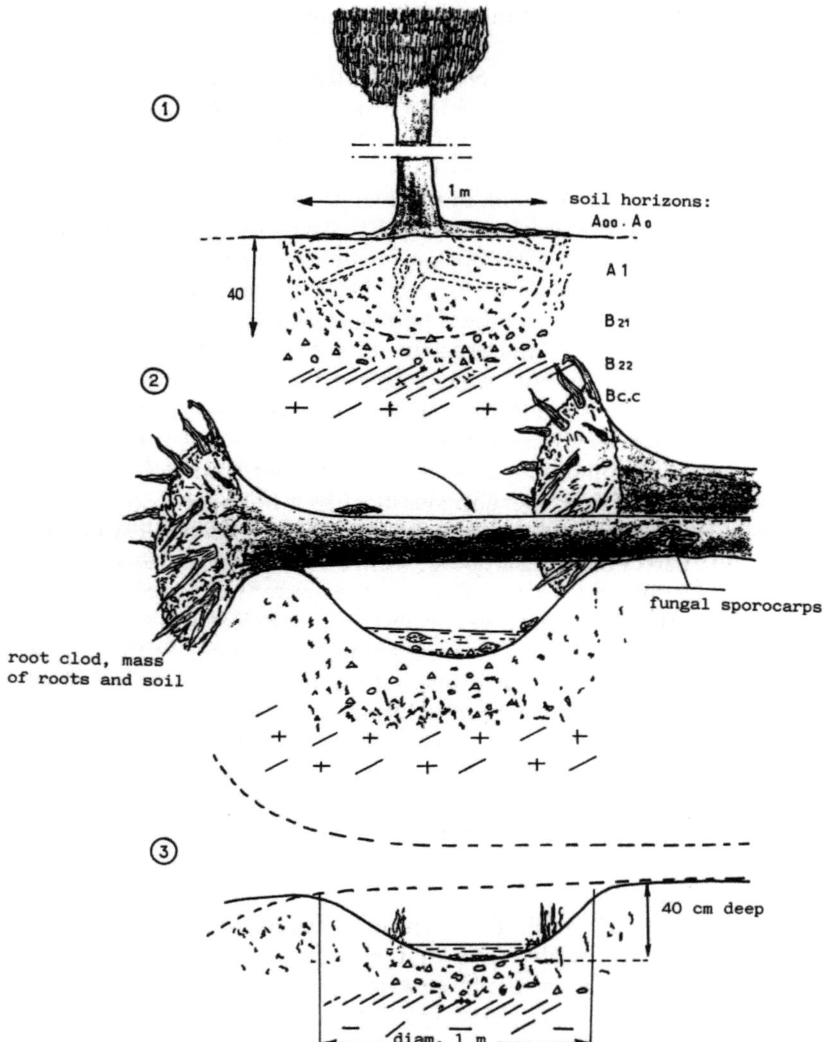

Fig. 5.24. Example of a root-floor effect in French Guyana. Soil horizons numbered as *A*- and *B*-subdivisions (cf. Fig. 5.20). The B_{22}-horizon is an impermeable hardpan. If uprooted, the root clod of the tree slips over this floor, leaving a hole that is very persistent, accumulates water and is unsuitable for tree germination. Only some monocots (Rapateaceae) become established. The Creole name for such holes is "djougoung-pété" ("jugung-paytay"). *1*, *2* and *3* steps in the process. (Blancaneaux 1973)

Fig. 5.25. Traces of the ripening of marshy soil by root activity. The hollow cylinder is ripened soil. In the cylinder, there once was the root of a plant. Around the cylinder: unripened and badly structured soil. Kourou, French Guyana, 1968. (Photograph by R. A. A. Oldeman)

Indeed, water-tables are rarely constant unless kept so by hydrological engineering. They form moving root floors, that are otherwise indispensable for many forest plant species, because these need the water ascending from the groundwater-table through the soil capillaries. Schmidt (1977) experimented with European poplar clones (*Populus*) in controlled soil environments within containers. He could explain some unexpected growth patterns as follows (Schmidt p. 159): "The roots find themselves in a container of 110 liters volume and a diameter of 44 cm. The groundwater-table is maintained at about 17 cm above the bottom. The capillary ascension in the sand is about 7 cm. This means that in only one layer of some 10 cm deep there is the right mixture of soil particles (sand), water (i.e. nutrient solution) and air that is needed for root growth and function. In this case only about 15 liters of soil, one seventh of the container volume, was suited to root growth and only this layer showed dense root occupation. Poplars do not form roots below the water-table (...) and above the dense root layer there were only a few very thick roots." (free translation from the German, RAAO).

An example of the friend-enemy relationship between groundwater and roots has been provided by Derkman (1979). He studied the influence of ontwatering of non-polder soils in the Netherlands by the pumping of drinking water on the survival and mortality of beech (*Fagus sylvatica*), after two particularly dry summers. For a more thorough, broad and theoretical treatment of these questions the reader is referred to Kozlowski (1981). The architecture of the rhizosphere in its lower parts largely depends on these relationships between water-table, capillary water and architectural root potential (cf. Fig. 5.19).

The combination of water-table, precipitation and soil profile leads to still another set of conditions that determine rhizosphere architecture at the eco-unit level. The interaction between soil and water in soil formation (pedogenesis) will be left out here, and soil structure will be considered as a given constant over the period considered.

Drying out of the soil may quite drastically affect the soil volume that is accessible to roots. One factor implied is the "wettability" of such soils as contain water-repellent layers, where roots are handicapped because infiltration rates can be 25 times slower than in a similar wettable soil. This has been studied by DeBano et al. (1979) in the Californian chaparral. Forest fires and controlled burning considerably change the spatial proportions between wettable and repellent soil horizons (Fig. 5.23C), at the cost of the litter layer that is burned off. Such phenomena are very important for the evaluation of the consequences of zero-events in these kinds of forests.

Wetting the soil excessively causes changes in soil cohesion to occur between soil particles. In simpler terms, soil turns into mud in extreme conditions. In many climates, however, such conditions occur with a predictable regularity during the rainy seasons. Rain may operate directly in very flat terrain with some kind of microrelief (cf. Raad 1981, in Sudan). The effects are more often indirect and are linked to the rivers and creeks. These fill up during the rainy season and temporarily inundate valleys and larger or smaller flood-plains, such as the Brazilian varzea in the Amazon basin, or the German Auen along the Rhine and Danube. Similar sites can be found on large surfaces along most large river systems, from the American Mississippi to the Siberian Jenissei or the African Zaire river.

Zero-events in these regions with soils showing fluctuating cohesion in a predictable way are largely determined by these fluctuations. If minimal soil cohesion is very weak, this leads to a marked increase in frequency of such events because uprooting is facilitated. Eco-units then do not become very old. Figure 5.26 gives two examples of such forests, one in the humid tropics and the other in Europe.

Fig. 5.26. Examples of forest on soils with fluctuating cohesion, **A** "Pri-pri" (French Creole) or "Várzea" (Brazilian); note the scarcity of high trees (*Triplaris surinamensis*; Polygonaceae) and the abundance of low vegetation, often *Tabernaemontana* shrubs (Apocynaceae), translating a high frequency of zero-events in the wet season when soil cohesion is minimal. French Guyana, Crique Calebasse, 1968. (Photograph by R. A. A. Oldeman). **B** Salicetum (willow forest) built as a large pioneer eco-unit on soils with fluctuating cohesion along the Rhine in France. Canalisation of the Rhine has fixed this soil and this forest, so that biostasis is now reached over a large surface. Marckolsheim, Alsace, France, 1979. (Photograph by J.-M. Walter)

Fig. 5.26

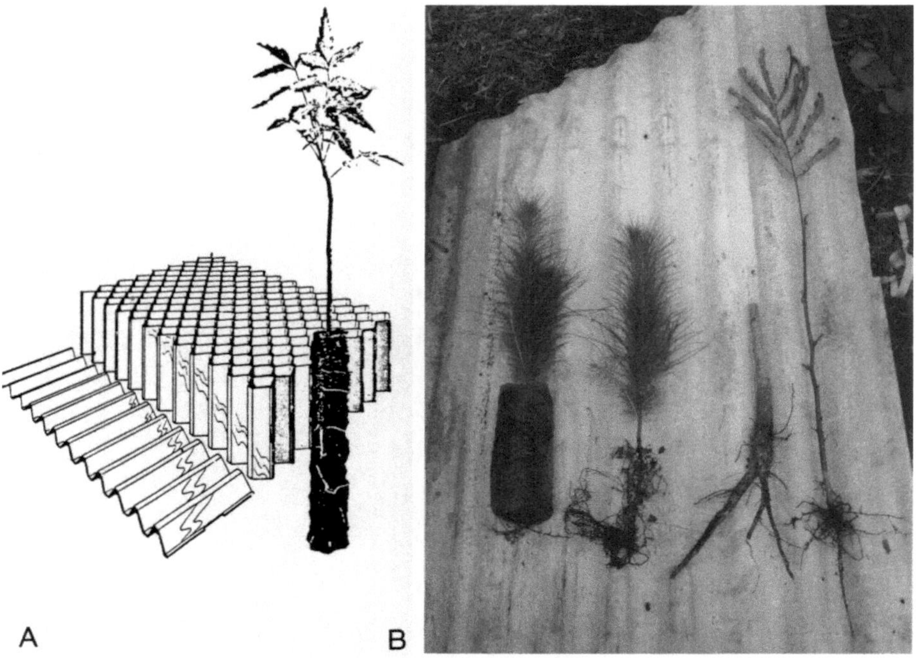

Fig. 5.27. Tools causing root-floor effects and their consequences. **A** Winstrips show air-pruning of roots below the soil-and-root cylinder in each vertical hole, and also air-pruning if roots penetrate between the corrugated sheets. (Courtesy Van Niftrik Industries). **B** Four different root systems from a nursery in Costa Rica, *from left to right* roots from plastic bag, note torsion of taproot (*Pinus caribaea* var. *hondurensis*), naked root seedling, also note root torsion caused by plastic bag (*P. oocarpa*), "stump" with pruned stem and roots (*Tectona grandis*, teak) and "stripling" pruned aboveground only (*Schizolobium parahybum*, a legume with root nodules). (Photograph by D. Lemckert)

Contact with dry air also blocks root growth. This property is used in devices for "air-pruning" of seedling roots, e.g. in plant containers of the winstrip type. These containers consist of strong plastic sheets that resemble corrugated iron, bound against each other in several layers so that the vertical cylinders that are formed can be filled with soil and seeded (Fig. 5.27A). They have no bottom. The taproot, that in other containers (e.g. plastic bags, Fig. 5.27B) spirals against the bottom and rarely if ever gives rise to a new, reiterated taproot after planting, ceases growth in winstrip containers when it meets the air below and resumes growth when planted out. Plant containers in general are the most extreme root floors that exist. If left too long, the root system that has grown in them never recovers completely and is retarded in occupying the surrounding soil volume after planting. This is accentuated with the use of a planting auger, because this tool more or less compacts the walls of the planting hole, and compaction creates a "vertical root floor". Van den Tweel (1980) studied Douglas fir that barely had roots outside the planting holes 3 years after planting.

A classical error in tree planting, when holes are made by seesawing a spade in the soil, is the spade being allowed to turn around a line that lies higher than the end of its blade. The top of the spade then makes an "air chamber" that is not visible from above, but that most often remains open under the planted tree seedling. This space filled with air prunes the taproot effectively, and the trees generally keep only a superficial root system throughout their life. Holes made with a spade in this manner are often also too narrow, whereas their walls are compacted. The root system then develops in only one vertical plane, defined by the original position of the spade. If at a later age the tree receives gusts of wind from a direction perpendicular to that plane, it is easily blown over.

Many urban areas today are built on polluted and badly structured soil that has been accumulated and levelled, i.e. compacted with heavy machines and that is covered later by paved road-surfaces in which openings have been left for trees. It may happen that these trees have access to no other soil than their original planting clod during their whole life-cycle. The surrounding environment is dry, compacted and chemically inimical and constitutes in fact a closed root chamber (L.K. Wiersum, P. Schmidt, W.Th.M. Smits, pers. commun.).

5.3.4 The Edaphon and its Compartments

Soil occupation by the roots of an eco-unit follows a strategy that is closely adapted to all described properties of the subterranean environment. A hypothetical image will be sketched later of the developing rhizosphere architecture during the four phases of innovation, aggradation, biostasis and degradation. However, the subterranean part of forest ecosystems, beginning with eco-units, largely depends for its functioning on the other forest components, the soil flora and fauna. Our impression is that these components are more immediately vital to eco-units than the non-arborescent components above the ground. Their role in the interactions in the soil compartment of the eco-unit is so vital as to be indispensable in the short term, whereas above the ground the absence of non-trees has more long-term consequences. An indication of this state of affairs is the fact that European forests have survived long centuries of strange and artificial wildlife and game management (cf. Mayer 1981), whereas two decades of pollution affecting the organisms in the rhizosphere have led to quite extended forest dieback (cf. Smits 1984; Bosch 1985; Termorshuizen and Schaffers 1987).

Edaphon is the name given traditionally to all organisms that live between and with the roots in the rhizosphere. The majority of the species in the edaphon are so small that their action radius does not go beyond one eco-unit, in which they live. That is to say that their penetration into new or other eco-units happens just as in seed plants. Their propagules (spores, eggs) are present in the propagule bank or immigrate there, and are mobilised if and when the eco-unit development is ready for them.

The term **rhizosphere** has been used in this text in order to indicate the **whole subterranean part of the forest,** in other words the subterranean compartment. For researchers that study groups of soil components other than higher plants, this term is used to indicate the roots alone. Still others (e.g. Marks and Kozlowski 1973) consider the environment surrounding the roots as being the rhizosphere.

It is impossible to discuss the other soil organisms without a functional diagram of this part of the forest ecosystem. Figure 5.28 unites some of the illustrations given by Lavelle (1984) in this context. This researcher distinguishes several compartments: the rhizosphere (roots), the drilosphere (earthworms), the fungi (no special name, probably included in the rhizosphere and "microorganisms"), the N-fixers (no special name). The coining of a special term "drilosphere" indicates the emphasis placed upon earthworms by this author. Other authors (e.g. Bongers 1985; Tamis 1986) add the nematodes as important components, but do not create a special compartment named for them. And what about insects, of which only the termites are mentioned in Fig. 5.28? Brauns (1976), PUDOC (1982), or Szujecki (1983, 1987) give them much more weight.

In Section 5.5.10 the notion of "connexus" (Konnex; Brauns 1976) will be discussed as a means to distinguish functional compartments of interacting species in forest ecosystems. A connexus is not limited to a taxonomical group, and therefore seems a more flexible concept than taxon-bound compartments in the analysis of both eco-units and sylvatic mosaics. In Sect. 5.5.6, some attention will be paid to synusiae, groups of organisms showing the same biological type (e.g. epiphytes).

For the present, the diagrams by Lavelle (Fig. 5.28) are sufficient to pinpoint the large groups of subterranean forest components in a functional context. For Europe, a beautiful and accessible book on the ecology of soil organisms has been written by Brown (1978). In her well-illustrated work, many research methods and identification keys for soil organisms can be found. Richards (1974) provides an excellent discussion of regulation and development of the subterranean part in ecosystems.

5.3.4.1 Mycorrhizae

Mycorrhizae are fungi living in symbiosis with plant roots. The term is derived from the Greek μυκος (mycos, toad-stool) and ριζα (rhiza, root).

Figure 5.29A and B shows mycorrhizae that grow on the outside of roots and hence are called **ectomycorrhizae**. Figure 5.29C and D shows fungi that live inside the tree root and its cells. Such combinations are called **endomycorrhizae**. On the roots of Indonesian dipterocarp trees, Smits (in press) found a third type that combines properties of both others, in an organisation type that had not been observed before and which he called **amphimycorrhizae**. The finer distinctions between mycorrhizal types and other terms can be found in the enormous literature on the subject, accumulated over a century. According to Redhead (in Mikola 1980), Treub already described vesicular-arbuscular mycorrhizae (VAM) on sugar-reed roots in

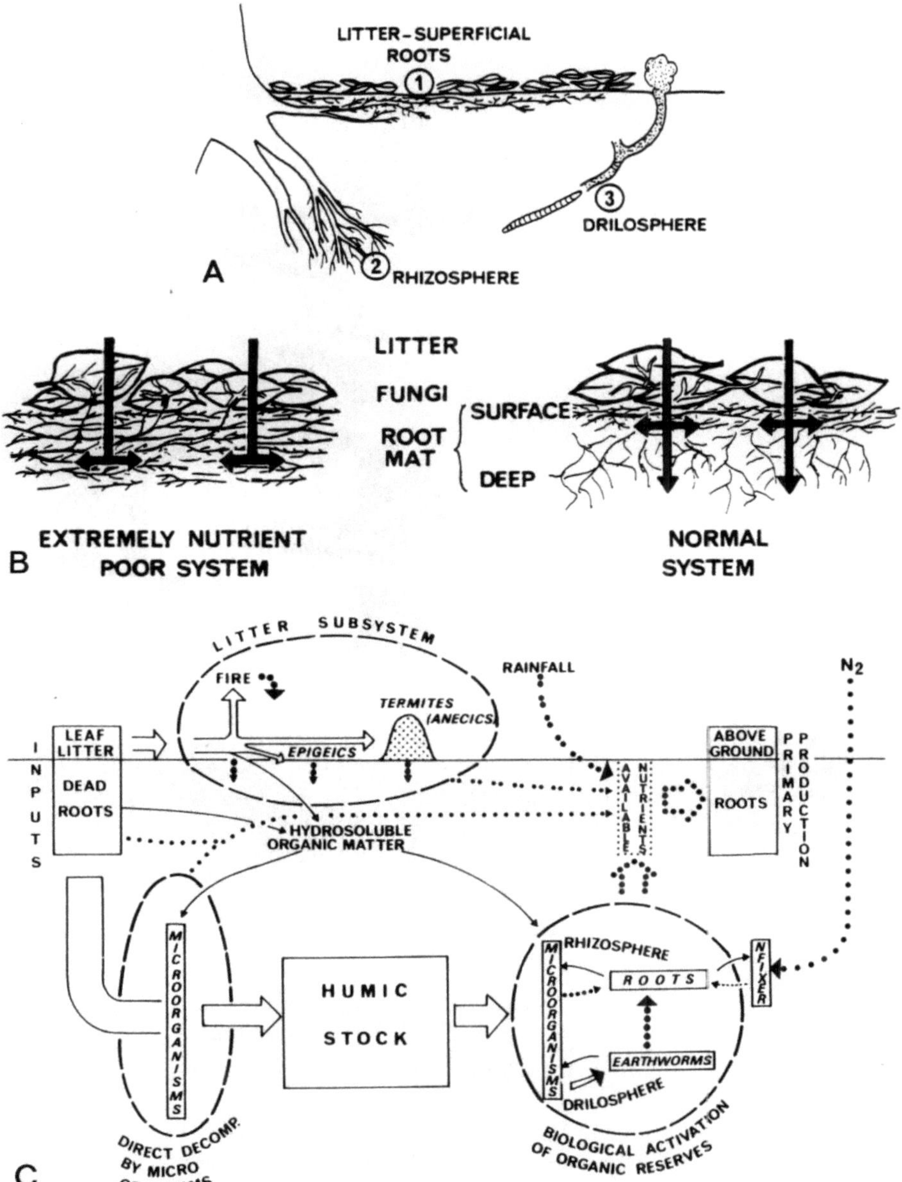

Fig. 5.28. Example of functional diagrams of humid tropical subterranean forest compartments. **A** Three regulating functional subsystems. Drilosphere = earthworm compartment. Note particular meaning of "rhizospere" by this author. **B** The litter-cum-superficial-root subsystem on soils of poor (to the *left*) and average (to the *right*) fertility. **C** Carbon and nutrient cycles in humid tropical soils (also cf. Fig. 5.78). *Solid lines* fluxes and stocks of organic matter. *Dotted lines* nutrients. Note that not all forest soil components are included (see text), and that there are two different cycling mechanisms, at the surface and deeper down. (Model and figures from Lavelle 1984)

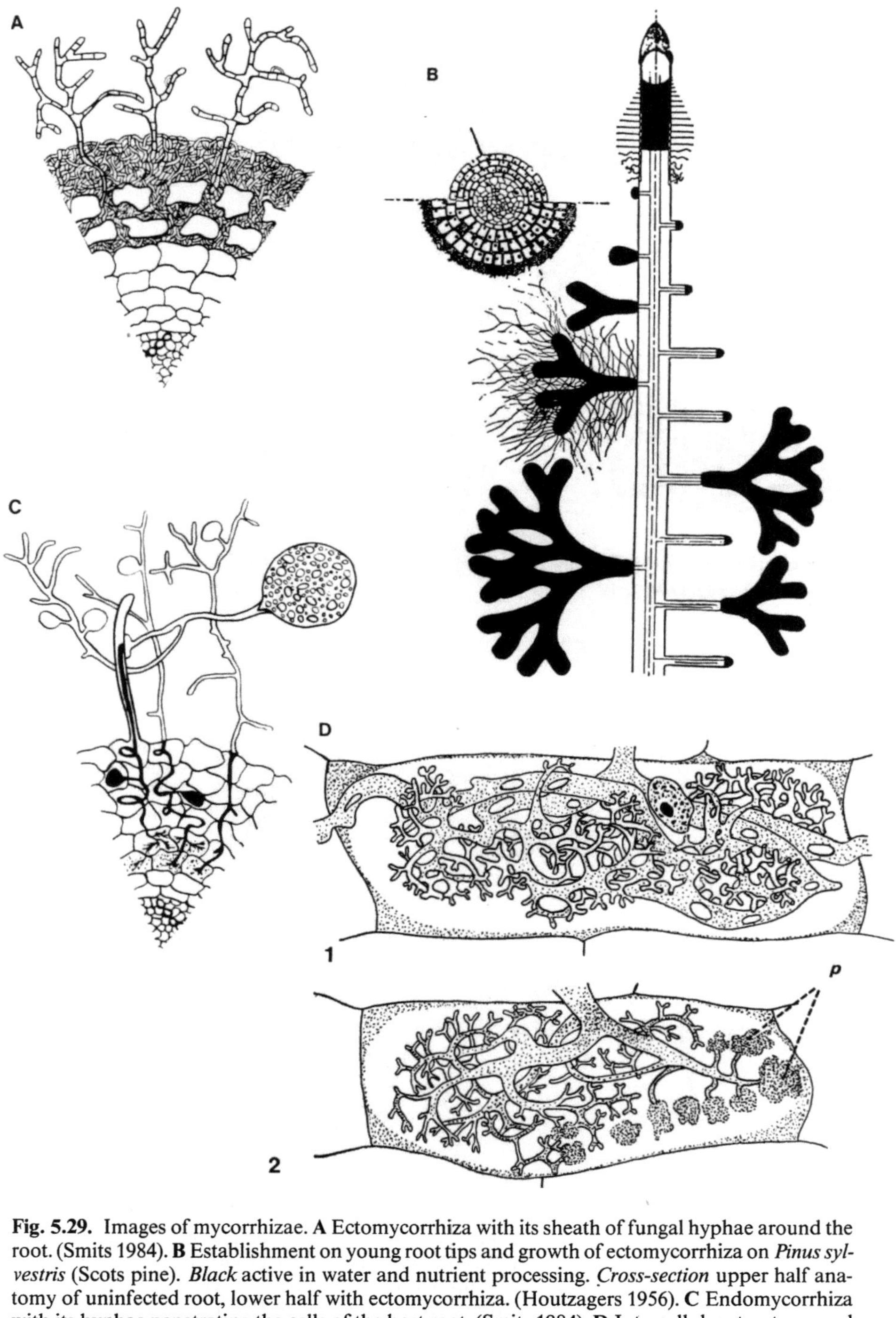

Fig. 5.29. Images of mycorrhizae. **A** Ectomycorrhiza with its sheath of fungal hyphae around the root. (Smits 1984). **B** Establishment on young root tips and growth of ectomycorrhiza on *Pinus sylvestris* (Scots pine). *Black* active in water and nutrient processing. *Cross-section* upper half anatomy of uninfected root, lower half with ectomycorrhiza. (Houtzagers 1956). **C** Endomycorrhiza with its hyphae penetrating the cells of the host root. (Smits 1984). **D** Intracellular structures and their development in endomycorrhizae in *Sequoiadendron giganteum* (D1) and *Allium sphaerocephalum* (D2). (Boullard in Pesson 1974)

Java in 1885, the same year when Frank coined the term "mycorrhizae" and stated that they were natural and not pathogenic.

For the tropics, Mikola (1980) united much useful information in his book, and for the temperate region much information may be found, for example, in Boullard (in Pesson 1974) or Marks and Kozlowski (1973). We now know that about 95% of all land plants have mycorrhizae (Smits 1984).

Mycorrhizal fungi have no chlorophyll, and receive photosynthates from the tree they live on. In return, they provide products and services to that tree. Some important interactions can be enumerated as follows.

- The hyphae of the fungus (cf. Sect. 4.7) increase the water-absorbing root surface manifold and may increase water absorption by trees by a factor of 100% or more.
- In the hyphae, a stock of nutrients is available to the tree, and in the tree there is one for the fungus. Both organisms buffer each other.
- The fungus can mobilise elements in the soil that the tree cannot absorb otherwise.
- The fungus occupies the space in and around the roots that can otherwise be occupied by pathogenic fungi, and so protects the tree.
- Mycorrhizal fungi can produce antibiotics that protect the tree-partner against pathogens. Both spatial and chemical protection mechanisms are important in seedlings.
- The fungus may absorb heavy metals in polluted soils, accumulate and immobilise them in the hyphae so that they do not reach and poison the partner tree (Morselt et al. 1986).
- The fungus often synthesises special nitrogen compounds, such as vitamins or amino acids that the roots do not produce, so that trees of such species show deficiency symptoms if mycorrhizae are lacking, e.g. short internodes or yellowish leaves.

The presence or absence of mycorrhizae in higher plants may be a matter of life and death. Some fungi, such as *Amanita muscaria* (fly-fungus) in the temperate regions, or *Russula* sp. in the humid Asian tropics, may be symbionts of many different tree species. Others, such as the partners of dipterocarp trees in certain regions of East Kalimantan (Smits 1982), or some temperate species mentioned by Mason et al. (in Fanta 1986) are quite specific for one tree species at least on the same soil type.

The importance of mycorrhizae for tree life was discovered accidentally when importing Christmas trees (*Picea abies*) in Australia. This species could not be grown on that continent until soil was imported from Europe with living Christmas trees. This soil contained spores of the right fungus that was introduced in Australia in this involuntary way. In Suriname, the introduction of *Pinus caribaea* from Central America was no success until pines with mycorrhizae were placed in the centre of the nursery beds, as shown in Fig. 5.30. Inoculated trees of *Pinus merkusii* were used in the same way when this species was introduced from the island of Sumatera to Java (Indonesia).

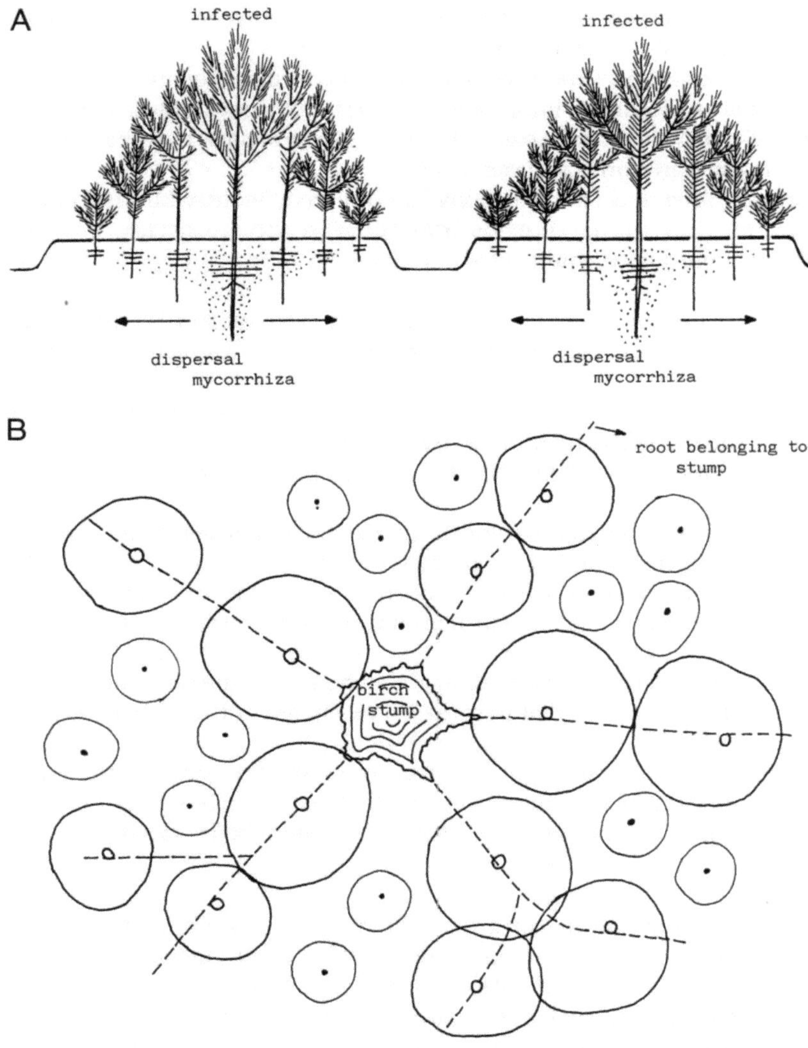

Fig. 5.30. Inoculation and dispersal of mycorrhizae. **A** Pines (*Pinus* sp.var.), if introduced in a region where their fungus partners do not occur, are grown in nursery beds with one tree in the middle that has been imported with its root clod, containing mycorrhizae. **B** Sown *Pinus sylvestris* on a clearcut (large eco-unit), next to a stump of birch (*Betula verrucosa*). The trace of surviving mycorrhizal birch roots can be seen by the distribution of better developed, inoculated pine seedlings. (Drawn after a description by Boullard in Pesson 1974)

Pritchett (1979, p. 173 ff.) describes the condition for optimal development of the symbiont on the tree. There has to be much light, and a not too fertile soil. One may assume that tree photosynthesis under unlimited light quantities produces many assimilates that descend with the sap-stream (Figs. 3.30, 3.32), but that these cannot be used up by the root system alone in a somewhat poorer soil environment. These carbohydrates are exudated by the roots, used by the fungus, which in its turn enhances the performance of rhizosynthesis by providing extra biosynthates to the tree. It is emphasised that this hypothesis may seem logical, but has not been checked experimentally, and therefore is unproven. Another relevant factor is that mycorrhizae become established on young, growing root tips (Fig. 5.29B), of which many can be formed by strongly photosynthesising plants, and which might be more numerous in poor than in rich soils.

For the development of an eco-unit it is important that different fungus species often occur on the same tree species in a sequence: a tree forms mycorrhizae with different fungi during its life cycle (Mason et al. in Fanta 1986). In experiments with Scots pine (*Pinus sylvestris*), young individuals of which were observed in perforons (Fig. 3.18A) by Bosch and Smits (unpubl.), a sequence of 15 different types of mycorrhizae was observed within 1 year. Given the age of the pines, this would place this fast sequence in the innovation phase (Fig. 5.4).

The first development phase in large eco-units on inhospitable sites, such as industrial slag heaps or eroded soils, shows very hardy innovation species in fungi as in other organisms. One example is the "miracle" fungus now commercialised for afforestation of such sites, *Pisolithus tinctorius* (Marx and Artman 1979). Fungi of the genus *Pisolithus* are now widely used all over the world, also in the tropics, to inoculate trees for afforestation of badlands. It should be emphasised that *Pisolithus* sp. clearly belongs to the innovation phase, and that its persistence in later development phases would be a symptom of an unhealthy, poor ecosystem with severe species deficiencies. In normal eco-unit development, the innovation species disappear and are replaced by others, as mentioned earlier.

Finally, mycorrhizae most often prefer acid soils. The white strands of their mycelium can be observed in the F-layer of mor-humus, a humus type for which these strands are a diagnostic symptom (Fig. 5.20). In the acid soils of the tropical rainforest, they occur in the litter, where they may extract more than half of the soluble carbohydrates from freshly fallen leaves in a matter of hours.

The fact that mycorrhizae are a form of symbiosis occurring particularly in the poorer, acid soils makes authors such as Pritchett (1979) or Armson (1977) recommend caution in applying chemical fertilisers in silviculture. The addition of such substances can render the environment unsuitable for the fungi. To a lesser degree, this also applies to nurseries. The cost of the fertiliser in wood production estates should then not only be compared to the benefit of increased yield, but also to the losses due to impaired mycorrhizal functioning.

In forest regeneration, the character of the propagule bank determines the array of available fungus species. Figure 5.30B shows how spontaneous growth of European pines (*Pinus sylvestris*) was enhanced if germination took place above an old birch root, that remained for some time a reservoir of mycorrhizal fungi. On the other hand, Smits (1982) describes the elimination of many fungi from the propagule bank in carelessly exploited dipterocarp forest in Asia. The mycelia and spores die by exposure to high soil temperature (Smits et al. 1987). In both cases, the fungus components of the propagule bank largely determine what kind of eco-unit will develop there. We remind the reader of the statements on lock-and-key mechanisms (Sect. 5.15). If either the tree species or the fungus species is missing, this mechanism does not function, and a different part of the propagule bank is mobilised.

In weakened trees that cannot feed their fungal partner any more, it may happen that the mode of symbiosis changes to parasitism. The fungus then lives at the cost of the tree and helps to weaken it further, until either death or recovery follows. Paraphrasing an American saying, one might state in this case: "no pay, no cure" (Bosch pers. commun.).

5.3.4.2 Other Fungi

Other fungi also play important parts in the life of an eco-unit. Some species are specialised in the decomposition of litter. Such plants are **saprophytes**. Saprophytes always depend for photosynthetic carbohydrates on the dead organic matter that they consume. They have no chlorophyll themselves. Many saprophytes are indeed fungi, but some families of flowering plants contain pale, whitish and diminutive saprophytic species with an analogous way of life. In the tropical American rainforests there are examples in dicots (Gentianaceae) and Monocots (Triuridaceae, Burmanniaceae). Because litter formation is an eco-unit property, the presence and species composition of saprophytes are eco-unit properties, too.

The role of fungi as **pathogens,** i.e. organisms that receive more from their partner than they yield in terms of matter, energy and quality of substances, is very important. There is, however, a gradual and by no means irreversible transition between saprophytism and parasitism (facultative parasites), as there is between mutualism and parasitism (see Sect. 5.3.4.1). An example of a facultative parasite is the cosmopolitan genus *Armillaria* (*A. mellea* is the honey fungus), that may be very frequent in a quite healthy forest and become a root-rot disease in weaker stands. We harvested several good meals of the edible sporocarps of the honey fungus in one healthy stand managed according to "nature-conforming" (Naturgemässe) methods in Schweinsberg, Hessen (see Hasenkamp 1976). But James (1982, p. 124) says about *A. mellea*: "Fungus spreads either by spores or by black underground growths similar to boot laces in appearance, and known as "rhizomorphs". These enter through roots, and form masses of felted white mycelium which frequently kills the tree". A case study of *Armillaria* rot in New Zealand *Pinus radiata* plantations shows the source of the infection to be left-over tree stumps (Van der Pas 1981).

Many handbooks cover these forest components in much more detail (e.g. see PUDOC 1982; Cosijn et al. 1983). In the present context, that of eco-unit development, it is only emphasised that diseases and epidemics depend on the nature of the interactions between components (cf. Sect. 5.6.8). **These interactions are properties of the eco-unit,** not of the participating species that only are the tools of interaction.

In silvicultural design, the balancing of components is too often limited to "species choice", by which is meant the choice of tree species. We propose to include the choice of all other necessary forest components. This could be the means for considerable improvement in silviculture.

An example of extended species choice is given by Gremmen (in PUDOC 1982, p. 58). Another European rot fungus than *A. mellea* mentioned above is *Heterobasidion annosum* (better known under its old name, *Fomes annosus*). *H. annosum* lives on the roots of many tree species, which it weakens so much that death or windthrow occur. Its source of epidemic expansion are the stumps left after thinning or harvesting. Gremmen discovered another, non-pathogenic fungus that occupies the same niche in the stumps, *Peniophora gigantea*, and that does not spread to living roots. Inoculation of stumps with a spore suspension of *P. gigantea* prevents epidemics by *H. annosum* to a large extent. In fact, this means that *P. gigantea* is included in the species choice.

5.3.4.3 Bacteria

Bacteria, especially **nitrogen-fixing species** are very important eco-unit components. Richards (1974, his Chap. 9) introduced the subject in more detail than can be included here. Nitrogen-fixing bacteria that form root-nodules with tree roots are best known from the Leguminosae family. Legumes always associate with *Rhizobium* bacteria. To call a halt to the spreading belief that only legume plants, and all legume plants, form root nodules, we cite Sprent et al. (1986), who reviewed legume symbiosis with *Rhizobium* bacteria and mentioned nodule-less legumes, e.g. *Parkinsonia*, a tree from desert borders often used in semi-arid agroforestry systems. Sharifi and Rundel (1986, see Fig. 5.19) studied the non-nitrogen-fixing *Cercidium floridum* in the Sonoran Desert, California. Furthermore, there is a list of non-legume genera that form root nodules with *Frankia* bacteria (Richards 1974), i.e. *Coriaria* (Coriariaceae, small trees, Andes, Mediterranean), *Cercocarpus, Dryas, Purshia* (Rosaceae), *Comptonia, Gale, Myrica* (Myricaceae), *Alnus* (Betulaceae), *Casuarina* (Casuarinaceae, Australian and Asian pioneer trees now spread world-wide in the tropics), and five shrub genera from the order of Rhamnales.

Nitrogen-fixing pioneer trees are often used for afforestation on degraded soils (National Academy of Sciences 1979). In Costa Rican and other agroforestry systems, *Erythrina* species (Papilionaceae) with root nodules are planted between other crops, such as coffee and cocoa, which then do not need nitrogen fertilisers. In Costa Rica, *Erythrina poeppigiana* is pollarded (cf. Fig. 3.23) twice a year. According to Budowski (pers. commun. 1979) the roots also die back and the newly formed roots produce a new stock of nodules afterwards. The use of alders (*Alnus* sp. pl.) in temperate regions and in the Central and South American mountains as an "auxiliary tree species", planted between commmercial tree species is wide-spread. It has been applied in the afforestation of new Dutch polders, where alder was mixed with poplar (*Populus*) in order to accelerate the ripening of the soil (Fig. 5.25) and to fix atmospheric nitrogen. This technique accelerates and enriches the innovation phase development of such large, artificial eco-units as poplar plantations on virgin soils.

Many of the free-living nitrogen-fixing bacteria belong to genera such as *Azotobacter, Clostridium, Pseudomonas* or *Beijerinckia*. There may be a division of work between different species. Some fix atmospheric nitrogen, others process the fixed nitrogen in other N-compounds. According to Richards (1974), this is the probable source of nitrogen taken up by mycorrhizal fungi, which cannot fix nitrogen themselves but may acquire it through bacterial mortality in the soil. Nitrogen processing in ecosystems is largely biotic. Large nitrogen losses at clearcut, as mentioned by Bormann and Likens (1979) might be due to heavy bacterial mortality and washing out of the highly soluble organic nitrogen compounds. According to Becking (1961), there is a geographical distribution pattern of nitrogen-fixing species: *Beijerinckia* would be more tropical and *Azotobacter* rather temperate, for instance.

Except for bacteria, blue-green algae and lichens fix atmospheric nitrogen, too. If, as a world-wide average, the amount fixed per hectare per year is taken as 10 kg, the relative contributions of different organisms would be of the following order (Richards 1974):

- lichens, abiotic events (thunderstorms)	less than	0.5 kg
- free-living heterotrophic bacteria	less than	3.0 kg
- nodulated non-legume plants plus woody legumes	from 2 to	4.0 kg
- non-woody legumes	less than	0.5 kg
- blue-green algae	from 2 to	4.0 kg

The cycles for nitrogen, carbon and nutrients in general are given by Brown (1978, her Figs. 9.2, 9.3 and 9.4; see Fig. 5.78), including the organisms that process these substances at every step. But these processes certainly change quantitatively and qualitatively with the development phases of an eco-unit. Szujecki (1983, p. 6) probably refers to the aggradation phase when he states: "Symptoms of inhibition of pine stand growth on afforested fields (in Poland, RAAO) become most pronounced in the period of the tree-top formation, when their requirement of nutritional substances, and particularly nitrogen is the highest." As yet, no studies are known to us concerning the microbial species composition and the quantity of nitrogen fixed in each development phase. Studies on the impact of ammonia pollution on forests and its effect on mycorrhizae, however, do not point to a different reaction per phase (not mentioned in Bosch 1985, for instance). But ammonia pollution leads to the opposite of the Polish nitrogen scarcity, i.e. to nitrogen saturation. Both extremes may block every development in every phase.

Nitrogen is the most discussed but is not the only element processed by bacteria. There are bacterial species that are specialised in chemical handling of nearly all important elements, including sulphur and phosphorus.

Just as fungi, bacteria can interact with higher plants in different ways, including as pathogens. An example is the willow disease that is scourging European *Salix alba* stands and trees for some 10 years, and that is caused by the bacterium *Erwinia salicis* (PUDOC 1982). The life cycle of this bacterium and its dispersal mechanism still remain quite obscure. It yields an example of a component of large pioneer eco-units that cannot be effectively coped with, because its role in the eco-unit is not understood.

5.3.4.4 *Soil Fauna*

The soil fauna is extremely diverse in species composition and ensures a large array of ecosystem functions. For soil animals the following classification is often used, referring to their size:

```
Microfauna                 Mesofauna                       Macrofauna
--------------------------------------------------------------------------
                 +                                   +
                 +                      ---  insects ---------
                 +                                   +
                 +          ---     false      ---   +
                 +                  scorpions        +
                 +--    collemboles ----------       +
                 +                                   +
                 +              ---  spiders  --     +
                 +                                   +
               - +--   mites    ---    ---  centipedes -----
                 +                                   +
                 +                      ----  millipedes -----
                 +                                   +    -- insect
                 +                                   +       eaters
               - +- tardigrades    --   woodlice     +----
                 +                                   +
               - +- wheel   ---    --       snails, slugs -------
                 + animalcules                       +
--- protozoa --  +                    --       potworms ----
                 +                                   +
               - +- nematodes ---                    +-- earthworms --
                 +                                   +
                 +                                   +
--------------------------------------------------------------------------
                 +                                   +
                 +                                   +
+                +    +     +     +                  +   +   +   +
0.02 log mm      0.2  0.5   1.0   5.0                10  20  40  80 mm
--------------------------------------------------------------------------
```

The data and names in this synopsis have been been compiled from Richards (1974) and Brown (1978).

If we know, for instance, that during Szujecki's work (1983) on pine forests in Poland 180 Arachnoidea species (spiders and two families of allied, eight-legged Arthropoda) and at least 100 species of Carabidae beetles were caught and studied, that in tropical forests Lavelle (1984) cites densities of between 35 and 700 earthworms per m^2, and that Brauns (1976, cf. his Fig. 879) in his second volume gives 876 illustrations of the more common insects in the forests of Northern Germany, the dimensions of the problem of analysing the soil fauna are somewhat exemplified. It is quite impossible to generalise in describing the functions of all these animals one by one, or the interactions in which they are involved.

For instance, Lamont et al. (1986) studied the case of a marsupial *(Bettongia penicillata)* that shelters in legume thickets in Australian Eucalypt forests. These animals burrow, and localise and consume the subterranean sporocarps of mycorrhizal fungi associated with the trees. The spores are dispersed with the faeces of the marsupial: five different mycorrhizae appeared in the laboratory on *Gastrolobium* trees inoculated with these faeces. Another example of complicated interactions is given by Bongers (1985, p. 1), who writes: "One kilogram of soil on the average contains some 30,000 nematodes. They consume bacteria, algae, lichens, fungi, higher plants, other nematodes and other soil animals such as oligochaetes, protozoans and collemboles. They are attacked or consumed in their turn by bacteria, fungi, protozoans, mites, tardigrades and larger soil organisms."

When treating some examples of connexus, or rather of eco-unit compartments in which numerous interacting organisms are considered, some more order will be brought in this apparently all-but-incomprehensible richness (Sect. 5.5.10).

One more useful classification, however, may be presented here, because it is made following the diet of the animals. Their diet places them in a certain compartment and in a certain role, although not in a detailed way. Soil animals may be:

- **Phytophages,** that eat living green plants (sucking insects, larvae that chew and eat roots, certain termites and ants that use plant material for fungus cultivation, one category of nematodes);
- **Detritivores,** that eat dead organic matter, in which they may feed on the other organisms present in that matter (representatives of nearly all major taxonomic groups given earlier); one of the roles of these animals is fragmenting of litter and humus;
- **Carrion feeders** or *necrophages*, that eat dead animals. Only few animals can digest proteins (some nematodes, earthworms and slugs); most often this function goes to saprophytes;
- **Coprophages,** that eat faeces of both larger animals and the small soil animals themselves (springtails, nematodes, but here also saprophytes take a large part);
- **Microbivores,** that eat fungi and protists (important are the fungivorous nematodes, certain beetles and fly larvae - see also the above example of the marsupial);
- **Carnivores** or **predators,** that eat living prey animals (beetle and fly larvae, ants - cf. Collinet et al. in Guillaumet et al. 1984, p. 113 ff. - many mites, nematodes

that eat other nematodes, other small worms and protozoans); this category includes **parasites** that eat but do not necessarily kill larger animals.

It is a good exercise to try and fit these animal behaviour patterns and functions into Fig. 5.28, inserting a new set of arrows. This rapidly complicates such relation diagrams. Numerous very complicated diagrams of this kind have been published, but are quite difficult to read back. They will be introduced in Section 5.6.

Many of such relations can be found in detail in Dickinson and Pugh (1974). A more recent account was published in the Vakblad voor Biologen in December 1986.

The usual theory formation in zoo-ecology (for forest insects see Grison and Géri, in Pesson 1974, p. 219 ff.) is of little help in linking their development to the systematic development of eco-units or ecosystems in general. No distinction is made between eco-units and mosaics (stands and forests), for instance. In fact large areas are preferred for the stochastic inventory of single insect species, so as to establish overall population change in the form of mean numbers of individuals against time. The time considered is climatic time, i.e. time as linked to seasons and weather. Ecological time, linked to forest development, has only rarely been considered and then most often indirectly, by referring to bio-regulation of populations. Given the fact that the silvatic mosaic is the direct reflection of climate and weather over a number of years, it is no wonder that system regulation, which occurs on the hierarchical level of eco-unit systems, is averaged out, and that a whole school of theoretical ecologists does not take it into account (e.g. see Pianka 1974; Den Boer 1977; Wolda in Sutton et al. 1983).

Publications that give some insight in the systems analysis of the rhizosphere as a developing compartment are rare, and this is certainly so in the field of soil zoology. Brauns (1976) gives several figures that pinpoint the spatial localisation and functional role of insects that play a part in litter decomposition, and particularly concern the sequences in dead wood (see his pp. 19 ff.), standing or fallen (cf. Fig. 5.23A). Animal-wise information is available in Jeník (1979). A direct connection between the development phases of a forest and the species composition of the entomofauna has up till now only been published by Szujecki (1983), for Polish pine forests on old forest soils and on former agricultural soils. His master diagram is represented here as Table 5.2. In this table, the innovation phase plus the beginning of the aggradation phase have been united as "thickets and plantations", the rest of the aggradation phase is divided into two periods (20 to 30, and 30 to 40 years old), and the end of the aggradation phase plus the beginning of biostasis are indicated as "timber stands maturing". Knowing this, Table 5.2 can be linked to Fig. 5.4 and yield insight into one case of subterranean eco-unit development.

5.3.5 Phasic Development of Eco-Unit Root Compartments

Subterranean forest eco-unit development now may be briefly sketched as follows, taking into account the uncertainties that remain.

- **The innovation phase** shows a root layer in which other plants largely dominate tree seedling roots, the tree seedlings showing a large array of strategies in which they do not necessarily root deeper than herbs, as is often believed. If mycorrhizae, formed in this phase by hardy fungus species, are deficient, early root rot ("damping off") may be caused by other fungi. Depending on the amount of organic matter left behind after the zero-event, saprophytes and hardy detritivorous soil animals are present, with their predators. Phytophagous, root-consuming nematodes and insect larvae would be numerous in this environment where root development is intense and therefore the amount of young, soft root biomass is high. The forest microclimate has not been established as yet in the innovation phase, and macroclimatic influence is still strong, so that high fluctuations of populations may occur, including epidemics and pests. Rhizosphere architecture is fluid and shows relatively short-lived patterns.

- **The aggradation phase** is structured by the architectural development of the roots of the surviving trees. The early mycorrhizal fungi are either eliminated

Table 5.2. Sequences of trophic macrofauna groups during different Polish pine forest developments, on agricultural and forest soils. (Szujecki 1983)

Timber stands	Stable meso-hygrophilous macrofauna associations with prevalence of phyto- and zoophages	*Heteroptera* *Lepidoptera* *Strophosomus*	*Athous subfuscus* *Empididae* *Dalopius marginatus*	*Nothiophilus biguttatus* *Othius myrmecophilus* *Myrmica ruginodis* *Myrmica lobicornis* *Tapinocyba pallens*
		Clearcut and reforestation		
Plantations	Xerothermic communities with prevalence of phyto- and zoophages	*Byrrhidae* *Tenebrionidae*	*Tetramorium* *Nopoiulus fuscus* *Bibionidae* *Dendrobaena*	*Bradycellus* *Amara quenseli* *Mycetoporus brunneus* *Cardiophorus*
Thickets 30–40 years old	Unstable regeneration communities of forest macrofauna with exchange of soil phytophages to saprophages	*Sericus brunneus* *Strophosomus* *Sericus brunneus* *Strophosomus*	*Nopoiulus fuscus* *Bibionidae* *Sciaridae* *Dendrobaena* *Dalopius marginatus* *Dolichopodidae* *Nopoiulus fuscus*	*Quedius boops* *Calathus erratus* *Prosternon tessellatum* *Calathus micropterus* *Othius myrmecophilus* *Sipalia circellaris*
Stands 30–40 years old	Stable macrofauna communities with prevalence of saprophages	*Heteroptera* *Strophosomus*	*Dendrobaena* *Athous subfuscus* *Tipulidae* *Dalopius marginatus*	*Calathus micropterus* *Othius myrmecophilus* *Myrmica ruginodis* *Empididae* *Myrmica lobicornis*
Maturing stands		*Heteroptera* *Strophosomus*	*Athous subfuscus* *Tipulidae* *Sciaridae* *Geotrupes stercorosus*	*Calathus micropterus* *Othius myrmecophilus* *Myrmica ruginodis* *Myrmica lobicornis* *Empididae*

Direction of forest macrofauna stabilization →

Table 5.2. (continued)

Arable land	Synantropic fauna of open land	*Selatosomus* *Agriotes*	*Lestremiinae*	*Atheta angustula* *Calathus ambiguus*
			Afforestation	
Thickets and plantations	Pioneer xero-thermic communities with predominance of phyto- and zoophages	*Selatosomus* *Agriotes* *Scarabaeidae* *Tenebrionidae*	*Tetramorium* *Lestremiinae* *Allobophora* *Geotrupes vernalis*	*Calathus erratus* *Myrmica rugulosa* *Philonthus nitidulus* *Oxypoda togata*
		Selatosomus *Brachyderes* *Agriotes*	*Orthocladiinae* *Tetramorium*	*Oxypoda togata* *Prosternon tessellatum* *Calathus erratus*
Stands 20–30 years old	Unstable communities of period of macro-fauna exchange	*Brachyderes*	*Dendrobaena* *Orthocladiinae* *Bibionidae* *Lasius alienus*	*Calathus micropterus* *Othius myrmecophilus* *Prosternon tessellatum* *Empididae*
Stands 30–40 years old	Mesohygro-philous comm-unities of stabilized forest macrofauna	*Strophosomus*	*Dendrobaena* *Orthocladiinae* *Geotrupes stercorosus*	*Calathus micropterus* *Myrmica ruginodis* *Othius myrmecophilus* *Sipalia circellaris*
Stands timber maturing	Stable meso-hygrophilous macrofauna communities with predominance of sapro-phages	*Strophosomus*	*Athous subfuscus* *Tipulidae* *Sciaridae*	*Calathus micropterus* *Othius myrmecophilus* *Sipalia circellaris* *Myrmica ruginodis*
		Strophosomus	*Geotrupes stercorosus*	*Empididae* *Tapinocyba pallens*
		Strophosomus *Heteroptera*	*Athous subfuscus* *Dalopius marginatus*	*Notiophilus biguttatus* *Othius myrmecophilus* *Sipalia circellaris* *Empididae* *Tapinocyba pallens*

Directions of succession

Direction of succession in second forest generation unknown – expected to be as on forest soil

with their partner plants if these were innovation species, e.g. herbs, or replaced on tree roots by other fungi of some mycorrhizal sequence, like the one described by Mason et al. (in Fanta 1986). The litter-decomposing soil flora and fauna are gradually shifting from decomposer complexes on gross litter, i.e. fallen trunks, tree stumps, towards fine litter decomposers. Phytophagous animals and parasitic fungi on roots become rarer and more specialised; there might be a shift to grazers on mycorrhizae, such as nematodes. The architecture of the rhizosphere takes shape. If above the ground there is a lower structural set (Fig. 5.4B), the roots of these smaller trees or shrubs penetrate the organic layers extensively, whereas the potential trees still may have to invest matter and energy in the positropic parts of their root system, depending on the soil and the tree species. The underground accumulation of dead root mass, the subterranean equivalent of litter, starts up with the death of the roots of all eliminated innovation species. This accumulation attracts the decomposer guild and their associated predators. Root mortality and decomposition in deeper soil horizons starts later. At the end of the aggradation phase "a quite complete separation of above- and underground litter decomposition systems" (Lavelle 1984, p.14) may have developed, as is illustrated in Fig. 5.28C.

- **The biostatic phase** is one of replacement of lost organs in individuals of large-sized species, and replacement of lost individuals in populations of small-sized species. There is no reason to doubt that this also occurs below the ground. Depending on the tree and shrub component species, one superficial layer or more and also deeper root layers have been occupied, the root territory of each component being delimited more or less like the crown space aboveground. It is too early as yet to speak of "subterranean structural ensembles", because no data are available to map them. The final species composition of soil fauna and flora should be reached here, with the "end species" of species sequences. But the species that are absent certainly are present in the propagule bank that is built up, in the form of spores and seeds. This may or may not be the case for the soil animals: some eggs may be quite long-lived, and span several seasons; others may disappear. The biostatic phase is resilient, i.e. it can correct with some architectural, floral and faunal adjustments all of the most common impacts from beyond the eco-unit like drought, animal grazing and frugivory or thinnings, up to a certain degree. If such impacts cannot be filtered and are heavy, such as accumulating pollution, resilience may not be enough. Biostasis becomes unbalanced, and a premature decay of the eco-unit starts. Rhizosphere architecture in the normal state of the biostatic phase, however, seems to be persistent and layered. Although soil homogenisation by animals (earthworms, termites, millipedes, collemboles) is to be reckoned with, it may be slow or blocked in the biostatic phase, so that the root layers are maintained. An indication in this sense is the fact that, from about the middle of the aggradation phase on, only one species of earthworm that is confined to the litter is left in the Polish pine forests studied by Szujecki (1983).
- In **the degradation phase** the root systems of decaying or damaged trees become underfed. Parts die and start decomposing, whereas from above gross litter fall (dead branch wood) increases. Subterranean eco-unit architecture degrades, like that above the ground, and niches open up for many of the decomposers, or even for innovation species. The process is very heterogeneous (see Oldeman, in Cosijn et al. 1983) and the said niches are formed in pockets, here and there in the degrading eco-unit. Abrupt and steep changes in architecture and species population structure may occur. This anticipates the innovation phase. The belief that degrading eco-units are sources of pests and epidemics is certainly unfounded if interpreted as a general rule. But the overall instability of the innovating eco-unit, that favours outbreaks of pathogenic events, certainly is prepared in the degradation phase. The zero-event may well be discovered to be of reduced importance below the ground and span several seasons (see Sect. 5.2.6, reorganisation phase).

This sketch is rough and incompletely proven. It may serve, however, to indicate research orientations concerning patterns and dynamics in the subterranean compartment of stands (eco-units). It can also serve as a provisional framework for silviculturists, in order to include the rhizosphere in silvicultural design. The recent fashion to include "dead wood management" in forest management (e.g. Cosijn et al. 1983) may have led to some aberrations, such as the artificial creation of dead wood compartments in forests in the aggradation phase. But it is also indicative for the real need to consider the root and litter zone more carefully. So are the studies on pollution impacts on forests (e.g. Ulrich in Ulrich and Pankrath 1983; Bauer 1985). Everything points to the view that forest dieback starts with the small components, especially belowground, and ends with tree mortality. The recent development of a nematode-based method, that gives genus composition, according to ratios between phytophagous, fungivorous and predatory genera, as a symptom for "forest vitality" (Tamis 1986), stresses this point.

5.4 Examples of Eco-Units of Different Size and Nature

Some examples of eco-units are presented in the following pages. These case studies will serve to show some of the facts at the basis of the architectural eco-unit model given before, and to furnish some precisions that will allow us to treat the species composition and production ecology of eco-units in a better way (see Sects. 5.5 and 5.6). For more examples of cases and processes, the reader is referred to Pickett and White (1985).

5.4.1 Large Eco-Units

Large eco-units are presented in different states, from different regions and originating from different zero-events.

5.4.1.1 Treeless Innovation Phase near Montpellier (France)

This innovation phase is illustrated by Ndiaye's analysis of a weed vegetation in Montpellier (1977). Two cases are presented in Figs. 5.31 and 5.32, which represent fragments of transects being respectively 2.5 m and 10 m long. Their different length and width have been determined according to the rule-of-thumb formulated by Oldeman (1974a), that a block or belt transect should have its width chosen between one and two thirds of the upper height of the vegetation considered. Neither of Ndiaye's transects presents tree seedlings; but they show an image of the behaviour of the non-arborescent components in many innovation phases. Examples of the architecture of herbaceous components can be found in Figs. 4.3 and 4.4.

The importance of the root zone in the innovation phase is evident. Tree seedlings develop strategies to keep alive in this densely occupied subterranean volume. One example has been studied by Bosch (1984), who analysed root competition be-

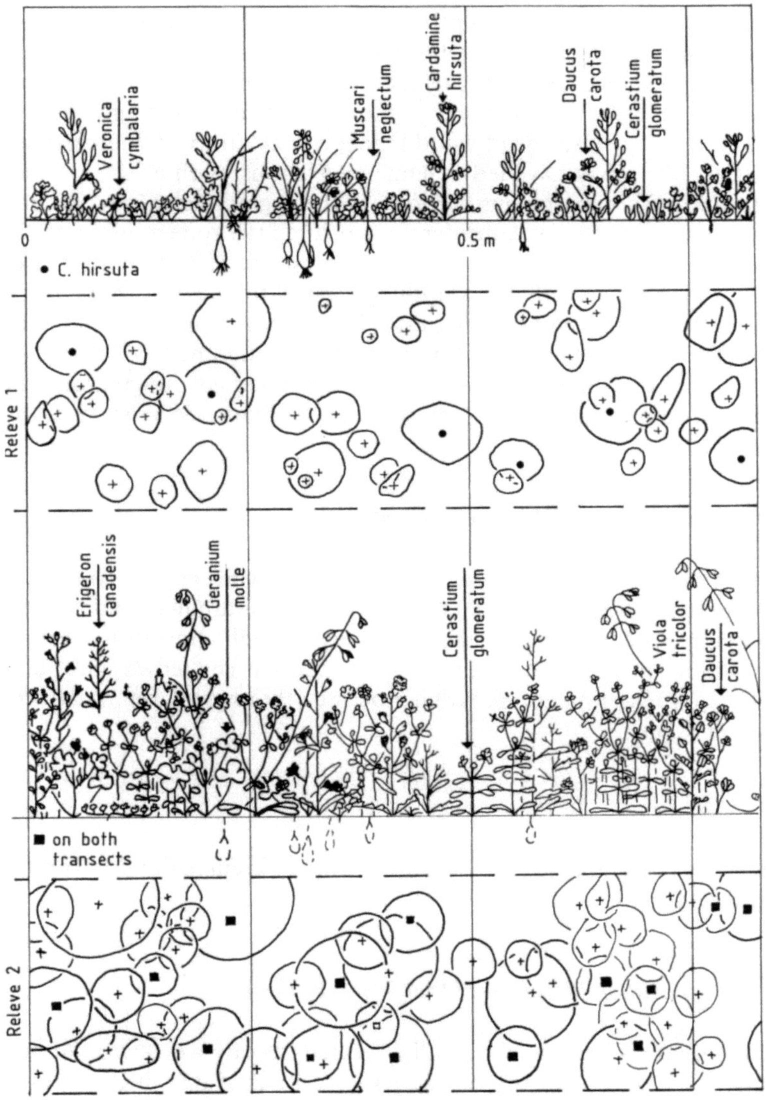

Fig. 5.31. Innovation phase development on weeded ground, Montpellier, after Ndiaye (1977). Part of larger transect. Two months and 17 days lie between releves *1* and *2*

tween a European oak species (*Quercus robur*) and a frequent herbaceous component of clearcut surfaces (*Chamaenerion angustifolium*) in perforon root-boxes (Fig. 3.18). Oaks start germination with the formation of a long and deep taproot and retard their growth aboveground until their root system is well established. *C. angustifolium* handicaps the further growth of oak roots, but the tree seedlings are established and many survive, notwithstanding reduced growth rates, possibly due

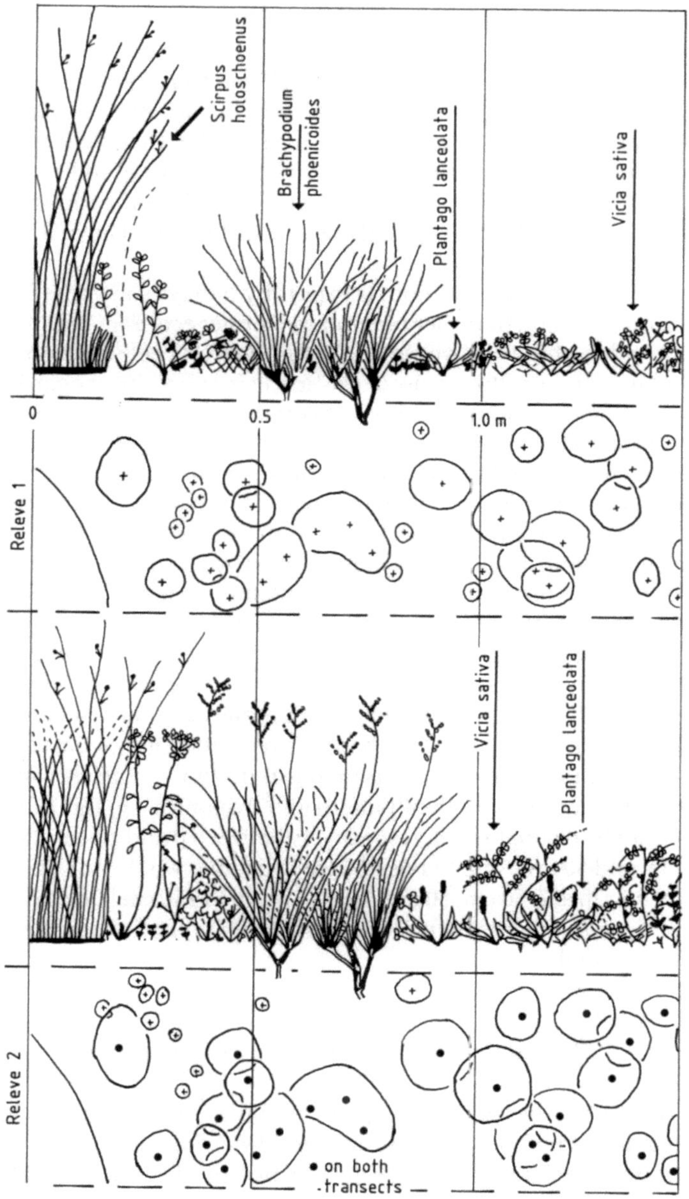

Fig. 5.32. Innovation phase development on fallow, Montpellier, after Ndiaye (1977). Part of larger transect. Three months between releves *1* and *2*

to allelopathic chemicals from the weed. The root pattern development of both species has been mapped by Bosch (1984). Such early root formation has also been observed by us in a tropical species, *Bertholletia excelsa* (Lecythidaceae, Brazil nut) in cultivation at the ORSTOM-Centre in Cayenne. Examples of such seedlings may be found in De Vogel (1979), next to other root architectures that probably result in the rapid saturation of a superficial volume of soil by means of a dense net of diatropic brachyrhizae (also see Ng, in Tomlinson and Zimmermann 1978). This last strategy is herb-like, but there also are herbs that have very deep roots (e.g. Figs. 4.3 and 4.4).

The comparison of both transects by Ndiaye shows the importance both of the zero-event and the inheritance of the former eco-unit. Figure 5.31 represents an innovation phase that would result from a zero-event of weeding a soil rich in organic matter. In fact, this transect was made in a bed full of garden weeds after initial weeding. The innovation phase here is built by immigrated species and bulbiferous species that probably were still left in the soil, the key-and-lock mechanism acting in favour of plant species that are nitrogen-demanding. Figure 5.32 was made after a weed vegetation on an abandoned lot of badland, on a shallow calcareous clay soil in the outskirts of the city. Here, the zero-event probably included mechanical soil levelling or soil accumulation some years earlier, and the propagule bank has been at the origin of hardy pioneer plants, some surviving underground from year to year and others invading the places between them every spring (immigration). The presence of grasses in this second transect is noteworthy.

Finally, if these transects concerned a vegetation of tenfold size, the question would have to be raised whether this is one eco-unit or a mosaic of small eco-units. Seen from above, such a herbaceous vegetation may seem to be one rather uniform carpet. But the analysis of its architecture shows patches with a completely different development, both of pattern and dynamics. On figure 5.31, different patches could be distinguished, respectively between 0 and 20 cm, 20 and 40 cm, 40 and 55 cm, and 60 to 80 cm. In Fig. 5.32, such patches would lie respectively between 0 and 20 cm, 20 and 50 cm, 50 and 70 cm and 70 to 150 cm. Do large eco-units exist in nature, or is early fragmentation the rule? This question will come back in the following examples.

5.4.1.2 Planted Early Aggradation Phase (the Netherlands)

The start of the aggradation phase in some large artificial eco-units has been architecturally analysed by Van Raffe and Roelvink (1986). Figure 5.33 shows samples of three such units with a reconstruction of the crown development of their Douglas fir trees (*Pseudotsuga menziesii*). The three stands were part of block 16 (Speulder-en Sprielderbos, the Netherlands), situated on holtpodzol soils (see De Bakker and Schelling 1966), and planted respectively at 3 x 3 m, 2 x 2.5 m, 2.5 x 2.5 m. This is reflected in the codes for these stands, that were studied in order to understand the effect of planting distance on stand architecture and development. Stand 16holt (3 x 3) covers 2.5 ha, stand 16holt (2 x 2.5) occupies 2 ha and stand 16holt (2.5 x 2.5) grows on 2.5 ha.

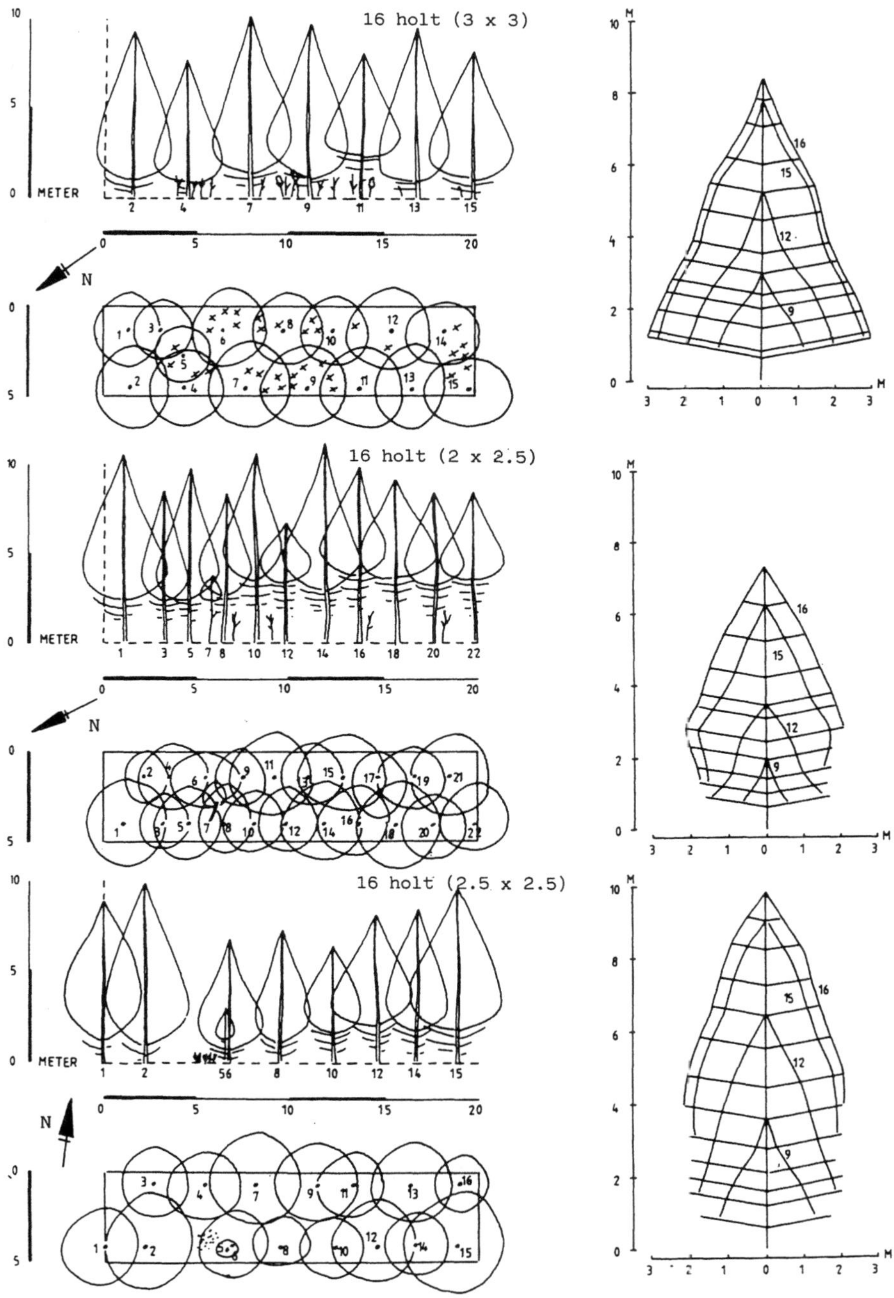

Fig. 5.33. Beginning aggradation phase in large, planted Douglas fir eco-units with reconstruction of crown development. (Van Raffe and Roelvink 1986)

Both the tree stand and the eco-unit are even-aged in the two last stands. But trees and eco-unit do not have the same age. The stand 16holt (2 x 2.5) is 12 years old, but was planted with trees of 4 years old, hence the trees are 16 years old. Before planting, the herbaceous flora - mainly ferns - was removed, so eco-unit development indeed started with all vegetation components of one age. Stand 16holt (2.5 x 2.5) is 10 years old with 16 years old trees, for the same reasons and including initial removal of shrubby *Prunus serotina.* Only stand 16holt (3 x 3), that was planted on a large windthrow of 2.5 ha without clearing of the forest floor, has an even-aged Douglas fir compartment (16 years), is 11 years old (planting), but may contain many other plant components of different ages. These also may include such trees as birches (*Betula verrucosa*) that are not present, however, on the transect.

The zero-event here is artificially homogenised in the two stands with weed removal. This can be seen on the transects, because non-arborescent plant components are decidedly rarer than in stand 16holt (3 x 3), notwithstanding the fact that canopy closure has occurred some years before, so that the innovation plant species are in the process of being eliminated. Homogeneity of the vegetation, visible in the architectural patterns of the stand, is here artificially brought about with the aim of wood production in a predictable way. Two steering methods have been used to achieve this:

- The use of planting stock that is older than the other plant components, altering the innovation phase to favour these trees and also in order to overcome the planting shock;
- the elimination of other innovation components, which is thought to enhance the advantages of the development of trees as forerunners, in two out of the three stands.

It has to be added that in stand 16holt (2.5 x 2.5) old trees have been left standing in 5-m-wide bands during some years (duration not mentioned) after planting, so that the young plantation grew in a slightly less illuminated and more humid environment. This resembles a stylised version of a zero-event leading to fragmentation of the eco-unit.

The application of such measures is clearly a race against time. The artificially impoverished propagule bank is certainly reconstituted again by immigration of propagules, both vegetable and animal. This may lead to proliferation of species from propagules that immigrated first and in large quantities. Pests, diseases and weed problems may then overtake the growth of the trees, because the unbalanced propagule bank contains no species to counterbalance them. At least one sylvicultural system for Douglas fir, the Auermühle system, consciously minimises weeding and site preparation (Brünig and Schneider 1975).

Finally, the style of these transects has to be noted. It is totally different from those by Ndiaye (Figs. 5.31, 5.32). This is not just a question of draughtsmanship. The different representations of reality, i. e. models, are made to answer different questions. Ndiaye aimed at the explanation of vegetational architecture in terms of plant component architecture (Figs. 4.3 and 4.4). Even the minimum of work and precision here required many details. Van Raffe and Roelvink aimed only at the diagnosis of the consequences of different planting distances, with little precision re-

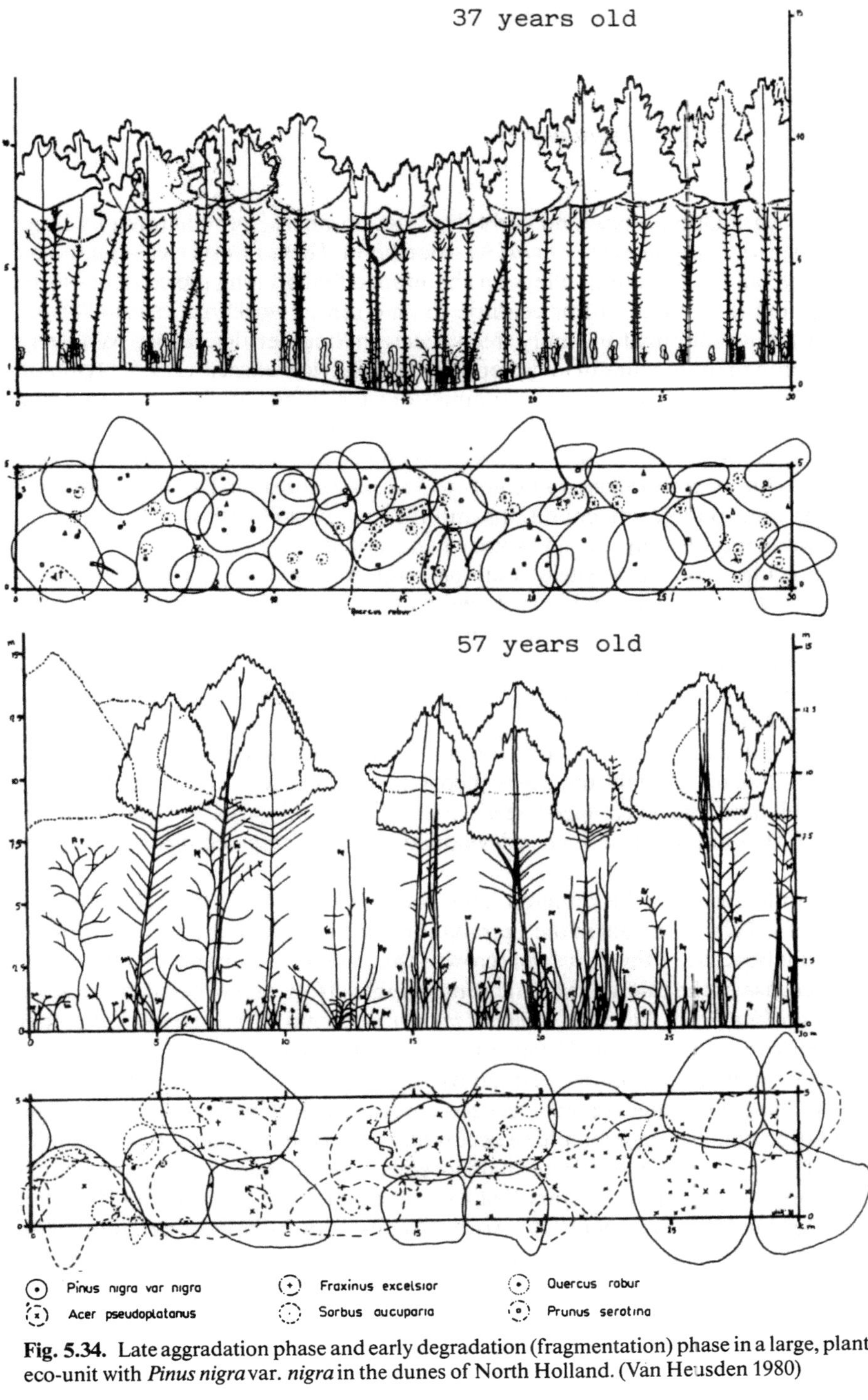

Fig. 5.34. Late aggradation phase and early degradation (fragmentation) phase in a large, planted eco-unit with *Pinus nigra* var. *nigra* in the dunes of North Holland. (Van Heusden 1980)

quired at this stage of research. Therefore their transects contain little detail. This aspect (also see Chap. 2) has to be borne in mind in the following examples.

5.4.1.3 Aggrading and Degrading Eco-Units (the Netherlands)

The late aggradation phase and early degradation phase are illustrated by large eco-units with *Pinus nigra* var. *nigra* (Austrian pine). These forests were planted on pure dune sands, or in dune valleys on abandoned fields, along the coast of North Holland in the first half of this century. In an effort to work in a "natural" way, which then was thought to imply tree species mixing per definition, oaks were mixed through these pines in low proportions. The two transects (Fig. 5.34), made by Van Heusden (1980) represent samples of such stands, each stand of several hectares large and on soils in which some humus formation has taken place only since plantation, at an age of 37 years (top) and 57 years (bottom). The fact that these trees have barely reached a height of 15 m after more than half a century indicates the inhospitable site conditions and the lack of adaptation of this species, which was a recently introduced exotic when planted.

Taking into account the site conditions, reinforced by periodic thinning which in the late 1970's was probably switched from low thinning to the rather heavy crown thinning now indicated in the management plan, and carefully observing the two development stages represented here, it seems that these eco-units just reached biostasis before starting to fragment. Late fragmentation is a diagnostic symptom of a degradation phase.

The architectural pattern shows a quite homogeneous stand that can be analysed by classical growth and yield theory in forestry because of the homogeneous soil and the even-aged trees. Such forests can also be simulated by means of such models as are now developed in production ecology (cf. Mohren 1987).

The undergrowth of the youngest stand contains many young broadleaved trees and shrubs. This is unusual in the aggradation phase. It has occurred here without any doubt because of the thin *Pinus* crowns that let pass an important amount of light, because of the thinnings that accentuate this effect, and because of periodic attacks on *P. nigra* by the insect *Tomicus piniperda* (common pine shoot beetle) that indirectly causes defoliation. In fact, this favours the management aim of transforming these stands into broadleaved stands. But the presence of these young trees is a signal that the propagule bank has been constituted and mobilised already. The preparation of the innovation phase occurs already in the aggradation phase of the stand! The build-up of a propagule bank in soils that have been without forest during centuries is an important story that will not be told here (but see Van Heusden 1980; De Coo 1981; Szujecki 1983). Artificial constitution of a balanced propagule bank is a concept that brings afforestation and reforestation projects far beyond mere tree planting.

Fragmentation of the pine forest shown here (Fig. 5.34, below) is also precocious with regard to probable natural development, and has been helped by quite

Fig. 5.35. Innovation phase with early fragmentation on a large eco-unit, 1 year old, in Tonka (Suriname). *1, 2* are *Cecropia* species, *8* is *Goupia glabra.* (Hagg and Pel 1983)

heavy thinning. The innovation phase of smaller broadleaved forest eco-units, that follows when such fragmentation proceeds further, has been shown already in Fig. 5.5, made in the same forest. The role of *Acer pseudoplatanus* (maple) and some surviving oaks (*Quercus robur*) is important; for the first species see Bonsen (1984), whose work contributes to the understanding of this forest component in this situation.

Finally, we note that the phases shown are not images of the same stand at two ages, but images of two stands deemed to be comparable in that the youngest is thought to develop towards a forest such as the older one. Because two eco-units are *never* totally identical, this introduces an element of uncertainty that cannot be avoided in most forest research. Projects that are followed over more than 10 years are rare (but see for instance De Graaf 1986; Bormann and Likens 1979; the SWECON project). The best check on eco-unit development models,

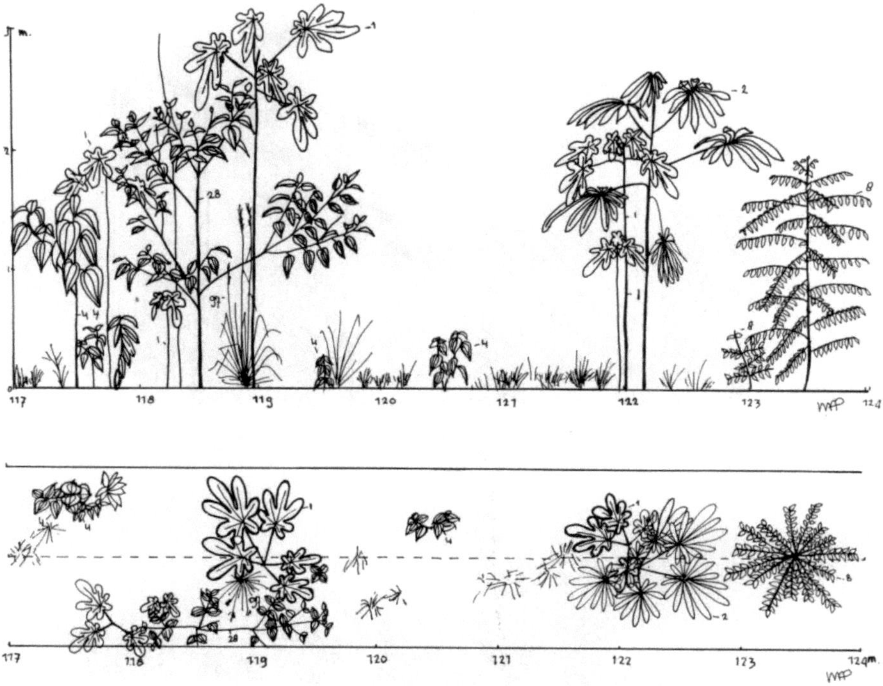

Fig. 5.36. Innovation phase with early fragmentation on large eco-unit, 3 years old, at Kabo Emplacement (Suriname). Species as Fig. 5.35. Note hanging leaves because of drought stress. (Hagg and Pel 1983)

then, is to follow comparable eco-units with an age difference of 5 to 10 years. After 10 years of research, predictions for every younger eco-unit, made at the beginning of the research project, can be checked and it can be verified whether they have reached the same state as their older counterparts 10 years earlier or not.

5.4.1.4 Late Innovation Phases, Humid Tropics (Guyanas)

Early development phases of large-sized tropical eco-units have been studied with particular thoroughness in the Guyanas during the last 12 years. This has led to a series of articles, theses and reports, of which we mention Ketelaars and Budelman (1976), De Foresta (1981), Pel (1983), Hagg and Pel (1983), Lescure (1978, 1986), and for the analysis of the seed bank Holthuijzen and Boerboom (1982), Prévost (1981, 1983 and other publ.) or Sabatier (1983).

Figure 5.35 shows a sample out of a 50-m-long transect (Hagg and Pel 1983), made in 1-year-old secondary forest on a hand-cleared experimental plot of 1 ha, with a deforested border strip of 25 m wide around it, and surrounded by primary

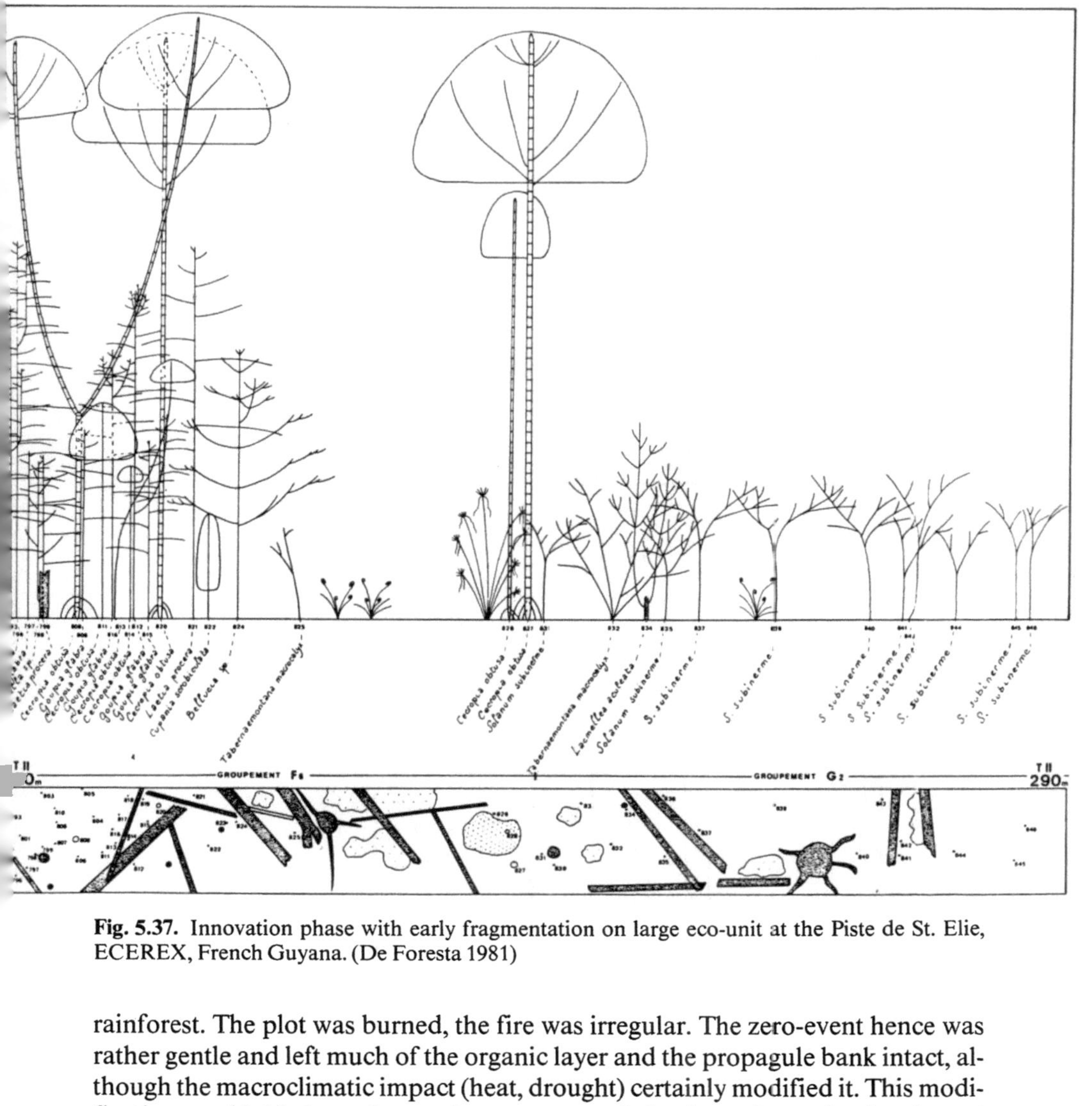

Fig. 5.37. Innovation phase with early fragmentation on large eco-unit at the Piste de St. Elie, ECEREX, French Guyana. (De Foresta 1981)

rainforest. The plot was burned, the fire was irregular. The zero-event hence was rather gentle and left much of the organic layer and the propagule bank intact, although the macroclimatic impact (heat, drought) certainly modified it. This modification was highly irregular too, because so much dead wood remained and acted as a climatic filter. In Europe, Koop (1981a, b) mentioned the same heterogeneousness in natural or semi-natural zero-events over larger surfaces.

Figure 5.36 represents a sample out of a 159-m-long transect (Hagg and Pel 1983), made in 3-year-old secondary forest at Kabo Emplacement (Suriname) on a mechanically cleared large terrain, on which the organic forest remains have been stacked by bulldozers on rows 60 m distant from each other, after which agricultural experiments have been done. The primary rainforest is near the transect, so that reconstitution of the propagule bank by immigration has certainly occurred. This brutal zero-event differs greatly from that at Tonka.

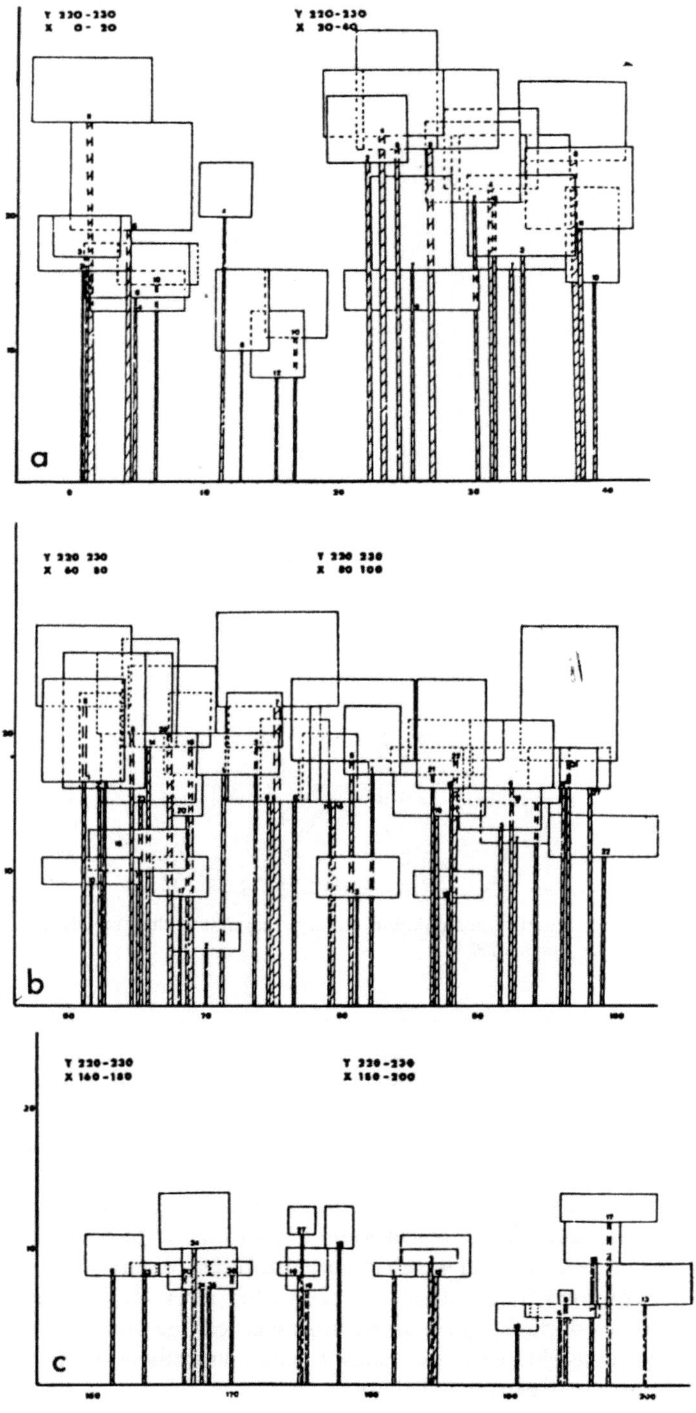
Y 220-230
X 0-20
Y 220-230
X 20-40
a
Y 220 230
X 60 80
Y 220 230
X 80 100
b
Y 220-230
X 160-180
Y 220-230
X 180-200
c

Fig. 5.38

Figure 5.37 shows a sample out of a 532-m-long transect, one out of two transects of which the other one is 448 m long, made by De Foresta (1981) in 3-year-old secondary forest at the ECEREX experimental project, Piste de St. Elie (French Guyana) on terrain that had been subjected to mechanical exploitation of the pulpwood-harvest type (procédé ARBOCEL), over 10 ha, that were enlarged experimentally to 25 ha later. The surface was fired afterwards by local inhabitants who wanted to profit from such large fields for shifting agriculture. The primary forest was farther away from the transect than from the Suriname transects. For Pel (1983), the ECEREX vegetation corresponded most closely to that at Kabo Emplacement (Fig. 5.36), and she mentioned the comparable initial treatment, in our terms the comparable zero-event. In Suriname, 8 transects in different situations were studied, two of which are exemplified here. Pel (1983) remarked that the secondary vegetation studied in French Guyana contained ca. 150 (flowering) plant species on 2 ha, whereas in Suriname ca. 240 plant species were present on 0.58 ha.

All samples presented here correspond to the other documents of the authors cited, in that the innovation phase and the early aggradation phase are extremely heterogeneous, as remarked upon by both Pel (1983) and De Foresta (1981). We would prefer to speak of early fragmentation or fragmented innovation on these surfaces (cf. Oldeman 1983b, his Fig. 3b). Pel (1983, p. 40) states that "... probably, the beginning of succession is very diverse and is determined for a large part by random events". This certainly is true. But in Chapter 2 it was mentioned already, that events on a much higher hierarchical level (e.g. climate, human society) and with low frequency are random events at a much lower level (e.g. eco-units), even if such events could be predicted at their own level.

One basic assumption, made on the basis of seed-bank research (see authors above, and cf. Vazquez-Yanes in Gomez-Pompa and Vazquez-Yanes 1976; Gomez-Pompa and Del Amo 1985; Janzen and Vazquez-Yanes in Hadley 1988) is that in undisturbed forest mosaics the seed bank may not be completely homogeneous, but it still has a predictable potential (cf. Fig. 5.3). If a zero-event were really homogeneous, the resulting large eco-units that are examined here would be much more homogeneous in architecture and species composition than they are. Three imponderables seem to increase random heterogeneousness in large eco-units (cf. Ketelaars and Budelman 1976):

- The spatial disposition of the organic remains of the earlier eco-unit or mosaic cannot be predicted. This may create a welter of small-sized surfaces that are the **key** to different **locks** in the propagule bank. This information is known and used

Fig. 5.38. Transects and scientific questions. Examples from the MAB project in San Carlos de Rio Negro, Venezuela. *a* Yevaro type; *b* Cunuri type; *c* Bana type. The rectangular crown model demonstrates length and width, not architecture, of the living crown. Compare with Fig. 5.37, where crown architecture is roughly indicated. Both transects are measured documents, providing spatial patterns for computer-simulated dynamic system models regarding processes such as litter turnover, water and nutrient cycling, biomass formation and microclimate. (Brünig in Furtado 1980)

in Amerindian shifting cultivation to choose the optimal planting spots for different crop species (Haxaire 1976).

- The spatial impact of the zero-event on the propagule bank in many cases cannot be predicted unless total destruction occurs (cf. Figs. 5.34 and 6.7, or Garwood et al. 1979). Irregular fires (e.g. the examples given here), a sun-and-shadow pattern that leads to local extinction of fungus propagules (cf. Smits 1982), the tracks of bulldozers and skidders that crush seeds and animals locally, will lead to a fragmented innovation phase or at least increase unpredictable heterogeneousness. Kilbertus et al. (1980) gave an account of the effects of deforestation on the soil microflora in the ECEREX project (Figs. 5.37 and 6.11).
- The physical impact of the zero-event on the soil cannot be mapped until after the event. This renders the soil heterogeneous as a germination environment for higher plants: compaction by the caterpillars of bull-dozers or skidders, local cooking of the soil in fire hearths or patterns of emergence of the mineral soil after uprooting of trees also are heterogeneity factors (cf. Fig. 6.11). In Fig. 5.36, the effect of mechanical soil degradation is visible in the drought symptoms of the underdeveloped trees with hanging and wilted leaves (Pel 1983).

In all cases shown (Figs. 5.32, 5.35, 5.36) where the zero-event resulted in particularly inhospitable environments with important reconstitution of a propagule bank on impoverished soils, grasses play a part. Grass invasion indeed is a symptom of particularly drastic zero-events, from total domination by *Imperata cylindrica* (alang-alang, cf. Sect. 4.6.1) on degraded lands in tropical Asia (e.g. Webb in Furtado 1980) to the consequences of pollution in European forest regions (Fig. 4.11B), and weed problems in afforestation projects all over the world. In some regions this may extend also to bamboos, e.g. *Guadua* species in the Andes.

The temperament of forest trees will be treated in Sections 5.5.11 and 6.2.4.2. Two examples of pioneer trees are exemplified on the transects shown here and may be mentioned already. These are the *Cecropia* species and *Goupia glabra*. Neither can tolerate much shade and survive (see Schulz 1960). But *Cecropia* species are short-lived, estimates ranging between 15 and 40 years, whereas *G. glabra* becomes a giant tree with hard wood that can still be traced back after several eco-units have developed, one after the other, below it. *Cecropia* has many ecological counterparts (e.g. African *Musanga, Macaranga, Harungana*; Asian *Macaranga, Anthocephalus*; European and Northern Asian *Betula, Salix, Populus*; North American *Rhus* are examples). The ecological equivalents of *G. glabra* also are many (e.g. Asian *Tectona grandis*, teak; European and American *Quercus*, oaks; African *Khaya, Entandrophragma*, mahogany).

Finally, we note again the different styles of the transects by Hagg and Pell (Figs. 5.35, 5.36) and De Foresta (Fig. 5.37), linked respectively to scientific approaches centred on species composition (Pel) and production ecology (De Foresta). This last approach, referring to biomass, abstract volumes of vegetation and production figures is most purely represented by graphical models of forest architecture such as those which are produced in this context by Brünig (e.g. in Furtado 1980). Figure 5.38 shows this kind of transect.

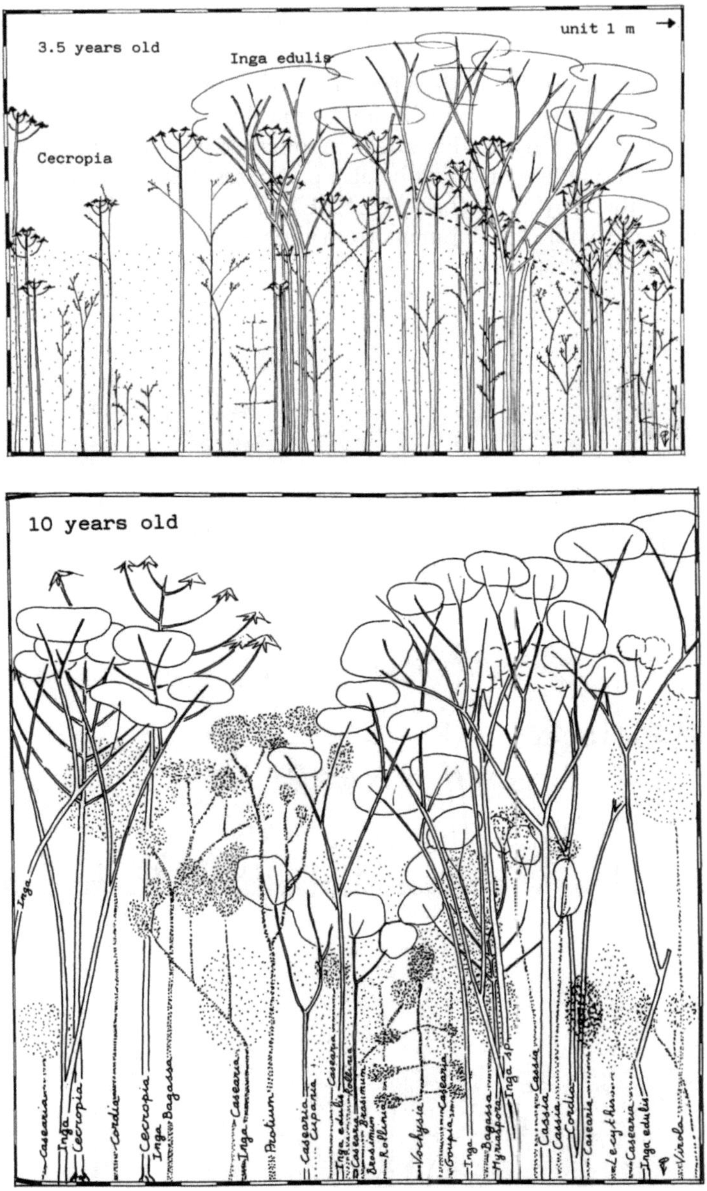

Fig. 5.39. Aggradation (3.5 years old) and biostatic (10 years old) phases in fragments of a large eco-unit growing on an abandoned shifting agricultural field on the Upper Oyapock River (Trois Sauts, French Guyana). *Dotted volume* shrubs and abandoned crop plants. (Lescure 1986)

5.4.1.5 Late Aggrading and Degrading Phases (Oyapock, Fr. Guyana)

Aggradation and biostasis of fragments of large-sized tropical American eco-units, resulting from shifting cultivation by the Wayapi Indians on the Upper Oyapock River (French Guyana) have been represented here (Fig. 5.39) from the study made by Lescure (1986) in which he also compares these secondary forests with those nearer the Guyanese coast (ECEREX, De Foresta; Fig. 5.37). Lescure's study covers the whole development, from agricultural field to reconstitution of the "primary" rainforest.

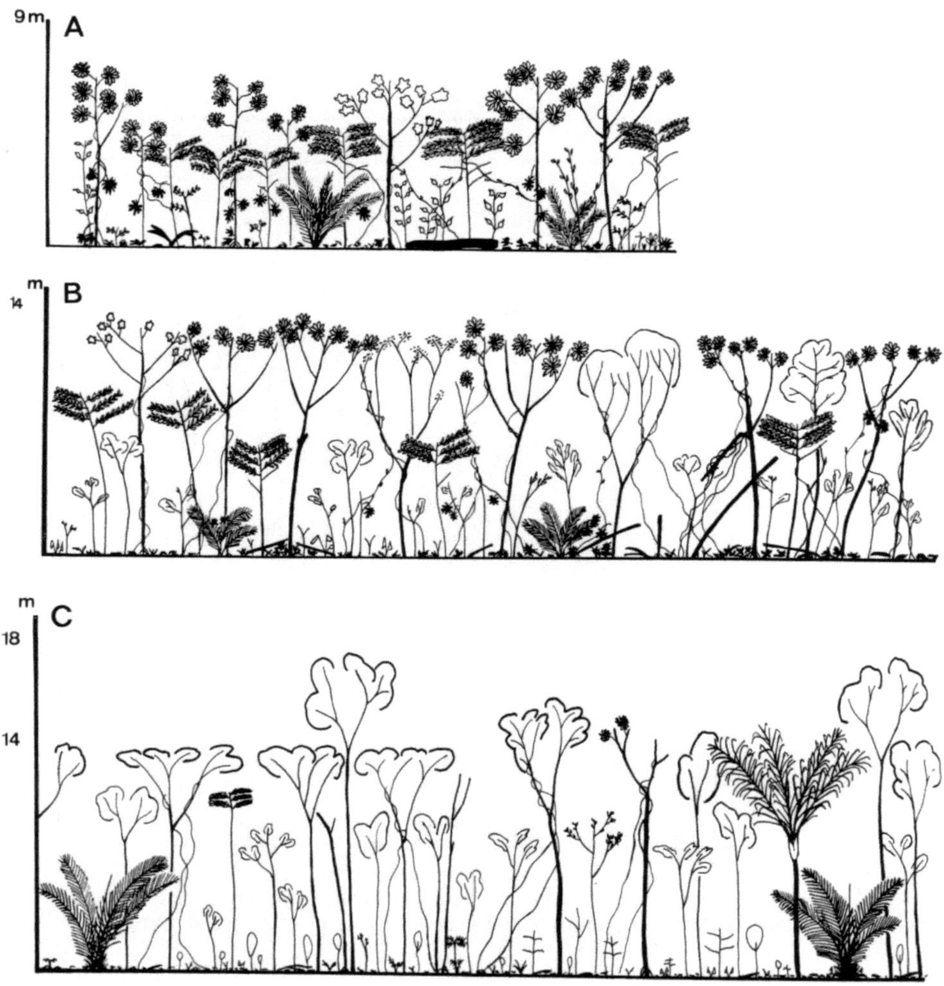

Fig. 5.40. Aggradation phase (**A**), biostatic phase (**B**) and situation after or during fragmentation (degradation) (**C**) of a large eco-unit, schematically rendered after observation of several cases near Manaüs, Brazil. (Kahn 1982)

In Fig. 5.39, attention is drawn to the development of a large eco-unit that was fragmented early in its innovation phase. The modification of the propagule bank here is local and might be exceptional. *Inga edulis*, the mimosaceous tree with Troll's model that forms the highest canopies on Fig. 5.39, has fruits of which the arils (*arillus*, fleshy tissue surrounding the seed) are a sweet delicacy that the Indians like to eat. In their fields, the seeds are spat out around the place of the meal and in these spots, *Inga edulis* starts building an eco-unit from which *Cecropia* is gradually eliminated, or in which it participates only as a rare component (Fig. 5.39, below to the left). Because Lescure still considered this whole vegetation as one large eco-unit, and because no trees of the present could be distinguished according to their architecture, Lescure (1986, p. 56) proposed a new subsystem in such large eco-units: the overtopping pioneer ensemble. This would function like a structural ensemble (see Sect. 5.2.4), but one that is not constituted by "regular" trees of the present.

In fact, the 10-year-old forest in Fig. 5.39 (below) may also be looked upon as showing *three* eco-units. On both sides, there are two biostatic phases with one structural ensemble consisting mainly of *I. edulis* trees. In the middle, there is a very small eco-unit in its aggradation phase, made up by more tolerant Guyanese tree species that are still not those found in the "primary" forest. After their temperament with slower growth rates and higher shade tolerance than pioneers, these have been called late pioneers or sciaphilous pioneers (e.g. by Oldeman 1974a). We will encounter these trees again in Section 6.2.4.2. Their ecological equivalent can be found the world over, and Douglas fir (*Pseudotsuga menziesii*) might well be one of their temperate-region representatives.

5.4.1.6 Large Tropical Eco-Unit Development (Manaüs, Brasil)

Large eco-unit development as a whole may be considered in the humid tropics. Tropical secondary eco-units develop very fast: the former examples go from the innovation phase to the degradation phase, fragmented or not, in some 15 to 25 years. Kahn (1982) gave three schematical scale drawings of this development in secondary forests near Manaüs (Brazil). The patterns that he found are represented in Fig. 5.40.

Contrary to the developments observed in the Guyanas, there is a homogenisation of an already somewhat fragmented aggrading eco-unit (3 years) when it attains biostasis (15 years). In fact, Fig. 5.40A and B may be closely compared to the basic and very abstract pattern diagrams in Fig. 5.4B and C, including the early formation of a low structural ensemble (genera indicated by Kahn: *Vismia, Casearia* and *Bellucia*). He indicates the zero-event in a general way as agriculture, which after being abandoned is soon followed by invading short-lived shrubs (*Solanum*), between and above which the forest of Fig. 5.4A develops. At 30 years (Fig. 5.4C), the structural ensemble made up mainly by *Cecropia* has disappeared or is disappearing. It looks as though fragmentation of the large-sized eco-unit is taking place at that moment.

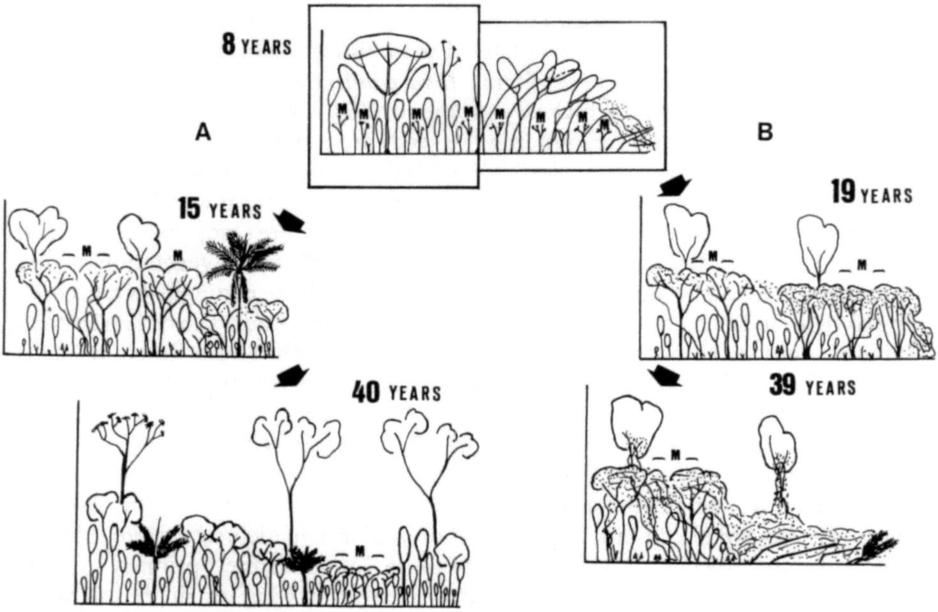

Fig. 5.41. Development and fragmentation of a large-sized eco-unit in Taï, Ivory Coast, schematically drawn. *M* = *Myrianthus* sp., a small moraceous tree building the first lower structural ensembles. (Kahn 1982 ex Guillaumet et al. 1984)

In terms of processes, not of biomass or species composition, there is a continuous series of possibilities: fragmentation occurring in the innovation phase (France, Suriname, ECEREX French Guyana, Figs. 5.32, 5.35, 5.36, 5.37), fragmentation starting in the aggradation phase (North Holland, Trois Sauts - French Guyana, Figs. 5.34, 5.39) and fragmentation starting in the degradation phase (Brazil, Fig. 5.40). It cannot be ascertained with certainty which is "the" natural process, because most of the examples shown involve human intervention in one phase or another. This emphasises the need for more wilderness studies as baseline studies (cf. Franklin 1981).

5.4.1.7 Eco-Unit Development in the Côte d'Ivoire

Kahn (1982, 1983, in Guillaumet et al. 1984) also analysed the development of large eco-units after shifting agriculture in the Tai region, Côte d'Ivoire (West Africa). Two alternative developments are schematically represented in Fig. 5.41. The aggradation phase (A-B) behaves very much as in some of the neotropical examples. Only the mechanism of fragmentation is different. In this case it is liane development that changes the *processes* on part of the surface, so that the definition of an eco-unit does not apply any more to this heterogeneous forest fragment, because this definition includes the same surface, zero-event and development. But even

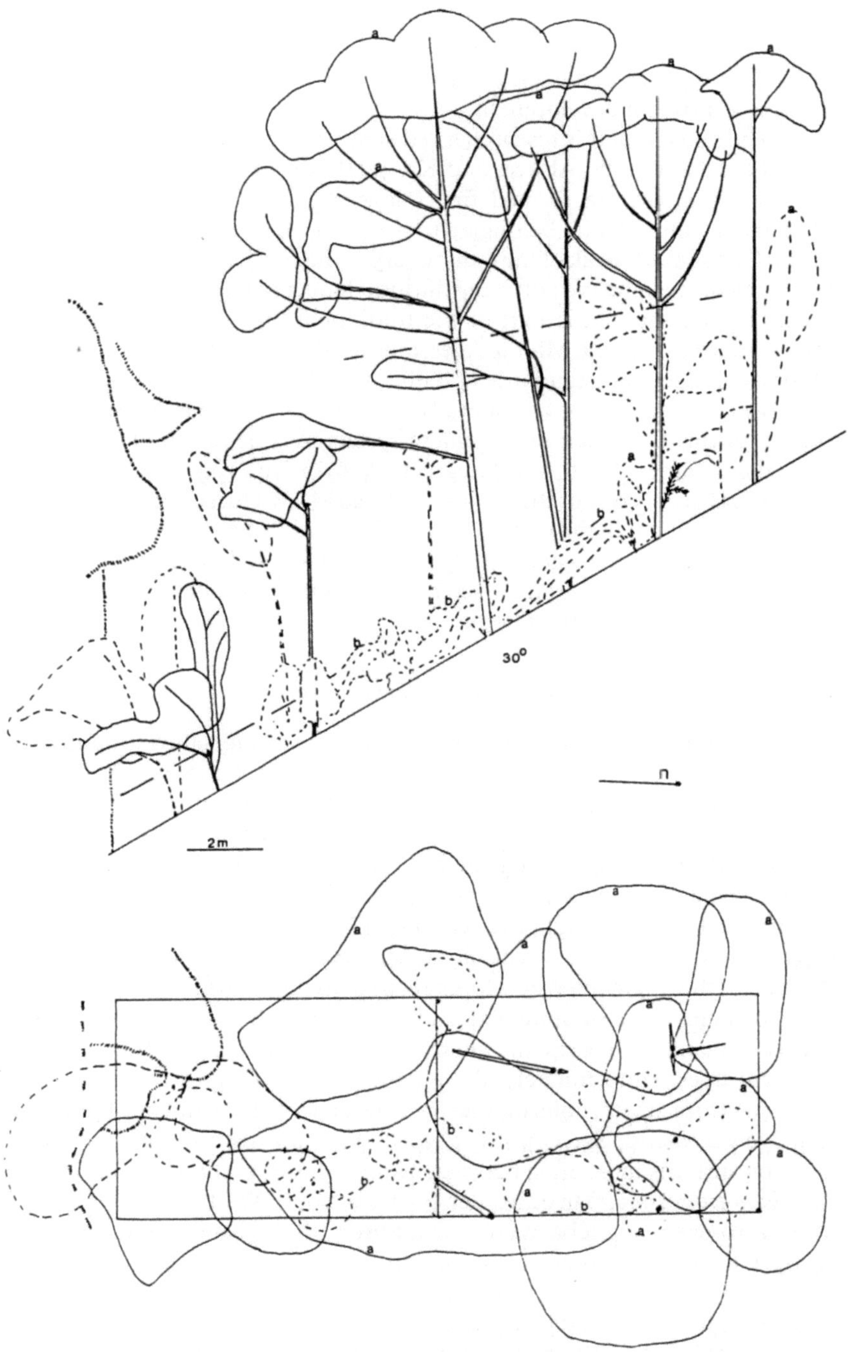

Fig. 5.42. Fragmentation of a large, 10- to 12-year-old eco-unit after shifting cultivation near Wau, Papua New Guinea. *a Lithocarpus* sp. *b Acalypha* sp.; *dotted* potential trees/shrubs; *solid lines* trees/shrubs of the present; *shaded* rooted beyond transect. (Louman 1986)

the 15-year-old phase at the left is fragmented, because the palm-dominated surface shows a different development.

It is to be noted that this researcher rediscovered, around the same time as Van de Winckel (1980) in the Alsace, and Koop (1981a) in Germany, that the lower structural ensembles are formed first ("layering from below"), the small tree species in the Côte d'Ivoire belonging to the genus *Myrianthus* indicated as *M* on Fig. 5.41. One should speak of the rediscovery of the species with this temperament, germinating in the light and surviving during a longer or shorter period when overtopped, because the fact is part of the traditional knowledge of farming communities. At least since the Middle Ages, they used such species in village forests of the type "coppice with standards" ("taillis sous futaie", "middenbos", "Mittelwald"), which now might be considered as agroforestry systems. More detailed descriptions of this European management system may be found in Lanier (1986), Mayer (1980), James (1982), Van Miegroet (1976) or Houtzagers (1956).

Figure 5.41 is only one of the results of a large MAB-Unesco pilot project on the forested region of Tai in the Côte d'Ivoire. The results of 8 years studies including social sciences and an ecosystem study covering many soil, floral, faunal, and soil-biological data have been summarised by Guillaumet et al. (1984). A complete bibliography can be found in their book.

5.4.1.8 Fragmenting Eco-Unit in Papua New Guinea

Fragmentation of a large eco-unit after traditional shifting cultivation near Wau, Papua New Guinea, is illustrated by one of the figures by Louman (1986). This represents an eco-unit of 10 to 12 years old, that has clearly split up already into two or three smaller ones: a biostatic one to the right, a degrading one in the middle (firewood harvest?) and an aggrading one just visible to the left (Fig. 5.42). Louman arrived at a schematical analysis of the silvatic mosaic.

For tropical Asia, we might cite Ashton (in Tomlinson and Zimmermann 1978) who published a transect of 4-year-old secondary forest in Pasoh Forest, Malaysia, that confirms the fragmented development pattern outlined on the earlier pages. R. Huc united many data for Sumatera (Indonesia), in the way of ECEREX research in French Guyana, with particular attention to the architecture and development of component trees. Unfortunately, these data, that again agree with the image of large humid tropical eco-units that seems to be general, have not been published.

For New Zealand, Six Dijkstra et al. (1985) published the development sequence, on surfaces most often between 3 and 10 ha, of natural terrace rimu forest (*Dacrydium cupressinum*) in Saltwater Forest in South Westland (Fig. 6.50B). The possible zero-event might be massive windthrow, the measured stands not having been logged. The transects of what these authors call "senility/regeneration growth phase", "early competition growth phase", "competition growth phase", "homeostatic growth phase" and "late homeostatic growth phase" show that in this case the general composition of the propagule bank is the main determinant. In the first two phases there is a tendency to early fragmentation, but in the "competition growth" (late aggradation) phase, forest architecture is quite homogeneous over an 80-m-long transect. From this stage on, this large eco-unit behaves as in Fig. 5.4. It is

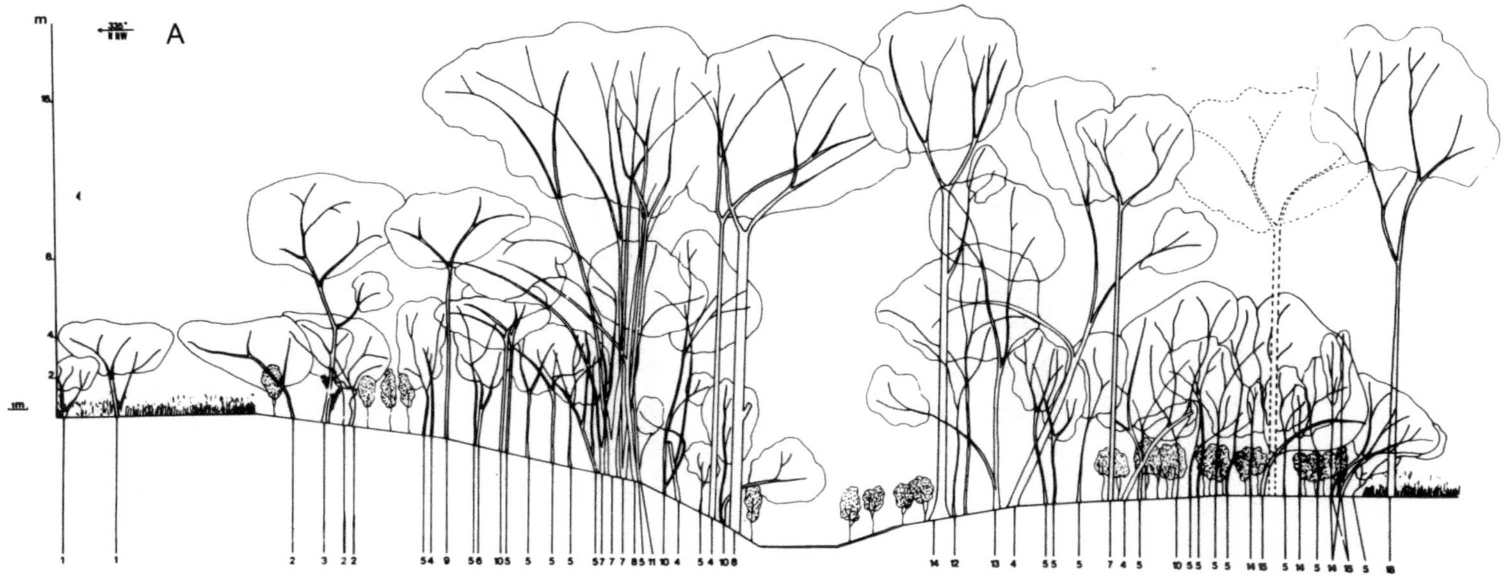

LEGENDES

1 *Piliostigma thonningii*
2 *Allophylus africanus*
3 *Holarrhena floribunda*
4 *Mimusops kummel*
5 *Drypetes parvifolia*
6 *Lecaniodiscus cupanioides*
7 *Dialium guineense*
8 *Cynometra megalophylla*
9 *Eugenia salacioides*
10 *Pancovia bijuga*
11 *Diospyros abyssinica*
12 *Cassipourea congoensis*
13 *Cola laurifolia*
14 *Manilkara obovata*
15 *Antidesma membranaceum*
16 *Lonchocarpus sericeus*

Fig. 5.43A
(Fig. 5.43B see page 252)

1. *Diospyros mespiliformis*
2. *Malacantha alnifolia*
3. *Lecaniodiscus cupanioides*
4. *Canthium multiflorum*
5. *Afraegle paniculata*
6. *Khaya senegalensis*
7. *Drypetes floribunda*
8. *Mimusops kummel*
9. *Monodora tenuifolia*
10. *Kigelia africana*
11. *Oncoba spinosa*
12. *Tamarindus indica*
13. *Manilkara multinervis*

Fig. 5.43. Intermediate-sized forest eco-units from the dryer regions of the Côte d'Ivoire. Both biostatic phases. **A** Gallery forest with five eco-units in the Guinean region near Lamto. **B** Dense, dry tropical forest with two eco-units in the Sudanese region, near Ouango-Fitini. (Devineau 1976 for **A**; 1982 for **B**)

possible to think that synchronous ageing of all these trees makes for a phase of general onset of decay, which renders the whole eco-unit vulnerable to storms. The large-sized eco-units would be adapted in this way to the macroclimate.

5.4.2 Intermediate Eco-Units

Intermediate-sized eco-units, of which the architecture of the biostatic phase is determined by several (e.g. five to ten) trees of the present cannot be traced back easily in published forest transect drawings. If and when they seem to occur in such documents, there is always an ambiguity. It is not easy to find properties that they have in common, their developmental history is seldom clear and there often is more than a hint of interference by site heterogeneousness.

Still, a whole category of silvicultural systems is based upon the concept of such forest units as management units. The zero-event here is induced by harvesting trees in limited groups, creating an innovation phase that has certain desired properties and components (tree species). They go under names such as "group system" (James 1982), "Femelwald" (German, see Mayer 1980) or "groepenkap" (Dutch, see Van Miegroet 1976, or Houtzagers 1956). The French "futaie par bouquets" seems to be limited to the use of such intermediate-sized groups in places with promising seed or seedling richness, within stands of coppice with standards to be converted into high forest (Cochet 1977, p. 156 ff.). The fact is that such systems work, and that they create artificial intermediate-sized eco-units that stay alive. This at least is experimental proof of the existence of a set of forest components that, after certain key-events, can be mobilised from the propagule bank to build such eco-units. A short reconnaissance of examples from literature will serve to illustrate some intermediate-sized eco-units and try and read the marks left by their development.

5.4.2.1 Biostasis in Drier Climates (Côte d'Ivoire)

Two examples represent biostatic phases, and originate from the dryer regions of the Côte d'Ivoire (Devineau 1976, 1982). The first one (Fig. 5.43A) is a gallery forest along a stream near the Lamto Station, close to its confluent with the Bandama river, and adjacent to the marsh forests on its banks. This gallery forest is one of the forest types in the "pre-forest district" in the Côte d'Ivoire (defined by Aubréville 1938 ex Devineau 1976), later classified as "semi-deciduous forests of the mesophilous sector of the Guinean domain" by Guillaumet and Adjanahoun (1971 ex Devineau 1982). The region now is largely covered by savannas that develop into forest once more when protected against fire (see Valkeman 1985).

The second example (Fig. 5.43B) is a forest that belongs to the dense, dry forests of the Sudanese savanna region (sensu Guillaumet and Adjanahoun l.c.) in the north of the Côte d'Ivoire, near Ouango-Fitini in the National Park of Bouna, where such forest relics still exist. Both forest formations mentioned were defined by Devineau by frequent tree species, according to phytosociological criteria. For the trees of these regions see also Geerling (1982).

In the first example at Lamto, the gallery forest clearly shows four intermediate-sized eco-units from left to right. The second and the fourth show two structural ensembles, and the first and third only one, the upper canopy. Even without knowledge of the development of these biostatic phases, the topographical context revealed by the altitude of the soil surface with reference to the river water is clearly linked to these eco-units. They could be named tree savanna (see grass cover), intermediate site, riverbank forest and (maybe) marsh forest. The tree savanna certainly stretches further to the left of the transect. As a preliminary explanation one might think tentatively about "intermediate-sized eco-units on intermediate-sized sites". More generally, this situation may arise on many toposequences, i.e. sequences of sites linked to relief and hydrology. Toposequences have been a key element in most forest ecology studies in the Côte d'Ivoire (cf. Beaudou et al. 1978; Vooren in prep.; Fig. 5.69).

As suggested by the descriptions due to Devineau (1982), who does not use this concept, however, fragmentation of a large-sized pioneer eco-unit is the probable cause of the development of such groups in the Sudanese forest type that can be observed in Fig. 5.43B. This may be compared to Figs. 5.35 to 5.39, showing early fragmentation in the humid American tropics. Other architectural data on these rare and beautiful dry tropical forests around Ouango-Fitini, published in the same article by Devineau (1982) suggest, but do not prove conclusively, that a large-sized eco-unit, constructed mainly by *Anogeissus leiocarpus* (Combretaceae, ecological equivalent to *Cecropia*?) may be broken up into intermediate-sized or small eco-units built by tree species such as *Afraegle paniculata* (ecological equivalent of *Inga*?) and *Khaya senegalensis* (Meliaceae, African mahogany, ecological equivalent of *Goupia*?).

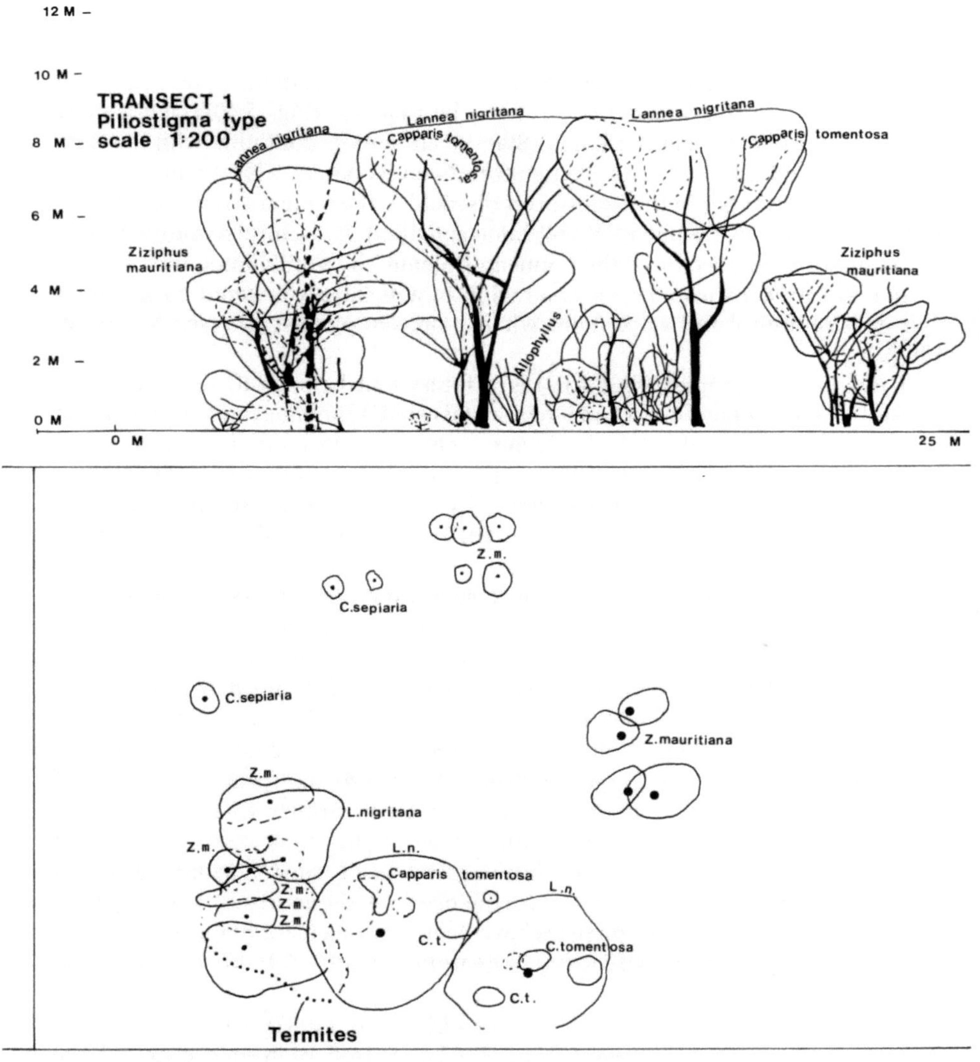

Fig. 5.44

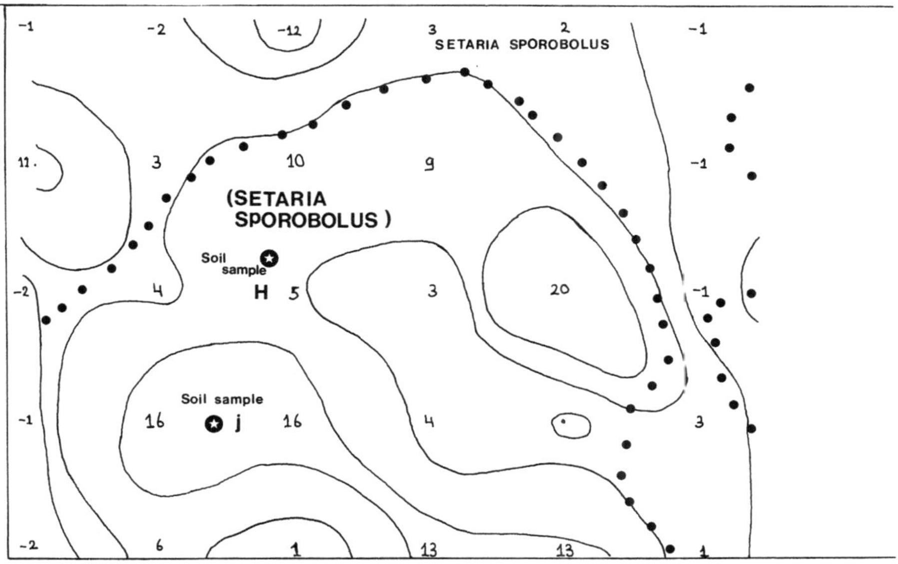

Fig. 5.44. Part of a transect in a semi-natural tree savanna on clayey impermeable soil with microrelief (*below*) in cm and termite activity determining intermediate-sized eco-unit occurrence and development. Upper structural ensemble built by *Lannea nigritana*; biostasis. Grasses may reach 1.5 to 2 m high. Bor Forest, South Sudan; · · · ·: grass; — : altitude in cm. (Raad 1981)

5.4.2.2 Biostasis in Savanna Forest (Bor, Sudan)

The biostatic phase of an intermediate-sized eco-unit in a still drier climate on the hard, impermeable clay soils of the tree savannas near Bor, South Sudan (Fig. 5.44), is taken from a report by Raad (1981). One has to look at the crown map and the map of the microrelief with altitudinal differences in centimeters (!), and one has to know that these savannas are subjected to frequent fires, in order to understand this intermediate-sized eco-unit, built by *Lannea nigritana* (Anacardiaceae), in the midst of grass vegetations that may reach heights of 1.5 to 2.0 m in the wet season.

Here again, there is a certain correlation between intermediate-sized sites that are some centimeters higher than the surrounding country, and intermediate-sized forest eco-units. There seems to be no link between zero-event and eco-unit size. The biostatic eco-unit shows only one structural ensemble, the upper canopy, at a height of ca. 8 m. Forest components other than trees have a decisive influence on the development processes. Termites cause the microrelief to be heightened in many places, and the isolated forest eco-units in the grass savanna often coincide with places of termite activity. These termites loosen the soil and may make it hospitable for the germination of lianes, which interfere with eco-unit development much in the same way as in Fig. 5.41.

As has been emphasised by Raad (1981), silviculture in these regions is a challenge to the forester, who is faced with a quite inhospitable environment for forests, a situation that is worsened by man-made fires and, in some regions, by excessive cattle grazing. The fragment of the transect shown in Fig. 5.44, one out of four transects in these regions, each 100 m long, represents a part of the region with only *light* human influence. It is the closest to the natural vegetation that Raad could find around Bor.

5.4.2.3 Aggrading Phase in Marsh Forest (Flevoland, the Netherlands)

An aggradation phase of intermediate-sized eco-units is shown in Fig. 5.45A, representing one of a series of short transects in the only truly natural forest of the Netherlands, after De Coo (1981). When the South Flevoland Polder was pumped dry in 1967/1968, in the deepest parts the water level came close to the ground level at the moment that wind-blown *Salix* seeds from the "old land" invaded the "new land". The humid clay represented an optimal germination environment for these pioneer species, that constituted a new propagule bank together with herbaceous species, mainly monocotyledons (reed, *Phragmites australis*, Gramineae), Typhaceae (*Typha* sp.) or sedges (Cyperaceae, *Scirpus* sp.) and also families like Compositae (*Senecio*, *Aster*), Cruciferae (cress, *Rorippa islandica*) or Chenopodiaceae (*Atriplex*, *Chenopodium*). The propagule bank was fed by wind-blown seeds and homogenised by the currents and the vertical movements of the remaining water level. In Southern Flevoland, *Phragmites* was sown out by plane.

De Coo and Wolffs (1980, ex De Coo 1981) established the architecture of the main components of the spontaneous forests that grew up in this way, and analysed the 11-year-old willow forests that existed in 1979. The age of the eco-units is of course well-known: in 1967, pumping was begun and in spring of 1968 the seeds could germinate. Two factors are responsible for the development of intermediate-sized eco-units in this young environment. First, there is a site factor. The sea-bottom was not flat, and the position of the water level as related to the new soil level created intermediate-sized sites. Second, the relative ecological efficiency of reed strategy and willow (*Salix*) strategy was not the same on each of these sites. Here again, the example shows the early functional fragmentation of a quite homogeneous propagule bank by a heterogeneous site.

5.4.2.4 Biostasis at the Forest Limit (Colombian Andes)

Figure 5.45B shows a vegetation that is not strictly speaking a forest, but that belongs here as an extreme example of woody plants germinating and growing in mini-sites determined by very small-sized site patterns in the Colombian páramos, the vegetation that covers the Andes higher than some 3300 m above sea level. Cleef (1981) analysed this vegetation very thoroughly. The *Espeletia* species that take the form of a small tree (Compositae, Tomlinson's model, also see Fig. 5.23B) is present in positions that are co-determined by the microrelief and the propagule bank. In other páramos, like some observed by us in Ecuador, they may sometimes form groups that are more or less comparable to the intermediate-sized eco-units ob-

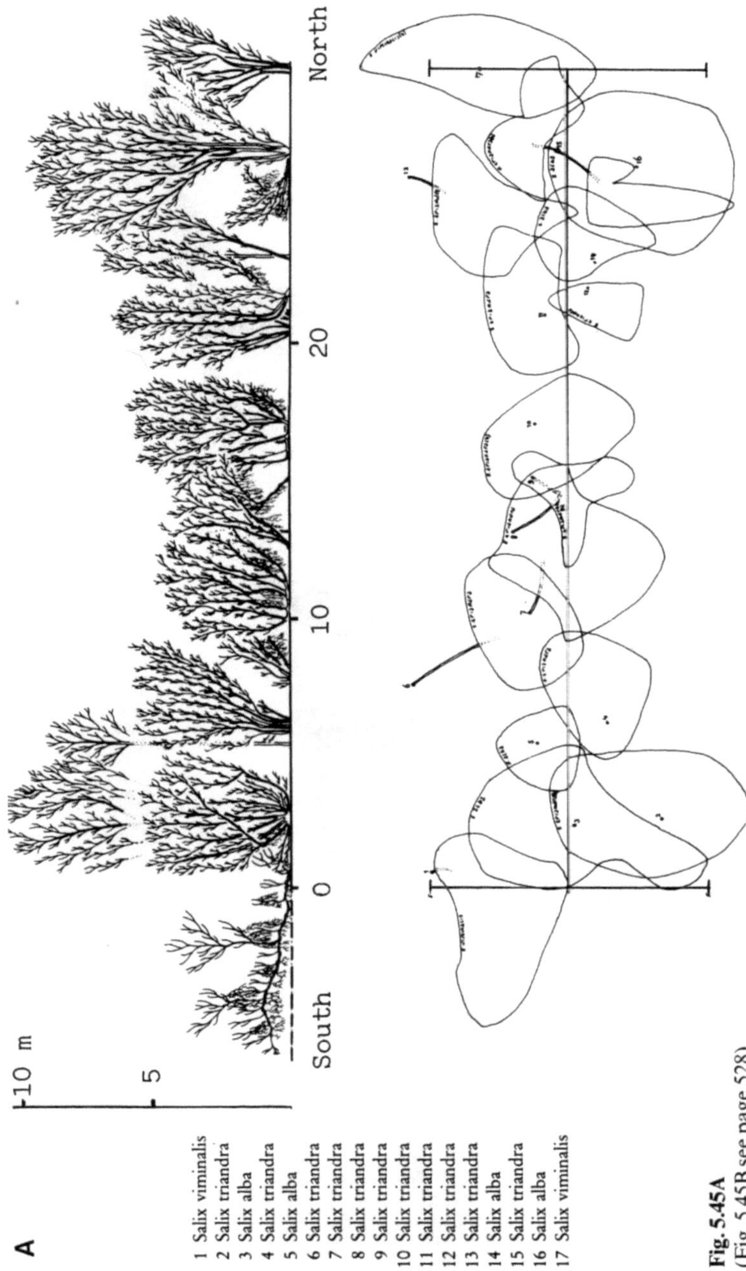

Fig. 5.45A
(Fig. 5.45B see page 528)

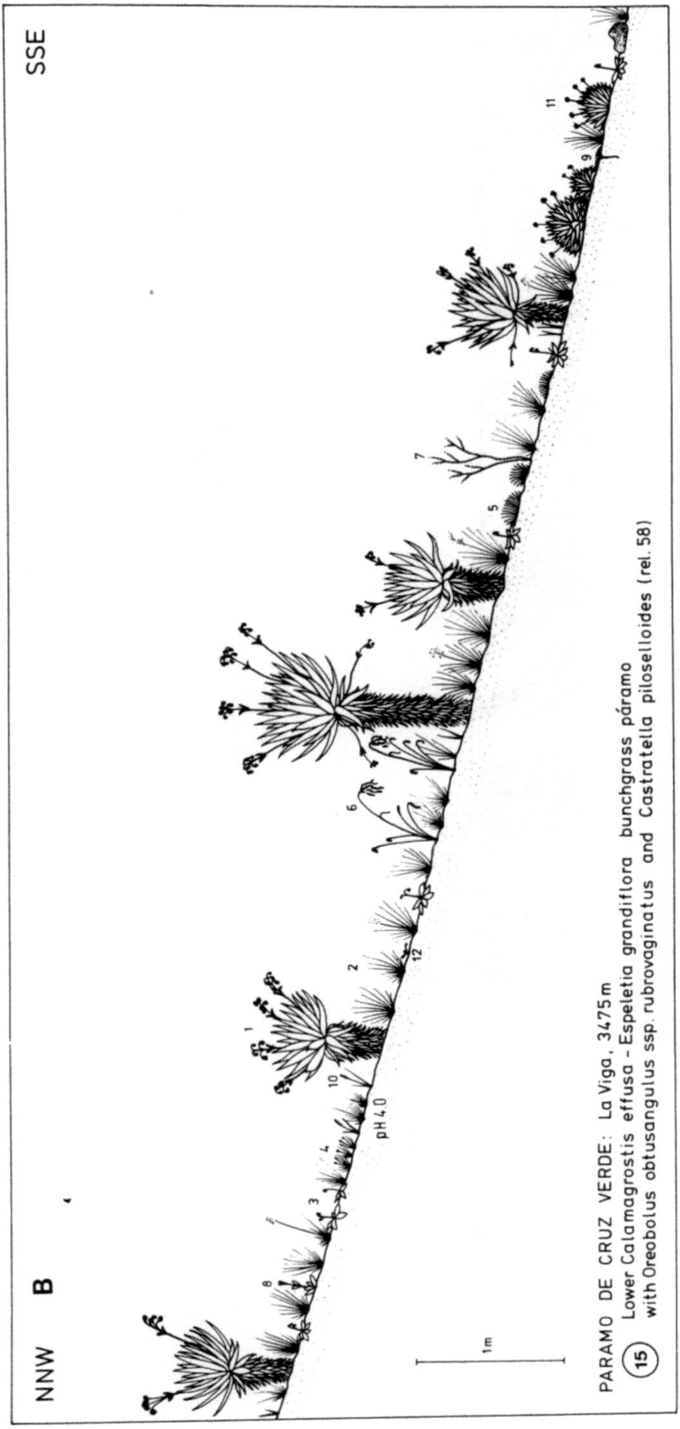
NNW
B
SSE
1m
pH 4.0
1
2
3
4
5
6
7
8
9
10
11
12
PARAMO DE CRUZ VERDE: La Viga, 3475 m
(15) Lower Calamagrostis effusa - Espeletia grandiflora bunchgrass páramo
with Oreobolus obtusangulus ssp. rubrovaginatus and Castratella piloselloides (rel. 58)

Fig. 5.45B

served by Raad in the Sudan (Fig. 5.44), here in an Andean environment that is also inhospitable for forests.

In the Andes, eco-unit development is determined largely belowground in the deep, peaty soil that may contain several hundreds of percent of its own dry weight in water. Figure 5.45B should be studied both in comparison with Fig. 5.20, in order to understand the influence of the meters-deep litter and peat layers, and in comparison with Figs. 5.31 and 5.32 in order to understand such vegetations with and without tree-like components. If in the páramo shown here there are forest eco-units, they are minimal in size, and as such should have been treated among the examples of small eco-units given later (Sect. 5.4.3). However, they illustrate so well the influence of quite small site factors on the fragmentation process that they are inserted here in the text.

5.4.2.5 Pine Forest Development in the Appennines (Italy)

This example shows a *Pinus nigra* forest (Fig. 5.46) on the Western flank of the Italian Appennines not too far from Perrugia, on calcareous soil in a locality called Poggio Miolo, that was studied by Amorini and Gambi (1976). These authors mention that the forest was planted in 1885 with both varieties of *P. nigra* (Austrian and Corsican pines), that imperfect clearcut was practiced during both World Wars (around 1916 and in 1942/1943), and that except for these harvests the forest development has been spontaneous. This study may usefully be compared with Van Heusden's work on plantations of the same species in the Dutch dune region (Fig. 5.34).

Amorini and Gambi (1976) published five transects of 20 x 2 m each. Three of these are shown here. The first one (Fig. 5.46A) shows an even-aged stand of 60 years old, that has certainly established itself on a clearcut after the First World War. On this site, *P. nigra* grew much better than on the Dutch dunes, but the same phenomenon of fragmentation of the larger eco-unit into smaller ones may be observed in both cases in the undergrowth. Figure 5.46B yields one example of the very small eco-units that may result from such fragmentations (after the Second World War), and Fig. 5.46C shows intermediate-sized eco-units, in the innovation phase (left) and the aggradation phase (right). In order to distinguish this last case, Amorini and Gambi use the term "struttura biplana" (two-level structure), to indicate the architecture of the whole mosaic shown (see also Fig. 6.9).

The species richness of the woody forest components should be noted, as well as the lens-shaped architecture of the intermediate-sized aggradation eco-unit, which is a floristically richer and somewhat larger version of Figs. 5.5 and 5.6A. The devel-

◄

Fig. 5.45. Intermediate eco-unit and forest limit. **A** Aggrading *Salix triandra* in spontaneous willow forest, after zero-event of drying the Southern Flevoland polder (the Netherlands), Lepelaarsplassen. (de Coo 1981). **B** Mini-tree groups (*Espeletia grandiflora*, Compositae) in Colombian Páramo vegetation; forest development mainly below the ground. (Cleef 1981)

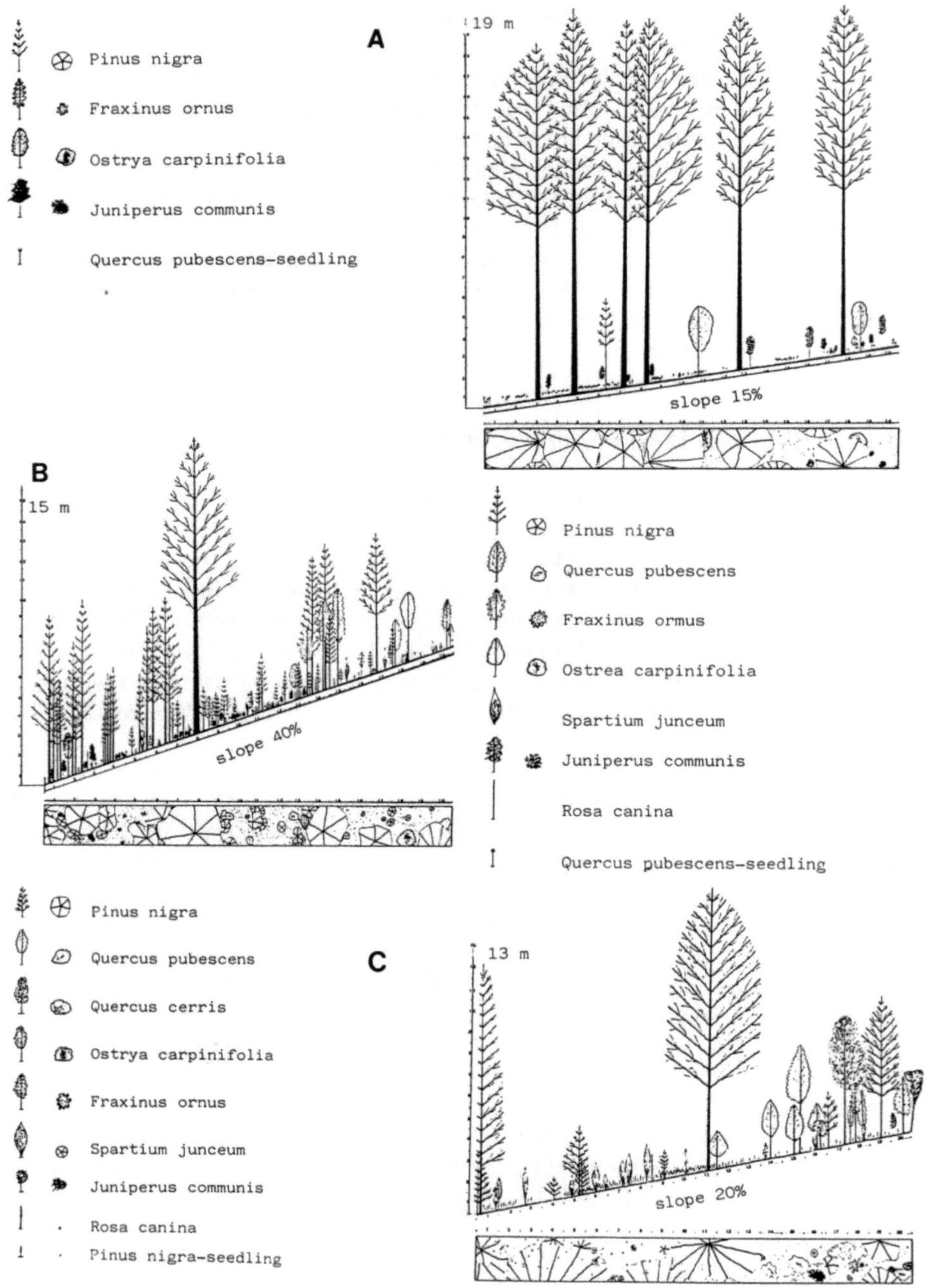

Fig. 5.46. Fragmentation of a large-sized eco-unit built by *Pinus nigra* on the west flank of the Italian Appennines. **A** Large, 60 years old. **B** Uneven-aged, in small groups. **C** Uneven-aged, intermediate-sized groups. (Amorini and Gambi 1976)

opment of intermediate-sized eco-units here can be explained by a combination of the heterogeneousness of the zero-event, which is emphasised by Amorini and Gambi, who point to the irregular disposition of the branches and other wood left after harvest, and the composition of the propagule bank, emphasised by these authors when they mention competition between trees and other plants in the innovation phase ("rinnovazione" in Italian). No mention is made in their study of other forest components than the flowering plants.

5.4.2.6 Early Aggrading Phases in Alsatian Oak Forest (France)

Two development phases, an early and a later aggradation one, in an Alsatian oak forest near Bischenberg, France (Fig. 5.47) have been published by Hoff (1978). They again illustrate, this time in a broad-leaved forest built by *Quercus pubescens,* characterised by Hoff as "low coppice with many openings", that even a propagule bank containing reiterating meristems (that yield coppice shoots) may lead to a forest mosaic that is early fragmented into intermediate-sized eco-units.

In the Alsace, this oak species occurs at the northern limits of its geographical area, which stretches much further to the south. It therefore does not reach impressive sizes. This is one reason why an eco-unit of 10 m diameter (Fig. 5.47B, right) is here indicated as "intermediate-sized". In forests built by higher trees, such eco-units would be simply small-sized, and if their difference in height and their limits were so subtle as in Fig. 5.47B, these would disappear a few years after this stage. In any case, they would become indistinguishable, and the whole forest

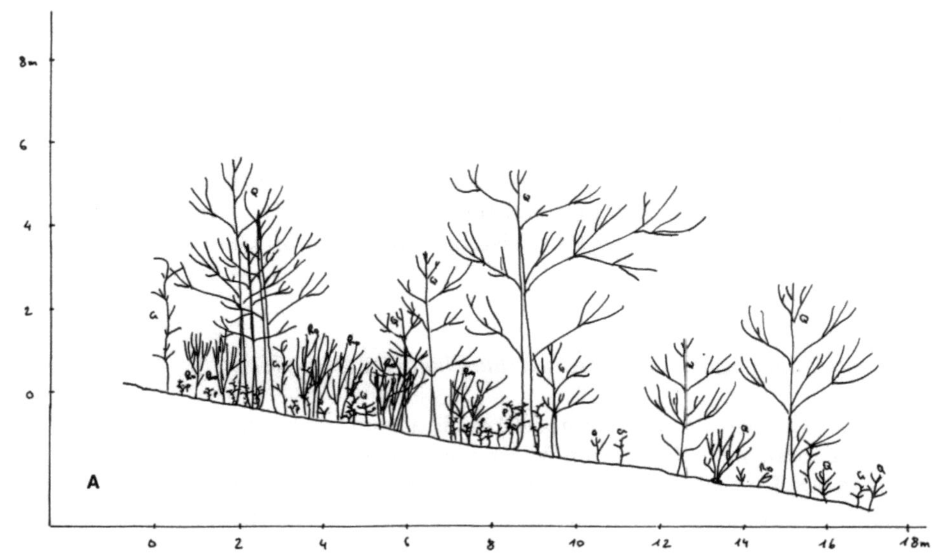

Fig. 5.47A (Fig. 5.47B see page 262)

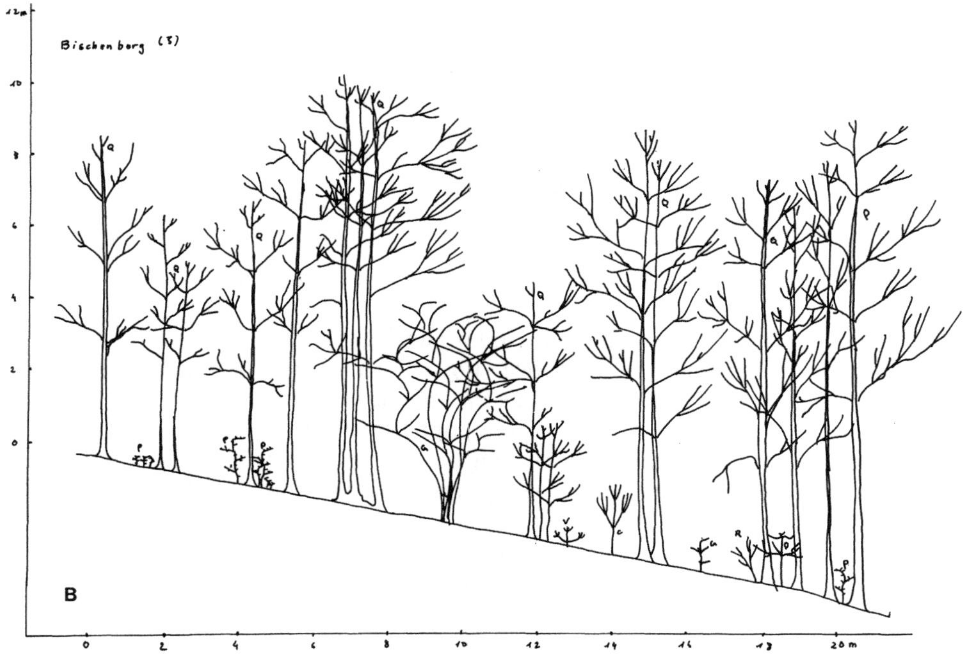

Fig. 5.47. Early (**A**) and later (**B**) aggradation phases in a low, open coppice built by *Quercus pubescens* at Bischenberg, Alsace, France. Note small size and relativity of the qualification "intermediate-sized" for the eco-unit in **B**, to the *left*. Relative size of eco-units is often defined by the ratio between diameter and height of neighbouring eco-units. (Hoff 1978)

would look like one large eco-unit, in which maybe traces of its fragmented past could be found back (see Vandeursen and Wisse 1985, for such architectural traces in high broadleaved Alsatian forests).

This is the inverted process of fragmentation. We will come back upon it in later paragraphs under the name fusion of eco-units, meaning the formation of larger eco-units by converging development of neighbouring smaller ones, or other processes. In both fragmentation and fusion, many forest components other than flowering plants may play a decisive part. In Hoff's article, these are not mentioned.

5.4.2.7 Late Aggrading Phase near the Rhine (the Netherlands)

A late aggradation phase, or the start of biostasis, in an intermediate-sized eco-unit a little south of the Rhine from Wageningen, on heavy river clay soils has been published by De Gelder (1981), and is shown in Fig. 5.48. The matrix of the whole forest, one of the few small relicts of broadleaved forests in these agricultural regions of the Netherlands, was an oak forest (*Quercus robur*) that was first planted around the year 1900. During the last World War, the estate to which this forest be-

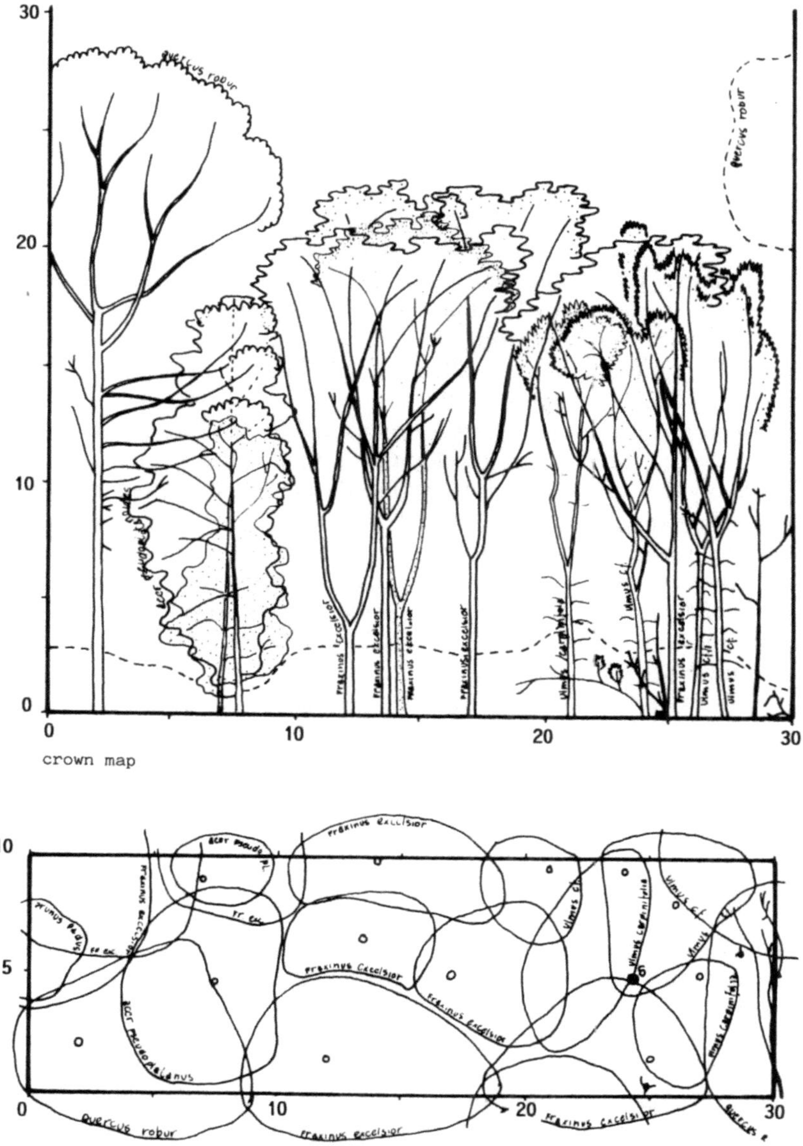

Fig. 5.48. Late, aggrading, intermediate-sized eco-unit in broadleaved forest on river clay near Hemmen, Betuwe, the Netherlands. Zero-event: bombing impact during Second World War. *Above* side-view; *below* crown projection map. (De Gelder 1981)

longs came under artillery fire and aerial bombardments in 1945. The intermediate-sized eco-unit shown in Fig. 5.48 grew up after a zero-event that was the impact of a shell or bomb. All trees have the same age: 33 years in 1978. Between 1987 and 1989 the elms became diseased and died.

Several features make this eco-unit interesting enough to present it for study. Its overall physiognomy is still lens-shaped. Different tree species have built this eco-unit: Ash (*Fraxinus excelsior*) in the centre, elm (*Ulmus carpinifolia*) on one side of the periphery and maple (*Acer pseudoplatanus*) on the other. The orientation of the transect is not given by its author, but there almost certainly has been an effect of exposition to the climatic factors, sunshine, wind and rain, that differred from one side of the periphery to the other (north/south). The border layers of the eco-unit are well visible on the drawing: the leaf canopy above, the deep crowns at the sides filter the inputs and outputs of the system. No data are available, as usual, on the eco-unit limits below the ground.

According to De Gelder, who also made a vegetation analysis of the flowering plant species, this eco-unit development (term not used by that author) agrees remarkably well with natural vegetation development as predicted by vegetation scientists like Westhoff and Den Held (1969). Although the zero-event was decidedly human in origin, and although its effects on the soil environment must have been the equivalent of deep ploughing, the propagule bank seems to have been capable of a "normal" reaction all the same. A comparison with broadleaved forest along the upper Rhine in the Alsace (Walter 1974; Van de Winckel 1980; Vandeursen and Wisse 1985) confirms this.

This example shows the origin of an intermediate-sized eco-unit by late fragmentation of a biostatic eco-unit of larger size. Most often, this is considered to be the "normal" case for such eco-units ("groups"). Our short survey of examples has already disclosed that this remains to be seen.

5.4.2.8 Eco-Units in Old-Growth Douglas Fir Forest (USA)

A mixture of decaying, biostatic (?) and aggrading eco-units in Douglas fir forests in the west of the United States is shown in Fig. 5.49. The document is one of a series established by Kuiper (1984). Publication of the whole series was achieved by Kuiper in 1988. The forest is described as an "old growth stand of over 1000 years old". In fact, what is meant is that Douglas fir trees of over 1000 years old occur in the uneven-aged tree populations covering the strip of 130 x 10 m that is shown in the transect and crown map. According to the architectural state of a 60-year-old unmanaged second growth Douglas fir eco-unit (cf. Kuiper 1986), and an old growth stand of over 400 years old, both from the same region (Cascade Mountains) and both included in Kuiper (1984), the eco-units shown in Fig. 5.49 are the result of late fragmentation of a large-sized Douglas fir eco-unit.

The pattern of fallen trunks on these three transects, and the size of these logs, show that fragmentation has started before the age of 400 years and is still not entirely finished at 1000 years. At the first millennium, intermediate-sized eco-units built by some ten large trees, nearly 70 m high at the left and 35 m high at the right of Fig. 5.49, have originated from the prolonged fragmentation process. Because architectural symptoms to establish the states "of the present", "potential" or "of the past" (cf. Fig. 5.4) are not known in conifers, the development phases of the two intermediate-sized eco-units in Fig. 5.49 cannot be diagnosed with precision. The one to the left (10–35 m on the baseline) might be a late biostatic phase, and the one to the right (85 to 115 m on the baseline) might still be aggrading, when looking at such empirical features as overall crown regularity and crown decay patterns (cf. Sect. 5.2.4.2).

The same dilemma presents itself as in the Alsatian case (Fig. 5.47). In the Alsace, small surfaces may bear a relatively "intermediate-sized" eco-unit, whereas in

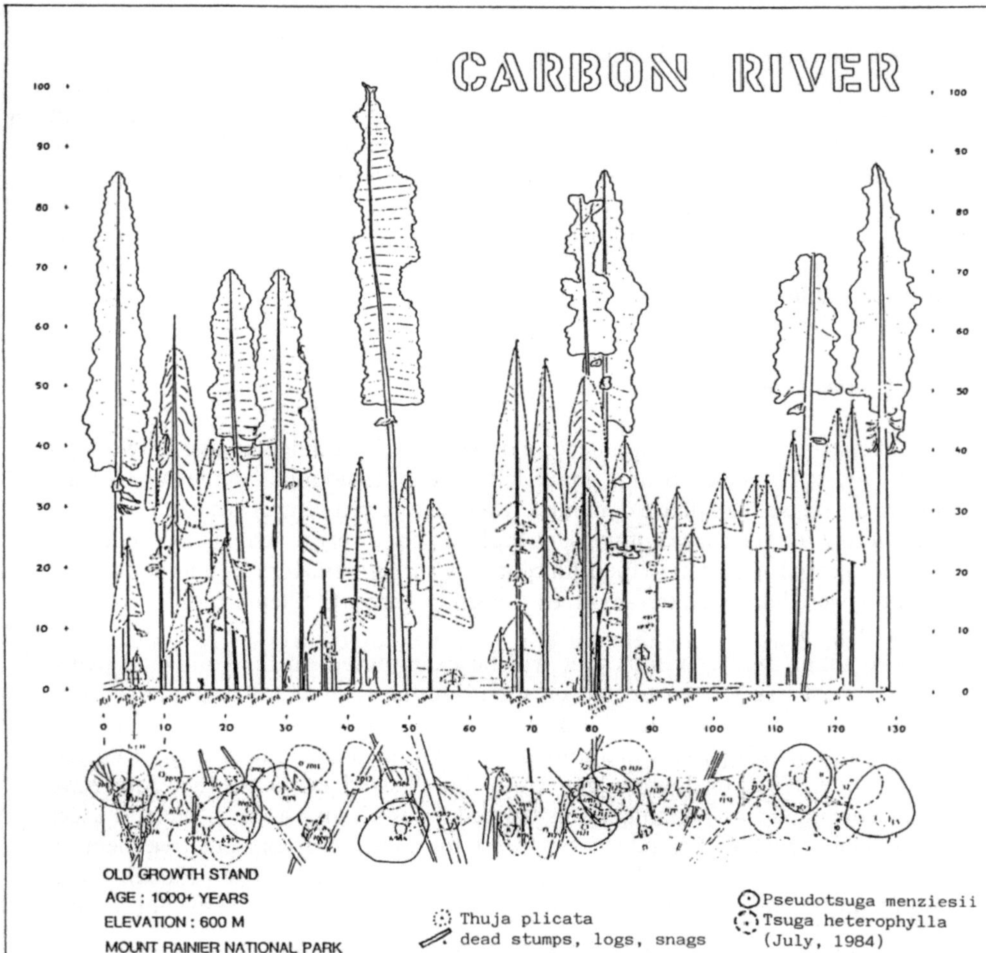

Fig. 5.49. Late biostatic (ca. 10 to 40 m, baseline) and possibly late aggrading (ca. 85 to 115 m) intermediate-sized eco-units resulting from fragmentation of an over 1000 years old Douglas fir eco-unit in the Western United States. (Kuiper 1984, 1988)

Fig. 5.49 larger surfaces are occupied by eco-units that look small in relation to the surrounding degrading eco-units that reach heights up to a 100 m. This dilemma is not new. In European forest studies it has been solved by expressing the diameter of a regeneration surface (= eco-unit) as a multiple of the average height of the surrounding forest (also see Collins et al. in Pickett and White 1985, p. 221). Limiting values between "small-scale" and "large-scale" forest regeneration were established according to microclimatic criteria. As long as the surrounding forest prevents the total predominance of the macroclimate on the regeneration surface, this is judged to be small. In North-Western European forests, and depending on the forest type, an eco-unit diameter of two to three times the average surrounding tree height defines the limiting size, below which it is judged to be "small" (among other authors, see Koop 1985; also in Fanta 1986).

We will come back to the relative status of an eco-unit as a sub-system of a forest mosaic. The comparison of Figs. 5.43 to 5.49 teaches us, however, that neither the absolute surface, nor the (potential) number of trees of the present in an eco-unit are useful criteria if forest eco-units are considered as to their role in complete ecosystems. A tree group in an extended savanna may then be conceived as a large eco-unit, whereas a similar group of trees in the midst of a 100-m-high forest can be seen as small. But for the explanation of the interior organisation of eco-units both the absolute eco-unit size and the number, architecture and size of tree components are indispensable data.

5.4.3 Small Eco-Units

Small-sized eco-units, the architecture of which is determined by one or two trees of the present in their biostatic phase, have been studied much in the last 10 years under the name "gap phase dynamics" and have been a central issue in European silviculture for at least a century. This tradition has a very practical basis: many of the European forests left in the 19th century were farmer's forests, and quite a few of these were harvested tree-by-tree ("selection system", "Plenterhieb", "uitkap", "Jardinage").

The small-sized eco-units that will be exemplified on the following pages have almost never been treated as such. During their innovation phase, they have been described as regeneration groups within the forest, most often linked to "gaps" in the canopy (for literature see Vooren in prep.; Shugart 1984; Pickett and White 1985; or Leibundgut 1982). Small-sized eco-units therefore have been considered as phenomena in "mixed forests", and as such we will meet with them again in Chapter 6 on forest mosaics. Their present treatment as living systems is not an alternative to the usual "mixed forest" model, but a supplement: *both* approaches are needed in forest ecology and silvicultural design.

Because these small eco-units were rarely considered in their own right, most documents show a forest mosaic of which they are parts. Hence some overlap with Chapter 6 is unavoidable. The documents published by other authors should not be mutilated to the point of isolating very small parts with the only aim to fit them into another approach. The only reason for reproducing fragments will be technical. Often, published transects are too large to print on book-sized pages. A choice has then to be made between reducing their size into incomprehensibility or reproducing them partly. This choice has also been made, to the best of our ability, for the preceding examples (Figs. 5.31 to 5.49).

5.4.3.1 Eco-Unit Development in Bialowieza (Poland)

The origin of two innovation phases of intermediate to small-sized eco-units in the natural forest of Bialowieza, Poland (Fig. 5.50) has been reconstructed by Derkman and Koop (1977) and Koop (1985). The zero-event in both cases was a chablis (cf. Sect. 5.1.3). Figure 5.50B shows how the mosaic of small-sized biostatic eco-units was fragmented within as little as 20 years. The plot had been chosen as a representative hectare for phytosociological studies in the International Biological Programme (IBP), because of its floristic homogeneity (cf. Falinski 1977).

However, the forest architecture in 1964, as reconstructed by Koop (1985; Fig. 5.50B), was not so homogeneous. From left to right there are four small-sized eco-units, one intermediate-sized, one with broadleaved trees (*Carpinus betulus*), and two small ones. They all seem to be in early to late biostatic phases. In 1977, the chablis on the left have already opened a place for a lens-shaped intermediate-sized eco-unit that is in the beginning of the aggradation phase. This eco-unit replaces two or three small ones, still existing in 1964. In 1984, the chablis formed by both *C. betulus* trees replaces their own eco-unit by an equivalent one. The architecture of *Carpinus betulus* (Troll's model, see Fig. 3.2) has not been rendered on the drawings of the young trees in Koop's transects.

Many details of the processes involved can be found in Koop's study of Northern German semi-natural forests (1981). This is particularly important for an aspect that can be seen in Fig. 5.50, i.e. the heterogeneousness of the situation after the zero-event, with dead wood irregularly spread over the surface and mounds-and-

Fig. 5.50A (Fig. 5.50B, C see page 268)

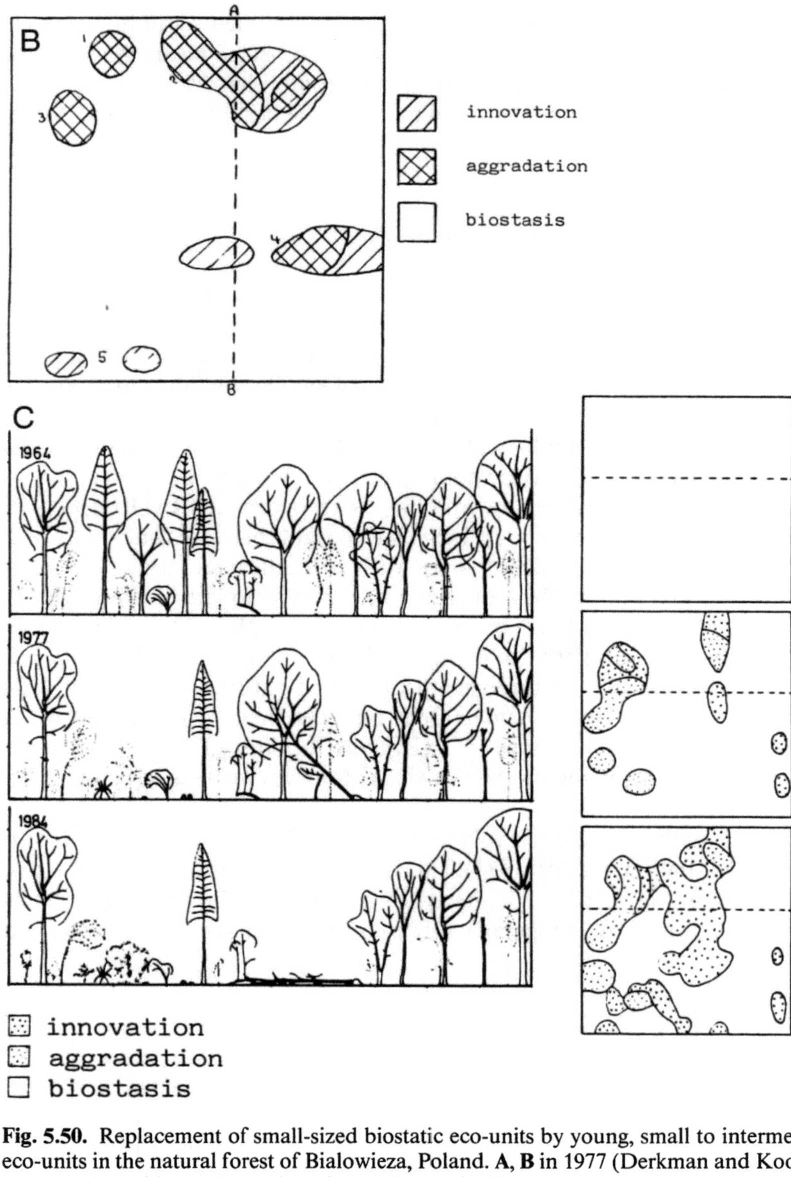

Fig. 5.50. Replacement of small-sized biostatic eco-units by young, small to intermediate-sized eco-units in the natural forest of Bialowieza, Poland. **A, B** in 1977 (Derkman and Koop 1977); **C** Koop (1985), with reconstruction of 1964 after Falinski's data

pits of mineral soil, appearing with a pronounced microrelief at the places of uprooting. The behaviour of the potential *C. betulus* (4th from the left, Fig. 5.50A) points to another important feature of small eco-units. This is the pronounced and deep border zone, with extended microclimatic gradients (Figs. 5.2, 5.59B), between the small eco-unit and the surrounding forest.

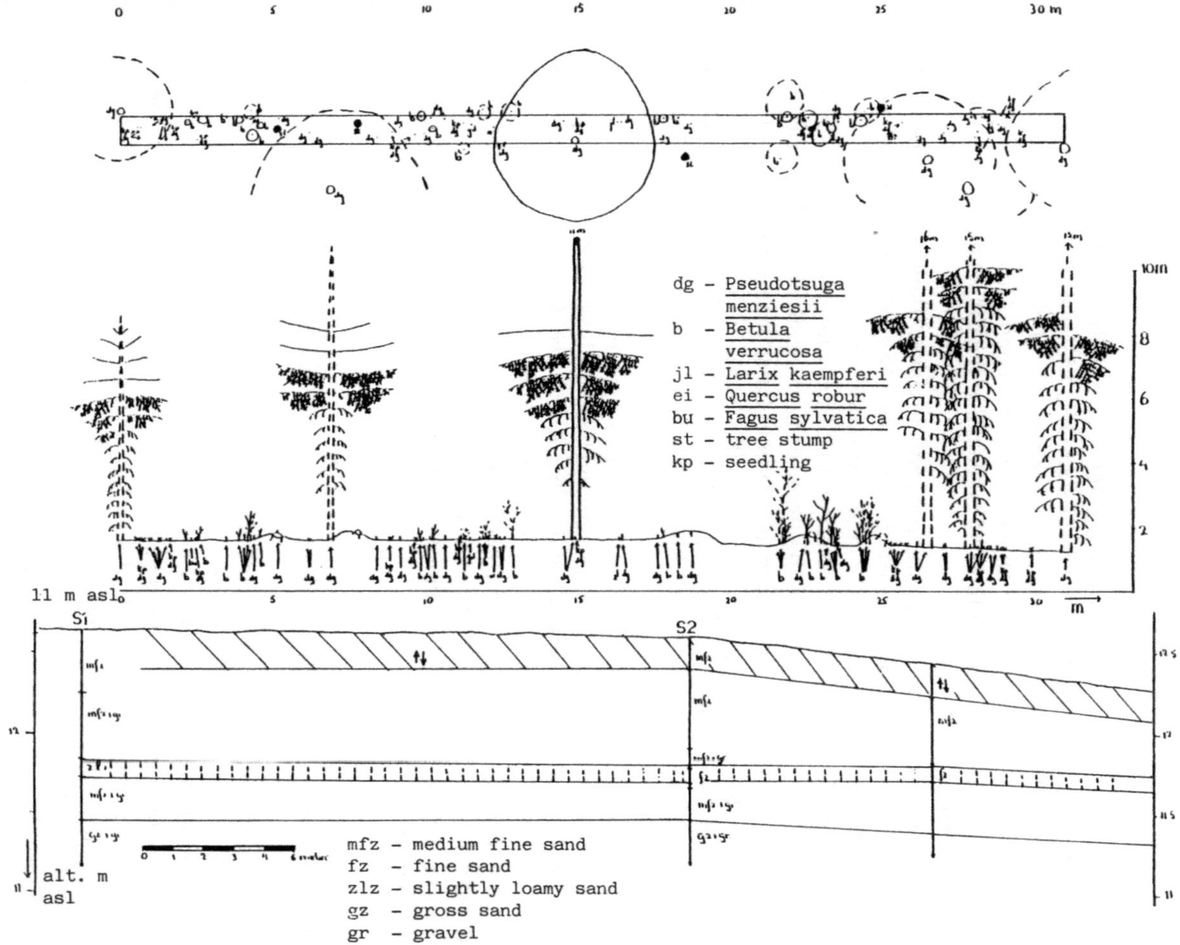

Fig. 5.51. Small-sized innovation eco-units in windthrows in a Douglas fir stand in Verwolde Estate, Eastern Netherlands. Note soil profile. (Projectgroep 1979)

5.4.3.2 Small Windthrow in Douglas Fir Plantation (the Netherlands)

The innovation phase of small-scale eco-units opened by windthrow in Dutch Douglas fir plantations on the Verwolde Estate (Fig. 5.51) was studied by a project group in 1980. The transect reproduced here is part of a series concerning eco-units of increasing size, taken from the Appendix to their report. On larger surfaces, replanting of Douglas fir and weeding had taken place, but the small unit given here was spontaneous.

This document contains a soil transect, which makes it unusual because soil data are too rarely vertically mapped in forest regeneration studies (but also see Fig. 5.58). The microrelief due to the chablis (uprootings) is clearly visible. To the left of the transect, where a little slope occurs and where the modified soil layers induce a different water budget, young trees seem to germinate preferentially on these mineralised, somewhat heightened mounds. Later behaviour of a small eco-unit is often determined by such mini-site differences in the innovation phase, when further development is prepared. This may make it quite difficult to obtain homogeneous wood plantations by using natural regeneration processes, if such is the aim of the manager.

The project group studied one animal compartment of the different eco- units. This was the group of carabid beetles (ground beetles). Both species composition and numbers of these animals differed with eco-unit size. In the limited study at Verwolde, it was impossible to link these insects to functional properties of the eco-units. For carabid beetles in the Netherlands see also Den Boer (1977), who considered whole vegetation mosaics in his study, under the name of habitat types, and therefore emphasised the importance of population change due to abiotic and human factors.

5.4.3.3 Eco-Unit Development in Tropical Rainforest (Suriname)

One rare document represents the development of the tropical rainforest mosaic in Suriname over 10 years after light selective logging. This is shown in Fig. 5.52. The situation shortly after logging was mapped by Sterringa (1971). Ten years later, the same sample plot was mapped again by Van Bodegom (1981). These kind of documents are at the basis of the 20-year series reconstructed by De Graaf (1986) for a similar forest, subjected to several chemical thinnings that increase the relative number and the growth performance of marketable tree species, to be harvested again 20 years after the first harvest in a "polycyclic silvicultural system".

The 50-m-long transect (Fig. 5.52) represents half of the total transect. In 1969, three eco-units could be distinguished (Fig. 5.52A): one in an early aggradation phase (0–10 m on baseline), the second in a later aggradation phase (10–30 m on baseline) and the third one in its biostatic phase (30–50 m on baseline). In 1979, the same small eco-units can be observed (Fig. 5.52B). In the first one, a left-over unhealthy tree (2853) has broken, and although the eco-unit has not grown in height, its tree crowns have expanded and entered into competition. Note, for instance,

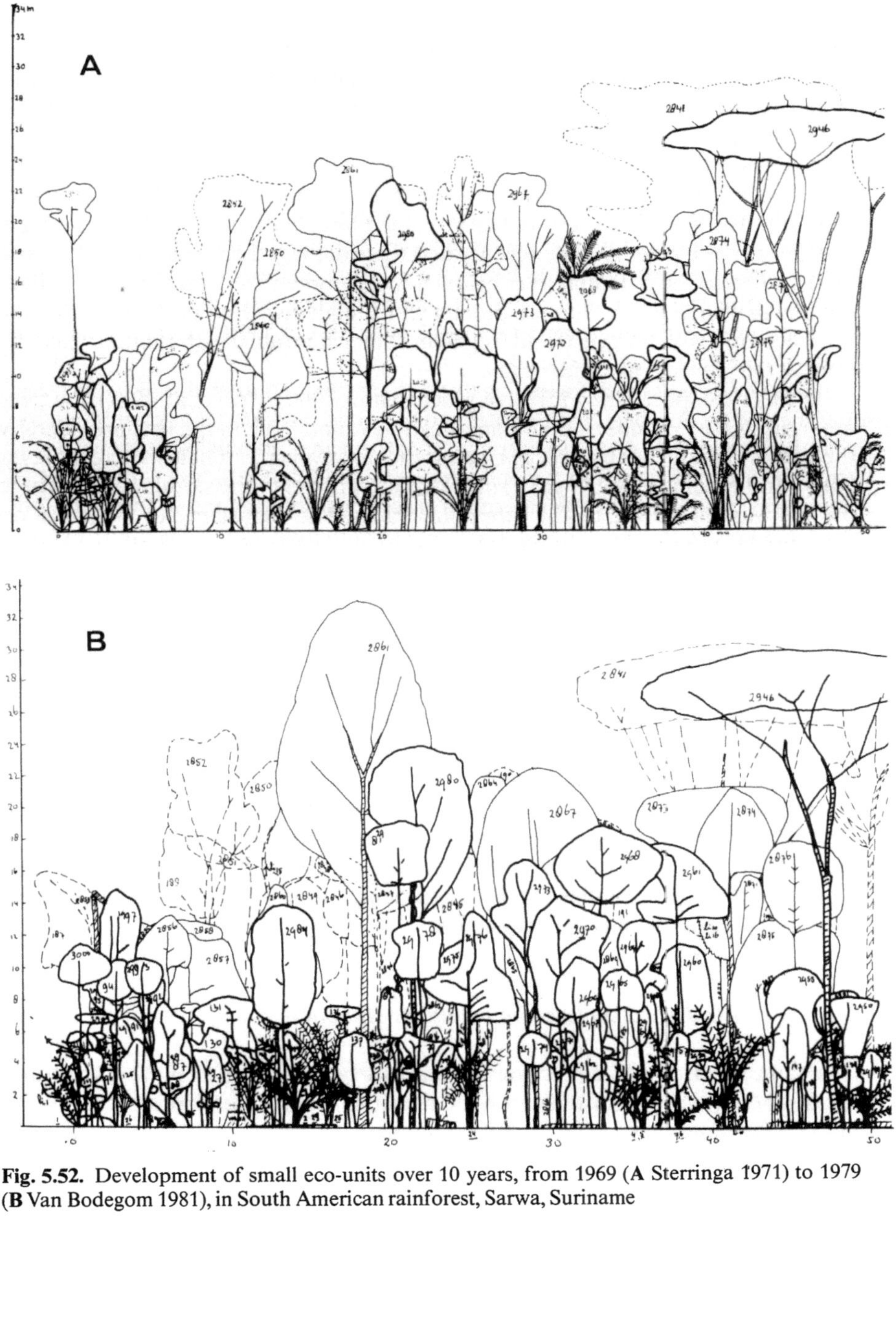

Fig. 5.52. Development of small eco-units over 10 years, from 1969 (**A** Sterringa 1971) to 1979 (**B** Van Bodegom 1981), in South American rainforest, Sarwa, Suriname

tree 2997, which has both expanded and overtopped its neighbours. In the background, between the first and second eco-unit, an aggradation phase is developing fast. In the second eco-unit, tree 2861 has taken an indubitable lead, having grown from 24 to 34 m in height in 10 years and so is starting to form a one-tree structural ensemble. This eco-unit is entering from the late aggradation phase into the early biostatic one. Between the second and the third eco-unit, tree 2867 has started to do the same, but this development is at least several years behind that of the second eco-unit. And the third eco-unit has remained largely in the same biostatic state. The trees of the present have regenerated their damaged crowns (trees 2946 and 2841), whereas the potential trees below have asymmetrically profited by the light coming in from the left. One can predict with a high degree of probability that the third eco-unit will disappear and fragment somewhere at 50 m (baseline) when the trees of the present enter into decay.

In this very species-rich forest (e.g. see Schulz 1960; Mori et al. 1987), the trees are indicated by numbers: their names can be found back in the lists provided by the authors cited. The *Astrocaryum* palms that form an apparent lower crown layer (structural ensemble; see Fig. 5.4) in reality are limited to the skidder roads where the soil was compacted by the caterpillar tracks. The persistence of the palms on these tracks prevents broadleaved trees from germinating there, whereas the compacted soil is a pessimal germination environment. This illustrates the vulnerability of tropical rainforests. Even a light logging operation changes forest development. The severity or mildness of the operations to be tolerated in forest management depends on the management aims, that may or may not include the sustained survival of the forest type in question. From a silvicultural viewpoint, managers should be warned if they take too many risks in this sense.

One has to note the extreme flexibility of the development of the small-sized eco-units in tropical rainforests such as this. It is difficult to predict just which trees are going to take the lead in eco-unit construction. And border influences (cf. Fig. 5.59B) are very strong, as illustrated by the asymmetric development in the eco-unit between 30 and 50 m on the baseline. Still, each of these eco-units shows a recognisable development phase and development. It is no figment of the imagination to look at them as distinct living systems.

5.4.3.4 Eco-Units in a Dipterocarp Plantation (Indonesia)

The same kind of eco-unit development can be found in tropical Asian rainforests. Many, very precise data on the humid and the dryer forests of the Western Ghats (India) have been united in the thesis by Pascal (1983). In the printing space available here, they cannot be reproduced without harming them. The same is true for the architectural studies on the Indonesian forests in Sumatera by Laumonier (1980). Both authors give a wealth of data, illustrations and literature that should be consulted by every forester and silvologist interested in this region.

Our example of a small-sized eco-unit from tropical Asia is taken from Torquebiau (1984). He described a man-made dipterocarp agro-forest from Sumatera, In-

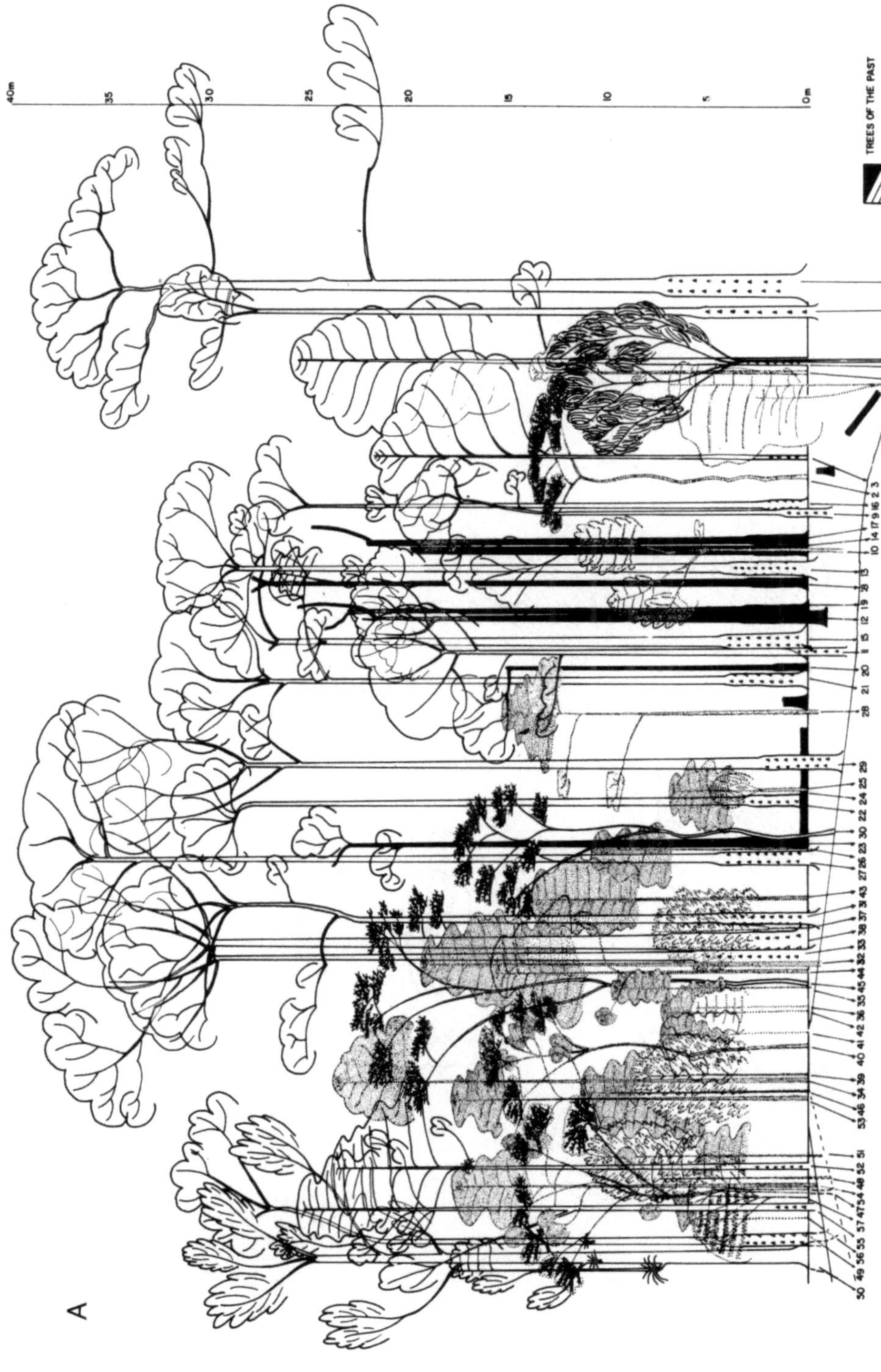

Fig. 5.53A (Fig. 5.53B see page 274)

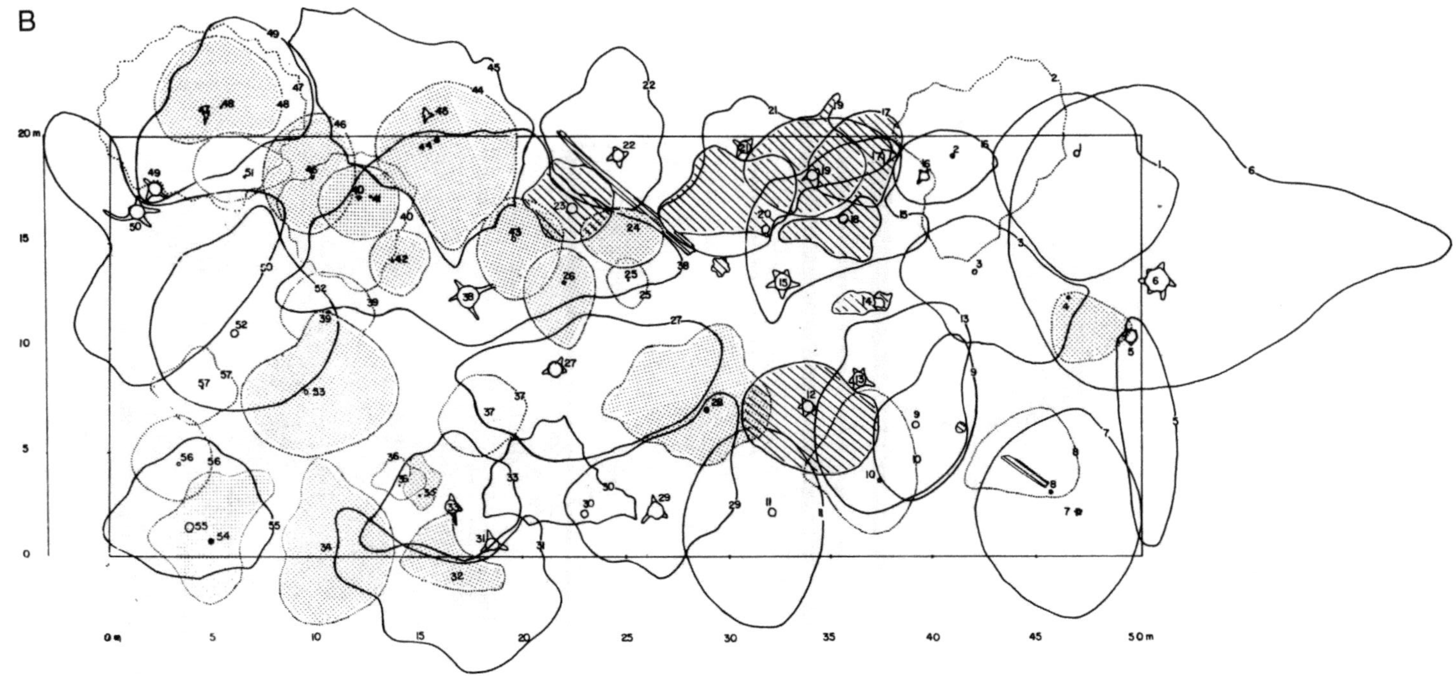

Aglaia domestica 25
Aglaia dookkoo 7
Durio zibethinus 10, 50
Garcinia parvifolia 36
Gnetum gnemon 8
Parkia speciosa 2, 30, 40, 45, 47
Plectonia glabra 56
Shorea javanica 1, 3–6, 9, 11–24, 26–29, 31–35, 38, 41–44, 46, 48, 49, 52–55
Syzygium aromaticum 37, 39, 51
? 57

Fig. 5.53. Small-sized aggrading eco-units, built by *Shorea javanica* under older eco-units with the same species, in a man-made dipterocarp forest (note traces of resin-tapping, tree bases). Krui, Lampung, Sumatera, Indonesia, 1982. (Torquebiau 1984)

donesia (Fig. 5.53). The trees which are cultivated in a semi- natural system are *Shorea javanica*, on the stem base of which the traces of resin tapping are to be seen. Resin ("damar mata kucing") is the cash harvest of villagers. A small-sized eco-unit, lens-shaped, and in the beginning of the aggradation phase, can be observed to the left on the transect. This is quite representative of the behaviour of many dipterocarp eco-units in tropical Asia. They slowly grow up below the seed-tree, the centre of the lens shape being close to its trunk. When the seed-tree finally shows dieback of the crown and decays, the young potential trees start growing and build an eco-unit that successively passes through its development phases, until only one large tree of the present remains in biostasis as a one-tree structural ensemble.

According to Smits (1982), who studied dipterocarp regeneration in several regions in Kalimantan, this eco-unit pattern can be explained by other forest components than trees, i.e. the mycorrhizal fungi. He found numerous dipterocarp species (e.g. species of *Shorea* and *Dipterocarpus*) which form mycorrhizae with rather specific fungus partners per site (also see Smits et al. 1987). The closer to the base of the seed tree the germination of the typical dipterocarp fruits (Fig. 5.54) takes place, the higher is the chance that the seedling finds the right fungus. Combined with considerable shade tolerance, this causes small-sized eco-units to have

Fig. 5.54. Winged fruits of *Shorea* spp. (Dipterocarpaceae) on the litter, which is invaded by fungal hyphae even at this fresh stage, and very near the surface. If these fungi belong to the mycorrhizal partner of the seed tree itself, the seeds stand a good chance of yielding a surviving seedling with the right mycorrhizae. Central Kalimantan, Indonesia, rainforest near Sangei Riang, December 1982. (Photograph by R. A. A. Oldeman)

an architecture that is very different from those growing up in gaps, because their centre is not in maximum sunlight (cf. Figs. 5.5. and 5.6A) but in the maximum concentration of the right kind of spores or mycelia. The zero-event is the death and decay of large standing trees, and here the meaning of "reorganisation" is very evident (cf. Sect. 5.2.6).

Once this is known, at least five small-sized eco-units can be distinguished in Fig. 5.53, if the young one is included. In the canopy there are, from left to right, one eco-unit of 30 m high, a second one of 40 m high, a third one of 32 m and a last one between 20 and 25 m. Their life-span is considerably shortened because the trees are stressed by resin-tapping. If they lived longer, these small-sized eco-units might show *fusion* and form a mosaic that could barely be distinguished from a large, even-aged eco-unit, except by refined measurements. For the architectural development of dipterocarp trees see Fig. 3.8 (after Edelin 1984).

5.4.3.5 Eco-Units in Tai Forest (Côte d'Ivoire)

Vooren (1986, and also in prep.) gives an example of small-sized eco-units in Taï Forest, Côte d'Ivoire (Fig. 5.55). From the left to the right, and also considering the map, there is one decaying eco-unit in the background, a second young aggrading one in the foreground, a third, late aggrading one with the lower structural ensemble already formed in the background, a fourth, degradation phase next to it to the right and a fifth, biostatic one in front to the right. And all these eco-units can be found on a surface of 35 x 20 m! If all eco-units were so small in this forest, there would be over 30 per hectare. It is not astonishing that Aubréville, when doing research in these African forests in the 1930's, published the often-cited idea of "tree-by-tree replacement" forming a mosaic (Aubréville 1938).

The zero-events leading to the development of such mini-eco-units are linked to tree mortality. Vooren (1986) mentions that in this forest about 33% of the fallen trees were uprooted (chablis), which is quite high if compared with some other figures. These are, for instance, 17 to 25% on Barro Colorado Island, Panama; 22% in Makokou, Gabon, according to different authors ex Vooren (1986). But it is quite low if compared to the statement by Mutoji A Kazadi (1977) that in French Guyana, Saül region, chablis are the main mechanism of tree mortality. Trees broken at the stem base after fungal attack in their basal parts or root system (rot) accounted for the other 67% of tree falls in Taï (volis). Quite often, this basal decay is linked to previous or later crown decay (see Fig. 5.55 and Sect. 5.2.5).

Even in the Tai example, the eco-units and their development phases can be distinguished. The whole transect, however, also shows the importance of border effects. No single small eco-unit can even be conceived in this forest without considering its neighbouring units or, rather, the whole mosaic. A mixed stand model is here as useful as, and maybe more practical than the model of a mosaic built by eco-units. In a mixed stand model (e.g. examples in Sects. 5.4.3.7, 8 and 9), the subsystems of the mosaic are not eco-units, but trees. It remains to be seen whether the

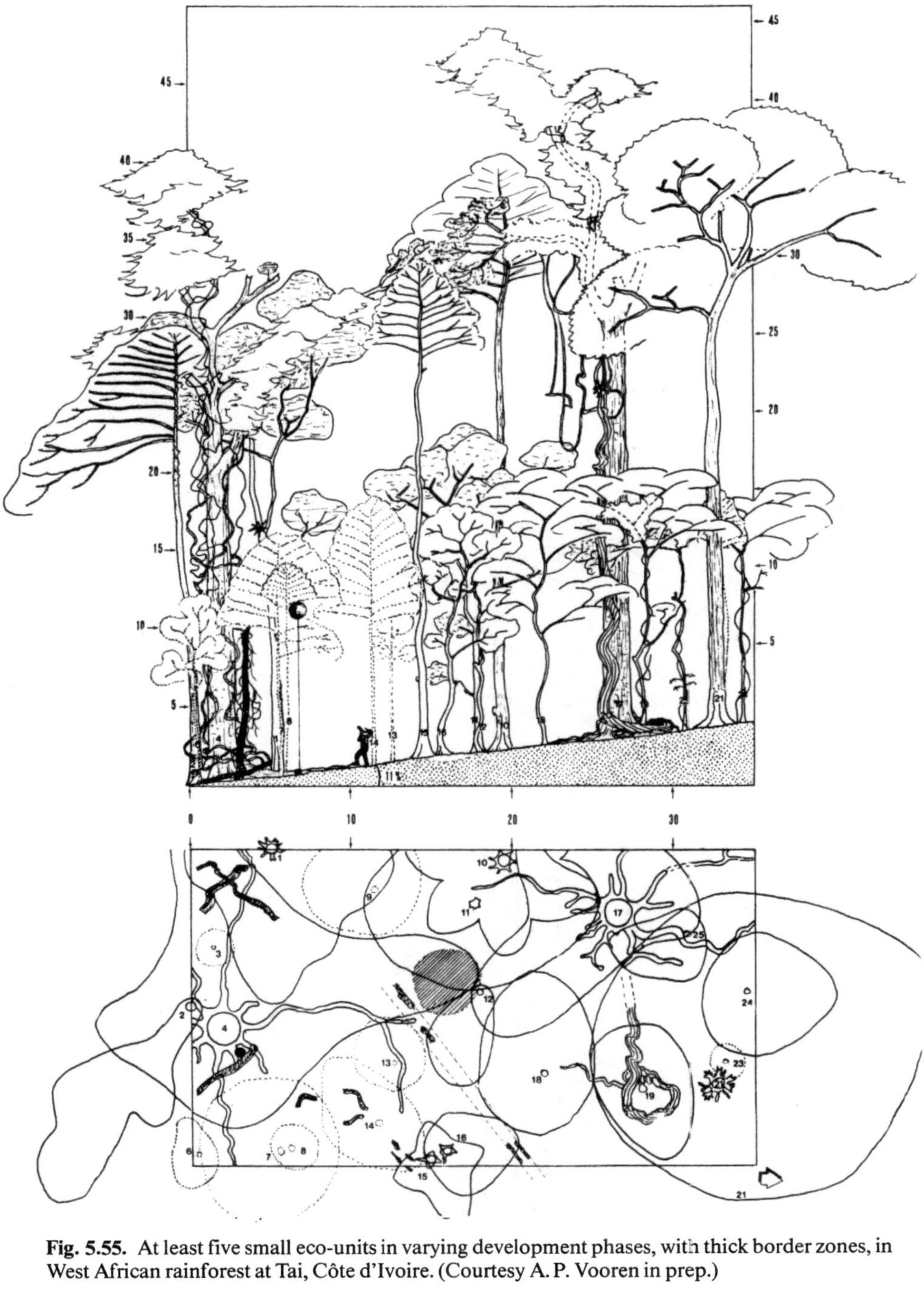

Fig. 5.55. At least five small eco-units in varying development phases, with thick border zones, in West African rainforest at Tai, Côte d'Ivoire. (Courtesy A. P. Vooren in prep.)

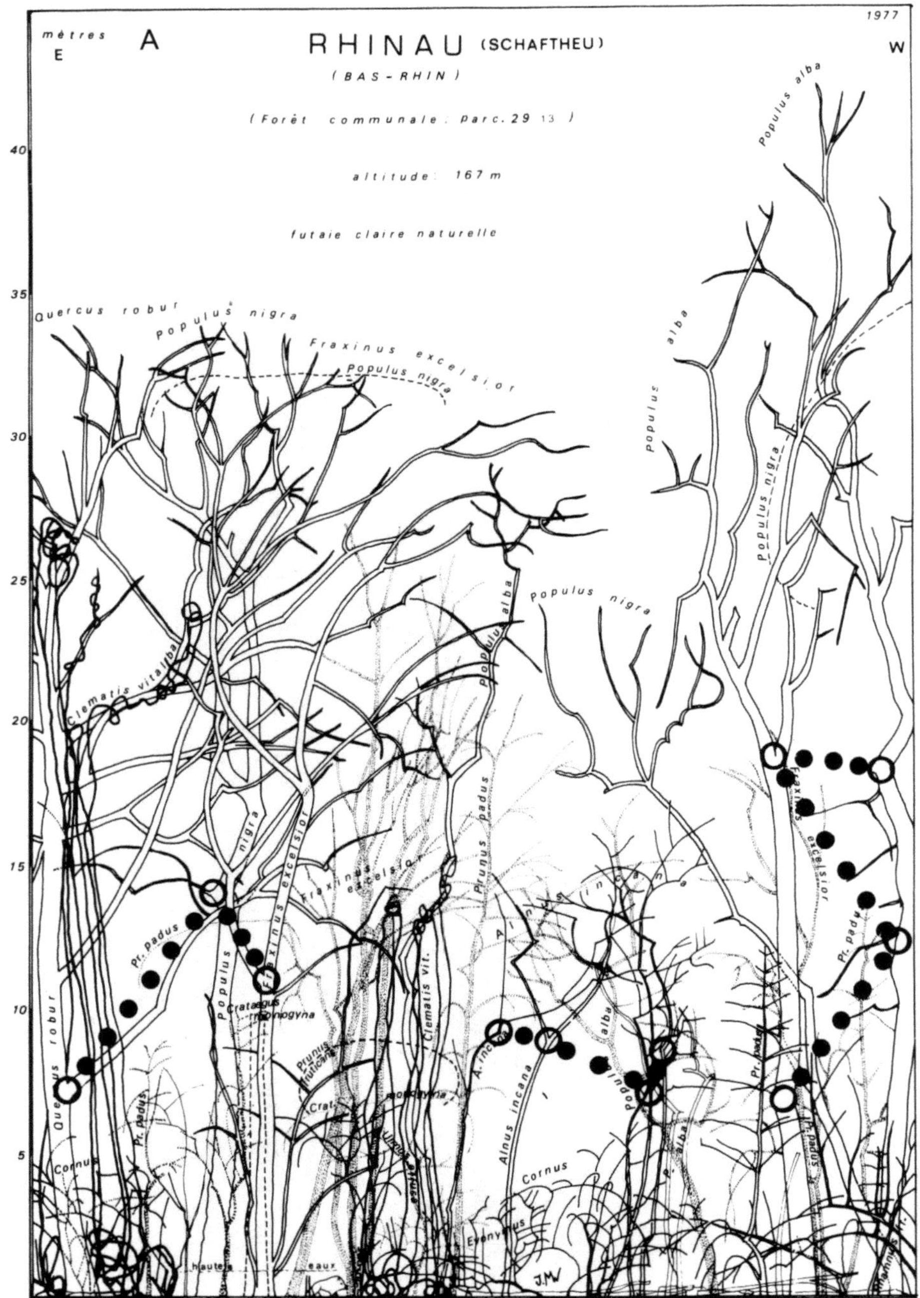
1977
mètres
E
A
RHINAU (SCHAFTHEU)
(BAS-RHIN)
W
(Forêt communale. parc. 29 13)
altitude: 167 m
futaie claire naturelle
Quercus robur
Populus nigra
Fraxinus excelsior
Populus nigra
Populus alba
Populus alba
Populus alba
Populus nigra
Populus nigra
Populus alba
Clematis vitalba
Pr. padus
Populus nigra
Fraxinus excelsior
Fraxinus excelsior
Prunus padus
Alnus incana
Fraxinus excelsior
Pr. padus
Quercus robur
Crataegus monogyna
Prunus fruticans
Clematis vit.
A. incana
Alnus incana
P. alba
Pr. padus
Cornus
Cornus
Euonymus
hauteur eaux
J.MW
40
35
30
25
20
15
10
5
5
10
15
20
25
30

Fig. 5.56A

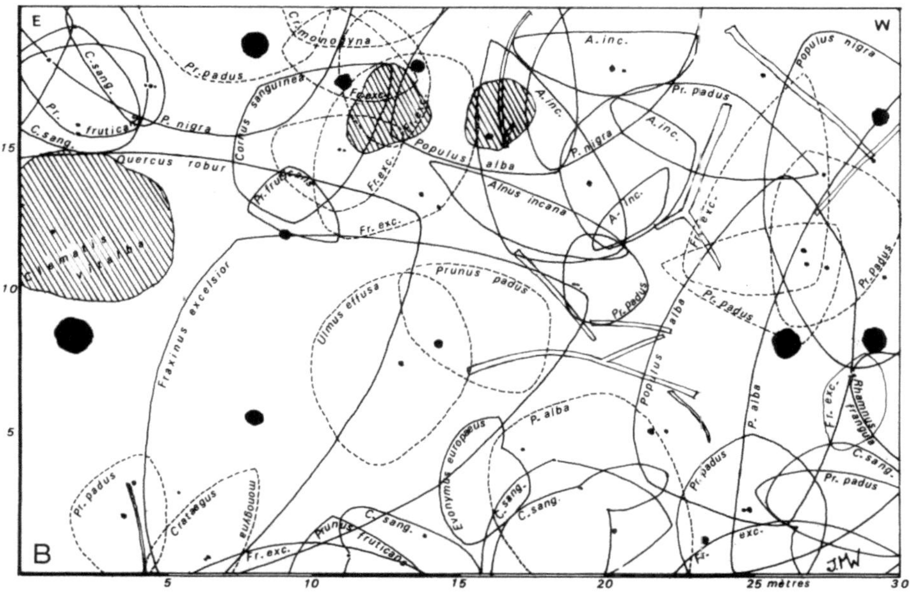

Fig. 5.56. Several small eco-units in an Alsatian broadleaved forest in the valley of the Rhine at Schaftheu (France). Note liane, *Clematis vitalba*, in border-zone. (Courtesy J.-M. Walter, University of Strasbourg)

mixed stand model can also account for the ecological functioning of non-tree components, especially small ones, without losing spatial and temporal resolution (see Sect. 2.3.2).

5.4.3.6 Eco-Units in a Mixed Hardwood Forest (France)

In order to rid ourselves of the idea that mini-eco-units such as those shown in Fig. 5.55 are an exclusive feature of the tropical rainforest, the present example (Fig. 5.56) concerns an analogous and by no means abnormal situation in the forests in the valley of the Rhine in the Alsace (Walter 1982). With the comments given above as to the previous example, the reader will easily be able to analyse the transect at Schafheu in the same way. The presence of lianes (*Clematis vitalba*) at the left of the transect should be noted.

Analogous examples can be found in other temperate regions, such as the Appalachian Mountains in the United States (see Oosterhuis et al. 1982). The documents provided by these authors have not been analysed in terms of eco-units, but can be easily interpreted in that way.

5.4.3.7 Eco-Units in Mixed Central European Forests

Figure 5.57 yields two examples of the many excellent studies on forest pattern and architecture that have been published in Central Europe since 1766 at least (see Fig. 2.3). More complete literature lists may be found in Mayer (1980), Koop (1981a, b, 1989), and Leibundgut (1982).

Figure 5.57A shows a mosaic with many small eco-units, the zero-events of which were windthrows, in a natural forest reserve in Austria at an altitude of between 1800 and 1900 m above sea level in the Alps of Niederösterreich. Zukrigl et al. (1963), who are the authors of a large series of transects out of which the present example has been taken, consider this forest as a "Verjüngungsfase" (innovation phase, see Table 5.1), which it is if one considers the mosaic. Zukrigl et al. compare the mosaic to the silvicultural "shelterwood system" (Schirmschlag, coupe d'abri) which consists in leaving a certain number of large trees standing after harvest, in order to shelter the young trees, to provide seeds, or to do both.

The Alpine forest type shown here is poor in tree species and is built by beech (*Fagus sylvatica*), silver fir (*Abies alba*) and Norway spruce (*Picea abies*). Because the temperament of each is different, they follow each other in elegant cycles of

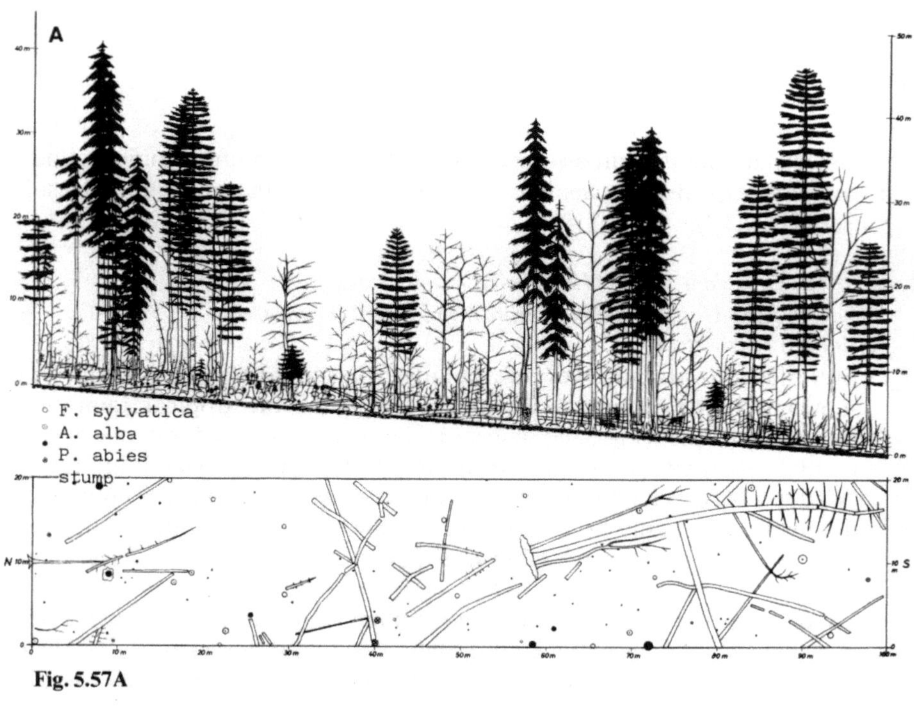

Fig. 5.57A

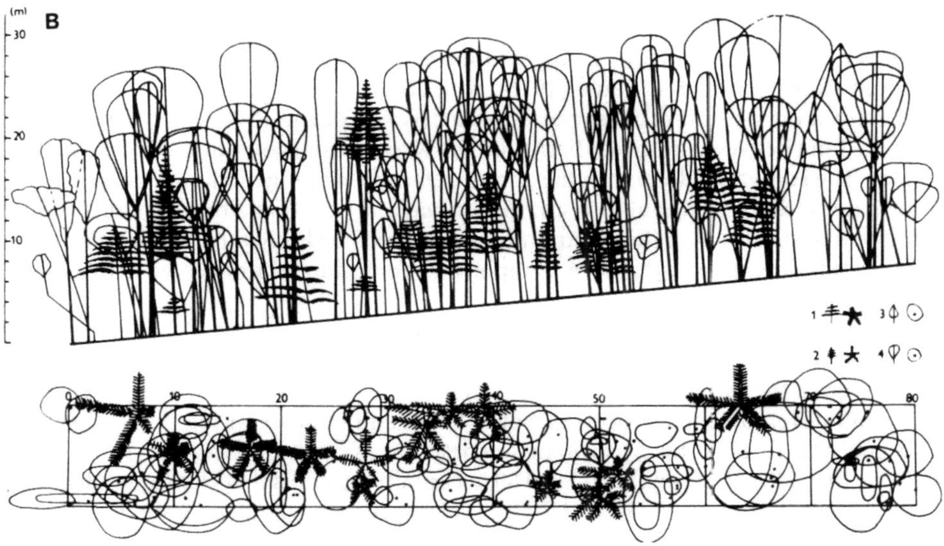

Fig. 5.57. Forest mosaics with small-sized eco-units in Central Europe. **A** *Fagus-Abies-Picea*-forest at ca. 1800 m asl in Austria, after Zukrigl et al. (1963). **B** Richer type of analogous forest at ca. 800 m asl in Czeckoslovakia, after Korpel and Paule (1976). *1 Taxus baccata*; *2 Abies alba*; *3 Acer* sp.; *4 Fagus sylvatica*. Note that innovation and aggradation are predominant in **A**, biostasis in **B**

eco-units which have their innovation phase in the light or light shadow (spruce), in the shadow (beech) or in heavy shade (fir). In the given transect all beeches are still potential, considering their architecture. The remaining firs and spruces (gymnosperms!) seem to be quite healthy still, in their state of the present, although they may be weakened by their more or less sudden isolation, exposing them to weather and biotic impacts. There is a large number of small broadleaved eco-units in different moments of the aggradation phase, and some (to the left) in the innovation phase. The older eco-units, built by gymnosperm trees, are less easy to diagnose. In every case, they form minimum-sized eco-units. Mulder and De Waart (1985), in their architectural study of the Austrian Plenterwald, in which tree-by-tree mortality is replaced by tree-by-tree cutting, had difficulties in establishing the limits of such coniferous eco-units.

These difficulties are real, and are due both to the lack of architectural criteria to determine the state of a needle-tree (potential, of the present, of the past, cf. Fig. 5.4), and to the large influence and importance of border effects between eco-units. Especially in these forest types, small eco-units blend into each other. It may seem to be a paradox, but especially these coniferous forests built by numerous small to very small eco-units were the least conducive to the discovery of the eco-unit concept and also the least apt to be easily analysed in eco-unit terms. At another hierarchical level, an analogous statement can be made for tree architecture. A comparison between Gatsuk et al. (1980) and Hallé et al. (1978) may illustrate that here also, other climatic zones favour other ways of analysis.

Figure 5.57B shows another Central European forest mosaic, with small eco-units that seem to be mainly in aggradation and biostatic phases. The transect was taken from Korpel and Paule (1976), who studied forest dynamics and species composition in a nature conservation area between 800 and 900 m above sea level, at the limit between the Great Fatra and the Kremnitzer Mountains (Czeckoslovakia). The study was done with the aim to determine the regeneration mechanisms of natural relictual populations of *Taxus baccata*, in a forest that contains some more tree species than the Austrian example, the Czeckoslovakian one being at a lower altitude. *Acer* sp. ("Ahorn", maple) may for instance be observed in this transect.

This transect gives an idea of the later development into a broadleaved forest of young mosaics such as shown in Fig. 5.57A, although the different geographical position of the two transects precludes more than a general correspondance in the overall developmental processes. Some traces still indicate in Fig. 5.57B that the beeches and maples have grown up as components of small eco-units. The blocks from 10 to 30 m (baseline) and 70 to 80 m contain eco-units that clearly are distinct and small. But between 30 and 70 m, fusion between small eco-units has taken place. The whole block may well behave henceforth as one large eco-unit, the origin of which can only be suspected by careful analysis of the inner patterns, which do not correspond to the architecture resulting from the development shown on Fig. 5.4.

The only coniferous tree components that are present are the most shadow-tolerant species (*Abies alba, Taxus baccata*). The silver fir (*A. alba*) certainly is potential but the yew (*T. baccata*) probably occurs both as potential trees and trees of the present (also see Koop 1981a). Both species are able to form structural ensembles, at different heights. Their distribution over the transect already announces where they will build small eco-units when the broad-leaved trees decay. The same phenomenon occurs in the oak-maple forests of New England, where *Tsuga canadensis*, in groups or alone, survives the oaks, maples and birches (cf. Hallé et al. 1978, their Fig. 93; Oliver 1978; Oliver and Stephens 1977).

The development cycles in such forests, as analysed by the Central European authors, will be treated in Section 6.3.3.

5.4.3.8 Eco-Units in Coniferous Japanese Mountain Forests

An example of small eco-units, composed by 30 m high *Abies sachalinensis* and *Picea glehnii*, overtopping and smothering at least one small eco-unit built by *Betula ermanii* and maybe *Sorbus commixta* (Fig. 5.58) is taken from Kojima (1983). This forest is situated in Kitami District, Hokkaido, Japan, and is part of some natural forest relics in the mountains. Plot 8213, shown on the transect, grows at an altitude of 1050 m above sea level. It is part of a series of 22 such plots which are carefully analysed by Kojima and convey a beautiful image of the coniferous forests of this region in Northern Japan.

Most of these plots show a mosaic of small eco-units in different phases of development. Some also represent intermediate-sized eco-units. The dynamics of the

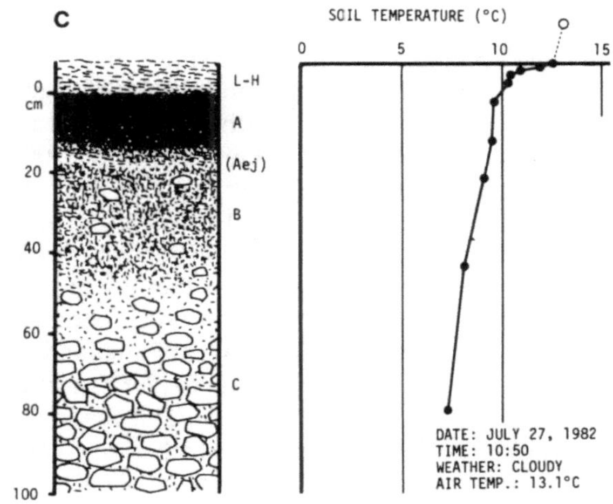

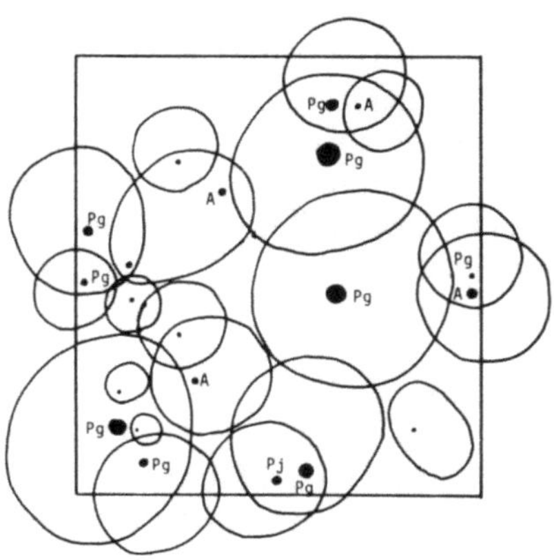

Fig. 5.58. Small-sized eco-units in aggradation and biostatic phases in an *Abies sachalinensis* forest at 1050 m asl in Kitami District, Hokkaido, Japan. **A** Side-view and crown projection; **B** h/d relationship for tree species mentioned; **C** soil profile and temperature gradient. Note soil data (thick litter) and floristic richness as compared with Europe (Fig. 5.57). (Kojima 1983. Reference line h = 100d added). *A Abies sachalinensis*; *Pg Picea glehnii*; *Pj Picea jezoensis*; unmarked *Betula ermani* and *Sorbus commixta*

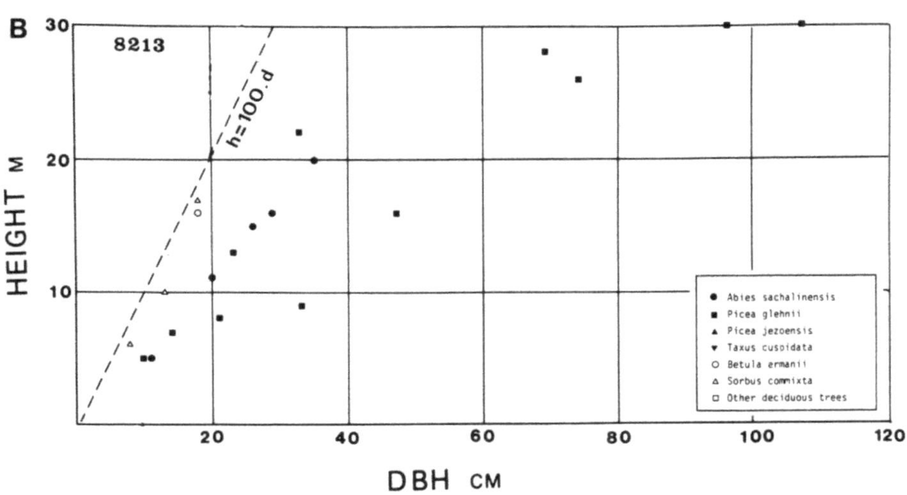

forest seem to be energised by zero-events in the form of chablis or volis (Sect. 5.1.3), but also by standing tree mortality, as testified by the photographs in Kojima's publication. Figure 5.58 shows the graphs that have also been united for each of the other plots, in addition to tables of vegetation data (species of higher plants and bryophytes, mosses, with their cover) and soil properties.

We have added the reference line h = 100.d on the h/d graph (Fig. 5.58B), so that it may be compared with Fig. 2.44. This comparison strengthens the suspicion that for conifers the constant 100 does not apply as it does for many broadleaved trees (cf. Sect. 3.7). The question needs more study.

The soil profile (Fig. 5.58C) does not show a clearcut root floor (cf. Sect. 5.3.3), which is confirmed on similar profiles for other plots on which the visible root fragments are drawn. The thick layer of litter and humus is to be noted. It is about 7 cm deep, but this differs from plot to plot. It seems to be thinner notably when aggradation phases are predominant. This fits in with the processes of litter formation during the development of British pine stands (Ovington 1965). Thinner humus layers also occur in plots where large oaks (*Quercus mongolica*) are present. The thickest layers occur under high, biostatic coniferous eco-units. This points to a feedback mechanism inside the eco-unit, between the crown canopy and the rhizosphere. Indeed, at first sight it is reflected by species composition and cover percentage of the non-arborescent plant components (cf. also Ovington 1965).

Soil and vegetation data are also available in the studies from Austria and Czeckoslovakia in Section 5.4.3.7, but they are given in another way, that is less closely linked to the plots examined, and depends rather on known phytosociological classifications and site correlations. These cannot be discussed in the short space available for the present examples, but they are eminently useful.

5.4.3.9 Eco-Units in Fragmenting Oak Forests (Austria)

Forests dominated by oaks are among the important forest types of the world, and occur in North and Central America, the Indonesian mountains, Europe, China and Japan, to mention some areas. The example given here is situated in Austria in the famous Wienerwald (Vienna forest) in a natural oak forest reserve called Johannser Kogel (Fig. 5.59A). This whole forest area has been studied by Mayer and Tichy (1979). Only two transects out of ten are reproduced here. As in the former Central European examples, vegetation and soil data are included in the study by Mayer and Tichy, according to known classifications and correlations.

These oak forests, built by *Quercus petraea* and *Q. cerris*, are quite old: an age of 200 to 300 years is mentioned for the trees. The large eco-units, which may have been the result of long-ago planting or fusion of smaller natural units, are now in fragmentation. Figure 5.59A(1) shows the start of this process ("Frühe Terminalphase" of the mosaic) and Fig. 5.59A(2) represents a later phase ("Späte Terminalphase"; see Table 5.1). One small eco-unit starts aggradation in the early terminal oak phase (see crown projections in Figures). It is composed by a group of *Carpinus betulus*. These groups occur in much larger sizes under the later degradation phase of the oak forest.

Carpinus betulus (common hornbeam), like beech (*Fagus sylvatica*) is a species that may germinate and survive in rather deep shadow. If, under an older biostatic eco-unit, the undergrowth is sparse, both beech and hornbeam may already form sparse aggradation phases, composed by meagre trees, below the older eco-unit. These two species grow according to Troll's model (Fig. 3.2), whereas oaks show Rauh's model. All axes being plagiotropic in Troll's model, leaves can been pushed easily in a lateral direction to intercept rays of through-falling light, each year again (see Fig. 5.61). In Fig. 5.59A (2), at the right side, there also is a small aggrading eco-unit built by maple (*Acer campestris*). This tree grows according to Rauh's model and is somewhat less shade-tolerant. On the same figure, to the left of the middle, there is a one-tree biostatic eco-unit with *Carpinus be-*

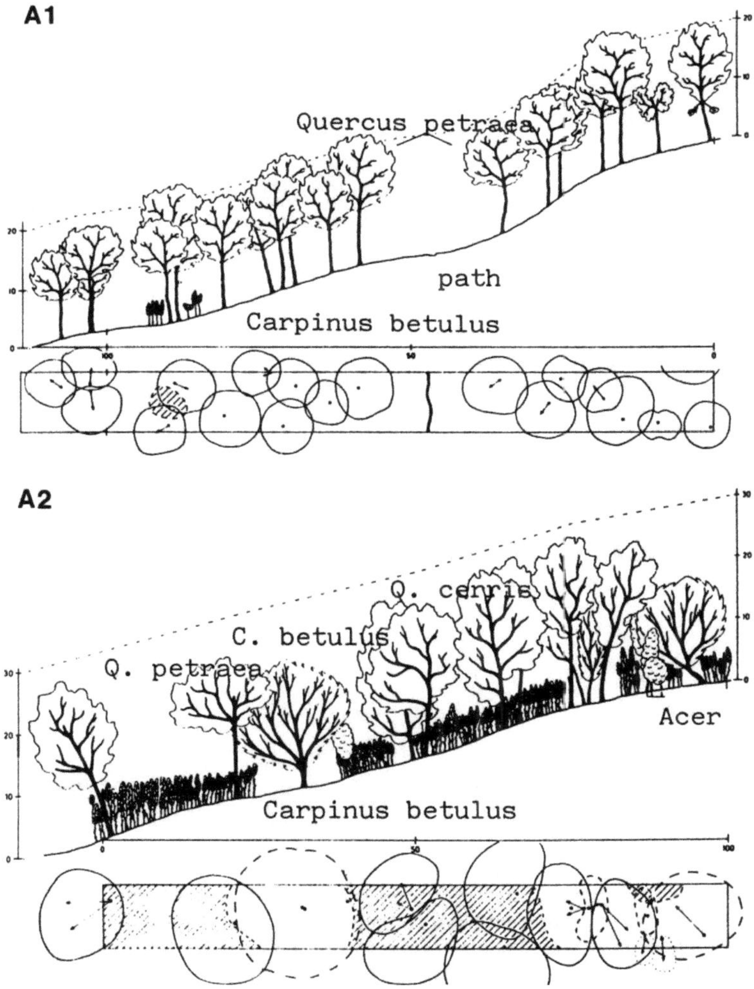

Fig. 5.59A1, A2 (Fig. 5.59B see page 286)

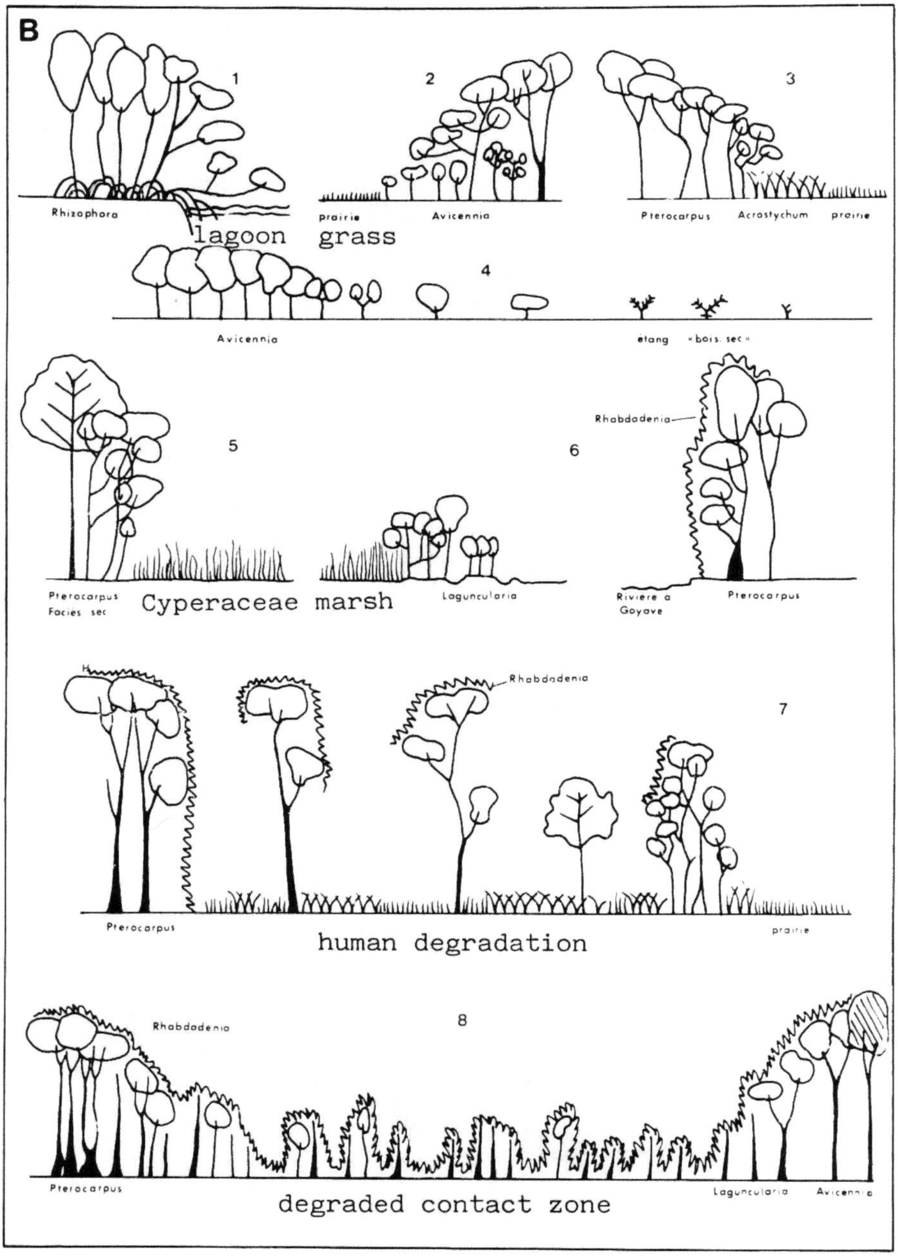

Fig. 5.59. A Two development phases of large biostatic (**A1**) and degrading (**A2**) *Quercus-Carpinus* eco-units in the Wienerwald, Austria; fragmentation yields place to smaller aggrading eco-units built by *Carpinus* and *Acer*, which will enter into fusion later. (Documents from Mayer and Tichy 1979). **B** Different types of eco-unit border zones in Antillean mangroves and marsh forest, some being determined by site factors, others by biotic or human factors. Note lianes. (Lescure 1980)

tulus. The species coexists with the neighbouring oak eco-units and does not invade them.

The future of this forest mosaic appears to be evident. The oaks will die one after another, a process that may well take some more centuries because oaks are long-lived (Fig. 5.6B). Wherever the oak eco-unit fragments, parts of the aggrading hornbeam eco-unit will develop fast. Oaks are gradually eliminated, and because they do not germinate and become established under the dense shadow of trees with Troll's model, the whole mosaic will later be mainly composed by *Carpinus betulus.* One of the main factors that comes into play here again is the influence of bordering eco-units.

5.4.3.10 Eco-Unit Borders (Antilles)

The only example of the architectural analysis of eco-unit borders known to us is represented in Fig. 5.59B. It is taken from Lescure (1980), and shows the border zones found by this author in Antillean mangrove and swamp forests. It is not too far-fetched to show tropical border-zones among paragraphs on temperate examples, because these particular tropical forest types are handicapped by the site and hence are not much richer in species and structures than the European examples. The architecture of the border zones is determined mainly by orthotropic growth models (Attim's model in *Rhizophora*) and plagiotropic axes (Troll's model in *Pterocarpus*), with *Avicennia germinans* probably showing a more complex architecture with metamorphosis (cf. Figs. 3.1 and 3.2). This metamorphic growth pattern might be comparable to temperate region birches (*Betula*, cf. Fig. 3.9).

Several of the border zones represented in Fig. 5.59B are linked to site conditions, certainly the cases 1 to 4 and maybe also 5. Case 5, however, shows the sealing of an eco-unit limit by the asymmetric crowns of the smaller trees, which are built both by unilateral branching and reiteration patterns (cf. Oldeman 1972). In all other cases, this sealing of limits, made by human interference in cases 7 and 8, is operated or strengthened by lianes of the species *Rhabdadenia biflora* (Apocynaceae). In Fig. 5.56, traces of lianes bordering an eco-unit can be read on the map (*Clematis vitalba*) and traced back to the transect of the Alsatian forest at Schaftheu. As a matter of fact, liane curtains in most forests of the world that grow on not too inhospitable sites fulfil some sealing function between eco-units that differ much in height (see Fig. 4.6). This function is an eco-unit property, not a liane property.

The thickness and architecture of these border zones may be such that they cover the whole of a small eco-unit which they surround. The microclimatic gradients, both vertical and horizontal, within larger eco-units have a steady pattern. This is established at canopy closure and changes with eco-unit development until biostasis is reached. Such gradients are absent from small eco-units with thick border zones. In fact, each small eco-unit influences its neighbours to such an extent that a complex pattern of gradients arises which can more easily be understood at the hierarchical level of the mosaic (cf. Oldeman in press 1989). This transfer of ecological functions from one organisation level to the next higher one (cf. Sect. 3.8.3) will be considered again in Chapter 7. But it announces itself already when considering eco-unit borders.

5.4.3.11 Fusion of Small Eco-Units in Beech Forest (the Netherlands)

The size of eco-units becomes quite obscured in such examples as were announced already by Figs. 5.57 to 5.59. The forest in the sandy centre of the Netherlands, in the Speulder- and Sprielderbos, contains well-studied examples of the temporary coexistence of oak (*Quercus robur*, some *Q. petraea* and hybrids) and beech (*Fagus sylvatica*). According to Dutch standards, the sites are not bad for forests, but in comparison to oak sites elsewhere in Europe they are poor. Hence the oaks do not grow fast and tall. Figure 5.60 shows the invasion by beech in oak plantations, as observed by Van Dort et al. (1979).

The sequence of events is that large, planted eco-units of oaks (Fig. 5.60A), which are still in a late aggradation phase, show an undergrowth that is a young aggrading eco-unit, or a series of smaller eco-units, built by spontaneously immigrated and germinated beeches in a young aggradation phase. For some time, the complex of two interwoven forests behaves more or less like a biostatic phase with two structural sets, potential beeches below mixing with a layer of some shrubs of the present and potential trees of smaller species (e.g. *Sorbus aucuparia*). This may be compared with Fig. 5.4C.

But soon, some beeches start growing faster, overtaking the oaks. Because the growth rate of the different beech trees is different, some start mingling with the oak crowns already, whereas others remain behind (Fig. 5.60B). The beech forest below is breaking up into small eco-units, some of which have already killed the oak trees by outshading them. In this stage, both the oak eco-unit and the beech eco-unit are fragmented.

In a third phase (Fig. 5.60C), all or almost all oaks have been eliminated, and if there were small beech eco-units before, they now have entered into fusion and form a new, composite large-sized monospecific eco-unit. On better sites, many oaks survive and may tower above successive beech forests for centuries (e.g. Koop 1981a). But in the forest shown in Fig. 5.60, beech is there to stay and to regenerate itself as soon as large enough openings come into being in decaying and fragmenting degradation phases of beech. The undergrowth of the beech forest on this kind of site is empty of other flowering plants.

This process and these patterns have been explained in terms of the above-ground architecture of oak and beech by Hilgen and Van den Brenk (1984). They studied pairs of oaks and beeches at different ages in the Royal Forest Estates near Apeldoorn, on comparable sites and close to the stands studied by Van Dort et al. (1979). Three of such pairs are shown in Fig. 5.61. Figure 5.61A was drawn in an even-aged oak-beech mixture, sown in 1949. Figure 5.61B and C were established in converted oak coppice, where stump shoots had been left and oak saplings had been planted between them in 1926. The processes, of which momentary snapshots are shown, hence were not

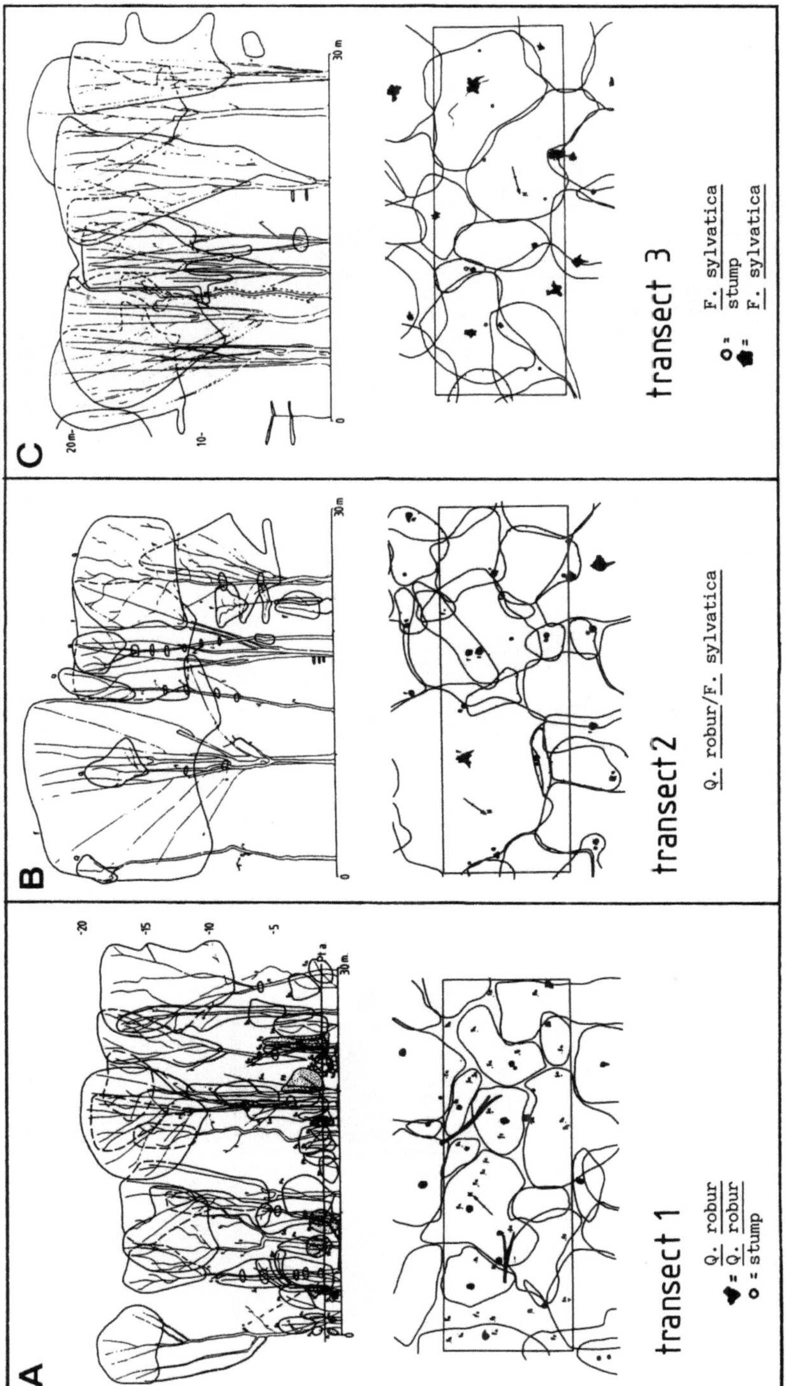

Fig. 5.60. Fragmentation of a large *Quercus robur* eco-unit **A** by invasion from below by small *Fagus sylvatica* eco-units, which emerge one after the other (**B**) until, after fusion, they form a large-sized *F. sylvatica* stand (**C**). Speulder-en Sprielderbos, Veluwe, the Netherlands. (Van Dort et al. 1979)

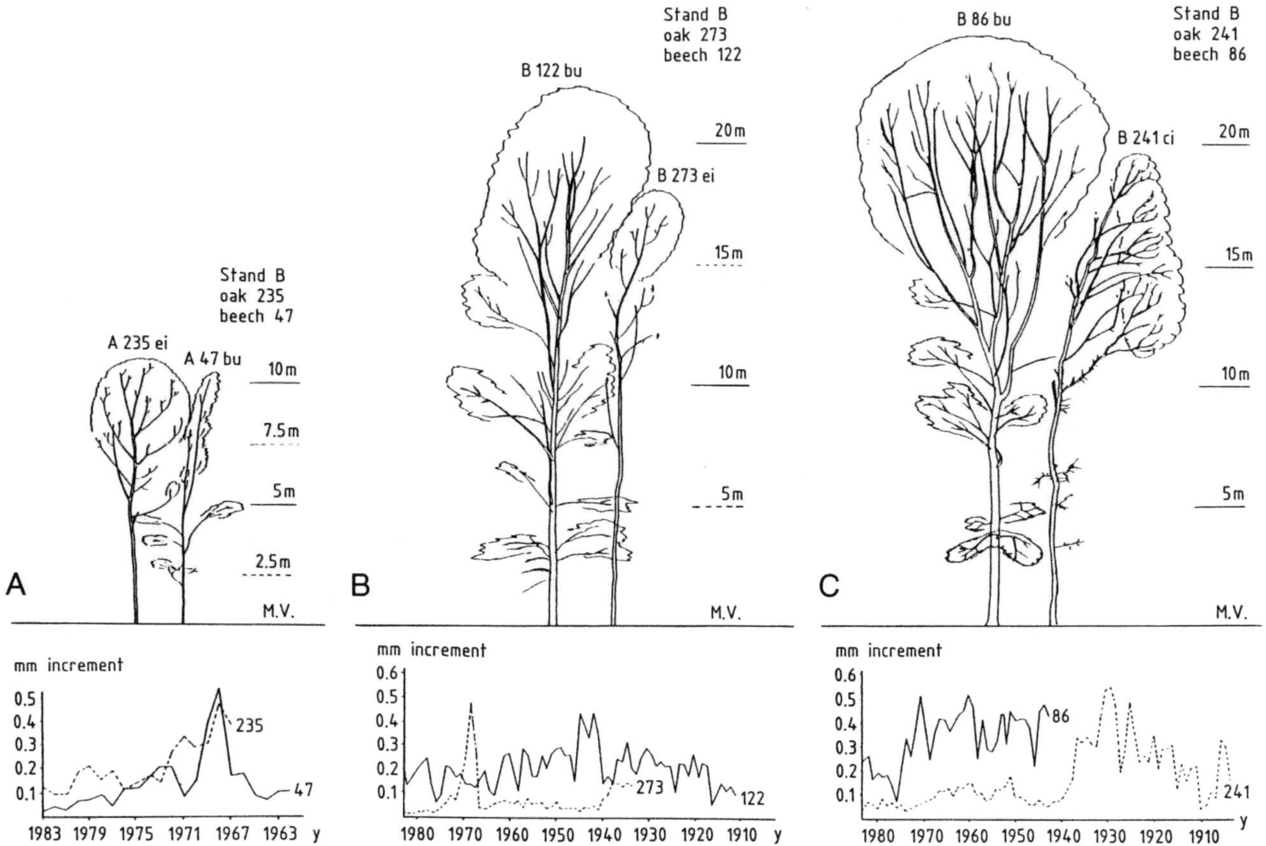

Fig. 5.61. Explanation of Fig. 5.60 in terms of tree components. Three stages of elimination of *Quercus robur* (*ei*), Rauh's model, light-demanding, somewhat slower growth, by *Fagus sylvatica* (*bu*), Troll's model, shade-tolerant, higher growth rate. Note growth rings below. **A, B, C** successive stages. Royal Domain near Apeldoorn, the Netherlands. (Hilgen and Van den Brenk 1984)

completely identical. They are sufficiently comparable, however, to understand how the combination of temperament, architecture and growth rate favours the beech.

The oaks (*Quercus*) grow according to Rauh's model (Fig. 3.2). This model is completely built by orthotropic axes. Moreover, oaks do not tolerate shade: outshaded branches die off. If overtopped, an oak can form only horizontal branches reaching out to the light by building a series of orthotropic axes that form a sympodium (Fig. 3.3b). The base of each axis grows out, the top rises up and bears the leaves. To bridge 1 m of lateral extension, an oak has to build ca. 2 m of axes, because about half of each grows vertically.

On the contrary, beech grows according to Troll's model (Fig. 3.2). This model is completely built by plagiotropic axes that may rise under a steep angle, but extend their tips horizontally as soon as there is light. Because beech is shade-tolerant, this strategy allows it to form long branches on the lower part of the trunk and to intercept light that reaches this level. And for lateral extension over 1 m, beech has only to build 1 m of plagiotropic axis, a monopodium or a very efficient sympodium (Goderie 1984; Roloff 1985).

Entering into competition with oak, a beech forms both horizontal axes beneath to catch sunflecks, and near-vertical axes at the top that slip in between oak crowns (Fig. 5.61A). As soon as it is higher than the oak, the beech forms spreading axes above it, and the oak starts an evasion growth to the side (Fig. 5.61B), its lower branches dying back. When beech crown expansion continues, the oak crown becomes very asymmetric (Fig. 5.61C) and finally weakens so much that the tree dies. The tree ring analysis below the figures shows how cambial activity reflects this crown behaviour.

This competition is an eco-unit property, not a tree property. Below the ground, an analogous process goes on, although it is architecturally unknown. Van Dort et al. (1979) have shown experimentally and by compiling data from literature that the absence of herbs and shrubs below the resulting beech forest is not due to the lack of light, but to the very dense, horizontally spread root systems of the beeches, between which there is literally no room left for roots of other plant species. If there is green undergrowth, it consists of mosses which are rootless (see Sect. 4.7). Van Tol (1979) confirmed this experimentally by isolating 1 centiare (m^2) of soil with trenches that cut through the beech roots. On the isolated centiare, beech seedlings became established, whereas there was barely seedling survival in the surrounding forest.

Such eco-units, once this is known, can also be explained in terms of more and other eco-unit components: mammals (Van Vuure 1985), birds (see Komdeur and Vestjens 1982) and maybe fungi and nematodes. Beeches structure the whole eco-unit niche configuration of food availability, light and shadow, moisture and already occupied physical space. It is convenient to start this chain of explanation with the trees, that form the skeleton of the eco-unit.

5.4.3.12 Fused Small Eco-Units in Japanese Beech Forest

In the context of his study on the beech forests of the Northern Hemisphere, Peters (1985) studied Japanese forests. One of his transects is shown in Fig. 5.62. The large *Fagus crenata* eco-unit, observed at 1180 m above sea level in Nishitani, Toyama Prefecture, Honshu, has been analysed both by the methods developed by Kojima (1983 - see Fig. 5.58) and Oldeman (1983c - see Fig. 5.4). Peters' text hence combines floristic, soil and architectural data.

The presence of a small *Betula ermani* eco-unit, included in the large beech eco-unit is to be noted (block Q2). The lower structural ensembles are continuous throughout the transect, and are species-rich for the biogeographical zone concerned. According to Kojima's floristic analysis, a B-1 layer (shrubs and trees of 2 to 5 m tall), a B-2 layer (low shrubs and trees under 2 m tall) are present. Moreover, these Japanese forests contain a grassy bamboo layer composed by *Sasa* species, and two species of lianes. Figure 5.62 does not differentiate potential trees versus small trees and shrubs of the present. It is remarkable that no potential trees occur between these layers and the upper structural ensemble, the upper crown canopy.

According to Peters (1985), who estimated tree ages by taking increment cores from the trees in the transect, there are several age classes among the Japanese beech trees, the oldest ones being around 200 years old. Birches (*Betula*) and others would be younger. This suggests an interplay between large eco-unit fragmentation, and later fusion of the younger and smaller eco-units. The processes have been described in earlier examples.

Beech forests, the world over, are among the best examples to study such balances and interactions between development processes of eco-units. This is due to the fact that they are both relatively simple floristically, and show gradients of floristic richness from very species-poor to rather species-rich. An enormous amount of data has been gathered on beech forests, which can be traced back through the reference lists of the authors cited in the previous examples. Beech species are of very different economic importance in the zones where they occur. For instance, in Europe, *Fagus sylvatica* is of great value for wood production, whereas *F. grandifolia* in North America is most often considered as a valueless weed tree. The comparison of the world's beech forests, however, yields both ecological insight and silvicultural knowhow that can scarcely be found in an easier way in other forest types, being, like beech forests, built principally by species with a tolerant temperament.

Eco-unit architecture and development which are implemented by trees showing a combination of shade tolerance, Troll's architectural model and potentially high growth rates occur in many other forest types with a much more complex mosaic architecture. In the Leguminosae, Troll's model is preponderant among forest trees (Oldeman 1989a in press). The tropical rainforests of Africa and the Americas contain vast forest regions in which leguminous tree species are dominant. The principles of beech eco-unit development as presented on the pages above hence are apt to illuminate a much larger array of eco-units, belonging to very different forest types. Offermans' study (1986) of the architecture of *Piptadeniastrum africanum* (Mimosaceae) in the Tai region (Côte d'Ivoire) is a case in point.

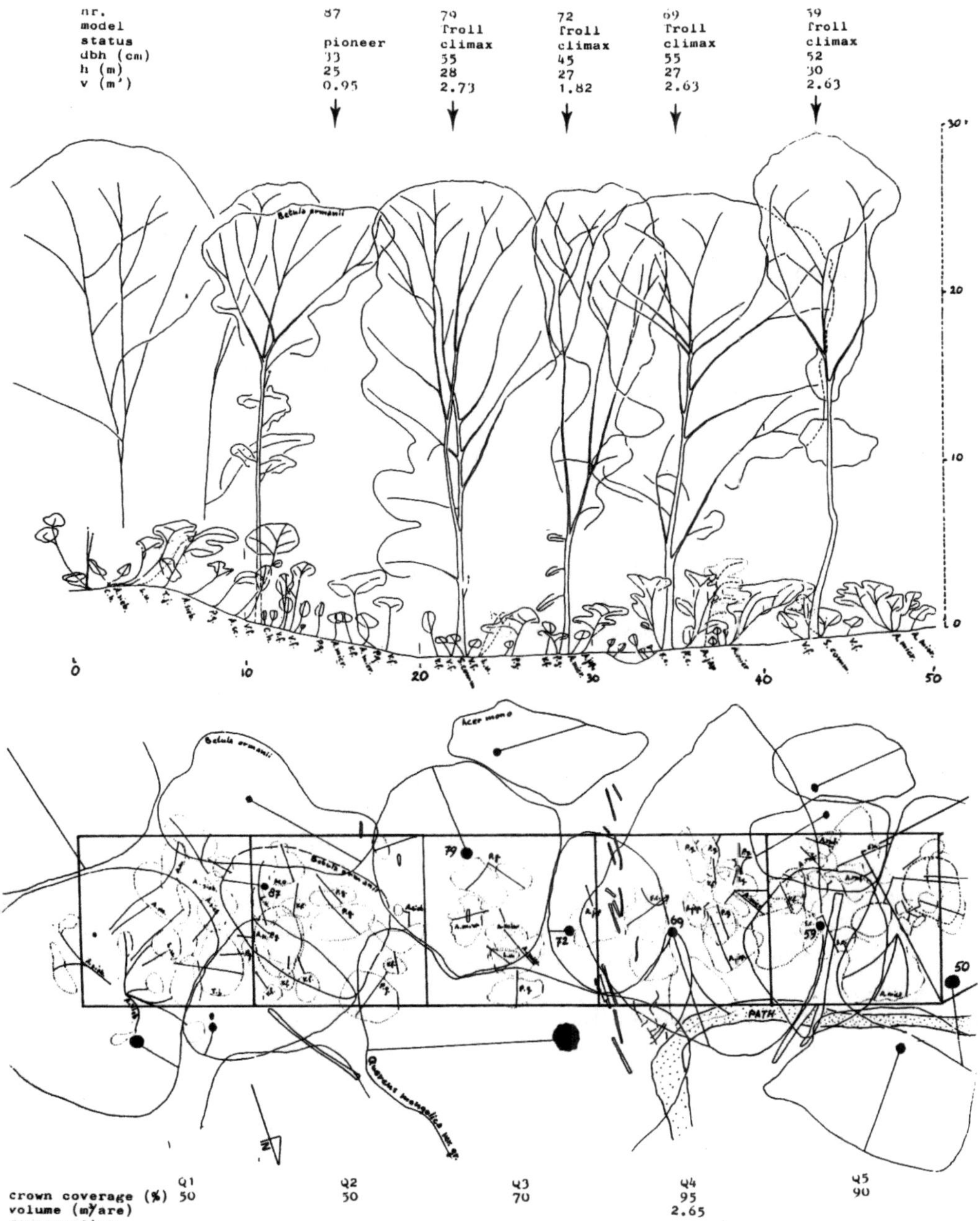

Fig. 5.62. Large Japanese beech eco-unit (*Fagus crenata*, Troll's model), including a small *Betula ermani* eco-unit. Oldest beeches 200 years. Natural forest relict, Nishitani, 1180 m asl, Toyama Prefecture, Honshu. (Peters 1985)

5.4.4 *Why Were These Examples Chosen?*

The above examples of eco-units differing in size and geographical origin have been chosen out of a large array of available material. They are meant to illustrate several points made in the present book. Sometimes technical reasons led to their exclusion. Technically, documents that were too small or too large or that required too many different graphs and tables to clarify their meaning, could not be fitted in without damaging these results of good research. In compiling other scientist's work the risk always is present of involuntarily presenting their intentions askew, and this should not be compounded by technical factors.

The first criterion for the choice among the remaining, numerous examples was a sufficient spread of very different forest situations to be shown. This is indispensable in order to introduce a step of intermediate abstraction between the very abstract diagrams presented in earlier paragraphs (e.g. Figs. 5.4. or 5.16) and the very complex nature of the forests themselves. Contrarily to a photograph, a transect drawing is an interpretation leaving out many details and using graphical codes, like all maps. However, the examples show maps of very specific forest situations, which can be directly checked in the field by repeating the observations. The validity of more general diagrams can only be checked by searching for exceptions, indeed by using case studies like those given above.

The second criterion for choosing examples was their bibliographical background. If no overriding reasons existed, e.g. if there was more than one good example depicting a certain forest situation, the choice was made according to the literature references attached. In this way, the reader of the present book can follow up other work and theoretical lines related to specific examples given here.

The third criterion for the choice of examples was bringing together forest studies with different styles and according to different schools of thought. Many excellent but different methods exist for tackling scientific questions related to forests. As stated in Chapter 2, these methods should be considered as complementary approaches, each being optimal to provide answers to special sets of questions, rather than choosing one among them as the road to "the" truth. In the next paragraph (5.5) many of them will be discussed, both among the widely spread methods and those used in more restricted research circles, among the classical and the more recent ones.

The fourth criterion was the illustration of a methodological dilemma which is paramount nowadays in international scientific networks, like the Man and the Biosphere programme (MAB-UNESCO), the Forest Ecosystem Research Network founded by the European Science Foundation (FERN), the subject groups of the International Union of Forest Research Organisations (IUFRO), or the International Union of Biological Sciences (IUBS).

This dilemma can be simply stated as a choice between either standard methods, to be used world-wide, or methods to the taste of each individual scientist, which often yield results that cannot very well be mutually compared. The above examples show the heterogeneity of methods used nowadays. However, they also point towards a solution of the methodological dilemma. This solution lies in harmonisation of the methods used, rather than their standardisation. Harmonisation requires every new concept to be linked to the existing ones in such a way that translation of the results from one set of concepts to the other becomes possible. This is the case with the examples shown in the preceding pages, simply because they all have in common being horizontal or vertical forest maps, which are directly comparable as to their basic traits once their scales are adjusted.

The whole set of examples therefore points to the very practical possibility to choose a method apt to yield answers to specific scientific or practical questions, without losing coherence within the silvological or silvicultural field.

5.5 Systems Analysis of Forest Eco-Units: The Compartments

5.5.1 *What is a Compartment?*

The concept of compartment is used here in a more general way than is sometimes done. For instance, the term component is used in German literature for what is here meant by compartment (cf. Ellenberg 1973). The word compartment has been coined in production ecology, where it is used sometimes in a very loose way, and sometimes in a very precise and restricted way (e.g. Ernst et al. 1985). In the present text its meaning can be circumscribed as follows:

- **we will consider a compartment: any set of abiotic and/or biotic subsystems or components in a living system, that is delimited as a recognisable, functional ensemble.**

In the preceding pages (Sect. 5.2.4), the set of potential trees (potential set), the set of trees of the present (set of the present), divided in structural ensembles, and the set of the trees of the past (set of the past) were all examples of compartments, as was the rhizosphere in the sense used here (see Sect. 5.3.4).

The circumscription used above leaves such latitude for definition of every compartment, that compartments in a model should be very well defined. This is done on purpose, because compartmentalisation is a scientific artifice in order to interpose auxiliary hierarchical levels between main levels such as those distinguished in Table 2.1. It is of considerable importance that this can be done freely, but such auxiliary levels should have a generic name so that we may indicate exactly what kind of analysis is used. In an eco-unit, every compartment lumps together a few or many components or even organs (leaf compartment, for instance).

Compartmentalisation within models has contributed considerably to the understanding of ecosystems in general. The compartments being used often or appearing very meaningful will be discussed in the following pages. It is to be noted that not all authors see every one of these categories as a compartment.

5.5.2 *Populations as Compartments*

Populations are compartments including all individual organisms of one species, here considered per eco-unit. Populations are very important compartments and have been central in ecosystem analysis for more than a century. Good insight into contemporary thought on populations can be gained from Freysen and Woldendorp (1978). Pianka (1974) approached population biology from a zoological viewpoint, and Harper (1977) wrote a standard text on plant populations, followed in 1986 by Begon et al.

The species concept will not be discussed here. The interested reader should leaf through the last 25 annual volumes of the journal TAXON to gather many of the ideas that concern the particularly elusive species concept, of which no watertight definition exists. In silvology it is enough to understand that organisms belonging to one species have roughly the same morphology, by which they are distinguished in identification keys, and the same genetical build-up, so that they can beget progeny among them.

Ecology concerning one population in a central role, and the rest of the ecosystem as its environment, is often called **autoecology** (litterally: self-ecology). When the interaction between different populations is considered and analysed, this is frequently indicated as **synecology** (literally: together-ecology). The approach in the present text is evidently synecological. This is not the same in many classical silvicultural studies, where very often one (useful, tree) species was considered in an autoecological way. In yield tables, the production class (Bonität) is taken to reflect the possibilities of the site, i. e. the complex of soil, climate, hydrology and other organisms surrounding the crop population. Several important site factors are sometimes considered in their own right, because their impact on the crop is so evident, e. g. soil fertility or populations of pathogens and parasites.

The emphasis on autoecology of crop plants is a classical feature of agricultural thought. Site factors may be artificially strengthened. For instance, soil fertility is increased by fertilisers. And other "factors" are weakened, such as the populations of weeds, parasites or pathogens, which are reduced by all sorts of techniques, from hand-weeding to chemical eradication. This approach has led to considerable success. It is proven by the fact that both the readers and the author of these lines are at least reasonably well fed, clad and provided with commodities such as a wooden chair. This way of thinking has pervaded forestry quite deeply. For instance, James (1982, p. 55) states that "silviculture is the commercial cultivation of trees for the production of timber".

From the viewpoint of the present book, the correct (French) name for this activity is **ligniculture** (literally: wood cultivation), whereas **silviculture** (literally: forest cultivation) encompasses a much larger array of products and services, going from the conservation of genotypes in natural forests to the protection of agricultural land against erosion, to recreation and to the growing of forest products other than wood alone. In other words, if an autoecological approach to crop trees is justified up to a certain limit in ligniculture, multiple-use silviculture requires a more synecological analysis.

The limits for justifying an autoecological model in any system whatsoever are reached when it is exaggerated to the point of considering isolated compartments, such as crop/pest/pesticide/soil fertility, while forgetting the ecosystem as a whole. Agriculture of annual plant species gets away with this exaggeration longer than forestry, because failures due to an incomplete model express themselves as misharvests that usually are written off as incidents. But a forest ecosystem, however simple it may be, lives at least some tens of years and if it is to survive it needs many more components than an agricultural field. If populations of forest organisms are neglected in a silvicultural model, they largely have the time to disrupt or to change forest development earlier or later. And there are many such populations.

The models that aim to take into account many, if not all populations contain the important categories of **syntaxonomic** and **phytosociological** models. The roots of such models were the theories from the beginning of this century by Clements (e.g. 1936, USA), Braun-Blanquet (e.g. 1934, France-Switzerland, or Zürich-Montpellier) or Cajander (e.g. 1930, Finland). These models established that there are repeated patterns of combinations of different populations (**taxa**), and one method calls these patterns syntaxa. Syntaxonomy in a wide sense is the basis for forest typology in many parts of the world (e.g. Mayer 1986; Kojima 1979, 1981). The limits it traces in a forest vegetation are different from those in architectural analysis (cf. Figs. 5.1 and 5.7).

The feature, but not the term that both types of analysis have in common, is the propagule bank. If this is natural, it contains all species that are taken into account in syntaxonomic models. If its whole potential were realised completely, over a homogeneous region, one might count the species number over an increasing surface. This is often done in natural vegetations and these counts lead to species-area curves. Such curves (Fig. 5.63) show a fast increasing number of species over the first ares (10 x 10 m), whereafter the number of added species per are starts to decrease. After x ares, the total number remains constant and the surface of x ares is called the minimal area on which this vegetation can exist completely.

This works more or less well in the temperate regions, where the species richness is not excessive. In 1956, at the Unesco Symposium in Kandy (Sri Lanka), where Van Steenis (1956, in UNESCO 1958) launched his much-cited biological nomad theory, the question was debated and settled de facto for the tropical rainforest: its species richness is so high, that no satisfying way can be found to determine a minimal area. However, a recent book by Hommel (1987) successfully uses phytosociological methods in delimiting land units favourable to *Rhinoceros* occurrence in one region of Java (Indonesia). The forests concerned were not the most richly developed tropical rainforests, but were still much richer in species than most temperate ones.

That populations may be considered as compartments in a vegetation is maybe most clear from the concept of indicator species. Certain species indicate by their presence, their abundance, or both that the abiotic site factors show certain (chemical or physical) properties or values. Spore plants, for instance, may be very precise pH indicators, that react to changes in the environmental acidity of the order of one decimal behind the comma (see Sect. 4.7). Pioneer species generally indicate abundant light, or pronounced mineralisation of the soil. The importance of nitrogen-indicating species, such as nettle (Europe, *Urtica dioica*) has become largely recognised in regions where nitrogen-pollution occurs. Such populations form compartments that play a functional role in the processing of H^+ ions (acidity), nitrogen, or light. It is implicit in Ulrich's

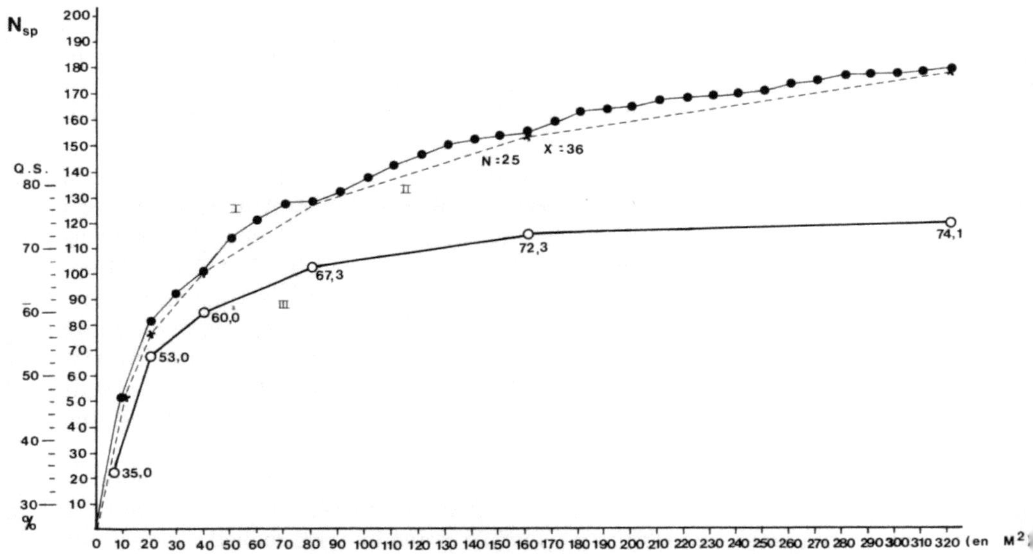

Fig. 5.63. Example of a species/area curve from the Yapo forest, Côte d'Ivoire. (After Mangenot 1955). Curve *I* represents the counts of species numbers on successive surfaces of 10 m^2 each. The "jumps" in this curve indicate transitions from one micro-site to another or one eco-unit to another (Oldeman 1978). Curve *II* is a theoretical curve, calculated under the assumption that after the first 100 m^2, each doubling of the surface adds 25 ± 2 species ($N = 25$) and that the Williams diversity index $\alpha = N/0.693$. Curve *III* represents the quotient of similarity (*Q.S.*) between successively surveyed plots, also according to a Williams formula. As a syntaxonomical association, this forest has been indicated as a *Diospyro-Mapanietum* by Mangenot, who considered all vascular plants, and predicted a plant species number of 290 to 300 per hectare, and 700 on a hundred thousand hectares of this association. This is one of the rare examples of the application of phytosociological methods in the tropical rainforest in such a thorough way

analysis (in Ulrich and Pankrath 1983) of forest acidification that the forest development phase plays an important role in the balance between assimilation and mineralisation that determines the pH of the site (see Sects. 5.3.1 and 6.2.5.2). This also links the presence of indicator species to eco-unit development.

Instead of single species that indicate certain ecosystem properties, finer analyses use complexes of several species, whose coexistence is indicative of these properties. The largest complex to be found, of course, is the set of all populations on one site. This is most often named after dominant species, in a system that has much in common with plant taxonomy. Just as species, genera, families, orders are used in classifying organisms, associations, alliances, orders and classes are the names of such hierarchical classification units in vegetation science. A critical discussion of the possibilities of population-based classifications of ecosystems has been published by Barkman (1985). He remarked for instance (p. 126): "In fact, flora and fauna cannot be separated within ecosystems, but alas, most ecosystem types are based only on plants, even on green plants alone and often on vascular plants exclusively. Mosses and lichens are frequently left out or mentioned incompletely. Certain researchers in forests even observe nothing but woody plants and also neglect the herbs. In this way, the types become ever less natural and give ever less information, both floristically and ecologically". This state of affairs is more or less reflected in Figs. 5.31 to 5.62. The species or population analyses, if present in the original publications, have been mentioned but not reproduced by us.

The trouble with a complete population analysis, even in a simple eco-unit such as shown in Fig. 5.34, is the processing of the complete information on all populations present. As mentioned above, this has been a major handicap in the tropical rainforest. A look at Akker and Groeneveld (1984) demonstrates the difficulties encountered, even including limited groups of insects only. The elementary compartments that are monospecific populations therefore have been lumped into compartments that contain many species. This, of course, is what humans have always done when using such concepts as "trees", "bugs", "herbs", "weeds", "climbers" or "game".

5.5.3 Biological Types (Life Forms) as Compartments

Biological types, as defined by Raunkiaer in 1934 (not cited) are in fact such categories of plant species, with the difference that they are objectively distinguished by the behaviour of their meristems during inhospitable seasons. For instance, a tree is a phanerophyte, together with all other plants that keep their meristems visibly above the ground in winter or drought. On the contrary, plants that survive such seasons with their meristems below the ground are cryptophytes (literally "hidden plants"). We will not use this classification further. It has been developed with considerable precision by adding plant sizes and other criteria, and it is useful to know the terms so as to recognise them in other publications. Instead of saying that a forest is dominated by phanerophytes, however, we will write about trees, shrubs or climbers.

Except for the "tree compartment", "herb compartment", and such series of biological types based essentially on meristem behaviour, plants have been separated in more kinds of population groups. These have in common some important ecological characteristic, e.g. leaf size: megaphylls have gigantic leaves (e.g. some palms), and nanophylls have dwarf leaves.

Because populations are defined by the interfertility of their members, the ways and means by which their offspring is begotten and dispersed has always been considered as one of the most important characteristics of populations. One of the convenient and ecologically useful sets of compartments is the following one.

Plants may be pollinated by some factor. They are then indicated by an adjective that contains the factor and ends on -philous, derived from the Greek word for friend, φιλος. The dispersal of fruits or seeds is indicated in the same way, but the words end by -chorous, χορευειν being Greek for dancing. The terms then are as follows:

Factor	Pollination	Dispersal
wind	anemophily	anemochory
animals (outside)	zoophily	exozoochory
animals (inside)	–	endozoochory
water	–	hydrochory
gravity	–	barochory

To these general terms, specific ones have been added whenever an author deems it necessary, for instance cheirodactylophily (pollination by bats), or loxodontochory (dispersal by African elephants). We will not use such specialised terms, but again the terminology is mentioned, so that it may be recognised in literature.

A complete manual for the description of vegetations in such compartments, and others, based on different combinations of plant populations that have some character in common, has been published by Godron et al. (1968). Usually, the numerical expression of the proportion between such compartments is used for calculations. It is indicated by the term spectrum as in "life form spectrum" (= biological type spectrum), "leaf shape spectrum", "dispersal spectrum", or "coverage", which is the spectrum of proportions of the ecosystem's ground surface covered by the leafy crown of each plant, summed per compartment or class.

This last property of a forest ecosystem is important in growth and yield studies for tree crops in managed forest stands. The percentage canopy closure of such stands, indicating the proportion of the soil covered by the tree crown projections, shows what part of the maximum production is reached. Between certain limits, the leaves or needles at the sides of the crowns will compensate for an imperfect crown coverage, but if the trees become too widely spaced, production of the stand decreases, because light interception lessens. Again, there is considerable agricultural bias in such values, which leads to the neglect of the rest of the forest ecosystem if used indiscriminately.

Once all these compartments are counted, the numbers can be processed by a computer. This is how analysis of such a multitude of data is done nowadays (cf. Wiegers 1985). The development of these analytical methods can already be found, for instance, in the proceedings of the International Symposia of the "Internationale Vereinigung für Vegetationskunde", published since 1959 by J. Cramer in Vaduz (Liechtenstein, Europe), in German. A good bibliography including the French authors, who developed methods for computer-aided ordination of population data, can be found in the thesis written by Cheviron (1985), shortly before his untimely death in a traffic accident.

We will come back later to Cheviron's work (see Fig. 5.86), which was a breakthrough in the integration of population-based and architectural analysis. First, some visual representations of plant populations with or without distinguishing compartments with different properties will be shown here. They have to be studied in order to evaluate their possibilities and limitations in providing answers to certain scientific or practical questions.

5.5.4 Examples of Population and Life Form Analysis

Figure 5.64 is a forest transect made by Grubb et al. (1963) in Ecuador, and concerns a montane forest at Borja, 1710 m above sea level in the Andes. The numbers concern the species occurring in this forest. No architectural codes indicate trees of the present or the past, or potential trees. No eco-units have been indicated, although they can be seen to exist on the transect, which may be called a physiognomical profile diagram, contrary to the analytical transects such as shown in Fig. 5.53. Physiognomy literally is "the understanding (γνωσις) of the nature of the body (φυσις)".

Some idea of the spatial biotopes of species may be derived from physiognomical profile diagrams of forests, of which many have been published (cf. Richards 1952; Ashton 1964; or the examples cited by Rollet 1974). Most of these profile diagrams are limited to trees of a diameter at 1.30 m (dbh) over a certain limit, e.g. thicker than 5 cm or 10 cm. They show no other forest components except, sometimes, a particularly large liane (Fig. 5.64), although a more complete plant species inventory is often added in the form of tables or computerised lists or matrices. This is the case also in Grubb et al. (1963), who analysed such compartments as epiphytes, monocotyledons (see Sect. 4.6), and the leaf size spectrum for several Ecuadorian forest formations. This article, as well as the ones published later on these forests, contains a worthwhile bibliography.

Figure 5.65 shows how to let the eco-units disappear in a reconstructed homogeneous mosaic. This artifice was invented by Holdridge et al. (1971). Two

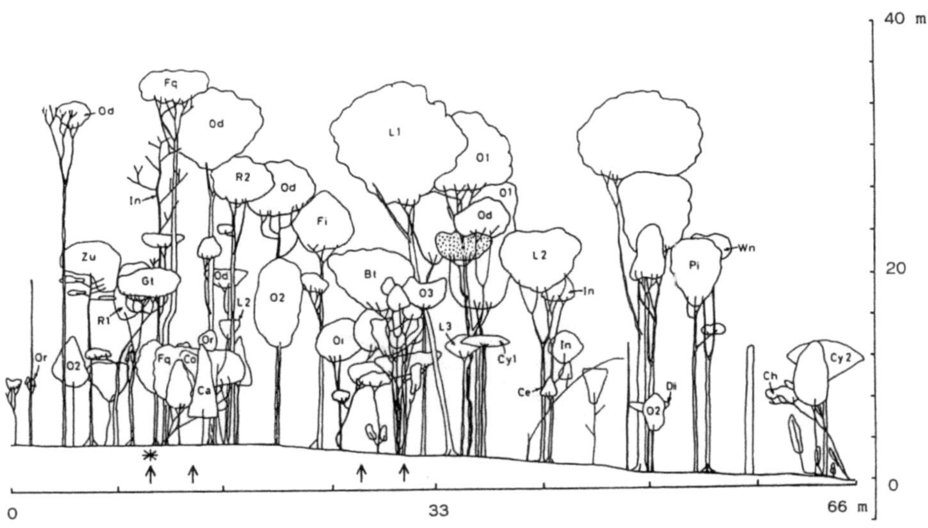

Fig. 5.64. Forest transect with emphasis on species populations and natural volumes. Example from Ecuador, Borja, at 1710 m asl in the Andes. *Arrows* and *star* refer to following articles by these authors. Scales approximately translated from feet into the metric system. (Grubb et al. 1963)

elements of observation are needed: a transect drawn to nature (Fig. 5.65A) and a frequency list of the species found in the minimum area, which is found with methods such as illustrated in Fig. 5.63 (Fig. 5.65B). "Idealisation" of the transect, i.e. transformation of Fig. 5.65A into Fig. 5.65C, is done by the following steps:

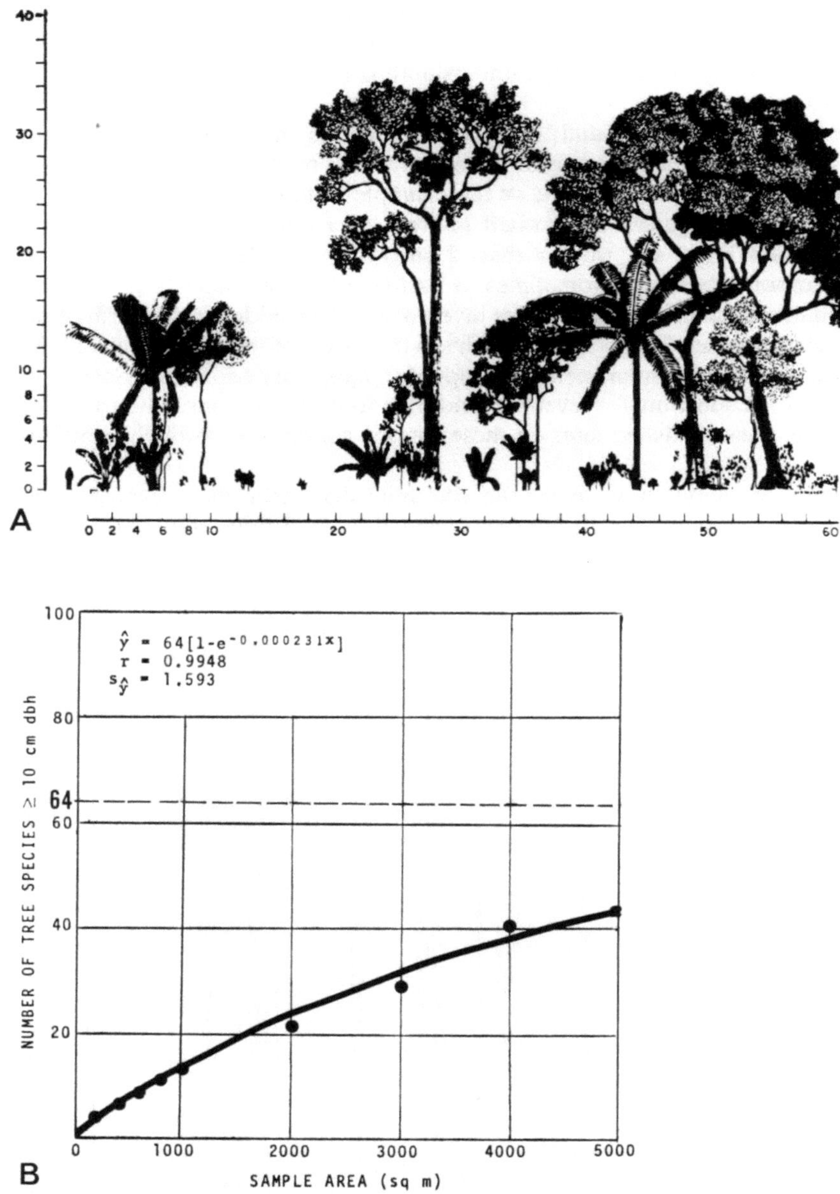

Fig. 5.65A, B

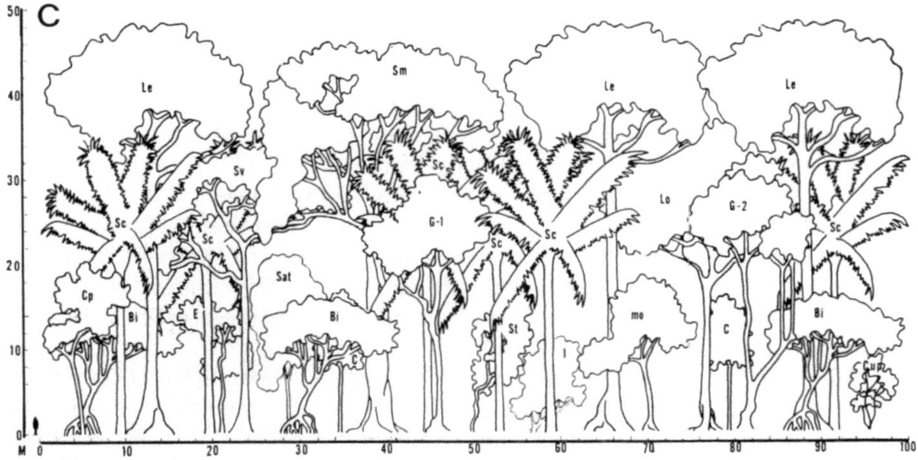

Fig. 5.65. The use of a "standard profile" (**A**), showing natural volumes, and species population data (**B**, and non-reproduced table) for the establishment of an "idealised profile" (**C**) which in fact is a visualised graph of species frequencies, showing "ideal" representatives of each species. (Holdridge et al. 1971). Costa Rica, Barranca, Pacific coastal plain, ca. 40 m asl

- the most frequent species are chosen in the species list, and their numerical proportions are determined: on 100 trees, there are 12 of species a, 9 of species b, 7 of species c, and so on;
- the best-developed tree of each of these frequent species is copied from the standard profile (Fig. 5.65A) and outlined as a schematical drawing;
- these schematic tree drawings are spread over a baseline as long as in the original "standard" profile, in frequencies that correspond to those from the species list.

The result is a visualised graph of species frequencies (Fig. 5.65C). In this way it is indeed idealised, because it represents its authors' ideal of a floristic model, which does not take into account the growth of the forest components and the biotopes in which they live. These biotopes are linked to certain eco-units, but the eco-units have disappeared from the "idealised" forest profile diagram. This is very clear from the position of the palm species *Scheelea rostrata*, which grows up in the sunlight, in innovation phases, or survives after being overtopped in aggrading eco-units (Fig. 5.65A; see also Fig. 5.15), but which in the idealised forest model grows in the deep shade cast by higher trees (Fig. 5.65C). Furthermore, it is to be noted that in Fig. 5.65C the potential trees have disappeared and, with them absent, there are no forest dynamics left. Holdridge et al. (1971) claim that this idealised forest may be recognised in the field. However this may be, and probably the trees can be recognised only and not the forest, these "idealised" profile diagrams would be misleading if they were used to demonstrate either the temperament or the growth behaviour of the trees. Neither can they be used to show forest development, because they depict a forest architecture that, if taken literally, could not be built by these plant components. Holdridge et al. (1971) indeed use these idealised diagrams for purposes of descriptive classification only, and do not claim to use them otherwise. Readers of their publication should be well aware of these facts.

Among the data-set for the forests studied by Holdridge et al. (1971), complete soil analyses take a dominant place. In several cases they also include a forest profile according to the methods of Dansereau and Lems (1957), which are illustrated here by a case from the Netherlands studied by Wiegers (1982, 1985) and concerning birch forests (*Betula pubescens*) on marshy sites near the former IJsselmeer coast, between Emmeloord and Wolvega. Figure 5.66 shows the codes invented by Dansereau and Lems and the structural forest diagrams of a birch forest in which these codes are used. Such drawings are sometimes called "danserogrammes", after the author of the method.

Wiegers (1982) studied the same birch forest with a 10-year interval (Fig. 5.66B). The codified compartments are distinguished according to a refined version of the biological type (life form) classification described in Section 5.5.3. No eco-units can be distinguished with any degree of certitude, but Wiegers showed shifting in clustering of life forms during the 10-year development. No natural height scale is used in the danserogrammes, but a "layering principle". Layer boundaries are represented as occurring at normalised height intervals, as in a grid (also cf. Godron et al. 1968), and layers are named after the non-scientific terms of biological types: moss layer, herb layer, shrub layer, etc.

A considerable literature contains this kind of danserogrammes, established for quite numerous forest types, including tropical ones. It is therefore useful to recognise and understand danserogrammes. As in all forest models, the array of questions that can be answered in a simple way, by such documents, has its limits. The abstraction of the danserogramme bears on the natural forms and volumes, which are not considered. The species making up the compartment of each biological type are generally detailed in tables or are computer-ordinated (also see Wiegers 1982). Animals are not considered in a systematic way, although they might conceivably be "hooked up" to this vegetation model. The field work and processing require the handling of huge amounts of data, and hence cost efficiency of such research should be carefully monitored. Danserogrammes answer questions on population structures, but barely do so for ecosystem structures. They give a good idea of the changes in plant populations during general forest development, but do not explain them in a functional way except for the general meristematic strategy per life form.

The next example shown does not only represent plants, but has been analysed to unite tree populations, insect populations and eco-unit development phases in one image. This is the study by Akker and Groeneveld (1984) in a tropical rainforest in Ouro Preto, Rondonia, Brazil. Table 5.3 shows the insects, divided into compartments that are very similar to those distinguished in the soil fauna (see Sect. 5.3.4.4). These compartments are the following ones.

- **phytophagous** animals which eat plants;
- **predatory** animals which eat other animals;
- **saprophagous** animals which eat dead organic matter (detritivores);

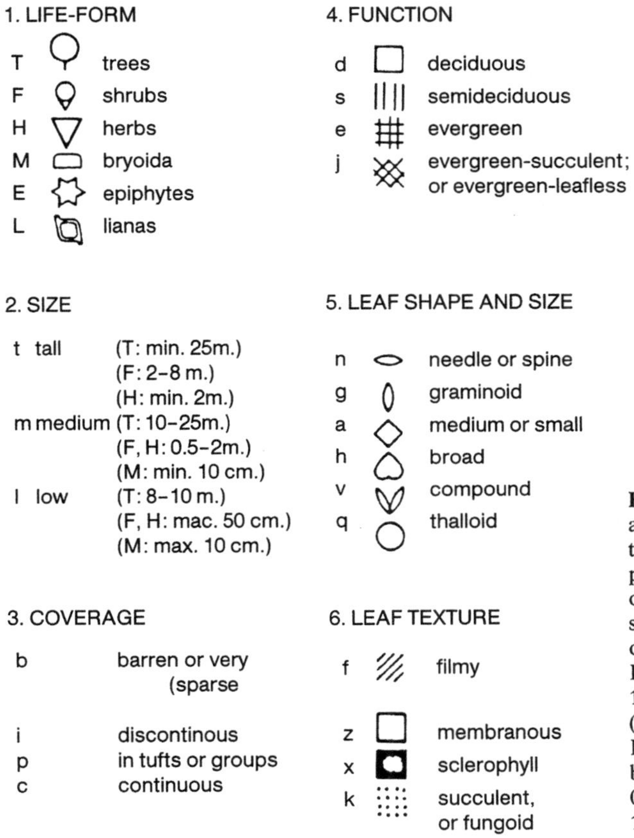

Fig. 5.66. "Dan serogramme", a type of transect indicating the occurrence of biological plant types. **A** Original version of codes as published by Dansereau and Lems (1957). **B** Use of the code in a transect of a Dutch birch forest between 1970 and 1980 by Wiegers (1982), also compare Fig. 5.52. Height classes or layers: *1* below 0.1 m; *2* 0.1 to 0.5 m; *3* 0.5 to 2 m; *4* 2 to 6 m; *5* 6 to 10 m; *6* 10 to 25 m

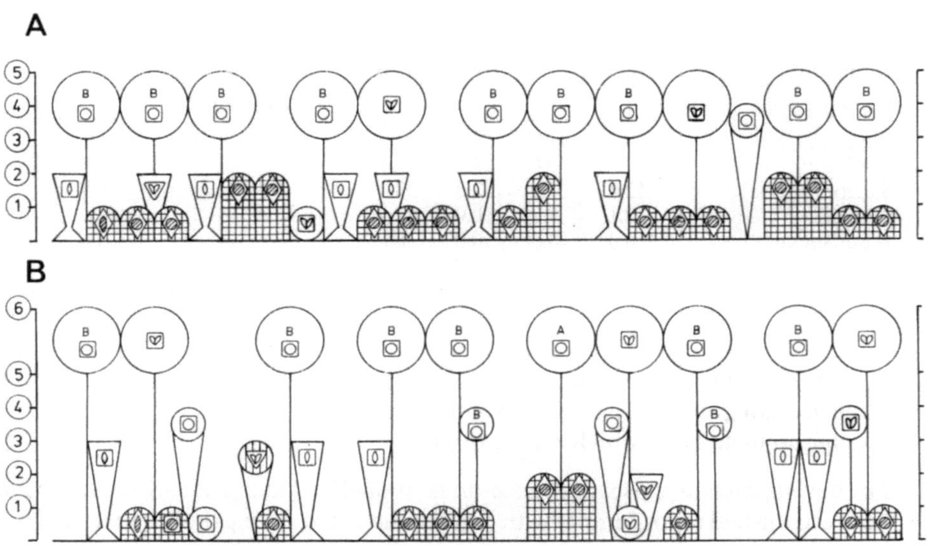

Table 5.3. Percentages of insect food compartments in different eco-unit development phases of a transect (20 x 200 m) of tropical rainforest in Ouro Preto, Rondonia, Brazil, by Akker and Groeneveld (1984). Columns represent different types of insect traps

Catching device	Mal-aise trap	Pitfall	Sweep net	Sticky trap at 1.80 m	Sticky trap at 8.00 m	Mal-aise trap	Pitfall	Sweep net	Sticky trap at 1.80 m	
Chablis						Steady state I				
Phytophagous	20.3	5.0	35.6	43.4	61.6			34.7	31.7	
Predatory	9.3	0.3	2.2	10.1	12.3			6.9	3.7	
Saprophagous	43.9	2.5	11.1	13.7	5.1			15.3	23.2	
Parasitic	7.7	0.3	0.0	16.7	4.3			12.5	19.5	
Bloodsucking	4.6	0.2	0.0	1.2	1.4			2.8	0.1	
Omnivorous	14.2	91.7	51.1	14.3	15.2			27.8	20.7	
Boundary chablis						Steady state II				
Phytophagous		6.8	34.2	29.4		22.3	18.7	57.1		30.0
Predatory		0.9	7.9	4.7		7.1	2.4	7.1		16.7
Saprophagous		6.5	15.8	32.7		47.5	16.1	7.1		26.7
Parasitic		0.8	10.5	18.7		6.3	1.0	5.4		10.0
Bloodsucking		0.0	0.0	0.5		10.5	0.0	0.0		10.0
Omnivorous		85.0	31.6	14.0		6.3	61.8	23.3		6.7
Dying stage										
Phytophagous		12.8	46.6	25.4	11.8					
Predatory		2.8	5.5	3.0	5.9					
Saprophagous		48.9	6.8	26.1	29.4					
Parasitic		0.4	15.9	17.1	23.4					
Bloodsucking		0.4	5.7	0.0	0.0					
Omnivorous		34.7	20.4	28.4	29.4					
Old chablis										
Phytophagous	24.6	8.6	27.9	31.5	30.0					
Predatory	6.2	1.0	7.0	7.6	0.0					
Saprophagous	32.7	1.0	9.3	26.1	0.0					
Parasitic	8.1	0.7	27.9	16.3	30.0			sticky trap		
Bloodsucking	8.9	0.2	0.0	0.0	0.0			between		
Omnivorous	19.5	88.4	27.9	18.5	40.0			20–35 m		
Growing stage						Everywhere				
Phytophagous		18.7	41.9	50.0	46.9				30.9	
Predatory		0.6	0.0	22.2	3.1				5.9	
Saprophagous		12.8	6.5	27.8	28.1				19.7	
Parasitic		0.1	3.2	0.0	12.5				19.7	
Bloodsucking		0.0	0.0	0.0	0.0				1.3	
Omnivorous		67.8	48.4	0.0	9.3				22.4	

- **parasitic** animals which eat parts of living prey animals or plants;
- **bloodsucking** animals, a subgroup of the parasites;
- **omnivorous** animals, which eat everything.

The developmental phases of the eco-units studied by Akker and Groeneveld have been established in conformity with Fig. 5.4, the criteria being the numbers and developmental stages of the trees (of the present, potential, of the past). We have conserved these authors' names for development eco-unit phases in the table; they are translated here into the terms used earlier in the present text:

- chablis (zero-event has occurred recently);
- old chablis (innovation phase);
- boundary chablis (border zone: cf. Fig. 5.59B);
- growing stage (aggradation phase);
- steady state I (biostatic phase, example I);
- steady state II (biostatic phase, example II);
- dying stage (degradation phase).

The third factor in Table 5.3 is the type of insect traps used. These are indeed selective instruments, each of which catches a different category of animals. For instance, pitfalls catch insects walking on the forest floor, whereas all others catch flying insects. And a sweep net cannot be brought inside dense vegetation, where a sticky trap can intercept flying insects. It is important to remember that all insect inventories are always biased because of the traps, none of which can catch all species, and some of which disturb the character of the environment. This is the case for instance with light traps, operated in the darkness, which attract insects to a lamp. It also occurs when fogging tropical forest canopies with insecticides (Erwin in Sutton et al. 1983), leading to all insect species being caught but without yielding information on the mini-biotopes where they live within such canopies (cf. Fig. 5.11). The human eye itself is biased, because some insect species are better camouflaged than others, so that the ratio between insects seen and insects missed is different per species.

Nothwithstanding the incertitudes that remain, Table 5.3 gives an impression of the insect compartments inside the development phases of this forest in Brazil. This is known for no other tropical forest, for the moment. Phytophagous insects dominate in the chablis, the aggradation phase and the biostatic phases studied. Saprophagous insects do so in both the degradation phase, which was to be expected after the fall of much dead wood, and the border zone, situated between chablis. Parasitic insects predominate in the innovation phase. In the aggradation phase, there are no bloodsuckers: this would point to an absence of warm-blooded animals in this phase, as was found in Europe by Van Vuure (1985). Except for spatial differentiation of the insect compartments in the development phases of the eco-units, there was a marked shift between the dry and the wet season. This would be a first confirmation of what was stated earlier, i.e. that macroclimatic regulation (sensu Den Boer 1977) occurs at the silvatic mosaic level, whereas biotic regulation occurs at the eco-unit level (cf. Sect. 5.3.4.4).

Akker and Groeneveld (1984) stated that most omnivorous insects were ants, in enormous quantities, which are difficult to analyse. It should be remembered that ants cannot be studied as populations of individuals, but should be regarded as colonies. Part of the biological functions of individual insects are transferred to the higher organisation level of the colony (for termites, see Marais 1938). Moreover, such colonies remove some of the mobility of the species, which becomes "rooted" in a certain territory and influences it profoundly (cf. Cherrett in Sutton et al. 1983).

Is a colony of Hymenoptera, the insect order that includes ants, bees, wasps, a population or an individual? In any case, it is a living system that has many properties added to the sum of its components, the insects. This has profound consequences for forest protection in silvicultural systems.

5.5.5 Compartments in Classical Vegetation Science

The examples of species populations, biological types and synusiae are part of an ecosystems hierarchy that has been developed since Lorenz (1858, according to Barkman 1973). This may be termed the "classical vegetation science" approach. Table 5.4 compares it with the systems hierarchy proposed in Table 2.1. In this way, many terms are shown that should be recognized by any reader consulting the rich literature accumulated in this field. The Handbook of Vegetation Science Series, edited by Whittaker, summarizes most of this work. It should always be used when existing case studies are consulted for special purposes and for a forest or forest region if these studies are expressed in these terms, which were developed by many authors from all over the world. This often is the case (e.g. Mayer 1986).

Referring to Table 5.4, the provisional formal definitions of three important system compartments are given here after Barkman (1970). In each of these, there is a careful distinction between the concrete system (e.g. an oak tree or the Taï forest) and its abstract nature (i.e. the class that they belong to, a species, *Quercus robur*, or a forest type, tropical moist leaf-shedding forest, respectively). Referring to Chapter 2, one might also read this as "system" and "abstract model".

- A phytocoenose (abstract: coenon) is a concrete volume of vegetation with uniform structure and with internal floristic differences that are smaller than the differences with the surrounding vegetation, or which are a clear consequence of (secondary) environmental differences created by the vegetation itself or the associated fauna (Barkman 1970, p. 103).
- A microcoenose (abstract: microcoenon) is a spatially delimited part of a phytocoenose inhabiting a determined microhabitat, and which differs in its species composition from the rest of the phytocoenose (Barkman 1970, p. 103/104).
- A society (abstract: synusia) is part of a microcoenose (or if no microcoenoses can be distinguished, part of a phytocoenose), formed by plants belonging to the same vegetation layer and not differing essentially in periodicity and method of exploitation of the environment (Barkman 1970, p. 105).

These definitions, and Table 5.4, clearly show the approach as being different from architectural analysis. This is most clear at the eco-unit level (Table 5.4, right hand). Eco-units are not only part of the biotic community, they are complete ecosystems in which the abiotic parts should also be considered. Strata, or layers, in classical vegetation science are considered as properties of the whole ecosystem (at the silvatic mosaic level) in which they are made up by stratocoenoses (Table 5.4). In architectural analysis, strata or layers are considered as eco-unit compartments which are very closely linked to and inseparable from development of the eco-unit (sets of the present and the past, potential set, structural ensembles;

Table 5.4. Compartments and hierarchy in classical vegetation science, according to Barkman (1973, his Table 1, modified) and the present book (cf. Table 2.1). For authors of different concepts see Barkman (1973). - - - - conceptual links

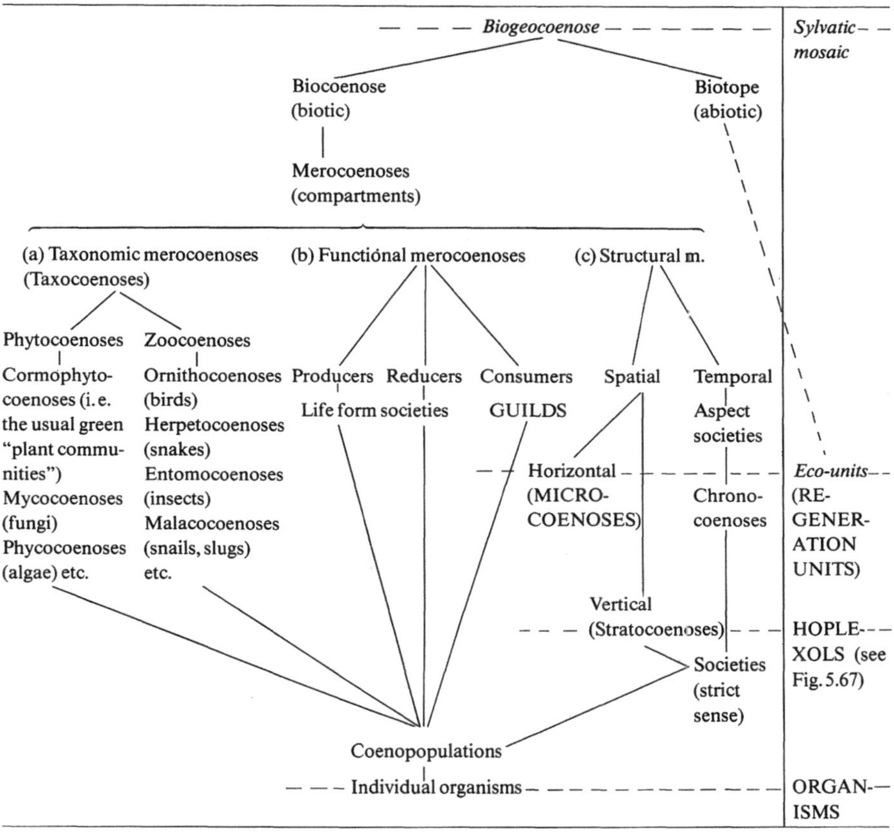

Fig. 5.4). In vegetation science, the "aspect societies", as linked to time and development, are groups of organisms that are linked to seasons mainly (Barkman 1970, 1973). The temporal aspects in architectural analysis are linked to whole eco-unit development, or may be considered for special compartments. Only in this last case they may be compared with the chronocoenoses from the table. Architectural analysis therefore takes into account at the outset the dynamics of the forest considered, whether this is at the level of the eco-unit or the silvatic mosaic. Spatial aspects are distinguished, but not separated from temporal aspects. The patterns of forest architecture are always linked to the processes that occur in the forest. There is a constant interaction between the process that is the cause of the development of a certain structure, and the structure which causes an ongoing process to proceed in a certain direction, and at a certain rate.

This may seem to be a highly theoretical argument. Therefore it is stated explicitly here that it has both scientific and practical importance, because:

- silvicultural treatments always start by modifying structures and patterns in order to change growth processes in a forest;
- there is no sense at all in the often-raised question whether processes or patterns are primordial in forest studies, because the one cannot be conceived without the other and they cannot be explained except by each other.

The theoretical background of this last statement, which implies considerations of randomness and structure (cf. Jensen 1987) and aspects of scale and hierarchy, will not be treated further here. It is sufficient to understand that "process-oriented" researchers think in terms of sources, flows and sinks of protons, atoms, molecules and energy, whereas "pattern-oriented" researchers tend to see the forest in compartments equal to or larger than organisms. One might say that the link between the two is, that "patterns show pathways of processes". This should be kept in mind when reading Section 5.6 on production models.

5.5.6 Synusiae as Compartments

Synusiae are compartments which are less numerous and more comprehensive than populations. The etymology of the word is unclear: συν is Greek for "together", and usia probably is a German transcription of ουσια, meaning essence, being, nature or even household. A synusia contains plants that together use a same ecological resource, according to a similar strategy? This is not the way in which they are defined in literature (see Sect. 5.5.5). Synusiae each contain all plants in an ecosystem that belong to the same biological type, and hence occupy rather well-defined spaces in the vegetation, their seasonality also being taken into account. Some authors use the concept of **synusia** as an equivalent of layer. A tree layer may be called, for instance, a synusia of megaphanerophytes.

The power of resolution of synusiae-based forest models should not be underestimated. Barkman (1985) wrote a draft for a determination key for European forest types, based mainly on the use of synusiae that are recognisable in the field, and which therefore can be used as diagnostic symptoms. All depends on the precision and clarity of the definition of each biological type. In earlier work, Barkman developed synusiae that are both so precise and so clear that their presence or absence can be used to pinpoint very precise properties of the ecosystem, such as the acidity of the substrate (see Sect. 4.7).

With this degree of precision, synusiae start to resemble other compartments, which have been defined from another viewpoint, that of production ecology. Please note that the image of a forest in terms of "trees", "herbs", "lianes", "arthropods" and so on, remains general and offers a poorly focussed explanation of that forest. Foresters, for instance, cannot work without at least dividing the trees in categories with different temperaments, e.g. "light-demanding", "shade-tolerant", and so further. Between the rough-and-ready compartment named "trees" and its very complex inside composition by many species populations, there are intermediate compartment sizes which are both more practical than the detailed build-up of all tree populations, and more precise that the large agglomeration of populations meant by "the trees".

It is interesting and relevant to note that the compromise reached seems to be at about the same level of precision in very different methods. Synusiae arrive at a degree of differentation such that the definition of the biological type (subtype, subsubtype) starts to reflect the ecological production mode of the species quite accurately. Plant morphology is not only used as an indicator for function, but at the same time it is used in identification keys to define species. From the other side, production ecologists note how an organism feeds, what are its size and biomass, and what are the calculated relations between inputs, outputs and interior processing of energy and nutrients (see Sect. 5.5.8).

5.5.7 Hoplexols and Transdisciplinary Language

Between 1975 and 1977, a group of ORSTOM researchers involved in the interdisciplinary study of the Taï forest ecosystems (Côte d'Ivoire, West Africa) agreed that the geologists, soil scientists, botanists, zoologists and other participating scientists used sets of concepts that were so different that communication between them became at best superficial. As a matter of fact, the results of "interdisciplinary studies" in the preceding 10 or 20 years generally took the form of a book with chapters on climate, geomorphology, soil, ecology, botany, zoology, hydrology and sometimes sociology, preceded by an "integrating introduction" written by an unhappy team leader who had to create one coherent image of a region (or its forests) based on data and models that had been developed independently of each other, each resting on a scientific tradition which sometimes went back more than a century.

The research team in the Côte d'Ivoire, which was involved in the MAB-UNESCO pilot study in Taï Forest (see Guillaumet et al. 1984), decided that they needed a new "transdisciplinary" language in order to communicate. The first necessity, they judged, was an ecosystem description that everybody agreed on and which could be used as a base-line for all further studies in whatever scientific field. The aspect that most of the participating scientists saw as a common denominator in their respective descriptions was the layer: the stratum from plant ecology or geology, the horizon in soil profiles, or the synusia of syntaxonomy.

After three seminars, in Paris and Montpellier in 1977 and at the Institut de Géographie Tropicale in Abidjan in 1978, the first version of a transdisciplinary language was published by ORSTOM (Beaudou et al. 1978). Any reader interested in methodology is strongly advised to read this small volume, of which only an incomplete outline can be given in these pages. Unfortunately, the whole team did not stay together in the field long enough, so that no complete regional ecosystem study according to this method is available as yet. The description of the vegetation compartment was made explicit by Guillaumet and Kahn (1979) and used in the analysis of tropical rainforests on both sides of the Atlantic by Kahn (1982; also see Fig. 5.40 in the present text).

This may explain why this language was not widely adopted in ecosystem analysis. The vocabulary may seem rather unwieldy and difficult to memorise. But this did not prevent the wide utilisation of the US Soil Classification System, which has a vocabulary that is at least as difficult and repulsive. And if we compare this with the vocabulary of classical vegetation science, of which Table 5.4 gives only an outline, but which has gradually accumulated hundreds of unfamiliar terms, there would be no reason not to learn the transdisciplinary language for mere reasons of commodity. On the contrary, these courageous efforts to develop interdisciplinary communication should be strongly pursued. Of course, it can only be expected to serve as a tool in research. Kahn (1982) understood this well and published his results in current French, leaving the formal transdisciplinary diagnosis in an Appendix.

Figure 5.67 gives an example of the analysis and description of two forests in Côte d'Ivoire (Taï region, MAB-UNESCO project, also see Guillaumet et al. 1984). One is a primary forest, the other a "humanised" forest, and the figure is taken from Beaudou et al. (1978). Hoplexols are more or less horizontal layers, observed volumes delimited by the plants themselves (Kahn 1982). The contents of each hoplexol are the leaf masses of maximally expanded tree crowns (paliphyse) or of potential tree crowns (prophyse), woody stems (stylagé), non-grassy herbaceous

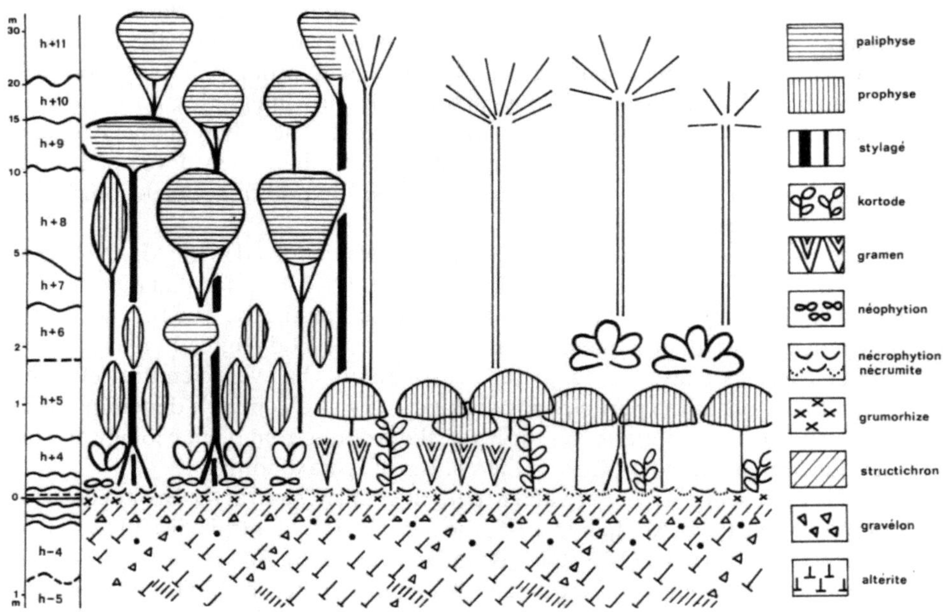

Fig. 5.67. Schematic example of forest analysis in terms of the transdisciplinary language. Situation inspired by forests in the Taï region (Côte d'Ivoire, W. Africa), in intact forest (*left*) and forest converted by humans (*right*). *Left column* rough indication of hoplexols (layers) with their successive numbers, above and below the ground. *Right* column contents of hoplexols, visualised, with their terms (translation see text). Detailed hoplexol description omitted, for example see Fig 5.69. (After Beaudou et al. 1978)

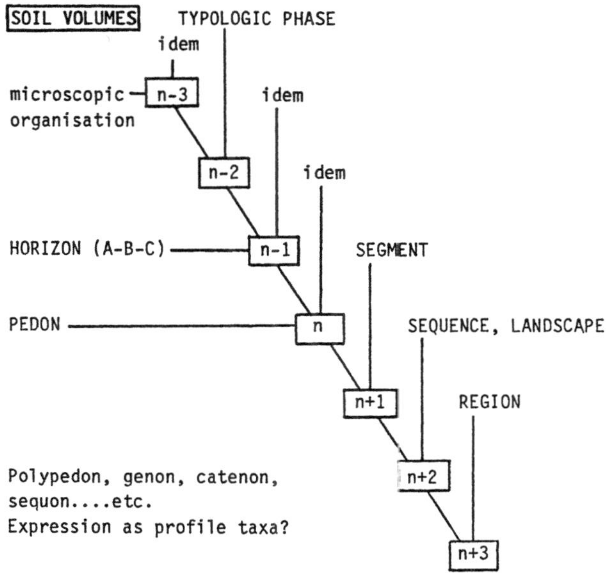

Fig. 5.68. Hierarchical system for soil analysis, built like Table 2.1, by a sequence of systems/subsystems. According to Chatelin in Beaudou et al. (1978), from whom this figure is taken, the analysis at different levels is expressed in current soil science by languages that have little in common

phytomass (kortode), grass biomass (gramen), seedling biomass (néophytion), litter (necrophytion), decaying litter (nécrumite), organic-mineral soil with dense roots (grumorhize), mineral, loose, crumbly, porous, homogeneous bright yellow or red soil (structichron), another soil horizon (gravélon), mother material (altérite). The arrangement and mixture of the contents in each hoplexol are quantified in well-defined classes, the proportion of volumes filled by air and by biomass of different kinds are given by the use of differentiating terms, prefixes and suffixes. Examples are "branches present but not abundant" (phase dendrigée), "volumes of plant mass and air mass of about the same importance" (stomaphique), or "leaf mass on tree crowns in transition between the potential state and the state of the present" (propaliphyse).

This list of examples may seem quite arbitrary, but in reality all terms and concepts rest on a logical combination of syllables, are united according to a well-defined grammar, and refer to a precise hierarchy. This is illustrated by the synoptic Fig. 5.68, taken from the analysis of the underground part of the ecosystem.

Figure 5.69 shows the result of soil analysis along a toposequence (hilltop/slope/valley). It should be compared with Fig. 5.20. One can then appreciate how finely the soil analysis in terms of hierarchical systems can be done in this way, each small block of "transdisciplinary words" containing the compacted information that normally would take a long paragraph and be less strictly formulated. These descriptions are not translated here for that reason. The figure itself shows the great difference of the superficial soil layers in different places along the to-

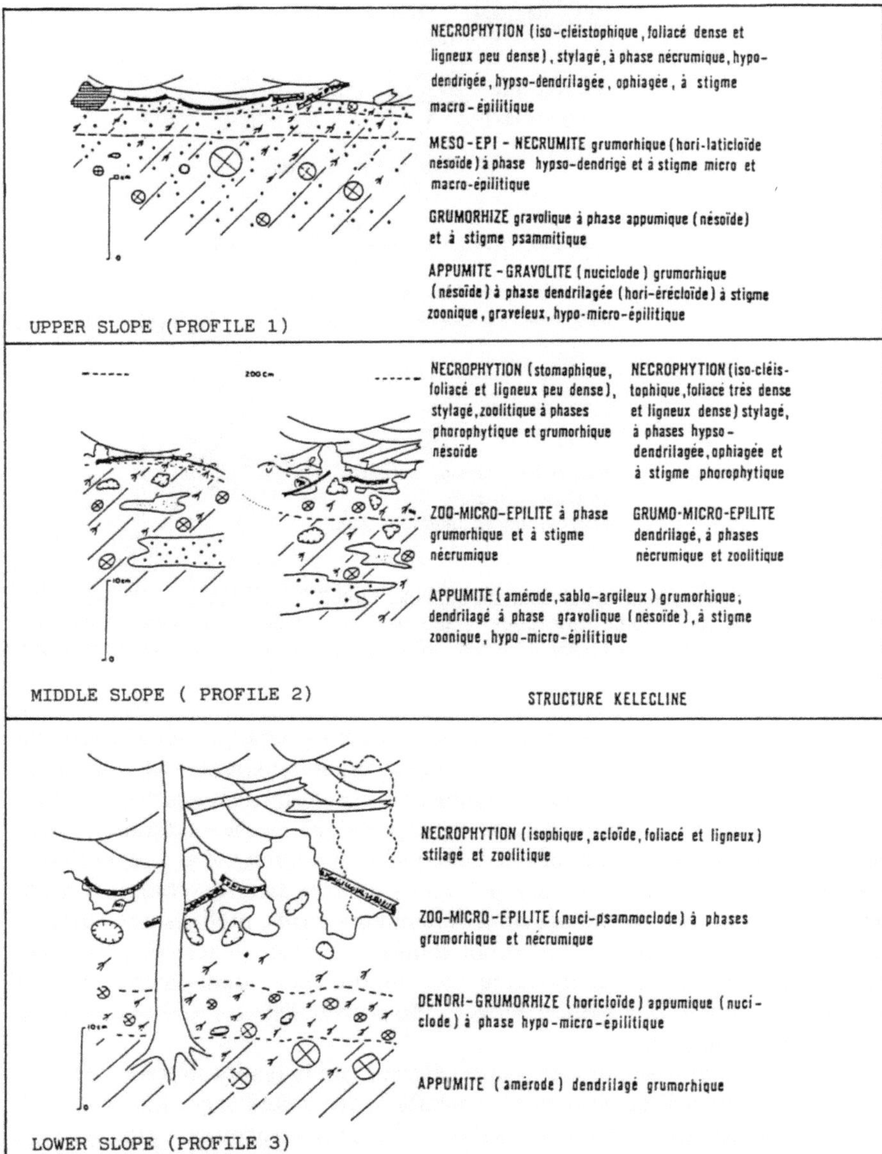

Fig. 5.69. Example of the transdisciplinary language being adapted to the analysis of soil hoplexols over a toposequence. Compare with the underground parts in Fig. 5.67. Each short, diagnostic description replaces a whole paragraph of text but is left untranslated here (see for explanation Filleron and Richard in Beaudou et al. 1978 pp. 31 ff., from whom this figure was taken; also Richard 1989). Note litter and soil differences over the toposequence, indicating different eco-units (for temperate region cf. Fig. 6.47)

posequence (for French Guyana, also see Turenne 1977). Litter decomposition and organic matter turnover, as has been said earlier, are eco-unit properties. The different soil structures given in Fig. 5.69 therefore indicate the presence of different types of eco-units along the toposequence. This fact will come back in Chapter 6 on silvatic mosaics.

Here, it is useful to compare the hoplexol approach with other hierarchies. In Table 5.4, **hoplexols** are clearly related to **stratocoenoses.** But contrary to these, hoplexols are defined by architectural criteria, not by species (or life form) composition. On the other hand, like stratocoenoses, hoplexols are considered as subsystems of a whole forest ecosystem, a biogeocoenose, and not of a smaller ecosystem like the eco-unit. Finally, the **eco-unit** is clearly related to the microcoenose in Table 5.4. But contrary to both microcoenoses and hoplexols, eco-units are subsystems of whole ecosystems because they always include abiotic components. Moreover, eco-units are not defined by species or life form composition, but by architectural criteria which resemble hoplexols at a smaller scale in the hierarchy: set of the present, potential set and set of the past (Fig. 5.4).

Both the eco-unit, compartmentalised in such sets, and the hoplexol, in which the state of components also has a dynamic connotation (e.g. paliphyse = foliage of the present; prophyse = foliage of potential crowns) may be used in establishing predictions for future forest dynamics. But these predictions refer to different hierarchical levels: the eco-unit itself (through the sets) and the sylvatic mosaic (through the hoplexols), respectively. Figure 5.55 represents a forest transect from the Taï Forest, where the transdisciplinary language was born. In the discussion related to that figure, the point was already made that in such cases, with very small and numerous eco-units, a "mixed forest model" may be easier to use. This is exactly what happened in the transdisciplinary studies. The concept of the eco-unit is vaguely suggested, because the authors (e.g. Guillaumet and Kahn 1979) state that hoplexols should be distinguished in "homogeneous units of vegetation". With very small eco-units, this may lead to a confusion if it is not clearly indicated whether homogeneousness is sought in a mosaic model ("eco-unit mosaic") or in a mixed-forest model ("tree-by-tree mosaic").

It remains to be seen how the transdisciplinary language will develop in the future. If it does, it may grow into a conceptual bridge between architectural models and production models, because the content of each hoplexol also clearly separates the assimilating and conducting organs, leaf mass versus stem mass. Production models are nowadays simulated with computers, and the transdisciplinary language also is becoming more and more computerised (Richard 1989). The way to an easier mutual translation of pattern models into production models and vice versa may be paved by research in this sense, combining eco-units, hoplexols and flow models. One promising possibility here is that computers may be programmed to receive an input in such artificial languages, but to print out the results in everyday language destined for practical use by professionals such as agriculturists and foresters.

After this excursion into very complex and abstract compartmentalisation in forest ecosystem models, one statement in the introduction (Sect. 1.2) is once more strongly emphasised. Silvicultural design has to be comprehensible for field personnel, and hence all traces of scientific, abstract tools should have disappeared from such designs.

5.5.8 Guilds as Compartments

This term is derived from the mediaeval organisations of people in one trade, e.g. a bakers' guild, a smiths' guild or an architects' guild. All people with the same profession belong to the same guild. In forest ecosystems, all organisms doing the same job, independently of their architecture, belong to one ecological guild. And the job of a component species involves the intake of matter and energy, its processing and the output of other forms of matter and energy, in a certain way. The ecological term is nutrient and energy turnover.

One might say that a guild is a synusia defined according to other criteria, but that these morphological and ecophysiological criteria meet in the middle. On the other hand, guilds have been defined for animals. However, they may be easily extended to compartments of either plants or animals, or mixed ones (Fig. 5.28). The guild of litter decomposers, for instance, contains insects, fungi and bacteria next to earthworms. Just as the guild of bakers may be divided into one of bread bakers and another one of pastry bakers, the large guilds often can be divided into smaller ones (Fig. 5.28). Their size is determined on a strict need-to-know basis, i.e. as a function of the scientific question asked.

The guilds of forest components are important in production ecology. As far as we know, they have never been linked to eco-units or regeneration units, but were always distinguished in arbitrary forest blocks that were taken to represent some forest type.

The most primitive way of defining guilds was in fact the distinction between "useful" and "unuseful" groups of organisms. Plants or animals may be classed as "edible" or "inedible", and the distinction between "crops" and "weeds" is another example. These are authentic guilds, because they are groups of plants and animals that are or are not doing the job of including human beings in the turnover cycles. "Timber trees" are a guild as much as "game", "cattle" or "medicinal plants". The concept of guilds therefore is of the utmost importance in all forms of agriculture and forestry. In silviculture, the dispensable or indispensable character of guilds as forest ecosystem compartments, and the possibilities of guiding their composition and behaviour are among the most important subjects. For Barkman (1985), this might mean that a utilitarian classification of ecosystems in such cases is preferred to a truly objective classification based on all properties of ecosystems. It is to be considered that there is a way out of this dilemma. This consists of including *Homo sapiens* and his social systems as compartments in the eco-unit analysis. We think that this is the only way to follow if sustained silvicultural systems are to be designed. Yield tables then become part of forest ecosystem analysis. The idea may sound strange at first to foresters and other conservationists alike, if formulated in this way. As a matter of fact, it implicitly underlies several existing silvicultural schools that claim "to follow Nature".

5.5.9 *Cohorts as Compartments*

With this concept, we move to still more dynamic models and their compartments. The compartment that is called cohort contains all individuals of one species within one ecosystem that are of the same age (e.g. see Beukema and Nienhuis in Bakker et al. 1985).

In all approaches to forest ecosystem analysis, this is an extremely useful concept. For population dynamics of major or rare components, the size of each cohort and the intervals between their ages are valuable variables to calculate birth or germination rates and rhythms, and mortality. By vegetative multiplication, one plant may form a whole cohort. This distinction between cohorts resulting from contemporaneous germination and from vegetative multiplication plays an important part in plant population analysis according to the methods of Harper (1977; also see Begon et al. 1986).

In production ecology, the presence of cohorts is of the utmost importance. The production curves shown in Chapter 3 (e.g. Fig. 3.38a) start from the moment of germination. If one eco-unit contains several cohorts in one of its determinant plant species, each cohort will be at another point in time on these curves, a value which is also influenced by the interaction between cohorts. This interaction codetermines the shape of each curve.

The architectural analysis of eco-units might often be considerably refined if one could pinpoint the presence of cohorts. Sometimes, this is possible in a rough way, for instance if there are relict trees from a former eco-unit, i.e. an older cohort participating in the development of a new eco-unit after a zero-event. In this development, there usually are younger cohorts involved, too. In the humid tropics, where the chronological code of wood structure cannot be read as yet, because growth rings sometimes do not exist and in any case may not be year rings (see Sect. 3.5.6), usually it is impossible to trace back the presence of cohorts.

The phenomena of fragmentation and fusion between eco-units, exemplified earlier by beech forests (Sects. 5.4.3.11 and 5.4.3.12) are closely linked to compartmentalisation by cohorts. Each beech eco-unit contains one cohort, all cohorts have fused into a larger unit that cannot be easily analysed as a mosaic, at least when all its eco-units are in their biostatic phase.

Finally, cohorts are of great practical importance. The difference between even-aged and uneven-aged forest stands in silviculture and forest management is one of cohort design and use. This is true for mixed stand silviculture as well as for pure stands in which only one tree species is grown in monoculture. Silvicultural practice always favours the appearance of clearcut cohorts because of the regular periodicity of treatments and wood harvests. De Graaf (1986) can be interpreted in this sense.

5.5.10 *Connexus and the Ecological Interaction Fabric*

The concepts that are reviewed in the present section seem to have been a side-line in ecological research, because they cannot be found in the main ecological textbooks that were consulted. A thorough treatment is given by Brauns (1976, vol. I, pp. 9–16), for those who read German. The concept of **connexus** ("Konnex") is to be seen in the same line of ecological thought that yielded the biogeocoenotic models (cf. Table 5.4), and its authors therefore called it "coenotic connexus" (Fig. 5.70).

Friederichs (1930 ex Brauns 1976, p.9) defined the biocoenotic connexus as follows.

"A biocoenotic connexus, within the interaction fabric of a biological community, as limited to each practically observed case, represents both the direct relationships and those close indirect relationships that can be grasped, between animals and other animals, between animals and plants and between plants and plants."

This German definition has been quite freely translated into English here, because the conceptual frameworks of both languages make any direct translation impossible. (Original text: "Biozönotischer Konnex" in dem "Verknüpfungsgefüge einer Lebensgemeinschaft" – man versteht darunter im einzelnen praktischen Fall die direkten und die näheren, greifbaren, indirekten Beziehungen der Tiere zu anderen Tieren und zu den Pflanzen und dieser untereinander")

The term connexus is the Latin form of "Konnex". It is composed of the prefix "con", meaning "together" and the noun "nexus" (plural also "nexus"), meaning "knot". The connexus is then the way in which organisms are knotted together in the network of an ecosystem. Many ecologists, cited by Brauns (1976), have visually represented such connexus for specific small ecosystems, or for compartments of such systems. One example has been given here in Fig. 5.70.

This does not represent a complete ecosystem, but one compartment. Its authors wanted to represent, for Hungarian oak stands, the ecological relationships that are vital for the parasitic and predating ichneumonid wasps that catch, and lay their eggs in the body of, herbivorous insects, which feed in their turn upon the forest plants. On the figure, the wasps are indicated by fat-drawn ovals, the host plants by fat-drawn rectangles, and the plant-eating insects by thinly drawn rectangles. As a matter of fact, the compartment defined by the "wasp-connexus" is here compartmentalised again as a food web with guilds of primary producers (plants), secondary producers (herbivorous insects) and tertiary producers (parasites/predators) – see also Section 5.3.4.4.

It is certain that the study of Fig. 5.70 leads one to understand much better how this compartment functions. The investment in a long, thorough and expensive research project yielding such a diagram pays back handsomely in increased insight, with scientific as well as practical rewards. The latter consist in the possibility to evaluate quite precisely the ways and means for biological control of oak pests, and to compare them with the impact of chemical treatments.

However, there are certain limitations to such diagrams, which have been enumerated by Brauns (1976) quite candidly:

- they cannot be made after literature data, but should only rest upon observations in a real forest, because otherwise they will represent nothing more than possible

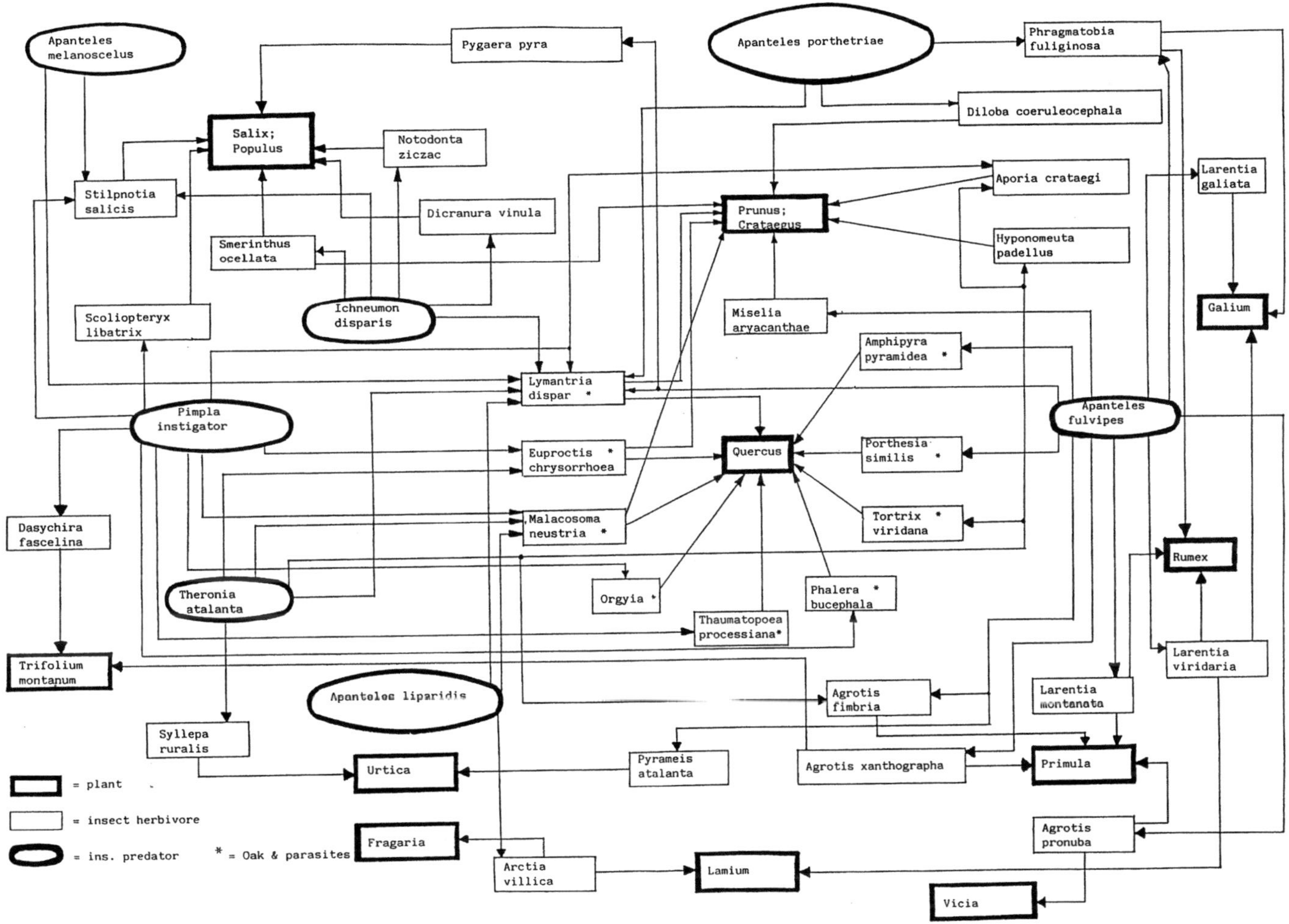

Fig. 5.70. Coenotic connexus in Hungarian oak forest. (See Tischler 1955 after Györfi 1951 ex Brauns 1976; modified)

interactions between plants and insects that may not even be present in the forest;
- the interactions are not complete, because only feeding is considered and no other interactions, e.g. from breeding and sheltering ecology; however, these are more or less linked to the basic necessity for food;
- there are no inputs and outputs represented between this compartment and its environment, neither within the ecosystem nor with other ecosystems; the difference between stationary and migrating insects defines one of these input/output relationships;
- the feeding relationships are not weighted, i.e. there are neither estimates of biomass and nutrient transfer, nor mortality impacts in relative terms (e.g. "*Ichneumon disparis* reduces *Lymantria dispar* populations by 10%");
- the visualised connexus is only an instant picture, because many of these insects are seasonal and, more important still, show quite distinct feeding habits in each of their life stages, from larvae to imago. This follows from the flexibility of insect, like fungus, life-cycles (see Sects. 4.7 and 4.8.5).

It may be added that the lack of a clear ecosystems hierarchy increases the confusion concerning the validity of connexus diagrams. Brauns (1976) spends several pages in discussing "biotope-bound" communities, "border communities", such as forest borders, "small surfaces of horizontal differentiation" of forests, like 6- to 8-year-old beech stands in mountainous forest districts and "cut-out parts of forest ecosystems". An approach that uses the eco-unit and its border zones as the standard, and adds input/output relations with the surrounding eco-units would clarify these matters greatly. This is one of the reasons why the eco-unit, as a hierarchical level, has been proposed in the present book.

Figure 5.70 is the simplest connexus given by Brauns (1976). When considering the more complex ones, like his example of decomposition of litter and dead wood, it is abundantly clear that the greatest risk in using this approach lies in a resulting model that becomes unmanageable, because it contains too many variable, small subsystems. Here we are again in the centre of the problem of compartmentalisation. Connexus such as those of Fig. 5.70 refer to species populations as their basic compartments, and it has been said earlier (Sect. 5.5.2) that the processing of all information on all populations, even in one eco-unit, cannot be managed. The Akker-Groeneveld model (see Table 5.3), from Brazil, is an example of more synthetic insect compartmentalisation.

The splendidly illustrated book on tree and shrub insects of the Canadian prairie provinces by Ives and Wong (1988) indirectly illustrates how complex connexus are. It contains a no less than 45-page index listing observed interaction points between each woody species and the more common local insects. Canada, due to its climate, cannot be considered as one of the most species-rich regions of the world. Moreover, the publication by Ives and Wong gives these ecological interactions rather as an aside, its main concern being field identification of Canadian insects, with some emphasis on crop damage.

The reason to introduce the connexus concept here, however, is not merely methodological. In the sections on production models, much simplified relation diagrams will be presented that are used for calculation of biomass, energy and nutrient flows and accumulations. The question to be asked quite practically is whether in such simplified diagrams information is eliminated that is essential for ecosystem analysis. One could, for instance, simplify the connexus in Fig. 5.70 as shown in Fig. 5.71.

With adequate measuring and estimating methods, this could be the basis for biomass, energy and nutrient turn-over models, in which numerical values are calculated. What information has been sacrificed from Fig. 5.70 in order to make such useful calculations?

We have seen earlier that biotic regulation of populations is linked rather to the eco-unit level than to forest mosaic or abiotic levels, where abiotic regulation plays an increasing role (Sects. 5.3.4.4; 5.5.6). If the connexus concerns a stand or a natural eco-unit, the information may define this regulation much better than other models. In the concrete case of the Hungarian oak forest of Fig. 5.70, the statement that increased numbers or biomass in the wasp compartment will lead to less damage caused to the oaks by pests in fact is a useless generalisation.

Even without weighting the mutual interactions between wasps and pests, one can easily see that two polyphagous wasp species, i. e. *Pimpla instigator* and *Apanteles fulvipes,* will exercise a more general pressure on many pest populations. On the other hand, if one pest is considered in particular, such as *Euproctis chrysorrhoea,* which causes havoc in European oaks from year to year (PUDOC 1982) much pressure would be brought to bear on this insect by wasps with a more restricted diet, in particular if their prey species that feed on other plants (*Prunus, Populus,* for instance) are absent. In that case, biotic regulation would depend upon wasps such as *Ichneumon disparis* or *Apanteles porthretriae.*

Again, the precision needed in analysis, which depends largely on the way in which compartments are chosen, is determined by the questions asked. No model is "better" than any other in itself, they merely address different questions. For the assessment of production and production losses, the simplified version of Fig. 5.71 is quite sufficient. But if the question is, how biotic regulation works, and how alternative options may be chosen to modify this regulation in a desired way, some connexus diagram like Fig. 5.70 is the optimal model now available.

We consider connexus to be compartment models of a kind that is too little known and under-used. For the tropics, we searched in vain for such documents in order to assess, for instance the ecological role of monkeys (cf. Van Roosmalen 1980: no connexus-like data), elephants (cf. Alexandre 1978a or Laws et al. 1975; only lists and descriptions), birds (cf. Van Dorp, in Gomez-Pompa and Del Amo 1985), or lianes (cf. Sect. 4.4). All kinds of statements are made concerning such forest components, but they cannot as yet be substantiated by visualised connexus diagrams.

5.5.11 Tree Compartments and Tree Temperament

Because trees are the determinant components of forest ecosystems, their compartmentalisation deserves a section of its own. Since time immemorial, trees have been classified in a sort of compartments that have both spatial and temporal aspects.

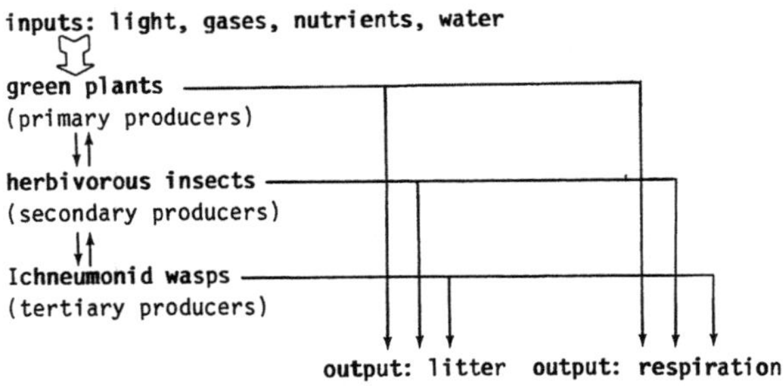

Fig. 5.71. Much simplified version of Fig. 5.70. Could be expressed in feedback terms in a way comparable with Fig. 3.32B

Trees have most often been considered according to their adaptation to the conditions they meet during their development. Most traditional classifications consider soil and light requirements, with emphasis on the latter. The germination requirements most often are treated apart. For the humid tropics, the subject covers an important part of the report by Hadley (1988), which is the basis for a MAB-UNESCO book on rainforest regeneration (in prep.).

The classical term in European forestry to indicate the requirements of trees is temperament (cf. Lanier 1986). Usually, trees are first considered as belonging to one out of three temperament classes:

- **shade-tolerant** trees may germinate and survive in even heavy shade until they have the chance to obtain enough light to accelerate their growth and penetrate in the canopy. The sequence is: shade first, light later.
- **semi-tolerant** trees may endure some shade during germination and establishment and survive quite some time until the full light allows them to grow into the canopy. The sequence is: half-shade first, light later.
- **light-demanding** species require light for germination, establishment and further growth; when outshaded by other trees, they die. The sequence is: light first, light later.

Regarding Fig. 3.1, one may note the temperament of a fourth category of trees which had disappeared from forestry handbooks with the management form known as "coppice with standards" (Mittelwald, taillis-sous-futaie), and which we propose to call:

- **late-tolerant** trees and shrubs, which require much light for germination and establishment, but which, when overtopped, survive in the shade during long to very long years, sometimes centuries (e.g. *Taxus baccata*). This has been mentioned already in Fig. 5.4 and in Section 5.2.3. The sequence is: light first, shade later.

These classical temperaments rest on the concept of extended forest zones with crown canopies of high density or low density, creating shade or half-shade, alternating with larger cleared surfaces, where much light is available. This is a forest model without eco-units, considering forest areas as Aubrevillean mosaics built by adult trees forming crown strata (see also Sect. 6.2.3.1). Geometric and numerical models of the light régime under forest cover in this model have been proposed, e.g. by Alexandre (1982a, 1984) in the Côte d'Ivoire. If eco-units or "gaps" are taken into account, the temperaments have to be distinguished in another way.

Therefore Shugart (1984) gave a mirror image of these properties by the indication of the "gap requirements" that would be identified here with the zero-event which trees need to start their life-cycle. His four "roles" of trees are (Shugart 1984, p. 120, his Fig. 5.5):

- role 1 : produces gap/requires gap;
- role 2 : produces gap/does not require gap;
- role 3 : does not produce gap/requires gap;
- role 4 : does not produce gap/does not require gap.

The requirements from classical forestry are here combined with tree mortality patterns. These roles are an important element in the "patch dynamics" models that are the cornerstone of Shugart's forest succession model (also see Pickett and White 1985).

The mineral soil requirements of trees are most often either listed or given as part of their silvicultural description. Van Goor et al. (1974) represent these requirements in the form of a matrix. Vertically, they list the Dutch soils, horizontally there are 18 tree species that can be grown in the Netherlands, and for every soil type their general growth performance in $m^3\ y^{-1}\ ha^{-1}$ is indicated by a colour code. This matrix allows the reader to see immediately whether a tree species is either a specialist on certain soils, (e.g. Douglas fir, *Pseudotsuga menziesii*, does well on 9 out of 28 soils, not advised on 19 others), or a generalist on many soil types (e.g. balsam poplar, *Populus trichocarpa* and hybrids, does well on 19 out of 28 soil types). The reader is reminded that a high growth performance is often a sign of healthy trees (Sect. 3.5).

Germination and establishment of trees are often considered according to a checklist containing numerous factors. One example is given by Shugart (1984, p. 62). Such factors include the presence or absence of certain seeds. This is an altogether different set of criteria which belongs to dispersal ecology and which has to be eliminated when considering the temperament. What is left, are typically those features that are the consequence of the nature of the zero-event (cf. Sect. 5.2.5.3). One could classify the "germination temperament" as follows:

- requirement for mineral soil/requirement for organic soil/requirement for ash;
- requirement for water/requirement for humidity/- but scarcely for drought;
- requirement for light/half-shade/shade - as said earlier;
- requirement for high/low/intermediate temperature;
- requirement for symbionts, e.g. mycorrhizae/no symbionts required;
- escape from seed predators required/not required;
- requirements for certain wavelengths of light, activating the germination through photopigments in the seed coat (e.g. Vazquez-Yanes in Gomez-Pompa and Vazquez-Yanes 1976).

If we add to such lists that trees often have the flexibility to change their requirements per site, sometimes because they are different genotypes within the same species, and that factors may compensate each other, which is often the case with light and soil fertility, it appears that the old difficulty with compartments is back. How far may one go into detailed requirements, in other words how many compartments may one create if the forest model is to remain manageable? The affair becomes still more complicated if one realises that trees start their life in a herbaceous life form, the seedling, meeting one environment, then develop through the potential phase in a second environment and finally have to cope with the canopy conditions when they have become trees of the present.

Many studies have been done in the last 15 years which illustrate parts of this problem. Bormann and Likens (1979) wrote a chapter on it for the Eastern United States. Mexican work on the seed physiology of light-demanding woody genera, such as *Urera, Trema, Cecropia, Piper* and *Eupatorium*, with extensive bibliographies, has been published by Vazquez-Yanes and Orozco-Segovia (1987a, b) and Orozco-Segovia et al. (1987). Alexandre (1978b) described and explained the behaviour of the light-demanding *Trema guineensis* in the Côte d'Ivoire, and of the shade-tolerant *Turraeanthus africana* from the same country (1977). He also studied small sapling populations of many tree species with different temperaments in the Tai forest in the Côte d'Ivoire (Alexandre 1986). The chapters by Bazzaz, by Bawa and Krugman by Janzen and Vazquez-Yanes in Hadley (1988), and the bibliography of that report also allow an up-to-date entrance in this field of research.

The simple tolerance ratings mentioned at the beginning of this paragraph (also compare Bormann and Likens 1979, p. 106) and the tree/soil tables remain very useful, simple approaches of the problem. Moreover, there is large consensus on their usefulness and most people can understand them right away, so that they are good and practical communication tools. As all classifications, they have been used in several more or less detailed versions (e.g. Mayer 1980).

In Section 5.2.2, distinction was made between initiation species, characterizing the initiation phase in **eco-unit** development and disappearing in the following phases, and the often-used terms of pioneer species, late secondary species, climax species, or "pioneer" versus "primary" species (T.C. Whitmore in Hadley 1988), which refer to **succession.** Succession was clearly stated to be a sequence of eco-units replacing eco-units (see Sect. 5.1.7), and **not** species replacing species within a developing eco-unit.

Now, trees are the only components which, by definition, determine forest eco-unit functioning from the initiation phase to the phase of decay. Tree species cannot be initiation species, because the same set of trees remains present in all eco-unit development phases. The question of tree temperament can therefore be linked only to succession, and be rephrased as: what temperament corresponds to what kind of eco-unit in a mosaic?

This leads to a comparison with the concept of ecological strategies of species, which is now gradually disappearing from literature as a central point of interest because the analogy between species and army strategies proved to be less than perfect. Begon et al. (1986), for instance, simply refer to life histories. An unverified story says that, at a botanical congress in Leningrad in the 1960's, an officer of the Red Army rose and stated that botanists as a matter of fact were not talking about biological strategy, but about tactics. However this may be, and with all lack of perfection, the ecological strategy of trees is a concept that is useful enough to order our ideas on temperaments.

The strategy of a species is conceived to be its programme of behaviour, developed by evolution and under selection pressure from its environment, to cope with its environment over its whole life cycle. Because a strategy belongs to a species, not to one individual tree, reproduction is an important concept. If reproduction is not efficiently ascertained in its strategy, a species will become extinct because of a lack of succeeding young generations.

The direct environment of a tree is the eco-unit which it helps to build and in which it grows up. In the following pages, the temperament of trees will first be placed in the context of general plant strategies, also relevant for the other plant components of forests treated in Chapter 4. After having done so, we will consider the temperament of trees in different kinds of eco-units.

One definition of species strategies is based upon the reproductive strategy of animals, later extended to plants. This rested on the first mathematical analysis of population numbers and their change with time by Verhulst (1848), more than one century and a half ago. His work was lost from view for a long time, and was rediscovered in the 1920's (Lotka and Volterra ex Pianka 1974). His formula is:

$$\text{Verhulst's formula} \quad \frac{dN}{dt} = r \cdot N \cdot \left[\frac{K - N}{K}\right],$$

in which *N* is the number of members of a population, *r* is the potential growth factor of that number, and *K* is the maximal number that can live within the local energy- and nutrient-budget.

This leads to the definition of two extreme strategies (e.g. see Pianka 1974; Hallé et al. 1978):

- *r*-strategies are based upon maximisation of population growth. This implies maximisation of the number of seeds, spores, eggs at the cost of energy or nutrients used for maintaining the parents. These organisms therefore are small, short-lived and numerous. This strategy can only be followed when there is plenty of energy and the nutrient supply is high, otherwise the mortality will be so high that most of the species is wasted. Open environments with few competing species therefore are the places where this strategy is most successful.
- *K*-strategies are based upon filling all the spaces where the organism can live, i.e. upon building specialised organisms. This implies a complex organisation of the parents, at the cost of the energy and nutrients to be used for seed, spore or egg production (cf. Fig. 3.30). This strategy is most successful when living space is limited and chances for survival are small, due to competition from other organisms in crowded environments. Closed environments with many competing or interacting species, like forests, therefore are the places where this strategy is found.

One can find paragraphs on strategies, resource allocation and budgetting and on strategy and competition by several authors in Bakker et al. (1985).

Pianka (1974) gave a table of properties for animals displaying both strategies. This table was formulated for trees by Hallé et al. (1978), and is reproduced here as Table 5.5 in a slightly modified form.

The expression of the reproduction and mortality pattern is illustrated by *Acer saccharum* (Aceraceae, sugar maple), which is a North-American tree species growing up in large eco-units with mixed tree populations. The environment is rather open, but becomes crowded fast (cf. Figs. 5.5. and 5.6A for a pure *Acer* stand of another species). This makes the sugar maple a *r*-strategist in its early reproduction, and a *K*-strategist in later forest development phases. The seedling numbers and mortality in Fig. 5.72 illustrate this.

This figure further shows a very general phenomenon. This is the strong mortality occurring among seedlings of forest trees the world over, and in all kinds of eco-units. There is nothing as ephemeral as a seedling carpet in a forest. One cannot judge about the future of a tree population in the seedling stage, because such carpets, growing "like hairs on a dog" as the old forestry saying goes, may have disappeared the next year (cf. Vooren 1979, in the Côte d'Ivoire). The habit of judging the future of forests by the seedling population comes from temperate silvicultural sys-

Table 5.5. Some characteristics of *r*- and *K*-selected trees. (After Hallé et al. 1978)

Character	*r*-Strategy	*K*-Strategy
Climate	Variable, unpredictable, macroclimate	Stable, predictable, (forest) microclimate
Biotope	Open communities or niches; "ecological vacuums"	Crowded communities with few free niches
Tactics	Colonisation	Infiltration
Mortality	Low when young, later high, or diagonal curve against time	High when young, later low during a long time, exponential curve against time
Reproduction	Early, not often, abundant	Delayed, often little by little, later sometimes abundant (mast years!)
Population size	All-or-nothing; changeing with time, below *K*	Always some, constant in time, at or near *K*
Population growth	High maximum rate *r*	Gradual, low *r*
Population density	Species-dependent, often high	Competition-dependent; often low
Competition within species and with others	Variable, often slight	Usually heavy
Individual growth	Fast	Slow
Individual size	Rather small	Large
Individual lifespan	Short to rather short	Long to very long
Production	Fast, quantity	Slow, quality
Wood	Often soft and light	Often hard and heavy

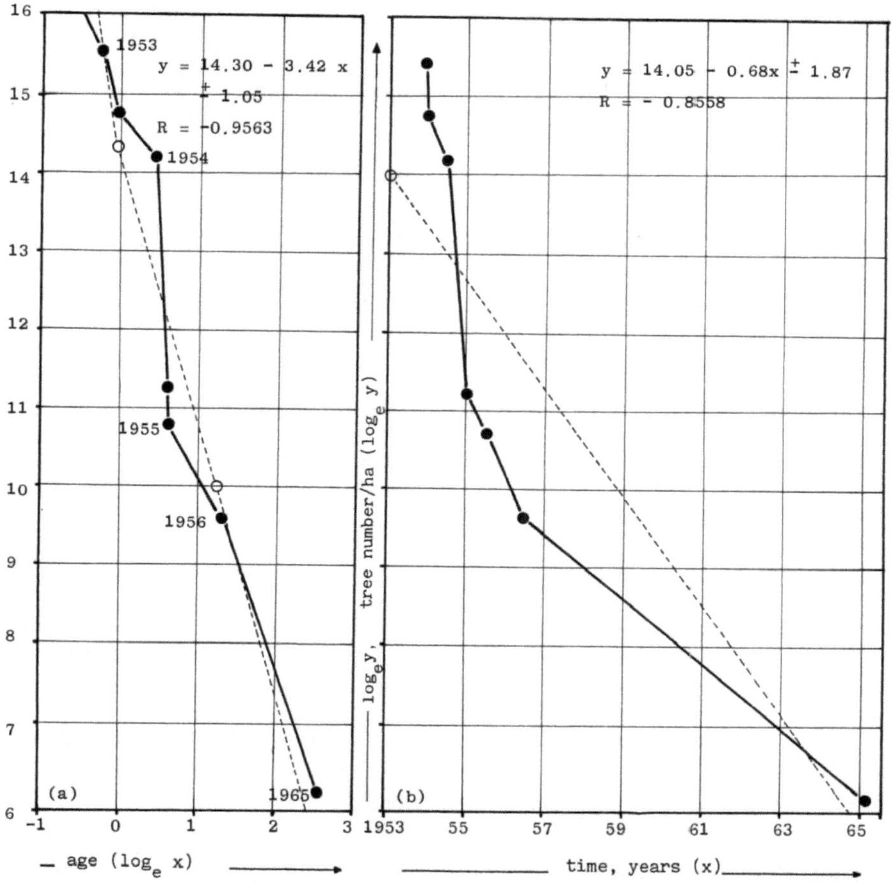

Fig. 5.72. Age class distribution of seedlings in *Acer saccharum* (sugar maple, Aceraceae) in *Acer/Tilia* stands in Wisconsin, USA. *Stippled line* linear regression. Note seasonal mortality. No observations available between 1957 and 1965. Nothing is as ephemeral as a seedling carpet. Compare this development with Table 5.5, on *K*- and *r*-strategies (also with Table 5.6). (After Hett 1971)

tems, where such populations are managed, i.e. first protected and then brought back to the desired numbers according to the management plan. This reduction of the population is called thinning or, in early development stages, tending. In plantations, these densities are reached by using normalised planting distances (e.g. 2 x 2 m; cf. Fig. 5.33), applied generally in a square or a triangular planting arrangement leading to a certain planting density per hectare.

The tree populations under management are not only reduced in number, step by step with each thinning, they are also selected as to their properties by the choice of the right seeds for the purpose and by eliminating the trees with undesired properties during each thinning operation. Please note that selection on pure economic criteria, such as free trunk height (desired for maximum marketable wood yield) or branchiness (undesired for lowering the wood quality) has to be done in such a way that the remaining trees are also the best components for survival of the

Table 5.6. Some characteristics of competitive, stress-tolerant and ruderal plants. (Grime in Freysen and Woldendorp 1978). Compare with Table 5.5

Characteristic	Competitive	Stress-tolerant	Ruderal
1. Morphology of shoot	High dense canopy of leaves Extensive lateral spread above and below ground	Extremely wide range of growth forms	Relatively small stature, limited lateral spread
2. Leaf form	Robust, often meso-morphic	Often small or leath-ery, or needle-like	Various, often mesomorphic
3. Litter	Copious, often persistent	Sparce, sometimes persistent	Sparce, not usually persistent
4. Maximum potential relative growth-rate	Rapid	Slow	Rapid
5. Life-forms	Perennial herbs, shrubs and trees	Lichens, perennial herbs, shrubs and trees (often very long-lived)	Annual herbs
6. Longevity of leaves	Relatively short	Long	Short
7. Phenology of leaf production	Well-defined peaks of leaf production coinciding with period(s) of maximum potential productivity	Evergreens, with weakly defined patterns of leaf production	Short period of leaf production in pe-riod of high poten-tial productivity
8. Phenology of flowering	Flowers produced after (or, more rarely, before) periods of maximum potential productivity	No general rela-tionship between time of flowering and season	Flowers produced at the end of tem-porarily favourable period
9. Proportion of annual production devoted to seeds	Small	Small	Large
10. Response to stress	Rapid morphogenetic responses (root-shoot ratio, leaf area, root surface area) maximising vegetative growth	Morphogenetic responses slow and small in magnitude	Rapid curtailment of vegetative growth, diversion of resources into flowering
11. Photo-synthesis	Strongly seasonal, coincid-ing with long continuous period of vegetative growth	Opportunistic, often uncoupled from vegetative growth	Opportunistic, coinciding with vegetative growth
12. Storage of photo-synthate, mineral nutrients	Most photosynthate and mineral nutrients are rapidly incorporated into vegetative structure but a proportion is stored and forms the capital for expansion of growth in the following growing season	Storage systems in leaves, stems and/or roots	Confined to seeds
13. Palatability	Various	Low	Various

remaining stand. They should give it as much resilience as possible against pests, diseases, mechanical damage and fire as well as an optimal niche structure for the needed non-tree components in each development stage.

In other words, a tree population will not survive without problems in whatever eco-unit, if its temperament or strategy is hampered. In nature, this can lead to a species being eliminated by selection. In silviculture, risks are heightened.

The distinction between *r*- and *K*-strategists has been refined with more criteria. Next to competition leading to selection, a criterion of stress-induced selection has been added. Grime (1978 in Freysen and Woldendorp), limits the stress concept to abiotic causes. The resulting compartments among plants in general are competitive, stress-tolerant and ruderal. In this classification according to strategies, ruderals show many points in common with *r*-strategists, and both other categories are variants on the *K*-strategy. Trees are claimed to adopt mostly competitive strategies. Grime's table is reproduced here as Table 5.6.

If we follow Grime's concepts, trees may be seen as following a ruderal strategy when young (invasion of degrading or innovating eco-units), a competitive strategy when in the potential state, and a stress-tolerant strategy once they become trees of the present in the canopy, which is an environment under high climatological stresses. This corresponds roughly to the life-cycles of tropical forest trees as described in 1903 by Schimper. The stress-tolerant behaviour of trees of the present fits in, for instance, with the periodicity of flowering which has, according to Grimes' table, "no general relationship with season". This would apply to mast flowering, at first sight (cf. Sect. 6.2.4.8).

Finally, we will link these strategies with the environment in different eco-unit types. We have seen earlier (Sect. 5.4) that large eco-units have a tendency to fragment. They are followed by intermediate or small eco-units. If these enter into fusion at a later age, germination occurs either in their shade on large surfaces, or in smaller eco-units when some zero-event causes late fragmentation. As a matter of fact, here one has the three temperaments we started with:

- light-demanding trees in large eco-units;
- half-tolerant trees in small eco-units;
- shade-tolerant trees under large, older eco-units or fused eco-units.

Large eco-units have always been considered as the beginning of succession, after large-sized and intensive zero-events like clear-cutting, fire or storm. Trees which grow in such eco-units are therefore called pioneers (starting a new forest) or secondary species, the disappeared forest being considered to have been primary. Sometimes, as in Table 5.7, authors consider pioneers to be shrubs and secondary species to be trees. Referring to Chapter 4 on non-arborescent forest components, this is not a consistent way to analyse temperaments, because it confuses life-form or biological type (tree, shrub, liane etc.) with temperament. Species germinating in smaller eco-units, after smaller-sized and/or less intensive zero-events, are considered either **early successors** to the pioneers, or late secondary tree species. Finally, **late successors** germinate either in very small eco-units or under biostatic or late aggrading phases in larger or fused eco-units. Late successors are called **climax**

Table 5.7. Characteristics of tree compartments in different eco-units that are part of successional processes in humid American tropical forests, inspired by Budowski (1965). Note distinction between pioneer shrubs and early secondary trees

Eco-unit	Large	Large, fragmented	Intermediate	Small (Fragmenting?)
Woody components	Pioneer shrubs	Early secondary trees	Late secondary trees	"Climax trees" (Successors)
Height	5 to 8 m	12 to 20 m	20 to 30 m, rarely up to 50 m	30 to 45 m, sometimes up to 60 m
Species	1 to 5	1 to 10	30 to 60	100, or little more
Geographical area species	Very wide	Very wide	Wide, including drier regions (Budowski's rule)	Usually limited, quite some endemics
Growth rate	Very fast	Very fast	Dominants fast, others slow	Slow to very slow
Lifespan dominants	10 years	10 to 25 years	40 to 100 years sometimes more	100 to 1000 years, perhaps more
Tolerance (shadow)	Very intolerant	Very intolerant	Young tolerant, later intolerant	Tolerant, except in phase of the present
Seedlings dominants	Very scarce	Nearly none	None, or many with high mortality	Rather abundant
Seed or fruit size	Small	Small	Small to intermediate	Large
Viability seeds	Long, latent in soil	Long, latent in soil	Short to intermediate	Short
Leaves in dominants	Evergreen	Evergreen	Often caducifolious	Evergreen
Wood	Very light	Very light	Light to medium hard	Hard and heavy
Diameter	Thin stems	Over 60 cm	Sometimes very thick	Including very thick trunks
Other components				
Seed dispersal	Birds, bats, wind	Birds, bats, wind	Mainly wind	Gravity, mammals, birds
Epiphytes	None	Few	Numerous, few species	Many species and life forms
Lianes	Abundant, herbaceous, few species	Abundant, herbaceous, few species	Abundant, few big ones	Abundant with large woody ones
Shrubs	Many, few species	Some, few species	Few	Few, many species
Grasses	Abundant	Few or many	Scarce	Scarce

species if one claims succession to reach an end stage, called climax. Climax theories will be referred to in the next chapter.

Table 5.7 is inspired by Budowski's table in an article which condensed his thesis and was published in 1965. His data are from Costa Rica, and their interpretation can be widely extrapolated in the humid tropics. The form of this table has been modified, so that its links with architectural forest analysis become apparent. Moreover, the order of the original table has been changed, so as to distinguish the woody components and other features, sometimes including interactions such as seed dispersal which are linked to silvatic mosaics rather than to eco-units (cf. Sect. 6.2.4.5).

Budowski (1965) formulated a general rule that, upon checking in the field, seems to be general. Oldeman (1983b) amended its formulation, so that it may be applied world-wide. The amended version is given here.

Budowski's rule: **any plant which plays a pioneering role in hospitable environments shows a geographical distribution which includes more inhospitable environments.**

As can be seen in Table 5.7, Budowski formulated this rule when stating that pioneer or secondary woody species of the humid tropics generally have a geographical area that includes drier regions. The same rule applies, however, in the temperate regions when considering pioneer species, e.g. birch (*Betula* sp. pl., Betulaceae). This pioneer of temperate regions also occurs in polar or boreal regions with much colder climates. The same can be said for larches (*Larix* sp.pl., Pinaceae), pioneers of the temperate regions which extend their area into quite inhospitable high mountain regions of Europe and Asia.

For the temperate regions, two examples of tree species will here be tested against the listed characteristics of Table 5.7.

Scots pine (*Pinus sylvestris*, Pinaceae) is well known as a pioneer tree. Its architecture remains true to Rauh's model (Fig. 3.2; Edelin 1977) for a long time. On better sites it may grow over 20 m high (late secondary), but it has a very wide distribution, including inhospitable, cold regions (early secondary). It shows a fast growth rate relatively to the other trees of these regions (late secondary) and lives much longer than 100 years, but it is a coniferous tree and lifespans are usually longer in the north than in the tropics (late secondary). It is quite intolerant to shadow (early secondary) and its seedlings have difficulty in establishing themselves under the parent trees, and do so in unpredictable ways (does not correspond to table; perhaps late secondary). The wood is rather light (late secondary), the seeds are intermediate-sized with not too long a viability period unless dried (late secondary). Foliage characteristics cannot be compared with this tropical table. Neither can seed dispersal be compared, because of the fauna being completely different. The position of Scots pine in this table, in conclusion, is not straightforward. This also is the case with N-W-European oaks (*Quercus robur, Q. petraea*), which combine characters of pioneer species with properties of late successor trees (cf. Koop 1981a).

On the contrary, European beech (*Fagus sylvatica*, Fagaceae, Troll's model) clearly seemed to be a late successor species, although its geographical area is quite large. However, when planted in the new Flevoland polders in the Netherlands (cf. Fig. 5.45A), on incompletely ripened soils and amongst the abundant grasses, the plantations grew well in an unexpected and decisive way. Did the grass vegetation protect the beech in its vulnerable young stages? Van Miegroet (1976, vol. 1) cites the behaviour of European ash, which reputedly is a half-tolerant tree species with late secondary characteristics, as having behaved like a late successor in the dense shade of coniferous stands in Belgium and Switzerland, where it survived in a suppressed state for up to 40 years.

These contradictory data are mentioned here intentionally. The reader, when judging the temperament of a tree species for whatever silvicultural or management aim, should be aware that the lists in existing handbooks should never be followed blindly. Douglas fir may display a different temperament in Europe than in its native Western United States, and mahoganies (*Swietenia, Khaya, Entandrophragma* and other Meliaceae) may show a quite different temperament when grown on other continents in the tropics, or even in other regions of their continents of origin. *Eucalyptus* is an Australian genus of the Myrtaceae, the temperament of which has been selected under heavy pressure from zero-events due to fires. This leads to special temperaments which, when defined well and used in mathematical simulation models, allow the prediction of resulting forests (see Shugart 1984, pp. 99–101).

In current tropical literature, the temperament of trees cannot always be traced easily. Here, tree species are often mapped over considerable tracts of forest (e.g. Van Roosmalen 1980 - Fig. 2.2; Hubbell and Foster in Sutton et al. 1983) and their distribution is statistically analysed as to clustering (e.g. Thorington et al. in Leigh et al. 1982). Indirectly, one may then suppose that tree species occurring in large groups would be pioneers. Our considerations on beech forests (Sect. 5.4.3.11) are a warning against such suppositions, unless carefully verified, because groups may result also from fusion of small eco-units built by a tree species with a successor temperament. In the tropics, this might apply to certain Leguminosae, or to the tropical Asian Dipterocarpaceae (cf. Fig. 5.53).

Finally, a preliminary documentation type is presented that is closely linked to the temperament of forest trees, to the connexus approach, to the nature of zero-events and to reproduction strategy. This is the silvicultural diagram (Oldeman 1983a, 1985; Oldeman et al. 1985). Figure 3.1 is a forerunner of the silvicultural diagram. After reading the preceding pages concerning the temperament of trees, the link between architectural deployment and adaptation to site conditions appears clearly in Fig. 3.1.

A silvicultural diagram, as shown in Fig. 5.73A, is an open-ended document, on which all environmental factors and organisms interacting with a tree of a certain genotype (species, subspecies, cultivar, clone etc.) can be indicated. Because the list would become very long, the importance of each interaction is weighed, and only the most important ones are included. No difference is made, in principle, between natural and human factors. If the price on the wood market is an important criterium for the maximum lifespan of a tree, this may be added, as well as some parasite that risks to wipe out the whole tree population at a certain age. Interactions are given in relation to the age of a tree, because often they occur only in certain developmental stages, e.g. seedling, suppressed potential tree, growing potential tree, tree of the present. The thicker the bar indicating a particular interaction, the more important it is at that age. This of course is a quantitative approach, because the number of millimeters indicating the thickness of the bar can be numerically expressed. The precision for the moment remains quite low, and the criteria for "importance" remain empirical in the preliminary studies done up till now (Oldeman et al. 1985). In the future, links with ecological production models will certainly make it possible to express the "importance" of an interaction as a number, e.g. biomass gain or loss at a certain age caused by that interaction. In a more global way, the principles to do so have been described for pollution stress by Rabbinge and

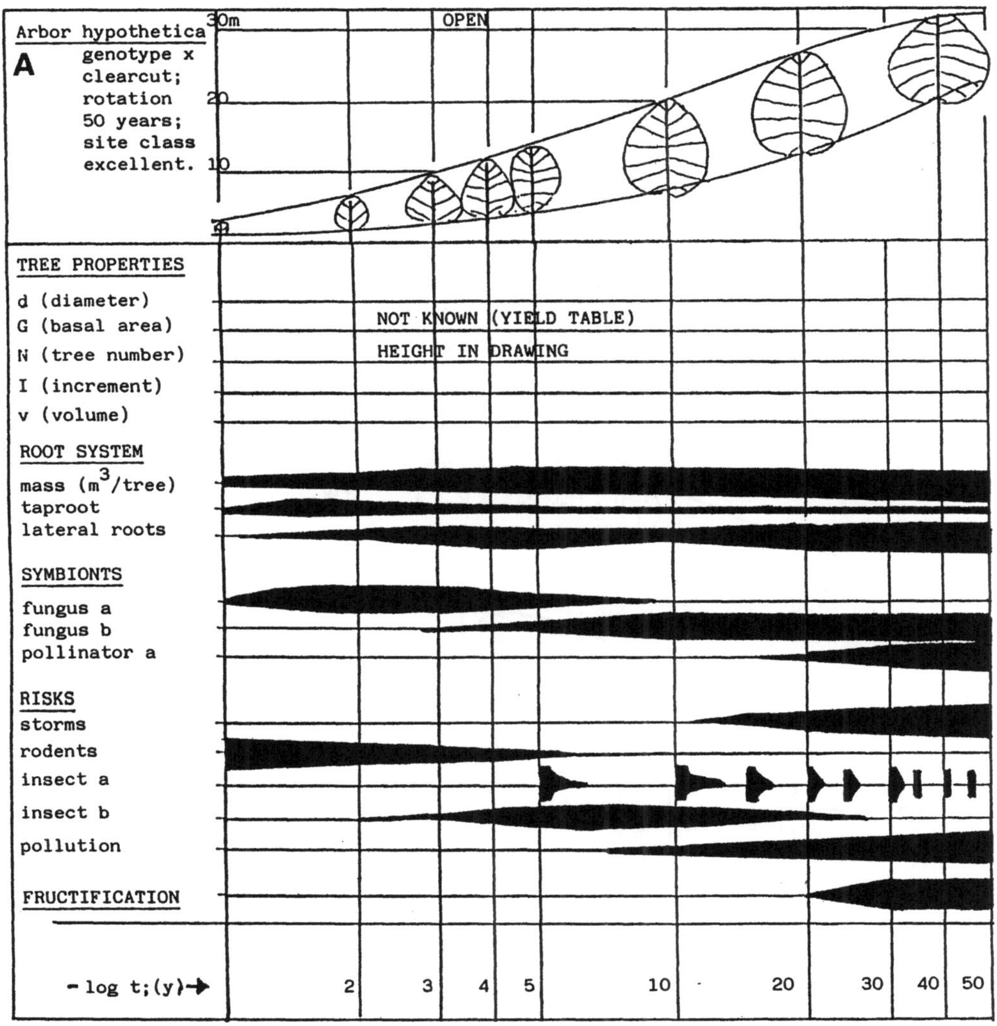

Fig. 5.73A (Fig. 5.73B see page 334)

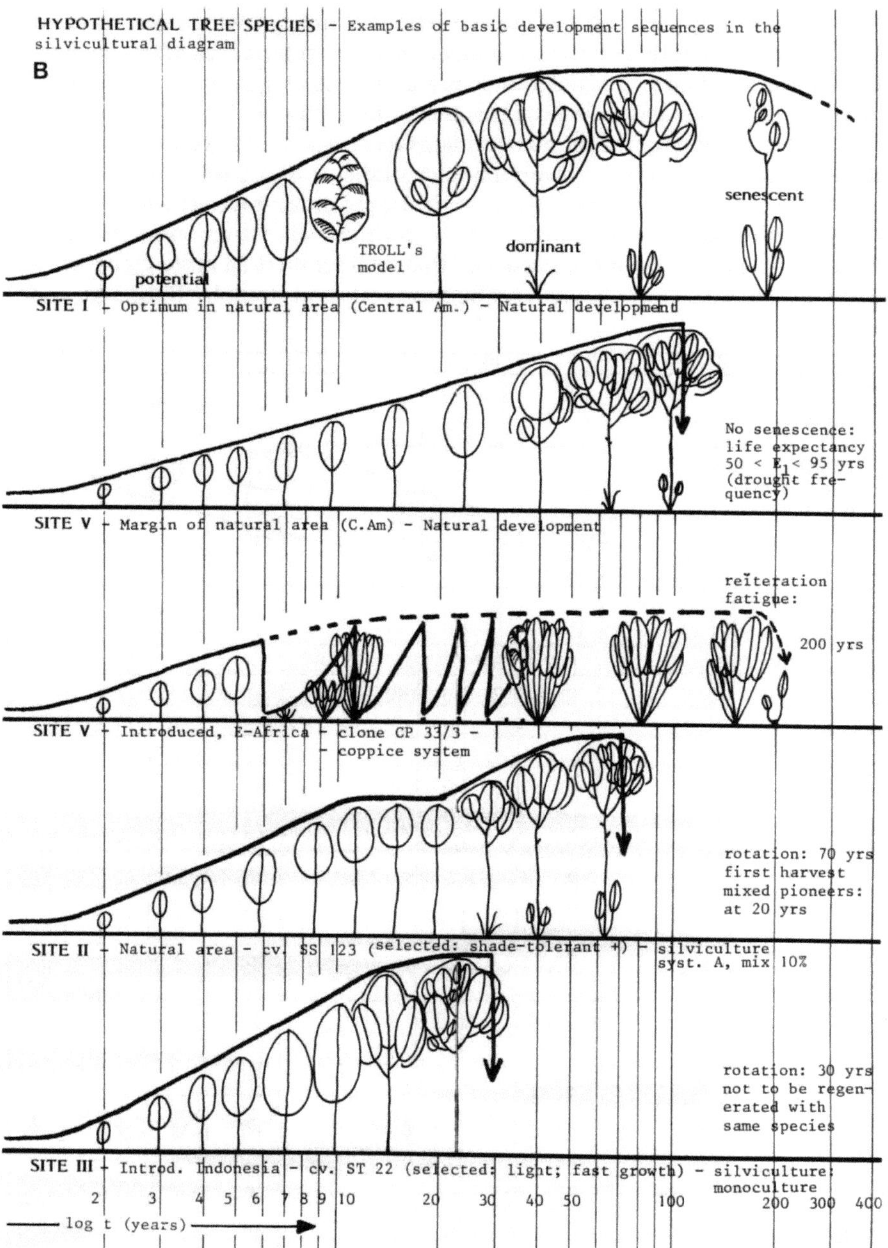

Fig. 5.73. Silvicultural diagrams. **A** Principle of the silvicultural diagram giving tree form, production data and main interactions against age. The diagram is open: one may add what is wanted. **B** Principle of diagrams for different genotypes, sites and silvicultural systems (natural base-line at top). Production data and interactions to be added. (**A** after Oldeman 1985; **B** from Oldeman 1983a)

Mohren (1984). The general principles of such models have been set out by Mohren (1987), who left out the biotic interactions intentionally. Another way to express interactions numerically would be an expression of lethality of an interaction for the tree, for instance by values referring to a lethal threshold. If production were reduced under a "critical growth rate", the tree would die (cf. Kohyama 1983).

The nature of the eco-unit concerned is here characterised by the silvicultural system. **Clear-cutting** systems, for instance, lead to large eco-units with an innovation phase under much light, drought, wind, direct impact of rain and, if mechanised, soil compaction to a certain degree and disturbance of the propagule bank. Another example is **selective silviculture,** which is not the same thing as selective exploitation. Selective exploitation is only a harvesting system, in which at best the logging agency considers that the forest will recover automatically after harvest (but see e.g. Johns 1988; Oldeman in Van Beusekom et al. 1987). Selective silviculture, on the contrary, creates a well-designed artificial late fragmentation process, leading to a certain type of small eco-units favouring the growth of well-selected tree species. **Coppice silviculture** still is another option. Figure 5.73B shows some of these options, starting with natural development in natural eco-units as a base-line.

The third condition considered in silvicultural diagrams, next to genotype and eco-unit, is the "site class". In forestry, sites are evaluated as to their productivity in terms of m^3 of wood per year per hectare. The principles underlying these classifications can be found back in handbooks on growth and yield. In Fig. 5.73B, site class I is the best, and site class V is the worst. Here again, an apparent link exists with ecological production models, but also with considerations of stress. One may consider that trees grow up under conditions of more stress if the site class is worse.

No perfect silvicultural diagrams exist as yet. Research is on-going to establish such documents for various species (e.g. *Acacia auriculiformis*, Schreuel and Stegeman 1986; *Pinus sylvestris*, Oldeman et al. 1985; Swart 1987; *Robinia pseudo-acacia*, Maas and Gubbels 1986a, b; *Faidherbia* (Acacia) *albida*, Van der Zandt and Sterck 1986; or the graphical basis for three *Shorea* species and *Eusideroxylon zwageri*, Suselo 1983). Once established, silvicultural diagrams are considered to become a communication link between silvological research and forest management, both for production and conservation objectives. If the researchers make the framework, the manager can add all specific factors encountered over the years in the special forest under management. This information can then be fed back into research.

Finally, the links between site quality and stress lead us to consider once more the different strategies, particularly *r* and *K* (Table 5.5), and the fact mentioned by different authors that tree species can change their strategy according to the site. Figure 5.74 gives a hypothetical image of what happens. It is well known from both yield tables and production ecology (Mohren 1987) that less biomass is formed on inferior sites which may be limiting in water or nutrients, or stressed by pests, diseases or pollution, or show a combination of these factors. Now, *K*-strategy implies the maximisation of biomass in a highly organised, large and long-lived organism, such as a tree. Moreover, the present author observed in several places with diminishing forest vitality that forest regeneration, i.e. the number and health of seedlings and saplings, was particularly vigorous. In broadleaved trees, this fits with Oldeman's observation (1974a) that abundant flowering and fruiting is linked to abundant, small-sized reiteration in large tree crowns, whereas Fig. 3.33 of the present book shows that reaction to stress leads to abundant, small-sized reiteration.

In conclusion, stress leads to shorter lifespans, less biomass accumulation, increased small-sized reiteration and hence increased flower and seed production. This in fact is a shift from *K*-strategy to *r*-strategy. Figure 5.74A shows the increase of biomass against time in *K*-strategists (left) and *r*-strategists (right). It is not necessarily the amount of biomass per species, accumulated after a number of years, that is different for these two types of plants. The difference lies in its distribution over either few large, long-lived organisms or numerous small, short-lived organisms. It is this distribution pattern that changes when *K*-strategists come under stress. The mechanism of seed production increases the number of trees which each contain less biomass and live shorter lives. This automatically means a different organisation and development of the eco-unit which they build, because smaller, differently grown and organised trees exert different regulation pressures, such as shading-, transpiration-, or interception-functions.

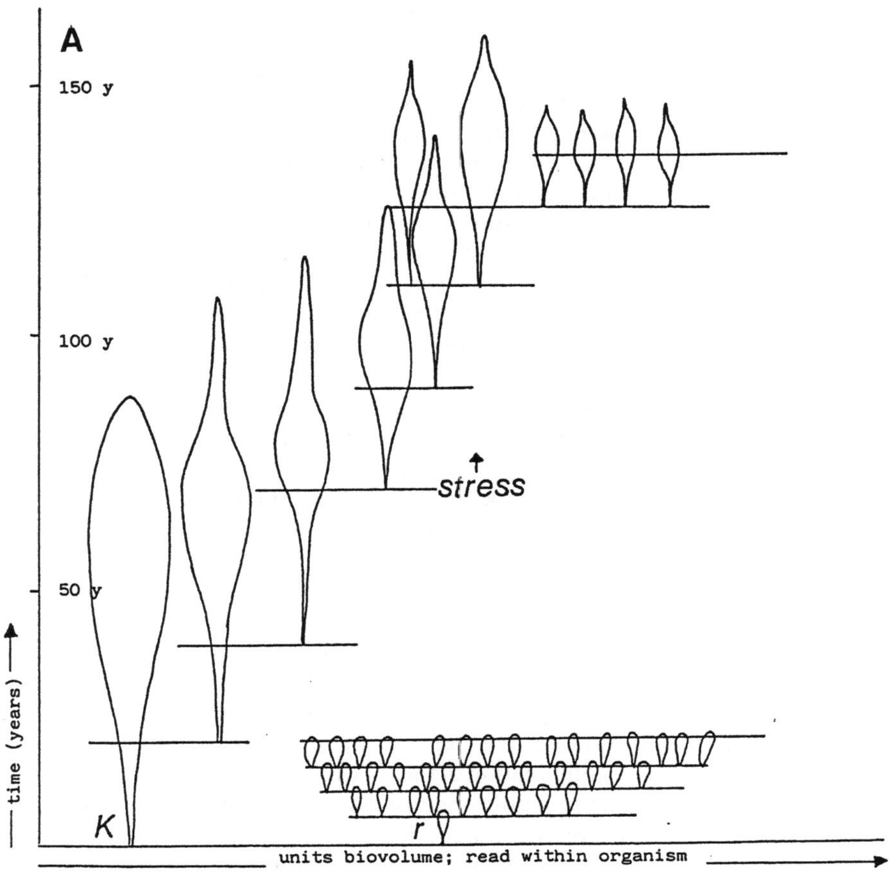

Fig. 5.74A

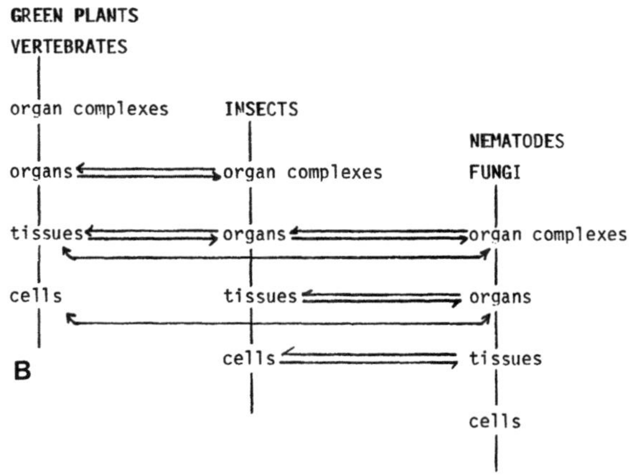

Fig. 5.74. *K*- and *r*-strategies, stress and interactions. **A** When tree species with a *K*-strategy come under chronic stress, from generation to generation they reproduce earlier and accumulate less biomass per tree; this convergence towards *r*-strategy may also be conceived as a selection pressure, e.g. in evolution coupled to climatic shifts or species migrations (cf. also Sect. 6.3.8). **B** Interactions between components of an eco-unit are not necessarily at the same hierarchical organisation level (cf. Table 2.1). Hence ecological regulation power of trees is important. Remember that an "organ complex" most often is an organism. See also text

Referring to Table 2.1, it now becomes interesting to pinpoint the hierarchical levels of interaction between different organisms, particularly between such *K*-strategists as trees, shrubs, larger birds or mammals and *r*-strategists such as insects, nematodes or fungi. An insect is an organ complex (Table 2.1). It feeds on leaves, or axes, or buds, which are organs (Table 2.1). In its turn, a nematode is an organ complex that is smaller than an insect, and it may feed either on an insect's organs or on plant tissues, whereas other nematode species may graze upon the whole organ complex of mycorrhizal fungi. Figure 5.74B represents these interactions graphically, although such a general overview may well have its exceptions because nature has no strict limits. It is interesting to note that "pests" (insects) and "diseases" (fungi, nematodes) are interacting with green plants at different levels. The distinction between pests and diseases therefore is a prescientific expression of an ecological rule. Furthermore, this diagram indicates the far-reaching effects that modification of tree strategies, due to anorganic stress, may have upon the biotic interactions within an eco-unit.

5.6 The Life Cycle of Forest Eco-Units: Production Models

The preceding consideration of the compartments of forest ecosystems, at the level of the eco-unit, has led us ever closer to the indispensable models concerning ecological production. This category of models includes the "process" models which focus upon turnover of matter and energy, and accumulation or breakdown of biomass. As such, they are the counterparts at the eco-unit level of the physiological models treated in Section 3.5 at the level of the tree organism (Figs. 3.28–3.37). For some remarks concerning the difference between the autoecological and the synecological approaches, which are valid particularly in production models, the reader is referred to Section 5.5.2.

5.6.1 Concepts

In ecosystem production models, the same features are used as in tree production models (Sect. 3.5.1), i.e. linear, surface or volume variables or mass (weight). These variables can be expressed against time. Turnover of volume or mass is expressed as the rate of modification somewhere in an ecosystem (space) against time, as compared to the modification somewhere else in the ecosystem over the same timespan. The aim of production models is the explanation of such variations by sources, flows and sinks of matter and energy between the one spot and the other. This requires the previous establishment of a relational or interaction diagram, representing the links between different ecosystem compartments.

These links generally are expressed as the carrying-over of amounts of energy, water and chemical elements. Submodels are therefore called "carbon cycle", "nitrogen cycle", energy cycle", "water cycle" and so on, the water cycle being important because many chemical substances are transported in watery solutions. This is true inside organisms, but it also accounts for a large part of nutrient turnover in ecosystems, for instance in eco-units.

The whole approach is based upon the classical notion of "food chains". We have not been able to find the origin of this concept, which is given as the base of ecosystems analysis in many handbooks (e.g. Odum 1975; Kormondy 1976; Ricklefs 1976; Beukema and Nienhuis in Bakker et al. 1985; Begon et al. 1986) or the splendid, illustrated "Pyramids of Life" by Reader and Croze (1977). Most probably, food chains and food webs have been considered to be self-evident by biologists, from Von Humboldt in the early 1800's onwards. According to Ellenberg (1973) the main notions for ecosystem analysis were available before 1914. However, the terms and definitions of today were coined much later.

In its most simple form, a food chain can be exemplified as follows:

- Pine consumes sunlight;
- insect consumes pine needles;
- bird consumes insect;
- fleas consume bird, or

- Young oak consumes sunlight;
- rabbit consumes young oak;
- man consumes rabbit.

This simple form is not sufficient to explain the transfer of matter and energy in a whole ecosystem. Chimpanzees, the large African apes, for instance, consume both plant matter and animals, and the same is true for the Asian orang utan (Rijksen 1978). Therefore, food chains are usually combined in food webs. We already saw an example of such a web in the section on connexus (Fig. 5.70). The diet of consumers is important to understand food webs, and the terms indicating different eaters can be found back in the section on the subterreanean part of an eco-unit (Sect. 5.3.4.4). More generally, green plants are called primary producers (with a few other organisms which do not depend on light), their consumers are called herbivores or secondary producers, the organisms that live by eating or feeding on these secondary producers are called tertiary producers, and so on.

It should be said that a considerable loss of energy and nutrients occurs between these trophic levels. As a rule-of-thumb, the efficiency of energy and nutrient use is about 10% (e.g. Asimov 1972) from one level to the next, 90% being lost. The rabbit uses about 10% of the energy and nutrients stored in the pine, the man uses about 10% of the storage in the rabbit. This may explain why very small organisms (pathogens) may make a larger organism very ill. On the other hand, it may also point to a considerable amplifier effect of inputs to larger photosynthetic organisms from smaller symbionts, e.g. in water input through mycorrhizae, if and when this rule-of-thumb works in reverse (see Smith and Douglas 1987).

If solar energy is considered to be bound in chemical substances by green plants, the whole cycle of nutrients may be conceived as an energy cycle (e.g. Ricklefs 1976; Vervelde 1975). This is a unifying concept which allows the building of quite simplified and manageable models. But it leaves out the character of the chemical substances which carry this energy, which may act either as neutral carriers, or as stimulants for processes, or as brakes. If excessive braking-down of a vital process takes place, the effect is toxic for the living system (see Sect. 5.6.8). Still other molecules act as stimulants or poisons according to their dosage. This question is raised here in order to show that these aspects generally are not included in production models, which usually are limited to the transfer of large quantities of specific and important elements (e.g. H, O, C, N, P, K).

Just as we considered the main sapstream (Fig. 3.30) in a tree, taking for granted the signalling, stimulating or braking system of phytohormones which is linked to special molecules in small dosages, we will limit ourselves for the moment to the general lines of the eco-unit production model (see Sect. 3.5.5).

5.6.2 Relation Diagram for an Eco-Unit

Figure 5.75 shows such a model in the form of a relation diagram. It refers to one eco-unit, contrary to most other diagrams of this type, which concern "average" forest ecosystems. Most data in the form of relation diagrams are assumed to obtain their "representativity for a forest type", not by establishing physical limits of living systems, but by choosing a "representative" part of a large forest surface through standard statistical procedures (but see, for instance, Berger and Berry 1988).

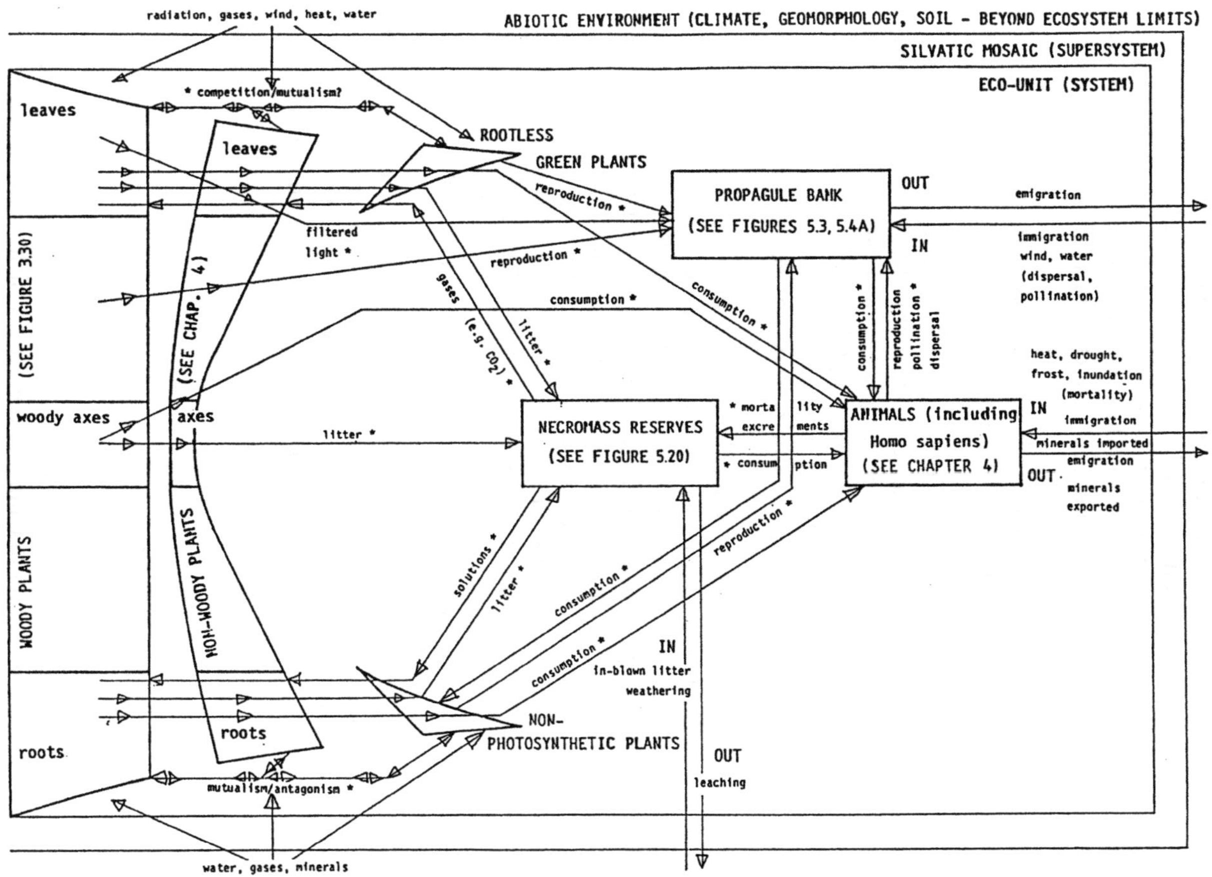

Fig. 5.75. Relation diagram for eco-unit production. Only compartments, flows and interactions proper to the eco-unit level are given, except in the compartments "woody plants" and "non-woody plants", which thus can be compared with other similar diagrams. Note stepped inputs and outputs, coming from the abiotic environment, but always selectively transmitted through the silvatic mosaic (surrounding eco-units). Interaction between necromass and propagule bank omitted. Eco-unit description: Sects. 5.1.4–6. For trees compare Fig. 3.30, through Table 5.8

The diagram in Fig. 5.75, on the contrary, is drawn for one eco-unit as defined and circumscribed in Section 5.1, with the five attributes of living systems (Sect. 5.1.5), i.e. a border layer, an internal communication system, a mass translocation system, the capacity for adjustment, and an information system for reproduction.

- The **border layer** is made up by the green leaf mass (phyllosphere) above, by the root mat (rhizosphere, including symbionts) and the root floor below and by stems and leaves at the sides.
- The **internal communication system** is represented by the interactions marked with an asterisk in Fig. 5.75. These interactions are authentic eco-unit properties, and are no properties of any one single participating organism.
- The **mass translocation system** is partly localised inside organisms, particularly trees, this part not being an eco-unit property but a property of organisms (see Figs. 3.30 and 3.32). Authentic eco-unit mass translocation is partly in parallel with communication lines and takes place through interactions such as a litter fall and decomposition (necrosphere: νεκρος, nekros in Greek = dead). Other interactions implied are consumption and symbiosis, i.e. food chains.
- The **capacity to adjust** lies in the feedback loops between the processes indicated by the arrows. For instance, if radiation increases by the elimination of neighbouring eco-units, all interactions coming from and going to the leaf compartment (phyllosphere) will change, with allocation of either more or less eco-unit resources to other compartments.
- The **reproduction system** of the eco-unit is the complex of interactions between all other compartments and the propagule bank. This is not mere reproduction leading to the stocking of this bank with propagules such as eggs, spores and seeds. For instance, it also includes regulation of germination processes by the quality of the light spectrum filtered through living leaves (Sect. 5.5.11) or by litter chemicals (Sect. 5.6.8), or the amount of propagules consumed by animals and fungi.

However, in order to make this diagram comparable with other such diagrams in literature, in which biomass most often is divided in the compartments "leaves", "small wood", "thick wood", "small roots", "thick roots" (see Fig. 6.54) and sometimes "animals", with the same division for the necromass compartment called "dead organic matter" which may be divided in "litter" and "organic soil matter" (or humus), such compartments are also indicated within the plant compartments on Fig. 5.75.

Processes indicated without asterisks are not eco-unit properties, but represent interactions with the surrounding forest mosaic or landscape. If a silvatic mosaic, built by many eco-units, is present, many inputs from the abiotic environment are filtered through the mosaic. The whole balance between immigration and emigration of animals, for instance, is determined by the nature of the surrounding silvatic mosaic. So is the immigration and emigration of propagules. The outputs of leaf and root compartments to the surrounding environment of the eco-unit have been omitted (but see Fig. 3.30).

Table 5.8. Comparison between relation diagrams for a tree (Fig. 3.30) and an eco-unit (Fig. 5.75)

Property	Tree	Eco-unit
Inputs/outputs	Through leaves/roots	Through propagules/animals/phyllosphere; less through rhizosphere/necrosphere
Primary production	Leaves/roots	Phyllosphere/rhizosphere
Transport system	Xylem/phloem	Necromass/animals
Direction large mass translocation	Bottom-up, in ascending sap stream	Top-down, in litter transfer and through necromass
Nature of mass translocation	Continuous liquid flow through different, contiguous tissues	Transfer chain, dry and liquid, through interrelating but often spatially isolated organisms
Regulation of mass translocation	Balanced feedbacks and phytohormone effects	Balanced feedbacks and chemical regulators (cf. allelopathy)
Reproduction system	Seeds (fructification, i. e. selective gene accumulation)	Propagule bank (selective propagule accumulation)
Information in reproduction	Carried by gene combinations	Carried by propagule combinations
Regulation of reproduction	Physiological feedbacks and regulators, shedding	Balances between propagule formation, longevity, consumption and germination/hatching stimuli
Dispersal to	Own eco-unit and surrounding mosaic	Surrounding mosaic (formation identical propagule banks elsewhere)
Defense	Feedback balances and chemical substances	Feedback balances and population shifts

Now let us compare the two ecophysiological relation diagrams, for a tree (Fig. 3.30) and for an eco-unit (Fig. 5.75). They have a few things in common, but their organisation is completely different. Table 5.8 shows this.

As in Fig. 3.32, which gives much simplified versions of Fig. 3.30, the detailed eco-unit diagram of Fig. 5.75 can be simplified. This reduces it to **rule-of-thumb** quality, but on the other hand it provides an easy way to make a provisional judgement as to the general effects of human or natural impacts on an eco-unit. Figure 5.76 represents this simplification.

The larger the arrows on this figure, the larger are the usual quantities of mass translocation through this type of interaction. As a concession to comparison with other published diagrams, the sapstream, which is properly speaking a property of an organism, is included in a much simplified way. The four compartments that are left in this figure do not include the propagule bank, just as reproduction is not mentioned in Fig. 3.32. The remaining compartments are:

- The **phyllosphere,** containing all green, photosynthesising mass in the eco-unit;
- the **rhizosphere,** containing all roots and soil (micro-)organisms active in non-photosynthetic chemosynthesis in the eco-unit;

- the **necrosphere,** containing all dead matter and the organisms that decompose it (decomposers), functioning as a transfer compartment ensuring transport;
- the **zoosphere,** containing all animals that feed on primary producers (herbivores) or on secondary producers (predators) and that ensure **transport** of eaten mass inside the eco-unit, but also between the eco-unit and the surrounding forest; moreover inside the eco-unit they often transport small amounts of important mass, such as pollen needed for pollination (pollinators) or seeds (dispersers) and so ensure transfer of information.

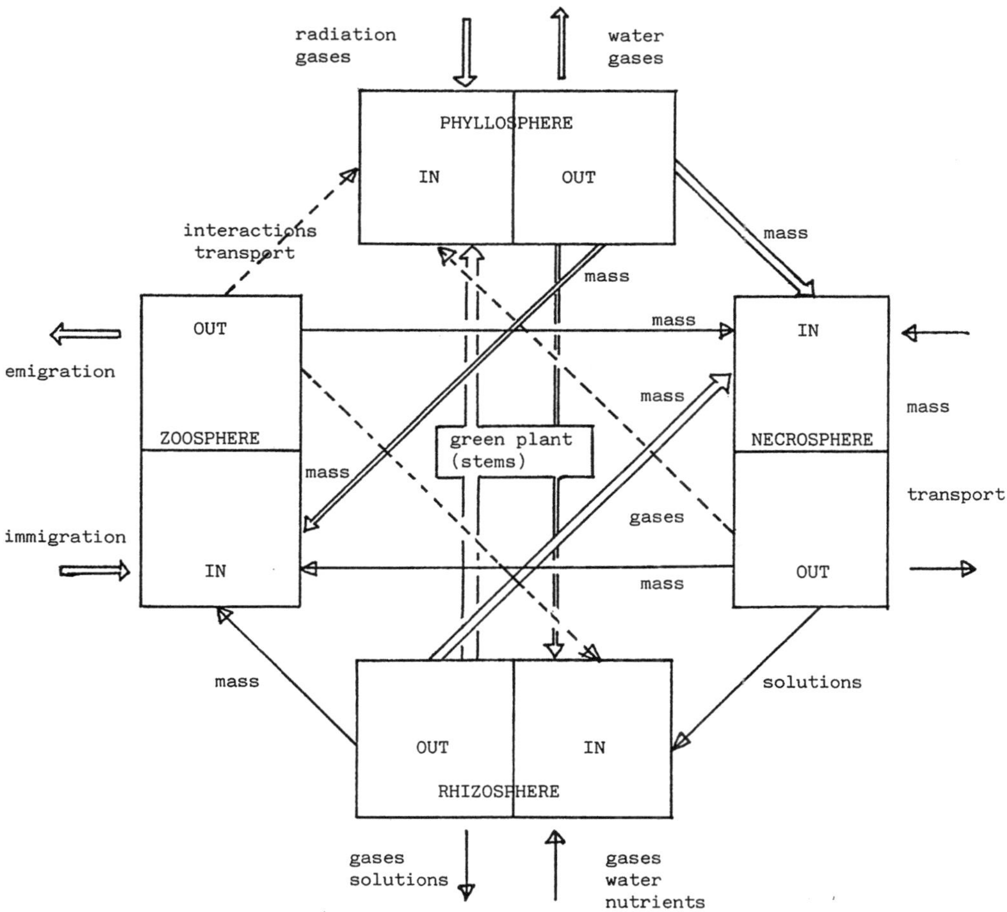

Fig. 5.76. Figure 5.75 simplified towards rule-of-thumb level. Like Fig. 3.32 for trees, this diagram makes it possible to establish a rough-and-ready diagnosis of the reactions of a forest eco-unit to impacts by natural causes or silvicultural measures. The factor to be assessed is introduced at the appropriate *arrow* (e.g. cattle: zoosphere, *IN*) or compartment (e.g. thinning: phyllospere and rhizosphere smaller, *OUT* smaller). The consequences can then be traced over the feedback loops to all other compartments, which are increased or decreased. No correction for internal regulation (e.g. impellors, Sect. 5.6.8) can be assessed without extra knowledge

5.6.3 Evaluation of Silvicultural Measures

Some examples show the kind of conclusions that can be drawn from this simple diagram in practical situations.

If wood harvesting is considered, the trees that are cut first are part of the necrosphere (dead mass), and the extraction of the wood from the forest is then an increased output from that compartment to the environment. This means that three compartments become smaller: the phyllosphere (crowns have disappeared), the necrosphere (trunks have been exported from the eco-unit) and the rhizosphere (roots die). Therefore, all outputs from these three systems will decrease. The direct effect therefore is a lessened primary production because the system is smaller and the inputs from the necrosphere also become smaller. Because of lessened inputs, the amount of animal biomass also will decrease. This may lead to shifts in population numbers and in species composition in both the zoosphere and the necrosphere. The more wood is harvested, the stronger will be these effects. For mortality patterns in trees, e.g. see De Graaf (1986) or Johns (1988).

If cattle are grazing in the forest, the impact comes from the zoosphere, whence outputs in the form of exported biomass (meat) are increased. The effects can also be traced through the feedback loops inside the eco-unit, just as with wood harvesting. The same can be done for the effects of hunting.

If chemical substances come in from the atmosphere, part of these increase the inputs in the phyllosphere, part of them pass through the phyllosphere and reach the necrosphere directly. Following the chain of effects, which is important for instance in the evaluation of pollution or forest fertilisation, can be done again by estimating the modifications in feedback. This will be quite different from substance to substance. But it will help not to forget any compartment of the eco-unit, in particular the necrosphere, which apart from trees is one of the major factors differentiating forest systems from agrosystems.

These examples show the usefulness of this kind of simple diagrams for a first, rough evaluation of a large array of management options or natural events, like the ice rains that covered some 40,000 ha of forest and all trees with a thick ice layer in the Dutch northern provinces in March 1987. But most often, a rough evaluation is not enough. The aim of many, if not most of the models starting with a relation diagram resembling Fig. 5.75 is the calculation of biomass production by organisms or populations (crops!) under different circumstances.

5.6.4 Biomass Production and Biogeochemical Flows

The first thing that is needed for such calculations is a numerical expression of the importance of each arrow, so that their respective values can be added or substracted. The most universal way to do so is the expression of these values in units of energy, kilocalories or Joules. Ricklefs (1976) calls this the "universal currency of ecosystems". Indeed, the amount of energy bound in, say, phosphate molecules, potassium chloride, water or chlorophyll molecules can be compared, whereas

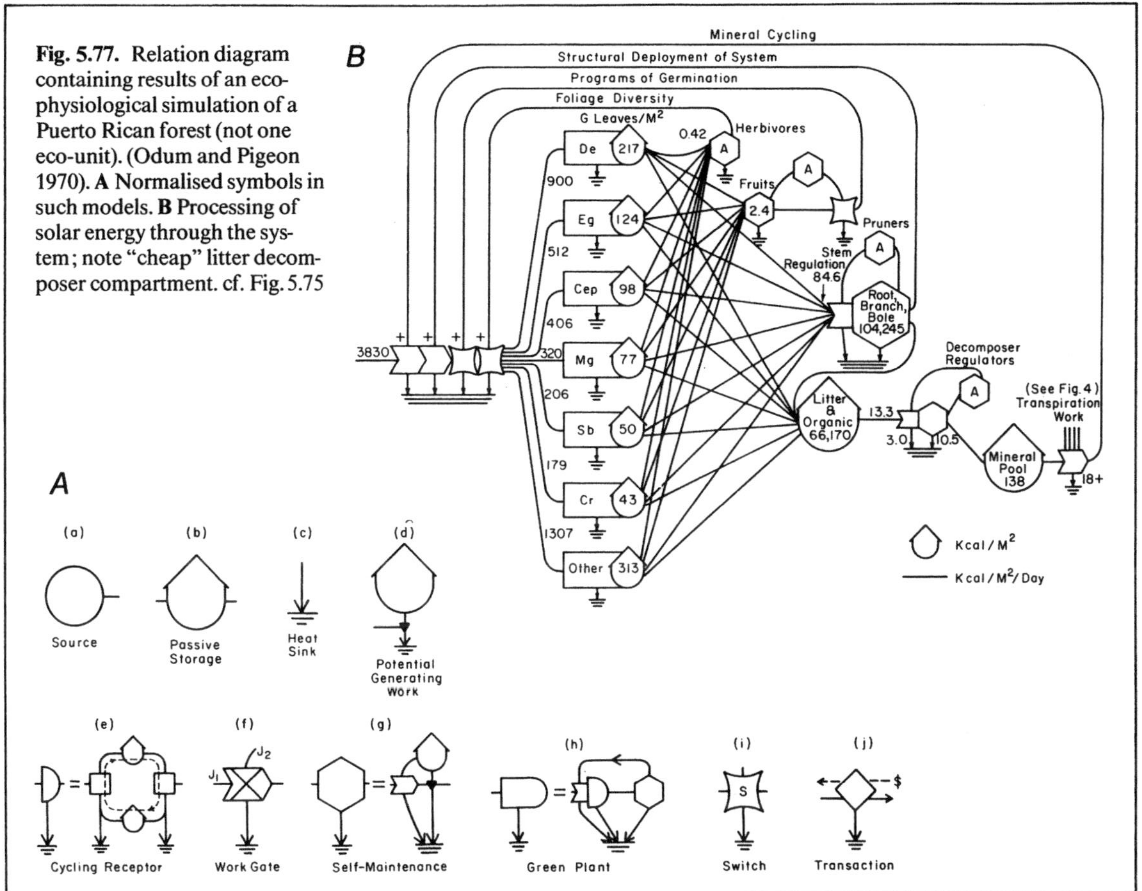

Fig. 5.77. Relation diagram containing results of an eco-physiological simulation of a Puerto Rican forest (not one eco-unit). (Odum and Pigeon 1970). **A** Normalised symbols in such models. **B** Processing of solar energy through the system; note "cheap" litter decomposer compartment. cf. Fig. 5.75

their other chemical properties and pathways through an ecosystem may be very different.

Figure 5.77 is an example of a forest ecosystem model from Puerto Rico, built according to these principles by H.T. Odum (in Odum and Pigeon 1970). Figure 5.77A explains the abstract symbols used in these kinds of models, indicating such features as storage places of energy, switches from one pathway to another or "transactions" where one process in one direction is paid for by a process the other way round. These symbols are used in numerous present-day publications. One may recognise the major compartments of Fig. 5.75, but they are arranged in another sequence. Again, the art of intelligent compartment choice is very important, as emphasised above. Important ecosystem functions, such as the work of decomposer organisms (right-hand, below in Fig. 5.77B) are "cheap" when comparing their "cost" with their "benefits". They consume only 13.3 kcal/m^2, this being 0.35% of the incoming 3830 kcal/m^2/day, in processing necromass (litter and organic) that stores 66170 kcal/m^2. In this figure, H.T. Odum has subdivided the compartment of primary producers (green plants) in six main tree species and "the others", whereas the phyllosphere (leaves) corresponds to Fig. 5.75 but is graphically placed outside the plants. The roots and axes are lumped together (root, branch, bole) in one compartment, also outside of and "preceding" the whole plants.

The system in which this analysis was made is no eco-unit. The authors specify that a prism of forest vegetation has been selected. The limits of the system considered hence do not correspond to eco-unit limits, and the values mentioned in Fig. 5.77B may or may not be averages over different eco-units. A visit to El Verde forest, where this study was done (see also Brown et al. 1983), suggested to the present author that indeed the probability is high that more than one eco-unit has been included. From the field notes on this half-day visit in 1985, the El Verde forest may be diagnosed as "located in a very divided landscape showing deep valleys, many rocks emerging from the soil. Damage of tree crowns by tropical cyclones is clearly visible. With its high rainfall, this is a tropical humid (moist?, rain?) lower montane forest. There is scarcely any organisation at the level of the silvatic mosaic, only interaction within a collection of eco-units, each of which grows on a different site fragment that is the main determinant of its properties". In terms of Fig. 5.75, the main input and output regulation of such eco-units is abiotic (outer border of the figure). The regulation by the silvatic mosaic (inner border of the figure) is reduced to more or less random interactions between eco-units.

Biogeochemical cycles are often considered instead of the "universal" energy cycling. Such cycles each represent the part of the general interrelations that concerns a particular chemical element, molecule or class of molecules. Dry plant matter contains more than 90% of carbon, oxygen and hydrogen (cf. Chéret 1987, p. 159 ff.). The largest input quantities in primary production, i. e. water and carbon dioxide, are also formed by these elements. Nitrogen is the next important element in quantity. Potassium and phosphorus, the well-known macro-elements in fertilisers, are important too.

Spore elements, or micro-elements have rarely been analysed in complete, specific cycles (but see Brown et al. 1983). Some 25 of them, among which

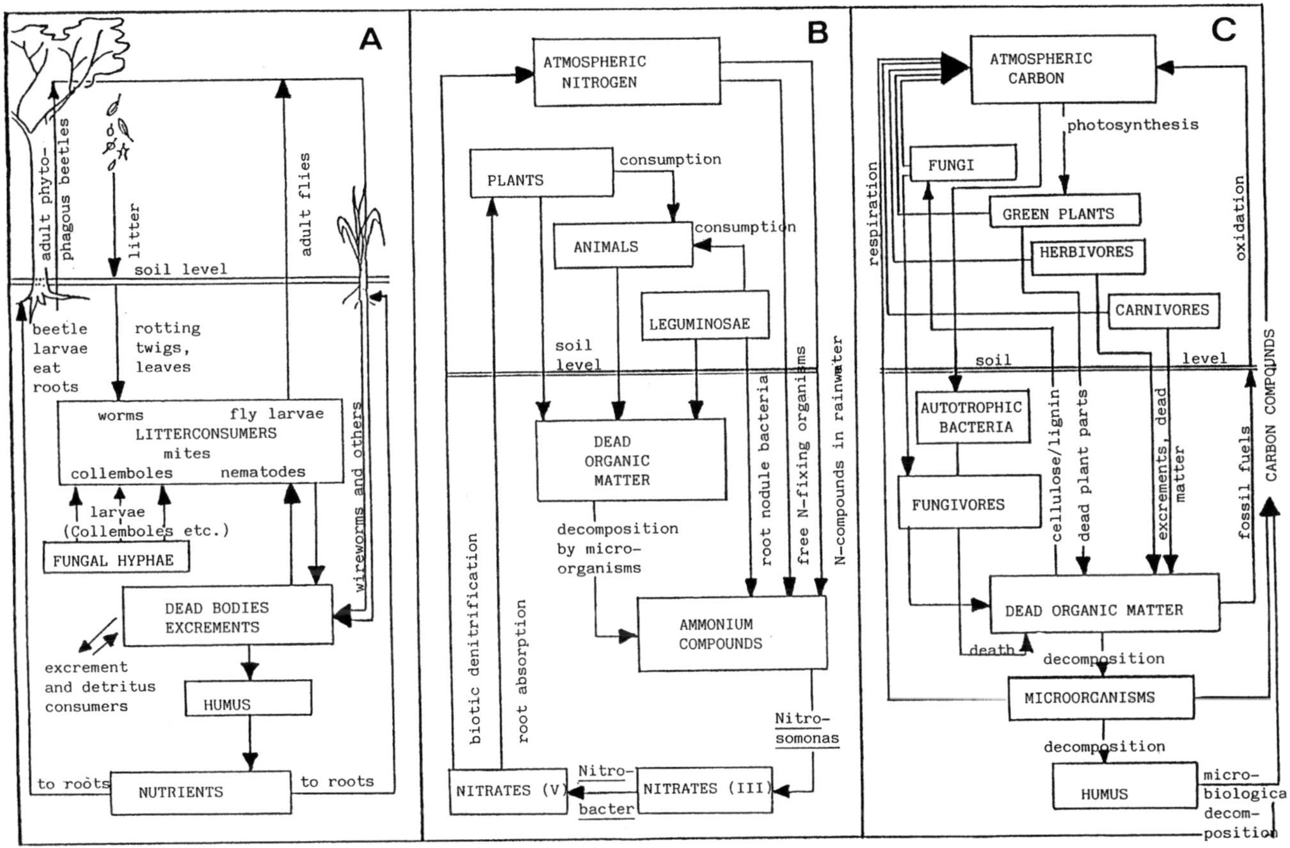

Fig. 5.78. Nutrient cycles represented as compartmentalised relation diagrams. **A** Litter decomposition. **B** Nitrogen fixation. **C** Carbon cycle. (Mainly inspired by Brown 1978, modified)

zinc, copper, magnesium, calcium and boron, are recognised as important in the functioning of forest ecosystems. They are most often considered as substances interacting with one of the major cycles. The most common of the latter are shown in Fig. 5.78. The humification cycle (Fig. 5.78A) refers to the decomposition of necromass with its inputs and outputs, i.e. to a large class of mostly organic molecules. The nitrogen cycle (Fig. 5.78B) has mainly respiratory energy inputs. The green plant compartment is divided in a different way than in Fig. 5.75, the Leguminosae being considered separately because of their role in symbiotic nitrogen fixation. The carbon cycle (Fig. 5.78C) is usually considered as the one authentic primary production cycle because its energy inputs are dominated by sunlight, which is processed by photosynthesis. In most handbooks on general ecology, different **nutrient cycles** are considered in more detail. A good text in German is that by Larcher (1976). It contains chapters on each of the main cycles.

The **water cycle** is another one that is often treated in detail. Considering, for instance, the sapstream (Fig. 3.30) or "transpiration work" in ecosystems (Fig. 5.77B, below, to the right), the importance of this cycle is evident. It is most often considered as a balance between inputs, throughputs and outputs:

- *inputs*:
  precipitation (e.g. rain, snow, dew, hail etc.).
- *throughputs*:
  interception by leaf canopies;
  stemflow, i.e. water that reaches the soil level along stems;
  throughfall, i.e. water that reaches the soil level directly or by dripping from the leaves;
  water uptake by root systems and rootless organisms, the latter being partially transmitted to the roots again (cf. mycorrhizae);
  storage in necromass and soil.
- *outputs*:
  evaporation from intercepting leaf canopies;
  transpiration through green plants (animals?) evapotranspiration;
  evaporation from other surfaces in the system;
  leaching, vertically through necromass and soil feeds streams;
  runoff, transversal through or over the soil and groundwater.

Depending on the scientific questions asked and very often upon the difficulties in designing measuring methods (e.g. see Leikola 1971, 1976 for Finland or Russell and Ewel 1985, for Costa Rica), these "water compartments" are established and various calculations are made. Advanced hydrological cycle and flow models have been published for instance by Mulder (1983), who worked in Coniferous dune forests in the Netherlands (see Fig. 5.34), and Bruijnzeel (1982), who analysed different man-made stands in Java, Indonesia. Dolman (1986, 1987) published interception models for broadleaved, temperate oak forests. Van Grinsven (in Van Breemen et al. 1988) made hydrological simulation models in polluted oak woodlands, using the SWATRE model as a basis.

The water balance of a forest ecosystem is essential for a number of reasons. Water is in the first place a carrier of dissolved or suspended matter. If the balance of the processes and storages cited above is shifted, for instance, towards high values of runoff, the water will carry the soil away. This means increasing erosion. Or, if insufficient water is available for uptake by organisms, ecosystem productivity is hampered, because the sap streams of plants cannot function optimally, or even not at all. This may easily happen if the soil has a low water capacity, as on sands (cf. Mohren 1987).

Water and nutrient uptake and their pathways through the roots in a Florida subsoil have been monitored by following the movements of introduced strontium by Van Rees and Comerford (1986). Moreover, water storage in the necromass compartment may be considerable. In high Andean ecosystems (see Fig. 5.45B), necromass has a very slow decomposition turnover, is accumulated in very thick layers and may stock as much as 300 or 400% of its own weight in water (Zebrowski 1975 pers. commun.). Degradation of such ecosystems entails high risk of necromass decrease, diminishing the water storage capacity strongly and causing irregular feeding of streams, which in its turn increases flood risks in the valleys. Afforestation of such high mountain lands with man-made forests is no solution for this problem of risks, because no forest ecosystem has been designed as yet that equals the water storage capacity of the páramo necromass. In this regard, tropical high mountain vegetations represent a dramatic example of the importance of establishing good models of the water cycles.

Acid substances in watery solutions are separated in a negatively charged "rest" and positively charged hydrogen ions. Because a hydrogen atom is built by one proton as its nucleus, and one electron as its outer peel, a hydrogen ion is one proton. In ongoing research concerning acid rain as a factor in forest dieback in industrial countries, the **proton cycle** is therefore more and more considered as an important chemical ecosystem compartment (e.g. bibliographies by Ulrich, in Ulrich and Pankrath 1983, or in Bauer 1985; detailed study by Van Breemen et al. 1988).

Energy cycles, biogeochemical cycles, water cycles or proton cycles are all conceived with the aim to simplify ecosystem models, so as to make them both manageable and apt to provide answers to specific scientific or practical questions. In the preceding pages, this need for optimising simplicity in ecosystem models has been emphasised over and over again, as it has been by many other authors (e.g. De Wit, in Vervelde 1975; De Wit et al. 1978). Now the questions asked traditionally concerned biomass, because biomass production determines the quantity of products that can be harvested (e.g. Spedding et al 1981; Mäkelä 1988). With the development of ecological science, it became apparent that biomass compartmentalisation and accumulation also have fundamental significance for ecosystem functioning. In forestry, volume and yield tables concerning growth and accumulation of standing woody trunk mass per hectare against time are the classical expression of biomass aspects, i.e. the wood to be harvested and the rate of its production in a certain stand of a certain tree species.

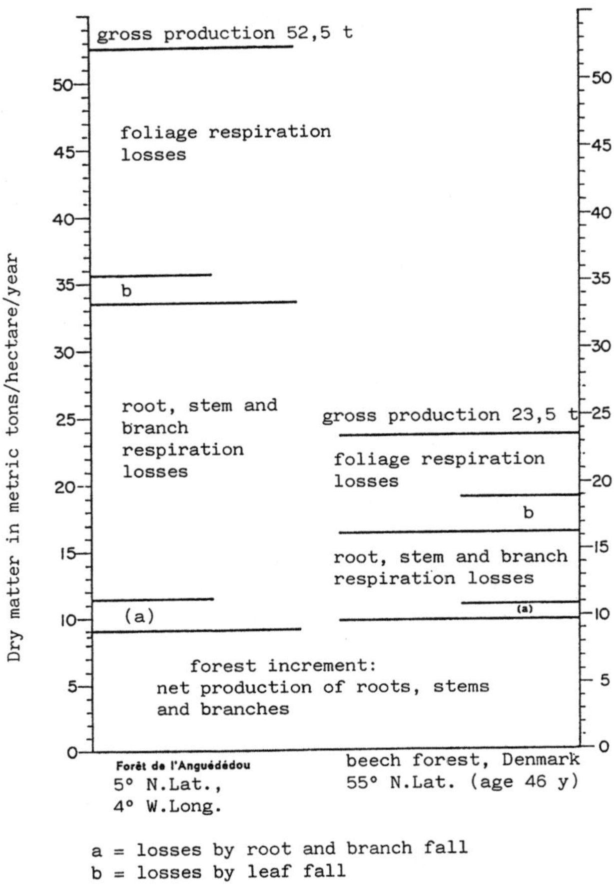

Fig. 5.79. Comparison of ecophysiological production in forests in West Africa (Côte d'Ivoire) and Europe (Denmark). Gross production, respiration losses and net production in metric tons per hectare per year. Values marked *a* are estimated, other values measured. (Müller and Jörgen Nielsen 1965). Note that average values per hectare probably do not have the same meaning for the two forest ecosystems, because the Danish forest is one even-aged eco-unit, the Ivorian one is a silvatic mosaic

5.6.5 Production Models: Some Principles and Examples

Linking biomass developments to physiological processes, such as photosynthesis, respiration and transpiration started over a century ago (for historical bibliographies see Wassink, in Vervelde 1975, or Müller and Jörgen Nielsen 1965). A document due to the latter authors is given in Fig. 5.79. It is one of the earliest comparisons between biomass production in a temperate forest ecosystem, beech in Denmark, and a tropical rainforest system, the Anguédédou forest in the Côte d'Ivoire which at that time still was an important tract of forested land. It is much degraded now. The conclusion from this figure is that, on the average, the net bio-

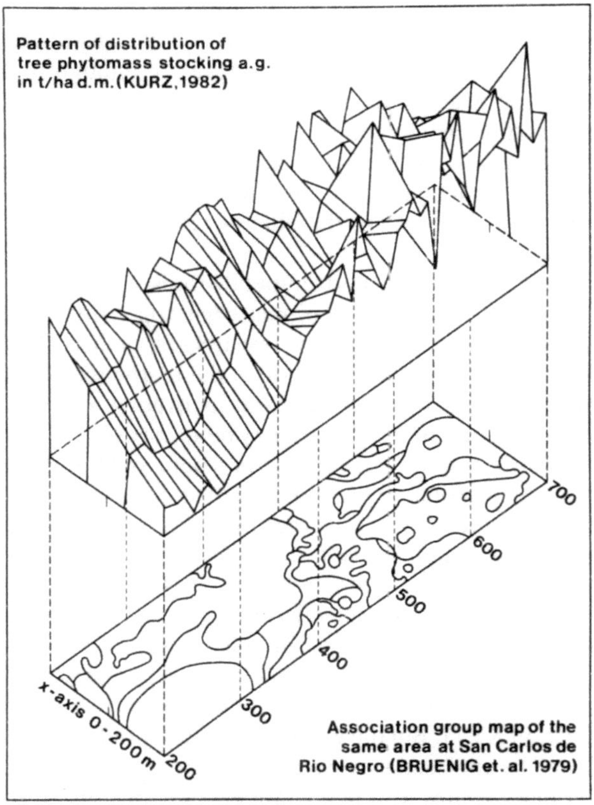

Fig. 5.80. Phytomass distribution in a rainforest in Venezuelan Amazonia, linked to a syntaxonomic vegetation map (below). Dry weight of trees over 1 cm diameter aboveground in 50 (X-axis) by 10 (Y-axis) m rectangles. Lowest tree mass 2.5 t/ha, highest 485.9 t/ha. The whole biomass and vegetation structures suggest a mosaic of eco-units in different development phases. (Brünig et al. 1982)

mass production in this tropical rainforest is slightly less than in Denmark. This is caused by much higher turnover rates in the tropics.

However, the Danish forest of 46 years old is probably one large planted eco-unit (cf. Figs. 5.60 and 5.62), whereas the Anguédédou forest as seen by the present author in 1963 resembled the Tai forest (Fig. 5.55). It was a mosaic of many very small eco-units in different phases of development. Therefore, the average values for biomass per hectare do not have the same meaning for these two forest ecosystems. In particular, the African data certainly include innovation and aggradation phases which have not attained their final amounts of mass and production as yet, and degradation phases in which decomposition (respiration) losses are high. This image of an Ivorian mosaic determining forest production patterns was confirmed in 1975 by Bernhard-Reversat and by Huttel (in Lemee 1975), after their research in the forests of Banco and Yapo.

Figure 5.38 is an illustration of the way in which researchers in Hamburg deal with this architectural aspect of the distribution of biomass, if statistical averages over a mosaic are not precise enough to answer particular questions (e.g. see Alder et al. 1979; Brünig et al. 1978, 1982). Figure 5.80 is a still more abstract representation of biomass heterogeneity in a tropical rainforest in Venezuela (from Brünig et al. 1982). The size of the associations mapped below in this figure suggests, but does not prove, that the units represented are eco-units. A book on this line of research has been published by Brünig et al. (1986).

For both tropical and temperate forest ecosystems, and without using a concept resembling the eco-unit, biomass heterogeneity was mathematically and numerically expressed by Japanese researchers, such as Hozumi and Shinozaki (1974) or Hozumi (1975). Their method is the mathematical comparison between the real weight of each forest plant and its average weight, calculated as total biomass divided by plant number. It may be seen easily that the three-dimensional, graphical expression of these numbers would yield a figure resembling Fig. 5.80. Their two-dimensional expression is called, by the authors cited above, the *M-w*-diagram, in which *M* is the average mass and *w* the real mass.

This was highlighted because the following category of production models, developed by teams of researchers such as those led by De Wit (cf. De Wit et al. 1978; Mohren 1987) or by scientists that can be found in Huxley (1983), e.g. Loomis or Connor, are mainly based on the analysis of average values. Cannell (in Huxley 1983) did take into account a certain heterogeneousness of shrubby or tree crops according to a system that may be compared to the silvicultural diagrams treated in Section 5.5.11. But the whole agroforestry field considered is so regularly built that it can be analysed as a system showing strictly localised oscillations around mean biomass values. Cannell's models may be conceived as intermediaries between Fig. 5.80 and the production models to be considered hereafter.

In such models, the maximum primary production of an ecosystem, AMAX, is determined by the amount of solar energy that reaches the surface of the earth at the place where the ecosystem is situated. At the equator this amount is maximal, at the poles minimal, expressed as luxes or joules per square meter. The differences in gross production in the two forests in Fig. 5.79 illustrate this. The second primary factor determining AMAX is the physiological efficiency of the plants in question: how large is the proportion of sunlight that they are able to process? Radiation and physiology together determine the potential CO_2-assimilation of the system. However, "radiation is a complicated, nonstationary process" (Lappi and Smolander 1984, p. 402). Calculation or mathematical expression of potential photosynthesis is easier said than done (Smolander 1984; Smolander and Lappi 1985). If all other conditions, such as the nutrient and water availability, and the biological environment, were optimal, the potential assimilation rate would be reached.

Simplification of the model has been attempted first in agricultural crop models, by using a hierarchy with two levels: the producing organs and the producing field. The latter by definition (see Sect. 5.1.7) is an eco-unit, although in agriculture it is treeless. But, in a short-lived system such as a maize field, the growth and development from sowing (innovation) to harvest (degradation) are determined by one

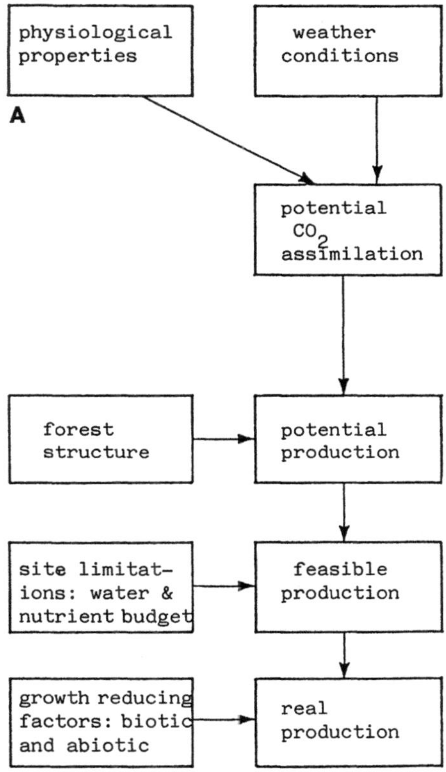

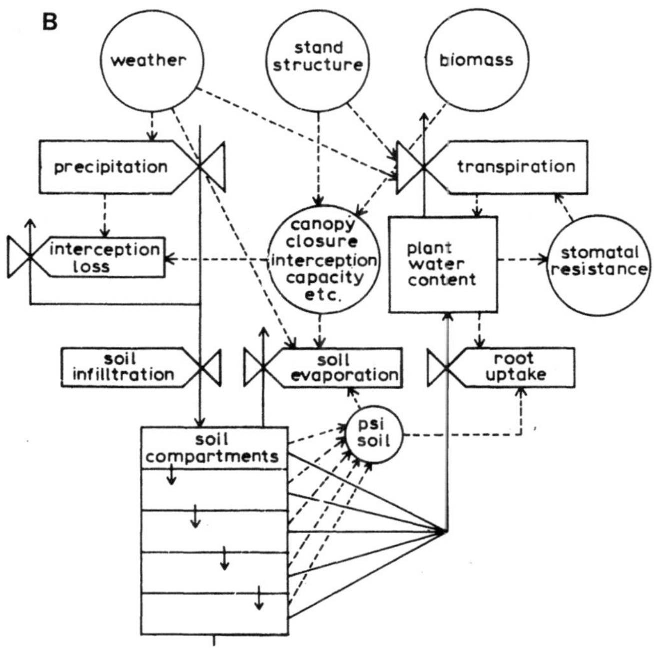

Fig. 5.81. Examples of relation diagrams for two-level forestry crop simulation models explaining stands through organs. **A** General flow diagram. Note that feedbacks are hidden in the compartments. Model is linear and reduces maximal assimilation (AMAX) stepwise to real production, applying forcing functions (*left side*). (Rabbinge and Mohren 1984). **B** Details of the hydrological part of such a model for even-aged Douglas fir stands in the Netherlands by Mohren (1987). *Solid lines* flows of matter; *dotted* relations between rate and state variables and between driving, intermediate and rate variables. Feedbacks made explicit

set of plants, so that the definition comes close enough. In the above hierarchy, the individual plant organisms are conspicuously lacking. These models resolve the problem by counting the plants, and allocating a mean production value to each. If there are notable differences from plant to plant, the production allocated to each individual can be a weighted average.

Figure 5.81 shows the steps followed in developing this kind of model, that can be simulated on a computer, e.g. by using programmes such as BACROS and PHOTON (De Wit et al. 1978, their Appendices A and B). Once the potential CO_2-assimilation is assessed from meteorological data and photosynthetic performance, the second-order modificator of the production is at the second hierarchical level, i.e. the vegetation as a whole. This is called "vegetation structure". In agricultural models it includes factors like crowding and mean angles of the leaves. For forests, leaf-canopy models were made by Horn (1971), who distinguished "monolayer" and "multilayer" crowns in a geometrical tree crown model that can be expressed in numbers.

Mohren (1987), in his production model for monocultures of Douglas fir (*Pseudotsuga menziesii*) in the Netherlands, approached the vegetation structure of these forests by much simplified geometrical crown models in which their architecture is taken into account to some extent, this tree species conforming to Massart's model (see Fig. 3.2). However it is expressed, the vegetation structure is only taken into account as far as it modifies the light interception properties of the leaf mass. Because neighbouring plants cast shade on each other, vegetation structure is considered as a factor that limits production to potential production level, which is less than potential CO_2-assimilation.

In the Brünig models, forest vegetation structure is also used to calculate a measure of canopy roughness, important in modelling wind regimes. Wind is influential both in gas exchanges in the forest canopy and in determining zero-events (windthrow). Philippe (1988) studied the influence of forest architecture upon vertical microclimatical gradients in French Guyana with high-definition airborne diffusiometers.

Once the potential production is assessed, the De Wit-models consider the availability of water and nutrients. If their supply is optimal, the potential production equals the attainable production. However, the inputs of water and nutrients may be either too scarce, as on unfertile soils or in drier climates, or too abundant, as in inundated floodplains or by nitrogen pollution. In both cases, these inputs act as stressors (cf. Lugo 1978). Water and each nutrient are often indicated as growth factors. In Latin, **the word "factor" means "maker".** Except for the presence of each factor in a sufficient amount, the proportion between these amounts also plays a role, because factors interact and one may block the other. This goes further than the statement on **one limiting factor** blocking the eco-unit to profit from the other, sufficient factors, because the time dimension should not be neglected.

If one takes into account that during the development of an eco-unit factors can be temporarily immobilised by becoming "encapsulated" in biomass, before they become available again through the litter cycle, it may be perceived that the growth and development of the eco-unit can show self-induced stress periods. Spectacular examples of this phenomenon have been described by Van Goor and other authors between 1950 and 1980 (see bibliography of Van Goor, Ned. Bosbouw Tijdschrift, 1984, **56**, 5:139–142). These case studies concern large areas that were reafforested in the 1930's and the 1940's with half a dozen tree species, on podzols that had been broken up in diverse ways and treated with a large array of different fertilizers from anorganic phosphorus to processed city waste containing many micro-elements such as zinc, manganese and others.

Even without taking into account the microbiological soil compartment, this indicates that fertilisation in forestry is less simple than in annual crops, because the amounts of nutrients in the soil fluctuate during eco-unit development over the years. This fluctuation should remain above the scarcity stress amount and below the excess stress amount of each nutrient. No model up till now can account for this very complex development involving dozens of chemical elements. Simulation and validation of these processes can easily become unmanageable. A forcing function is a set of instructions in a simulation model to force some imitated production process or nutrient flow in a certain way. Forcing functions may be used to account for micro-nutrient effects. Compartmentalisation of the model then is done in interacting main compartments (e.g. water, nitrogen, phosphorus, carbon and maybe potassium cycles) each compartment including its forcing functions imitating interaction with minor factors. This is an application of the black box principle (Sect. 2.4.1.1).

In such a way, the attainable production level is calculated. Biotic stress factors and those abiotic stressors that are neither nutrients nor water are finally introduced as forcing functions to simulate the actual production. Here again, a manageable model requires severe limitation to a few major stressors, most often major pests or diseases of crop plants.

The kind of questions that can be answered by these kinds of simulation models are rather broad and general. The answers are not very precise, but high precision is not required in land use policy decisions at national, regional or provincial levels. The sweeping generalisations that are the result are exemplified in Table 5.9.

Table 5.9. Potential and actual production compared in Dutch agriculture and forestry. (Rabbinge and Mohren 1984)

	Agricrop	Grass	Forest
Max. rate of photosynthesis (kg CO_2/ha/y)	30	30	30
Average growth rate (kg dry matter/ha/day)	200	110[a]	110
Growing season (days)	100	180	180
Dry matter production (kg/ha/y)	20 x 10^3	20 x 10^3	20 x 10^3
Economic production (kg/kg)	0.5 to 0.7	0.6	0.5 to 0.6
Net economic production (kg/ha/y)	10 to 15 x 10^3	12 x 10^3	10 to 12 x 10^3
Actual economic production (kg/ha/y)	5 to 10 x 10^3	7 to 8 x 10^3	2 to 3 x 10^3
(% of net econ. prod.)	50 to 70%	60 to 70%	20 to 30%

[a] Mowing causes incomplete canopy closure during part of the season.

This type of table allows a rough estimate of the influence of omnipresent factors, such as humans or polluting chemicals or sunshine. Conclusions therefore are broad and general, e.g. "it is urgent to improve silvicultural systems if a higher wood production is to be reached", or "the simulation of high amounts of atmospheric SO_2 by a forcing function shows the general decline of forest vitality over the territory, actual production being reduced by x percent". The table concerns the Netherlands, and may in no way be used for other regions in the world, not even the neighbouring countries such as Northern Germany or Belgium, where other inputs and outputs of the model will prevail.

But the same kind of model has been developed at a more detailed hierarchical level, the stand, in our words the eco-unit. The model for Douglas fir stands in the Netherlands by Mohren (1987) has been mentioned already. Hari et al. (1985), in a Finnish-Russian research project, modelled pine stands in these two countries. For a Polish Scots pine silvicultural simulation model see Bruchwald (1986). In such cases, much more detailed recommendations can be made, e.g. for thinning or fertilising regimes.

Figure 5.82 represents some features of the model built by Hari et al. (1985) and improved as to the carbon cycle by Mäkelä (1988). Two young Scots pine stands (*Pinus sylvestris*, Pinaceae, Rauh's model) were simulated. One is situated in Haapajarvi, Karelia, USSR, the other in Hyytiälä, Finland. They grow at about the same latitude, but differ in history, age, stand architecture (Fig. 5.82A) and soil type. Hyytiälä and Haapajarvi are about 600 km distant from each other. Contrarily to the Mohren model, in which crown architecture of Douglas fir was taken into account to a limited extent, the model by Hari et al.(1985), the systems concepts of which were announced by Hari et al.(1981), approaches light distribution in the stand by the analysis of hemispherical ("fish-eye") photographs (Fig. 5.82B). These photographs are used by other European researchers, such as Walter in the Alsace or Koop in the Netherlands. In Fig. 5.82B, the architectural model of the pines can be approximately recognised.

Many elements of the Finnish/USSR model can be understood in terms of sapstream models of trees (see Chap. 3.5.4), as illustrated by Fig. 5.82C, which derives the number of non-conducting tree rings in the wood, i.e. also the remaining rings that do conduct sap, from an allometric correlation with dead branch whorls. The model reposes on numerous allometric relations of this kind, closely related to tree architecture (cf. Oldeman 1974a) and to the pipe model theory, in which sapwood is considered as a growing net of water-conducting pipes (Shinozaki et al. 1964; Mäkelä 1988). Allometric relations of this nature that were found by Kaibiyainen et al. (1986) in Finland and Karelia (USSR) confirm this, because according to these authors (p. 26) these relations "are constant for pines growing under different ecological conditions" (also see Chap. 3.7). The elegant models established by Barreto (1987) on the basis of such Japanese and Russian "allometric constants" (cf. also Ross 1981) and using Portuguese data sets, also confirm the validity of this approach.

One of the resulting simulation graphs by Hari et al. (1985) is shown in Fig. 5.82D. It concerns needle mass increment over 14 years. Properties such as stem volume, tree number or mean degree of interaction between trees (cf. the Faber model mentioned below) are also simulated over different time spans, most

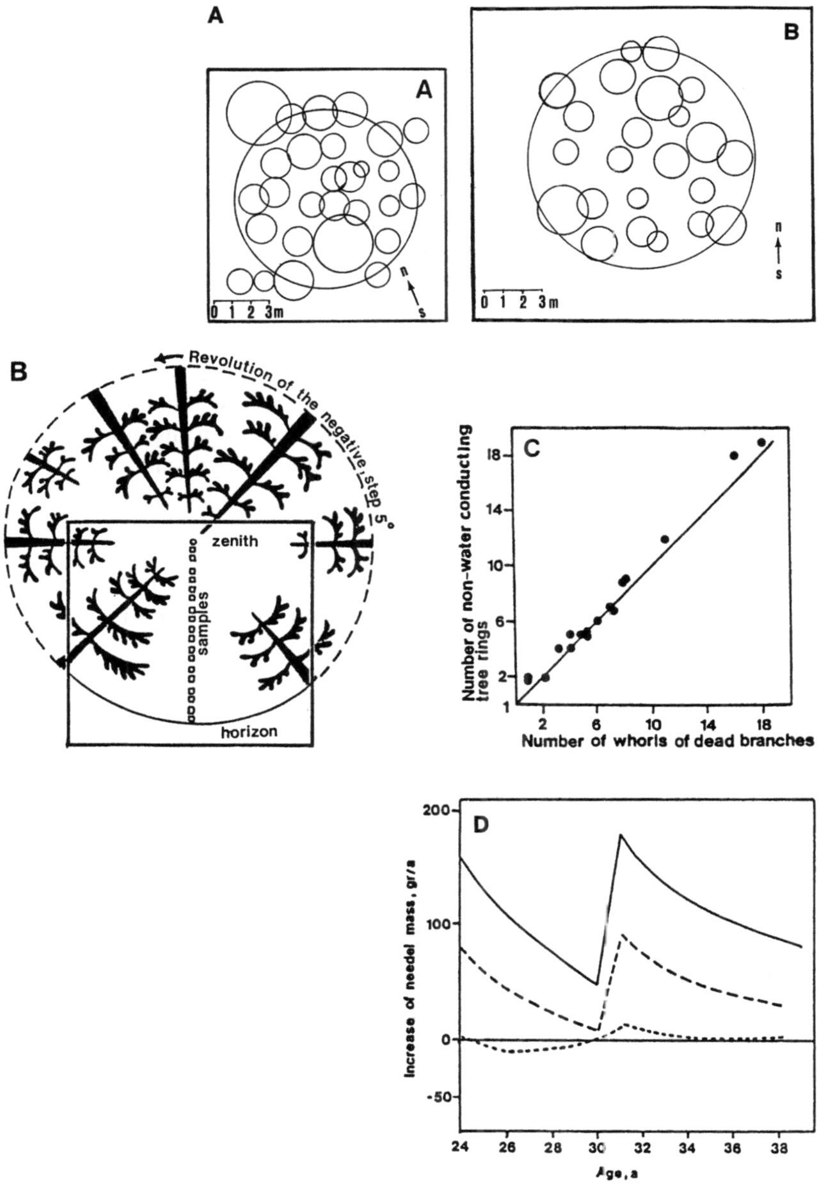

Fig. 5.82. Example of different steps in the analysis of a two-level stand model explaining stands through trees in forestry. Comparison between Scots pine stands in Haapajarvi (Karelia, USSR), random crown distribution, natural regeneration after fire, oldest trees 50 y, mean age 30 y, site "*Calluna* type" (**AA**) and in Hyytiälä (Finland), approx. regular crown distribution, sown after prescribed burning ca. 23 y ago, thinned to 3200 trees/ha, site "*Vaccinium* type" (**AB**). **B** Hemispherical photograph ("fish-eye") on which tree architecture is indicated, for estimating light regime by sampling (*squares*). **C** One of the numerous allometric relations (here dead whorls/non-conducting rings) used to assess sapstream and production. **D** Model *output*, exemplified by needle mass; minima: thinning. In the model of Fig. 5.81 needle mass is an *input* state variable. (Hari et al. 1985)

often a rotation of 80 years. The graphs contain no references to observed values in analogous stands at different ages contrarily to the curves simulated by Mohren (1987) for Douglas fir.

The two models differ fundamentally in that trees are included in the Finnish/USSR-model as organisms at their proper level, but are interpolated between the organ level and the stand level by Mohren. Figure 5.82D is the proof. Needle biomass is an input in Mohren's model as an estimated value determining photosynthesis, whereas it is an output in the other model. Neither model includes biotic or abiotic non-nutrient stressors.

When Mohren et al. (1986) formulate fertilisation recommendations in case of air pollution effects, this therefore implicitly rests on the hypothesis that litter decomposition and other nutrient cycles in the eco-unit will not change and that no major pests or diseases will occur as a consequence of shifting nutrient inputs. This is a classical viewpoint in European forestry and is linked to its agricultural origin. Tree species choice has commonly been at the basis of stand establishment, i.e. a crop-oriented approach. Sometimes during the present century this has included leaf litter properties for different tree species chosen. But it has always implicitly been assumed that the correct choice of tree species would be automatically followed by the healthy adjustment of all other stand components to the trees.

However this may be, it is more easily said than remedied. No manageable ecophysiological model of non-arborescent eco-unit compartments has been formulated as yet in such a way as to be compatible with silvicultural production models. Figure 5.77 represents an example of one first step towards such more complete models, although it does not refer to an eco-unit. It does include, however, a qualitative differentiation in species clusters (compartments), determining quantitative flows of energy and nutrients.

5.6.6 Crop Models and Gap Models

The long-term studies at Hubbard Brook drainage basin (NE, USA) for large clearcut eco-units (Bormann and Likens 1979), the synthesis of many studies on temperate and subtropical eco-units of differing sizes by Shugart (1984) and partly the rainforest studies in Suriname by De Graaf (1986), Jonkers (1987) and Poels (1987) take into account the theories of population biology (plants and sometimes animals), of production ecology (from classical growth-and-yield approaches to theoretical production models) and of eco-unit size (implicitly, or explicitly as in Shugart's book). The degree of mathematisation of the models is highest in Shugart, Bormann and Likens and Poels, who consistently use computer-aided simulation. The crop models described in 5.6.5 depict a large-sized eco-unit populated by a monospecific planted crop. Its extension to mixed-species forests without any concept resembling an eco-unit was made possible in the JABOWA model by Botkin et al. (1972 ex Shugart (1984), who gave a FORTRAN IV version of the derived FORET simulation model by Shugart and West 1977, in his Appendix).

The JABOWA model has given rise to a whole family of JABOWA-inspired simulation models. According to Shugart and Seagle (1985, with an 86-page Bibliography to which the reader is referred) the main members of this family are, in alphabetical order:

- BRIND (Shugart and Noble 1981), a model of Australian *Eucalyptus* forests;
- FORAR (Mielke et al. 1978), a model of Arkansas (USA) mixed *Pinus-Quercus* forests;
- FORET (Shugart and West 1977), a model of Tennessee Appalachian (USA) mixed hardwood forests;
- FORICO (Doyle 1981), a model of Puerto Rican (Carribean) Tabonuco montane rainforest;
- FORMIS (Tharp 1978), a model of Mississippi (USA) floodplain deciduous forests;
- JABOWA (Botkin et al. 1972), a model of mixed northern USA hardwood forests;
- KIAMBRAM (Shugart et al. 1981), a model of Australian "subtropical" rainforests;
- SWAMP (Phipps 1979), a model of Arkansas (USA) wetland forests.

The results of these models in terms of validated predictions at the forest mosaic level will be cited in a later chapter. Here, we are still at the eco-unit level. Therefore, the following paragraphs will summarise the basic unit of the listed models, i. e. the gap model.

These gap models are two-level, dynamic explanatory simulation models, the core of which is an abstracted image concerning the trees as components. They are two-level models, because they concern the gap system ("gap-phase", "patch") and selected components of the eco-unit growing in a gap, the trees, as subsystems. They are explanatory because the eco-unit development is explained by interacting tree growth. In all first versions, the non-tree components are abstracted. The impact of one such component, the fungus *Endothia parasitica*, the infamous chestnut blight which has almost exterminated the American chestnut (*Castanea dentata*, Fagaceae) in the Eastern US, has been indirectly assessed in the FORET model as a 100% chestnut mortality (Shugart and West 1977). Animal niches have been simulated at a landscape mosaic level (Shugart and Seagle 1985).

Figure 5.83A represents the flow diagram of the FORET model, as an example of the type of analysis. It is applied to circular forest surfaces of 16.29 m diameter (1/12 ha) and considers 33 tree species. The model follows JABOWA (Botkin et al. 1972 ex Shugart 1984) in simulating the annual change (time dimension, temporal resolution see Sect. 2.3.2) in biomass of a forest stand of the given size by calculating the yearly increment of each tree building that stand. This is the subroutine GROW. Loss of biomass by tree mortality is simulated by the subroutine KILL, also like in JABOWA, and would for instance allow to introduce a forcing function leading to 100% chestnut mortality. BIRTH simulates germination and establishment of trees, also following JABOWA, by yearly evaluation of the availability of mineral soil, the presence of leaf litter, the above-average climatological temperature (heat) and the density of deer and smaller mammals.

SPROUT is an original subroutine in FORET. It simulates the coppice-like sprouting of trees (cf. Fig. 3.21), which is an important way of vegetative propagation in many tree species of this region (also see Sect. 5.2.3, aggradation phase). Sprouting for the moment ensures vegetative survival of *Castanea dentata* in some regions, but the blight does not allow the sprouts to reach sexual maturity, so that BIRTH is empty in the model for this species, and GROWTH is aborted early. Shugart and West (1977) did not enter into these complications and just deleted *C. dentata* from the species list entered into the model for "forest without chestnut", without seeming to impair the model's performance by this simplifica-

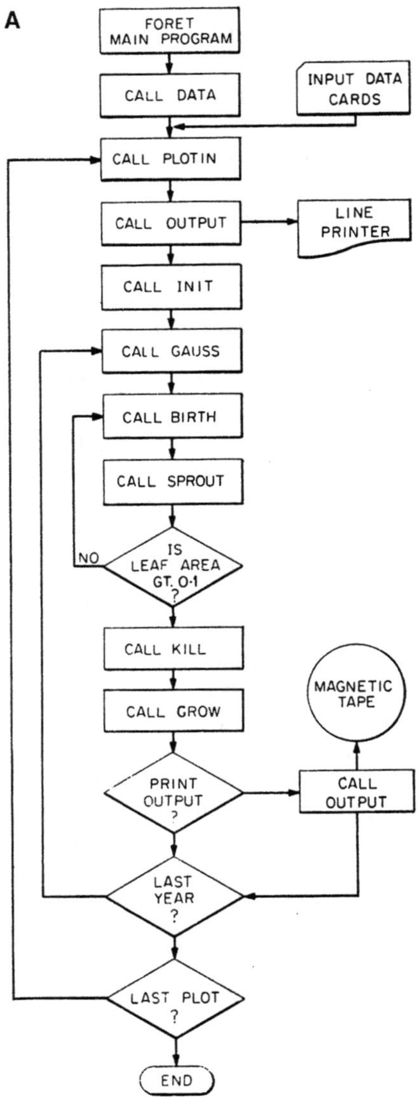

Fig. 5.83A

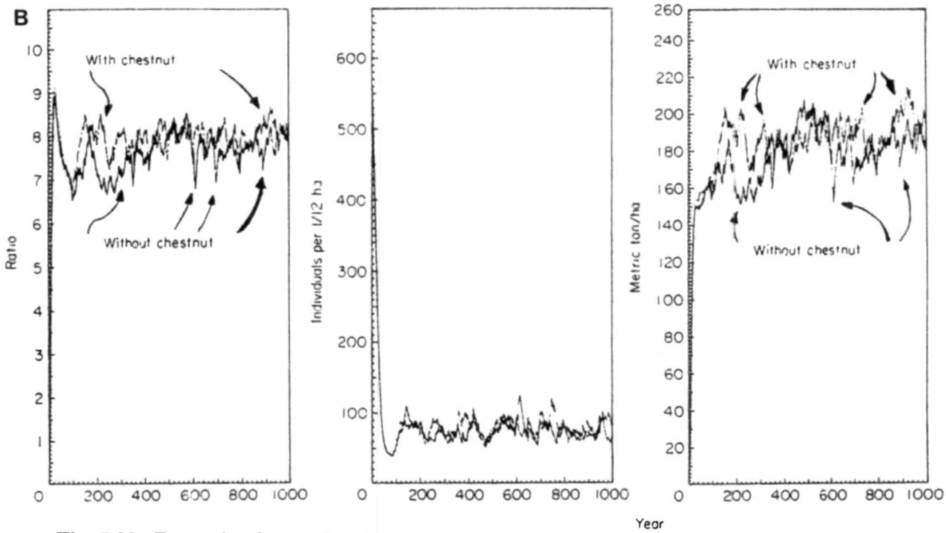

Fig. 5.83. Example of a relation diagram and some outputs from a two-level gap model, trees explaining the dynamics of mixed forest patches, different from eco-units by their artificial circular shape and normalised 1/12 ha size. Criteria both biomass accumulation and tree species; other forest components may be introduced as forcing functions by expanding the model. The model is a member of the JAWOBA simulation model family. **A** Relation diagram of the FORET model for Tennessee Appalachian mixed hardwood forests. *Rectangles* show computer subroutines, explained in the text. **B** Outputs of the simulation model: three indicators of overall forest dynamics over 1000 years: **(a)** Leaf area (m^2 leaf area/m^2 land area); **(b)** total stem number (trees with d over 1.27 cm per 1/12 ha); **(c)** biomass (t/ha). Graphs show mean values for 100 standard plots for each of 1000 years. *Dashed* chestnut present; *solid* chestnut extinct. *Arrows* indicate significant differences, the eco-unit build-up of the mosaic then being certainly different (see text). (Shugart and West 1977)

tion. GAUSS, also specific for the FORET model, is a subroutine introducing randomness. FORET thus is made into a stochastic model. Other subroutines (DATA, PLOTIN, OUTPUT, INIT) are service subroutines that do not influence the model itself and its interpretation. They replace equivalent JABOWA subroutines.

Some remarks have to be made in order to place the JABOWA family, as exemplified by FORET, in the context of silvological models in general. First the normalised surface of 1/12 ha shows that a pixel-like delimitation of each patch-system is followed. The simulated gap hence is not the exact equivalent of an eco-unit, which has natural limits. The performance of the model shows, however, that this artificial surface corresponds near enough to the natural one to allow reliable predictions to be made. Second, the tree-by-tree growth simulation is a different approach than in the Mohren simulation model, which is a stand model in which the share of each tree is calculated and allocated later (Mohren 1987).

In the third place, the stochastic approach is different from the structural approach in architectural models. In a spectrum of different models, stochastic behaviour of subsystems in a system as one extreme has always been the assumption in physical systems analysis. Structural, non-random behaviour of subsystems as the other extreme has been most often assumed in living systems research, at least at the organism level and below. However, both structure and chance events play their role in forest development (e.g. zero-events, Sect. 5.2.5.3). In silviculture it is essential to develop designs in which the two are correctly balanced (cf. Brünig in Wiersum 1984; Christensen 1987).

The stochastic bias in FORET is a consequence of the introduction of population biology according to the principles of the zoologists Andrewartha and Birch (1954, see Pianka 1974) although Harper's (1977, see also Begon et al. 1986) more structural botanical approach has also been assimilated to some extent. Statistical treatment of population data stems from those syntaxonomic approaches that repose on average frequencies of species over large surfaces. FORET as a matter of fact rests on the hierarchy represented in both right-hand columns in Table 2.1, although some consideration of structure has been introduced (cf. Shugart 1984, his Chap. 3).

Finally, the two-level nature of FORET, in which forest "patches" as systems are explained by organisms and their interactions, means that interaction among patches in the mosaic has been omitted. Shugart and Seagle (1985, p. 354) say this explicitly:" The less the dynamics of the landscape are a consequence of contagion among the mosaic elements, the more easily the computations of the model are performed". This is another difference with the architectural model, in which the eco-unit is conceived from the outset as a system within a supersystem (Fig. 5.75). This three-level set-up is a useful addition to pristine forest analysis, but it is also true that no professional forest manager would consider a stand without looking into its position among the other stands of his forest district. Eco-units are no islands.

Figure 5.83B shows some results from the FORET model by presenting simulated development of leaf area, stem number and biomass per 1/12 ha, with and without chestnut, over a millennium. The arrows indicate periods of between 25 and 250 years in which developments with and without chestnut are significantly different. These conclusions may be used as a cross-check, by an independent method, of the statement made earlier concerning Fig. 5.75, i.e. that adjustment of an eco-unit to changing circumstances is done by reallocation of resources to different component organisms. In the two cases simulated by FORET, trees are considered only, and their different numbers, leaf areas and especially biomass point to such reallocations. This may be formulated as: "during some developmental phases, fewer trees get more resources each in one case and more trees obtain less resources each in the other case, whereas the differing leaf areas suggest that different species may be involved." This last suggestion is borne out by Shugart and West's millenary simulation of biomass per species, not reproduced here. This whole analysis should be compared with the study by Kohyama (1987), who produced a gap-dynamics model for a warm-temperate rainforest in Yakushima Island (Japan), concentrating upon the analysis of diameter distribution among trees by using diffusion equations.

The synthesis by Bormann and Likens (1979), reached by development modelling of northern hardwood forests in the USA after a case study of several decades at Hubbard Brook, is given here as a second example of modelling largely based on JABOWA. There is the more reason to give some information about this research because in Section 5.2, on developmental phases, several terms of these authors were referred to. In particular, the notion of "aggrading phase" was adopted from this source.

Figure 5.84A shows how these phases are percieved by Bormann and Likens, on the basis of what they call "the biomass accumulation model". The authors specify that the "steady state" is defined in terms of biomass, species composition and biogeochemical functioning. They state (1979, p. 4): "We use the word Steady State with some trepidation, since ample evidence suggests that the condition we define above is at very best one of slow net changes or possibly a period of quiet antecedent to a cyclic revolution". Among their basic concepts, this state is included as a "shifting-mosaic steady state" and taking into account their precise definition of the zero-event (clearcut, see Fig. 5.84A) this points to eco-unit fragmentation processes as exemplified in Sect. 5.4.3.9. Also compare Fig. 5.83B.

The compartments of biomass in this model are represented in Fig. 5.84B, to be considered carefully next to Fig. 5.75. The most obvious difference is that all organisms other than green plants are lumped into one huge compartment "heterotrophs", literally: organisms that "feed on others". Green plants are "autotrophs", literally "those that feed themselves", a notion mentioned earlier under the name "primary producers". Figure 5.84C should be compared with Fig. 5.4. The JABOWA-simulated amounts of biomass are completely compatible with the architectural images of the innovation phase and the aggradation phase.

Altogether, the conclusions drawn by Bormann and Likens (1979) from the Hubbard Brook model, in which JABOWA was used as a tool to establish the main orientation, biomass accumulation patterns agree with patterns obtained by architectural analysis, if we accept that the shifting-mosaic steady state is the result of fragmentation of a large original eco-unit.

Among their conclusions, some require to be discussed here so as not to create misunderstandings.

- In architectural terms, the reorganisation and aggradation phases of Bormann and Likens represent one eco-unit (Fig. 5.84C), whereas their steady state, as suggested by these authors themselves, refers to a mosaic of eco-units, each of which might be treated according to one of Shugart's gap models. The lower average value of biomass for the mosaic is then caused by a somewhat oscillating proportion of small eco-units in the innovation and degradation phases. This also tallies with our conclusions about Fig. 5.79 and with the three-dimensional phytomass graph of Fig. 5.20. There is no conflict between the usual asymptotic biomass curves and the curve in Fig. 5.84A, if this last curve is considered as a sequence of asymptotic curves, each at different integration levels (Table 2.1), and the steady-state ones representing averages over many small eco-units.
- Bormann and Likens found that the use of photosynthesis/respiration ratios (P/R-ratios) as a simple index of ecosystem development has strong limitations. These limitations also seem to be linked to the comparison of one eco-unit (reorganisation + aggradation of Fig. 5.84) to a mosaic of eco-units (steady state), many of which have strongly different P/R-ratios because they are either very young or degrading.

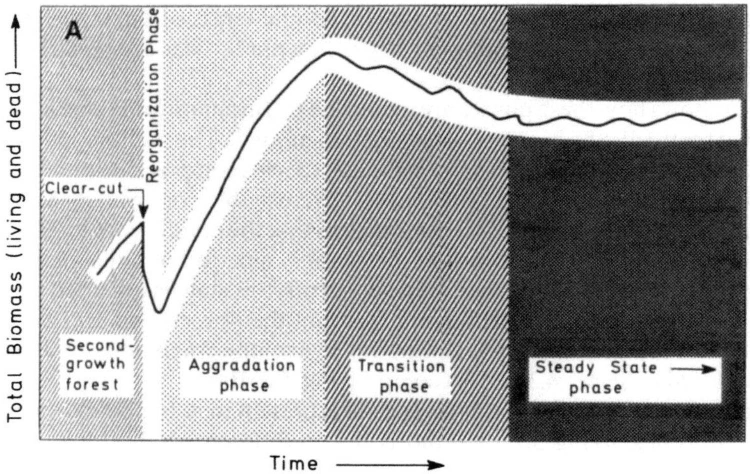
A
Total Biomass (living and dead)
Reorganization Phase
Clear-cut
Second-growth forest
Aggradation phase
Transition phase
Steady State phase
Time

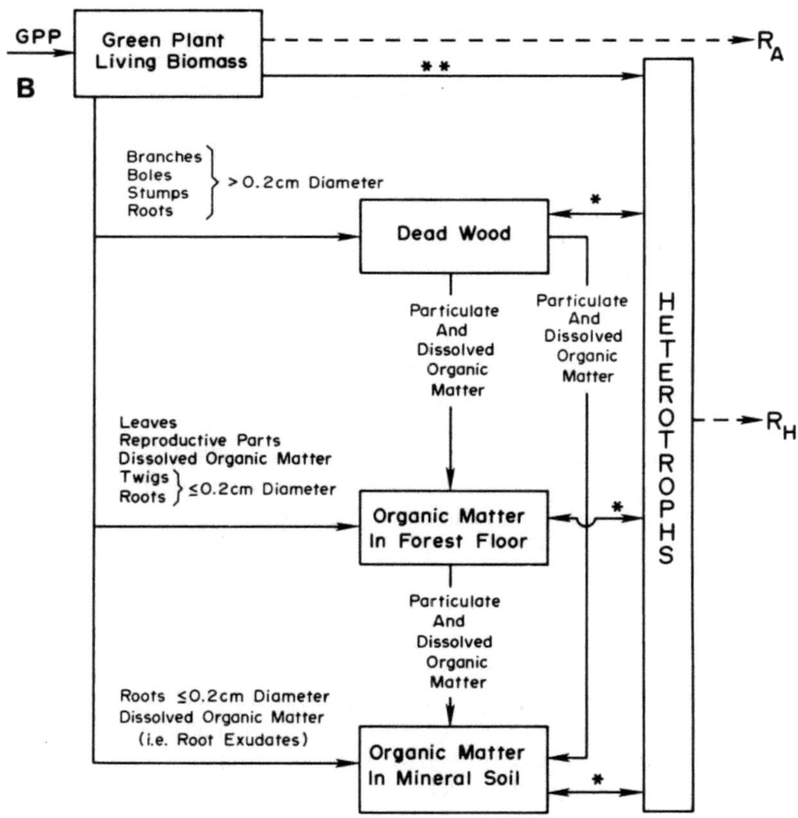
B
GPP
Green Plant Living Biomass
R_A
**
Branches
Boles
Stumps
Roots
> 0.2cm Diameter
Dead Wood
*
Particulate And Dissolved Organic Matter
Particulate And Dissolved Organic Matter
HETEROTROPHS
R_H
Leaves
Reproductive Parts
Dissolved Organic Matter
Twigs
Roots
≤0.2cm Diameter
Organic Matter In Forest Floor
*
Particulate And Dissolved Organic Matter
Roots ≤0.2cm Diameter
Dissolved Organic Matter
(i.e. Root Exudates)
Organic Matter In Mineral Soil
*

Fig. 5.84A, B

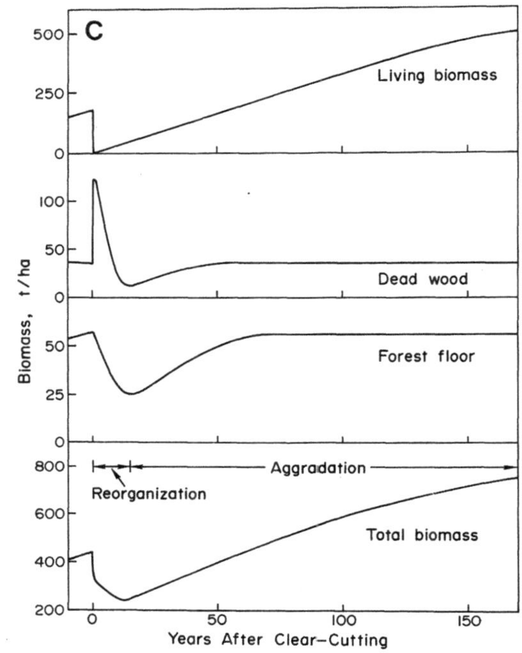

Fig. 5.84. Example of elements from the JAWOBA-based, two-level (tree/stand) simulation model of US mixed Northern Hardwood forest at Hubbard Brook. (Bormann and Likens 1979). **A** Biomass accumulation, the basis of the model (cf. Fig. 5.83). Note development phases. Authors denote steady state as shifting mosaic; transition is probably a fragmentation process (see text). Zero-event clearly indicated (clear-cut). See also Table 5.1. **B** Relation diagram for the model: compare with Figs. 5.75 and following. Non-tree components included (heterotrophs). **C** Simulated biomass accumulation in different compartments over more than a century for reorganisation and aggradation phases, i.e. before fragmentation and in one large, clearcut eco-unit

- Statements on habitat diversity, being maximal in the "steady state" indeed reflect its mosaic character (see discussion of non-arborescent components earlier; also Sect. 6.2.4) and do not conflict with the habitat (niche) diversification sequence sketched for one eco-unit (Sect. 5.2).

The most frequent zero-events in the region of Hubbard Brook have been carefully examined in a chapter apart, which is exemplary for those who want to improve their understanding of the driving forces of forest development. In their final chapter, Bormann and Likens (1979) list the conditions under which different types of wood harvesting and landscape management measures are optimal or pessimal.

The examples given above allow us to understand other systems-analytical models based on simulation. The model described by Bossel (in Brünig et al. 1986) for a Chinese-German MAB-project may profitably be used for comparison. This author uses the term "component" for what in the present book is called "compartment". His compartmentalisation clearly derives from the developments in ecosystem modelling in Germany, described in earlier publications, by authors such as Ulrich and Mayer (in Ellenberg 1973).

A beautiful British production model for forest development can be found in the elegant book by Cousens (1974). This author explicitly describes the basic principles defined by Watt (1947) in a pioneer article on development phases in a "standcycle" (cf. Fig. 5.4), which he studied in heathlands (Ericaceous vegetation). Watt is mentioned also by Shugart (1984) and, for tropical forests, by Whitmore (1982), as the originator of the idea of phasic forest development on patches. Because it is evident that such phases have been empirically distinguished in French and German forestry for two centuries at least, we take Watt to be the author who introduced the notion in ecology by extending it beyond human-made or managed stands.

Cousens book (1974) includes very precise instructions for measuring and data processing. His flow diagram for production assessment in a pinewood is a detailed version of Fig. 5.81A, including more compartments, and sets of instructions for human operators next to those for a machine (Cousens 1974, p. 66/67). His model includes non-arborescent components. It is mentioned here, because it explicitly rests on food webs, nutrient cycling and population biology, like the other models cited in Section 5.6.6.

The models summarised here are not the only ones having a biogeochemical base. But, except for some of the first ones discussed (e.g. Figs. 5.77–5.79), they have in common that they consider the limits of the ecosystem which is analysed by referring to stands, patches or gaps. In this aspect, they are the ones as close to the eco-unit concept as could be found. Another category of ecophysiological models, considering average values for mosaics, will be mentioned in a later chapter.

5.6.7 Computer-Linkage Between Architectural and Biogeochemical Models

Computer-simulated ecophysiological models should in the longer run be linked up with computer-aided simulation models of eco-unit architecture. These are now being developed in several places. Their main handicap is that the growth of individual plants can be simulated (Fig. 5.85) in quite sophisticated ways, that the volumes they successively occupy in the forest (growing space, "spatial niche", ecotope sensu Oldeman 1974b) can be simulated in an increasingly handsome manner (Fig. 5.86), but that the simulation programme for each plant is so complex that it would become very problematic to fit it into a complex ecosystem model, exerting variable behavioural pressures upon the plant model. One or both of the levels, the plant or the eco-unit, has to be simplified in order to create a manageable model in terms of computer time and memory.

Figure 5.85 represents one result of the stochastic programmes built by De Reffye (1979), followed by De Reffye et al. (1986, 1989). It simulates tree growth internode by internode, introducing stresses such as competition in establishing the "choice" of an internode: to grow or not to grow, to be either vegetative or inflorescential, to grow either plagiotropically or orthotropically. The figure represents Rauh's model, and in De Reffye's programmes, all architectural models and their reiteration can be simulated. The hierarchical levels used in his simulation are very far apart: the whole tree (organism) is explained by the behaviour of parts of organs, the internodes.

In Table 2.1 several auxiliary levels were introduced between the tree and its organs, because already when drawing by hand it becomes apparent that the internode-by-internode approach is very complicated. Internodes, together, form axes according to simple patterns, which are limited in number and are recurrent in nature (Stevens 1974). Such linear patterns have been simulated by Niklas (1986), who used them in computer-simulated evolutionary "war games" between hypothetical fossil tree species with different tridimensional, projected growth patterns, spectacular colours indicating the growth dynamics. This method has not been used as yet in ecological modelling. The mathematical principles of fractal scaling, as used in pattern research of biological systems, e.g. by West and Goldberger (1987), might also be of help to simulate the complex patterns of tree architecture in a simpler way. As a concept, fractal analysis would fit in with strategic tree growth principles, such as architectural gigantism versus repetition (Hallé 1986) or the transfer of architectural adaption from one organisation level to the next one (Oldeman 1989a in press).

The images of tropical agroforestry vegetations that have been generated by De Reffye et al. (1989) are collections of CAD-models of single trees and shrubs, presumably drawn into one image according to an agricultural planting grid. This is an enormous simplification in comparison with the complex build-up of even a simple forest eco-unit. Because indeed in agriculture this kind of simplification is

Fig. 5.85. Output of simulated architectural model of trees growing according to Rauh's model (see Fig. 3.2; e.g. *Pinus* spp., *Quercus* spp., *Cecropia* spp., *Khaya* spp. etc.). (De Reffye et al. 1989). Two-level stochastic model, explaining tree growth through the growth of internodes which may grow or not, be rhythmic or continuous, bear flowers or not etc. The difference between, e.g. free-standing and forest trees rests on the introduction of factors forcing this choice. This model occupies so much computer memory that it cannot be included in a simulated architectural interactive forest model. It represents in detail the distribution of biomass over different parts of the tree.

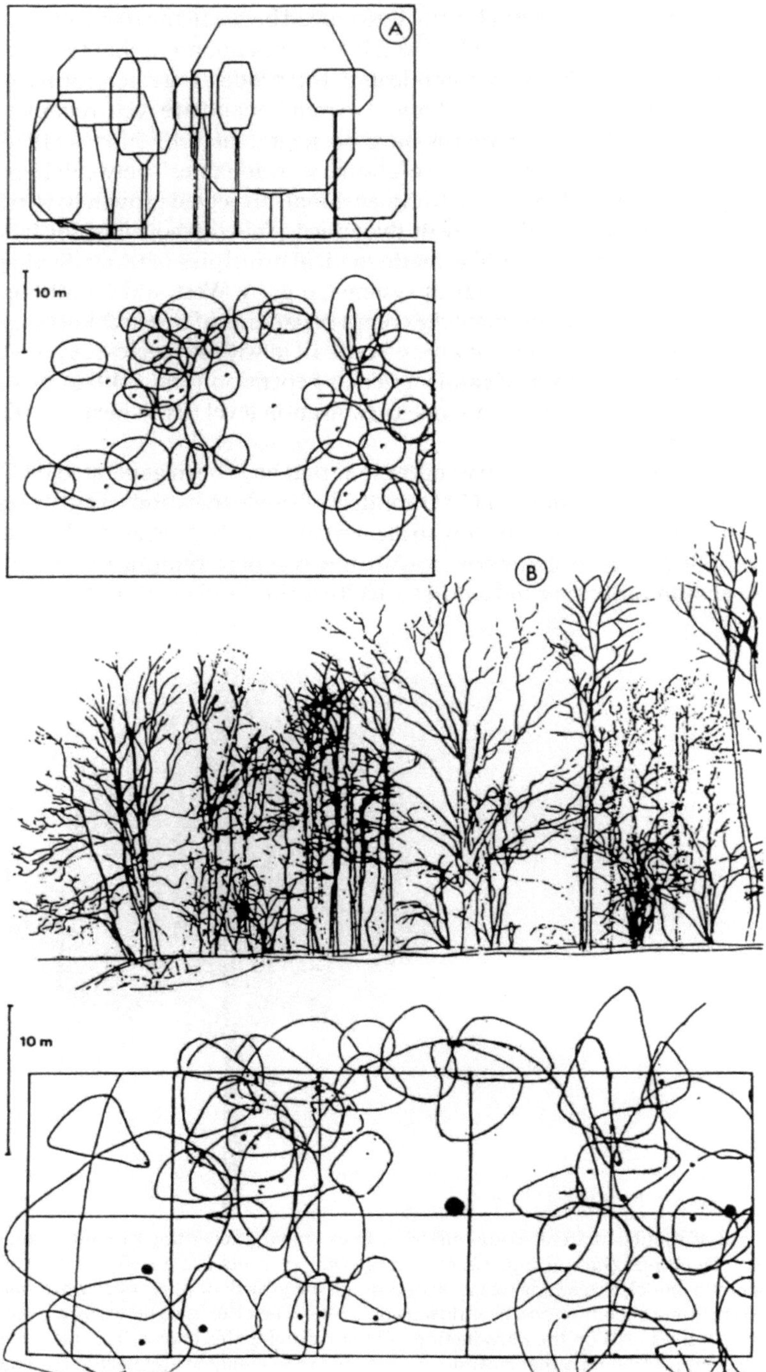
A
10 m
B
10 m

practiced in real life, the cited image looks realistic. However, to paraphrase Shugart and Seagle (1985, p. 354), there is little "contagion" among the subsystems, in this case the plants. In a real-life forest there is, in the form of multiple ecological interactions among organisms.

Figure 5.86, due to Cheviron (1985) shows the transformation of a transect measured and drawn by Van de Winckel (1980) in the Wyhlerwald, a flood-plain forest in the Alsace, into a computer-simulated transect that resembles Fig. 5.38. The loss of information is clear in this figure. For existing biogeochemical models, the eliminated information is unnecessary and such transformations therefore are justified (cf. Brünig et al. 1978, 1982; Fig. 5.38). On the other hand, Koop (in Fanta 1986), Koop (1987, 1989) starts by measuring eight coordinates for each tree, defining the tree base on the map, tree height, first fork and the outer periphery of the crown. The computer fits these together on a surface composed by parts of spheres, ovoids or cones (e.g. Mansoer 1986).

With the resulting image, calculations can be made or it may be transformed, for instance in order to reconstruct a fish-eye image of the forest canopy seen from below. Because the volumes delimited by such surfaces represent tree ecotopes in the sense of Oldeman (1974b), their growth or shrinkage with time can be simulated by using allometric functions. It should be emphasised that Koop uses these simulation techniques in parallel with hand-drawn forest transects and maps, in all cases where more detailed information might be or become indispensable. An interesting application of this simulation model is the reconstruction of remote sensing images. But as long as no simplified simulated architectural models of tree growth exist, the crown volumes in the forest model will remain empty. Another experiment with partial simulation of tree crown volumes can be found in Bücking et al. (in Fanta 1986).

Figure 5.87 shows another, two-dimensional simulation model of crown space (Fig. 5.87B). It is the application by Faber (1985) of his more general growing space model for tree crowns to a population of a poplar clone (*Populus x interamericana* cv. Rap) planted in an experimental Nelder design (Fig. 5.87A). As a matter of fact, this is a projected biomass accumulation model, because crown spaces are calculated as a function of basal area (Fig. 5.87C), corrected for competition by neighbouring trees as a function of reciprocal distances. The proportions are brought to realistic values by the use of empirical constants. These are different per tree species. The model has been calculated for monospecific tree stands exclusively up till now, with one exception. This is the study by Hilgen and Van den Brenk (1984) on mixed oak/beech forests (Fig. 5.61). These authors did not find a way to insert the tree crown architecture and dynamics in the Faber model.

Fig. 5.86. Transformation of a hand-drawn transect (**B** by Van de Winckel 1980) into a computer-drawn transect (**A**) by Cheviron (1985). Wyhlerwald, a river plain forest in the Alsace along the Rhine, France. The simulation has been made with numerical data read from the hand-drawn transect. Note loss of detail, which is justified if only crown volume is needed for adjustment to other models. Dynamics will be simulated accurately only if models such as Fig. 5.85 are simplified in order to "fill" the crown volumes (present figure) by simulated tree growth processes. cf. also Fig. 6.29 and Koop (1989)

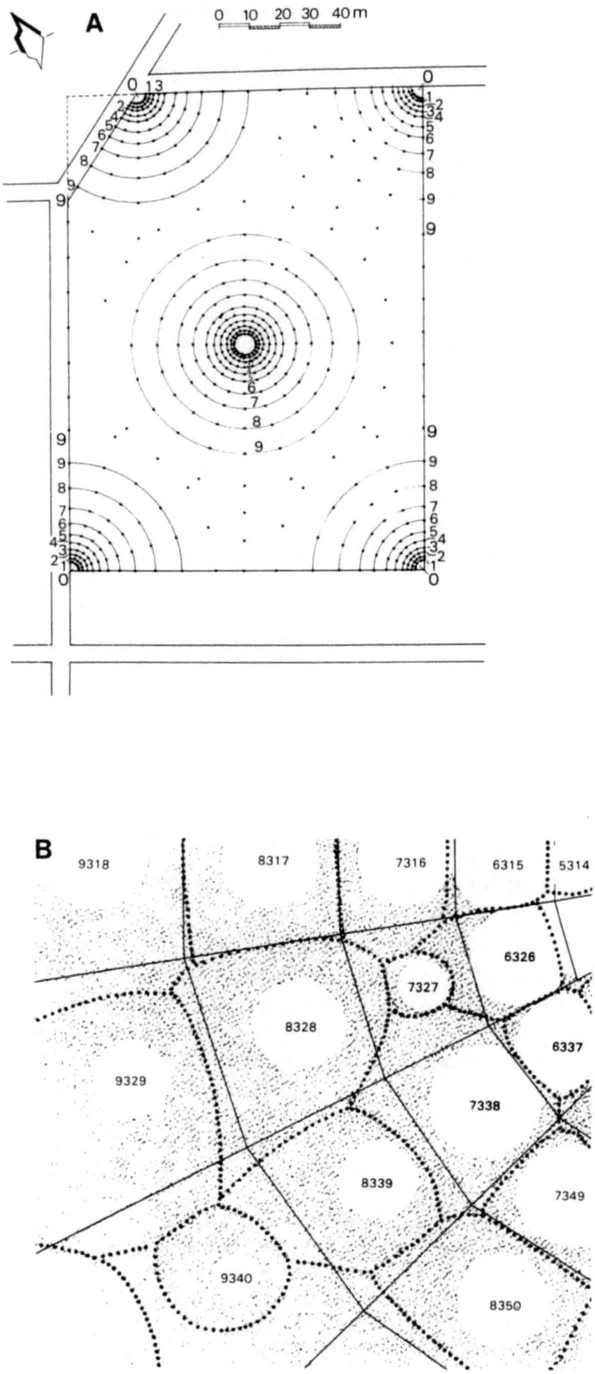
A
0 10 20 30 40 m
B
9318
8317
7316
6315
5314
6326
7327
8328
6337
9329
7338
8339
7349
9340
8350

Fig. 5.87A, B

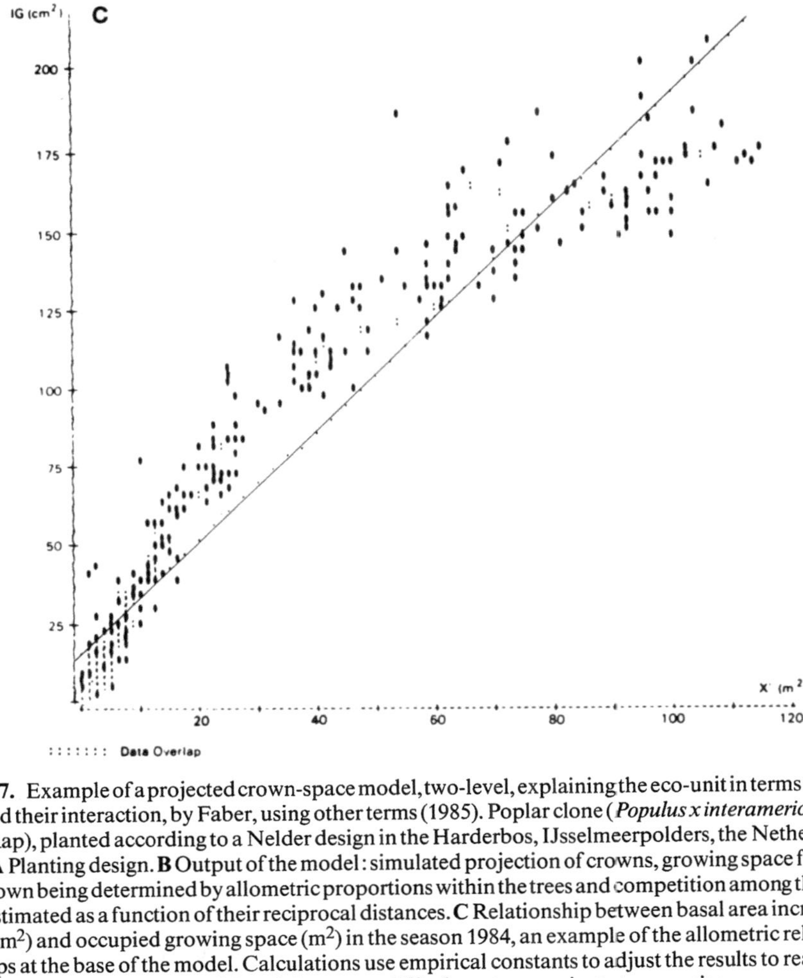

Fig. 5.87. Example of a projected crown-space model, two-level, explaining the eco-unit in terms of trees and their interaction, by Faber, using other terms (1985). Poplar clone (*Populus x interamericana* cv. Rap), planted according to a Nelder design in the Harderbos, IJsselmeerpolders, the Netherlands. **A** Planting design. **B** Output of the model: simulated projection of crowns, growing space for each crown being determined by allometric proportions within the trees and competition among the trees, estimated as a function of their reciprocal distances. **C** Relationship between basal area increment (cm^2) and occupied growing space (m^2) in the season 1984, an example of the allometric relationships at the base of the model. Calculations use empirical constants to adjust the results to realistic values. These have been validated. Constants differ from tree species to tree species

Therefore we are left, for the time being, without any widely known and accepted architectural model in which component architecture, were it only for trees, can be causally linked to eco-unit or stand architecture. This is a serious lacuna for different reasons.

- Tree vitality, and hence the condition of the eco-unit, cannot be appraised precisely by biomass accumulation alone (see Fig. 3.33). A two-level simulated architectural model, linked to a biogeochemical model, would enable us to explore the whole complex effect of stressors or growth factors on a stand, including its visible, architectural symptoms.
- It will not be possible to understand the precise effect of silvicultural measures such as thinning, or natural impacts such as wildlife feeding, on a stand unless

the architectural modifications can first be introduced in the two-level architectural simulation model, in order to assess their effects on energy and nutrient flows. In the meantime descriptive yield tables are all that is available.

- The precise architectural feedback between eco-unit architecture and tree crown architecture is an indispensable tool to understand niche formation for small organisms that live in the canopy. This not only applies to larger animals, such as the birds, for which the point is approximately proven by a descriptive matrix of bird diversity versus architectural canopy diversity (Komdeur and Vestjens 1982). It also concerns insects and cryptogamous plants (cf. Chap. 4), pests and diseases included.
- It will not be possible, if such a model is not built, to assess the effects of the location of energy, nutrients and stressors in the system. Location is linked to the system's architecture. This also applies to ecological impellors, chemical substances which block or favour growth at precise points, moments or surfaces in an eco-unit.

This last aspect requires a specific paragraph, because it leads to a series of quite neglected facts and insights with a most actual and important ecological and silvicultural importance.

5.6.8 Ecological Impellors, Poisons, Chemosilviculture

In organisms, substances exist that are transported from one subsystem to another, there to exert a driving force that blocks or favours growth or metabolic processes. To drive, in ancient Greek, is ορμαω (hormao). Therefore these driving substances have been called **hormones**. Hormones, in very small quantities as compared with the mass of a whole organism, ensure the orderly course of the vital flows and accumulations of energy, nutrients, water and biomass. More recent research (e.g. Si-Jiu Liu and Tillberg 1984) points to a strongly topophysical role of these substances, polarising cells, tissues, organs or organisms. Some plant hormones have as a task to kill and eliminate organs when the latter become superfluous or dangerous. Leaf abscission in green plants is an example, in which leaves are killed after having been emptied of their nutrients, so that the plant may live. The killing is driven by hormones of the group of abscisin and dormin, being sesquiterpenes chemically (Abercrombie et al. 1977).

Resistance factors, including "antibodies" that eliminate and kill aberrant body cells or immigrants in the organism are in another class of such substances. At the hierarchical level of the organism (living system) they act as mini-systems in the internal communication system that allows adjustment and survival. At the level of the cell, tissue or organ (subsystems) they are **poisons.**

The concept of poison is very anthropocentric. It comes from the Latin word "potio", meaning a (bad) drink. "Toxic" is derived from the ancient Greek word τοξον (toxon), denoting a bow for shooting arrows. Toxic substances kill organisms as if they were shot. The word poison has exactly such agressive, deadly and immorally warlike emotional connotations. Poison is bad. It is difficult to discuss poisons in a business-like way, and this negative impression nowadays is also attached to synthetically made killing substances such as biocides, pesticides, herbicides or arboricides.

Brünig (1983), in a visionary paper presented in Viet-Nam, made the following statement (p.3 ff.): "Biochemical warfare is closely associated with the evolution and distribution of species and ecosystems. On a global scale, the enriching of the atmosphere with oxygen by photosynthesising aquatic plants waged a fierce and effectively successful territorial war against the realm of the anaerobic organisms. Final victory came into reach when the oxygen level had risen sufficiently and the occupation of the land had become possible by plants developing and displaying in action a cybernetically regulated self-support water supply system. Chemical warfare continued on land and with time, devices became increasingly more sophisticated and extended to include chemical warfare between plants and animals." A beautiful account of a comparable viewpoint may also be found in the Gaia-theory (Lovelock 1987).

This image was developed by Brünig into a sketch of competitive chemical interactions between plants and animals in tropical forests, due both to direct chemical evolutionary competition and indirect effects of litter decomposition or soil (peat) formation. It points to the ecosystem functions of such chemical substances, in as far as systems analysis is emphasised in Brünig's text.

There are countless examples of such natural substances. Those that can influence human behaviour or cause human death generally are feared as natural poisons. Altmann (1979) listed and illustrated over 100 poisonous plants and animals from North-Western Europe: 15 poisonous toadstools (fungi), 59 poisonous green plants including medicinal plants, 17 poisonous arthropodes including insects, spiders and scorpions, 7 poisonous amphibians, mainly toads and 4 poisonous snakes (reptiles). Like in Leeuwenberg's (1987) "Medicinal and Poisonous Plants of the Tropics", of which the species list contains over 650 species, and the chemical list over 250 compounds or families of molecules, the useful and dangerous aspects are combined.

The fact that medicinal plants are included points to the effects of **dosage**: clearly there are lower dosages at which these natural substances are lethal at the cellular level, but regulatory and beneficial to survival at the organism's level. Increasing the dosage makes the same substance lethal for the organism, but in that case it regulates the ecosystem by influencing population density and biomass of the organism killed. This is also true if it concerns a human organism. In their field guide to medicinal plants of the Appalachians (US), covering more than 400 species, Krochmal and Krochmal (1984) tell their readers that self-medication with wild plants is highly risky.

Figure 5.75, being the core figure of Section 5.6, is useful in explaining the balance between regulation and toxicity. At the eco-unit level, chemicals interacting among animals, e.g. in a population of poisonous snakes reducing a mice population, belong inside the compartment "animals". Even substances that are described as hormones may do so. For instance, Wingfield et al. (1987) showed how the steroid hormone testosterone stimulates aggression in birds, in response to certain perceived threats, which most often emanate from other birds of the same species or another. Both substances, snake poison and the hormone complex of which testosterone is part, are chemically very complex and transmit information which is lethal at some hierarchical levels but regulatory at the eco-unit level.

The regulation of the arrow "plant consumption by animals" (Fig. 5.75) takes place in the same way. Hladik (1978) showed this by her chemical analysis of the food eaten by tropical arboreal folivores. Most often, the substances involved are named according to the image of a "chemical war between organisms" as sketched by Brünig, i.e. "defense substances" or "attractants". This is well illustrated by the extensive literature published by Janzen on the subject (also see Sect. 6.2.4.8). His overview of the physiological ecology of fruits and their seeds (Janzen 1983) includes the balance between "attractants", "nutrients" and "protective secondary compounds" determining fruit or seed consumption, or both, by animals. This balance would operate a powerful selection among animal consumers and seed dispersers. In the view of the present book, the selective balance between attractants and repellents or poisons would be the regulatory part of plant-animal interactions (information flow), to be considered apart from nutrient interactions (transfer of mass and energy). The most important balance, linked to the dosage of such substances, is that between toxic effects (lower hierarchical levels) and regulatory effects (higher levels).

The extremely complicated nature of many such interactions is exemplified by the case of a bruchid beetle, *Caryedes brasiliensis*, feeding on toxic seeds of the neotropical liane *Dioclea megacarpa*, a papilionoid legume of deciduous forests in Costa Rica (Rosenthal and Janzen 1985). The larvae of the insect detoxify L-canavanine and its catabolic derivative L-canaline, but also produce ammonia that they can use themselves, in a series of chemical reactions involving amino acids and other protein fragments. The fruits of the lauraceous *Ocotea veraguensis* of the dry forests in Guanacaste (Costa Rica), as a second example, are consumed by large vertebrates, they are toxic for mice and predated by one weevil, an insect species that is able to deal with the 12 neolignans, 11 bicyclooctanoids and one benzofuranoid present in parts of the fruits (Dodson et al. 1987). The regulation power of beetle or weevil is to be noted. Its absence or presence leads to quite different allocations of resources to the tree species of which they eat the seeds, as compared with the other components of the same eco-unit.

Hundreds of case studies have been published in this realm. A prudent book on fungus-insect relationships (Wheeler and Blackwell 1984) documents many, while emphasising the need for very careful taxonomic identification of the organisms involved. Still, listing all relations and all chemical molecules involved seems to lead to such a mass of data that no manageable ecological model can be expected to explain them. Are there any main lines in this welter of information?

One of such lines may be the grouping of substances involved in regulation. Snakes, which even more than poisonous plants have long been the symbols of medicine, surrounding the staff of Aesculapius, yield poisons that are derived from saliva and contain mainly proteins and enzymes (Morris and Morris 1965). Janzen et al. (1986) discovered potentially "defensive" proteins in the mature seeds of 59 species of tropical Leguminosae, most of which were from Guanacaste in the dry forest region of Costa Rica. Snake poison and these legume proteins at least have in common a destructive effect upon the red blood corpuscules of animals. L-dopa, a derivative of alanine, an amino acid, according to Bruns (in Wheeler and Blackwell 1984) is a substance that may intervene in consumer animal repulsion by *Strobylomyces floccopus* (Boletales), against the insects and rodents that might eat this fungus, as well as in several leguminous plants being known for antiherbivory.

A second "family" of molecules that are found back in plant-animal interfaces as regulators, not carriers of energy or nutrients, are aromatic carbon ring derivatives, like the neolignans mentioned above. They can be encountered in the next category of substances, which are regulators of plant-plant relationships or plant-necromass interactions.

This category of substances is allelopathic. They inhibit functions in plants other than those producing them. Aubréville (1938), who half a century ago touched upon the one-tree eco-unit concept by stating that there was mosaic-wise tree-by-tree regeneration of the West African tropical rainforest, remarked in the same publication that some tree species showed a circle around their base, within which no seedlings occurred from other tree species, and sometimes no seedlings at all. He mentioned the possibility of some chemical substance excreted by roots, inhibiting the development of other plant species. In 1967, Webb et al. confirmed this in the case of *Grevillea robusta* (Proteaceae), a tree from the Australian forests in S. Queensland and N. New South Wales which excretes a chemical that is toxic to its own seedlings. Such phenomena have been found to be true by other researchers, who coined the word allelopathy and allelochemicals (cf. Whittaker and Feeney 1971, or Rice 1984, both ex Kuiters 1987). In the 1970's allelopathy began to be extensively studied, one famous case being the South-East Asian grass *Imperata cylindrica* (alang or cogon) with its allelopathic effects on trees (e.g. Mendoza 1978).

There seems to be some confusion as to the limits of this concept. Some researchers use it for substances excreted by living tissues of plants, particularly by roots, whereas others include indirect effects which are transmitted through the litter compartment, by organic decomposition products. In all cases, however, allelopathy concerns inhibiting, not facilitating effects.

In Fig. 5.75, the regulatory functions of such chemical substances hence can be found back in two places. These are the arrows at the lower left, "mutualism/antagonism" between plant roots, and the arrow between necromass and roots, "solutions". Such solutions contain substances with a regulatory function. Again it is emphasised that no flows of nutrients and energy are referred to, but that these substances are information carriers either blocking or facilitating energy or nutrient flows.

Kuiters (1987) did specific and detailed research on phenolic acids in forest litter in the Netherlands, and their roles concerning the inhibition of seed germination, seedling establishment and herbaceous growth. We refer the reader to his book for a bibliography. Kuiters found that polyphenols occur in different forms and concentrations in the litter of different tree species. He indicated that there are distinctive growth-altering effects by each of these litters, that these effects change with time in function of the progress of decomposition, and that nutrient element uptakes by plants grown on different litter were altered (also cf. Maheswaran and Gunatilleke 1988). No straightforward correlations with biomass could be found. Next to polyphenols, other organic compounds such as terpenes, fatty acids and abscisic acid were each found to exert their own and specific influence on green plant growth. Moreover, these organic substances codetermine, with the mineral substrate, the kind of humus profiles that are formed (Fig. 5.20). Organic soil properties indeed are eco-unit properties that may not be averaged over whole forest mosaics. This was stated earlier on the basis of other data (Sect. 5.5.7).

The regulation by such substances of germination processes within a forest eco-unit has not been indicated in Fig. 5.75, but it would justify an arrow between the necromass and propagule bank compartments. Germination is most strongly inhibited by fresh litter, which in the temperate regions is shed in the fall. This amounts to a germination release after the winter, i.e. the inhospitable season. Therefore, this is a clear example of an interaction process, belonging to the eco-unit but to no one single component. The process enters into an interior communication circuit contributing to eco-unit adjustment to unfavourable circumstances.

These regulatory processes are not only important for forest innovation through their effects on tree seed germination and seedling establishment. They also profoundly influence herbaceous plants, the behaviour of which was particularly studied by Kuiters (1987). Soil microflora and -fauna are affected, in feedback with soil acidity (Fig. 5.20). For instance, in the acid tropical forest soils of East Kalimantan there is a special insect compartment able to decompose the abundant tannins from dipterocarp litter (Smits pers. commun. 1985), whereas Kuiters cites Rice and Pancholy (1973) as stating that in temperate forests tannins accumulate in forest soils because they are not easily broken down by micro-organisms, in contrast to phenolic acids.

The strategy of herbaceous plants in the undergrowth of forests shows a temporal unfolding which is regulated among other things by the temporal gradient of necromass decomposition. A tropical example of such processes is the study by Maheswaran and Gunatilleke (1988) on different litter decomposition sequences in two vegetation types in Sri Lanka. Herb strategy moreover includes a particular architecture (Chap. 4) and a metabolism which, according to Kuiters (1987), includes a low respiratory rate. In this particular microclimate (e.g. see Barkman and Stoutjesdijk 1987), a more careful study may show still other adaptations by herbs. In the humid tropics, for instance, Lee (in Givnish 1986), described several herbaceous species from the rainforest undergrowth that show variegated leaves, leaves with red or purple undersurfaces and surface sheens or iridescence. Colours correspond to non-chlorophyllic photosynthetic pigments, raising leaf temperature, or using such light wavelengths as are left over in the deep forest shade, or doing both.

Some sheens and iridescences have been observed to be due to "epidermic lenses", leaf cells that function as lenses concentrating the scarce incident light on its way to the photosynthetic cells (Fig. 5.88A). An unproven hypothesis could assume that in this season-less environment it would not be the time but the place of germination and establishment of such herbs that are determined by the spatial distribution of specific necromass substances. This is indirectly supported by Kuiters' statements about the different properties of mixed-species litter as compared to monospecific litter. The observations by Aubréville (1938), cited above, might also be explained by heterogeneous distribution of decomposing litter belonging to each species, instead of exclusively assuming active allelopathy by root exudates.

Again, if inhibitory, the organic substances that regulate germination are lethal or toxic at the level of the seed or seedling, but act as regulators in a communication circuit allowing for a healthy adjustment of eco-unit architecture and functioning by steering the reallocation of energy and nutrients.

The phenomenon is both extremely common and extremely complicated because of the huge numbers of substances involved. Brünig (1983) adds several examples from tropical Asia. *Gonystylus bancanus*(ramin, Thymelaeaceae, also see Jacobs 1988), a commercial tree from the perimeter of peat-swamps in Sarawak, shows particular growth responses in experiments on peat from *Shorea albida* (Dipterocarpaceae) forests, from which *G. bancanus* is otherwise absent. In the same region, introduced *Casuarina* spp. and *Dacrydium* spp. produce a litter so rich in non-hydrolysable tannins that native tree species germination is completely blocked. This would also point to an unbalanced soil fauna without tannin-decom-

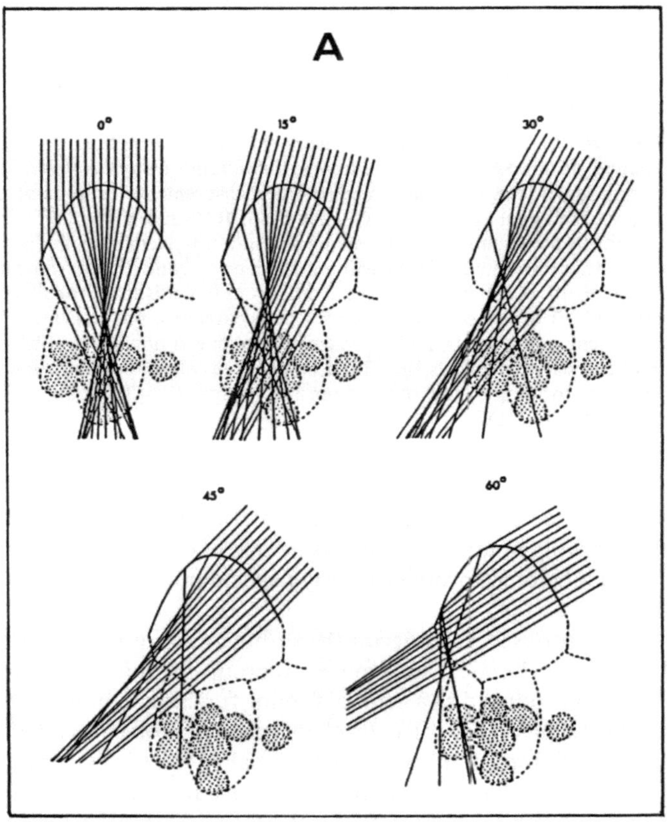

Fig. 5.88A (Fig. 5.88B see page 378)

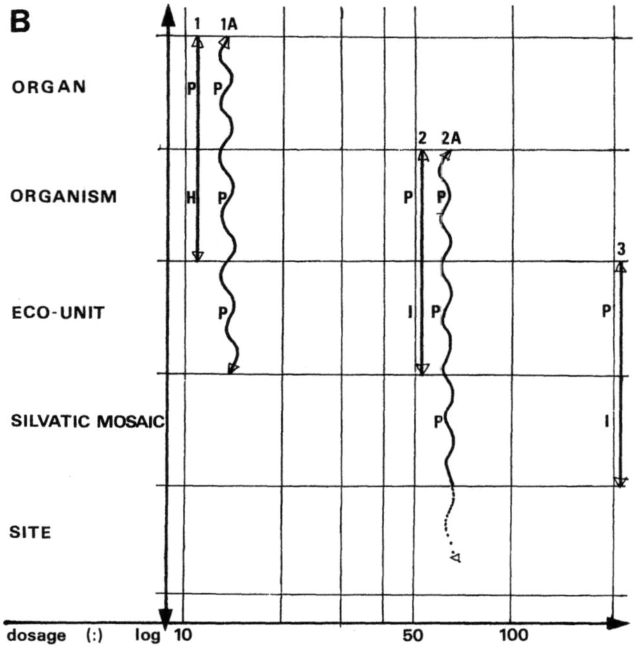

Fig. 5.88. Production strategies, poisons, hormones, impellors. **A** Undergrowth herb strategy, example of "epidermal lenses" in leaves of *Anthurium warocqueanum* (Araceae, S. American rainforest). Ray-tracing diagram as mean of five measurements for epidermal leaf cells under different angles of indident light. (Lee in Givnish 1986). **B** Visualisation of effects of fictional dosages of a chemically active substance at system level and subsystem level, in the hierarchy of Table 2.1, expressed as proportional to a logarithmic function. *1*, *2* and *3* one or more natural substances. A hormone (*H*) drives survival processes of organisms but poisons (*P*) or alters organ behaviour. An impellor (*2*, *3*) drives survival processes of ecosystems, but poisons (*P*) or alters behaviour of subsystems, e.g. organisms (*2*, cf. snake poison) or eco-units (*3*, cf. volcanic SO_2). *1A* and *2A* show synthetic active chemicals, broad action spectrum (wavy line), more or less toxic (*P*) at many levels according to dosage, width of spectrum and life-span (long life span: accumulation effects). Examples are dioxin, DDT

posing insects. The empty undergrowth volume in *Eucalyptus* plantations all over the world is notorious, and this genus, like other members of the Myrtaceae to which it belongs, produces many particular organic compounds such as ethereal oils.

It is proposed, considering the all-but omnipresent nature of eco-unit regulation by such biochemical substances, to group them in a chemical compartment that may be introduced as a general forcing function in biogeochemical models. In the living systems at organism level, this class of molecules is called hormones, as said in the beginning of this section. If we called the eco-unit regulators eco-hormones, this would perhaps lead to the misunderstanding that eco-units are considered as superorganisms (see Sect. 5.1.5). We therefore emphasise the non-organismic nature of the living system "eco-unit" by coining another word, derived from the

Latin verb "impellere" which means "to drive": impellors. Two parallel definitions would be the following ones.

Hormones are biochemical substances that implement the communication system in the organism and solicit growth adjustments, being beneficial to the organism in their natural dosages, but either toxic or tonic at the levels below the organism (organ, tissue, cell).

Impellors are biochemical substances that implement the communication system in the eco-unit and solicit growth adjustments, being beneficial to the eco-unit in their natural dosages, but either toxic or tonic at the levels below the eco-unit (organism, organ, tissue, cell).

The effects of impellors, as those of hormones, have two causes. The first cause lies in the chemical character of the molecule, which determines its longevity and which also drives a specific process. Such molecules may decompose rapidly or slowly and therefore act in the system during a short or a long period. The second cause is their dosage, shortages leading to such unhealthy growth phenomena as prolific growth and excesses leading to toxic effects at a higher level than their normal functional level. Figure 5.88B visualises this for natural substances (1 to 3) and broad-spectrum, long-lived artificial, or artificially dosed substances (1A, 2A).

The use of artificial chemical biocides is a case in point. Figure 5.88B shows how responsible use can be distinguished from abuse, at least according to the dosages applied. An example is the artificial substance 2,4,5,T (2,4,5,trichlorophenoxy acetic acid), which in small dosages is a growth stimulator in green plants, i.e. a phytohormone (cf. Ten Houten and Gysels 1984). As a phytohormone, this substance stimulates meristem and cambial activity. But in higher dosages, it becomes toxic for green plants, which literally grow themselves to death. In such dosages it is thus an impellor which kills or weakens treated plants and so regulates populations in an ecosystem. 2,4,5,T has been used in tropical silviculture in such dosages for chemical thinning, by treating such trees as would have to be weakened or removed for silvicultural reasons (e.g. Brünig 1983; De Graaf 1986; Jonkers 1987). In still higher dosages, it has been sprayed over whole vegetations as a weed-killer, or worse over whole forested regions in the Second Viet-Nam war (1963-1973; see Von Meyenfeldt et al. 1978; Brünig 1983), in a vitiated form under the name "agent orange". In such high dosages it kills whole forest mosaics and at that level it might be termed an **ecocide.**

It should not be forgotten that it is not the substance in itself, but its dosage and purity that determine its function as hormone, impellor or ecocide. In tropical silviculture, the early experiments with anorganic chemical thinning (see Bibliography in De Graaf 1986) and also those with organic substances included heavy overdoses of broad-spectrum substances which disrupted whole eco-unit developments. Our own observations in the Côte d'Ivoire in 1963 were that these substances in high dosages caused replacement of small eco-unit mosaics by large, pioneer eco-units (built mainly by *Musanga cecropioides*, Moraceae) and in other cases provoked fragmentation leading to developmental regression of eco-units and heavy invasion by lianes in the resulting, numerous (neo-)aggradation phases (cf. Sect. 5.2.3). Only in more recent years have dosages been determined, of the order of 500 g/ha/8 y or less, in which 2,4,5,T, having a narrower spectrum because of selective, tree-wise application, acts as an impellor regulating eco-unit (or mosaic) development, not as an ecocide.

The requirement of purity of the product is closely linked to the fact that each organic substance capable to act as an impellor has one or a few specific processes to regulate. The insecticide DDT is an example of a substance that is not specific enough to act as an impellor, because it kills all arthropodes and therefore disrupts ecosystem functioning in general. Moreover, it decomposes very slowly and therefore has a persistent action and even accumulates in unexpected places such as the eggs of insectivorous birds. Agent orange, the vitiated form of 2,4,5,T used for warfare, was fast and carelessly produced and therefore contained substantial traces of an industrial by-product, dioxin, which is a broad-spectrum biocide that acts in very small dosages and has a long decomposition period. Agent orange therefore automatically became an ecocide, the consequences of which may be expected to manifest themselves for long years to come, not because of 2,4,5,T but because of dioxin introduction into the chemical communication channels of the forests sprayed.

The use of chemical substances in silviculture has to be considered in a very sober way. Their application is nothing more or less than an alternative option replacing other silvicultural measures, executed by hand or using mechanical means. In the CELOS silvicultural system from Suriname (De Graaf 1986; Jonkers 1987), chemical thinning is an alternative option for mechanical thinning with chainsaws. The chemical option, if well designed, is much less disruptive within the forest ecosystems than the mechanical option. Chemical options are conceivable for the following measures.

- Chemical harvesting. This is not considered up to this day and almost certainly is not feasible.
- Chemical regeneration. As forms of chemical weed control in the innovation phase, the methods applied still rest on substances with too broad a killing spectrum, applied in overdoses by spraying. Better designs are urgently needed to eliminate the ecocidal features of the present methods.
- Chemical thinning. Some methods now approach a semi-natural prodecure, in which the substances applied act as impellors instead of ecocides. However, specificity is not yet guaranteed by the chemical properties of the products but only by application by hand to specific trees in the right dosage. The design still has to be improved.
- Chemical forest protection. Methods of chemical pest control in their vast majority are rather ecocidal, complicated by resistance and immunity build-up in pest populations and by transfer of ecological functions from one species to another. Studies on specific substances such as pheromones (e.g. De Jonge Mulock Houwer 1984) are expensive and rest on a problem-by-problem basis in too many cases. **Pheromones** are substances that transmit signals within an animal population, often as aerial suspensions ("smells"), e.g. sexual attractants. They certainly belong to the impellors localised inside the animal compartment (Fig. 5.75), but also influence the arrow "emigration/immigration".

The difficulties in designing chemosilvicultural measures have been severely underrated up to this day. Studies such as that by Kuiters (1987) clearly show that chemical measures in silviculture are not necessarily "anti-natural", because chemical impellors play a very important part in natural forest eco-unit development. In the same way, "nature-conform" ("naturgemässe") forestry as practised in Germany uses mechanical methods that are designed to imitate natural processes.

However, imitation or steering of the impellor system in eco-units still is in a very early and primitive stage. This is due in the first place to imperfect knowledge of the natural impellor system, which is extremely complex, involves the interactions of countless organisms amongst each other and with the necromass, concerns huge numbers of chemical substances, and is strongly linked to the biological soil and litter compartment. This compartment has too often been taken for granted in silviculture, provided that the right tree species were chosen.

Further research is needed if chemosilviculture is to become an adult technique in forestry, avoiding the ecocidal by-effects that accompany too many of the applications at present. Such research should emphasise the following aspects.

- Establishment of the chemical "families" of natural molecules that contain the most frequent and most active impellors, and may be linked to management problems.
- Introduction of these "families" as compartments or driving functions in ecophysiological simulation models, population models and architectural simulation models, these last ones being necessary to localise the spatio-temporal foci of chemical action.
- Isolation or design of molecules to be used in specific chemosilvicultural measures.
- Thorough experimental testing of such substances in many different eco-units.
- Establishment of instruction booklets for forest managers, giving simple explanations concerning the exclusive aims of the measure, the risks of wrong applications and the dosage, location and time to apply the specific product in a stand.

Real progress in this sense would require the foundation of a large, multidisciplinary international Institute for Chemosilviculture. In the meantime, extreme care is recommended in the use of chemicals for silvicultural purposes.

This section would not be complete without coming back to the fact that numerous impellors are substances found in medicinal plants and animals. These chemicals, in the correct dosage, are used to combat human diseases, an all-but hormonal use. It raises the same questions as in chemosilviculture with regard to chemical structures, dosages and synthetic pharmaceutical products. Plants and animals which produce such substances are to be always kept in mind, because they can yield valuable minor forest products in addition to wood. As has been stated in Chapter 1, these products were considered to belong to the major forest products in several traditional forms of land use, where wood was a minor product.

These resources should not be underestimated. Gomez-Pompa (1985a), for instance, cites 91 medicinal plant species from two villages in a semi-arid Mexican region alone. The numbers of tropical and North-American medicinal and poisonous plants mentioned above (Leeuwenberg 1987; Krochmal and Krochmal 1984) can be supplemented by many examples from other countries. The multiple traditional uses of 18 native plants of Central Australia in Yankunytjatjara plant use have been compiled by Goddard and Kalotas (1988) and for the whole of that continent Cribb and Cribb (1981) enumerated over 570 useful wild plant species.

Craker and Simon (1986) claim that the worldwide trade in herbs, spices and medicinal plants exceeds a billion (10^9) US dollars per year, in the light of which it is amazing that these plants are scarcely mentioned by Prescott-Allen and Prescott-Allen (1983) in their book on "genes from the wild". In the US strategy on the conservation of biological diversity (McPherson 1985) medicinal plants and animals are mentioned, however, in the very first paragraph of the introduction. Krochmal and Kologiski (1974) found about 30 medicinal plant species in a 30-ha loblolly pine plantation in North Carolina (USA). The present author encountered more than 30 such plant species in the region of Río Curaré in East Ecuador (Amazonian basin), pointed out by Wau Indians during 10 days of field work. These few examples should be sufficient to illustrate the enormous importance of such species, which should always be kept in mind in silviculture and agroforestry.

5.6.9 Transfer of Ecological Functions

In Section 3.8.3 some examples of transfer of biological functions were discussed. They included the transfer of functions from one subsystem to another in trees, e.g. photosynthesis operated by axes instead of leaves, and transfers from one hierarchical level to the next, e.g. from the flower to the inflorescence. In the present section, the transfer of ecological functions inside an eco-unit will be considered.

These transfers, or functional shifts are very important to understand. Still, they have not been considered as such in silvological studies or silvicultural systems. We already met with an example in an earlier part of this text. This was the reallocation of resources in an eco-unit after light storm damage or thinning, with the result that productivity shifts from the tree compartment to the insect compartment. This is perceived as an insect pest by forest managers.

A case-history serves us to assess what exactly happens during such processes. The ecology of the woodwasp *Sirex noctilio*, an important pest in Australian and New Zealand's *Pinus radiata* plantations, as described by Taylor (1981 in Kitching and Jones) yields a good and well-documented example (Fig. 5.89). It is chosen also, because *P. radiata* stands are well-defined artificial eco-units. The transfer processes go as follows.

Generally in less vigorous *P. radiata* stands, the vitality of which is low as a consequence of low quality soils or over-dense tree populations due to thinning delays, considerable proportions of trees emit impellors in the form of monoterpene hydrocarbons attracting the female wasps after they swarm and are fertilised. The impellors are emitted due to a highly respiring, weakened cambium/phloem system (cf. Fig. 3.30).

- The transfer is initiated by impellors, which are attractors for the wasps.

Female wasps drill a hole through the bark and depose their eggs and the spores of a fungus, *Amylostereum areolatum* in the sapwood, as well as a mucus that is toxic to the wood, another impellor (Fig. 5.89A). The larvae grow together with the fungus and feed on it.

- The transfer is implemented by poisoning the sapstream tissues, diminishing wood production, accompanied by an impellor-initiated increase in insect biomass production.

Now, among the biological predators on *Sirex noctilio*, a nematode was found that is at present being used in biological control, *Deladenus siricidicola*. Its life cycle is represented in Fig. 5.89. It is attracted by a combination of the fungus, yeasts (principally *Saccharomyces*) and bacteria, living on the frass left behind by wasp larvae in the wood. According to Taylor (in Kitching and Jones 1981) these produce attractants.

- The transfer was continued by impellors, being attractants to nematodes (or others).

In function of the presence and frequency of *S. noctilio* larvae, a certain proportion of the nematode population eats the fungus (Fig. 5.89B, above) and another proportion eats the wasp larvae (Fig. 5.89B, below).

- The continued transfer was implemented by consumption of woodwasps by nematodes.

Finally, the nematode and bacterial biomass produced arrives in the litter compartment, directly if the tree was killed and collapses to the ground together with its foreign biomass load, or indirectly if the tree recovers and processes this biomass by scar formation. In this last case, it will be incorporated in the litter when the tree dies later.

- The last transfer always is passive and goes to the litter.

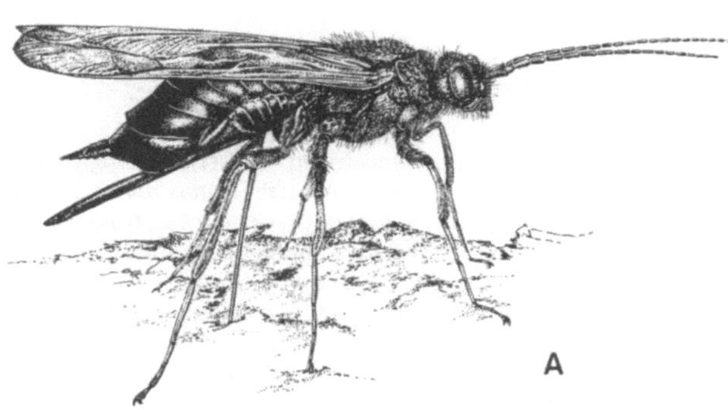

Fig. 5.89A (Fig. 5.89B see page 384)

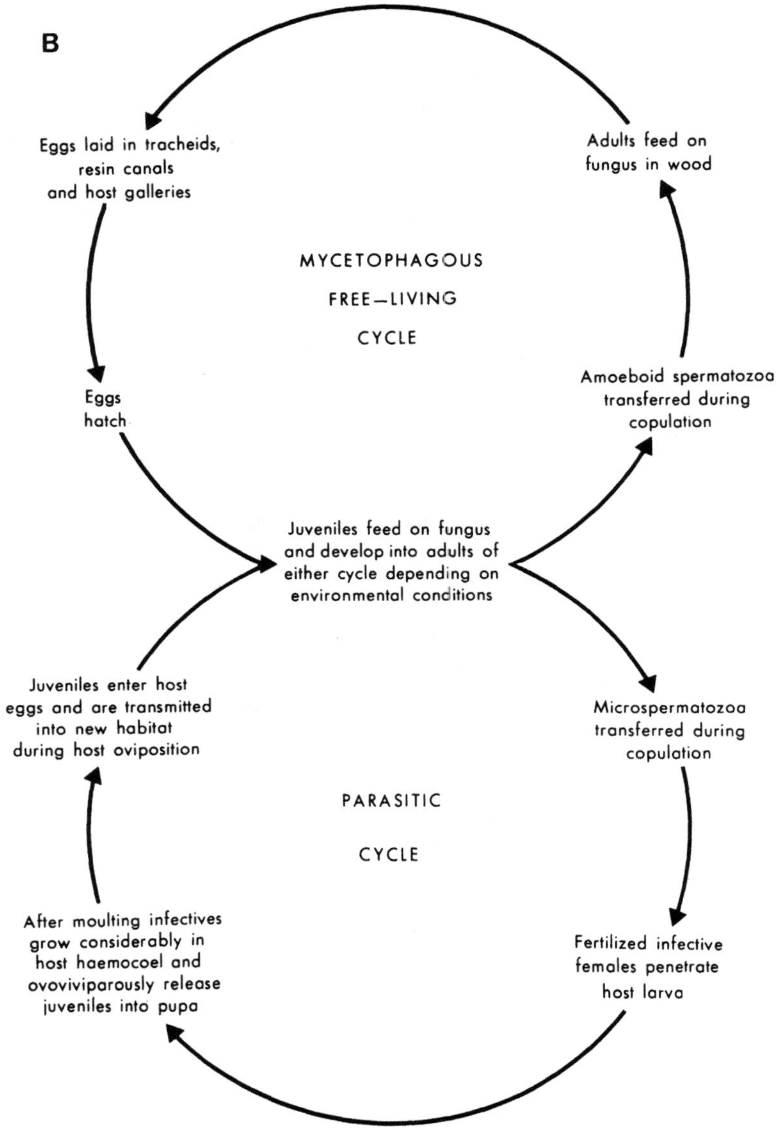

Fig. 5.89. Example of an impellor-driven transfer of ecological functions in an eco-unit, the production function being transferred from trees (*Pinus radiata*, Pinaceae, Rauh's model, Australia, New Zealand) to the parasitic woodwasp *Sirex noctilio*, from the woodwasp to the parasitic nematode *Deladenus siridicola*, all produced biomass ending at last in the litter cycle. Example simplified to illustrate some main transfers only. **A** *Sirex noctilio*, adult female ovipositing through pine bark, to which she was attracted by a chemical impellor. Spores of the fungus *Amylostereum areolatum* are deposited with the eggs. **B** *Deladenus siridicola* is activated by attractants (impellors) to follow in part a fungus-eating cycle (*above*), in part a larvae-consuming cycle on *S. noctilio* (*below*). The result is a reallocation of biomass production among forest components. (Taylor in Kitching and Jones 1981)

This final transfer is like the earth-wire attached to an electric machine that evacuates excess voltage to the earth.

This case story has been considerably simplified, because other wasp predators and several processes have been left out. Enough details have been given, however, to understand how impellor-driven processes lead to transfers of the production function from one component to the other. Such a chain of transfers finally balances out, all excess matter and bound energy ending up in the litter. The whole set of processes has then led to the adjustment of the eco-unit by a reduction of the tree compartment and through oscillations in the amounts of biomass in other compartments.

The reduction of the tree compartment means the death of numerous trees, which is often contrary to the aims of the forester. There are several options to avoid this happening. The first one is a design of the stand such that these accidents are minimised and if occurring after a climatic incident, that they are readily buffered (e.g. Brünig in Wiersum 1984; Brünig and Schneider 1975; Leibundgut 1970). This implies optimal tree species choice and planting schemes for the site, and an adapted and flexible system of treatments that is consistently followed.

The second option is the introduction of forest components that predate upon the organisms consuming tree biomass, such as the nematodes mentioned above. These predators ("natural enemies"), if inoculated upon the eco-unit, will dampen the oscillations in biomass of secundary producers and so reduce the decrease of primary biomass in the trees. Predators have to be brought in in the right dosage. The biocenotic connexus discussed earlier (Fig. 5.70) shows that there often are many possibilities, which have to be carefully assessed before choosing the optimal form of biological control.

The third option is the direct killing of the pest that feeds on the trees. This has to be done chemically, because it cannot be done cheaply by hand or mechanically. Now chemical pesticides may seem simple and efficient means to reach that goal, but the environmental conditions that were faced by the eco-unit through production transfers are not changed by pesticides. Moreover the latter are nearly always too little specific to act as an impellor. They trigger many processes with unpredictable outcomes by killing many more species than the target pest. Debach (1974) gives many documented cases and reports many experiments showing "how to use pesticides for fostering pests" and "how to make pests out of formerly rare and innocuous species". An interesting figure, well supported by literature, is cited by Debach: less than 1% of those phytophagous insects that are potential pests ever achieve the status of even minor pests. Their chances to do so are rather enhanced than reduced by pesticides.

As a matter of fact, if an eco-unit has to adjust, this will somehow implement a transfer of production towards members of one guild (see Sect. 5.5.8). If this is a tree-consuming guild of some sort, the extinction of one species will be followed by the activation of the next guild-member species. And if natural enemies of the pest species are eliminated, or even killed at a somewhat higher rate than the pest species, the pest outbreaks become both more frequent and more severe. The transfer of production functions then becomes unchecked, because the chain of transfers, from primary producer organisms through consumers towards the litter, has been broken off at the shackle of the eliminated species. Production runs wild at the last remaining shackle, i.e. the last surviving species receiving transfer signals.

A second type of functional transfer is that of functions linked to the sapstream. They have been met in earlier parts of this text, but it is useful to pinpoint them here very clearly as functional ecological transfers between eco-unit components. Root anastomosis (Fig. 3.16C) implies the flexible transfer of absorption of soil solutions or stocking of reserve nutrients to that member of the group of mutually grafted trees that is optimally suited to process them. In some cases, this may mean that small members of the grafted tree group are kept alive in the shadow and can start growing if something happens to the crowns of the big trees. In other cases, the small members may die in helping the big ones to reach the canopy. Mycorrhizal mutualism (see Sect. 5.3.4.1) is a second example. It includes the transfer of a large proportion of the water absorption capacity from the green plant roots to the fungus, as well as several nitrosynthetic processes and protection functions.

In these cases, particularly if only specific functions are transferred, the question may be raised of the limits of the individual organism. Does a tree with its multiple mycorrhizal fungal partners constitute a population, two populations or one individual with new characteristics, in which the fungus represents an organ? Probably, coherent models could be built according to every one of these system concepts, so that none of them is totally wrong. There just is no clearcut limit between an organism and a population, as we saw earlier in reiterated crowns (Sect. 3.8.1).

Independently of such a model, however, the fact of these transfers of function is of the utmost importance both to the silvologist and the silviculturist. Taking them into account, it becomes clear that no forest eco-unit can survive without such transfers. Silviculture is not merely choosing and cultivating trees. At every moment of the rotation, the accompanying species to which tree functions have been transferred have to be cared for, too. They include not only parasites, pathogens and mycorrhizae, but many of the other organisms mentioned in Section 5.3.4. and Chapter 4.

During an excursion to the Auermühle forest in Germany (Brünig and Schneider 1975), we observed isolated, old, healthy pines *(Pinus sylvestris)* that had been sparsely left over after a storm 10 years earlier. The manager remarked that after this storm, fortunately, there had been an attack on the surviving pine crowns by *Tomicus piniperda*, the common pine shoot beetle known as a major pest, so that needle transpiration was reduced and the roots could adapt to the new, isolated position of the trees. The mechanisms for reducing transpiration in the tree had been transferred to the insect. Of course, *T. piniperda* remains a pest, and in all such cases of interaction production functions are transferred from Scots pine to this animal. But an open eye for the transfer of functions makes it possible to judge each case on its merits, as an eco-unit mechanism for adjustment to the direct environment. This may or may not be in accordance with the aims of the manager.

A third type of transfer of functions is that from an older to a younger cohort. This is a general escape mechanism in species under stressed circumstances, one example of which certainly was the post-war human baby- boom of the late 1940's in Europe and the USA. Survival functions of a tree compartment are transferred in this way to a new eco-unit. Figure 5.74 gave an impression of this mechanism which, if stress remains constant, selects new cohorts with life strategies becoming more *r*-oriented. In stressed forest eco-units, this transfer of all life functions to younger cohorts may become apparent as abundant natural regeneration, i. e. dense tree seedling carpets. This in its turn may lead to a next shift in the production function, from the seedlings and saplings to abundant herbivorous populations that eat them.

There are more types of functional transfer within eco-units than described above, such as litter accumulation shifted towards the funnel-shaped crowns of S.E.Asian or South American palms in the undergrowth of tropical rain forests (cf. De Granville 1978). They will not be detailed here, but deserve a whole book. In Section 6.4.4, some remarks will be made on the transfer of ecological functions in silvatic mosaics.

5.7 From Eco-Units to Silvatic Mosaics

Eco-units occupy a relatively large part of this text. This is necessary in order to present a coherent image, in which data, models and case-studies are sufficiently numerous to convey the meaning of this concept, that has never been completely exlained before. It is deemed an important concept, both for the better delimitation of forests as living systems at this hierarchical level and for silviculture. Stands are the foci of human impacts on forests, and the innovation phase (regeneration) largely determines the success or failure of forest management and conservation plans. The notion of the eco-unit is a unifying concept, that may be used as a tool for comparing regeneration phenomena that up till now have been treated separately, for building more comprehensive simulation models and for facilitating compartmentalisation in all studies.

6 Silvatic Mosaics

6.1 What Are Silvatic Mosaics?

6.1.1 Elements of Definition

In the preceding chapter, silvatic mosaics have been mentioned already as the living systems of which the eco-units are the subsystems. When explaining eco-units as systems in terms of their subsystems (components or compartments), it was impossible to do so without referring to the environment of these eco-units (cf. Fig. 5.75). When further considering eco-unit development and degradation, processes such as fragmentation or fusion could not be explained without at least mentioning a mosaic concept, even if it were not completely explicit. Anyhow, it has been clear that the eco-unit fragmentation concept is quite different from the "fragmented forest" of Harris (1984), who considers landscapes with forest "islands". It has been stated above that Shugart and Seagle (1985) look at mosaics as the sum of island-like forest "patches" and consider, for the time being, that "contagion" amongst these patches is a complicating factor (Sect. 5.6.6). This "contagion", in terms of interaction among eco-units in a mosaic, will be the main element of explanation used in the present chapter.

Because the term "mosaic" has been used in quite a number of meanings (see Sect. 5.1.1; also Harris 1984), Oldeman (1983c, but for final definition see Sect. 6.1.6) coined the word "silvatic unit", which he defined as: "*. . . the minimum surface at any one site needed for all architectural growth phases of all locally possible eco-units.* In the original definition, the concept of "chrono-unit" was used, which is omitted here because it is not strictly needed. We now think that the epithet "silvatic" is clear enough to use the term "mosaic" instead of "unit" in a very clear sense. This has the advantage of using a comparison that everyone can understand immediately, i.e. that between an artistic mosaic and a vegetation mosaic. It also shows that the model given here belongs to a family of mosaic models, going back at least to Aubréville (1938).

Table 2.1 shows the place of a silvatic mosaic in the systems hierarchy used in this book. Its subsystems are eco-units. We know of no existing, coherent forest model in which eco-units are lumped together in compartments. Some tendency to do this is maybe implicit in categories such as "pioneer stands", "second growth", or "small gaps" which, however, are never systematically treated in groups as subsystems. Koop (1981a, see Figs. 5.1 and 5.7) for European forests or Torquebiau (1981, 1985, 1986) for Mexican and Indonesian ones, use eco-units directly without grouping them in larger compartments, even if they do not employ the term. The supersystem within which the silvatic mosaics are subsystems is an abiotic site mosaic (climate, soil, water: cf. Table 2.1). This is a non-living system, some remarks on which are united in Sect. 6.5. Site mosaics will emerge earlier in the present chapter, however. This is due to scale problems, linked to the minimum homogeneous site area needed to host one complete silvatic mosaic. This problem was mentioned here and there in Chapter 5 (e. g. Fig. 5.43 and others). Just as for the eco-unit, this leads directly to a consideration of the limits of silvatic mosaics.

6.1.2 Spatial Limits and Gradients. Succession

For a survey of the concept of limits, the reader is referred to Fig. 5.2 and Section 5.1.2. Clearcut natural limits, which can be directly observed, rest on the presence of short and steep gradients. When gradients become more gradual, limits are broadened into transitional zones. These can be made visible on abstract drawings or maps.

In the above definition of a silvatic mosaic, it has been characterised by its eco-unit composition. In this, it differs from the usual forest types, the typification of which rests on species composition. But both concepts share the use of abiotic patterns in establishing geographical limits between different silvatic mosaics or forest types. This is a prickly question, on which much has been written by vegetation scientists. The dilemma common to all is determined by two considerations: precision and gradual change.

One very precise study is that by Koop (1981a). In a semi-natural forest in North-Western Germany he found that a difference in depth of about 10 cm of a clay horizon determined the occurrence of different syntaxonomical forest types, accompanied by different silvatic mosaics. This author, also basing his statement upon a broad literature survey, wrote (Koop 1981a, p. 62, translated): "An important building block in model formulation is the smallest surface upon which all stages and phases, representative of forest development by forest self-regulation, are perpetually represented. This surface is the **minimal structural area.** The temporal aspect of the minimal structural area is difficult to evaluate.... The size of the minimal structural area depends on the degree of internal dynamics, i. e. on the climatic, hydrological and soil factors." (bold face by translator).

Mangenot (1955), in a much less precise way, linked syntaxonomical forest types (cf. Fig. 5.63) to soil maps of the Côte d'Ivoire, West Africa, that had been established by Leneuf. Here, only broad soil classes and climatic classifications were considered. Limits had the precision of any general soil map of a large country. His forest types ("associations") therefore certainly include different silvatic mosaics according to smaller site differences inside the mapped soil and climate regions. This may be one of the reasons why, as discussed before, syntaxonomical methods were pronounced to be unsuitable for the tropical rainforest regions (see Sect. 5.5.2).

Hommel (1987) repeated the same kind of study in Ujung Kulon, a nature reserve in West Java, Indonesia. His "land units", established by the most recent and precise research techniques taking into account geomorphology, hydrology, soils, altitude, human impact and climate, correspond to syntaxonomical vegetation types with neat limits and precise characteristics. He did not consider the silvatic mosaic and its build-up by eco-units. Ellenberg et al. (1988) investigated the forest island Pirang (Gambia, W.Africa) according to a similar approach but with a rapid diagnostic method. This forest also is species-poor (ca. 60 woody species) compared with mainland rainforest, but the animal diversity, which was particularly studied, could be provisionally accounted for. Finally, the precisely analysed silvatic mosaics published by Torquebiau (1986) are linked to a few general site properties only and do not include the fauna.

These examples all rest upon the thought that there are natural, recognisable site limits between distinct forest types or silvatic mosaics. Kessell (1979) approaches the situation from the other side. If there are limits, forest types can be classified. If, on the contrary, there are only continuously changing environmental gradients, the vegetation changes slowly with distance, without being really the same anywhere ("ecotones"). According to this author, there is no conflict between the two viewpoints, because in the complex web of environmental gradients there are always discontinuities, forming limits. Both classification and gradient analysis therefore are valid entries into forest mosaic research, which is part of landscape ecology. The main thing is to build an objective model (see Chap. 2) of the environmental gradients, to which all approaches can refer. In Kessel's view, this kind of model has to be mathematically built, and this determines all steps of research from the collection and analysis of preliminary data onwards (Kessel 1979, p. 12). This could lead to the suspicion that problems are sought that conform to the method, instead of looking for methods to solve specific problems. But Kessel's whole text contradicts this. It is always oriented to the simplest and cheapest way to help solving precise forest management problems with the desired degree of accuracy.

Figure 6.1 gives a systematic overview of silvatic mosaics on more or less "hospitable" sites. Site hospitality may depend on one or a set of environmental properties: steep slopes are less hospitable than gradual ones, too dry or too wet sites less so than intermediately humid ones, barren soils less so that fertile ones and so on. Bearing in mind what has been said on limiting factors in Section 5.6, even a fast reconnaissance of a site (cf. Ellenberg et al. 1988; Oldeman and Van der Meer in prep. 1990) may allow us in many cases to "place" it in the series of Fig. 6.1 and to assess what will probably be the eco-unit composition of the local silvatic mosaic. The figure is compatible with both the classification and the gradient approach. Along a gradient, many intermediary silvatic mosaic states will be found between each pair of states shown in the figure.

It is necessary to elaborately discuss the limits of silvatic units, because it has to be absolutely clear that such limits are much less easily perceived than eco-unit limits. This is due to the fact that at the site level we find the meeting point of two completely different processes: the biological growth processes that determine the silvatic mosaic, and the physico-chemical distribution processes of solar energy, air masses, precipitation and geological matter that result in a site mosaic. The essential point to be understood when considering the limits of silvatic mosaics therefore is that they are living systems within near-permanent, non-living limits, which may only be shifted if and when the living system modifies the physico-chemical system (cf. Fanta 1985, with extensive bibliography; Fanta 1986, pp. 5–103).

Such interactions between the vegetation and its abiotic environment have determined many scientific approaches to vegetational development up to this day, and coloured nearly all. In combination with the wide-spread use of the hierarchy at the right-hand side of Table 2.1, which uses the analysis of species populations and their biology as its criteria, one tenet of most ecosystem studies can be formulated as follows: "A population of organisms, upon its disappearance or extinction, leaves its abiotic environment changed, so that populations of other organisms invade it". This set of processes so clearly exists and is so common in vegetations, that its modelling has dominated ecology for over nearly a century, under the name of succession (e.g. Begon et al. 1986, p. 610 ff.). However, contrary to eco-units and silvatic mosaics, populations have neither spatial dimensions nor definite lifespans.

Succession theory has lead to the construction of powerful and abstract scientific models. However, the present chapter on silvatic mosaics is based on another type of ecosystems analysis, in which ecosystems are considered as living systems with spatial and temporal dimensions, according to the architectural column at the left-hand side of Table 2.1. The two approaches provide answers to different scientific questions in a complementary way.

They show the following basic differences in outlook. Succession, as analysed by means of population models, builds ecosystems of all sizes and lifespans out of populations, and therefore always falls back on a type of "if..., then" analysis. For example, if there is a forest fire, then populations x, y and z will be reduced by 90%. This analysis confronts population developments with each other and in interaction with stochastic, exogenous impacts acting as driving forces. Chance events are very important, and statistical analysis therefore is a preponderant method used in such vegetation models. The basic unit is a dimension-less population, which is regarded on an arbitrary pixel, being "representative" of the landscape, e.g. per centiare (m^2), per are ($100 \, m^2$) or per hectare, depending on the size of the organisms. This approach is well exemplified by the FORET model that was summarised above (Sect. 5.6.6), i.e. a model that rests on populations, uses a stochastic analysis, and considers circular pixels of 1/12 ha (for statistical approach see Shugart 1984, p. 162 ff.; Jensen 1987 or Berger and Berry 1988).

Succession, as studied by means of architectural analysis, maps the landscape in site classes ("land-units" of land evaluation science, e.g. see Hommel 1987, or Van Beusekom et al. 1987), maps each site as a mosaic of different eco-units, maps each eco-unit in the most convenient compartments, and arrives at species populations as the bearers of certain functions and structures in these compartments. Therefore it always falls back on a type of "when and where..., then" analysis. For example: when and where a forest mosaic is in a certain development state, then fuel accumulation is high and fire risks increase. Predictable processes in the living system condition the effects of stochastic risks, so that mapping and cybernetics come first, statistics come second. The special kind of succession that takes place in forests, with all its sub-processes such as eco-unit development, fragmentation or fusion of eco-units and exchanges between eco-units has been termed by Oldeman (1974a, 1983c) silvigenesis, literally the "form-taking of forest", in a bad combination of a Latin and a Greek word.

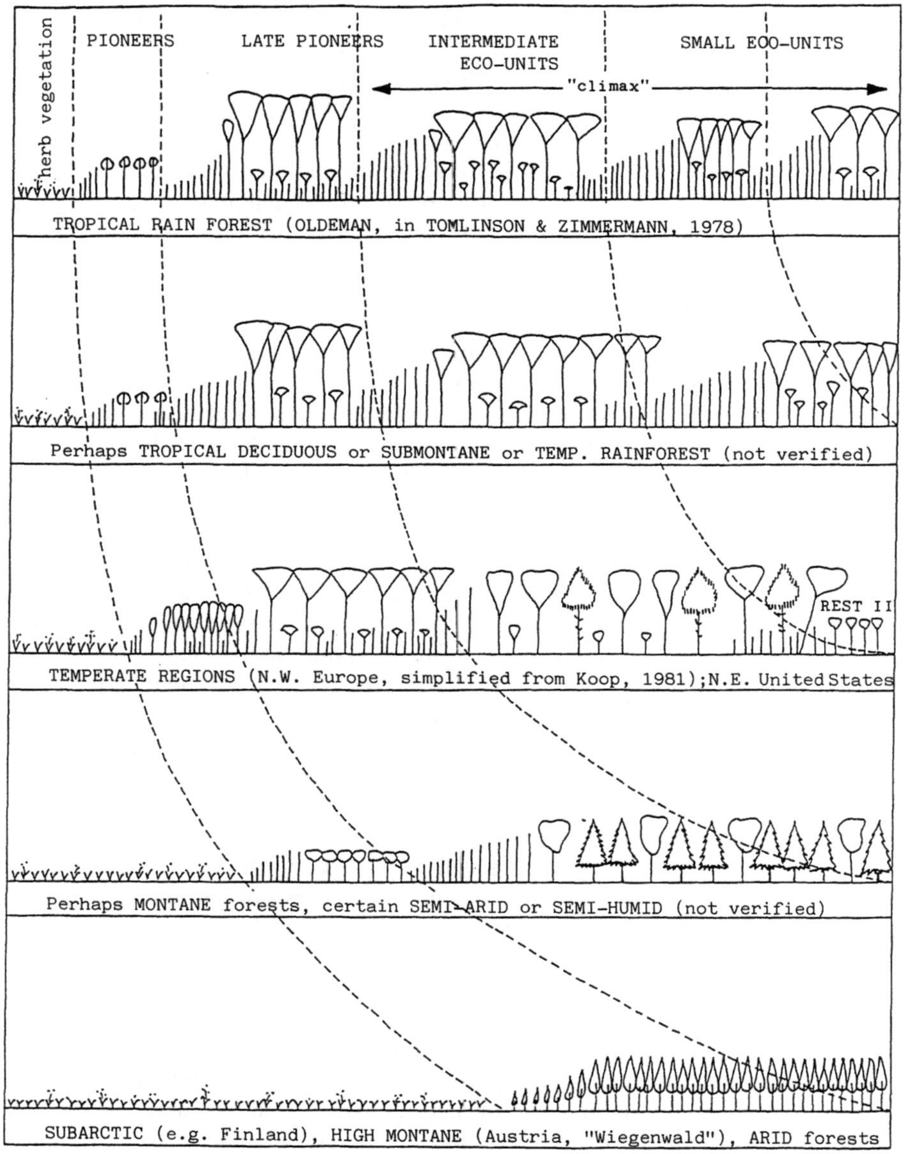

Fig. 6.1. Spatial distribution of eco-units in silvatic mosaics along a "site hospitality" gradient. Soils and geomorphology or inadequate land use may cause the site to "descend" along this scale. Mosaics reflect the result of lifespan versus risks of eco-units in each mosaic, also cf. shifting-events (Sect. 6.1.3). For wood production, pioneers, late pioneers and sometimes intermediate eco-units are most interesting, later eco-units often are too slow-growing. This figure is not an exact graph but the visualisation of a principle. (After Oldeman 1980)

It is emphasised that population research could profitably orient its choice, amongst the hundreds of thousands of populations that can be studied, by using architectural analysis to pinpoint key species, and that the other way round architectural analysis can only predict the development of ecosystems precisely if the behaviour and biology of key populations are known. Ecophysiological models would form the link relating the processing of energy and matter through populations to the functioning of ecosystems. Therefore it is quite urgent in these times, while widely different models blossom, that scientists with different approaches agree on a way of compartmentalisation of (forest) ecosystems that maximises the plus-value of each special study.

6.1.3 Temporal Limits and the Shifting Event

Figure 6.1 assumes that the proportion of each kind out of all different kinds of possible eco-units in all their developmental phases will be approximately constant for a given homogeneous site, occupying a large enough surface to include all these eco-units. But eco-units grow. If no outside factors were acting, that regularly create innovation phases (Fig. 5.4) of every kind of eco-unit of such a mosaic, it would lose the younger phases and the oldest ones would die. We have seen in the former chapter that the trees of large eco-units do not all die at the same time, and that natural tree mortality therefore leads to eco-unit fragmentation. This would happen in a "super-hospitable" environment. Mortality of isolated standing trees or small, standing, groups as the **only** regeneration factor in a silvatic mosaic leads to an eco-unit composition of very fine texture, the vast majority of all eco-units being one-tree eco-units (Fig. 6.2A).

This kind of "small-scale" mosaic is sometimes considered by extremists among nature conservationists to be the "ideal" virgin forest. This opinion is based on the misunderstanding that factors like fire, storms and other high-energy, large-surface events are "catastrophic", which is a notion borrowed from agriculturists or city dwellers. It is for this reason that such terms as "catastrophe" of "disturbance" were rejected (Sect. 5.1.3).

Most probably, this kind of silvatic mosaic which is only regenerated by gradual mortality of standing trees exists nowhere on Earth. Shifts from older to younger eco-units always are partially caused by outside impacts. Let us consider the other extreme, i.e. the regular but not too frequent occurrence of impacts like a fire every 30 years, a cyclone twice a century, or the well-documented case of tropical earthquakes (Garwood et al. 1979). Their interval should leave time for a fast-growing tree to reach its state "of the present" (Fig. 5.8), or at least its sexual maturity when it produces viable seeds. In that case, the silvatic mosaic will have the extreme configuration shown in Fig. 6.2C. This would be the "ideal" target forest for some extremists among loggers, who want seas of harvestable wood only.

The concept that we are approaching here would be termed zero-event if the eco-unit level were considered. At the level of the silvatic mosaic, it shifts the eco-unit composition towards a greater or lesser proportion of larger or smaller eco-units. It is emphasised that no static conditions, permanent except over millennia, such as geomorphological slopes are meant here, but dynamic **events** that occur in a short time and directly influence the eco-unit build-up.

Natural or artificial events that cause the eco-unit composition of a given silvatic mosaic to change as to its spatial and temporal pattern will be referred to as **"shifting events".**

Usually, silvatic mosaics undergo the impact of different kinds of shifting events, with different intensities, extents and periodicity. The higher the intensity and the wider the extent, the more an event is a shifting event and not merely a zero-event affecting one eco-unit. If a rough classification of shifting events were attempted according to their effects, one might distinguish, for instance:

- buckshot effects, hitting the mosaic at short and regular intervals in many small surfaces with high intensity, e.g. lightning, whirlwinds, gusts or selective cutting;
- sweeping effects, hitting the mosaic rarely but over its whole surface and with high intensity, e.g. hurricanes, fires, earthquakes or clearcutting;
- interactive effects, hitting the mosaic all over in a "soft" way and interacting with the weak spots on the site in causing eco-units to disappear, though not always immediately, e.g. waterlogging after heavy rain, pollution, or thin firewood harvesting;
- biotic effects may act in the same ways. For instance, tree-by-tree mortality causes buckshot effects, large pest outbreaks may be sweeping (Fig. 6.3) or epiphyte accumulation in large tree crowns may in the end cause trees to topple over because of their weight when saturated with rain water, an interactive effect.

6.1.4 Silvicultural and Ecological "Normalcy"

In most forest regions, buckshot effects occur frequently, sweeping effects occur rarely and interactive effects have little regularity if at all. A "normal" silvatic mosaic, subjected to a "normal" regime of shifting events, would look somewhat like Fig. 6.2B.

This "normalcy" is not to be confused with the concept of "normal forest" as used in forest management. A simple and maybe the most essential definition of this silvicultural notion is given by James (1982, p. 76): "Normal forest. A forest in which all age classes are appropriately represented". The key word in this definition of course is "appropriately", and in lignicultural plantations or many other forest management systems the meaning is quite simple, i.e. "appropriately to the desired wood harvesting frequency". In a plantation forest with a rotation of 48 years and a desired harvesting every 3 years, the normal forest contains stands of 0 y, 3 y, 6 y, 9 y,... 48-y-old in exactly the same quantities, e.g. 5 ha of each. Each year, the wood can be harvested over 5 ha in age class 48, which are then replanted and shift to age class zero.

Although this is an authentic mosaic composition determined by planted, contemporary eco-units, it is generally not perceived as being different from **tree** age classes. The usual treatment of data is so much inspired by population analysis that

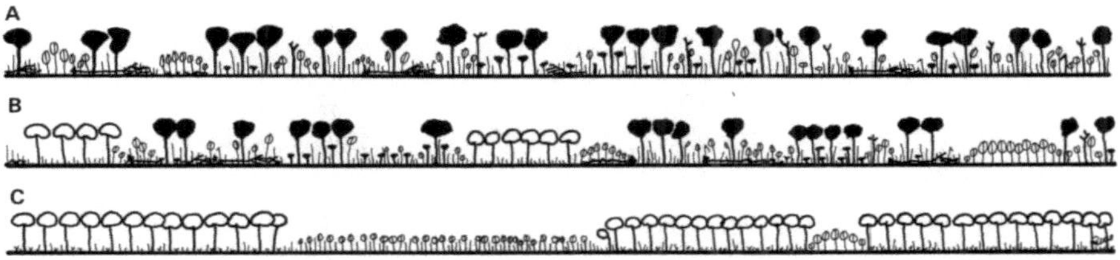

Fig. 6.2. Shifting events and forest mosaics. **A** Natural mortality of trees, and maybe some rare small event from outside (gust, thunderstorm) yield a mosaic of numerous small, one-tree eco-units of all ages, sometimes considered as "the" natural mosaic, but which is exceptional. **B** Shifting events of different nature and frequency, as is most usual in forests all over the world, yield a mosaic with numerous eco-units of varying sizes, architecture, age and composition. **C** Preponderance of sweeping, high-energy shifting events yield a mosaic of few, large-sized and fast-growing eco-units of different ages, as usual in extreme climates or after fires or volcanic eruptions. Note that Fig. 6.1 gives a temporal overview, whereas this figure displays a distribution of eco-units in space

a stand is perceived as a population (French: "stand" = "peuplement") and age as the mean age of the members of such a population.

This is still more evident in selective cutting systems (Plenterbetriebe, uitkapsystemen, jardinage), in which each management stand contains many one-tree eco-units (cf. Fig. 5.57 for an impression of such mosaics). Each stand is conceived, however, as a mixture of trees in different age classes. Here, the number of trees in each age class has to be such that every harvesting year (e.g. every 15 years) enough trees have arrived in the oldest class to make harvesting worthwile. Because there is high mortality in the youngest classes, and low mortality in the oldest, tree numbers in the youngest classes are highest. Taking the diameter as an approximate parameter for age, which may be far from the truth (see Sect. 3.6), a "normal diameter distribution curve" is calculated.

Figure 6.4 illustrates the comparison (from Chéret 1987) between the real and "normal" curves for a selective cutting system in the French Pyrenees from 1895 to 1962. It shows that in no year the management aims have been met. This was caused by a storm in 1910, a natural shifting event that counteracted the silvicultural measures. The latter are artificial shifting events pushing the mosaic towards the "normal" state as defined by the managers. Half a century later, this counter-action by one, rare, sweeping event still had not been overcome. Nowadays, pollution plays the role of a factor working against management aims by causing unwanted shifts in the mosaic (Chéret 1987, in other terms).

This example shows that calculated forest "normalcy" for management purposes need not at all correspond to the natural state of the silvatic mosaic to be managed. If silvicultural "normalcy" differs too much from silvigenetic "normalcy" at a given site class (land-unit), management aims are unrealistic. This is easy to say, but less easy to correct. The storm in 1910 in the above example could not have been predicted by the then existing climatological science, and such events with a periodicity counted in tens of years even now cannot be anticipated with climatological models (Wartena, pers. commun. 1984).

It is the reversal of the question that will bring us nearer to a solution.

Fig. 6.3. Example of a sweeping shifting event. Result of drought and forest fires in East Kalimantan, Indonesia, 3 or 4 years afterwards. The whole mosaic has died and, as reported after an earlier drought by a 19th-century traveller, "points to heaven with fleshless, white fingers". Note the installation phase underneath, already in full fragmentation (cf. Figs. 5.35-38, 6.6). Longnah, moist river plain forest near the Mahakkam River, December 1986, courtesy Inhutani II. (Photograph by R. A. A. Oldeman)

Instead of considering the effects of shifting events upon a silvatic mosaic, let us consider the properties of such mosaics that make them either vulnerable or resistant to such effects. This will lead both to new insights concerning the characteristics of silvatic mosaics and to the means to apply and use these characteristics in silviculture. This standpoint was formulated earlier by Brünig and his school (e. g. Brünig et al. 1982; Brünig and Schneider 1975; Brünig in Wiersum 1984). These researchers have chosen as their prime approach an ecophysiological model with structural aspects (cf. Figs. 5.38, 5.80). In the following pages, the prime approach will be architectural. Syntaxonomical and ecophysiological aspects will be added later, as in Chapter 5.

6.1.5 Silvatic Mosaics Are Living Systems but Not Organisms

In the same context as for eco-units (Sect. 5.1.5), the attributes of silvatic mosaics will be examined in order to test whether mosaics may be modelled as living systems or not. Attributes of living systems were:

- The presence of a filtering layer between the inside and the environment of the system;

- the presence of an internal communication system;
- the presence of an internal mass translocation system;
- the capability to adjust to changing exterior circumstances;
- the capability to reproduce.

An eco-unit can be convincingly analysed as if it were such a living, non-random system. The issue is less clear in the case of silvatic mosaics, because so many of their properties are determined by the direct contact with the abiotic environment both above and below the ground. If a silvatic mosaic is a living system, its consideration as such should in any case be marked by its being one of the longest-living on earth, with the least possibilities to adopt an autonomous form.

- **The filtering transition layer** between a silvatic unit and its environment is built by the interacting eco-units that form it. As a superficial analogy, it may be compared to a tissue built by interacting cells. Its horizontal filtering layers are shaped by the phyllosphere, the green upper layer of its eco-units, and the rhizosphere below. To understand that these are indeed two mosaic layers, and not the mere sum of eco-unit compartments, it is sufficient to consider the fact that no eco-unit phyllosphere or rhizosphere can be formed without interaction with the surrounding eco-units. This interaction is a silvatic mosaic property, not a property belonging to any one eco-unit.

The vertical borders of a silvatic mosaic most often are not neat, because the site limits that define them are transitional border zones (see Sect. 6.1.2). Around a silvatic mosaic, there is often a zone in which the eco-unit composition changes more or less gradually towards another mosaic composition. It can best be made visible by careful mapping, e.g. from detailed aerial photographs of from the ground (cf. Koop 1989). Figure 6.5 visualises some aspects of mosaic borders. It should not be forgotten that land consisting of a mixture of small-sized site areas will be covered by an interwoven complex of bits and pieces of different mosaics, sometimes lonely eco-units. In such a case, the level of the silvatic mosaic is so underdeveloped that analysis should explain the site system in terms of eco-unit subsystems directly, dropping the mosaic level altogether (cf. the Puerto-Rican El Verde forest, Fig. 5.77 and text).

- **The internal system of communications** in silvatic mosaics, like in forest eco-units, passes mainly through chains of organisms. Contrary to eco-units, in mosaics these organisms are mainly animal. In the preceding chapter, we saw that the faunistic composition of an eco-unit changes with its development, and that most animal species which are not microscopic have territories exceeding the size of one eco-unit. In very simple mosaics, such as European beech (*Fagus sylvatica*) or pine (*Pinus* sp.pl.) forests, communication is partly left to chance, e.g. anemophily (wind pollination), but seed dispersal as a means of transfer of genetic messages often is assumed by frugivorous birds. In the tropical rainforest, animals are involved in pollination, seed dispersal (cf. Van Dorp in Gomez-Pompa and Del Amo 1985) and "germination messages".

An example of the last effect is the germination of seeds in the particular environment of the excrements of herbivores or frugivores, transported from one eco-

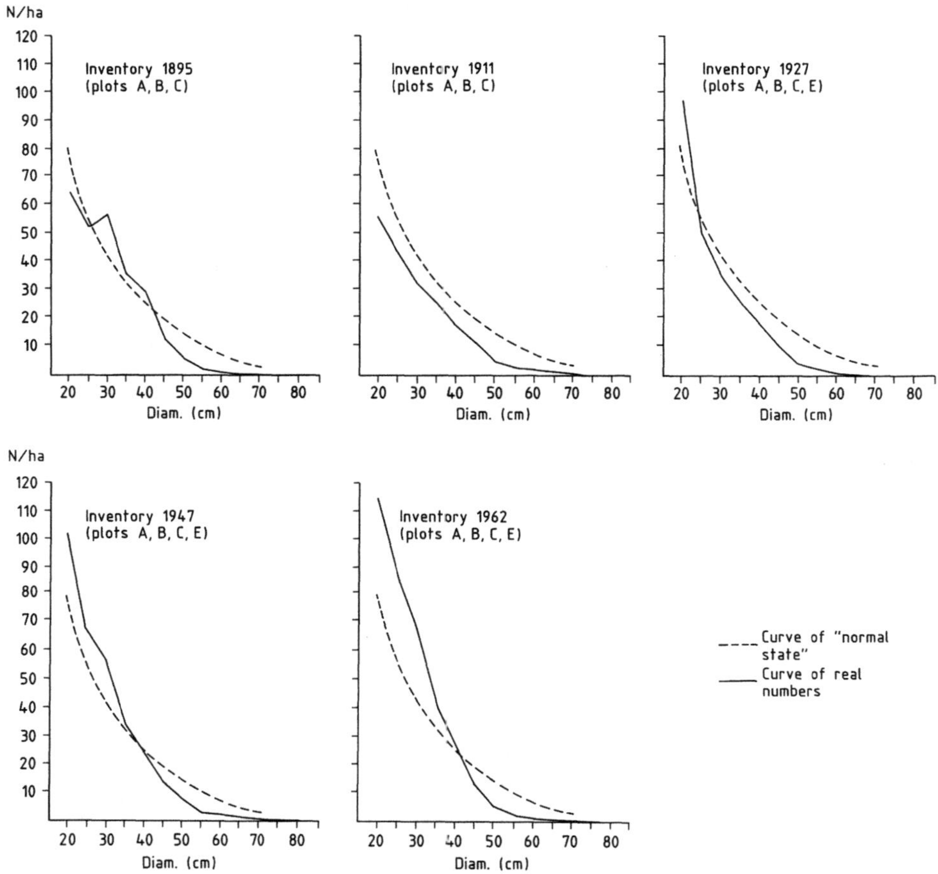

Fig. 6.4. Normalcy. Change in stem number (*N*) per ha for each category of diameter (*DIAM*) over the period 1895 to 1962 in a selection cutting system in the Communal Forest of Cazeaux-de-Larboust (Pyrenees France). If diameter is taken for an approximate parameter of age, it may be seen how a storm in 1910, combined with insufficient wood harvesting, influenced the composition of the intended fine-grained mosaic in this *Abies alba* forest until 1962. The "normal state" indicates the diameter distribution in the planned management mosaic, allowing regular harvesting of the thickest trees. The conflict between the natural and the management mosaic is clearly visible in this case, as is the influence of rare, heavy shifting events like storms. (Chéret 1987)

unit to another. Again, this is not a property of any particular organism or eco-unit, but a mosaic property, i. e. the interaction between secondary producers and seeds in transport and germination. A beautiful example is that of the interaction between fungus-eating pigs and spore dispersal, linked again to either favouring or hindering certain green plant species elsewhere, in a *Quercus ilex* forest near Béziers in the French Mediterranean region (Genard et al. 1986, with important bibliography; see also Sect. 5.3.4.4). The role of **impellors** as chemical messengers is also important in the silvatic mosaic (e. g. pheromones) and physical messages, such as light-and-shadow patterns or bird songs (Chappuis 1971, 1974–1985) play an important part in transmitting signals between neighbouring eco-units in the mosaic.

– **The internal systems of mass translocation** are the easiest to show. They depend upon food chains by moving organisms, i. e. animals along marked trails, following messages ("trap-lining"), or in a more erratic searching pattern. Mass translocation also depends on litter movement between adjacent eco-units, with its inherent transport of minerals and organic matter. The transport of propagules (eggs, seeds, spores) should not be underrated in terms of mass. Hladik (pers. commun. 1981) found monkey-less forests near Libreville (Gabon) with carpets of hundreds of kilograms of fruits per hectare on the ground, whereas this whole quantity had been translocated by monkeys in comparable rainforests further inland. The yields of forest fruits, such as blueberries (*Vaccinium myrtillus*, Ericaceae) or acorns (*Quercus* sp.pl.) in North-Western European forest, which are eaten by humans or cattle but in natural circumstances by wild animals, also amount to hundreds of kilograms of biomass to be translocated per hectare per year.

– **The adjustment to environmental change** is so perfect in silvatic mosaics that it seduced many researchers to explain forests in terms of purely environmental, random impacts. Still, in natural forests, the situation oscillates around a predictable eco-unit composition of the mosaic, as calculated by Torquebiau (1981, Fig. 6.49) for the Mexican rainforest in Los Tuxtlas (Veracruz). Depending on the shifting event, resources are more or less allocated to certain kinds of large or small eco-units, and all organisms are available that have to participate in building them. Figure 6.2 shows some of the states around which a silvatic mosaic may oscillate, and Fig. 6.4 in fact shows the difference between an unadjusted mosaic (silvicultural "normal" state) and an ecologically adjusted mosaic (real state). In population terms, adjustment is implemented by opening niches for certain species, meaning that **the right kind of** resources are made available to them. In architectural terms, certain kinds of eco-units are innovated that contain such niches.

– **Self-reproduction of silvatic mosaics** is highly hypothetical. Experiments to reconstruct forest types on comparable sites outside their original continent, like in the famous Arboretum of Tervuren near Brussels (Belgium), never have the complete set of original organisms available. The result, if large enough to contain a complete silvatic mosaic, which is not the case in Tervuren, is a completely new kind of mosaic in which components of the two continents are mixed, e. g. European mycorrhizal fungi on trees from the Rocky Mountains in the Western USA, the seeds of which are dispersed by European squirrels and birds, building their nests in American shrubs.

Silvatic mosaics in formerly deforested Western European regions (e. g. Great Britain, the Netherlands) often show this character. They display a mixture of components from all over the world, a superficial resemblance to the prehistorical local forests because of the native tree species (but often not genotypes), and nearly always a completely artificial eco-unit composition (cf. Fig. 6.31). Nowhere in the world do forests remain that are completely untouched by humans. And humans always dis-

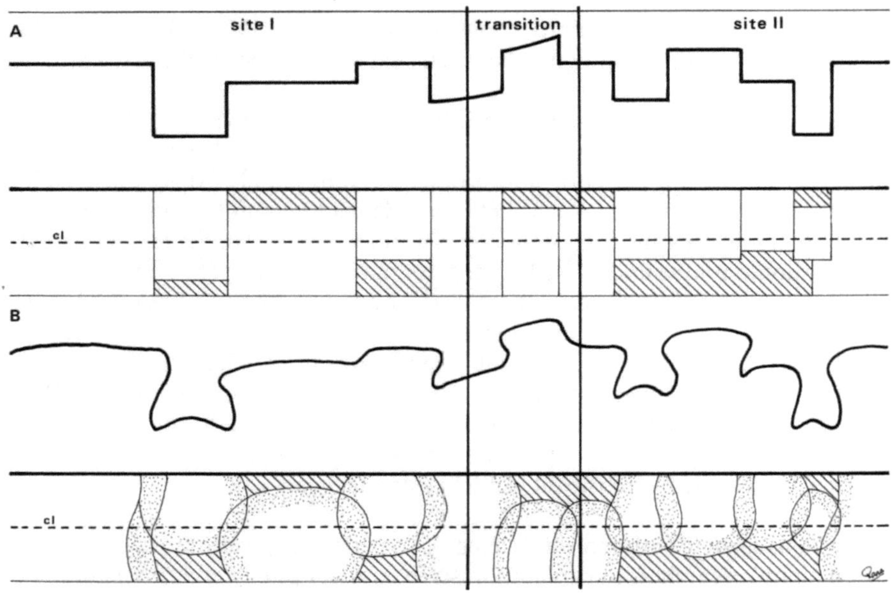

Fig. 6.5. Sum of eco-units versus silvatic mosaic. **A** Non-interacting patches, on two sites with transition, pixel-like representation. **B** Silvatic mosaic with eco-units, natural borders, interaction zones (*dotted*). *Hatched* not represented on vertical profile contours; *cl* central line of the imagined transect, excluding the hatched forest parts

perse species with them to be used for food, decoration, medicine, building, painting, weaving, weapons, and other utilisations. Therefore, hypotheses on the self-reproduction of silvatic mosaics could nowhere be put to the test for at least a thousand years. For the Mexican, Central American and Pacific South-American rainforests, see Jacobs (1988). These forests are profoundly marked by their pre-Columban utilisation and conversion by neotropical civilisations that are extinct now.

Still, on all-but primaeval sites in faraway forest regions the following hypothesis might apply. We have seen (Sect. 5.1.5) that reproduction rests on the presence of a collection of relevant genetical information in the propagule bank, in which the vast majority of information-bearing propagules die and in which the remaining information is activated by a key event. Now, if a land-unit of site class A were completely denuded by fire or bulldozers, and if other silvatic mosaics on site class A had survived not too far away from the denuded land, the following could happen. Animals and wind would start to build a new propagule bank, in which the pioneer and innovation species adapted to A-class sites would have the advantage. A large sized pioneer eco-unit would grow up, probably with early fragmentation occurring because of random distribution of tree seeds and occurrence of micro-sites (cf. Sect. 6.2.2.4). The fragmented large eco-unit would preferably attract the animals from other class-A site silvatic mosaics because the plants of these mosaics would be most common. This would narrow down the number of species arriving in the propagule bank, because dispersers from site class-B forests would more and more shy away from the new class-A mosaic. In the end, the composition of the propagule bank there would converge with that in the class-A forests.

The hypothesis can be developed further into a comparison between eco-unit reproduction and silvatic mosaic reproduction (Table 6.1).

This table suggests an explanation of the often-claimed redundancy within guilds containing many different member organisms ensuring the "same" forest functions. It may well be that what seems redundant at the eco-unit level is less so at the silvatic mosaic level. This again has to remain hypothetical for the moment. However, some of the case studies cited below contain facts that would fit in with such an explanation.

6.1.6 Behaviour of Silvatic Mosaics

In comparison with other viewpoints, the silvatic mosaic is situated approximately at the level of "forest types" in vegetational studies (Sect. 6.1.2) and "forest districts" in forest management planning (Sect. 6.1.4). If silvicultural design ("Waldbauliche Planung" in German) often concerns large eco-units, such as large timber-production stands, in more refined systems such as selection silviculture (Plenterbetrieb, jardinage) they include the complexity of silvatic mosaics. A clear circumscription of a silvatic mosaic hence is useful, so that no doubt can exist as to what kind of complexity is addressed.

We consider to be a silvatic mosaic: every forest surface that is situated in one continuous volume with the same resource regime of climate and soil, which shows the same complex of silvigenetic dynamics, resulting in an eco-unit composition that oscillates around a specific state and determining the architecture and ecological functioning as long as the resource regime remains unchanged.

It is to be noted that the time dimension is defined according to site change. This leaves room for the well-established interactions between vegetation and site from classical succession theory. Moreover, no "moment of birth" of a silvatic mosaic is indicated. This would only be possible if reafforestation occurred, and would therefore define artificial silvatic mosaics and a few known natural ones only. In most cases of existing natural silvatic mosaics it is completely impossible to define a moment of origin, even approximately. One therefore has to be content with an "eternal forest" image, in which not the beginning but only the end of the system can be considered, arriving when site conditions change beyond previously recognised limits.

Taking into account the geological history of forests (Flenley 1979, for the tropics), it is clear that theses like "the tropical rainforest has existed for millions of years" are only to be trusted as very general statements. As a matter of fact, sites have changed so often and so thoroughly, even during the last 50,000 years, that different "rainforest" mosaics must have succeeded each other. These did not necessarily resemble each other, except sometimes in the main lines of their organisation. There certainly were site changes that indeed exterminated some mosaic, after which a new one was born. But the data are so few and far away in time that no precise analysis is possible.

Notwithstanding this uncertainty in natural forest mosaics, there are so many instances of contemporary sweeping effects that silvigenesis will be described in the following pages from the naked land on, as if a new mosaic were born. The usual, existing mosaics which may remain near their specific near-constantly oscillating state during centuries to come, will be found as one silvigenetic (successional) state which is represented by the "ecuilibrium stage" (see Sect. 6.2.4, also for letter c) in the overview of silvatic mosaic development.

6.2 The Life Cycle of Silvatic Mosaics: Architecture

6.2.1 Definitions of Eco-Units as Subsystems

The compartments which indicated the developmental phase in an eco-unit were sets of trees, i.e. the set of the present, the potential set and the set of the past (Fig. 5.4). In silvatic mosaics, the subsystems are whole eco-units. They will serve us to establish a diagnosis of a silvatic mosaic in the same way. Eco-units may be (Oldeman 1983c):

- appearing if they are represented mainly by innovation or aggradation phases;
- dominating if they are mainly represented by biostatic phases;
- surrendering if they are mainly represented by degrading phases.

As to size and architecture, eco-units are roughly classified as follows (cf. Sect. 5.4):

- pioneer eco-units, if large-sized, fast-growing and tending to a component spectrum maximising the share of **r**-strategists (cf. Table 5.5);
- late pioneer eco-units, if intermediate-sized, rather fast-growing and tending to have more **K**-strategists among their components;
- successor eco-units, if small-sized, slow-growing and tending to maximise the **K**-strategists among their components.

All three classes of eco-units, which can be subdivided again if one wants to do justice to the enormous array of different cases in each class, should in any case be distinguished according to their longevity:

- short-lived eco-units live some decades, and sometimes are said to be composed of "weed-trees";
- average-lived eco-units live some centuries, and may correspond to what is most often found in natural forests in which the average lifespans of trees are usually between one and two centuries (cf. Bech 1983);
- long-lived eco-units may live from two centuries to over a millennium (cf. Mulder and De Waart 1984).

The connection that is usually made between "pioneer" and "short-lived", "successor" and "long-lived" (cf. Whitmore in Hadley 1988) is here abandoned. There are well-known instances of very long-lived pioneer eco-units, even if they survive in fragments, such as teak stands in Asia (*Tectona grandis*, Verbenaceae) or oak forests

Table 6.1. Comparison between eco-unit reproduction and hypothetical silvatic mosaic reproduction

Stand size (ha)	Test 1: Can an individual tree grow to its maximum size?	Test 2: Does gap-phase replacement occur?
0.01	No	Yes
0.02	No	Yes
0.04	Yes	Yes
0.08	Yes	Yes
0.20	Yes	Rarely
0.40	Yes	No

in Europe (*Quercus* sp. pl., Fagaceae), and several tropical tree species are suspected to play a role in such long-lived pioneer eco-units (Hallé et al. 1978, p. 383 ff.).

Contrary to the four phases distinguished in eco-unit development, five stages will be distinguished in silvatic mosaic development. Such mosaics are so long-lived that only short intervals of their development usually can be observed during one human lifetime. The image shown here is reconstructed, like all other images that have emerged from different ecological theories at this level, concerning "patch mosaics" (e.g. Shugart and Seagle 1985), "successional dynamics" (e.g. Fanta 1982) or "natural potential vegetation" (e.g. Jovanovic et al. 1986).

The five silvigenetic stages of mosaic development are an installation stage, a precuilibrium stage, an ecuilibrium stage (ec refers to eco-unit), an elimination stage and a stage of collapse. They are shown schematically in Fig. 6.6.

6.2.2 The Installation Stage

All silvatic mosaics that start developing on a large, empty site of at least several tens of hectares begin with an installation stage (Fig. 6.6A). This may happen after a particularly heavy, sweeping shifting event such as an earthquake, the volcanic eruption of Mount St. Helens (see Sect. 5.2.5.3), the forest fires in East Kalimantan (Indonesia) in 1982/1983, in West Africa in the same years or in the Mediterranean region every year. In most cases that we observed, a propagule bank must have been present and active rapidly, except on some heavily eroded and hardened tropical red soils where some sparse grasses were the only visible organisms during several years (Rorota, Mt. Mahury, French Guyana, 1966, pers. obs.), or on sterile, shifting wind-blown sand dunes (Tunesia, Sahara border; Veluwe, the Netherlands) where nothing grew prior to their fixation.

During the early part of the installation stage of the mosaic, all eco-units are in the innovation phase, i.e. they form an appearing set of eco-units, which may be easily mistaken for one large eco-unit in its innovation phase. A silvatic mosaic in its installation stage contains only appearing eco-units, which is a criterion to recognise it in the field (cf. Fig. 6.3).

The horizontal architecture (texture) of the silvatic mosaic can be distinguished early by those who know what to look for, and more easily when the appearing eco-units enter into their aggrading phases. As a matter of fact, no installation stage that we observed ever could be interpreted as one, very large innovating or aggrading eco-unit, with the exception of land that had been artificially replanted with trees after site homogeneisation (Fig. 6.7). Fragmentation normally starts very early in the mosaic's installation stage. Like all mosaic phenomena, this too is due to the confrontation between an already more or less heterogeneous site and the inputs from the surrounding mosaics. After the studies in West Africa reported on by Alexandre et al. (1978), a beautiful, experimental case study from French Guyana is to be found in De Foresta et al. (1984). A recent study on tropical forest reconstitution in Gabon, after shifting cultivation, is due to Mitja and Hladik (1989). They showed fragmentation to occur, site properties determining dynamics, and the role of temperament in "demographic groups" of plant species.

Hurricanes were not mentioned among the shifting events that sweep the mosaic back to a new installation stage. Their effects are to be considered in a later phase, because the biomass left behind in the field causes such heterogeneousness that it does not lead to the appearance of pioneer eco-units, the more so because the propagule bank is left intact (cf. Koop 1981b). But even fires and topsoil removal (e.g. earthquakes, bulldozers) still leave a microrelief, or unequally distributed organic matter or ash, which form microsites. Only tillage may erase most of the site heterogeneousness (Fig. 6.7).

Still, structural micro-heterogeneousness usually persists in a more or less attenuated form, e.g. the outcrop of rocks, the presence of filled or open brook beds of 1 or 2 m wide (erosion gullies), other forms of small slope gradients, and traces of former land use. This heterogeneousness is confronted by the build-up of a new propagule bank in a partly structured, partly stochastic way. The experiments by Szujecki (1983, see Table 5.2) have demonstrated at least one case in which tillage could not erase all traces of the former land use pattern. In stands planted on ploughed former farmland, the development of the entomofauna was more different from a typical pine forest entomofauna than in stands planted after pine harvesting and soil tillage.

6.2.2.1 Biotic Remanence

Biotic remanence in the propagule bank is the first structural factor determining the character of the installation stage. There is a degree of remanence, from 100% when the propagule bank remains intact, which does not concern the installation stage, to a small fraction that "colours" the installation stage of a new silvatic mosaic. The fact that the seed bank is the part of the propagule bank that has received most attention up to this day certainly has obscured remanence effects due to the microfauna and microflora.

This may explain why unexpected pests and diseases befall extended monocultures in the tropics that are planted on cleared and burned rain forest lands. Not only is this due to the local biotic remanence, but also to selective remanence. Selection is due to the impacts on the land that is cleared. Selection pressure by topsoil

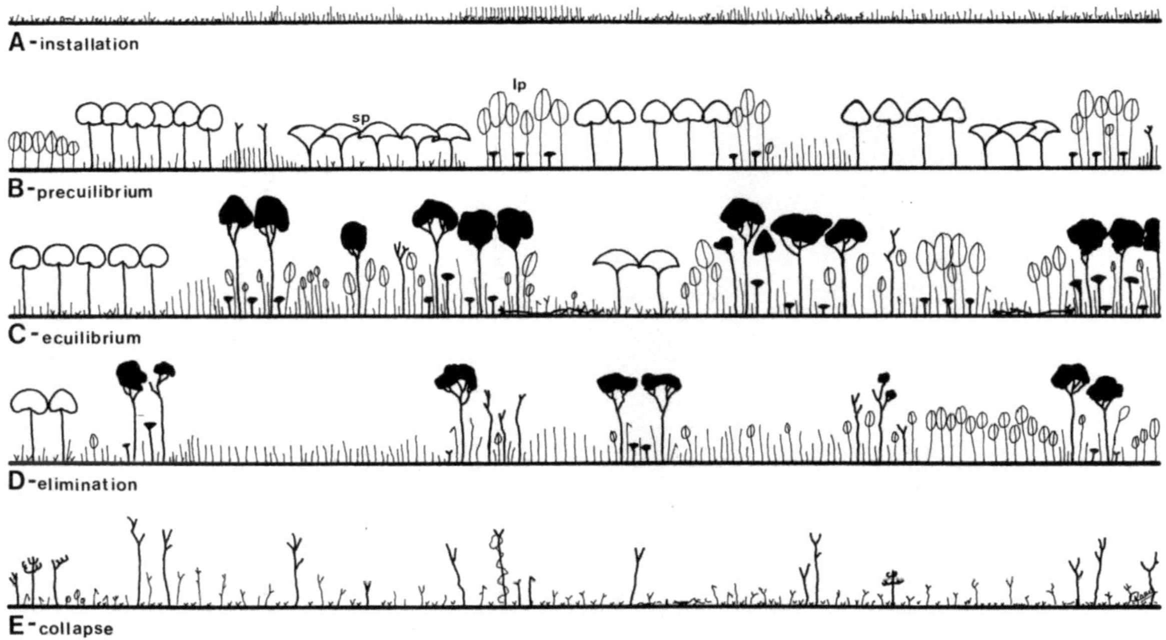

Fig. 6.6. Development (succession) of a silvatic mosaic. *Installation* (**A**) on completely cleared land. *Precuilibrium* (**B**) with dominating pioneer eco-units. *Ecuilibrium* (**C**) with mosaic in balance with natural shifting events. *Elimination* (**D**) with many eco-units swept away by rare shifting event or beginning stress. *Collapse* (**E**) under permanent stress. The *ec* in ecuilibrium refers to the first letters of eco-unit

Fig. 6.7. Artificial installation phase, homogenised by ploughing (*foreground*), *Pinus sylvestris* plantation, large pioneer eco-unit (*background*). These forests were studied by Szujecki (1983). Poland, Stara Brda, April 1987. (Photograph by R. A. A. Oldeman)

removal favours another set of propagules to survive than tillage or burning. But in all cases, the surviving organisms are innovation species (Sect. 5.2.2), which are hardy, short-lived generalists. There may also be a "deeper" remanence of organisms that belong to later development phases of eco-units or to later eco-unit types, and which remain inactive in the propagule bank until their time comes in the form of an appropriate key-event.

6.2.2.2 Biotic Reach of Propagules

Biotic reach of propagules is the second structural factor determining the character of the installation stage. If the dispersal agent is abiotic, e.g. wind or water, the weight and architecture of the propagules determine this reach. Massive, heavy fruits or seeds, containing little or no matter that is edible for animals, cannot travel far. Wings or air spaces make their reach longer, and very slight size and weight, such as in fungal spores or *Eucalyptus* seeds, allow for a reach so extended that these propagules can be said to be omnipresent (e.g. see Van der Pijl 1969).

If the dispersal agent is animal, the biotic reach of the propagules coincides with the distances that can be bridged by the animal species concerned. Alexandre et al. (1978) emphasise that woody pioneer plants in West Africa are mostly dispersed by birds, i. e. they travel far. Although it does not concern seed dispersers directly, Fig. 6.8A illustrates the vast differences in biotic reach amongst animals. This figure concerns the predator fauna that finds its refuge in hedgerows in North-Western Europe, and that was studied for species conservation purposes in a communal landscape plan (Van den Berg and Saris 1985). Such hedges correspond to forest borders surrounding cleared lands, which are considered here. The biotic reach of predators reflects that of their prey animals, which often are propagule dispersers.

The dispersal agent can also be interactive, like dispersal by wind or water of litter fragments or other organic remains containing arthropode eggs, larvae or spores. The organisation of such ecosystem fragments goes straight from existing eco-units to other silvatic mosaics. Therefore it does not belong to any one species that yielded the litter, or the eggs, or the spores, but to an organised structure that is a fragment of an eco-unit, and is transferred by mechanisms working at mosaic level.

6.2.2.3 Abiotic Impacts and Neo-Mosaics

Abiotic impacts are the main stochastic factors determining the character of the installation phase. Their importance increases with the size of the natural or artificial land clearing, and with the distance to comparable or other silvatic mosaics. Figure 5.45A shows an example of a forest on new polder grounds in the Netherlands, out of reach of nearly all biotic dispersal vectors. This forest probably was the result of a random hit, because willow fruiting, the right wind and the right water level at the site all three occurred at the same moment. All forest organisms within biotic reach at present are exotic plant species in the newly built town of Almere and the birds that live in gardens, parks and urban forests (De Coo pers. commun. 1979). The omnipresence of fungal spores was attested by the presence of a Canadian fungus species in planted forests in the same polder region.

One general conclusion would be that in such cases wholly new types of silvatic mosaics will develop with time, something which cannot be observed as yet in the installation phase. The formation of such neo-mosaics does not seem to be the general case, however. It more or less corresponds to primary succession as considered in vegetation science if completely virgin soils (emerging atolls, new lava layers, polder soils) become covered with vegetation. Secondary succession is the concept for forest regrowth on land from which a former vegetation has been cleared. The difference between the two, in our terms, is biotic remanence which is zero in primary succession.

The architecture of an installation stage is determined by the interaction between the pattern of the propagule bank and the pattern of microsites within the general site. This architecture (Fig. 6.6A) has the following characteristics.

6.2.2.4 *Fragmentation*

There always is some degree of fragmentation. The descriptions by Boerboom (1974) of the development of forest on large, cleared sandy terrains in Suriname give a clear picture of strong, early fragmentation. In plantation forestry, almost all site preparation and tending measures aim at counteracting it, especially its effects on the tree compartment. Such measures include tillage, fertilisation, seeding or planting a monospecific crop and thinning individual trees that stray too far from the average production norm, being either too small or too large (for many aspects see Wiersum 1984). Early fragmentation phenomena in the installation stage of a mosaic have been illustrated at eco-unit level in Figs. 5.35 to 5.37. It is emphasised once more that eco-unit fragmentation does not correspond to Harris' fragmented forest (1984). He considers landscapes being fragmented in "biological islands" being in a different state.

6.2.2.5 *Eco-Unit Composition*

All eco-units are in the innovation or early aggradation phase, i. e. they form an appearing set of eco-units. This means that the installation stage is rich in flora and fauna components that belong to innovation phases, particularly herbivores and phytophagous insects, self-supporting and lianescent herbs, generalist mycorrhizal fungi, and special fern species. If the installation stage is considered to be the equivalent of the Ivorian "stage with dominance of herbaceous plants . . . and shrubs . . . in forest reconstitution after shifting cultivation" (Alexandre et al. 1978), these

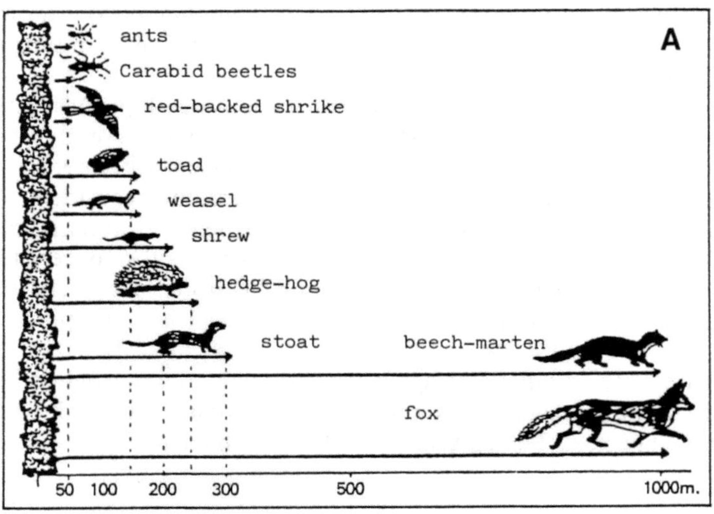

Fig. 6.8A

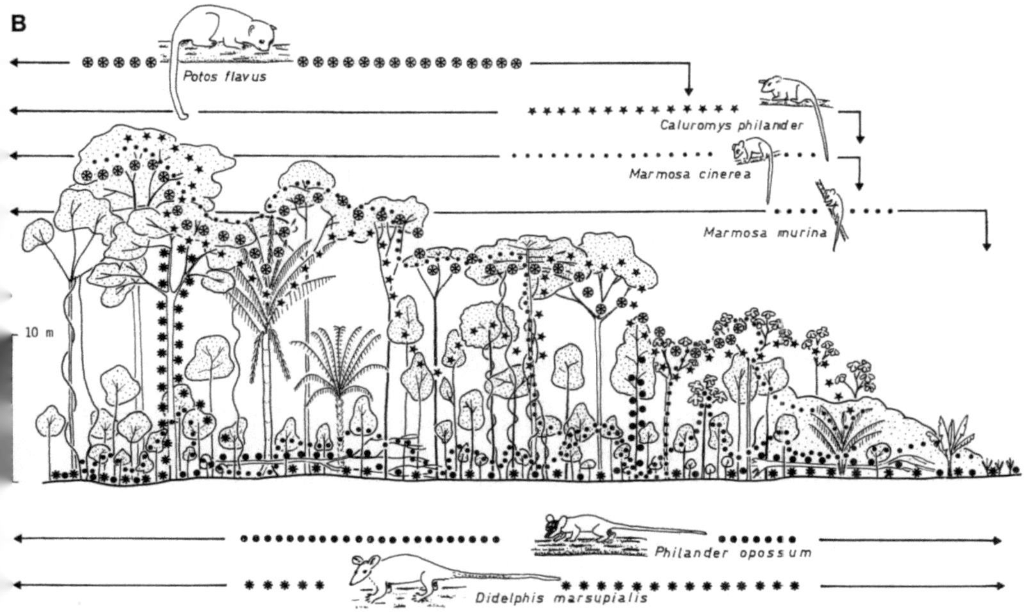

Fig. 6.8. Animals and silvatic mosaics. **A** Biotic reach. Distances reached by animals from a N.W. European hedgerow, equivalent to a forest border. Predators are the "shadow" of herbivores and fructivores dispersing seeds into mosaic installation stages. (Inspired by Müller 1981 ex Van den Berg and Saris 1985). **B** Schematic reconstruction of the itineraries of five marsupials and the kinkajou in a young, precuilibrium mosaic. Eco-units older *from right to left. Arrows* indicate limits of species. French Guyana. (Charles-Dominique et al. 1981)

principles of eco-unit composition seem to be implicitly confirmed. The local inhabitants of the Oubi tribe (SW Côte d'Ivoire) have their own name for it ("piri").

All eco-units at first are pioneer eco-units of some sort. If they are not large-sized in the absolute sense, they at least are so in the relative sense comparing their surface to the height of the surrounding mosaic (cf. Fig. 5.47). Their tree compartment, at least during the installation stage of the mosaic, is ordained by fast-growing tree species. Depending on the speed and pattern of the propagule bank build-up, seedlings of later mosaic development stages may be present below the pioneers in an early phase. This does not seem to be so for animal components, which prefer certain eco-units to others and will not visit mosaics that do not include their preferred biotope (Fig. 6.8B). In terms of the silvatic mosaic this means that the communication system and the translocation system amongst eco-units are different per development stage (Sect. 6.1.5).

The proportion between short-lived and long-lived eco-units in the installation stage may be very different from one mosaic to another. This is a determinant of mosaic behaviour during centuries to come. Still, it is seldom recognised. For instance, the pioneer mosaic that was taken to be "the" climax vegetation in North-

Western Europe during a long period of scientific research, the Querco-Betuletum (see Van Dort et al. 1979), can be interpreted to be a mixture of rather short-lived pioneer eco-units built by birch (*Betula verrucosa, B. pendula*; Fig. 6.29) and long-lived pioneer eco-units built by oaks (*Quercus robur, Quercus petraea*). During the installation phase, this mosaic may have the appearance of a tree-by-tree mixture not unlike the ones depicted in Figs. 5.35 to 5.38. But the further behaviour of the silvatic mosaic is strongly determined by the early death of the birch eco-units and the persistence of oak eco-units, showing a late fragmentation pattern in the pre-cuilibrium stage.

6.2.2.6 Apparent Monotony of the Installation Stage

The monotonous image of the installation stage hides its potential diversity. This should be clearly stated, because "secondary vegetation" (capoeira, belukar, brousse, in the tropics) is often represented as a poor, monotonous forest type having lost all the rich attributes of such forests as grew on the same site earlier. This image can only be the result of too casual observation. Such casualness probably is based on the wrong impression of "déjà vu" in artificial, homogenised, fast-growing timber plantations. This prejudice is widely spread and taught, maybe because most of the teachers and writers originate from countries where they have been exposed to the view of timber plantations from their childhood on, adopting it as an unconscious standard.

The vocabulary of "disturbance", "catastrophe" and "damage" is tainted by this preconception. Hence it was rejected in Section 5.1.3. As a matter of fact it is astonishing that the fragmented nature of secondary forest has not been seen earlier. Even as late as 1978, Alexandre et al. state (p. 267): "If the **monospecific or near-monospecific nature of woody secondary stands** appear to Aubréville (1947) as the fruit of chance, it still is incontestable that certain **environmental conditions** are of first importance, like the character of the soil, its impoverishment by succeeding cultivations, or the original forest type. And although the component species have extended, sometimes showing pantropical areas, reaching their ecological optimum and their **tendency to gregariousness** will lead to the **dominance of one or the other species.**" (translation and bold face R.A.A.O). In this statement, the lack of an intermediate hierarchy between the abiotic environment and the species is to be noted, resulting in unexplained, descriptive items like a "tendency to gregariousness". Somewhat more advanced views can be found in recent books, such as Leigh et al. (1982) or Muséum National (1986), both for the tropics.

In European literature, the preponderant attitude is polarisation between "nature-conform" ecologists and foresters, rejecting large-scale phenomena as anti-natural or unnatural, and management-oriented forestry authors who accept and promote homogeneous, large-sized young forests. Balancing-out of these views is starting by some authors, such as Fanta (1982, 1986) and others. For summaries of good recent research data see also Andersson et al. (1988). However, it will still cost much work, time and teaching to modify the opinion that "secondary vegetation is bad (or excellent)" and to replace it by a more objective image. This should not mix up indiscriminately such notions as forest development and succession, human aims and preferences, or ecological wholeness and degradation.

Monotonous sound patterns both by day and by night, and a lack of striking colours in large-sized installation stages of silvatic mosaics contribute to the impression of homogeneity. The meaning of sounds and colours in the communication systems that make silvatic mosaics into integrated, living systems will be discussed later. Sound was passingly mentioned in Chapter 4. In a large-sized tropical installation stage, striking sounds are the screeching of numerous phytophagous insects belonging to hardy, **r**-selected species, e.g. from the class of Orthoptera (including grasshoppers, locusts, crickets). At night there is the humming of parasites on warm-blooded herbivores, their predators and humans, e.g. Diptera such as mosquitos and sandflies. Colour is mainly expressed by herbaceous plant species or somewhat later by shrubs and climbers, such as Melastomaceae or Rubiaceae species. But there is a clear pattern of sound and colour that strongly differs from that in later development stages, and which has been very imperfectly analysed.

6.2.2.7 Stability of Mosaics Versus Eco-Unit Fluidity

Contrary to the innovation phase of an eco-unit, which shows a fluidity of pattern, process and species composition that can be easily influenced (cf. Sect. 5.2.1), the installation stage of a silvatic mosaic cannot be influenced so easily. The cause of this resistance to change is the buffering interaction amongst the eco-units, which are the subsystems of the mosaic. Once the eco-unit composition is established, further fragmentation processes seem to be compensated by hardy, omnipresent innovation species migrating through the mosaic and adjusting the behaviour of all eco-units to the mould of the mosaic. Such species may be trees, but may as well be other organisms, particularly pathogens or parasites. In other words, it is the fluidity of eco-unit development itself that guarantees the inflexibility of their mosaic, because interaction between eco-units is open and erases the effects of small or somewhat larger diversifying forces. This is linked to the difference between shifting events and zero-events (Sect. 5.1.3; Sect. 6.1.3).

An example of quite wide-spread intervention has been analysed by Lette (1981) in Colombia. He studied the effects of firewood cutting by local inhabitants on large surfaces that had been clearcut earlier and were covered by a pioneer forest in the end of the installation stage. Only such large, sweeping impacts as generalised fuelwood harvesting can shift the balance of installation stages by introducing the same type of change in all eco-units present. Other examples are the giant campaigns of forest tree seeding from airplanes in China around the middle of the 20th century, and application of chemicals from the air. The principles of installation stage establishment in a dry forest area of Costa Rica, to be restored to its original state, have been described in other terms by Janzen (1986a).

6.2.3 *The Precuilibrium Stage*

After the installation stage, we distinguish a long precuilibrium stage. The term, just as "aggradation" (made by Bormann and Likens 1979), does not figure in dictionaries and hence is new. The letters **ec** in **p**r**ec**uilibruim refer to **ec**o-units which are not yet in dynamic equilibrium with each other and the site. Architecturally, the precuilibrium stage starts with the closure of the mosaic canopy.

6.2.3.1 *Canopy Closure: Layers, Stufung and Schichtung*

This closure of the canopy is a process unlike that in eco-units, where a horizontal canopy layer, however unstable it may still be, screens off the forest patch from its environment. This is a form of horizontal closure, layering, strata formation, storey building or whatever terms have been coined to describe it (Fig. 6.9A). In population models, the strata (also see synusiae or hoplexols, Sects. 5.5.6, 5.5.7) include all plants and divide them up in life forms (cf. Sect. 5.5.3), e.g. upper tree layer, lower tree layer, shrub layer, herb layer and moss layer. Figure 5.4C, like Fig. 6.9A, defines layers in biostatic phases only and uses the compartment of fully grown plants (set of the present) as its only criterion. This set is divided again in structural ensembles (cf. Oldeman 1974a; Hallé et al. 1978).

Canopy closure in silvatic mosaics is vertical (Fig. 6.9B) and occurs when neighbouring eco-units (not trees) of different heights completely cover the soil.

This kind of canopy closure has long been known in forestry. It is indicated in German as stufung (pronounce ['shtüfüŋ]), each stufe, litterally "step", including all upper tree crowns at a certain level. We will use this specific term in this sense here. Loose use of the word, seen sometimes to include crowns within the forest (e.g. the two shrubs of the present at the left of Fig. 6.9B), will not be followed, and other meanings or derived terms (e.g. "Stufigkeitsmass", Mayer 1980), not referring to forests, are not taken into account in the present text. To be complete, the German term for layering, schichtung (see Fig. 6.9A), is also mentioned.

As soon as the canopies, prepared by the eco-units in the installation stage, have closed the mosaic canopy by stufung, the precuilibrium stage is reached. As described in Section 5.2.3, the canopy of eco-units may be closed by plants other than trees, e.g. shrubs or lianas. However, these processes also contribute to mosaic canopy closure by stufung. Only the height of each stufe depends on the process. Examples of stufung can be found in Figs. 5.31 to 5.62 as far as they concern more than one eco-unit. Stufung is the consequence of fragmentation processes.

Because "stufe" also means a step in a process, it is closely linked to the concept of "normalcy" in managed forest stands (cf. Sect. 6.1.4). In this case, each younger step in time (stand age) is taken to represent a lower step in canopy architecture. In this double meaning, stufung is used both in pure stand clearcutting systems and in selective cutting silviculture, concerning tree groups or trees within a stand. In the latter case, one stand contains a fine-grained mosaic, in the first case more stands form a coarse-grained mosaic together. The whole concept therefore is really linked to eco-units of all sizes and the silvatic mosaics they form. It is as rich in meaning as the French "chablis" (Sect. 5.1.3) or the Dutch "ontwatering" (Sect. 5.3.3), and such concepts are better not replaced by imperfect translations in other languages, which never do them justice.

The three concepts most often mixed up in forest ecosystem studies are "layering" in the vegetational sense, "layering" in the architectural sense and "stufung" as an implicit meaning. Mean values over all eco-units in the same stufe in a mosaic too often are not distinguished from mean values established over the same layer in one eco-unit. Figure 6.10 exemplifies how this kind of layers were thought to occur earlier. It represents a forest fragment in Sri Lanka (Holmes in UNESCO 1958), in which we may now note a transitional zone between two eco-units, each partially visible at the left and right. Alexandre (1984) explained such architectural features in the Côte d'Ivoire by the light-and-shadow patterns cast by both trees of the present and the past, without distinguishing stufung and schichtung. Not surprisingly, he concluded that (his p. 23): "...Taï forest, although not layered, is definitely structured". For the eastern and northwestern US, Oliver (1980) followed a kindred reasoning, describing stratification in even-aged, mixed stands as "vertical sorting by species". He made no distinction between strata and stufen.

This is the reason why layering is extensively discussed here. Confusion may lead to dead-alley modelling of forests. Dead-alley models do not lead to explanations or answers needed to understand or manage forests at precise silvological levels. The "idealized forest" of Fig. 5.65, for instance, breaks off the hierarchy (Table 2.1) at the eco-unit level, because all superior levels are reduced to artificial eco-unit images with layering and without stufung. The dead end then is the eco-unit level, because the "idealized forest model" cannot be used to explain production and species demography in the real forest concerned and on the surfaces considered, i.e. in a mosaic.

Probably, confusion about layering also leads to blindness for the structured systems of communication and translocation which are implemented by animals. Usually, animal species populations are tallied according to height levels without taking into account mosaic architecture. For instance, Rees (in Sutton et al. 1983) made population counts of hemipterous insects in the rainforest of Sulawesi (Indonesia), based on the contents of traps situated at normalised height levels. The result of such studies, of which many have been made, is the superposition of an artificial geometrical pattern over the forest. This pattern of abstract layers has something in common with stufung, because it goes from one end to the other of a mosaic, but also has something in common with layering, because it traces storeys within eco-units.

Populations counted in this way produce a species composition model including plants and animals from all phases and all sizes of eco-units. Again, the whole silvatic mosaic is in fact treated as one, average eco-unit. Sometimes this image covers various subsites of one site type also. Such models completely mask the communication and translocation patterns, because they average away an architecture (texture) with eco-units that are attractive to animals and others that are not. This architecture determines animal routes, from one attractive eco-unit to another, avoiding the unattractive ones, i.e. it builds the structure of communication and translocation channels. The small, vertical "rainforest maps" given in the

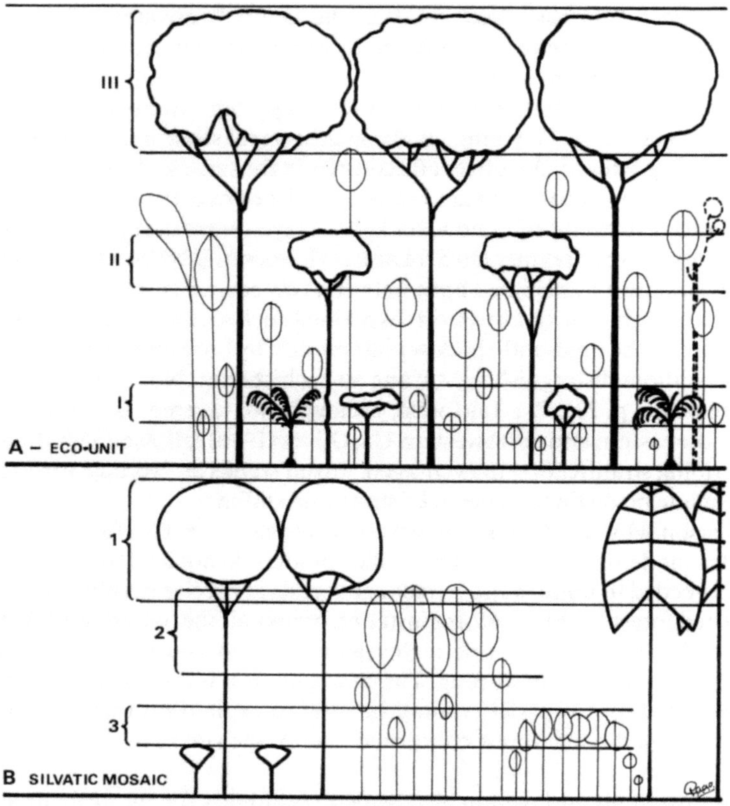

Fig. 6.9. Canopy closure in forest systems at different hierarchical levels. **A** *Layering* (horizontal closure) by different structural sets, containing trees of the present (*III, II, I*) at eco-unit level. **B** *Stufung* (vertical closure) by different eco-unit canopies (*1, 2, 3*) at the silvatic mosaic level; note exclusion of small trees of the present at the *left.* In current literature, these different principles of closure are often confounded, particularly in zoological publications referring to "storeys" in "the" forest

upper right corners of the pages treating animal groups in Ayensu (1980) exemplify the use of a visualised average-storeyed forest image. Such images can be useful as a very general information source, indicating to us whether to look up or down in the forest to find a certain animal species. But they again are dead ends for explanatory models at the silvatic mosaic level.

In this context, a very fitting statement was made by Gasc (in Muséum National 1986, p.97): "Total diversity has no meaning, each species must be looked at separately, taking account of some features of its life history which may be related to the structural characteristics of the microhabitat". This researcher derived his statement from a study on the herpetological fauna associated with a Guyanese rainforest palm species from the lower reaches of the forest, *Astrocaryum paramaca.*

Fig. 6.10. Example of an older "layering" concept in which no distinction is made between *stufung* and *layering* (see Fig. 6.9). The limit between two contiguous eco-units would be visible on aerial photographs at ca. 8 m from the *left*, where stufung is clearly occurring. This profile diagram, made in wet evergreen forest in the Kottawa Forest Reserve (Sri Lanka), relies on implicit methodological criteria concerning average phenomena and figures over larger forest surfaces, for which this profile diagram is judged to be representative. This eliminates eco-units. (Holmes in UNES-

6.2.3.2 Eco-Unit Composition Throughout the Precuilibrium

The development of a stufed precuilibrium mosaic can go through numerous steps, which are different per site and per forest type. The principles are the same the whole world over (Whitmore 1982), but the variations on the theme are without number, and no classification or typification exists as yet because the matter has rarely been studied from this viewpoint. Some cases will be illustrated later in this chapter. In general, the step represented in Fig. 6.6B can be said to have the following eco-unit composition.

- larger, short-lived pioneer eco-units dominating, long-lived pioneer eco-units appearing, and (depending on biotic remanence), some late pioneer or successor eco-units appearing. Except for the short-lived pioneer eco-units, all others are still in the aggradation phase, a set of appearing eco-units. All eco-units involved participate in stufung, as in the next sequence.

Some of the next steps will be the following

- large, short-lived pioneer eco-units surrendering (degradation, fragmentation), long-lived pioneer units dominating, late pioneer eco-units appearing, some small successor eco-units appearing. This is one of the first possible steps between Fig. 6.6B and C. The exact version of this step is determined by the local regime of shifting events and the site possibilities, including the build-up of organic soil contents and propagule bank. This is also the case for the next steps.
- long-lived pioneer eco-units dominating, late pioneer eco-units appearing and dominating, small successor units appearing in a more or less rich array of sizes and architectural patterns.
- long-lived pioneer eco-units dominating and surrendering, late pioneer eco-units dominating and surrendering, small successor eco-units appearing and dominating.
- few long-lived pioneer eco-units dominating and surrendering, late pioneer eco-units mainly surrendering, small successor eco-units appearing, dominating and sometimes surrendering. Fusion among neighbouring small eco-units becomes more frequent (Sect. 5.4.3.11).

The mosaic now only needs some rare but regular local shifting event, maybe once every 30 years or every century, for the appearance of scattered pioneer eco-units and with them the stage of Fig. 6.6C, the ecuilibrium, is reached.

In order to visualise silvatic mosaic architectures at the beginning of the precuilibrium stage, Fig. 6.11 shows the images of an 8-year-old, artificial, homogenised ash plantation on virgin polder soils in the Netherlands next to a 3.5-year-old secondary vegetation on non-homogeneously cleared soil in French Guyana. Even in the homogeneous Dutch situation there is a clear tendency to fragmentation, although it is accentuated by the presence of two genotypes (*Fraxinus excelsior* cv. Westhoff and *F. excelsior* cv. Eureka). Nuesink (1980), who did this study, made a complete analysis of component ash architecture. Maury-Lechon (1982), who published the Guyanese image, clearly indicated soil heterogeneity, due to treatment after clearing, and different herbaceous components in each fragment of the pioneer mosaic.

The steps in the precuilibrium stage, some of which have been shortly diagnosed above, vary in different silvatic mosaics on different sites. They may also appear in slightly different orders in time in the same kind of mosaic. Therefore, the only term used to unite all these successive steps remains "precuilibrium stage" which has the degree of abstraction of "juvenility" in organisms. Comparing the "juvenile" period of an insect with that of a bird gives an idea of the profound differences that may exist between a "precuilibrium" stage in sites such as the Siberian taiga or the Amazonian hylaea (also see Fig. 6.1).

Still, the development processes driving silvatic mosaics thoughout the precuilibrium stage all obey to a common set of rules, some of which are mentioned by J.-P. Lescure (1986). These rules are made visible by the following architectural symptoms.

- The number of eco-units increases.
- The proportion of small-sized, long-lived successor eco-units increases.
- The cumulated surface of small eco-unit innovation phases ("chablis") in the mosaic increases.
- The surface covered by the set of appearing eco-units decreases.
- First the surface covered by the set of dominating eco-units increases, later the surface covered by surrendering eco-units.
- There is an increasing frequency of eco-units in fusion (see Sect. 5.4.3.11).
- The total length of linear limits between eco-units increases, as a measure of the total increasing surface under transitional conditions. These limits may be defined according to different criteria (e.g. Bongers and Popma 1988).

These symptoms are in part visible in Fig. 6.12, taken from Lescure's study (1986) on forest succession on abandoned agricultural fields of the Wayampi Indians on the Upper Oyapock river, near the village of Trois Sauts (French Guyana). His terms, which are not completely congruent with the silvatic mosaic concept, have been replaced by the terms used in this book, because otherwise confusion might arise between eco-unit development and silvatic mosaic development. Please note that no fragmentation has been indicated in this case, except in the pioneer eco-units, and that Lescure's model shows the other kinds of eco-units to come up through the pioneer vegetation, from an appearing set composed of aggrading eco-units already established below the pioneers. This corresponds to the "even-aged" image given by Oliver (1978) for northern mixed hardwood forests in Massachusetts (USA).

The reason for this deviation from the development path sketched in Fig. 6.6 is that in neither case had the propagule bank been destroyed and hence in both a high degree of biotic remanence existed. Figure 6.6 was made to represent the case of an insignificant amount of biotic remanence, on totally cleared terrain, and with a newly built-up propagule bank. In the case of high biotic remanence, i.e. of truly secondary succession, the biomass and plant (tree?) numbers have been schematised against time in Fig. 6.13A, also established by Lescure (1986) for his case study. These kinds of curves have been experimentally confirmed by De Graaf

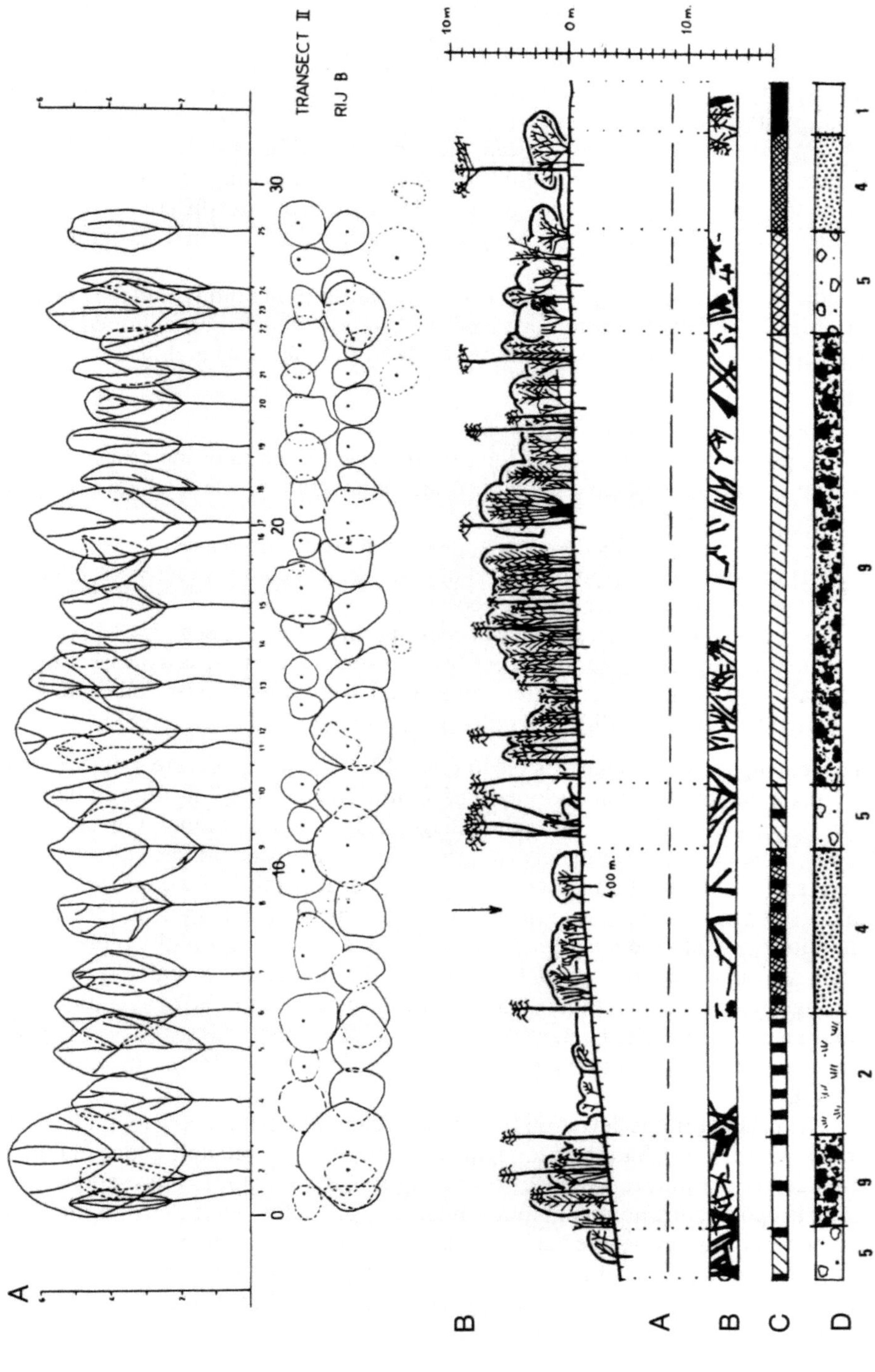
A
TRANSECT II
RIJ B
B
10m
0m
10m
400m
A
B
C
D
5
9
2
4
5
9
5
4
1

(1986) in Suriname and Weidelt (1984) in Indonesia and the Philippines. For the temperate regions, they are remarkable complements to the curves by Bormann and Likens (1979, see Fig. 5.84), which have been discussed above. A comparable curve, but without the highly significant oscillation at the right, had been published in 1967 by Kira and Shidei.

Finally, Fig. 6.13B shows the frequency of successive species being the main woody components ensuring the architectural and biomass development in Lescure's study area. This figure shows that there may well have been some build-up of the propagule bank during the earlier phases, unless seedlings and saplings of the species appearing later were present below the earlier ones. In any case, Lescure's figure does not contradict the model of plant life cycles along a successional gradient as presented by Gomez-Pompa and Vazquez-Yanes (in Gomez-Pompa and Del Amo 1985), reproduced here as Fig. 6.14. It is to be emphasised that neither Fig. 6.13B, nor Fig. 6.14 are based upon a clear distinction between eco-unit level and mosaic level, although Lescure (1986) refers to eco-units.

6.2.3.3 The Precuilibrium, Animals and Fungi

Contrary to the aggrading phase of an eco-unit, out of which many niches for animals are eliminated (Sect. 5.2.3), the precuilibrium phase shows an enrichment in animal species of its eco-units, and at the same time a lesser mean density of each species over the mosaic. The case is illustrated for small rodents and marsupial rats in the rainforests of French Guyana and Gabon, in a comparative study by Guillotin and Duplantier (in Muséum National 1986). Some examples of the gradual enrichment in both plant species from non-pioneer eco-units and animal species that disperse them, principally birds and bats, have been documented for the mosaic development studied very thoroughly in French Guyana (cf. Figs. 5.37, 6.11) by De Foresta et al. (1984).

Another example for the neotropics is presented by Koptur (1984a, b). She studied the interactions between Costa Rican species of the pioneer tree genus *Inga* (Mimosoideae, Troll's model) and the complex of pollinators, which include insects and hummingbirds, as well as their interactions with ants and phytophages. She proved the importance of long-distance transport of pollen, enhancing genetic

◄

Fig. 6.11. Two silvatic mosaics in an early precuilibrium stage. **A** East Flevoland, Netherlands, plantation of ash (*Fraxinus excelsior*, Oleaceae) on very young soil, homogeneous, biotic remanence insignificant. Age 8 years. Note fragmentation of the mosaic due to genetic factors: two cultivars of ash were used. (Nuesink 1980). **B** Piste de St. Elie, French Guyana. *BA* Secondary forest, ca. 3.5 years old on mechanically cleared, compacted soil after fire. Important biotic remanence and microsite differentiation by organic matter and ash distribution. (After Maury-Lechon 1982). *Arrows* footpaths. *BB* diagram: dead wood distribution; *BC* diagram: soil disturbance (the *blacker*, the more compacted and: the *denser* the *hatching*, the heavier the fire); *BD* diagram: herbaceous flora. *1* to *10* below diagrams: "vegetation formations", in this case nearly equivalent to eco-units (most are early aggrading)

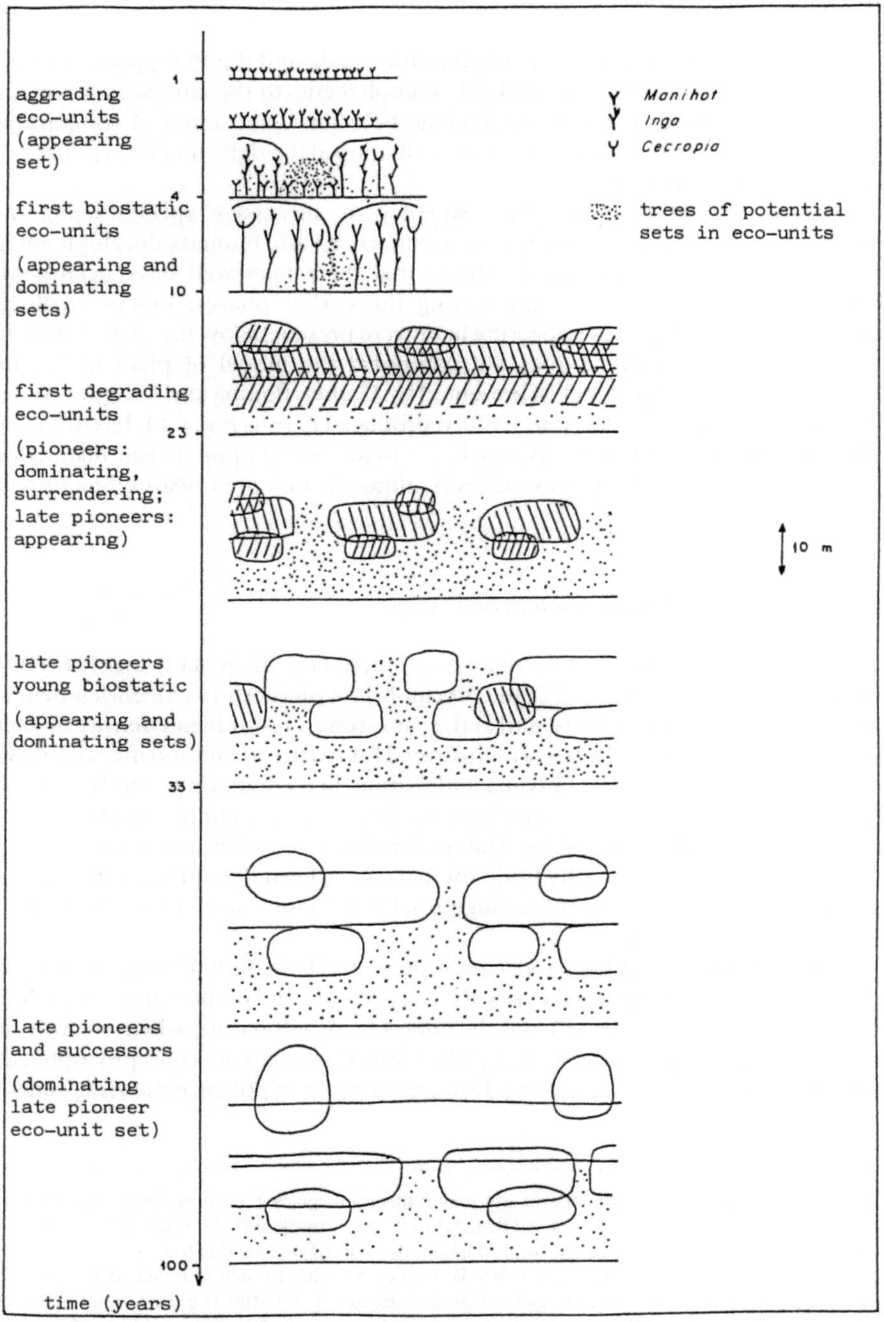

Fig. 6.12. Example of silvatic mosaic development after shifting cultivation along the Oyapock River (French Guyana). Highly diverse microsites, biotic remanence, human remanence (seeds from eaten fruits). Original terms replaced by terms used in the present text, otherwise taken from Lescure (1986)

diversity by enhanced fruit setting if compared with pollination from nearby pollen sources. The efficiency of ants in removing caterpillars and three other types of phytophagous insects from leaves also illustrates the importance and efficiency of animal transmission of signals and matter translocation in forest mosaics in the precuilibrium stage, where *Inga* plays an important role. The "defense" interactions are linked to mechanisms that are proper to mosaics, because the "attacking" organisms have ranges exceeding an eco-unit. The modification of the defense mechanism along an elevational gradient, i. e. a site gradient, shows that these communication and translocation systems are indeed changing with the change in silvatic mosaic caused by site change (cf. Janzen et al. 1976; Koptur 1985).

The role of ants in precuilibrium stages needs to be emphasised (see also Sect. 6.4.4). It is well known that many pioneer trees and shrubs (cf. *Cecropia* sp., Moraceae, or *Hirtella* sp., Rosaceae, S. America, Fig. 6.15, or *Macaranga* sp., Euphorbiaceae, S.E. Asia) are myrmecophilous. They live together with ants in a relation that may often be termed symbiosis. Leaf-cutting ants from Costa Rica, belonging to the genus *Atta*, cut leaf fragments all over tropical America and use them as a substrate to cultivate fungi in underground "fungus gardens". The ants have been experimentally shown to be highly selective in cutting the leaves of "successional" species (many pioneers; Blanton and Ewel 1985). This is part of the life strategy of these ants, in which short-term and long-term defoliation rhythms of tree species are balanced out in such a way that no trees are killed and the ant's resources are conserved (Cherrett in Sutton et al. 1983). The animals range far beyond any one eco-unit. This implies the transmission of information and nutrients between eco-units, which is a mosaic function. *Atta* is an important pest genus in tropical American agriculture, in other words in ecosystems that have an architecture and species composition radically different from natural silvatic mosaics.

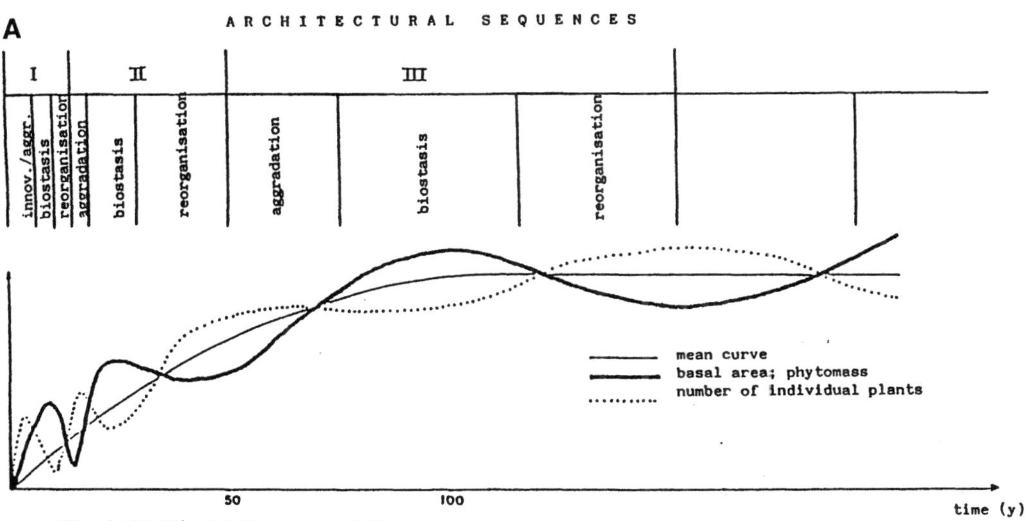

Fig. 6.13A (Fig. 6.13B see page 422)

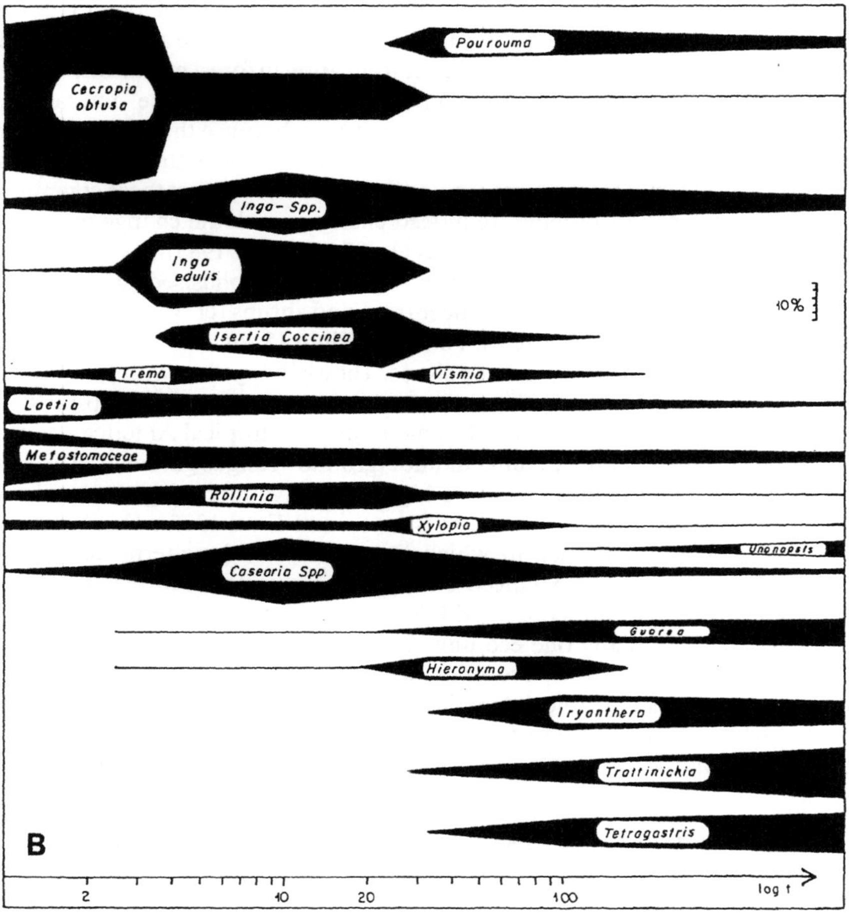

Fig. 6.13. Example of production parameters and species populations sustaining the silvatic mosaic development shown in Fig. 6.12. **A** Phytomass and plant numbers oscillate around a mean, asymptotic curve with mosaic development. Terms used here for development phases are those at eco-unit level, as seen by Bormann and Likens (1979). They replace the original terms by Lescure (1986), because their "steady state" and "reorganisation" as a matter of fact refer to mosaics and concern fragmentation phenomena. Values being expressed here per surface area (ha, a), mosaic phenomena only appear as variations of the average. **B** Estimated relative density of some taxa of trees during the development shown in Fig. 6.12. At about 30 years, the typical pioneer genera of northern South America are replaced by late pioneers; some intercalary genera are to be noted. This points to an impoverished seed bank after shifting cultivation, otherwise they might have germinated at the same time as the pioneers and waited below for these to die. Biotic remanence certainly reduced. (Whole figure Lescure 1986)

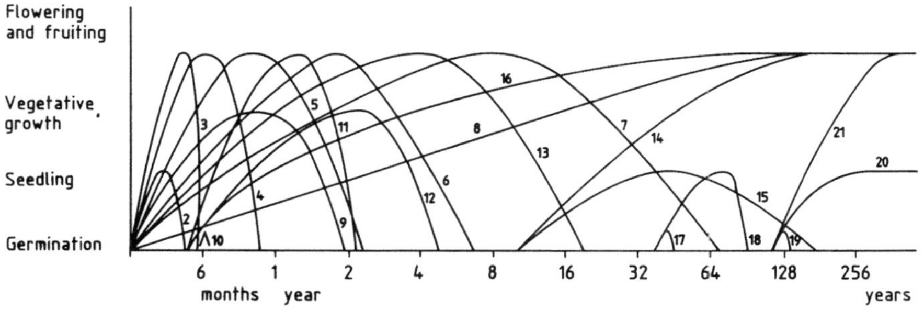

Fig. 6.14. Different life cycles of plants participating in silvatic mosaic building, which is neither even-aged, nor completely uneven-aged. Life cycles: *1* germinate and die; *2* germinate, produce some leaves and die; *3* complete life cycle in months; *4* annuals; *5* biennials; *6* life cycle less than 10 years; *7* secondary species living some decades; *8* long-lived pioneers living for centuries; *9* never reach reproductive state; *10* germinate shortly after pioneers and die fast; *11* annuals germinating in the installation stage; *12* germinate later than pioneers, never reach reproductive state; *13* short life cycle, less than 30 years; *14* long life cycle, germinate in later precuilibrium stages; *15* germinate in later stage and survive some years as seedlings; *16* germinate in early precuilibrium and live for centuries; *17* germinate and die in later precuilibrium stages; *18* germinate and live as seedlings or saplings in end of precuilibrium, then die; *19* germinate and die in ecuilibrium stage; *20* germinate and grow in ecuilibrium stage, survive and wait for shifting event to grow on; *21* germinate and grow in ecuilibrium stage, reach reproductive state after long life cycle ("primary species"). (Gomez-Pompa and Vazquez-Yanes in Gomez-Pompa and Del Amo 1985); terms in legend replaced by terminology used here

European carnivorous red forest ants of the genus *Formica* are often found in pine and spruce stands (Brauns 1976). These are among the types of eco-units that are often dominating in precuilibrium stages in many European silvatic mosaics. The ants are judged to be "useful" in forestry, because they consume herbivorous insects, among which there are "tree pests". Szujecki (1987) states that an average nest may consume around 2 million insects per year, of which up to 80% may be pest insects in pest years. In other years, the diet is much more diverse (Brauns 1976; Szujecki 1987) and includes the "honeydew" that is "milked" from aphids, which were in earlier literature considered as the ants' "cattle". The aphids generally are not feeding on trees, with the exception of some that live on the trunk and suck sap from the phloem (cf. Fig. 3.32), but most often occur on forest herbs and shrubs. The functions of such ants in silvatic mosaics are again to be placed in the processing of signals, e.g. "stop" signals for pests coming in from other eco-units, and translocation of biomass.

Animals can transmit powerful information for silvatic mosaic succession through the propagule bank (also cf. Sect. 6.2.4.6.). Janzen (1982), in one of his studies on the interactions between tapirs (*Tapirus bairdii*) and seeds in Santa Rosa National Park (North-Western Costa Rica), showed that there is a delicate balance between seed digestion and intact dispersal after roughly 10 days of intestinal treatment of seeds by tapirs. The tree species surviving transport are mimosoids with large, hard seeds, belonging to the late pioneers. The tapir therefore transmits information which will induce self-replacement of late pioneer eco-units, rather than accelerated appearance of a set of successor eco-units.

In the 90% of the secondary successional forest (precuilibrium mosaics, in our terms) included in the Santa Rosa National Park, Janzen (1986c) experimentally showed how the selection of the seeds in the propagule bank was carried further by seed predation by the spiny pocket mouse *Liomys salvini.* Janzen's title speaks for itself: "Mice, big mammals, and seeds: it matters who defecates what where". And still, this is not the end of the chain of silvatic information. Riding on the back of two mouse species, one of them being *L. salvini*, two new species of tineid moths were discovered (Davis et al. 1986), which possibly have a function in litter decomposition in and around their location of oviposition, the mouse nests. The precise mechanisms of these interactions cannot as yet be described like in the Australian case mentioned earlier (Sect. 5.6.9).

Whatever may be the precise contents of such transfers of information and translocations, they slow down stochastic developments and have a "silvistatic" effect. A comparable phenomenon with a very different mechanism has recently been described from Korup National Park in Cameroon (Newbery et al. 1988). The data of these authors suggest that aggregation of ectomycorrhizal fungus species, seedlings and adult trees of the genera *Tetraberlinia* and *Microberlinia* (caesalpinioid late pioneers) may lead to the favouring of groves (middle-sized eco-units, in our terms) of these tree species. This might be comparable with certain dipterocarp features from the Asian humid tropics (Smits in prep.).

6.2.3.4 The Precuilibrium and Silviculture

Finally, it is to be noted that most silvicultural systems are linked to and inspired by the installation and precuilibrium stages of local silvatic mosaics, especially in regions with a marked human population pressure. Some reasons for this state of affairs are enumerated below, and they explain many of the "agricultural" approaches in silviculture.

- Human presence with very few exceptions leads to forest clearings larger than 0.5 to 1.0 ha (cf. Haxaire 1976, French Guyana, Amerindian shifting agriculture). This causes an atypically high proportion of appearing pioneer eco-units in the mosaic, which may be seen as a regression along the precuilibrium steps mentioned above. The present author could observe this increased short-lived pioneer share from an airplane above an Amazonian silvatic mosaic in Eastern Ecuador (Río Curaré), with an estimated human population of 600 persons on 20,000 km^2.
- If human population density increases, the above effect is emphasised. Human forest clearing tends to increase the proportion of sweeping effects over shorter intervals in time than those caused by climatological and biotic factors. It should be compared with an increasing modification of the shifting events.

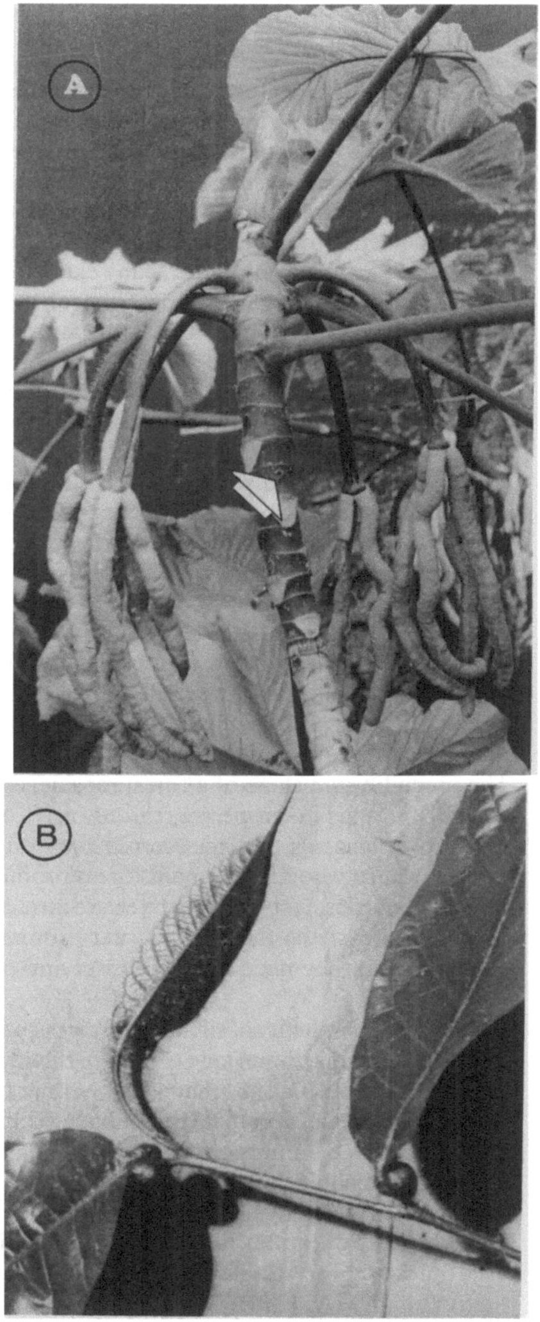

Fig. 6.15. Examples of the importance of ants as components of precuilibrium pioneer eco-units. Symbiosis between tropical pioneer trees and ants. **A** *Cecropia* cf. *surinamensis* (Moraceae, French Guyana, 1968); ants live in the hollow internodia which have a preformed entrance (*arrow*). **B** *Hirtella* sp. (Rosaceae; Approuague, French Guyana, 1968), ants living in special bladders at the leaf base (also cf. Fig. 4.14C). (Photographs by R. A. A. Oldeman)

- If human population reaches high densities, the silvatic mosaic disappears altogether in its original form. Two limits are then to be considered. The first one is reached when human awareness includes the necessity of the conservation and wise use of the forest relics, which usually are eco-units belonging in a precuilibrium stage for the reasons mentioned above. The second limit is reached when all or nearly all forest has disappeared and scarcity of forest products obliges human societies to develop reafforestation techniques. This entails the reconstruction of installation stages. Many existing silvicultural traditions are related to the first limit (Central Europe, France), many others to the second (Netherlands, Scandinavia in part, Great Britain, North Africa). The reafforestation traditions are closest to agriculture, because agriculture is the yearly repetition of installation stages without trees (e.g. Alexandre et al. 1978). The developing North American silviculture is certainly inspired by both limits of awareness and so is the Australian one.
- In most cases, silviculture came after the disappearance of the very knowledge of what the local natural silvatic mosaic looked like before strong human impacts changed it. This gap in silvicultural knowledge certainly is at the base of many approaches in silviculture with a forest concept that is precuilibrium in its essence. Wherever possible, this should be avoided, and silvatic mosaics with minimal human impacts should become the base-line of optimised silvicultural models and systems. This approach has been followed for a long time by Central European silviculturists, exemplified by Leibundgut (1970, 1982), Mayer (1980) or Van Miegroet (1976). Oliver (1980) points to the silvicultural risks in mistaking base-line forest developments for others.
- The precuilibrium stage of a forest mosaic, especially in its first steps between 50 and 100 years in temperate regions or 15 to 40 years in the tropics, is attractive also because its pioneer eco-units are often built by fast-growing pioneer trees. Because wood is the main forest product focussed upon by silviculture for two centuries, fast-growing trees are attractive for economic reasons. Many silvicultural systems therefore cluster around such tree species, building the eco-units that they need and so conserving precuilibrium-type processes and structures.
- But also the complexity of silvatic mosaics usually increases from the installation stage to the ecuilibrium stage (Fig. 6.6). Silvicultural systems are as complex as the particular mosaic to be managed. Simpler systems have long been thought to be cheapest, a question that cannot have such a general answer and that is still being debated. Complex and remunerative silvicultural systems within certain market and technical conditions have long been developed in Central Europe (Plenterwald, jardinage, selective forest, uitkapbos, bosque irregular). They are now sought again, after a lull of some decades, in the most complex of all forests, the tropical rainforests (cf. De Graaf 1986; Torquebiau 1986; Jonkers 1987; Wöll 1988). But there will always be a tendency to go back to the precuilibrium stage for reasons of simplicity.

6.2.4 *The Ecuilibrium Stage*

6.2.4.1 *Kaleidoscopic Balance in Eco-Unit Composition*

After the precuilibrium stage an ecuilibrium stage is distinguished in the developing silvatic mosaic (Fig. 6.6C). The word ecuilibrium, in this spelling, is new and indicates that the proportions of **eco**-units have reached an e**quilibrium** with the local regime of shifting events. In other words, for every eco-unit in the surrendering state, an analogous eco-unit in the appearing state is created. Earlier, we indicated the resulting forest as a kaleidoscopic mosaic (Oldeman 1974b), after the well-known children's toy, the kaleidoscope. Kaleidoscopes show mosaics of the same coloured glass patches, each at a different place after a shift.

The existence of an ecuilibrium stage presupposes that there is a regular regime of shifting events. In order to prove this, this regime would have to be predictable. Although it is widely known that it is not, and least of all for strong events that occur with intervals of 10 years or more (Wartena pers. commun.) such as hurricanes (cf. Emanuel 1988), floods, or droughts that lead to fires, this uncertainty is also widely overlooked. Not only is it rarely taken into account in forest management plans, but vegetation types, including forest types, are correlated with average, long-term climatological features only (e.g. Holdridge et al. 1971).

Few climatological maps exist that show the frequency and impact of shifting events, with the exception of some tropical maps given by Nieuwolt (1977) in his chapter on "Tropical Disturbances", in which he treats cyclones, thunderstorm frequencies and monsoon depressions. Even these few data go a long way to explain silvatic mosaic characteristics in the regions concerned, as we have rapidly checked for some regions in South America. In the next pages, the general inadequacy of climatological data formulation for explanatory models of silvatic mosaics will be mentioned again in the context of stresses (Sect. 6.2.5).

Figure 6.6C shows the architecture of a possible ecuilibrium stage of a forest mosaic in a rather hospitable environment, which might be a site somewhere in the upper half of Fig. 6.1. The disappearance of each one of the eco-units shown would be followed by the appearance in the mosaic of a new eco-unit of the same kind. Eco-units have different lifespans and architectures. The appearance of small eco-units is very frequent and that of larger eco-units is rather unfrequent. In comparison with the precuilibrium stage, the reappearance of a certain rather small proportion of pioneer eco-units is a sign that the ecuilibrium is reached, because they indicate the effects of those shifting events which are locally rare, but are "normal".

Because the shares of different eco-units in the silvatic mosaic have been stabilised, the flora and fauna are stabilised over the whole mosaic, too. Therefore it is in the ecuilibrium stage that species counts and calculations using mean numbers for populations over large surfaces have a characteristic significance for that particular silvatic mosaic. As a matter of fact, this developmental stage of mosaics is closest to what most vegetation scientists understand when referring to "climatological climax". For instance, Werger and Westhoff (in Bakker et al. 1985, p. 322), state: "The climax situation is stable if there are no disturbances from outside. This stability can be expressed by regular population regeneration (constancy), when structure in the strict sense is stable also, or, at another scale, in stable energy flow and matter turnover, showing periodical oscillations at the population level (cyclic change), accompanied in that case by periodical differences in structure in the strict sense." Figure 6.2A might show an architectural diagram of the first case mentioned by these authors, with only "natural" mortality of organisms determining change and without outside impacts.

The most precise analyses of silvatic mosaic architecture in ecuilibrium stages with minimal human impact that we know are those by Koop (1981a; see Figs. 5.1, 5.7, 5.50) in Northern Germany and Poland, by several authors in France (see Sect. 6.2.8.2), by Vooren (in prep.) in the Côte d'Ivoire, by ourselves in South America (e.g. Fig. 6.34) and by Torquebiau (1986, in Nooteboom 1987) in Indonesia. Pruša's standard work on the virgin forests of Bohemia and Moravia (Czechoslovakia) and his 12 years' monitoring data (Pruša, 1985, 1987) deserve special mention for their rich, very carefully prepared and as yet unparallelled data sets (Fig. 6.16C). Figure 6.16A, B represents such an ecuilibrium stage on 2.7 ha in Central Kalimantan (Indonesia), but it has to be noted that the site is not homogeneous, the forest covering several toposequences going from the top of small, steep hills to the swamps below (Van Reuler in Nooteboom 1987).

The mosaic analysis has been effected by characterising each eco-unit in terms of sets of trees (present; future or potential; past - cf. Fig. 5.4) or as a chablis (innovation). The results are drawn to scale in Fig. 6.16A, and the eco-unit composition is shown in Fig. 6.16B. This silvatic mosaic differs considerably from those analysed by the same author in Sumatera (Indonesia), particularly by the amount of dominating ("steady state") eco-units. The large, seemingly unbroken surface of biostatic eco-units coded 2H (this code indicating that the trees of the present have long free boles and shallow crowns, a sign of ageing: Fig. 5.8); is due to the fusion of many smaller eco-units, as we observed ourselves in this forest (also see Torquebiau in Nooteboom 1987, his Figs. 11, 12).

6.2.4.2 Tree Species and Their Strategies

Such studies go a long way to supporting the hypotheses by Orians (1982) on enhanced tree species richness due to the diversity of eco-units (this author uses the term "tree-fall") in such mosaics, which are rich in specialised habitats for seeds and seedlings that invade them. Precise data for Côte d'Ivoire chablis formation can be found in Vooren (1986). But a tree has to face, after its germination environment, a sequence of conditions prevailing in aggrading and biostatic phases, complicated still by interactions between its eco-unit and neighbouring ones. This requires trees to specialise along lines of great variety in life strategies, a subject studied among others by Oldeman and Van Dijk (in Hadley 1988). It is to be emphasised that the properties studied by all authors mentioned are linked to the complete mosaic, with habitats in one eco-unit being influenced by all surrounding eco-units. This entails a larger view than the temperaments of trees as considered earlier (e.g. Tables 5.6 and 5.7).

Figure 6.17 shows these strategies schematically. They are based upon the difference between "strugglers", establishing themselves in difficult conditions and struggling upward every time that a new, small eco-unit opens up nearby or around them, and "gamblers", saturating openings with seedlings and having to reach the canopy fast because they are unable to sustain biological competition for long. There are also intermediate strategies, and trees that change over from one strategy to the other when reaching certain forest levels. Criteria to distinguish these strategies are multiple, from tree architecture to leaf characteristics and reproductive behaviour. Van Dijk (1987) checked them in a Venezuelian cloud forest and Barkman (1988) presented a comparable view on two temperate *Quercus* species. A precise and important study on forest seedlings in Gabon was done by Miquel (1985), who described many and analysed their germination environment as a basic study in agroforestry. It is to be noted that Fig. 6.17 is not a classification, but a vis-

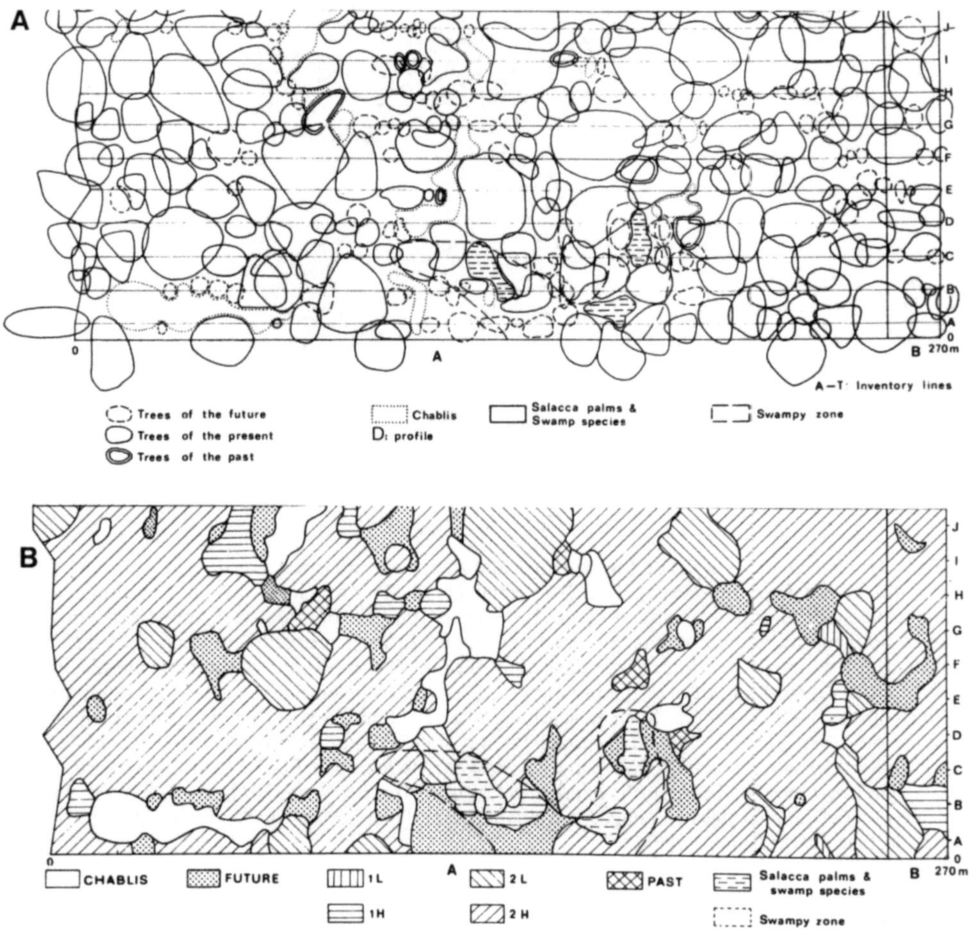

Fig. 6.16A, B (Fig. 6.16C see page 430)

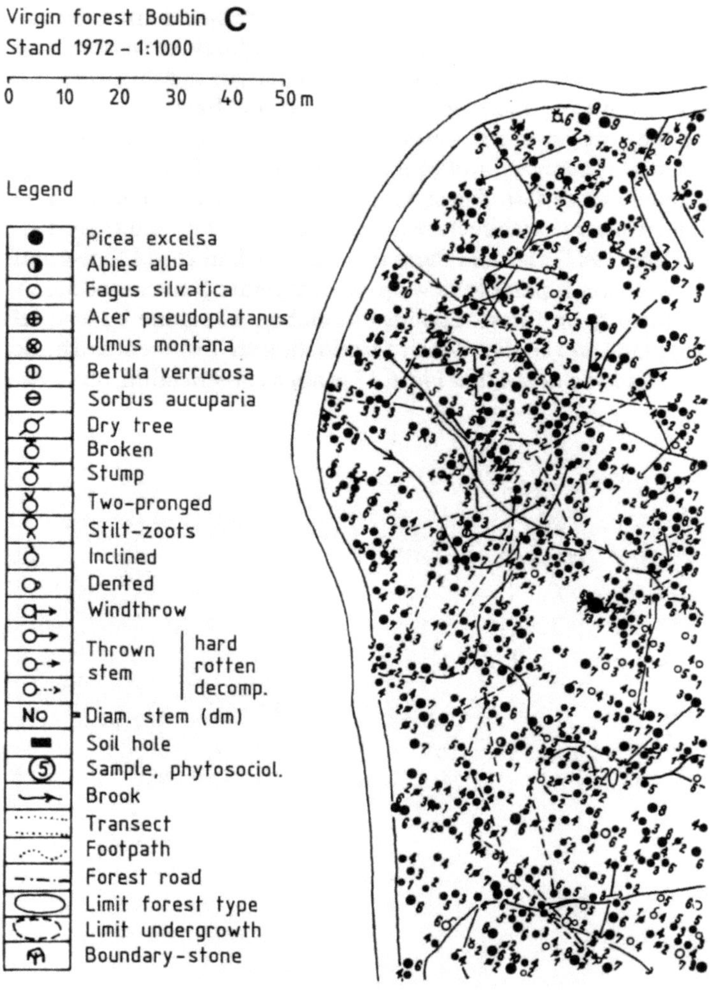

Fig. 6.16. Two analyses of the build-up of an ecuilibrium stage in a forest mosaic at Sungei Samba (Central Kalimentan, Indonesia), from Torquebiau (1986). This mosaic spans different sites in a toposequence (hill-top to marsh) and hence is in fact a composite of mosaics. **A** Analysis in terms of component tree compartments: potential trees (of the future), trees of the present, trees of the past. These define the eco-units distinguished in: **B** eco-unit composition of the same mosaic, in which four categories in trees of the present (*1L, 2L, 1H, 2H*) determine the canopy of different kinds of steady-state eco-units. Such mosaic images can be obtained by automatic pattern analysis of remote-sensing images with small enough "steady state" eco-units. Probably fusion of smaller eco-units has occurred, so that a finer pattern is masked. Also compare Fig. 5.1 (European mosaic). **C** Virgin forest mosaic in Czechoslovakia (after Pruša 1985). Criteria used indicate the zero-event quite precisely and so yield an image giving ecuilibrium dynamics rather than architecture (eco-unit surfaces; cf. Oldeman 1989)

ualised reference framework, on which many various tree life strategies can be situated with regard to their position in silvatic mosaic build-up.

6.2.4.3 *Succession and Silvigenetic Cycles*

At the eco-unit level, the biostatic phase may be quite prolonged and long-lived (Sect. 5.2.4). At the silvatic mosaic level, the ecuilibrium stage is very dynamic, and the architectural image hence is shifting yearly or seasonally. Figure 6.18 shows the "natural picture" of an ecuilibrium mosaic, reflected in the water of Río Curaré (Eastern Ecuador). The skeleton of this mosaic is built by hundreds of tree species with their different life strategies. These form tens of different eco-unit kinds, mainly of small sizes, as can be seen in the stufed mosaic canopy along and between the two river fragments on the photograph. Monitoring from the air or from space with images containing sufficient detail, such as provided by the 10 x 10 m SPOT pixels, coupled to automatic computer-aided pattern interpretation, might be the best way to arrive at an adequate analysis of the architectural dynamics. The analysis of aerial images obtained by radar backscattering may provide a still more refined "signature" of eco-units (Hoekman 1985).

Ground truth establishment is nevertheless still far from perfect. Webb started working on a cyclone-driven model of Australian rainforest canopies as early as 1958. His pro forma for the description of structural features in tropical forest typology (Webb et al. 1976) rests on different "canopy types", which are easy to assimilate with states of the canopy of certain kinds of eco-units. These canopy types, which were developed as part of Webb's "physiognomic classification" (1959) and later "structural types" (Webb and Tracey 1981) of Australian rainforests in Queensland, were brought together in a truly silvigenetic diagram (Webb 1977). The difference is that Webb's basic concept remained focussed on species populations, so that his structural types remained physiognomic, not being conceived as parts of a hierarchical system's analysis, like silvatic mosaics. The two approaches are compared in Fig. 6.19. Both leave open the possibility to consider a certain share of secondary forest as part of of the "climax" or ecuilibrium.

One of the tools derived from the Webb system is a dendrogram for the identification of canopies. Ohsawa (1981) did so in a parallel but not identical way for the mixed subalpine forests in Kitayatsu Mt., Honshu, Japan, with the aid of very precise forest regeneration maps indicating canopy gaps. In Ohsawa's dendrogram (his Fig. 11) the terms "pioneer", "building", "mature" and "degenerate" phase are clearly used for the whole mosaic, because in each the consequences of gap formation are discussed. One interesting feature mentioned by this author is gap closure by lateral encroachment. This is also indicated as "autogenic gap closure" (cf. Oliver and Stephens 1977). In terms of this book, it is the disappearance of an eco-unit in its innovation phase, i. e. early eco-unit mortality. For the explanation of the mechanisms of closure by expansion of neighbouring tree crowns, the article by Loehle (1986) on phototropism of whole trees is worth consulting.

Other Japanese work in search for such "forest regeneration patterns", in our terms mosaic dynamics, was done by Ohkubo et al. (1988) in primary Japanese

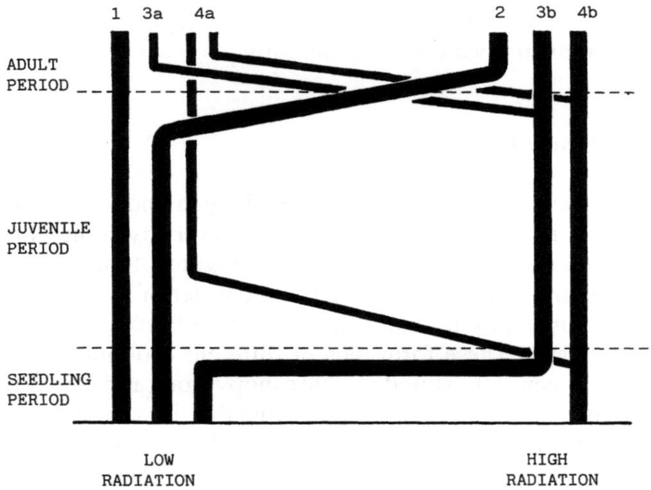

Fig. 6.17. Tree strategies as linked to silvatic mosaics, where microclimatic interaction amongst eco-units creates a welter of different conditions and gradients that require adaptation of the tree in each period of its life cycle. Tree behaviour is here linked to light (radiation) and reproduction. At the *left* (low radiation), there are the pure "strugglers" (*1*), at the right the pure "gamblers" (*4b*), which respectively give few but specialised seedlings or deposit their seeds massively to germinate in full light. Strategies *2*, *3a*, *3b* and *4* are examples of intermediate cases, in which behaviour changes when the tree reaches a certain level in the forest. (After Oldeman and Van Dijk in press). Compare with Fig. 6.14

beech forests (also see Sect. 5.4.3.12) and, in a way more closely related to architectural analysis, by Kojima and Peters (1984) in a Japanese birch forest in Hokkaido. Koop (1987, 1989) conceived a complete, computer-aided monitoring system for forest mosaics in general, named SILVI-STAR. With it, aerial images can be simulated from ground-truth data. If the system's promises hold true, automatic image processing of remote sensing images will be easily trained upon any regional or local forest mosaic with the help of this system of architectural simulation (cf. also Fig. 6.29).

One of the main handicaps for remote-sensing analysis of forest canopies as a means to understand mosaic dynamics is that such canopies are difficult of access. Over the last decade, this has been changing fast. In the tropical rainforest regions, Hladik and Hladik (1980) gathered data from photographs taken by a cable-balloon. Mitchell (1982) described climbing techniques and walkways through the tropical forest canopy, from tree crown to tree crown. In the Côte d'Ivoire, Vooren devised ways to obtain very detailed photographs of the canopy from an ultra-light airplane (Vooren and Offermans 1985). Detailed diagnosis of forest canopy vitality in Europe through false-colour remote-sensing images is exemplified in the magnificently illustrated book by Janse and De Boer (1987, pp. 110,126). The most recent technique was developed by the team of F. Hallé (pers. commun.), who, from 1986 on, land a "canopy raft" borne by a Montgolfière (hot-air balloon) on top of the forest, so scientists can work there during periods of weeks or months.

Fig. 6.18. Ecuilibrium stage of a silvatic mosaic on rio Curaré (Eastern Ecuador), as mirrored in the water. This "natural profile diagram" is to be compared with Fig. 6.6C. The parts of the mosaic between the river loops show the image of a mosaic such as analysed on Fig. 6.16, note configurations of chablis, appearing, dominating and surrendering sets of different kinds of eco-units. (Photograph 1975 by R. A. A. Oldeman)

For the moment, the analysis of mosaics along the lines of Fig. 6.16 may lead to time-series like those pictured in Fig. 6.1, in which different kinds of eco-units in different development stages are shown as if they succeeded each other from left (pioneers) to right (successor eco-units). This succession, as a matter of fact, seems to be an exceptional case in a process where "regression" to pioneer eco-units and "progression" to small-sized successor eco-units are balanced out, so as to make for a rather constant kaleidoscopic eco-unit composition of the mosaic. Figure 6.19A shows how shifting events, acting as different zero-events in various eco-units put in a "successional" order, affect the whole silvatic mosaic in a truly shifting architectural pattern of silvigenetic cycles. On different sites, with different climatic and pedological hospitality levels (Fig. 6.1), particular cycles are emphasised. If these are among the regressional cycles, the resulting ecuilibrium stage will be the equivalent of the level of one of the precuilibrium steps mentioned in Sect. 6.2.3.2.

This is shown in Fig. 6.20, depicting a natural forest mosaic in a somewhat drier part of the basin of Río Caroní in Venezuela. The shifting events include fire once every 40 years approximately, and intervening dry years. Many aggrading eco-units can be observed on this aerial photograph, and many pioneer trees can be rec-

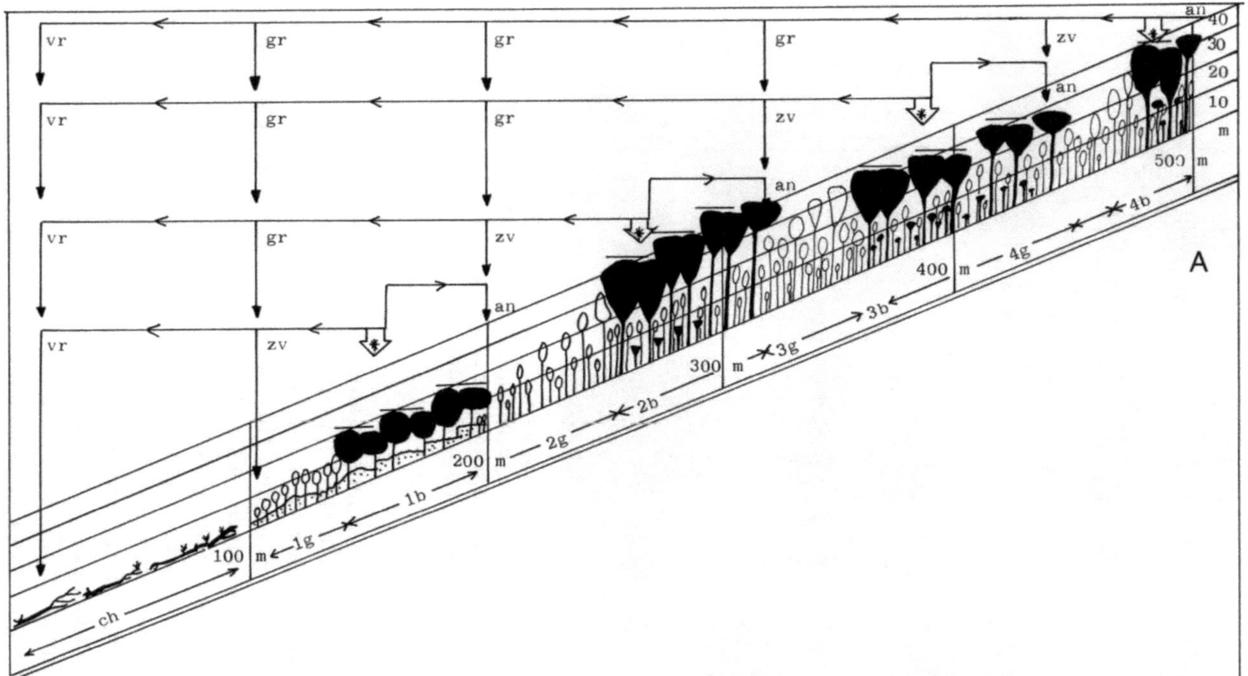

Fig. 6.19. Silvatic mosaic, silvigenesis, succesion. **A** Silvigenesis with four *kinds* of eco-units, i.e. *1* pioneer, *2* late pioneer and *3*, *4* different successors. Each eco-unit is either *growing* (*g* innovation + aggradation) or *not growing* (*b* biostasis + degradation). *Star in open arrow* shifting event. *Simple arrows* consequences, i.e. complete regression (*vr*), partial regression (*gr*), self-replacement (*zv*), or fragmentation (*an*). The older an eco-unit, the greater the risk of a zero-event. In the order from vr to an, such events have less energy and more restricted impact. The order of eco-units in the Fig. is that, often postulated, of succession theory, from large and short-lived to small and long-lived. This time sequence is probably rare in nature. Processes usually occur in an arbitrary order (silvigenesis). (Oldeman 1979)

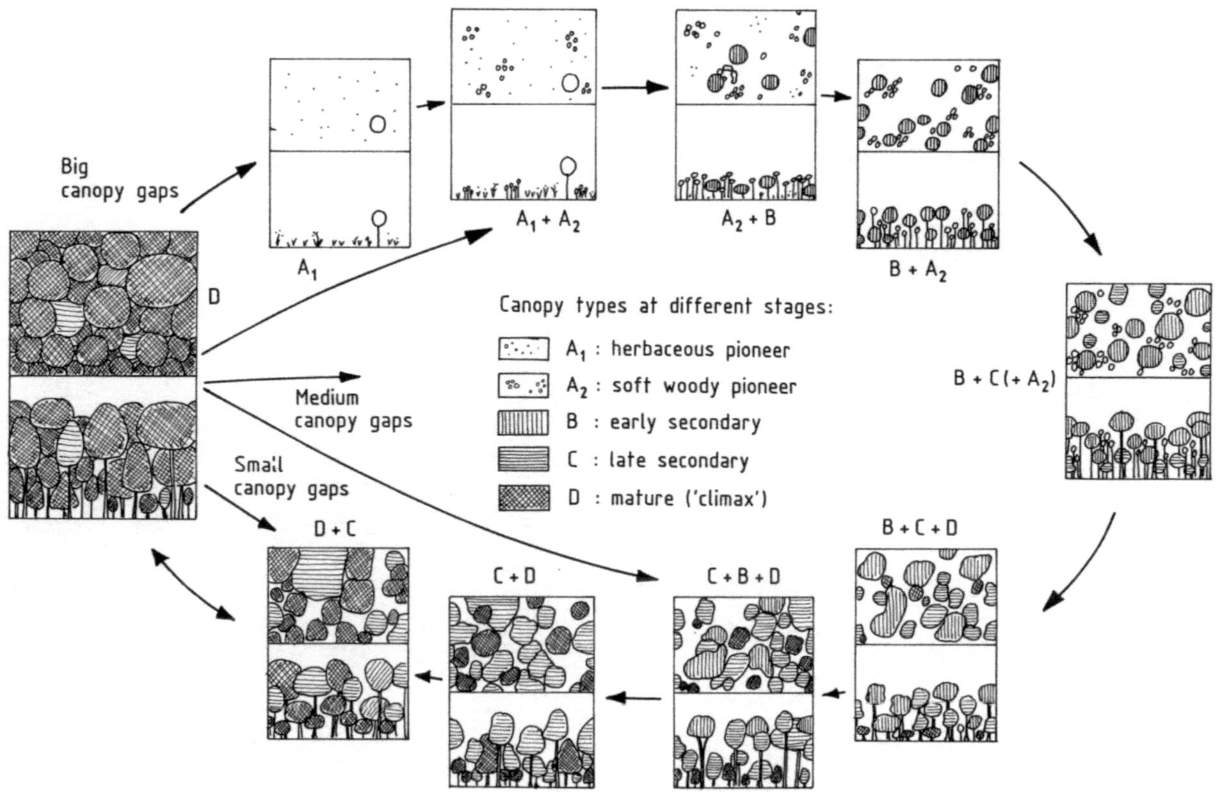

Fig. 6.19B. Progressive succession involving four "elementary canopy types" (*A–D*), following different-sized gap phases in a hypothetical, "undisturbed" rainforest in Australia. The dominant (tallest) canopy type is placed first at each diagrammatic stage. Note that the eco-unit level is not a fully-fledged system level in this model. (After Webb 1977)

ognised after their crown architecture (Rauh's model and Troll's model particularly; Fig. 3.2), which shows little reiteration in this biogeographical zone. Moreover, the lower left side of the picture, below the diagonal, shows that many eco-units are built by lianes and large herbaceous plants. These eco-units form a network without clear limits amongst them, which is a form of premature fusion of the participating eco-units in a blocked aggradation phase (cf. Section 5.2.3).

Indeed, "liane curtains" are a frequent phenomenon in vertically closed mosaic canopies (see Fig. 5.59B). Stufung is completed by these vertical leaf masses, as has been analysed by Caballé (1986) in Gabon and by Beekman (1981b) in the temperate riverside forests near the Rhine in the French Alsace. An early description in which the mosaic notion is not used, and which shows that different liane species fulfill this function in different forest types from Suriname is due to Van Donselaar (1970).

The other plant components of each silvatic mosaic are also often closely linked to the architectural dynamics of such a mosaic, on its specific site. Kahn and de Castro (1985), Kahn (1986a, b) and Kahn et al. (1988) established this to be the case for palm species in Amazonian forest mosaics, which therefore have a different species diversity in palms (Arecaceae). He states (1986b, p.214): "...one (forest type) is characterized by tall and broad-trunked trees up to 50 m in height and 2.5 m in DBH, and by tall arborescent palms; the other by more modest-sized trees up to 40 m in height and 1.3 m in DBH, and by abundant acaulescent and smaller arborescent palms.". He links this distribution pattern of different palms to different chablis ("gap") characteristics. In our terms, the architectural dynamics are different from one silvatic mosaic to another, and plant habitats in small-sized eco-units are largely determined by the interacting, surrounding eco-units of the mosaic.

Koop (1981a) established the same phenomenon for shrubs and herbaceous plants in Northern Germany in a natural forest mosaic close to the ecuilibrium state. He explained it by referring to environmental macro-gradients, characterising the whole mosaic composition, as opposed to microgradients affecting the distribution of the plants studied. In our terms, microgradients are closely related to eco-unit borders and mosaic dynamics. He thus established the fact that species judged to be indicators of a certain forest type (cf. Fig. 5.7) could occur in other phytosociological forest types if and when microgradients due to the mosaic structure created the right biotopes.

6.2.4.4 Large Animals and the Ecuilibrium

In considering the stabilised and stabilising role of animals in silvatic mosaics, let us start with the largest terrestrial species, the elephant. Van Wijngaarden (1985) very precisely studied the role of this species in open forests or wooded savannas in the Tsavo National Park in Kenya. In high population densities, elephants act as part of the shifting event régime, the silvatic mosaic being shifted towards a high proportion of tree-less eco-units (cf. Fig. 6.1, below). Van Wijngaarden concluded that a high cover by trees and shrubs was possible only at densities smaller than 1 elephant per 5 km^2, and that elephant mortality due to food shortage in dry years would occur at densities of over 1.3 elephant per km^2, this being the highest sustainable density.

Fig. 6.20. Ecuilibrium mosaic in a stressed environment near the rainforests in the Rio Caroni drainage basin (Venezuela). Stresses include regular fires, ca. 40 years apart. Note abundant network of linked eco-units with lianes and giant herbs below the *upper-left-lower right diagonal.* For details see Fölster or Huber (*in* INTERCIENCIA 1986; Fig. 6.60). (Photograph 1986 by R. A. A. Oldeman)

Moreover, the elephant density also determines the food sources and densities of other herbivores. An important element in his analysis is the finding that biomass production by trees and shrubs is highest in those species that are moderately consumed by herbivores, those that are either heavily attacked or not at all showing lower productivity. His "elephant-trees-grass-grazers-model" is so precise that computer simulation of the consequences of different management options is possible.

This function of elephants in drier African regions is quite different from the function of elephants in rainforests, like those in Gabon or the Côte d'Ivoire, where tree biomass formation is so fast and trees become so solid that elephant impact in shifting the mosaic becomes accessory (cf. Florence 1981) and their function as seed dispersers is enhanced, always in combination with the translocation capacity of biomass in their excrements (Alexandre 1978a). The impact of their consumption pattern in inducing productivity differences among food plants in Kenya has not been studied elsewhere, to our knowledge. It is a main factor in the action of other large herbivores, however, as is the case in Europe.

European densities of large herbivores (deer, aurochs, European elk, etc.) are artificially enhanced in most cases, for reasons of hunting or tourism, or both (cf. Mayer 1981). In his "ten ecological forest-wildlife commandments" this author

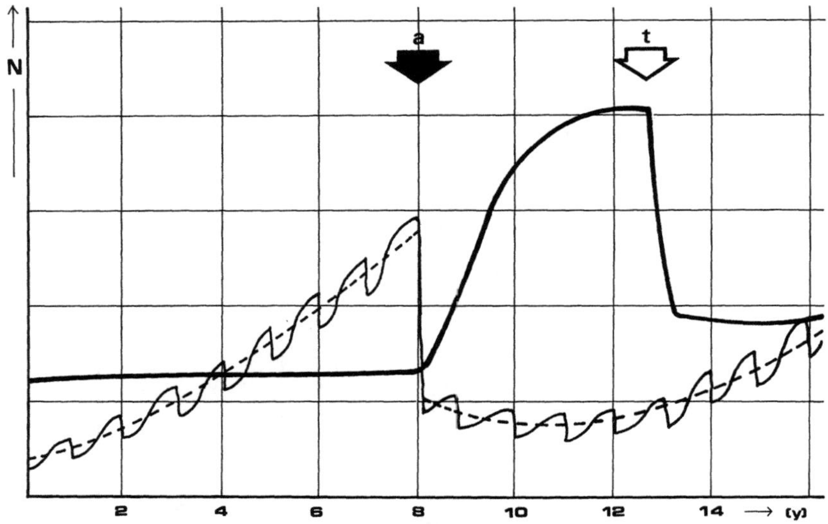

Fig. 6.21. The "large herbivore dilemma" in ecuilibrium stages. *Black arrow* shifting event causing high mortality in animals (*a*), whose population numbers are shown by the *thinner line*, reduced each winter below an average tendency (dotted). *Open arrow* shifting event causing decrease in trees over 5 m high (*t*), whose population number is indicated by the fat line. Event *a* may be a particularly severe and long winter, event *t* may be a hurricane. Both populations influence each other: many herbivores let few trees grow up over 5 m high, and many trees over 5 m high make for little accessible food for large herbivores. For forest mosaics other than the West-European ones, the 5-m limit may differ. For each site, there is an average, long-term balance between animal and tree shifting events. In some regions high forest mosaics with few large herbivores are the rule, in others, park-like tree savannas with big herds of large herbivores. Graph is a visualised principle and does not represent counted values, but see Berryman (1986)

pleads for the reestablishment of managed forest types closer to nature, with a high proportion of selection-managed forests (small-sized eco-units, in our terminology) and an animal density that can live in such mosaics without supplementary feeding in winter. In Europe, winters constitute a sort of "zero-event" for herbivores, many of which die when severe and prolonged cold spells occur. One of the reasons that this mechanism fails is that many animals are saved by wildlife management. They are fed in winter, and in extreme cases receive medical attention.

Figure 6.21 visualises how in principle the increases of animals and trees are determined by their interaction and by different influences upon that interaction. Heavy shifting events create many surfaces with young eco-units, on which herbivores feed abundantly and multiply. But once high mortality occurs in herbivores, many eco-units can aggrade above the consumption height level and the animals can no longer reach the tree crowns. Less animals can be fed on the remaining innovation phases. Feeding the animals, without enough counter-action by hunting, then increases their impact on the silvatic mosaic, which is kept in a precuilibrium stage with dominating non-forest eco-units, a wooded savanna.

The point is raised here both to understand the role of such large herbivores better, which is clearly one of signal transmission and mass translocation, and to bring into discussion an existing school of thought that considers high herbivore populations ("forest grazing") and the concomitant park-like forests as the acme of natural development and nature management. In the role of operators of shifting events, such as the elephants in Kenya, large grazers and browsers are thought to be a general forest component by the adherents of the "bulldozer herbivore theory", advanced by Kortlandt (1982). He concluded: "What botanists and foresters currently consider as a climax vegetation may often represent the final stage of a process of impoverishment resulting from the extermination of most of the large game". Janzen (1985, 1986d) reconstructed evolutionary big mammal loss from the Mexican desert and semi-deciduous Costa Rican forests in Santa Rosa (see Sect. 6.2.3.3.) by examining the plants showing traits apt to be useful in interaction with large mammals. He coined expressions like "defaunated big mammal vegetation" or "megafauna-free forest". In such a view, high closed forest mosaics might be due to underpopulation of large herbivores, and the natural silvatic mosaic would always be a "diversified mosaic landscape of savanna, woodland and forest" (Kortlandt 1982).

Figure 6.21 provides the answer to this "bulldozer herbivore" thesis. Some site factors determine both the frequency of shifting events and the growth performance of the trees. Other site factors determine animal mortality. Their balance determines the ecuilibrium architecture of the silvatic mosaic. On Kenyan sites, this is a wooded savanna with "bulldozer elephants", in Gabon or the Côte d'Ivoire it is a high, closed rainforest with numerous small chablis and "delivery boy elephants" (cf. Alexandre 1978).

Ongoing Dutch experiments in nature management concern the introduction of half-wild bovine races in artificial precuilibrium mosaics (e.g. planted pine stands) with the aim to increase the "natural value" of these woods. Such experiments can only prove that artificial enhancement of animal impact on forests is the equivalent of site degradation, in the sense of decreasing hospitality of the environment for forests (Fig. 6.1, below). This should not be confounded with nature conservation or management. It is the artificial creation of neo-natural mosaics with mainly recreational functions (cf. Sect. 6.2.2.3). This is by no means forbidden or dangerous, but it should be recognised as such and not be presented in other ways.

6.2.4.5 *The Propagule Bank and the Ecuilibrium*

In the ecuilibrium stage, the propagule bank has been built up towards its ultimate composition, with a general presence of hardy, long-lived seeds and other propagules and local, temporary presence of short-lived seeds and other propagules. Holthuijzen and Boerboom (1982) studied the long-lived seed bank compartment of *Cecropia* seeds in Suriname. Guevara Sada (1986) confirmed the prédominance of long-dormant pioneer tree seeds in seed banks in the interior of intact Mexican rainforests, which we suppose to be in the ecuilibrium stage. The seeds of *Cecropia obtusifolia*, their photoreceptivity and its reaction to interaction with dispersing animals were very thoroughly studied by Vazquez-Yanes and Orozco-Segovia (1986, 1987a, b). Different strategies of tree seeds in the North-Eastern USA are treated in the book by Bormann and Likens (1979).

The data provide a partly still hypothetical image of a rather uniform, species-poor basis of the propagule bank, consisting of hardy, long-lived propagules, with a shifting pattern of other component propagules that are shorter-lived. The Mexican authors and Janzen (e.g. 1986c) consider these patterns to consist of differently thinned and trimmed "seed-shadows" cast by trees on the seed-bank. If animal transport is lacking, seeds cluster around the seed plants or for animal propagules around an array of host plants (eggs) and only respond to shifting events that are close in time to their landing on the soil.

Some of the processes implied seem to be rather limited to particular eco- units, others extend to clusters or series of eco-units in the silvatic mosaic. Figure 6.22 shows the pollination and seed dispersal of certain *Ficus* trees from tropical rainforests after Verkerke (1986a, b). The pollination is effected by special wasps (cf. Janzen 1979), guided by impellors (Sect. 5.6.8) and which are so closely linked to the ripening cycle of the figs and so short-lived, that they can barely be supposed to operate outside the eco-unit occupied by the huge fig tree. On the contrary, the dispersal of the seeds is operated by birds, monkeys or bats (Orozco-Segovia and Vazquez-Yanes 1982; Fig. 6.22), which cover large distances inside the silvatic mosaic, consume the ripe and odorous figs and depose the sticky seeds with their droppings on tree trunks and branches. This indeed is a particular part of the propagule bank, because these seeds germinate high up on a host tree, form adventitious roots around its trunk and finally strangle it (strangling figs), as shown in the figure. Extreme as it may be, this example yields a clear image of the highly structured webs of interactions that make a silvatic mosaic function in its ecuilibrium stage.

One of the complicated examples involving migration of a tiny animal species, the Amazonian saddle-backed Tamarin, from one mosaic to another one ("vegetation types") has been studied by Terborgh and Stern (1987). The animal is both very small and rare. For 10 months it lives on fruits produced inside the mosaic, i.e. below the canopy. The other 2 months it gathers nectar from the flowers of *Combretum assimile*, a canopy vine in precuilibrium ("flood-disturbed") mosaics and from *Quararibea cordata*, a middle-sized tree below the canopy of an undefined mosaic. With a biomass of 5 kg per km^2, the tamarin thus caters for large territories in which it assumes communication functions (e.g. pollination) between eco-units and mosaics. These functions may seem quite out of proportion with its discrete, traplining existence menaced by food scarcity for 2 months every year.

These organised interactions have received the particular attention of Janzen (1967, 1970a, 1974, 1978), who among other things studied the synchronicity of flowering in species-rich Central American rainforests and the role of seed predators, mainly insects, on the reproductive success of high forest trees. In the temperate regions synchronicity of flowering (cf. also Sect. 6.2.4.8) and fruiting is induced by seasonal influences, such as cold winters. In drier regions, synchronicity of flowering and fruiting are bound to the climatic rhythm of the rains. It is therefore not astonishing that many authors, mentioned by Alvim and Alvim (in Tomlinson and Zimmermann 1978), have also tried to link tropical flower and fruit phenology to abiotic signals, particularly to climate, e.g. day-length. Janzen (1967) gave an ecological explanation for periodicity in flowering in tropical rain forests. This

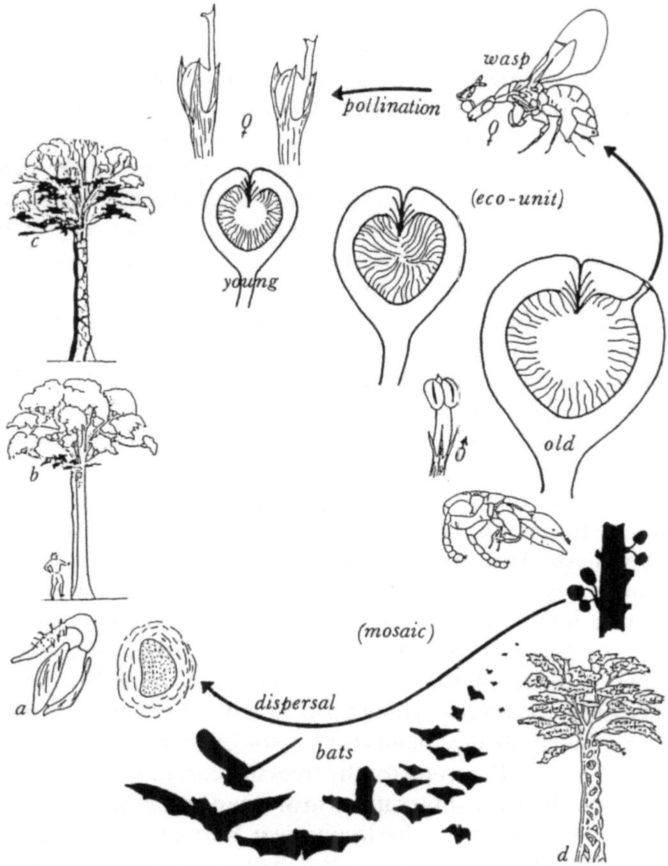

Fig. 6.22. Interactions regulating eco-unit composition of silvatic mosaics through the seed-bank. Example of strangling figs (*Ficus* spp. Moraceae), from the tropical rainforest. Note interactions mainly within eco-unit (wasps) and mainly amongst eco-units (bats). (After Verkerke 1986a)

explanation is linked to interactions amongst organisms and has to be interpreted at the silvatic mosaic level.

According to this author, if there is only slight climatic influence, a feed-back mechanism occurs, in which pollinator insects are more numerous in periods of abundant flowering, i. e. abundant honey and other food, and plants that flower in that period have higher chances to be pollinated because of the abundant animals. This enhances both the chances for off-spring of trees that flower at the "right" period and the survival chances of insects, birds or bats that have a life cycle keyed in to such periods, and their offspring. Such processes are indicated as co-evolution of species or groups of species, that become more and more adapted to each other (e. g. the case study on the tree family of Lecythidaceae in French Guyana, Mori et al. 1987). With Wheelwright and Orians (1982), we point to the fact that zoophilous pollination is a highly organised process, because the "target" of the animal is another flower of exactly the same kind as the one visited before. Regulation of these processes takes place by impellors (see Sect. 5.6.8) in the form of biochemical attractants manifest in odours, by colours and shapes (cf. Muséum National 1986) and by "payment upon delivery" in the form of food (Wheelwright and Orians 1982). Among pollinating animals there are specialists and generalists.

One curious observation concerning colours, by the present author, was the predominance of yellow flowers among coastal or semi-arid tropical tree species. According to P. Raven (pers. commun. 1988) this may be explained as follows. Most yellow flowers contain carotenoids which reflect ultra-violet and so produce "bee's purple", a term coined by Von Frisch to indicate a mixture of yellow and UV attractive to bees. Five or more families of bees would then pollinate these yellow mass-flowering trees, species diversity in bees being far greater in the dry tropics than in the rainforest, where mainly the genus *Trigona* and Euglossine bees dominate, at least in the New World (cf. also Mori et al. 1987).

Mori and Boeke (in Mori et al. 1987) state that the geographical distribution of lecythidaceous trees with zygomorphic flowers (bilaterally symmetric) falls within the range of Euglossine bees, pointing to mutual adaptation. They paid less attention to the distribution of species of Lecythidaceae with yellow flowers. It may be hypothesised that the distribution of colours in the flowers of plant species in a region is a symptom for gradients in the nature and diversity of pollinating organisms, a dominating hue indicating a dominating pollination mechanism.

As a practical consequence of this knowledge, Prance and Mori (in Mori et al. 1987) explain that the failure of plantations of the Brazil nut, the lecythidaceous *Bertholletia excelsa* reaching up to 50 m in height, may be tentatively ascribed to the lack of alternate host plants for Euglossine bees. The bees would have to pollinate the flowers of *B. excelsa* for the trees to bear fruit, but pure plantations lead to bee scarcity. Our own observations in an abandoned Brazil nut plantation, established in the 18th century by the Jesuits near Roura (French Guyana) showed high production of the surviving trees. But then, the plantation had long reverted to a semi-natural late precuilibrium mosaic since the Jesuits left around 1770.

The system of fruiting and seed dispersal in species-rich tropical rainforest mosaics has been studied in French Guyana by Sabatier (1983). He provides a broad and general insight in the factors determining the seed compartment of the propagule bank in silvatic mosaics the world over, because his complicated case study has been organised in such a way that processes may be omitted or simplified if necessary when they either do not occur, or are side-issues in a special mosaic to be studied.

Sabatier's study is based upon the "offer and demand" by component organisms being regularly present in the eco-units that compose a silvatic mosaic, mainly on the Arataye River (French Guyana). Figure 6.23 shows the ways in which trees belonging to an ecuilibrium mosaic present fruits to the dispersal agents. It therefore links together four hierarchical levels (Table 2.1): i.e. organs/organisms/eco-units/mosaics, of which the first pair is visualised, the second pair treated in the columns at the right side which concern the interfaces with animals and other seed vectors in the ecosystems. The complicated scientific terms have been left in because they have a reminder value.

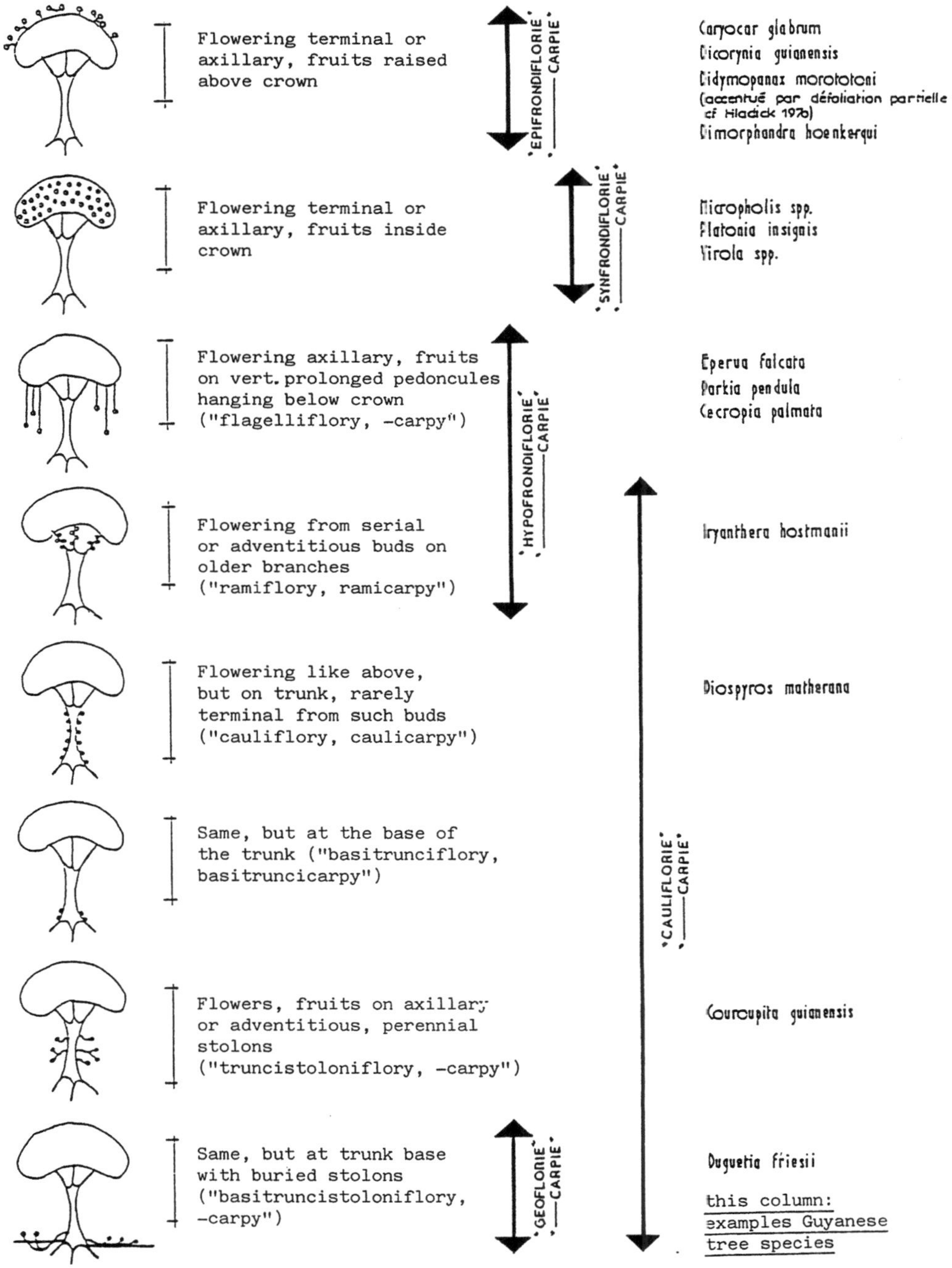

Fig. 6.23. "Offer" of fruits on tropical trees, for "demand" by dispersers. (After Sabatier 1983). French Guyana, tropical rainforest

The interface between fruits and dispersal vectors (e.g. animals, wind, water) has been more precisely characterised by Sabatier (1983) in a key, using criteria that are relevant to fruit availability and attractiveness for dispersal agents and which may be easily observed in the field. Contrary to most "learned" classifications of fruits, this classification is so basic for understanding interactions in silvatic mosaics that the key is reproduced here in extenso (Sabatier 1983, p. 45-46).

Physiognomic classification of fruits studied in French Guyana
Dichotomic key to the fruit types

- Fruits with at least one pulpy or fleshy tissue .... (1)
- Fruits with no pulpy or fleshy tissue at all .... (12)

1. Outer fruit cover dehiscent .... (2)
   Outer fruit cover indehiscent .... (5)
2. Fruit pulp or flesh surrounding dispersal units .... (3)
   Fruit pulp or flesh intimately linked to dispersal units .... (4)
3. Lively outer colouring of the fruit, contrasting with environment .... CLASS I
   No such colouring, fruit hidden in its forest environment .... CLASS II
4. Lively outer colouring of the fruit .... CLASS III
   No such colouring .... CLASS IV
5. Flesh or pulp outside, or outer cover soft .... (7)
   Flesh or pulp inside (outer cover hard, resistance more than 2) .... (6)
6. Lively outer colouring of the fruit .... CLASS V
   No such colouring .... CLASS VI
7. Pulp or flesh intimately linked to dispersal unit ("seed") .... (8)
   Pulp or chair surrounding dispersal unit .... (11)
8. Lively outer colouring of the fruit .... (9)
   No such colouring .... (10)
9. One or two dispersal units per fruit .... CLASS VII
   Several dispersal units per fruit (2 to 15) .... CLASS VIII
   Numerous dispersal units per fruit (over 15) .... CLASS IX
10. One or two dispersal units per fruit .... CLASS X
    Several dispersal units per fruit .... CLASS XI
    Numerous dispersal units per fruit .... CLASS XII
11. Lively outer colouring of fruits .... CLASS XIII
    No such colouring .... CLASS XIV
12. Outer peel dehiscent .... (13)
    Outer peel indehiscent .... (15)
13. Naked, lively coloured dispersal units .... CLASS XV
    Dispersal units with an appendice .... CLASS XVI
    Naked, not lively coloured dispersal units .... (14)
14. Dispersal units shot away by dehiscence .... CLASS XVII
    Dispersal units not shot away .... CLASS XVIII

15. Fruits with wing-shaped appendices. . . . . CLASS XIX
    Fruits with adhering appendices. . . . . CLASS XX

This key has been made after 400 species of which the fruits were studied. The term "dispersal unit" denotes one seed or a complex of seeds that are dispersed as one unit. Sabatier warns that there may be some more characteristics not covered by this sampling, such as different types of appendices than the two mentioned in the key. Nevertheless, we think that most essential plant/animal or plant/vector interface characteristics are included. In Europe, for instance, Oaks (*Quercus* sp.) would belong to class XIX/XX (appendices absent), beech (*Fagus* sp.) to class XVIII, and cherries (*Prunus* sp. pl.) mainly to class VII, denoting different dispersal strategies of the tree species and different interfaces with dispersal agents. It is extremely important both for questions of a silvological nature and concerning natural forest regeneration as a silvicultural method to have such a systematic view of the "dispersal offer".

Another key variable is the quantity of seeds produced in different zones of tree crowns. This information in principle allows calculation of the potentially available seed compartment in the propagule bank on a species-by-species basis. Such calculations have been made by Sabatier (1983) for some few species. Figure 6.24 shows why. He linked fruit production to crown architecture, in particular to the reiteration pattern (cf. Oldeman 1974a). The reader is also referred to Figs. 3.1, 3.4 and 3.8. The investment in time-consuming studies like Sabatier's is only justified if it leads to a reliable assessment of the number of fruits produced per crown in important species, on the basis of simple architectural parameters that can easily be observed and tallied in the field. It is not relevant for all fruiting patterns presented on Fig. 6.23.

6.2.4.6 *Birds, Propagules and the Ecuilibrium*

The "offer" of fruits, in certain quantities and qualities per eco-unit, and with their different fruit/vector interfaces which are often linked to tree and eco-unit architecture, is "demanded" by the animals that are the dispersal agents. Contrary to pollinators, such animals are "paid in advance" when they eat fruits or seeds (Wheelwright and Orians 1982). Many case studies have been made on the behaviour of such animals (e.g. see Richards 1952, 1970; Ayensu 1980). For monkeys in tropical forests, we already referred to studies such as those by Van Roosmalen (1980) or Rijksen (1978), and for elephants by Alexandre (1978). Many studies concern the role of birds as seed dispersers. Some of these are cited here as examples and entries in relevant literature.

Holthuijzen and Sharik (1985) analysed the role of birds in dispersal of the seeds of *Juniperus virginiana* (Cupressaceae) in South-West Virginia (USA). They found that eight bird species dispersed the seeds left on the trees after seed predation, and that each species ensured its own type of dispersal which could be either slow and sustained or rapid and instantaneous. Both fruit production and bird behaviour are seasonal in this temperate climate. The passage of the seeds through the avian digestive tract enhanced total germination by 1.5 to 3.5 times. If different digestion

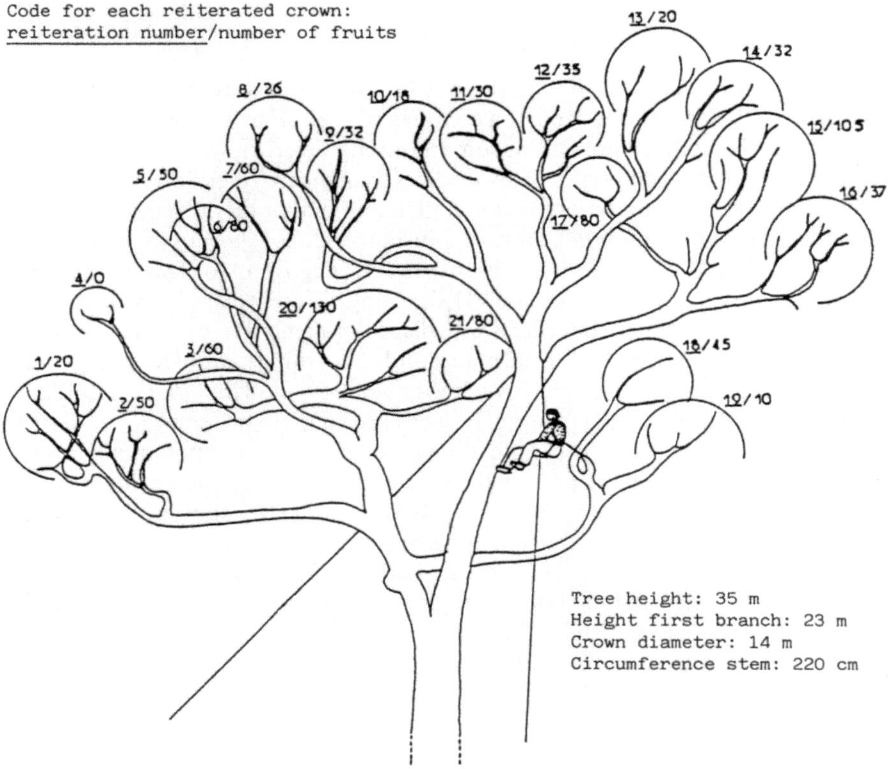

Total number of fruits counted; evaluation by random sampling (0.33):
a) (numbers 2, 10, 13, 1, 19, 16, 20) x 3 = 855
b) (numbers 14, 11, 10, 6, 16, 1, 4) x 3 = 651
c) (numbers 18, 9, 21, 17, 20, 2, 14) x 3 = 1227
d) (numbers 12, 21, 18, 15, 9, 6, 3) x 3 = 1311
e) (numbers 16, 2, 18, 20, 7, 14, 8) x 3 = 1140

Fig. 6.24. Estimation of total fruit production on a tree of *Couepia caryophylloides* (Rosaceae, French Guyana), by considering fruit production per reiterated crownlet (cf. Fig. 5.8). Estimates have little accuracy as yet, research method is slow and dangerous, but the principle of investing research time in finding simple architectural parameters for fruit production deserves attention. (Sabatier 1983)

periods, flight patterns and flight distances are taken into account, it can be easily understood why the propagule bank in these silvatic mosaics is abundantly stocked with seeds of *J. virginiana.*

Van Dorp (in Gomez-Pompa and Del Amo 1985), encountered this difference in bird behaviour among the ca. 50 bird species which he studied in a slightly disturbed ecuilibrium mosaic in the rainforest reserve at Los Tuxtlas (Veracruz, México). He wrote (Van Dorp as above, p. 356 ff.): "Seed dispersal shows two patterns. Seed density may decrease exponentially with distance to the seed tree. The majority of locally deposited seeds stay in a radius of 25 m. Within this belt, the seeds are aggregated rather than uniformly distributed. However, dispersal agents may bring in "foreign" seeds, from other trees that fruit elsewhere, and deposit these seeds in situ, so enhancing the seed diversity in the "seed shadow" of the tree mentioned.... Moreover, dispersal is strongly directional. Most seeds are dispersed downhill and/or in the direction of a high vegetational cover between the 5 and 20 m height levels. High cover percentage in these strata was interpreted as high availability of nesting sites for frugivorous birds. We suppose that there is a positive relation between the "availability of fruits" and the probability of foreign seed introduction in a particular site." (freely translated from the Spanish by RAAO).

This nesting behaviour of birds was studied by Brosset (1974) in Gabon (Fig. 6.25). He states that the camouflage of nests in the tropical rainforest differs from that in the temperate regions, where nests are often hidden under or behind leaves. Nesting is always largely determined by the reduction of risk. Brosset (1974) found that the position of the nests, as "mapped" on an architectural sketch of part of the silvatic mosaic (Fig. 6.25), indicated different risks. Of the nests located in small undergrowth trees, 70% were destroyed by predators or accidents, those built at the tips of twigs were destroyed for 50%, and those built over water for 35%. These risk factors for birds codetermine their part in the organisation of seed dispersal in silvatic mosaics, and therefore are another important feature of the ecuilibrium state discussed here.

The combination of such bird and plant factors in eco-units and in the silvatic mosaic is very far-reaching. As discussed in Section 5.2, the animal species composition per eco-unit development phase (innovation, aggradation, biostasis, degra-

Fig. 6.25. The efficiency of birds as seed dispersers, mass translocators and signal transmitters is linked to population survival, which is codetermined by nesting behaviour. Very different nesting spaces are available within the complex architecture of an ecuilibrium mosaic, like in the example from Gabon (tropical rainforest) shown here. "Nesting space classes" are mapped on an architectural sketch of part of the undergrowth of a mosaic: branch tip over empty space (*A*, *C*); liane tip over empty space (*B*, *D*, *F*, *H*, *I*, *N*); buttresses at tree bases (*E*); isolated shrubs in undergrowth (*G*, *J*) and nest constructed over water (*K*, *L*, *M*). Brosset (1974), from whom this figure is taken, names bird species in each of these cases in Gabon. He also points to the differences of nesting spaces per forest type

dation) is different. At the level of the silvatic mosaic, this is expressed as a distribution of animal species over the whole mosaic, according to its eco-unit composition. For birds, this has been confirmed by such authors as Schemske and Brokaw (1981) in central Panamá, who found that species richness was greater in "gaps" (innovation phases; appearing set of eco-units) than "forest" (aggradation, biostasis; dominating set of eco-units), but capture rates were nearly identical. Of 31 species with adequate sample sizes, these authors found five to have significant habitat preferences (either "gap" or "forest"). This would fit in with the hypothesis that the other bird species to a certain, undetermined extent are communication agents between different eco-units in the mosaic, both as seed dispersers and as mass translocators.

For a forest mosaic in the precuilibrium stage, Komdeur and Vestjens (1982) have designed a silvicultural system that will lead to high diversification of birds of song, gradually transforming some large pine eco-units on unfertile, sandy soils in the Netherlands. It is interesting to note that they propose to do so by what can now be recognised as eco-unit fragmentation, i. e. groupwise replacement of pines either by the same species at another age, or by broadleaved tree species. Their criteria were taken from a matrix linking occurrence and frequency of bird species to architectural tree and eco-unit features. Their conclusions are supported by a study on the influence of forest treatment on birds in France (Ferry and Frochot in Pesson 1974). The sole aim of Komdeur and Vestjens was bird species management.

Their study may be taken as an independent confirmation of the silvatic mosaic development stages of Fig. 6.6, precisely because these authors did not set out to prove it, but arrived at the conclusion that conservation management techniques should follow such a development. Gasc (1981) confirmed an analogous link between distribution and frequency of saurians and the eco-unit composition of rainforests in French Guyana. In line with his remark quoted earlier, he wrote (Gasc 1981, p. 316): "It seems essential to me, in order to understand well how niches and hence species are diversified, to understand the vegetation, being the main context of this environment, in its architectural and dynamic aspects. The divisions that were classically applied by zoologists to pinpoint forest parts corresponding to characteristic animal biotopes to us seem to be completely denuded of nuances".

The earliest serious studies on the role of sounds in the communication network established by birds in silvatic mosaics seem to have been published by Chappuis (1971, 1974–1985). He first analysed the acoustic frequencies of bird songs in Gabon, both in the rainforest and in savannas. The transmission of acoustic signals amongst forest birds proved to be in significantly lower frequency bands than in open ecosystems, as proven by sound frequency spectra established electronically for the songs of 78 bird species. As a general rule, the denser the habitat, the lower is the pitch of bird songs. The average sound frequency of small, shrill-noted birds and medium-sized species using middle frequencies is 1000 Hertz lower in the forest undergrowth than in the savanna. In families with low-pitched song this difference is ca. 300 Hertz.

It appears that a strong selection pressure exists against frequencies higher than 1500 Hertz in dense habitats. That this is linked to the efficiency of the transmission of signals follows from the fact that the forest undergrowth in Gabon is particularly rich in thick-cuticled leaves of 20 cm size and over, which absorb, disperse and block frequencies of over 1500 Hertz in a particularly effective way. Low-pitched bird songs therefore enhance the carrying power of acoustic signals in forests. This is confirmed by the human practice to signal one's presence by beating a stick against buttresses, because the low, beating sound carries much further than a human cry. Low-pitched drums also are used in the "jungle telegraph" of traditional forest dwellers.

Acoustic forest research was taken up for practical reasons at Nijmegen University (Martens 1986; Huisman 1987). These researchers first considered the properties of hedgerows, shelterbelts and forests as absorbers of traffic noise, before adding fundamental questions about acoustics and forest animal communication. They consider vegetational architecture and sound reflection by the forest floor as two variable factors to take into account next to foliar absorption and dispersion of sound waves. The basic significance and implementation of communication by sound among forest insects is passingly evoked in a theoretical context by De Jong (1988).

6.2.4.7 Significance for Nature Conservation and Silviculture

Up to this moment, these factors linked to animals have been taken into account nearly exclusively in nature conservation or management. The design of nature reserves, as linked with "patch dynamics", shifting silvatic mosaics in our terms, has for instance been considered by Pickett and Thompson (1978). For isolated, small, forest relics they refer to the school of island biogeography that was founded by Mac-Arthur and Wilson (1967; also see Andrewartha and Birch 1984). Most conservation planners consider "islands" consisting of sizeable forest mosaic fragments and consider island size and conservation (Zakaria-Ismail 1988). Shugart (1984) also refers to this theory when considering forest "patches" because he does not consider for the moment the interactions between eco-units ("patches") in a mosaic (cf. Sect. 5.6.6). The effect of isolation on the number of woodland birds in small woods in the Netherlands has been studied by Opdam et al. (1984). By multiple regression analysis they showed that, apart from eco-unit size ("patch area"), the amount of woodland near the forest "island" and the distance to extensive forest areas have a significant effect on the number of woodland birds. This is in agreement with our earlier considerations on the installation and precuilibrium stages.

However, the extended discussion on these interactions between eco-units, in which animals play a preponderant part, is also meant to emphasise that most silvicultural systems are not feasible without taking the animals into account. This is particularly the case in all systems depending on natural regeneration to a higher or lesser degree (cf. Mayer 1981), but it also includes many cases of conversion from one silvicultural system to another, in order to implement new management aims.

Up to this day, the world of foresters has often taken a somewhat negatively coloured view of these matters. On the one hand the field manager is proud of the wildlife in "his" forests. On the other hand he is inclined to resent the often occurring fact that fauna and game management are not in his hands, but in those of other persons, services or even ministries which are in charge of wildlife. Both management aims are nearly never harmonised (cf. the "bulldozer herbivore" school, Sect. 6.2.4.4) and therefore the wildlife that is a real source of joy may also become a forest pest, when for instance animal population densities are maintained that exceed the silvatic carrying capacity.

As for the smaller animals, from birds and lizards to insects and other arthropodes, foresters usually see them either as potential pests, or as aesthetically nice side-issues in their forests. The role of such animals in keeping the whole fabric of silvatic mosaics together by implementing ecological communication, mass translocation and seed dispersal amongst the eco-units is neither consistently taught, nor consistently respected in management. In tropical forests, local inhabitants are usually free to kill and eat as many small forest animals as they like. The effects of human overpopulation then cause severe breaks in the population balances of these forest animals, which will change the whole natural resource pattern of such regions in two or three human generations, even if no wood harvesting or forest clearing were involved. Again, the separation between forest management and wild-life management is counterproductive in these cases. And as soon as city dwellers start to hunt in an uncontrolled way for sport or commercially ("game restaurants", souvenirs) the impact on silvatic mosaic development becomes critical.

In order to avoid an excessive narrowing of the silvicultural possibilities all over the world, two ways should be followed.

- Wherever possible, forestry, wildlife and conservation services should merge and harmonise their aims and methods, including their publicity.
- All silvicultural systems that are designed should use all animal resources that are known and available to support forest development towards the "target forest type" desired by management. This includes among others the design of natural regeneration methods, of fertility schemes in which litter decomposition and fertilisation are never seen apart, of biological control of pests and diseases, and of (agro-)forest production of commodities other than wood, such as honey, fruits or silk, to mention a few.

In Section 6.4, on silvatic mosaic production, the interactions determining the dynamics and momentary architecture of such mosaics will be shown, as they were in Fig. 5.75 for eco-units.

6.2.4.8 Flowering and Fruiting Periodicity: Mast Years

Seasonality of flowering and fruiting was mentioned above. The rules which it follows have been critically reviewed by Sabatier (1983), who included the Janzen analysis (1967) and many others. Observed periodicity of flowering, fruiting and vegetative development of component plant organisms in general is induced by abiotic site factors that change rhythmically (warm/cold seasons; wet/dry seasons; periodical inundation) and biotic factors that are linked to interactions within forest mosaics rather than eco-units, at least in the ecuilibrium stage.

Mast fruiting is so called because in the mediaeval European oak and beech forests, the massive quantities of fruits produced in "mast years" served to "mast" (fatten) the domesticated swine of regional farmers. Because fruit formation drains away resources from the sapstream in trees (Fig. 3.32), it has to take place in the same year in which resources (light, water, nutrients) were plentiful, or 1 or 2 years later (Fig. 3.34). Mast years are several "intermast" years apart, from 1 or 2 years in some tropical leguminous trees (Janzen 1978 in Tomlinson and Zimmermann 1978) to 9 or 10 years in oaks (*Quercus* sp.) in Eastern France (Pardé pers. commun.), after which the whole tree population flowers and fruits simultaneously and abundantly.

Although climatic and hydric factors in a certain year have to meet the critical threshold for massive fruit production, the fact that whole tree populations are fruiting simultaneously according to some "internal clock" or "key event" (see also Janzen 1978) points to another cause determining the number of "intermast" years at a specific site. This, again, is linked to one animal compartment, the seed predators, which are often insects (Janzen 1978). With selected populations that flower "as one tree" every so many years, such tree species saturate the environment with seeds, so that predators are saturated also. On the seed crop, a high population of seed predators builds up, but the next year is an intermast year in which their offspring has no seeds to eat, so predator population density sharply drops again. Therefore, in the third or fourth year, the risk of predation again has decreased and the trees may fruit again.

Mast fruiting as an "escape trick" from predation is only one of the hypotheses that might explain it. However, it completely fits in with the image of the silvatic mosaic with its animal-driven communication and biomass-processing systems, that may be reinforced by coevolved properties of eco-unit organisms (see Sect. 6.3.8).

In the African humid tropics, Alexandre (1980) checked the yearly periodicity of flowering and fruiting for 201 forest plants in Taï forest, Côte d'Ivoire. He concluded (Alexandre 1980, p. 349):" A definite seasonality has been found, most trees bearing fruits in December-January, in the middle of the dry season. This is particularly the case for animal-dispersed species, by far the most numerous. The peak production of fruits by wind-dispersed species occurs a little later, at the end of the dry season, whereas mechanically dispersed (autochorous) species bear fruit at the end of the rains".

Many important tree families show species with mast flowering, e. g. the South-East Asian Dipterocarpaceae (cf. Janzen 1974; Oldeman and Fundter 1986; Smits in prep.), the temperate-region Fagaceae (such as oaks and beeches), many Leguminosae (Janzen in Tomlinson and Zimmermann 1978) and the South-American Vochysiaceae (cf. Oldeman and Fundter 1986). They may show distinct behaviour in each species, or appear to mast-fruit with the genus as a whole. This last case is exemplified by the dipterocarps, which are, however, less strict in this sense than was assumed earlier (Smits et al. 1987).

The phenomenon of mast-flowering is extremely important for natural regeneration silviculture. For instance, beech silviculture has been completely geared to it in the "shelterwood systems", gradually removing the trees of a stand and finally leaving enough seed trees behind for the expected mast year. For plantation forestry, mast fruiting means that seeds for nurseries are abundantly provided in some years only, with often long intermast periods when seeds in sufficient quantities are unavailable. If the trees are judged valuable in plantations, vegetative propagation methods from cuttings or **in vitro** have to be developed (Smits et al. 1987), for which thorough insight in reiteration phenomena is indispensable (Fig. 3.21).

6.2.4.9 The Ecuilibrium Stage: Summary Outline

Summarizing, the ecuilibrium stage of a silvatic mosaic is characterised by a kaleidoscopic eco-unit composition which is dynamically constant. Each disappearing eco-unit is replaced somewhere in the mosaic by another one of the same kind. Diversity in eco-units is highest in the ecuilibrium stage, and with it, species diversity. However, average population densities of any one species over a large surface tend to be lower than those of the species in eco-units involved in earlier or later stages (precuilibrium, elimination). The system of interaction between eco-units, implemented for a great part by animals, is most highly organised in the ecuilibrium stage. Being the most complex development stage of all, the ecuilibrium stage is the most difficult to manage.

6.2.5 The Elimination Stage

6.2.5.1 Stress and Eco-Unit Elimination

If an ecuilibrium stage comes under stress, an elimination stage of the silvatic mosaic results (Fig. 6.6D). The phenomenon has been mainly studied for stress caused by human impact, and even then it is not expressed in mosaic terms. In many publications on the impact of acid precipitations on forests in industrialised regions (e.g. Western and Eastern Europe; USA; parts of some humid tropical countries), scant attention is paid to the silvatic mosaic level. Both researchers and managers concentrate on forest components (trees mainly, sometimes animals and fungi), taking into account some of their interactions (e.g. in mycorrhizae), but without arriving even at the eco-unit level (stand level) in a thorough way. The ecosystem at stand level has indeed been considered in production models (cf. Ulrich in Ulrich and Pankrath 1983; Mohren et al. 1986).

The elimination stage has been named after the loss of certain kinds of eco-units, particularly the smallest and rarest kinds that depend for survival upon precise interactions with healthy neigbouring eco-units. If stress occurs, all eco-units become less healthy, and blocks of eco-units may be eliminated at the same time, their place being taken by larger, pioneer eco-units. As said earlier, these are hardier and more resistant, because their component species are hardy colonists of inhospitable sites (Budowski's rule, Fig.6.1. Table 5.7).

Elimination stages are best known to result from human settlement of a region. This has been documented by Jenik (1979) throughout his book. The fact alone that forest clearing for the most simple forms of agriculture concerns at least surfaces between half a hectare and one hectare (cf. Haxaire 1976) emphasises that larger eco-units are taking the place of smaller ones (see Sect. 6.2.3.4). However, forestry is concerned rather with building up forests than with their elimination.

Stress is any factor (stressor) that diminishes or disturbs vital processes in a living system. Its effect on that system is called strain. Two kinds of strain are distinguished. The first one is elastic strain, which is reversible, e.g. the symptoms of a disease. The second one is plastic strain, which is irreversible and is an equivalent of "damage", e.g. dieback of stands, malformation of trees or site degradation. These notions have been defined by Levitt (1980).

Looking at Fig. 6.6 again, the ecuilibrium stage appears in an uneasy balance between precuilibrium and elimination. If and when periodical sweeping shifting events occur, such as fires or hurricanes with a periodicity of 50 to 100 years, the ecuilibrium stage is set "back" several steps towards precuilibrium, but in the next half or whole century, the ecuilibrium stage is re-established. The rare, sweeping shifting events act as momentary, heavy stress factors causing elastic strain. This may be spectacular enough (see Fig. 6.3). But biotic remanence is high (cf. Koop e.g. 1981a), fragmentation occurs immediately and the ecuilibrium stage may be back in an amazingly short time.

Unpublished data from different personal sources suggest this period to be of four decades approximately in tropical rainforests. From Oosterhuis et al. (1982), it might be estimated that in some parts of the USA (Appalachians) this value would be of somewhat over a century, except for rare eco-units containing trees that that grow much older. In Massachusetts (Harvard Forest), we observed very old eco-units with 60-m-high *Tsuga canadensis* (Pinaceae, Eastern Hemlock), which had survived the periodical hurricanes in a deep valley, protected against the wind. Of course, this is a different site than the hilltops, and therefore it supports a different silvatic mosaic with a different ecuilibrium architecture.

For silviculturists, it is of the utmost importance to gather data on these rare but periodical events in the shifting event regime. According to the popular weather book by Pelleboer (1983), concerning extreme meteorological situations in the Netherlands over the last two centuries, this country may count with one heavy storm every 3 to 5 years, but not covering the whole territory. Every forest district might be hit once every 9 to 15 years, not counting the buckshot impact of local small storms. Extremely wet or extremely dry years would occur every 10 years on the average, sometimes in series of years. Such data suggest that rotations of over 30 years in the Netherlands are risky, taking into account that the first 30 years are in innovation and aggradation phases of eco-units (stands), which may be protected by measures and design, and which are still flexible enough to survive as a whole. Prolongation of rotations, which is a general aim of the long-term National Forestry Plan, therefore requires silvicultural systems that take into account and regulate the fragmentation processes that are to be considered normal after some 30 years stand age, and maybe earlier in some stand types (*Populus*; cf. Fig. 5.14). The average 6-year intermast period in several N.W. European broadleaved tree species suggests another interesting average, i.e. one especially hospitable year in six.

These movements of fragmentation and sweeping impacts increasing the proportion of pioneer eco-units in the mosaic are "normal" stress factors that cause elastic strain to occur. This is reversed in due time.

6.2.5.2 Long-Lasting Stress: The Ulrich Model

Long-lasting or permanent stress factors may, however, lead to plastic strain. This amounts to irreversible site modification. Most often, this is the equivalent of forest degradation. The elimination stage then, as a matter of fact, is a substitute mosaic of the former ecuilibrium stage. This will be illustrated by the changes due to acidification, as described by Ulrich (in Ulrich and Pankrath 1983). For the impact of wood harvesting on humid tropical forests see Oldeman (in Van Beusekom et al. 1987).

According to Ulrich, the following steps lead to a change in forest type, which corresponds rather closely to the one mentioned above.

- **Destabilisation I:** humus disintegration. Due to acidification of the soil, the last silicate buffers disappear and after that, the processes of phytomass production and mineralisation become uncoupled (see Sect. 5.3.2). This can be monitored by following the processes of nitrate and base cation leaching from the rooting zone, as an output surplus, because they are replaced by acid cations, such as aluminium or magnesium. Humus types change (cf. Fig. 5.20), the root systems of trees become shallow and soils in the long run may become compacted and waterlogged, with concomitant erosion. The change of humus type of course is accompanied by strong changes in composition, distribution and biotic activity in the edaphon.
- **Destabilisation II:** build-up of a decomposer refugium. During the first destabilisation, the mineral soil has become inhospitable or even toxic by the increased concentrations of acid metal cations. Soil-burrowing animals, such as earthworms, have disappeared. Organic matter in the soil contains more root litter than before, because rooting is more and more restricted to these organic zones. Organic matter above the mineral soil and between the mineral soil aggregates fills the only places accessible to roots and decomposers which find refugia there. The full result is a moder humus profile on an acidified A-horizon, and podzolisation. Podzolisation is a cyclic process, in which acidification causes profound vegetation changes, e. g. from forest to heather, to be followed by desacidification, for instance through grass vegetation. Even-aged, coniferous forest plantations belong to the acidifying vegetation types, according to Ulrich's data. Heavy thinning or selective cutting regimes, in our terms fragmentation in general, can lead to accelerated creation of decomposer refugia that may stabilise the vegetation again. If this does not occur, species with long-lived roots, such as trees, tolerate continuous cation acid production in the soil for a limited time only and then die.
- **Stabilisation II:** stability with low resilience. When the mineralisation rate equals the litter production rate approximately, proton excess "production" ceases and the soil below the silvatic mosaic reaches an uneasy balance which is not buffered any more by silicate minerals that can take up H^+ ions. The compartmentalisation, in spaces with organic matter, roots and decomposers versus mineral spaces that are inaccessible to these organs or organisms, is stabilised. No soil mixing by burrowing animals occurs any more, so that the soil compartments are

long-lasting, with a certain mutual compensation in ion contents by abiotic processes such as water flows. The roots, being confined to a limited, mainly superficial portion of the soil volume, cause the trees to become more vulnerable to drought, notwithstanding mycorrhizae. This situation can only be redressed by abiotic processes, such as weathering in the subsoil that may create new buffer minerals (silicates) for proton uptake. In this way the local ecuilibrium mosaic, Ulrich's stability range I, may be approached again to some extent.

According to Ulrich (1983 in Ulrich and Pankrath), excessive phytomass use by humans also leads to uncoupling of phytomass production and mineralisation, and hence to acidification. This is substantiated by studies on European forests having been used for wood production for centuries. It also provides a link between Ulrich's model and the evaluation diagrams of the consequences of wood harvest following different methods in the tropics, given by Oldeman (in Van Beusekom et al. 1987). Excessive wood exploitation will bring acidity problems to the tropics. Strongly and moderately acid ranges being typical of tropical rainforest and savanna soils (Young 1976), the European and North-American problems should not go unnoticed by tropical foresters and conservationists. Tropical forests, far from being immune to the threat of acid deposition or acidification by excessive use, are particularly vulnerable (cf. Poels 1987). Klinge's image of root functioning in Amazonia (1973b) corresponds surprisingly well to Ulrich's spatial configuration of range II stabilisation.

6.2.5.3 The Elimination Stage: Final Remarks

The architecture of a silvatic mosaic in its elimination stage tends to "slide down" the site hospitality range shown in Fig. 6.1, and can be recognised by increasing proportions of large-sized, woody pioneer eco-units and herbaceous eco-units (Fig. 6.19). If oscillations in development between the precuilibrium and the ecuilibrium stages are reversible and often revert faster than was thought, for the elimination stage the rule is different.

Elimination stages can be quickly established by careless exploitation or clearing of forest mosaics, yielding most often a considerable quantity of fast dollars, ecus or yens and costing little, because no sustainable use is considered. To reverse this trend, however, is much more difficult and expensive. In many cases, the site has been changed so drastically that restoration can only start in an early step of the precuilibrium stage, and will always lead to an ecuilibrium that is different from the pre-existing one, because the site has irreversibly changed.

6.2.6 The Stage of Collapse

If uncorrected, elimination stages are followed by a stage of collapse (Fig. 6.6E). Little needs to be said in this section. Photographs of formerly forested landscapes completely denuded by chemical deposition (e.g. parts of Central Europe) or by unchecked erosion (e.g. North America, Mediterranean region, tropics) are rife in the media and other publications. There is an enormous scientific and general literature on this subject. Here, only some few examples are mentioned, i.e. the classical "Topsoil and Civilization" by Carter and Dale (1955) and "Silent Spring" by Rachel Carson (1962), or "Climate, Soils and Vegetation" by Money (1965); Bauer (1985, mainly chemical degradation), Walls (1980, on desertification), and Morgan (1979, on erosion).

Collapsed silvatic mosaics, of which nothing was left but bare and exhausted or even poisoned soil, have often been confided to foresters. The last chance of restoration for lands that became useless and barren seemed to be reafforestation with hardy tree species. Large European regions have been recovered in this way in the past two centuries.

Collapse is a stage that can lead back to a new installation stage of a silvatic mosaic, with little biotic remanence if at all and a site diversity that is arbitrary with regard to the original situation. Platt, in summarising the results of the 1975 conference on "Recovery and Restoration of Damaged Ecosystems" in Blacksburg (Virginia, USA; in Cairns et al. 1977), recapitulated some basic, simple concepts, which are reproduced here and reworded as required by the present context.

- Existence of nearby propagule banks providing re-invading organisms;
- presence of vectors for the propagules (may be human);
- minimal hospitality conditions of the collapsed habitat after stress ends;
- residues of toxic matter below a critical threshold;
- acceptable physical and chemical water quality following stress;
- sufficient local organisms having formed resistance to the stress factors;
- some functional and structural similarity between the disappeared and the coming combinations of site and silvatic mosaic;
- availability of natural "cleansing capacity" (output mechanisms for stressing products);
- considerable distance between actual stress level and ecocidal stress level (cf. Fig. 5.88B);
- management capabilities for control and recovery of the damaged ecosystem.

This is a useful checklist for afforestation or reafforestation projects. Of course, it should be made to include measurable criteria under each heading. And in many cases, the species referred to will include exotics, whereas it is emphasised that in no case tree species may be considered alone, because the non-tree components play an extremely important part, especially in installation stages.

6.2.7 Rhizospheres in Silvatic Mosaics

Roots and the rhizophere are mentioned here at the silvatic mosaic level, rather as a reminder, because roots have rarely been studied at this level, if at all. There are no extensive root maps (but see Schmidt in prep., for Suriname, and some partial Central European data in Köstler et al. 1986), let alone belowground profile diagrams of representative parts of silvatic mosaics. Existing documents are exemplified in Sections 3.3 and 5.3. They concern individual trees or at best eco-units, like the Okoumé root map from Gabon shown in Fig. 3.16. Therefore it is impossible to determine the belowground interaction between eco-units with anything resembling accuracy.

Some empirical evidence points to the rapid invasion of the soil compartment in innovating eco-units by tree roots from neighbouring eco-units in the mosaic. Small eco-units can be closed by roots and crowns from surrounding eco-units before seedlings get a chance to become established. "Autogenic canopy closure" (cf. Sect. 6.2.4.9) might well start belowground. Of course, this is the effect sought in silvicultural thinnings, where no stand fragmentation is aimed at. Rooting space becoming available by removing thinned trees is to be used by specific other trees of the stand, which are destined to grow best until harvesting time. Thinnings maximise this effect right to the innovation threshold, but not beyond it.

Form and surface of regeneration surfaces therefore depend very much upon the growth dynamics of the trees to be cultivated, particularly their root dynamics. Line plantings in tropical rainforests, as documented by Bleijendaal (1979), are an example. This method, which would include all kinds of variants, e. g. wheel-spoke set-ups, was meant to be a compromise, sparing most of the existing forest, but enriching it with commercially interesting species. A planted line is a long and narrow innovating eco-unit. If it is too narrow, invasion by neighbouring root systems is fast, and planted trees are not able to expand their root system. This is particularly so with container plants, which need time for their roots to develop outside the planting clod. Finally, the canopy is closed by the crowns of neighbouring eco-units and the linear eco-unit dies young.

On the contrary, if the lines are too wide, the eco-unit behaves like a pioneer eco-unit, and planted trees have to compete with hardy pioneer trees and lianes. In both cases, the commercial tree species have little chance, unless the line has the exact width that is compatible both with the silvatic mosaic and with the exposition to sun and wind. A North-South line has the full sun at noon, an East-West line receives some sunshine in the morning and evening, whereas the situation is different again according to the toposequence considered. Lines on hill-tops, slopes facing North, South or West, or in valleys all present different exposure. Few line plantations in the tropics have been full successes, and it is certainly due to root dynamics at the mosaic level that many of them failed.

There may remain many unresolved misunderstandings as long as no precise research is done. For example, in the east of the Netherlands, farmers believe that poplars (*Populus* sp.pl., Salicaceae) from hedgerows are strong root competitors for the crops in neighbouring fields. In 1980, this was checked by a graduate project group of Wageningen University, who found that the diatropic roots of oaks (*Quercus robur*, Fagaceae) from the hedgerows penetrate much further outside these linear, narrow eco-units than those of poplars. Such data remain the results of case studies. They cannot as yet be generalised.

The effects of root anastomosis (Fig. 3.16) at the level of silvatic mosaics are all but unknown. They may have a significant effect in intermediate-sized eco-units consisting of tree groups belonging to the same species, and their expansion beyond this eco-unit wherever a contiguous innovation space opens up (cf. Newbery et al. 1988, discussed in Sect. 6.2.3.3).

Klinge (1973a, b) established root mass over tracts of Amazonian rainforest mosaics, but did not analyse his results in terms of eco-unit composition. He found that the percentage belowground phytomass decreases with increasing diameter classes of trees. Drawing conclusions from these data as to the possible strategy of seedlings, invading the soil before leaf-bearing axes invade the aerial space in innovation phases, fits in with his image but remains hypothetical until such conclusions are checked. In his publications, comparisons are drawn with many other root mass inventories all over the tropics, but these data are widely divergent. For instance, in vegetations classified as "tropical rainforests", the percentage belowground phytomass as compared with the aboveground phytomass ranges between 5% and 25%. Because analytical data on silvatic mosaics, or comparable vegetation units, are completely lacking for each of these cases, the only warranted conclusion is that such data are indispensable for correct interpretation of the figures cited. Such biomass studies have been united in a bibliographical study by Van Zadelhoff (1976), to the bibliography of which the reader is referred for the older and basic literature. Newer sources are available from Bruijnzeel (1982, 1986).

Data and a long international bibliography on the seasonality and abundance of forest floor arthropodes in tropical rainforests have been united by Pearson and Derr (1986). Their research concerned three Peruvian forest types, i.e. forest on terra firme (dry land), bamboo stands and flood plain forest. Seasonality in arthropod biomass fluctuation was well expressed in the first two forest types, less so in flood plains, whereas in all forests comparable circadian fluctuations occurred. These data point to different niche structures for arthropodes in the superficial litter and maybe in the organic horizons in different forest types, i.e. different silvatic mosaics. This remains hypothetical until checked.

Extremely interesting data on root tubers in Central African rainforests and neighbouring savannas were united by Hladik et al. (1984). Such tubers are organs of nutrient and energy storage. Contrary to woody tubers, such as those found in the Mediterranean *Erica arborea* (Ericaceae, Fig. 6.26), the Central African root tubers are fleshy. The authors write (Hladik et al. 1984, p. 283): "The different underground storage structures are considered as adaptive to the rain-forest environment. During the long period that most plants spend in the shade of the undergrowth (as the "potential set", see Sect. 5.2.4.1 – RAAO), the energy stored in tubers is more likely to escape insect and/or vertebrate predation than the aerial parts of other plant species. Accordingly, many tuberous plant taxa, both dicots and monocots, might well have originated in rainforest areas rather than in more seasonal woodlands or savannas."

According to their data, tuber-forming species belong to many biological types, of which the yams (Dioscoreaceae, *Dioscorea* sp. pl.) are conspicuous herbaceous lianes and important human food crops. The authors state that in many cases sexual reproduction of such plants is very effective, but that fragments of root tubers that have been left behind by consumer animals will give rise to a vegetatively reproduced young plant. In our opinion, this particular root strategy both serves the survival chances of individual trees (e.g. cassava!) or lianes (e.g. yams) and supple-

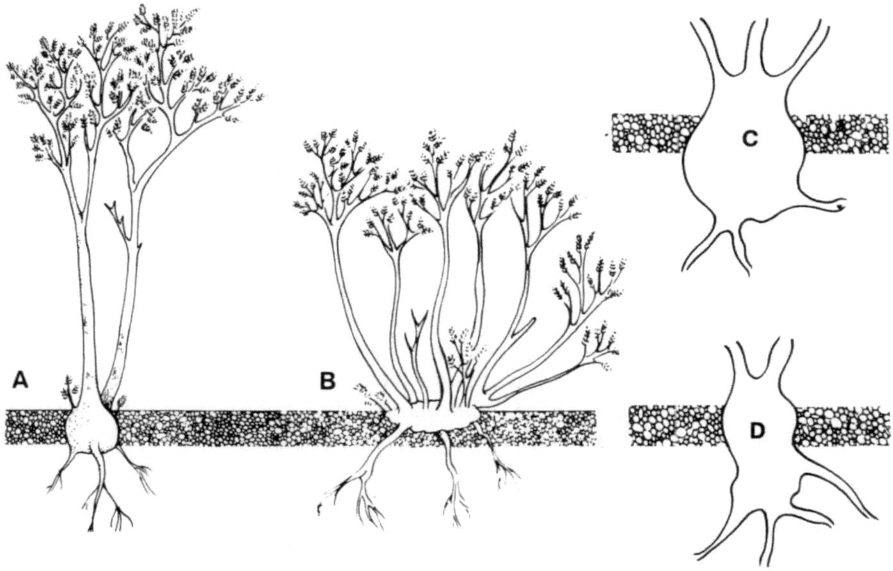

Fig. 6.26. *Erica arborea* (Ericaceae, Mediterranean region). Woody root tubers, probably reserve stocking organs like the fleshy root tubers in many tropical rainforest plants described by Hladik et al. (1984) from Central Africa. The root tubers of *E. arborea* are used in the manufacturing of tobacco pipes and are empirically distinguished as useless "male" (**C**) and useful "female" (**D**), formed by shrubs that can be distinguished by their architecture (**A, B**). (Alexandrian 1981)

ments the abundant but risky seed dispersal strategy with a little abundant but also less risky way of reproduction. This fits in with the extremely complex life environment provided by tropical rainforest mosaics, in which small eco-units in large numbers prevail (cf. Figs. 5.52, 5.55 and examples in Sect. 6.2.8). So many interactions are important in this mosaic, that one species may evolve various strategies along different lines.

No general picture of the rhizosphere in silvatic mosaics can as yet be provided from the above scattered data. Rhizosphere research at this level, which is expensive and time-consuming but rewarding, should be undertaken in the years to come and follow lines of ecological system analysis.

6.2.8 Examples of Silvatic Mosaics

Some examples of silvatic mosaics will be given below to visualise developments explained above. Because many aspects have been illustrated earlier in this book, the number of examples is reduced and references are judged to complete this section sufficiently.

None of the examples can be said to be totally without internal site differences. This is illustrated by Poels' study on site characteristics in Suriname (1987). It emphasises that silvatic mosaics, as the largest and most long-lived terrestrial ecosystems, usually overlap with smaller-sized abiotic site patterns. Emphasis in forest ecosystem analysis should be placed at the eco-unit level, where site heterogeneity usually is negligible. This also brings many models back to manageable proportions. As we will see later, the same effect is reached for mosaics composed by mini-eco-units (cf. Figs. 5.50 to 5.59), by considering "stands", i.e. management units that are generally located in a rather homogeneous site area. Models used often consider such mosaics as tree mosaics, not eco-unit mosaics, in other words as mixed stands (see Sect. 6.3.3).

The examples go from the temperate regions to the humid tropics. This more or less follows Fig. 6.1, from the bottom up, along a gradient of increasing site hospitality. This is loosely done, because there is no strict classification of world sites as yet, notwithstanding the "primary productivity maps" of the world that have been established since Paterson (1956). They developed further as more data and more sophisticated, computer-aided processing techniques became available (Walter 1968, 1973; UNESCO 1974, 1979; National Academy of Sciences 1975; also see Sect. 6.5). But such very general documents cannot descend to the level of each particular silvatic mosaic, where quite significant architectural change can be wrought by seemingly small site modifications in the soil compartment (cf. Fig. 5.69). Such modifications are rarely the result of mere chance, and most often are linked to abiotic distribution patterns, such as toposequences.

6.2.8.1 The Broadleaved Forests at Fontainebleau, France

La Tillaie and Gros Fouteau in Fontainebleau have been forest reserves without human intervention since 1861. These silvatic mosaics have developed in a late precuilibrium, or an early ecuilibrium stage dominated by small-sized beech eco-units. Their shifting event regime includes heavy storms, such as those in 1973. This forest has been thorougly studied over long periods by Professor Lemée's group of Paris University. Data and bibliographies can be found in Faille et al. (1984) or Van Baren and Hilgen (1983). The basic document is a complete map by Bouchon et al. (1973), on which soils, all crowns, fallen trees and other salient features are shown in that stormy year.

Figure 6.27A shows a transect in a small-scale silvatic mosaic, studied in La Tillaie on 1 ha approximately by Van Baren and Hilgen (1983). Although the site is heterogeneous, all soils are more or less podzolic: 36% on washed-out neopodzols, 19% on podzolic soils and 45% on podzols, according to Bouchon et al. (1973). Following the Braun-Blanquet classification, 77% of the forest belongs to the Fago-Quercetum, and 23% is transitional towards the Melico-Fagetum on better forest soils. Van Baren and Hilgen also studied a hectare on these better forest soil sites

Fig. 6.27. Near-ecuilibrium mosaic at Fontainebleau Biological Reserves (France), **A, B** in La Tillaie, **C** in Gros Fouteau. **A** Side view of transect on good forest soils, note proportion of biostatic eco-units, fusion. **B** Side view of transect on podzolic soils, note proportion of aggrading eco-units. **C** Eco-unit development after chablis by fallen beech (*Fagus sylvatica*), note invasion by neighbouring eco-units and passage of central eco-unit from innovation (1969, 1974) to aggradation (1982). (**A, B** Van Baren and Hilgen 1983; **C** Faille et al. 1984)

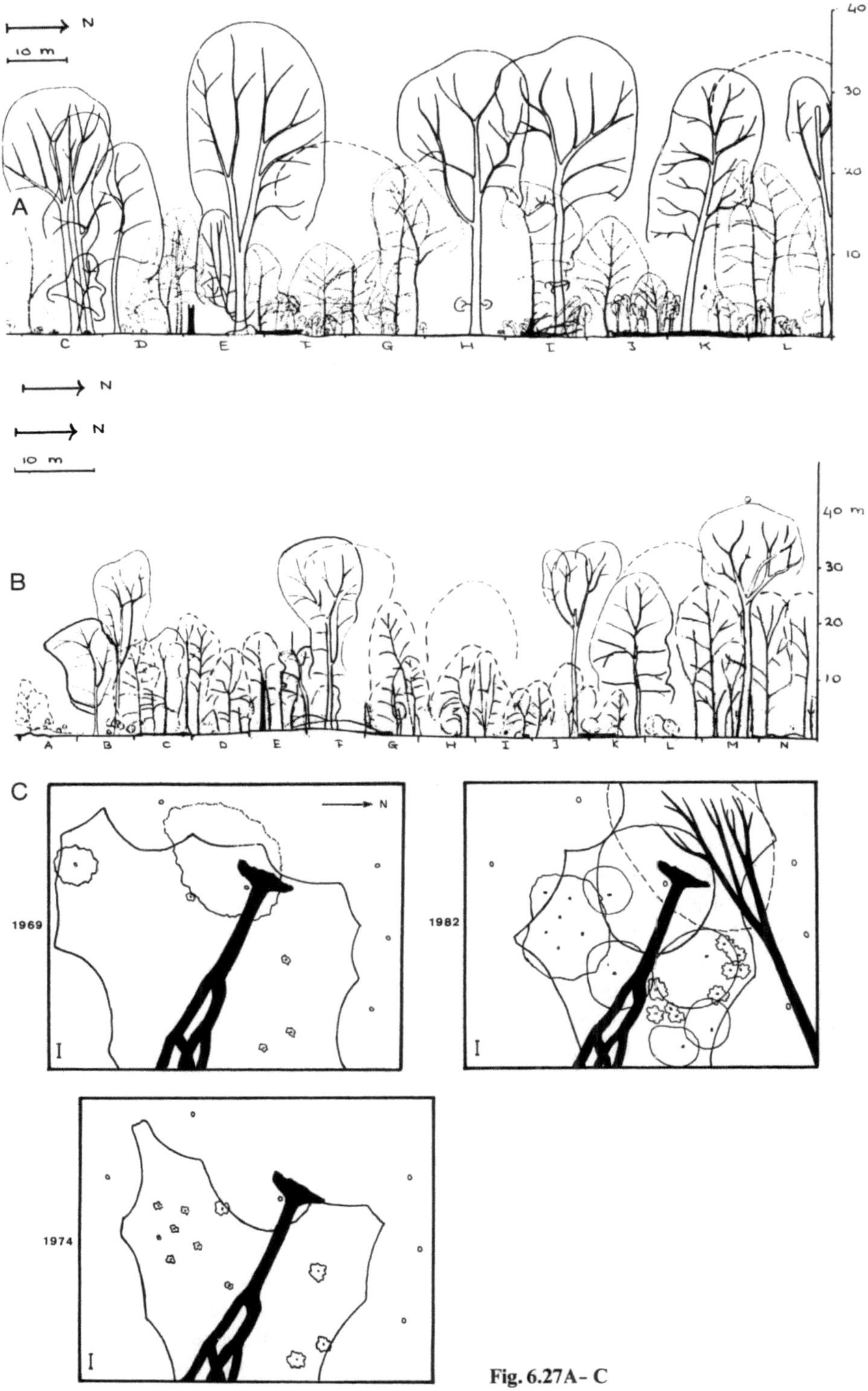
A
N
10 m
40
30
20
10
C
D
E
F
G
H
I
J
K
L
B
N
N
10 m
40 m
30
20
10
A
B
C
D
E
F
G
H
I
J
K
L
M
N
C
N
1969
1982
1974

Fig. 6.27A- C

(Fig. 6.27B). The difference in mosaic architecture is clear. On the less hospitable podzolic soils, dynamics are more intense and the share of the appearing set of eco-units is much higher than on the more hospitable soils, where eco-unit composition shows a high proportion of dominating beech eco-units. Eco-unit fusion is probable in the second case (see Sects. 5.4.3.11/12).

Figure 6.27C shows the detailed dynamics of eco-unit development in another part of the Fontainebleau forest, the biological reserve of Gros Fouteau, from the innovation phase in 1969 to the aggradation phase in 1982, when an adjoining innovation phase had recently been created by a chablis. The growth of surrounding high beech crowns into the gap is to be noted. If the gap had been smaller, no innovation phase would have occurred, but neighbouring biostatic eco-units would have shared the open space for lateral expansion (see Sect. 6.2.4.9). There is clearly a critical minimum size for innovation to happen. This defines the natural threshold between thinning and selective cutting in silviculture: selective cutting aims at the creation of new eco-units, whereas thinning aims at the use of opened spaces by the trees of existing eco-units.

Figure 6.28, after Van Baren and Hilgen (1983), represents changes in eco-unit composition between 1968 and 1982 for the silvatic mosaic on podzolic soils (Fig. 6.27A). It should be noted that the storms of 1973 have played the role of a rare but regular shifting event, so that the development over those 14 years was not continuous. This is visible in the biostatic eco-units. As many have disappeared from this phase as have grown into it. On the other hand, more innovation phases disappeared than were appearing. This indicates that, after the storms, the mosaic recovered fast. Provisionally, as long as it has not been checked, this might perhaps be a sign that the mosaic is in or near the ecuilibrium stage.

Other European studies on silvatic mosaics that have been left alone for half a century or more are due to Koop (1981a; see Figs. 5.1 and 5.7) in North-Western Germany, to Bartelink et al. (1987) in Lady Park Wood, Wales (Great Britain), where long-term studies have been done by Peterken's team (Peterken and Jones 1987), to Falinski (1977) or Derkman and Koop (1977) in the very old and nearly undisturbed giant forest reserve in Bialowieza in Poland (see Sect. 5.4.3), to Bücking (e.g. 1985) in the "Bannwald Waldmoor-Torfstich" in the lower Black Forest (Federal Republic of Germany), a book including animal studies, to Pruša (1985,1987; see Fig. 6.16C) in Czechoslovakia, to Zukrigl et al. (1963, Fig. 5.57) in Nieder-Österreich, and to Van de Winckel (1980; also cf. Figs. 6.44; 6.50A) or Walter (1982) in the French Alsace, where they studied the semi-natural forests along the Rhine (Fig. 5.56).

These studies have in common that the architectural element in methodology is strong. A general survey of "Urwald"-data in mountainous regions of Europe has been united by Leibundgut (1982), who is one of the principal authors to have given shape to the Central European School in silviculture, using the principles of natural forest development (see Sect. 6.3.3). Leibundgut (1959) wrote a basic article on the analysis of virgin forests ("Urwälder"). Except for the study in the Pyrenees (see Sect. 6.1.4; Chéret 1987), those by Mulder and De Waart (1985) in Austria and by Van der Burg et al. (1986) in the French

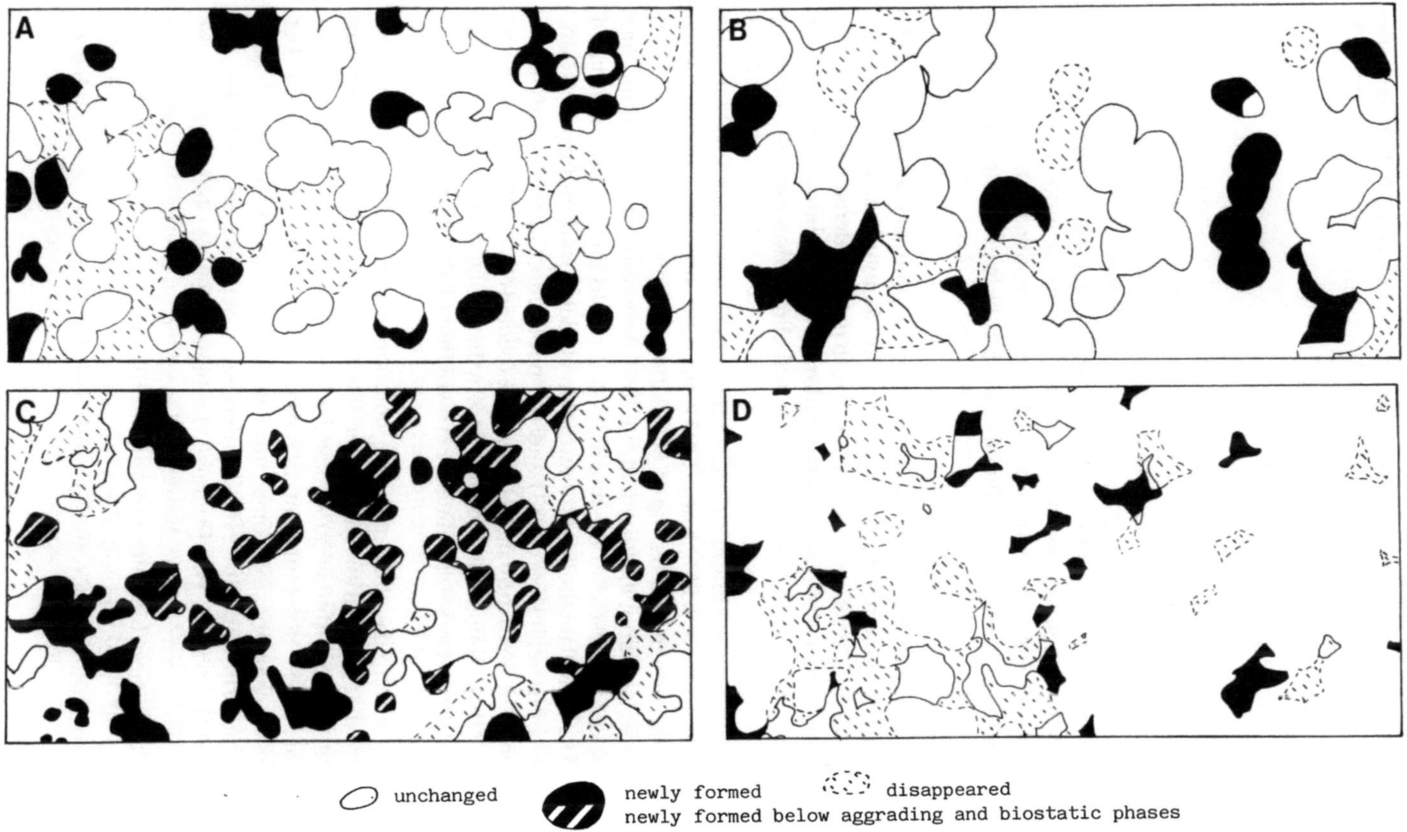

Fig. 6.28. Changes in the mosaic of Fig. 6.27B from 1968 to 1982. **A** Biostatic phase. **B** Aggradation phase. **C** Early aggradation and late innovation phases ("dense phases"). **D** Clearings (innovation, early). (Van Baren and Hilgen 1983), terminology modified to fit in present book

Vosges Mountains are mentioned here for their recent examples of imitated natural silvatic mosaics, managed by selection silviculture (jardinage, Plenterwald, uitkapbos).

6.2.8.2 *Pioneer Birch Mosaic, the Netherlands*

Koop (1983) provided the example of a fragment of a spontaneous young precuilibrium mosaic, which is representative for many sites in North-Western Europe with low fertility, often sandy soils, and podzolisation in a more or less advanced state. In the Netherlands, such podzols often have been broken up and mixed in reafforestation projects during the past century, but the main architecture of the precuilibrium stage remains about the same. With recent chemical site changes due to atmospheric deposition (pollution), such mosaic architectures become more frequent but should then be regarded as elimination stages.

The difference is in the presence of young oaks (*Quercus robur*, Fagaceae) below the large eco-units built by birch (*Betula pendula*, Betulaceae). This is a first indication of later fragmentation, involving long-lived pioneer eco-units built by oaks (see Sect. 6.2.1). Therefore, this forest still is to be considered in the precuilibrium context and not as the result of elimination of mosaic parts. Of course, the oaks may die precociously if stress increases, and elimination may still occur in this way.

Figure 6.29B shows the graphical, computer-aided simulation of this mosaic architecture (cf. also Fig. 5.86). Koop (1983, no page number) writes: "Work has begun on the design and testing of simulation models for prediction. With the techniques in question, existing patterns and developments can on the one hand easily be demonstrated to the manager in the field and on the other hand analysed with statistical methods and simulation models for scientific purposes.". His work has continued (Koop 1987) and the publication of the first dynamic architectural mosaic simulation models is to be expected (Koop 1989). The reader is reminded that these simulation models concern the ecological volumes that contain crowns, but cannot as yet simulate the crowns that fill these volumes (see Sect. 5.67).

In order to exemplify some non-tree components in this mosaic picture, Fig. 6.30 is added. It gives an analytical overview of the spaces in N.W. European forest mosaics where different moss species can be expected to occur. This figure therefore does not represent a mosaic transect drawn to scale, but rather presents schematically some unrelated eco-units on different sites (north and south slopes). However, border-zones between eco-units are not omitted but can be traced. For instance, the biotope on the left (dead leaves blown away) will occur at the border of large eco-units in the innovation phase, where they meet biostatic eco-units in the main wind direction. Stumps with accumulated litter around them are characteristic for small, artificial eco-units, the microclimate of which is determined by the surrounding, high eco-units.

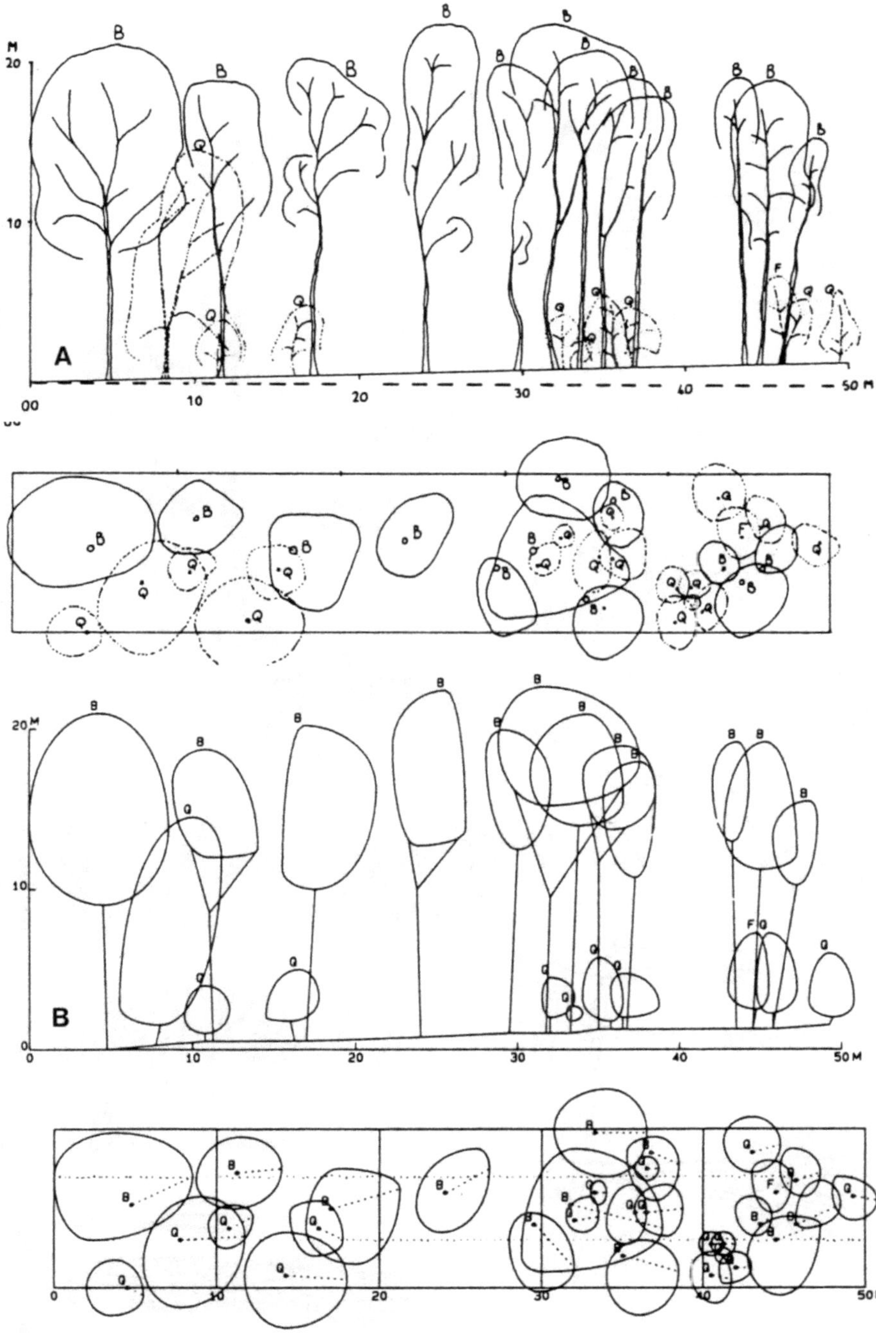

Fig. 6.29. Early preculibrium stage of silvatic mosaic on unfertile soils, common on podzols in North-Western Europe. Large pioneer eco-unit built by birch (*B Betula pendula*) with appearing smaller long-lived pioneer eco-units between-and-below, built by oaks (*Q Quercus robur*). **A** Sketch to scale, side-view and map. **B** Computer-simulated architecture of the same mosaic fragment (cf. also Fig. 5.86). Note that crown volumes remain empty, because crown architecure can-

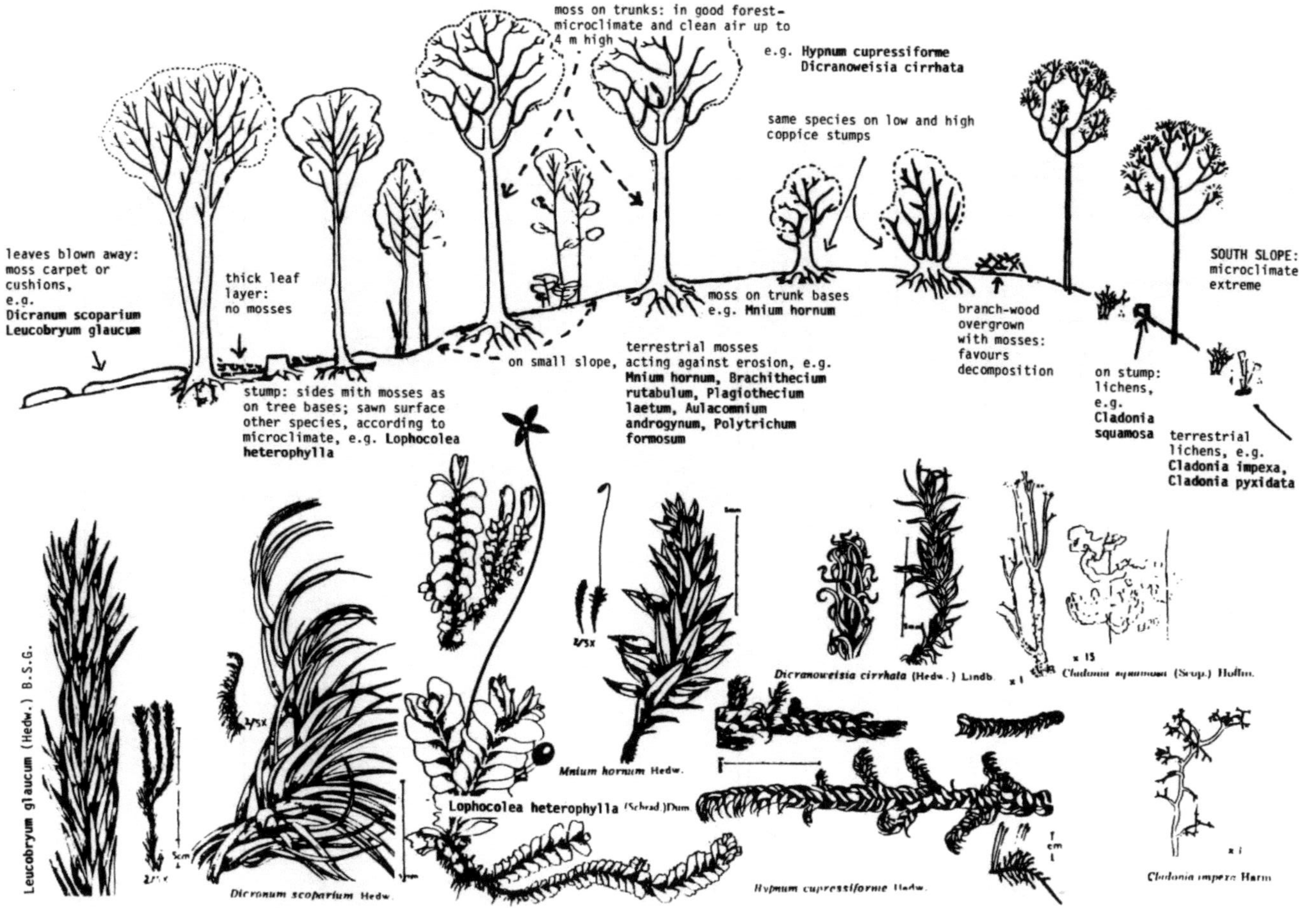

Fig. 6.30. Moss species and silvatic mosaic composition. Note that interaction between eco-units can be read in this figure, e.g. "leaves blown away" (border innovation/biostasis) or "stump" (small innovation surrounded by biostatic eco-units). Different mosses in different life spaces. (Courtesy A. L. Jansen, AUW-Phytopathology)

6.2.8.3 Neo-Natural Mosaic Fragment, the Netherlands

In their very detailed study of the forest vegetation on the Wageningse Bergrand, which is the slope of a sandy moraine, formed by glaciers during the Riss Glaciation and exposed to the south and the Rhine valley, Wilderink and Zakee (1985) analysed sites, history, floristic composition and architecture. Human history is the main determinant of the forest that exists there now. The soil has been worked more than once and many tree species have been introduced from abroad. Figure 6.31 exemplifies a moderate case of such mixtures, American red oak (*Quercus rubra,* Fagaceae) building a forest mosaic together with a European oak (*Q. robur*), birch (*Betula pendula,* Betulaceae - see Fig. 6.30) and beech (*Fagus sylvatica*).

As a matter of fact, this is a later step in the precuilibrium phase, following the one shown in Fig. 6.30, and enriched by red oak. It is clear that the exotic species creates its own, powerful eco-units (e.g. group at the left-hand side, Fig. 6.31A), that birch eco-units (to the left of the first group) are less high and that European oaks remain behind altogether (group before the birches, next to *Q. rubra*). Different life-spans and temperaments (cf. Barkman 1988a) of these species will make visible the fragmentation now masked by fusion and may allow the aggradation of beech eco-units. The whole canopy is fused now, except some little stufung on the left and right sides of the figure. The forest has not been tended except for recreational purposes, and the neo-natural development is a variant of local natural development.

Figure 6.32 shows the same mosaic fragment as a crown projection map. It is evident that some eco-units, such as the one crossing the middle, are in a more advanced state of aggradation than the others. The whole architecture announces a phenomenon that has been established by Van de Winckel (1980) in the Alsace, i.e. young aggrading eco-units in broadleaved, mixed-species European forests may form a network amongst and between eco-units that are in later development stages including biostasis (Fig. 6.44). This was shown for a natural Venezuelan mosaic with lianes in Fig. 6.20.

Wilderink and Zakee (1985) sketched the potential lines of silvigenesis for several natural and neo-natural silvatic mosaics at the Wageningse Bergrand. Their work provides implicit reasons to respect introduced species as fully-fledged components in silvatic mosaics, and shows no reasons to banish them. This of course does not contradict the need for "museum mosaics", in which historic forest types in a country are carefully reconstructed. There is such a thing as "historic nature". But in densely populated nations, the importation of exotic genotypes and species has gone on for many centuries and in high quantities. Hence no natural ecosystems can be imagined there in the future, in which such immigrated components play no important role.

In other parts of the world, original forests remain alive that are still largely conforming to ancient natural mosaics. Particularly if species-rich, such mosaics deserve to be conserved, before too many species become extinct because of excessive forest clearing or conversion. Species are the biological capital for many human generations to come, irrespective of the question whether we nowadays

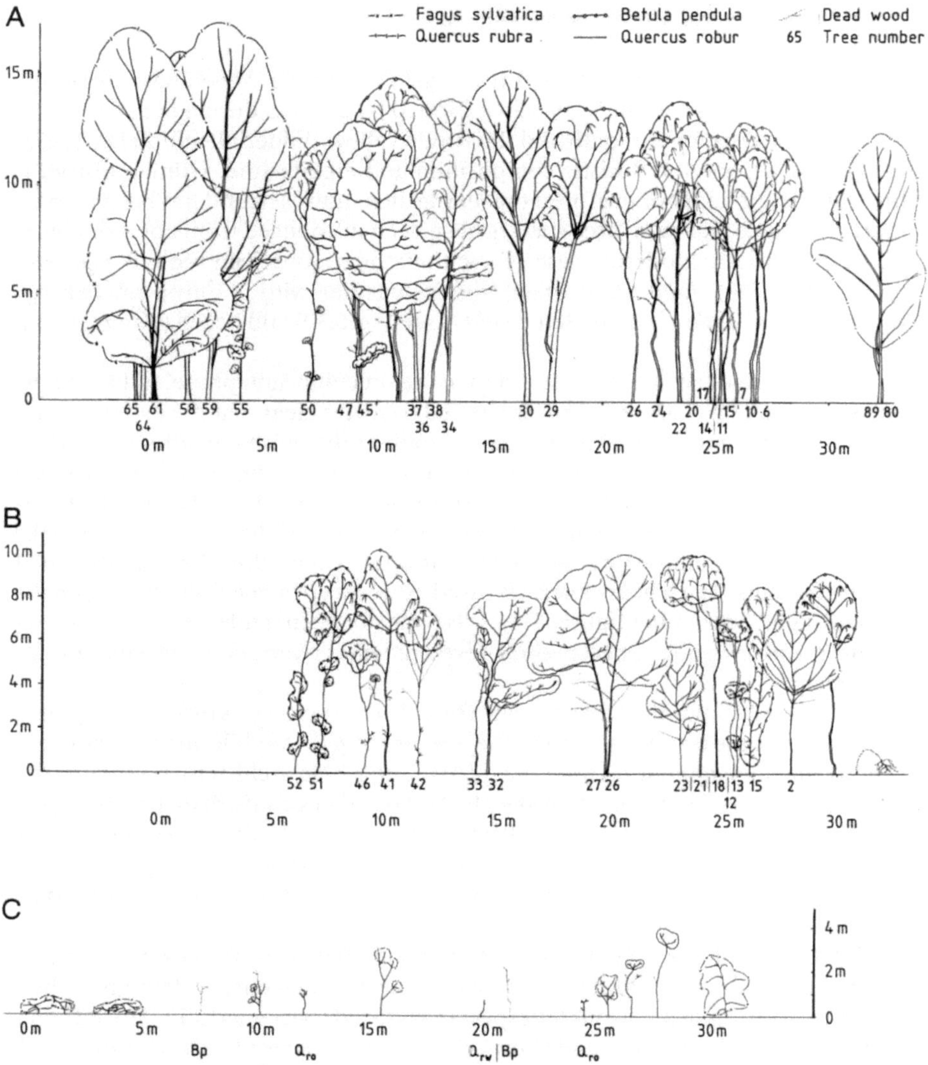

Fig. 6.31. Later precuilibrium stage of a neo-natural silvatic mosaic at the Wageningse Bergrand (moraine from glaciation). Side views of transect, in which *Quercus rubra*, American red oak, is an introduced component. **A** "High tree layer" (including older potential trees, some trees of the present). **B** "Lower tree layer" (mostly including potential trees: potential set). **C** "Shrub layer" (including few shrubs, more potential trees). Note eco-unit composition, e.g. *Q. rubra* late aggrading (*left side*), *B. pendula* late aggrading (next to *Q. rubra*, *behind*), *F. sylvatica*, aggrading (*right side*). This mosaic is entering in fusion between eco-units and shows little stufung. (Wilderink and Zakee 1985). See maps of this mosaic fragment, Fig. 6.32

Fig. 6.32. Maps of transect in Fig. 6.31. **A** Crowns "high tree layer". **B** Crowns "lower tree layer". **C** Ground layer and herbs. (Wilderink and Zakee 1985)

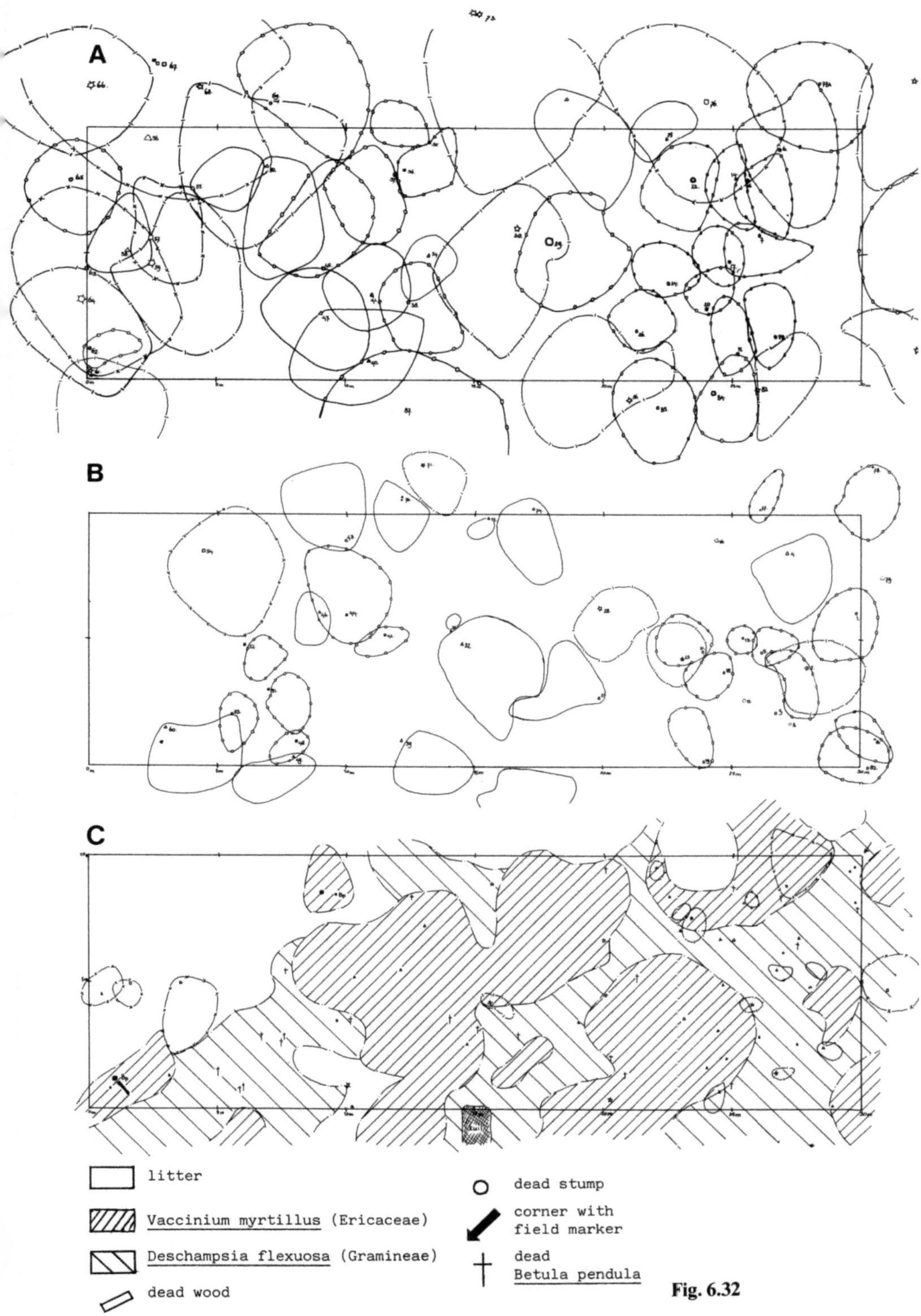
A
B
C
litter
Vaccinium myrtillus (Ericaceae)
Deschampsia flexuosa (Gramineae)
dead wood
dead stump
corner with field marker
dead Betula pendula

Fig. 6.32

consider them as pests, assets or luxuries. It is to be noted that ecosystem conservation and species conservation are two different management aims, which overlap in different ways on different sites.

6.2.8.4 Semi-Natural Mediterranean Silvatic Mosaics

The Mediterranean region in Europe is a transitional case between the temperate and semi-arid tropical regions. There are both cold winters and well-marked dry seasons. Mayer (1983) presented examples of mosaics in the mountainous regions around the Mediterranean. Probably lowland forests in this region have been used so intensively and have been cleared and converted so thoroughly, that nothing remains of the natural mosaics that once covered these lands. Human influence has been marked for over 3000 years. The total area of closed mediterranean mountain forests over 1000 m above sea level in the Mahgreb (Algeria, Morocco, Tunesia) might be around 5 million hectares only. The samples analysed by Mayer (1983) are among the least damaged that could be found. We consider them as later steps of a precuilibrium.

Figure 6.33A to C represents examples of cedar forests (genus *Cedrus*, Pinaceae, model with mixed axes, perhaps close to Champagnat's model). The two Turkish mosaics (Fig. 6.33A) certainly are in different stages, the upper one an earlier step and the lower one a later step in a precuilibrium stage. The difference in slope points to site differences, so that the ecuilibrium certainly is not the same in the two cases. The main species is *Cedrus libani*. In the upper transect, this is present in intermediate-sized, mainly appearing eco-units, whereas in the lower transect late fragmentation has left some dominating cedar eco-units, and is going to create space for broadleaved eco-units (*Quercus libani*). The role of *Juniperus foetidissima* (lower transect, on the left) is unclear. Maybe it constitutes a layer in an old, biostatic eco-unit.

The two transects in the mountains of Cyprus (Fig. 6.33B) also represent an early and a later step in a precuilibrium stage of a mosaic, in which *Cedrus brevifolia* is the main tree species. Traces of pioneer eco-units (*Pinus brutia*) are still visible in the upper transect, small broadleaved eco-units are appearing in the lower one. Analogous slopes indicate not too different sites, although their exposition is opposite (west and east).

The two Moroccan transects (Fig. 6.33C) show a development comparable to the Cypriot examples. The main tree species here is *Cedrus atlantica*, and the place of appearing broadleaved eco-units (*Quercus faginea*) is to be noted, as well as the giant cedar in the lower transect, which is so old that it has a local name (Cèdre-Gouraud). This is a clear relict of an old eco-unit, still biostatic. Cedars appear to characterise late pioneer eco-units in precuilibrium stages of all these mosaics. Site differences certainly are largely due to the water balance, because precipitation within the cedar area ranges from 600 to 2000 mm per year.

The examples from Algeria (Fig. 6.33D) concern different silvatic mosaics, both close to the ecuilibrium stage. In the upper one, *Cedrus atlantica* remains in the mosaic in modest proportions, whereas *Abies numidica*, some broadleaved

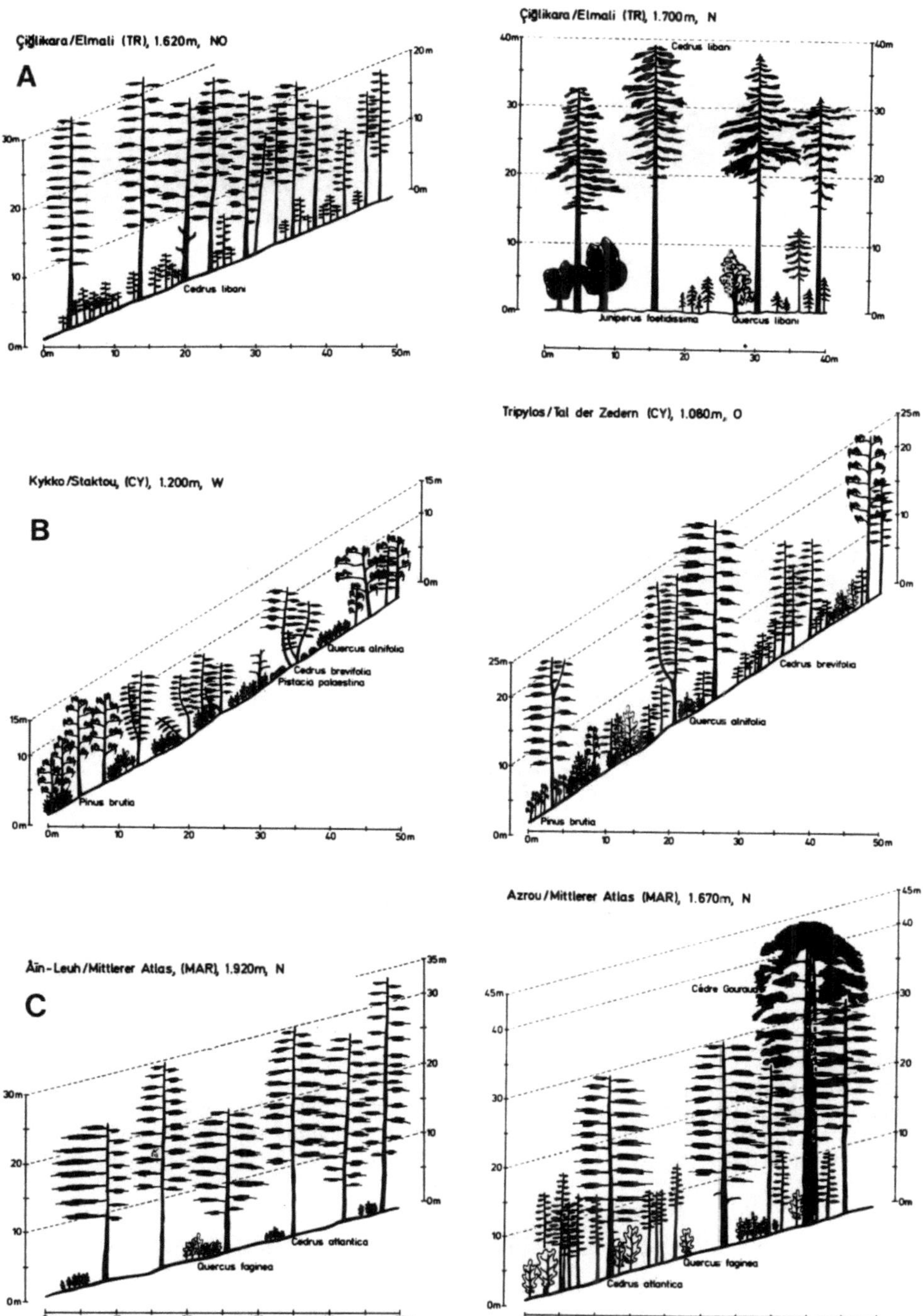

A
Çiğlikara/Elmali (TR), 1.620m, NO
Cedrus libani
Çiğlikara/Elmali (TR), 1.700m, N
Cedrus libani
Juniperus foetidissima
Quercus libani
B
Kykko/Staktou, (CY), 1.200m, W
Quercus alnifolia
Cedrus brevifolia
Pistacia palaestina
Pinus brutia
Tripylos/Tal der Zedern (CY), 1.080m, O
Cedrus brevifolia
Quercus alnifolia
Pinus brutia
C
Âin-Leuh/Mittlerer Atlas, (MAR), 1.920m, N
Cedrus atlantica
Quercus faginea
Azrou/Mittlerer Atlas (MAR), 1.670m, N
Cédre Gouraud
Quercus faginea
Cedrus atlantica

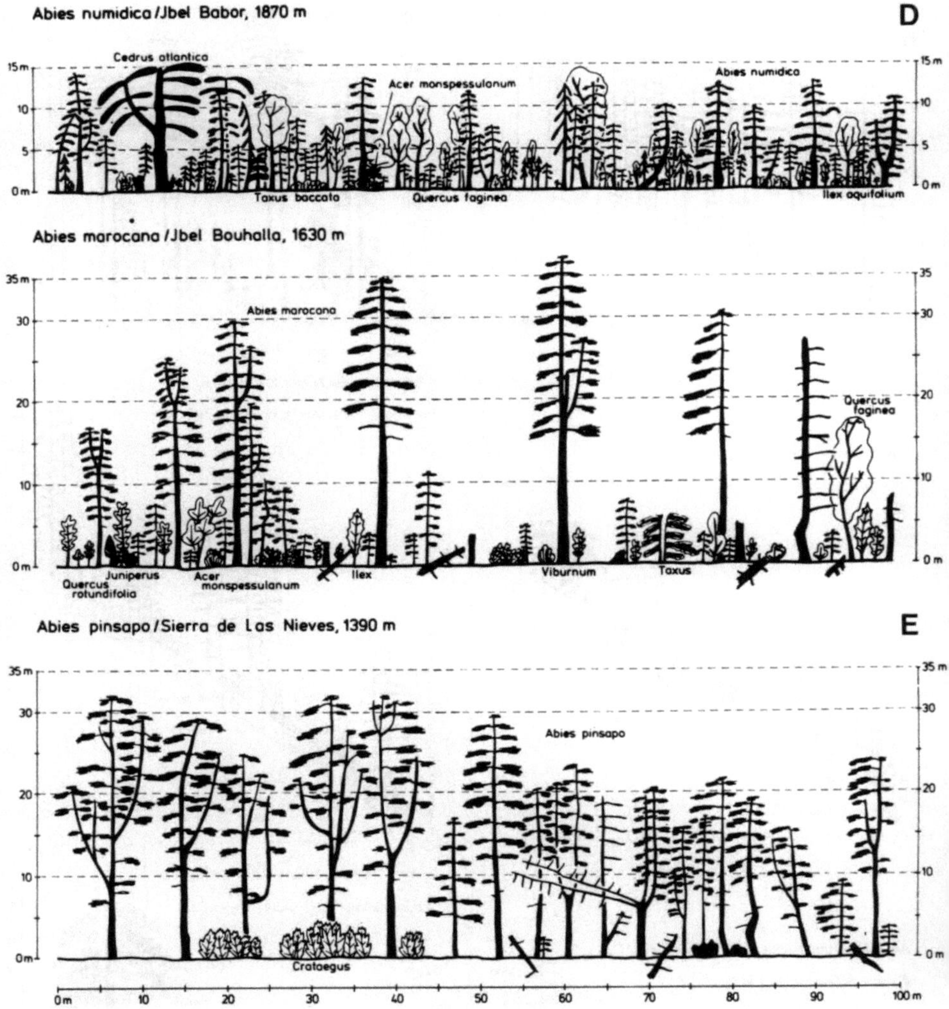

Fig. 6.33. Examples of mountain forests over 1000 m altitude, Mediterranean region. **A** to **C** main tree species of genus *Cedrus* (Cedar); **D** and **E** main tree species of genus *Abies* (firs). **A** Two pre-cuilibrium stages in Turkey. **B** Two precuilibrium stages from Cyprus. **C** Early and late precuilibrium stage in Morocco. **D** Two different mosaics, both near ecuilibrium, in Algeria. **E** Late pre-cuilibrium stage in Spain. (Mayer 1983)

species, and *Taxus baccata* (cf. Fig. 5.57) constitute many small eco-units in different development phases. The lower transect in Fig. 6.33D, showing an *Abies marocana* forest, represents a silvatic mosaic in a late step of the precuilibrium, as witnessed by the early aggradation phase of nearly all broadleaved eco-units. The small relictual area of *A. numidica* is drier than the precipitation-rich *A. marocana* area.

The Spanish transect (Fig. 6.33E) shows a silvatic mosaic with, at first sight, two intermediate-sized eco-units. The one on the left would be biostatic and the one on the right late aggrading, both seem to be built mainly by *Abies pinsapo.* Looking closer, however, both apparent eco-units are not so consistent in architecture as they should be (cf. Fig. 5.4). The one on the left side almost certainly consists of three to five eco-units in fusion. The set of shrubs (*Crataegus* sp.) is a possible trace of border-effects when the eco-unit filled by the *A. pinsapo* at 22 m from the left was young and aggrading. The reiteration pattern of the *A. pinsapo* crowns is to be noted. It gives the architecture of these eco-units a distinctly broadleaved physiognomy. In the aggrading eco-units on the right, the reiteration pattern suggests that many tops have been broken, and that the shifting event regime hence would contain regular storms.

Other studies from the Mediterranean region are scarce in so far as they concern forest architecture. The study by Van der Giessen et al. (1984) on the Pyreneen beech forests in Spain at Dehesa de Moncayo includes a transect covering the whole climate-and-soil gradient on a mountain side from 1100 to 1750 m above sea level. At significant points, detailed transects were drawn, showing the change in eco-unit and mosaic fragment architecture with changing sites. Research such as that by Kaplan (1984) in Israeli *Quercus ithaburensis* forests or by the team from Toulouse on the French side of the Pyrenees (e.g. Chéret 1987; Baudière et al. 1986; Genard and Lescourret 1984) has mainly been done according to other scientific methods (see Sect. 6.3).

6.2.8.5 Tropical Ecuilibrium Mosaic in the Andes, Ecuador

This part of a mosaic, published by Oldeman (in Bol et al. 1984), belongs to a forest on a spur in the Andes West of Quito at ca. 1900 m above sea level. It is part of a series of unpublished forest transects on the occidental side of the Andes, between ca. 3600 m and 400 m altitude. The site (Fig.6.34) is characterised by volcanic soils, high precipitations and is subjected to shifting events including frequent and violent winds.

The resulting silvatic mosaic in its ecuilibrium stage is situated at a hospitality level somewhat below the medium (Fig. 6.1), with a high proportion of dominating, short-lived pioneer eco-units (*Cecropia*) that form a network surrounding some rare intermediate to small-sized late pioneer eco-units. On the transect, one tree had been cut. At this altitude tree ferns are taking over from palms, as low components that ensure first closure of aggrading eco-units (left side).

Fragmentation occurs often and with violence. The amount of large dead wood on the forest floor over only 100 m transect length bears witness of this. Lianes (dotted) play a role in the stufung process, mainly on dead standing trees ("muertos"). The architecture and dynamics of this mosaic are encountered in many stressed forests in the whole world. But it shows its tropical species richness, which is considerable notwithstanding the rapid dynamics. Richness also is apparent from the other forest components, such as the herbs ("hierbas") and the epiphytes (see Fig. 4.8A/B, same location).

Fig. 6.34. Ecuilibrium stage of silvatic mosaic in stressed environment with violent storms (cf. Fig. 6.1) at 1900 m altitude in the Andes, Ecuador. Note network of pioneer eco-units with *Cecropia* sp., surrounding small late pioneer eco-units, and tree ferns replacing palms as undergrowth components. For epiphytes see Fig. 4.8A–B taken at the same location. (Published earlier by Oldeman in Bol et al. 1984)

Fig. 6.35. Ecuilibrium stage of silvatic mosaic of várzea forest on Rio Guama, near Bélèm, Brazil, stress by fluctuating soil cohesion caused by periodical flooding (cf. Fig. 6.1). Note overall mosaic architecture comparable with that in Fig. 6.34. Pioneer eco-units built by various palm species, appearing set of eco-units important, maybe one biostatic eco-unit with large tree of the present (*extreme right*). (Aubréville 1961)

6.2.8.6 Tropical Floodplain Ecuilibrium Mosaic, Brazil

In his study on the forests of Brazil, Aubréville (1961) gave a representative profile of a *varzea* forest (cf. Fig. 5.26A). Shifting events here are linked to **fluctuating soil cohesion** (Sect. 5.3.3), which is seasonal and caused by the river water level, rising during the rainy seasons. Stress factors and shifting pattern are very different from the Andean example, but the general silvatic mosaic architecture is comparable.

Instead of the Moraceous *Cecropia* trees in the mountains, palm species here form the pioneer eco-units, which can be quite extended (Fig. 6.35). Oldeman (1969) studied such large-sized pioneer mosaics composed nearly exclusively by pinot palms (*Euterpe oleracea*) in French Guyana. In the Brazilian example, more palm species are involved, the eco-units they build are smaller and broadleaved eco-units are present. Except for the large tree of the present at the left side of the transect (Fig. 6.35), which presumably is a legume according to its umbrella-shaped crown (Troll's model) and its huge buttresses, nothing points to broadleaved, biostatic late pioneer or successor eco-units.

The mosaic displays a high share of appearing eco-units, highly fragmented, and built by pioneer tree species mainly. The comparison between Figs. 6.34 and 6.35 shows that very different forests, under very different shifting event regimes caused by different stress factors, are built according to the same general architecture. At this hierarchical level, the silvatic mosaic, the recurrence of such patterns shows the same rule as for trees (Fig. 3.33). Whatever the stress factor, if it has about the same intensity and pattern, the living system reacts to it in essentially the same way. Elimination stages after permanent artificial water level changes in the temperate regions often show an analogous architecture, although they are built by different tree species.

6.2.8.7 Ecuilibrium Mosaic Fragments, Southeast Asia

The fragments shown are borrowed from Brünig (1985) and have been drawn in four forest areas in Brunei and Sarawak (Island of Borneo). The drawings include the soils, for which horizons and geological matrices are roughly indicated. The climates are linked to the successive altitudes. Each small transect is ca. 20 m long and barely shows more than one eco-unit, two or three at best. They therefore do not represent mosaic architecture as a whole. Consider, for instance, the numbers 31, 29 and 32, in Fig. 6.36. Number 31 shows a biostatic phase, number 29 two aggrading eco-units, and number 32 a biostatic eco-unit of another kind than 31. All three grow on comparable soils at identical altitudes. No eco-unit diagnosis has been made by Brünig. It is added here if necessary.

Comparing all fragments is less easy because no architectural graphic code allows to differentiate between potential trees, trees of the present and trees of the past. However, the fragments, except the isolated case in Bako National Park (upper right), clearly represent bits and pieces of ecuilibrium mosaics. This can be concluded from the size and architecture of the eco-units, and their interactive architecture. Number 30, for instance (middle left) shows fusion having occurred

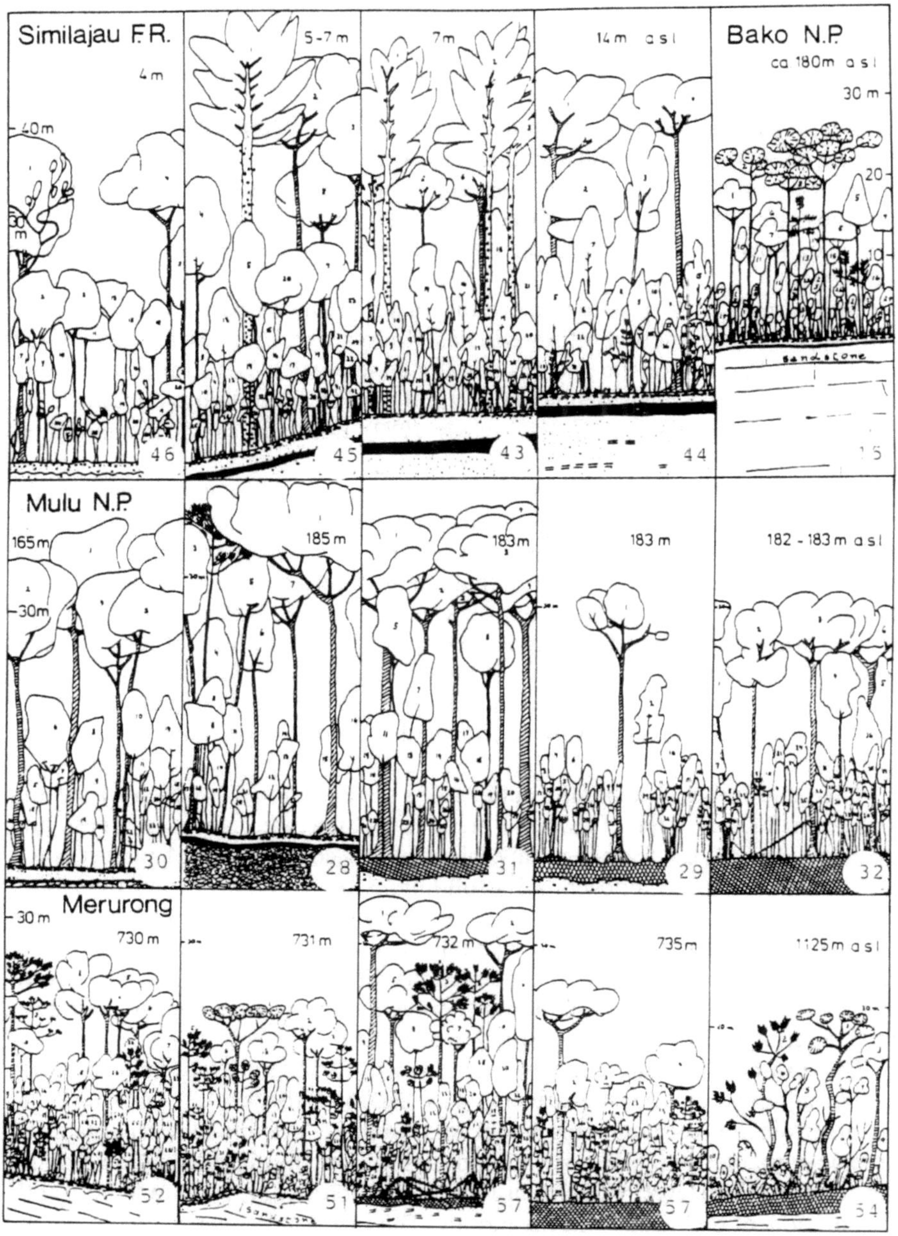

fragments of probable ecuilibrium stages of silvatic mosaics at different alti-
ent soils in Brunei and Sarawak, SE Asia. *F. R.* Forest Reserve; *N. P.* National
; 1985). Explanation in text

between two neighbouring eco-units, which can be distinguished both by their highest sets of the present and their undergrowth, showing traces of different developments. Number 46 (upper left) represents an aggrading eco-unit bordered by two biostatic ones, which are in different phases of maturity, as can be read from the crown depth of the trees of the present at both sides. All 15 cases can be analysed in this way.

Bratawinata (1986) provides interesting material for comparison with the stand build-up of mountainous rainforests in East Kalimantan, Indonesia (Island of Borneo), according to floristic and structural characteristics. His structural analysis contains transects which are not architectural but show forest physiognomy in the way that has been classical since Richards (1952). Eco-units can be distinguished, however, and compared with Fig. 6.36.

Brünig (1985) emphasises the change in tree stature and biomass density in the different mosaics, which according to him is not related to soil fertility. He writes (p. 13): "In Similajau, plot 46, on a fairly nutrient-rich, undeveloped Holocene terrace soil, tree height and biomass are less than on the flanks of the tertiary terrace with much poorer soil and single- or two-species dominance. Similarly, in Mulu National Park, the taller forest and highest biomass is not on the less nutrient-deficient lower terrace level, but in plots 28 and 31 with beginning peat formation and single-species dominance. The same feature is repeated on the Merurong Plateau.". Neither is there a straightforward relation with altitude, or with any other separate site factor.

The same author sketches the outlines of a systems model in which all site and vegetation factors interact: forest ("stand") architecture, species richness and diversity, canopy tree physiognomy, biochemical composition of plant tissues, micro-organisms (decomposers only), soil-humus, biomass as a nutrient filter, litter decomposition as a driving force, water availability and cycling, nutrient cycling, balancing and outflow, static and dynamic stability features, growth, increment and yield of trees as a result of the interactions within the system, and stand dynamics as a result of regeneration/mortality cycles and succession. Such models have been described in great detail by Brünig et al. (1986).

This example and the conclusions drawn by Brünig (1985) emphasise that neither shifting events nor sites are simple (cf. Sect. 5.1.3; Sect. 6.1.3). Looking at the factors taken into account in his model, this is a three-level model with organisms (trees in particular) explaining silvatic mosaics without the intermediate level of the eco-unit. The third level in the Brünig model is the abiotic site. It interacts both with the component organisms and the mosaic. The fact that eco-units can be distinguished without much trouble in Fig. 6.36 indicates that this level could be introduced in such models without changing them in principle. It would make them more manageable, particularly if eco-unit patterns could be read from remote sensing images (cf. Sect. 6.2.4.3).

The thesis by Djailany (1987) on the radiometric characteristics and structure of natural and secondary vegetations in Sumatera (Indonesia) shows that this still represents problems. These are principally due, in this first study on the try-out of parameters, to several of these proving to be useless for this aim. Further investments in studies on this subject are fully justified, given the potential gain in time, precision and model simplification that can be obtained. According to Djailany (1987), such studies on automatic vegetation classification should bear on the following aspects in priority.

- A fine-grained and simple classification of secondary vegetations according to the notion of "vegetation series";
- Ground-truth on every type in every series, and its radiometrical characteristics;
- Better sizes of ground-truth plots, of 10 ha at least, to be determined for each type (also see Brünig et al. 1986);
- Establishment of a "pattern dictionary" for the use of radiometric characteristics patterns as basic data.

6.2.8.8 Eco-Unit Interaction in Ecuilibrium, Indonesia

To make the preceding example more comprehensible, the detailed drawing of a forest plot in Sumatera, at Pasir Mayang, Jambi (Fig. 6.37), is presented here after Suselo (1983). It should be compared with Figs. 5.52 to 5.55. The map and side-view of the transect, considered together, show at least five small-sized eco-units. The numbers 5 on the map show young aggrading ones, around the number 2 there is an older aggrading phase, the number 1 is placed inside two giant crowns dominating one older biostatic eco-unit, and the black tree (tree 4) on the extreme right of the side-view certainly structures a small eco-unit entering into its biostatic phase. This eco-unit architecture is somewhat similar to that on plot 30 (middle left) in Fig. 6.36.

The young aggrading eco-units show that fragmentation and fusion are ever-continuing in the dynamics of this silvatic mosaic. Stufung therefore is clear. It cannot be read completely from the side-view, because the lower canopies in zones 5 (map) appear "below" the upper ones on the profile, whereas in reality they are situated "besides" them. Wide micro-climatical transition zones are apparent between the old biostatic eco-unit casting shadow on the young aggrading eco-unit to the north of it. On the contrary, sunlight penetrates from the south, over the young aggradation phase in the upper right corner of the plan, into the young biostatic eco-unit (black tree) to the north of it. This younger biostatic eco-unit is nearing the state of fusion with the older biostatic eco-unit to the north (number 1, map), after which only the crown depths of the trees in the upper structural ensembles will show the traces of different preceding eco-unit development. The older biostatic eco-unit has been damaged, large reiterated subcrowns having broken off from the tree of the present to the left (tree 42). This influences interactions.

This silvatic mosaic fragment shows that life spaces for non-arborescent plant components and for animals are many and varied. Litter build-up certainly is quite different from eco-unit to eco-unit, and with it the soil flora and fauna. Spaces in light and shadow alternate (cf. Fig. 6.58A; Koop 1989). For birds, a comparison may be made with Fig. 6.25, not forgetting that this represents an African rainforest. The life-spaces of monkeys may be defined according to the methods described by Hladik (1978; cf. Fig. 6.8B), which were applied by her in all tropical

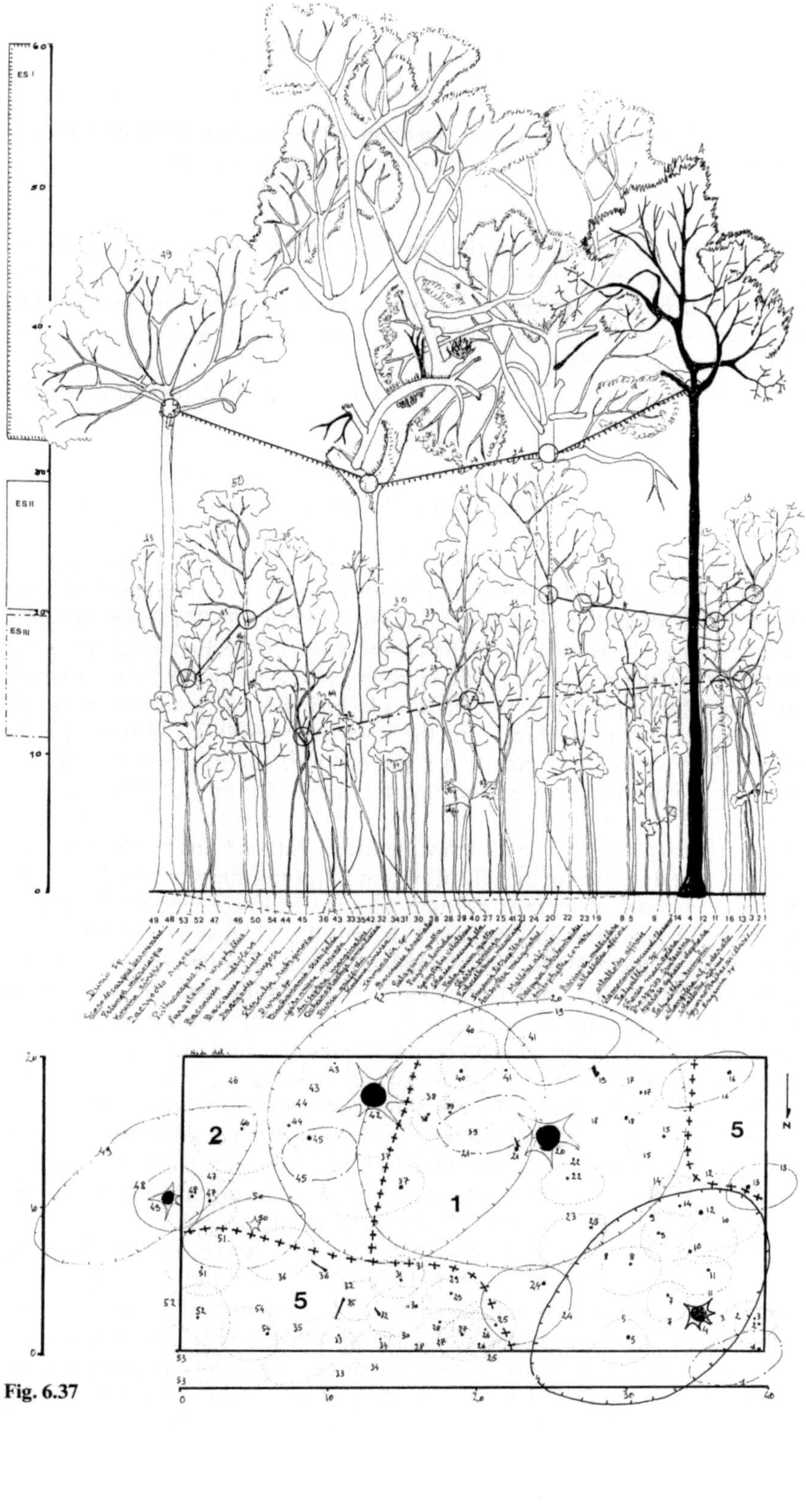

Fig. 6.37

continents (e.g. Gabon, French Guyana, Panamá, India). Terrestrial herbivores may be rare, because in this mosaic fragment canopy closure is complete, and tree crowns have grown out of reach of most such animals.

These interactions and life space structures are implicit in mosaic analysis in terms of eco-unit composition, such as represented in earlier examples. No detailed data, e.g. feeding behaviour or eco-unit architectural features, can be processed at the hierarchical level of the mosaic. But the architecture of the silvatic mosaic shows the variety and intensity of interactions among eco-units. This, again, clearly indicates the overall richness and variety, expressed as a pattern of deviations from mean values rather than as means themselves.

The zoological sources mentioned throughout this chapter indicate, by their approach, that silvatic mosaic analysis may finally make it possible to join together the concepts of animal territory and vegetational minimal area, in an integrated notion of minimal complete mosaic surface. For the moment research has stopped here. One great problem is the connection with site patterns, because the theoretical minimal ecosystem area probably always exceeds the maximal homogeneous site area. Hence on the one hand there is a question of components shared by different mosaics, and on the other hand an ever-increasing richness because mosaics on different sites interact amongst each other (Fig. 6.5; also see Terborgh and Stern 1987, discussed in Sect. 6.2.4.5.). The minimal area for complete forest ecosystem richness and variety would then include the land surface of our planet, as far as it is hospitable enough for forests. This is caused precisely by the fact that no single silvatic mosaic can reach its maximum richness and diversity, for want of sufficiently extended site surfaces. For the moment, we do not even have the right notions available to deal with these questions, not even Lovelock (1987). They are further complicated because of the irreversible changes that forests impose upon the original abiotic sites, a central notion in succession theory (cf. Sect. 6.1.2).

Other South-East Asian studies that contain numerous examples showing silvatic mosaics and eco-units from these regions are that by Hotta (1984), on the forest ecology and flora of Gunung Gadut (Gadut Mountain), West Sumatera, Indonesia, by Laumonier (1980) on the ecology and structure of Sumateran forests, and by Pascal (1983) or Reijneveld (1981) on the silvatic mosaics along a wet-to-dry site gradient in the Western Ghats, India.

6.2.8.9 Ecuilibrium Mosaic in African Rainforest, Zaire

The transect and map given in Fig. 6.38 are from Bosveld and Botterweg (1985), and are reproduced at a very reduced scale as compared to the original document. For this reason, the last 90 m of the transect had to be left out. At the original scale,

Fig. 6.37. Fragment with five eco-units in a probable ecuilibrium stage of the forest at Pasir Mayang, Jambi, Sumatera (Indonesia, after Suselo 1983). Details of eco-unit interaction make explicit similar principles as Fig. 6.36. Explanation see text; also compare Figs. 5.52–55

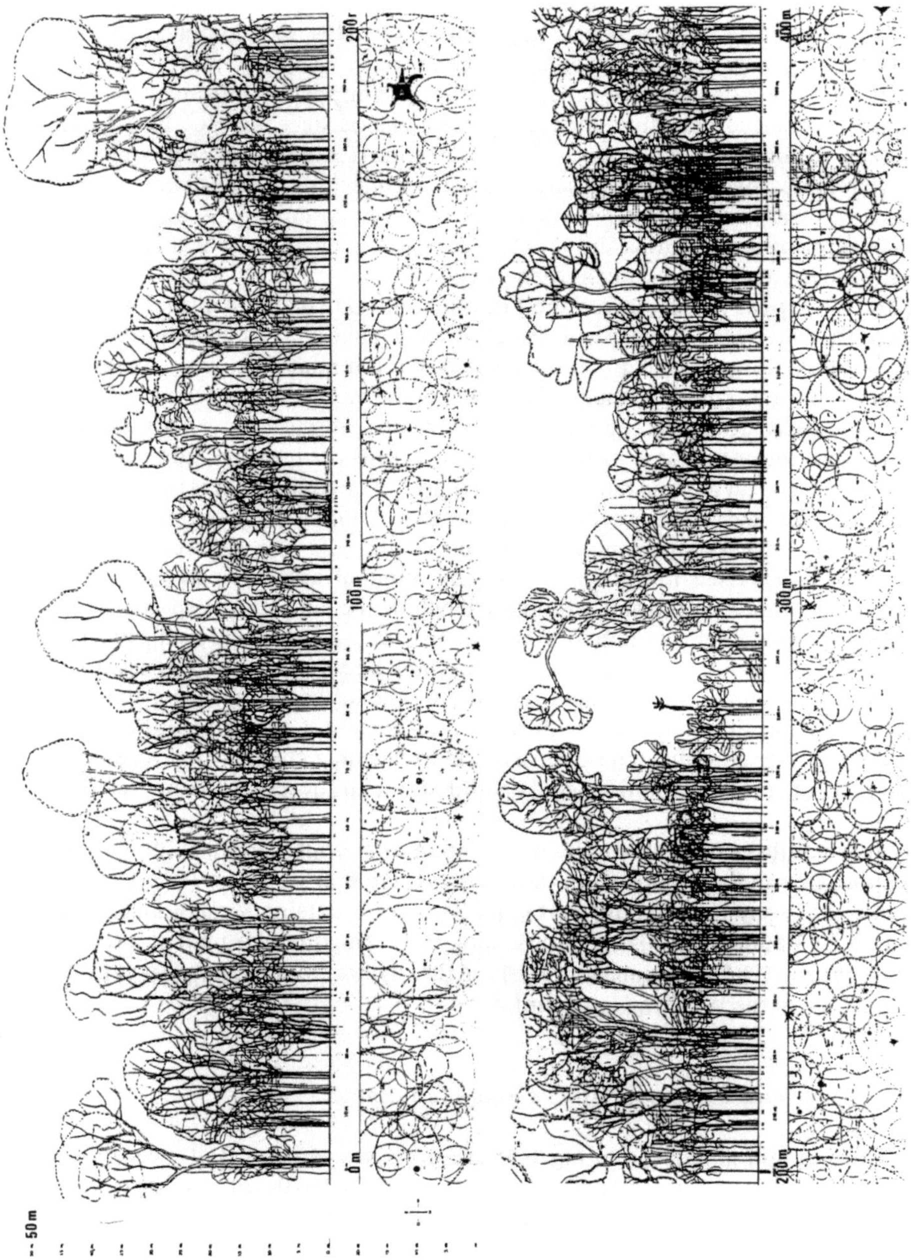

Fig. 6.38. Ecuilibrium mosaic, tropical rainforest near Yangambi (Zaire), after Bosveld and Botterweg (1985); explanation see text

the names of all component tree species can be read above the mosaic, and the architecture of each eco-unit can be analysed in detail. The authors did not make a formal silvatic mosaic analysis in their text.

This mosaic typically represents a near-complete silvatic mosaic with its eco-unit composition in the ecuilibrium stage. Animal territories are not considered, and for animal living spaces the same applies as in the preceding example (Fig. 6.37). The first 100 m show a complex of young biostatic and older aggrading eco-units, which are quite far advanced in the process of fusion. From the ground, these eco-units cannot be distinguished by the eye any more. From 100 to 200 m, the mosaic is mainly composed of early, biostatic pioneer eco-units with the exception of the giant *Entandrophragma cylindricum* (Meliaceae, African mahogany) at 190 m, which dominates a biostatic late pioneer eco-unit. The lower pioneer eco-units have light-requiring pioneer species in the upper structural set, ensemble, e.g. *Musanga cecropioides*, *Petersianthus macrocarpus* and *Ricinodendron heudelotii* (identification from Bosveld and Botterweg 1985). Between 160 and 180 m there is an aggrading successor eco-unit.

Between 200 and 300 m, fusion has almost invisibly welded together several eco-units into one ecological complex. Between 270 and 300 m this fragments again, as can be seen from the young aggrading eco-unit. This contains at least two potential pioneer trees. The broken stem of one of the trees of the past, having belonged to an earlier high structural ensemble, is still visible in the centre. Note the reiterated crown that has grown laterally into the opening above, and which belongs to the crown of a tree of the present at 300 m. Successor eco-units in different development phases, most often late aggrading, occupy most of the transect between 300 and 400 m. Note the dead crown (tree of the past) above the living crowns, near 400 m.

Stufung is evident over this whole mosaic fragment, and with it the intense interaction amongst the eco-units. The main lines of mosaic architecture as shown here indicate that indeed this mosaic functions as one, large, long-lived, living system. This is a mosaic with a most complex organisation due to eco-unit interaction, which is very dynamic and cannot be analysed in a manageable model by explaining it directly in terms of component species.

Management of such silvatic mosaics is very complex. Still, progress is being made in this sense. For recent advances see De Graaf (1986), Poels (1987) and Jonkers (1987) for tropical America, Vooren (1986, 1987, in prep.) for West Africa and Wöll (1988) for the Phillippines and Indonesia.

Some other publications on the subject that may yield valuable information are those by Vooren (1985) on the patterns of tree and branch-fall in West African rainforest, for South America by Herrera et al. (1986, Cuba, Sierra del Rosario), Mutoji-A-Kazadi (1977, French Guyana, on silvigenesis and aerial photography), Schröder (1977, on the undergrowth of the transects studied by Mutoji-A-Kazadi), and a general study on the dynamics of Amazonian terra-firme forests in a special issue of Experientia (Schubart and Walker 1987). Many transects from over the whole humid tropics can be found in Rollet (1974) and Oldeman (1974a).

6.3 The Life Cycle of Silvatic Mosaics: Populations and Succession

6.3.1 Concepts

Mosaic analysis in the past has been done without the notion of eco-unit, although other concepts such as microcoena (see Table 5.5; Sect. 5.5.5) or serules have been used. Serules are small seres. Seres are the "steps" in vegetational succession leading from a pioneer vegetation to a "climax" vegetation. Climax is a term derived from climate and means the state in which the vegetation is in permanent balance with the climate. The underlying hypothesis is that the vegetation itself slowly and unidirectionally modifies the soil substrate until the same soil prevails over large surfaces. These ideas, developed by Clements in the beginning of this century still largely determine ecological thought today. They have been changed and adapted, e. g. in Whittaker's "polyclimax" theory (1975) but both initial postulates remain intact, i. e. that converging soil development and succession lead from some form of pioneers to some form of climax vegetation.

Pioneers and climax, and the seres between them, are usually characterised by their species composition. The interaction between the abiotic site, i. e. a soil that allows a certain vegetation in a given climate, and species populations is thought to lead to a following, more mature sere characterised by a more advanced soil and another set of species populations. Within one sere, individuals of certain species germinate, grow up and die and their mortality and replacement occur in serules. This latter process has been analysed mathematically for trees by Horn (1971, 1975) in New England forests (USA). The models built according to these methods are not treated here extensively, because they can be found in many handbooks, for instance Begon et al. (1986).

Succession is a process comparable to silvatic mosaic development. Succession occurs over large surfaces, so species composition very often has been expressed as the average composition per hectare (or other surface unit) which is chosen in such a way that it is representative for the sere or climax considered. In succession theory, the abstractions concern subcommunities (such as eco-units) and climatic or irreversible abiotic soil changes. The assumptions are constancy, over considerable timespans, of both site and replacement of individual plants and animals by individuals belonging to the same species. The abstracted features have been fed into the models of succession as **correction factors** whenever they play an obvious role. The notion of indicator species is one of the tools for such corrections because such species indicate not only one association (sere or climax), but may also show variations (cf. Fig. 6.45).

In architectural silvatic mosaic analysis, on the contrary, the abstractions concern species replacement and site constancy or surface. The assumptions are eco-unit shifting as a driving process leading to an ecuilibrium stage which may well be ephemeral and is always in oscillation around a state that persists as long as site and climate show no important abiotic changes (cf. Figs. 5.6.1, 6.6). The abstracted features, if obviously important, can be fed into the models whenever necessary. An example is the silvatic mosaic in which exotic species are introduced, leading to neo-natural mosaics (cf. Figs. 6.31, 6.32).

In the present pages, several ecological approaches will be considered which are used extensively, and which have in common that they rest upon the assumptions and abstractions at the basis of succession theories. Here and there they will be interspersed with architectural approaches that may form bridges between the two methodologies.

6.3.2 Physiognomic, Descriptive Classifications

The simplest way to talk about climax forests in different climates is to name them. In Europe, with its poor post-glacial tree flora, these names are derived from (tree) indicator species according to one of the European phytosociological schools. One may speak about an oak-birch association (Querco-Betuletum) and the meaning will be completely clear (e.g. see Mayer 1986, on "The Forests of Europe"). In the species-rich forests of the tropics, such a nomenclature has been attempted (Mangenot 1955; Schulz 1960; Hommel 1987, for instance). But in general and preferably, a physiognomical approach was used first, representing tropical forests by their exterior appearance. Species composition can be added later.

One internationally agreed system of physiognomic classification of tropical vegetation formations has been formulated in Yangambi (Zaïre) in 1956. It has been discussed by Aubréville in 1965 in an article remarkably illustrated by N. Hallé (Fig. 6.39A/B/C). The main vegetation types of Africa were summarily defined, named and classified, and every type rested on studies published at that time. In Aubréville's article, this classification is extended world-wide. The illustration is without architectural analysis. Eco-units may be distinguished vaguely, but forest components like animals, grasses or lianes have not been forgotten by Hallé.

Figure 6.39D is from De Granville (1976) and was made along the Grand Inini River, a tributary of the Maroni in French Guyana. He writes, after having pointed out the difficulties in using tropical phytosociological studies in the field for the non-initiated user of such information (De Granville 1976, p. 23): "The transects established have a schematical character only: in no way are they true profiles resulting from architectural studies, on which plant density, form and dimensions and other such data are exactly reproduced. Our only aim is to visualise the presence of certain forest species, the most frequent ones according to our estimates, as a function of topography and soil." (freely translated from the French, RAAO). This kind of information bridges architectural studies and phytosociological studies, with a low degree of precision but with a high value for orientation in these forests. Figure 6.39E shows that more precision can be added. It concerns the presence of monocotyledons and ferns (cf. Chap. 4) in different biotopes which can be found back on elementary documents in the style of Fig. 6.39D (also cf. Fig. 5.15).

6.3.3 Physiognomy, Species, Dynamics: The Central European Model

The link between mosaic architecture and species composition has been established in a particularly careful way by Koop (1981a, cf. Figs. 5.1, 5.7) for late precuilibrium stages of North-Western German forests (cf. Sect. 5.4.3.1).

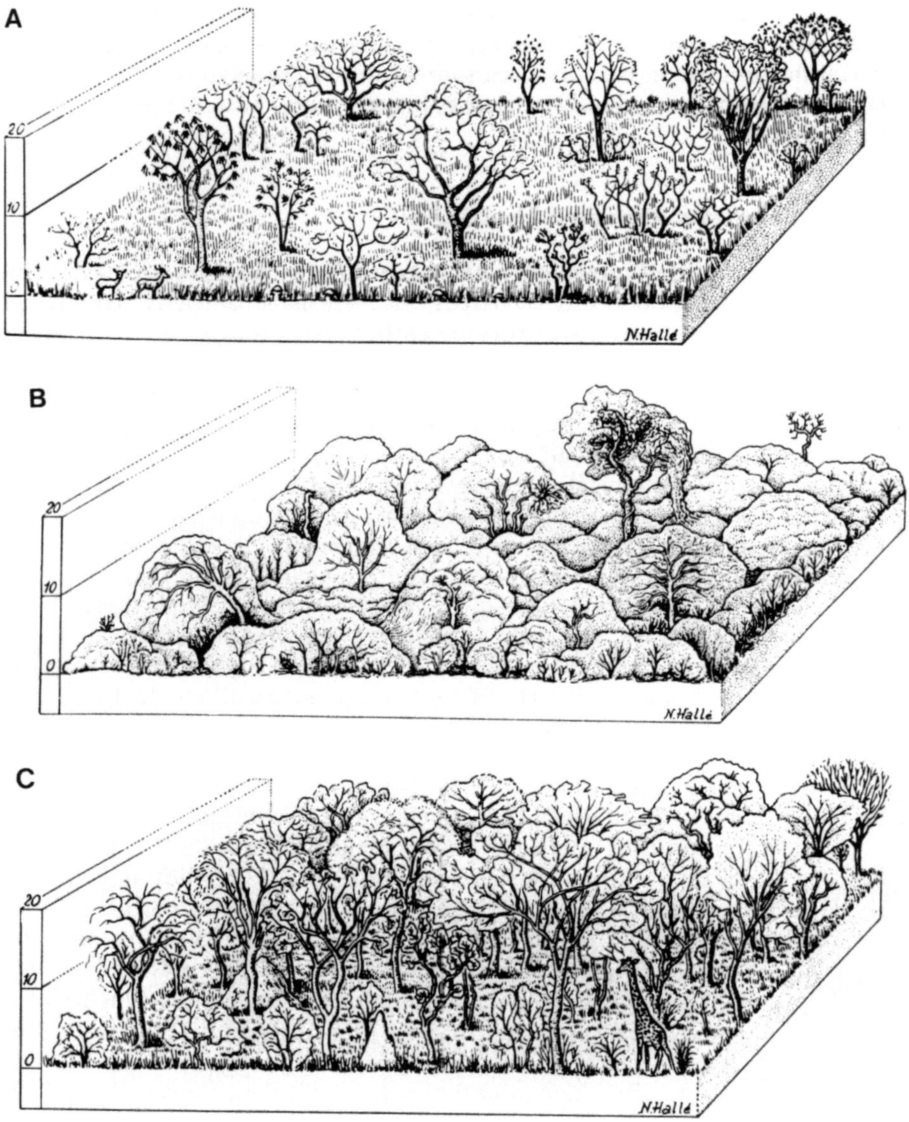

Fig. 6.39A-C

Carefully drawn transects of European forest mosaics on which tree physiognomy (not architecture) is represented in an attempt to pinpoint their position in such mosaics are quite old (cf. Fig. 2.3). They have been periodically used mainly by German-speaking foresters, and are exemplified here by a document from Köstler (1955, Fig. 6.40).

The theoretical development based on such data has been summarised by Leibundgut (1959). In the following quotation, tree species and their mixture are the very first criteria mentioned, which is consistent with the approach outlined above. Leibundgut (1959, p. 133 ff.) wrote: "The methodology for a virgin-forest analysis that can be interpreted statistically has to fulfill the following requirements in particular:

- Typical forms of structure should be recorded quantitatively and in a comparable way as to tree species mixture, layering (schichtung – see Fig. 6.9, RAAO), standing mass build-up, vitality, conversion tendency, and tree crown development (cf. Sect. 5.2.4.1, RAAO).
- Increment analysis should provide comparisons between different forms of structure according to statistically comparable data processing methods and following the characteristic criteria for these forms of structure.
- The analysis of regular increment processes should make possible the proven evaluation of its structural consequences and hence the deduction of the laws of natural virgin forest dynamics." (freely translated from the German, RAAO).

Leibundgut further defined "development phases" as linked to the development of "particular, chosen stands". He stated (p. 114): "We therefore define a phase as a structurally well recognisable development step within a well-defined forest community (association).". The reference to an association again indicates his phytosociological basis. He continues (p. 114): "We define a stage, on the contrary, as a development step within a developmental sequence of different communities (successional series).". In the present text, the same dictinction has been made, but with the difference that our "chosen stand" is an eco-unit, and that succession has been replaced here by sylvatic mosaic development. This means also, that the emphasis on statistics would have to be complemented by an emphasis on largely nonrandom mathematical simulation methods in architectural analysis (cf. Figs. 5.86, 6.29). The programme SILVI-STAR (Koop 1989) makes this possible.

Figure 6.41A shows Leibundgut's famous forest development diagram (Leibundgut 1978), which he limits to European forests. The upper four blocks ("optimal phase", "old age phase", "degradation phase" and "regeneration phase")

Fig. 6.39. D (Fig. 6.39E see page 488)

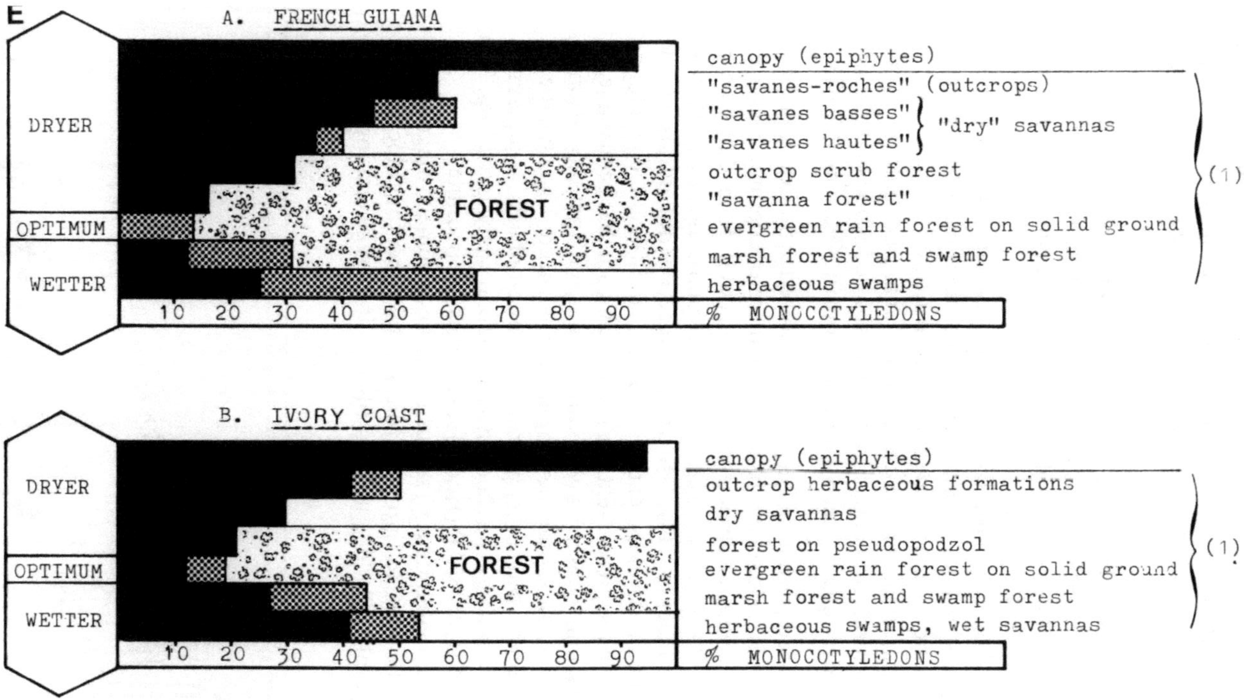

(1) Terrestrial plants only

Fig. 6.39. Ways to show forest physiognomy in order to determine forest types composed by plant and animal species (our level: silvatic mosaic). **A, B, C** African shrub forest, open deciduous forest and tree savanna, according to the Yangambi terminology. (Drawings by N. Hallé, from Aubréville 1965). Tridimensional form exceptional in contemporary literature. Note that eco-units are vaguely indicated by the artist. **D** Part of a stylised forest transect on the Grand Inini River (French Guyana), visualising overall forest physiognomy and localising the most remarkable and frequent plant species by symbols. Tool for first field orientation. (De Granville 1976). **E** Proportion of monocotyledons among main plant formations in French Guyana and the Côte d'Ivoire. Schematical, physiognomical indication of vegetation structure (De Granville 1984). For further explanation see text

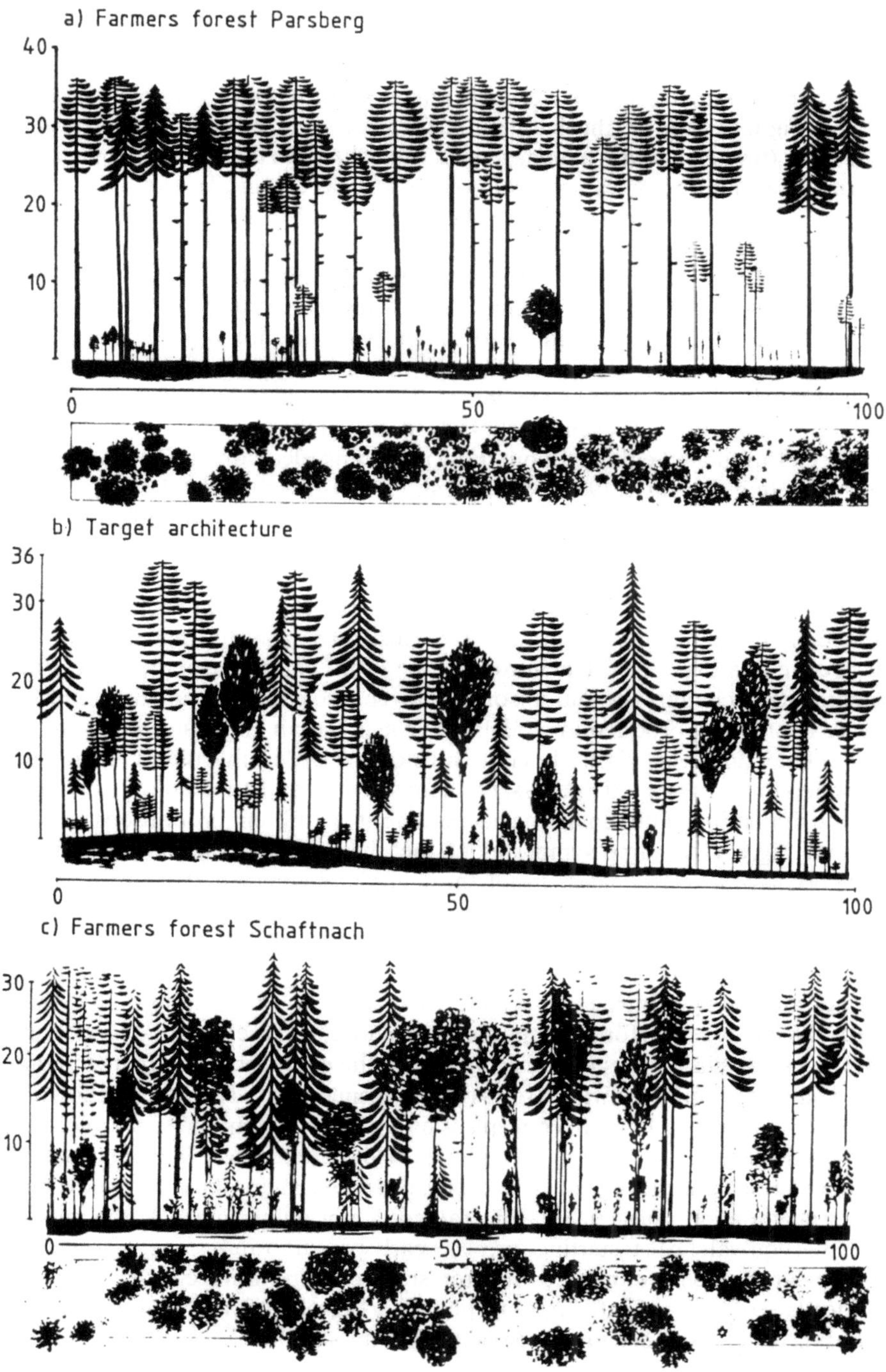

Fig. 6.40. Forest mosaics represented in terms of tree architecture, not eco-unit architecture, in the Central European school. (Example from Köstler 1955)

clearly correspond to forest development phases such as in Fig. 5.4, which is different from Fig. 6.41A only in that it is defined for an eco-unit, not a stand. It is emphasised that a (timber) stand is a management unit, although the term has sometimes been taken over in ecological work without testing its ecological meaning.

Four of the lower blocks in Fig. 6.41A correspond to silvatic mosaic stages, or successional stages. The "plenterwald-phase" is an old stand in fragmentation, the "starting forest with pioneers" is a pioneer community, the "transitional forest with pioneers and final forest species" is a forest community with both small patches from the "plenterwald" and larger patches with pioneer trees, and the "final forest" is a community in a later development stage, with all its remaining original patches. The whole image comes close to the hierarchical, architectural model presented in the present book. Indeed, this architectural model could not have been formulated either without these preceding ideas, or without its tropical viewpoints (e.g. Oldeman 1974a; Hallé et al. 1978; Torquebiau 1981,1985).

Our architectural forest model mainly differs from that of Leibundgut by considering ecological process and pattern independently from management-induced process and pattern. So the interposition of another hierarchical level became possible, i.e. the eco-unit. By virtue of this level, processes and patterns could be defined in a biological space-and-time frame instead of using the manager's values per hectare as a first entry. The introduction of the eco-unit level therefore quite fundamentally changes the outlook on species composition, relation diagrams for species interaction, production models, site classification and management models. The latter do not come first, but last. Silvology precedes silviculture. In numerical models this is expressed by a lesser role for statistics and a greater role for deterministic simulation, both remaining necessary but in different proportions.

From Fig. 6.41A, some symptoms of the changed outlook can be derived. There would be no difference between the degradation phase and the plenterwald phase, because late fragmentation starts during degradation. If the plenterwald phase is left out, an arrow between "degradation" and "transitional forest" could be added to show that fragmentation may lead to many kinds of mosaics, and the same is true for a second arrow between "degradation" and "final forest". No arrows from forest communities would lead to the "optimal phase" any more, because this belongs to one eco-unit. On the contrary, early fragmentation would have to be included with arrows from "regeneration". The result is shown in Fig. 6.41B. The equivalence of concepts is intended as: optimal phase = biostatic phase; old age phase = late biostatic phase; degradation phase = reorganisation (fragmentation) phase = plenterwald phase; regeneration phase = innovation phase; aggradation phase, taken for granted, is not represented. At the mosaic level, this equivalence is: starting forest = early precuilibrium stage; transitional forest = late precuilibrium stage; final forest = ecuilibrium stage.

The place of Leibundgut's "optimal phase" in Fig. 6.41A is ambiguous, because according to his diagram it may be a part both of the stand cycle and the forest community cycle. We have made a choice here by seeing it as the equivalent of biostasis (Fig. 6.41B), but according to Leibundgut's text he considers that both stands and communities may end up by forming high, closed forests. We accommodated this idea earlier by coining the term **eco-unit fusion** for the growing together of eco-units

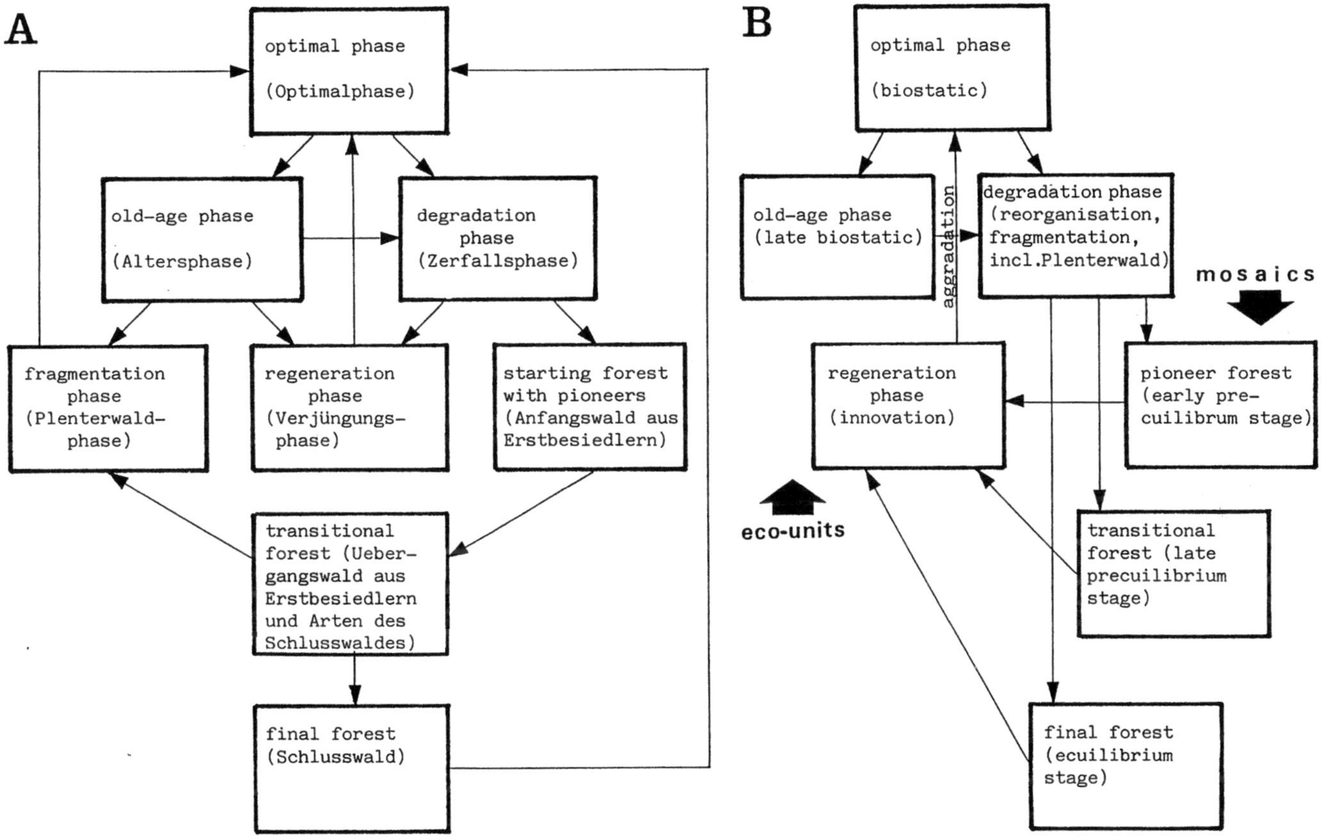

Fig. 6.41. Successional dynamics according to Leibundgut (1978) in **A** translated into eco-unit and mosaic terms (**B**). See text for explanation

from a mosaic with stufung, towards a situation of comparable layering of many neighbouring eco-units (see Sects. 5.4.3.11/12).

Figure 6.42A shows an example of the analysis of an Italian forest mosaic according to the approach by Leibundgut and others. Susmel (1979), to whom this figure is due, announces it as the schematical representation of an un-even aged forest structure originating from an even-aged one. The figure according to the author is the result of a numerical model taking into account species composition, aboveground biospace (determined by forest height, in this case 41 m), stand dynamics, regeneration as a tree replacement process, development stages (Fig. 6.42A: A to H), somatic-chronological structure (that would be called mosaic structure here), complex structure which is related to time, longevity of trees, unit number of trees per age class per hectare, tree biomass and annual growth and increment. The criteria from Leibundgut's list can be recognised; they are complemented by others. Susmel's simulation model, a software print-out of which is appended to his article, shows the fragmentation and fusion processes described above.

This model has been illustrated here, so as to demonstrate one of the many ways in which diagrams like Fig. 6.41A have been elaborated upon and quantified. Susmel's model has been validated in nature, but the author emphasised that age indications are an output of his model and have not been checked in the forest. Species composition in this mixed, natural *Abies alba* forest type is 50% fir (*A. alba*), 30% beech (*Fagus sylvatica*) and 20% Norway spruce (*Picea abies*). Species composition, stature (maximum forest height) and age vary with the site. For instance, the age at which the "first evolutive cycle" (precuilibrium in our terms) ends is indicated on the figure as 410 years, but may be 100 to 250 years on hospitable sites and run up to 2000 or 3000 years on very inhospitable ones, such as stony debris of landslides in the mountains. This all agrees with the general lines of silvatic mosaic development (Fig. 6.6), of which this model yields a parallel.

Figure 6.42B shows the way in which the Leibundgut forest development diagram was formulated earlier (Zukrigl et al. 1963; also Mayer et al. 1979). Here, no trace seems to be visible as yet of a distinction between eco-unit or stand development and mosaic or community development. The dotted cycle on the left side is to be noted, however. It indicates how from a "terminal phase" (Terminalphase), according to the rate of biomass decrease, the stand may pass through a degradation phase (slow rate, trees die standing), a regeneration or innovation phase (trees fall all together) or a plenter or fragmentation phase (some trees disappear abruptly, others keep standing).

This diagram very clearly indicates why Central European silviculturists often prefer selection cutting. This corresponds to fragmentation (plentering) and leaves a maximal standing mass per hectare ("Vorrat"), because only some few big trees per hectare are harvested per cutting cycle every 5 to 10 years. One of the principal management aims of plenter silviculture is indicated as "Vorratspflege", caring for the standing mass. Both the production of high-quality sawnwood and the sustaining of a forest with stufung (i. e. a fine-grained mosaic) are then ascertained, and the forest yields both high-priced wood and protection against flooding and avalanches, a typical form of perialpine forest management. If plentering has become an

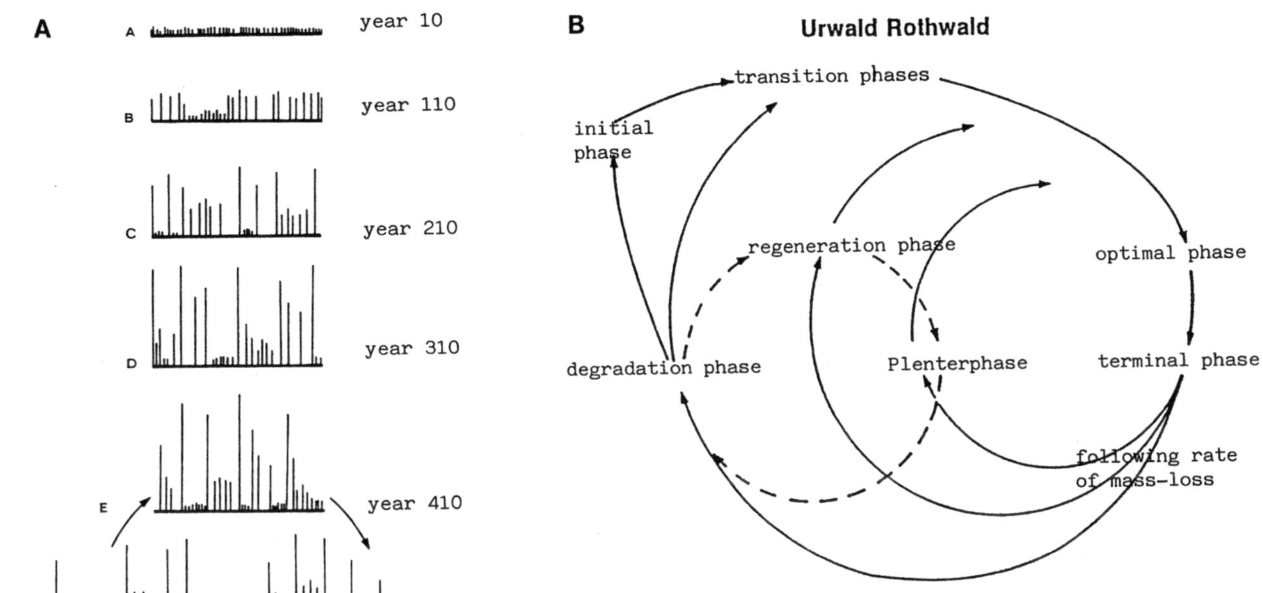

Fig. 6.42. Two succession models from the Central European School (e.g. Fig. 6.41A), other versions. **A** Simulation model after Susmel (1979), structural development of uneven-aged forest from even-aged beginning (*A* to *E*). Age calculated from model, may differ widely with site. Natural *Abies alba*-dominated mountain forest. **B** Cycles in *Picea abies-Fagus sylvatica-Abies alba* virgin mountain forest. (After Zukrigl et al. 1963 ex Mayer et al. 1979). Note that no difference is made between eco-units (phases) and mosaics (stages), contrary to Leibundgut (see text)

ideal of many European foresters, this is due to the viewpoint that it is also more "natural" than other forms of silviculture. As shown in Fig. 6.42B, this is not so in any absolute sense.

6.3.4 Successional Models Using Species Composition

For more detailed studies on these cycles, authors from this school use forest maps on which different development phases are indicated, and also follow the phytosociological approach. An example of this last case is shown in Fig. 6.43. It has been borrowed from the beautiful study by Fanta (1982) on the natural regeneration of Dutch forests on dry sandy soils. These soils are very inhospitable for forests. They have been reafforested for about one century with great trouble and mainly with the one species hardy enough, Scots pine (*Pinus sylvestris*, Pinaceae). If there is time enough and no pressure on the land by humans or animals, even blown and moving sands are fixed by lichens and mosses. Then, Scots pine can establish itself spontaneously. But this may take centuries.

This process, from the installation of Scots pine on, is the "tail" of Fig. 6.43. It has to be considered as primary succession or, in our terms, an authentic installation stage. The secondary cycles of forest development, once an oak-birch forest has been established (Querco-Betuletum), may still entail treeless stages as shown on the outside oval in the interrupted line, indicating heathlands with *Calluna vulgaris* and grasses. Such treeless interludes have been described also in other Dutch sandy biotopes, such as the coastal dunes near the Hague, where *Crataegus-Betula* woodlands may turn into open grasslands or cryptogamic vegetations (lichens, mosses) - see Van der Meulen and Wanders 1985. In other cases, no treeless interphases may occur and silvatic mosaic development leads to one of the inner cycles at the top of Fig. 6.43.

The way in which Fanta (1982) noted the tree species as occurring in the upper canopy or in the undergrowth of the successive communities of successional seres vaguely reflects the potential for fragmentation of eco-units. This potential is not very strong in these particular forests, because the main difference in the three principal tree species is not one of tolerance, all three being light-demanding, but in tree architecture and life-span: *Pinus sylvestris*, *Quercus robur* and *Betula verrucosa*. Comparing Fig. 6.43 with Figs. 6.1 and 6.6, the inner cycle in Fig. 6.43 may be conceived as a summary of ecuilibrium dynamics on these very inhospitable soils, seen from a species populations viewpoint.

One of the most thorough studies with links to these Central European succession models (Vredenbregt and Van de Winckel 1982), gives a broad ecological basis for a proposal to establish and manage forests on young clayey polder soils. These authors add other models to their tools for silvicultural design. One of these is the figure by Van de Winckel (1980) that is reproduced here (Fig. 6.44). It shows how, in different silvatic mosaics, different patterns of eco-units determine the architecture, the stufung and hence the species composition. Figure 6.44A shows biostatic eco-units emerging like islands above a network of aggrading eco-units, in the species-rich forests along the Rhine in the Alsace (cf. Fig. 5.56). On the con-

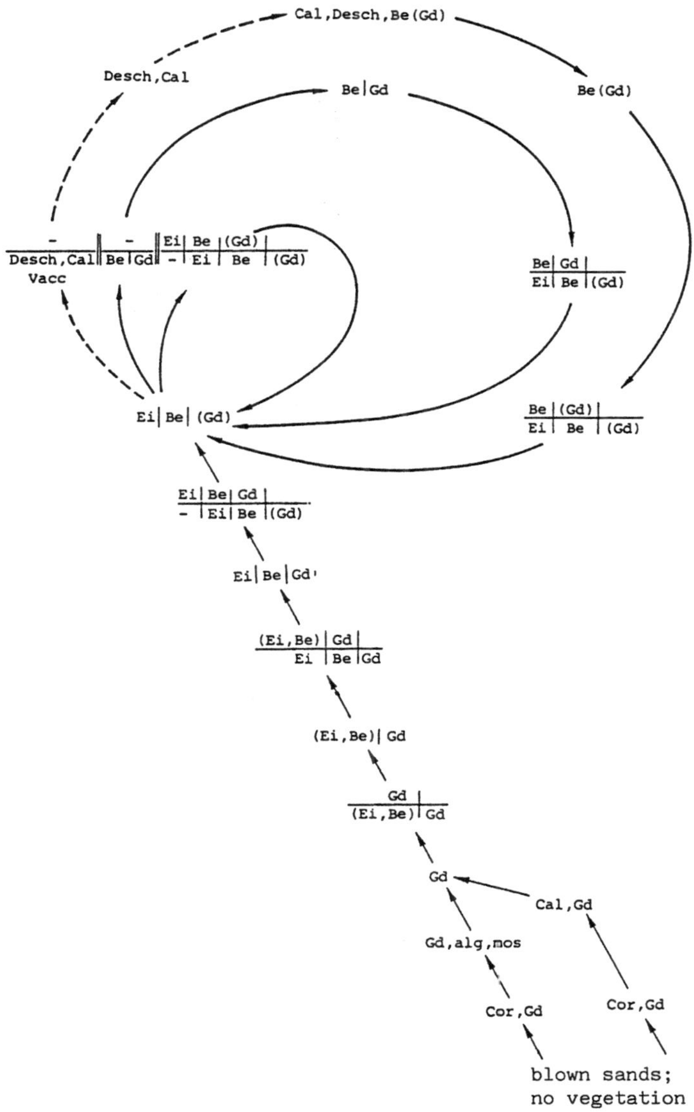

Fig. 6.43. Connection between succession diagrams (e.g. Figs. 6.41A; 6.42) and species composition, through indicator species for potential natural vegetation and its development through successive seres. Poorest variant *Querco-Betuletum* on Dutch sterile, blown sands and developing its own organic soil layer. *Thin arrows* primary succession; *fat arrows* secondary succession; *interrupted arrows* treeless secondary succession. *Cor Corynephorus canescens*; *alg* algae; *Pol Polytrichum piliferum*; *Cal Calluna vulgaris*; *Vacc Vaccinium myrtillus*; *Desch Deschampsia flexuosa*; *Be Betula* cf. *verrucosa*; *Ei Quercus robur*; *Bu Fagus sylvatica*; *Ei-hak Q. robur*, coppiced. Note two crown layers in some seres. Fanta (1982), from whom this figure is taken, says "theoretical model"

trary, Fig. 6.44B shows either a network of fused biostatic eco-units, or a large pioneer eco-unit starting fragmentation, with very small aggrading eco-units in numerous gaps. These schematic drawings should be compared with Figs. 5.31–42 and 6.27–38.

Figure 6.45 shows the example of a species composition analysis made by phytosociological methods (Van der Meulen 1979). It is a matrix listing all plant species encountered, their frequency and other characteristics expressed by a numerical or letter code, the site in which they were encountered inside the vegetation type and their role in the vegetation. Characteristic species for the whole formation are followed by indicators for variants and by companion species. The figure refers to a semi-arid part of the Western Transvaal bushveld in South Africa. Analogous matrices for European marshy forests can for instance be found in Wiegers (1985). Such documents concern mosaics. Their links with forest architecture are indicated in a very schematical way (Fig. 5.66; danserogrammes).

Such matrices nowadays are treated by computer, and many models exist to process the numerous data (e.g. Hermy 1984a, b). The results of data processed in this way are exemplified by an on-going case study in the coastal sand dune woodlands near the Hague (Van der Meulen and Wanders 1985), represented in a document by Van der Hagen (unpubl., Fig. 6.46). He used the programme DECORANA (cf. Hermy 1984a, b) to obtain the ordination of all species composition data along two axes with an unsharply defined meaning. The vertical axis refers to a combination of shrub vegetation and species poverty linked to grassification (sign of a fragmentation process?) and the horizontal axis indicates a complex of environmental factors related to drought. Each line of the graph denotes a permanent quadrat (PQ), monitored from 1952 to 1980 or 1981. Although Van der Hagen has some reserves about the weighting of the role of rare species in the DECORANA programme, development lines of the vegetation can well be read from such figures. The permanent quadrats are not large enough to contain more than a few eco-units, but have been chosen to be representative for the vegetation when the observations were started by Boerboom in the early 1950's.

6.3.5 Physiognomic Structure and Mosaic Architecture

Vegetation architecture cannot be read from such highly abstracted documents. Other phytosociological researchers included it in other ways than the danserogrammes mentioned earlier. The "vegetation pyramid" described by Bertrand (ex Baudière 1974) is shown in Fig. 6.47 from Baudière et al. (1986) for a forest in the Pyrenees. One may recognise the concept of strata or layers, applied to larger forest complexes, which probably are silvatic mosaics. These strata indicate larger trees, smaller trees, shrubs, low shrubs, herbs and mosses and may be considered to be predecessors of the hoplexols (Beaudou et al. 1978, Figs. 5.67–69), which have an architectural definition. In vegetation pyramids, the development of strata can be indicated as progression, regression, stability. In each stratum, the observed coverage is indicated. The method may be readily compared to the one employed by

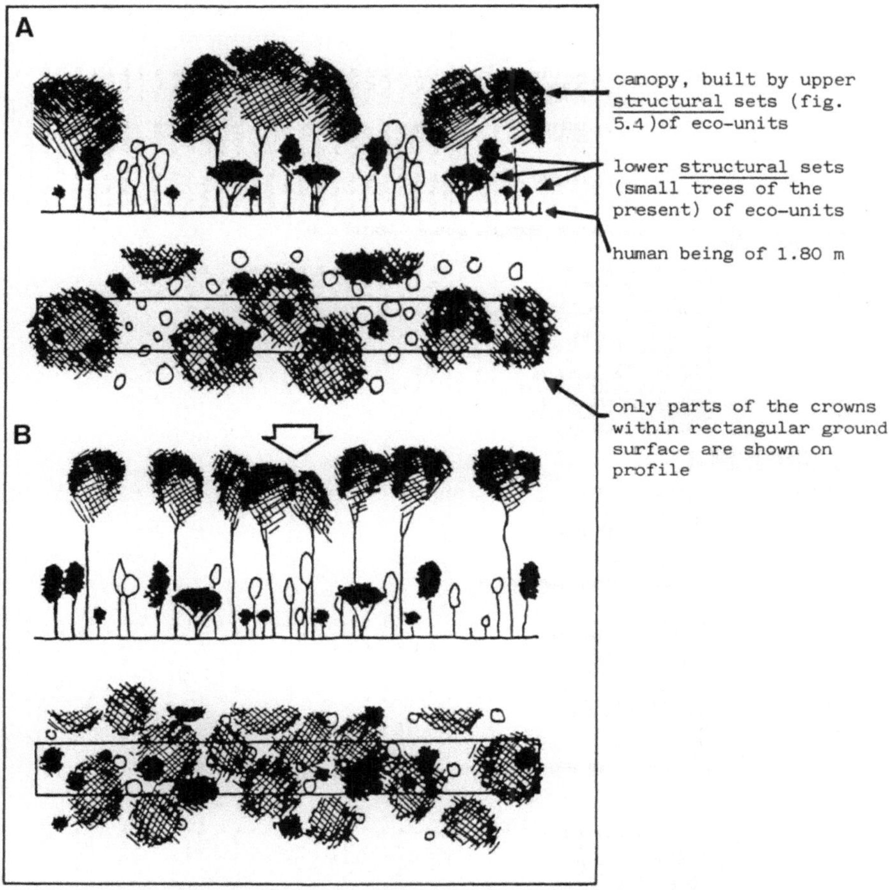

Fig. 6.44. Two patterns of biostasis and aggradation in eco-unit composition of a silvatic mosaic, explaining in part some particularities in Figs. 6.41–43. **A** Aggrading eco-units in a network surrounding biostatic eco-units like islands, observed by Van de Winckel (1980) in the Alsace. Note clear stufung and asymmetrical tree crowns forming vertical limits of biostatic eco-units. Authors think this pattern characteristic for (semi-) tolerant tree species, i. e. for intermediate precuilibrium stages. **B** Very small aggrading eco-units in a network of biostatic eco-unit phases, observed by Van de Winckel (1980). Biostatic units may be fragmenting, or fused in part; see traces of stufung below *large, open arrow.* Note diffused character of aggrading eco-units, because border layers amongst eco-units are asymmetrical (often below higher crowns) and large compared with the eco-units themselves. May be interpreted easily as "mixed forest model" instead of "mixed eco-unit model". (Vredenbregt and Van de Winckel 1982)

McEvoy et al. (1980) to describe a Southwestern Virginian oak forest in the Appalachians.

Such documents are rough, average sketches representative of larger forest mosaics, without considering their eco-unit composition. They are sufficiently precise to yield answers to many questions on these forests, e.g. in calculating the interaction between trees and seed consumers, such as insects, birds and mammals.

		1.1.1																		1.1.2							Sociability
	ALTITUDE	1	1	2	2	2	2	2	1	1	1	1	2	1	1	2	1	1	2	1	1	1	2	2	2	1	
	SLOPE	L	L	L	L	L	L	L	L	L	L	L	L	L	L	L	L	L	L	L	L	L	L	L	L	L	
	ASPECT	-	-	-	-	-	-	-	-	-	-	-	-	-	-	-	-	-	-	-	-	-	-	-	-	-	
	SUBSTRATUM	N	N	C	N	N	N	N	N	N	N	N	N	N	N	N	N	N	N	C	C	C	S	C	C	S	
	STONINESS	L	L	L	O	O	O	O	M	M	O	L	L	O	L	O	O	L	H	L	L	O	L	L	L	L	
	SOIL	C	C	C	C	C	C	C	C	C	C	C	C	C	C	C	C	C	C	C	C	C	C	C	C	C	
	RELEVE NR	151	168	087	253	031	144	291	452	456	484	216	180	175	506	147	475	510	159	220	231	215	089	290	288	225	
ACACIA TORTILIS - ARISTIDA BIPARTITA WOODLAND ON BLACK VERTIC CLAY																											
CHARACTER SPECIES																											
CS	ARISTIDA BIPARTITA	+	2	1	+	+	3	+	2	1	1	+	1	+	1	1	+		+								2-3
CP	HIBISCUS TRIONUM	+	+	+	+	+		+	+	+	+	+	+	+	+	+			+		+		+	+	+		1
CE	SETARIA WOODII	+		+	+	+		3	+	+	+	2	+	1	+		+	3	+				+				2-3
CE	ISCHAEMUM AFRUM	+	1	+	1	+	+	1		+	+	2	+	+		+	+							+			2
CE	SEHIMA GALPINII								2	1	1				2		+	+									2
CS	SUTERA LUTEIFLORA			+				+				+		+		+											1
CE	STRIGA FORBESII	+	+				+											+									1
CP	SORGHUM VERSICOLOR			+	R				+																		2
CE	CORCHORUS TRILOCULARIS	+	+				+																				1
CE	GLADIOLUS SERICEO-VILLOSUS		+				+						+														1
CE	ASPILIA MOSSAMBICENSIS	+	+		+																						1
CE	DICHANTHIUM PAPILLOSUM			+		2																					1-2
CE	DIGITARIA POLYPHYLLA					+	+																				2
SPOROBOLUS NITENS VARIANT																											
DIFFERENTIAL SPECIES																											
D(2,3,4)	SPOROBOLUS NITENS																			+	1	+	+	1		+	2
D(2,3,4)	LYCIUM CINEREUM																			+	+				+		1
D(2,3)	ANTIZOMA ANGUSTIFOLIA																			+	+			+			1
D(3,4)	MONSONIA ANGUSTIFOLIA																		+			+			+		1
D(3,4,5)	TRAGUS RACEMOSUS																			+			+				2
D(3,4)	TEPHROSIA PLICATA																				+	+					1
ACACIA TORTILIS - PANICUM MAXIMUM WOODLAND																											
CHARACTER SPECIES																											
CP(4,5)	ACACIA TORTILIS				1	+	+	1	1	2	1	1		+	+		1	1		1		2		2	2	+	1
CP(4)	BOTHRIOCHLOA INSCULPTA	+		+	+		+	+	+	+			+	+		+			+		+	+	+	+	1	+	2
CP(4,5,7)	ACACIA NILOTICA	2	1	+		1	1	+	+				1	2		2			+	+			1	+	+		1
CP(5)	ACACIA MELLIFERA					+			+	1	+			+	+					2	2	+	+	+		2	1
CP(4,5,6,9)	PANICUM MAXIMUM				+			+		+	+	+		+	+		+				1	+				+	2
CP(4,5,6,9)	GREWIA FLAVA	+		+	+	+								+		+	+			1	1		1	+			1
CP(4)	SEDDERA CAPENSIS			+	+	+									+					+	+	+	+		+	+	1
CP(4,5)	SCHKUHRIA PINNATA								+			+				+			+	+		+	+	+	+	+	1
CP(4,5,9)	ACACIA KARROO		1	1	+	1	2						+	+		1							+				1
CP(4)	CHLORIS VIRGATA		+	+																+			1			+	2
CP(4,5)	EHRETIA RIGIDA	+					+																	+			1
CP(4,6)	ABUTILON AUSTRO-AFRICANUM									+				+						+							1
COMPANION SPECIES																											
	NIDORELLA HOTTENTOTICA	+	+	+	+		+	+		+	+	+	+	+	+	+	+	+	+	+	1	+	+	+		+	1
	ERAGROSTIS CURVULA	2	+	2	1		1	1	+	1			1	2		2	1	+	2	+	1	1	+	+	+	+	2
	CYMBOPOGON PLURINODIS	+	+	2		+		+		+	2	+	+	+	+		2	+		+		2	+	1		2	2
	RHYNCHOSIA MINIMA	1	+	+	+	+	+			+	+	+	+	+	+	+		+				+	+		+		1
	ASPARAGUS SUAVEOLENS	+		+	+	+	+							+		+	+		+	+	+	+	+	+	+	+	1
	DIGITARIA ERIANTHA				+			+	+	+	+	+	+				+	+		+	1	+		+	+	1	2
	DIOSPYROS LYCIOIDES SUBSP. LYC.	+		+	+	+	+		+						+		+			1	+		+	+		+	1
	ALOE TRANSVAALENSIS			+	+	+								+		+			1	1	1		1	+	+	+	1
	ZIZIPHUS MUCRONATA			+		+	+	R		+				+	+	+				+	+		+	+			1
	EUPHORBIA INAEQUILATERA					1			+							+	+	+		+		+	+	+	+	+	1
	CRABBEA ANGUSTIFOLIA		+	+	+		+						+			+					+	+	+	+		+	1
	CORCHORUS ASPLENIFOLIUS			+						+	+					+				+	+		+	+	+	+	1
	KOHAUTIA LASIOCARPA				+	+	+			+				+		+			+		+			+	+		1
	ARISTIDA CONGESTA			+										+					+	1	+	+	+	+		+	2
	LANTANA RUGOSA	+			+		+							+		+			+			+		+	+		1
	ORTHOSIPHON SUFFRUTESCENS	+	1	+	1			+	+	+					+			+									1
	INDIGOFERA PARVIFLORA	1	+					1	+			1									+			+	+	1	1
	THEMEDA TRIANDRA	+	+		+		+						+			+	+						+				2
	COMMELINA ERECTA					+			+			+			+		+			+		+		+			1
	ACALYPHA INDICA	+	+	+					+					+		+			+				+				1

AND THE SPECIES OF APPENDIX II.1

Fig. 6.45. Example of a species matrix describing silvatic mosaics in terms of site factors and species composition in a part of the Western Transvaal Bushveld (South Africa). Semi-arid bush woodland, with characteristic species for the whole vegetation and differential species for variants. No possibility for analysis in terms of eco-units; frequencies and coverage percentages per species are averaged over mosaics. Sketched links with vegetation structure may be added by dansero-grammes (Fig. 5.66) or vegetation pyramids (Fig. 6.47), both for mosaics, not eco-units. (Table from Van der Meulen 1979)

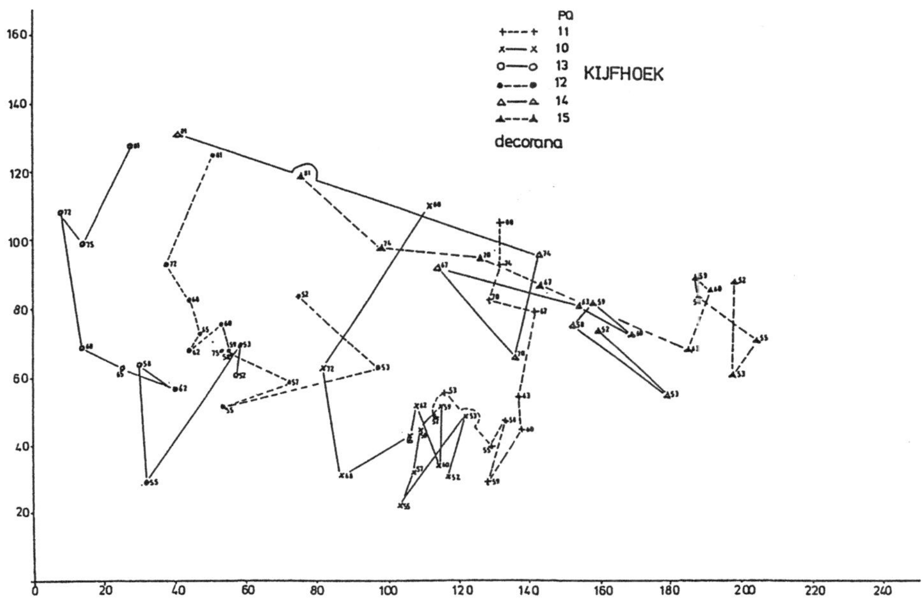

Fig. 6.46. Ordination of phytosociological data for the development of permanent quadrats (*PQ*) in the Dutch Kijfhoek forest reserve, coastal dunes near the Hague, between 1952 and 1980/1981. Low, stressed woodlands on poor sands, with treeless development stages of the mosaic (cf. Fig. 6.43). Ordination made with DECORANA computer programme. *Vertical axis* indicates composite criterion linked to shrub development, species poverty and grassification (treeless). *Horizontal axis* indicates composite criterion indicating factors linked to drought. According to Van der Hagen, the weight of rare species may in some cases distort conclusions from DECORANA ordination. Note that each permanent quadrat has been chosen according to criteria that are not linked to architectural development of the woodlands but to characteristic areas on which all species present may occur. (Unpublished figure, reproduced courtesy H. van der Hagen, DWL, The Hague)

This was done by Genard and Lescourret (1984) in a region of the Pyrenees, the forests of which have been classified by the methods of Baudière. Figure 6.48A shows the basis for calculation of the potential natural regeneration of pines in these forests.

Figure 6.48B shows a relation diagram for tropical dipterocarp rainforest in East Kalimantan (Indonesia), which might be quantified in the same way if the number of data needed would not exceed by much the number of data required in the simpler Northern forests. Of course, the interactions referred to in Fig. 6.48B are between trees and mycorrhizal fungi, not between trees and animals, but the analysis also goes straight from the component level to the mosaic level, without considering eco-units. This may be an error in this case (cf. Newbery et al. 1988; Sect. 6.2.3.3).

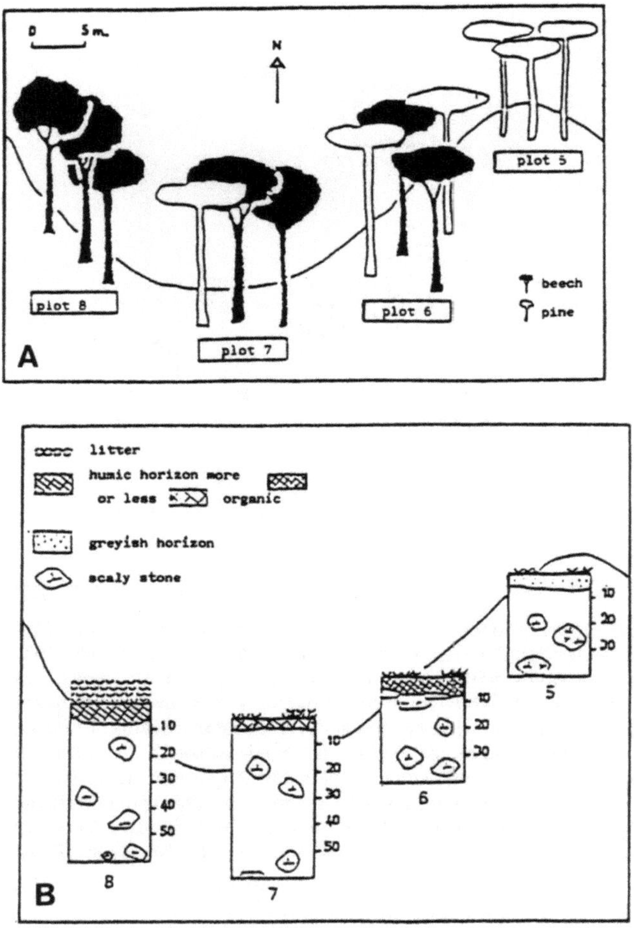

Fig. 6.47. Forest analysis based on site and species composition, average structures indicated by vegetation pyramids. Mont Coronat, ca. 1500 m asl., Eastern Pyrenees, France. (Baudière et al. 1986). **A** Forest over a toposequence on the north slope, note that plots are not eco-units but parts of a mosaic judged representative for a forest type. This is conformed by **B** soils over the toposequence (cf. also Fig. 5.69 for the tropics); because a different site allows a different silvatic mosaic to develop, the plots represent mosaics, not eco-units. **C** Vegetation pyramids showing the structure of the species composition by representing the percentage crown coverage per site, per height level, per species, for three of the four sites shown in **A** and **B**. The indicated strata are linked to biological types, as in danserogrammes (Fig. 5.66). *Numbers to the left* of each stratum: number of species present. The toposequence is short, distance between plots 5 and 8 being about 50 m (Baudière et al. 1986)

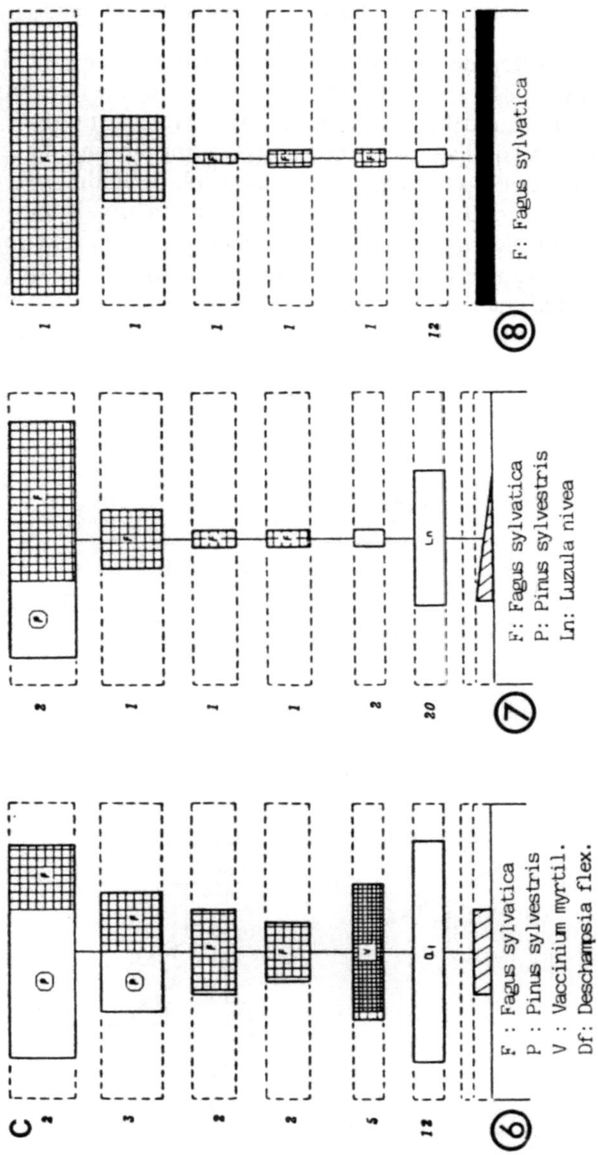
C
6
F : Fagus sylvatica
P : Pinus sylvestris
V : Vaccinium myrtil.
Df: Deschampsia flex.
7
F: Fagus sylvatica
P: Pinus sylvestris
Ln: Luzula nivea
8
F: Fagus sylvatica

Fig. 6.47C

6.3.6 *A Shifting Mosaic Matrix: The Torquebiau Calculation*

Forest development calculations have been made in a totally different way by Torquebiau (1981). He analysed the behaviour of the tree components, the eco-unit composition and the silvatic mosaic in the Mexican rainforest reserve of Los Tuxtlas (Veracruz). His diagnosis of eco-units uses the morphological inversion level, which is the more or less horizontal, undulating plane uniting all morphological inversion points (cf. Sect. 3.7; the first large living fork in broadleaved trees of the present, according to Oldeman 1974a,b). Torquebiau developed this notion further in 1985 and 1986 for Indonesian rain forests (see Fig. 6.16). For his Mexican forest area, he distinguished chablis or young vegetation without inversion level (innovation phase), zones with the inversion level below half-height of the forest (aggradation), with irregular inversion level oscillating around the mid-height level (degradation) and

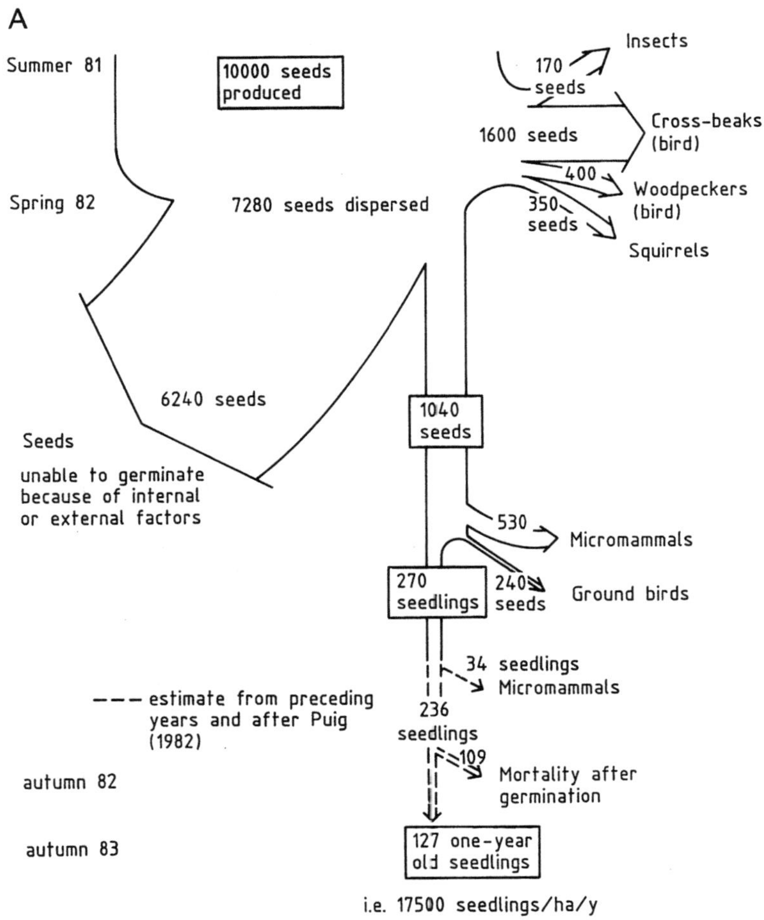

Fig. 6.48A

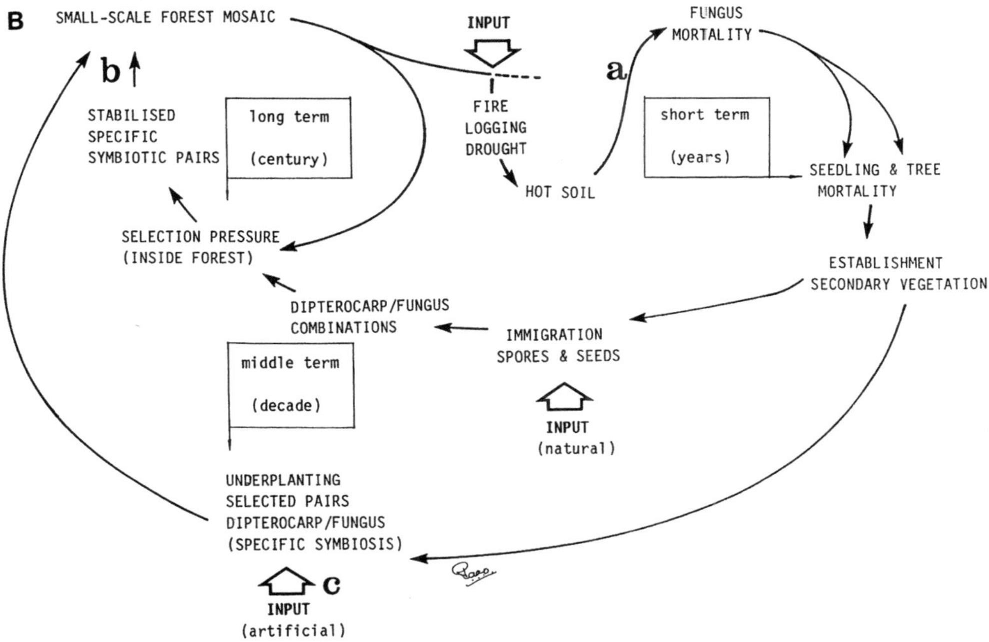

Fig. 6.48.B. Other forest components in two-level models explaining silvatic mosaics in terms of components. **A** Seed production by *Pinus uncinata*, consumption and germination in the forest of Osséja (Eastern Pyrenees, France), previously analysed as in Fig. 6.47. Note that the result here is expressed as seedlings/ha, and so is relevant for natural regeneration. Genard and Lescourret (1984). **B** Relation diagram for symbiosis between dipterocarp trees and fungi, selection pressure and plantation techniques (artificial) in tropical rainforest mosaics of East Kalimantan (Indonesia). This diagram may be quantified like Fig. 6.48A, but will require compartmentalisation to handle the numerous data. Like **A**, it shows a mosaic in terms of species, including non-trees. (Smits et al. 1987)

with more or less horizontal inversion level above the mid-height level (biostatic). The terms between brackets are added to establish the link with the present text.

Figure 6.49 shows the topography of the land (Fig. 6.49A) and the mosaic upon it (Fig. 6.49B), with tree dimensions indicated symbolically and mapped accurately. Torquebiau then planimetered the surface of each of his development phases on the map, and compared these surfaces with those found by other authors (e.g. Whitmore 1975; Mutoji-A-Kazadi 1977), who in other rainforests in the tropics had made comparable maps. He finally proposed a mathematical function using the rate of change-over from one silvigenetic phase to another, i.e.:

- the transformation of later phases into chablis, which is a function of chablis frequency;
- the transformation of chablis into later phases, which he defines as "everything that shows no clear traces of chablis anymore". This is a function of the rate of

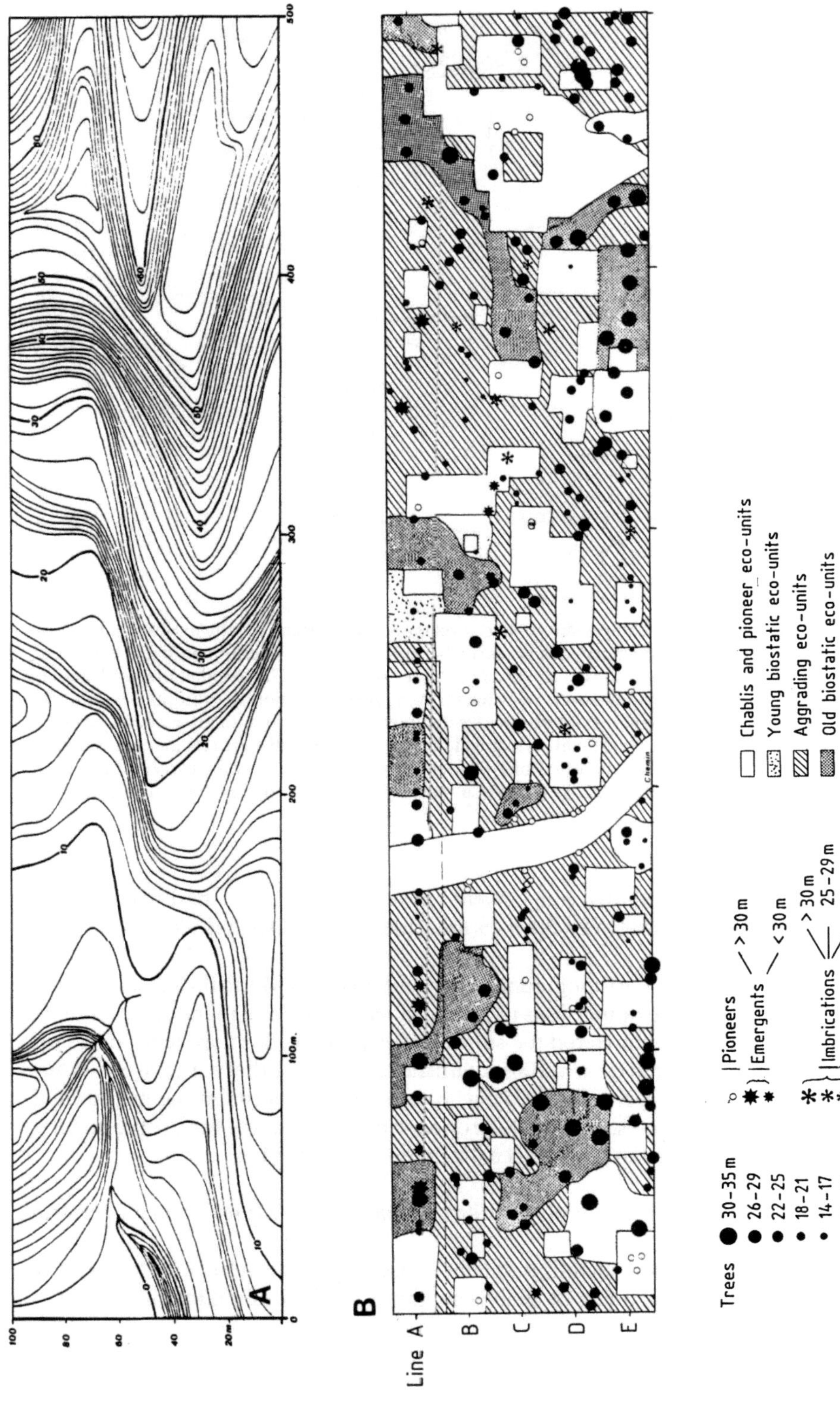
A
B
Line A
B
C
D
E
Chemin
Trees
30-35 m
26-29
22-25
18-21
14-17
10-13
Pioneers
Emergents
> 30 m
< 30 m
Imbrications
> 30 m
25-29 m
< 25 m
Chablis and pioneer eco-units
Young biostatic eco-units
Aggrading eco-units
Old biostatic eco-units

Fig. 6.49A, B

forest development (eco-unit development). This model can conceivably be made more precise by introducing more phases and more eco-unit kinds.

The result is a transition matrix. It simulates the development in time of both states from each other, starting with their share in the mosaic observed at a given moment, and extrapolating their reciprocal transformation rate observed over a certain timespan. This will be more precise when monitoring time of these processes has become longer. The matrix is a 2 x 2 matrix, with entries "proportion of chablis" (Ch) and "proportion of non-chablis" (F) at times t and t + 1:

$$Vt \qquad \begin{bmatrix} F\cdot t \\ Ch\cdot t \end{bmatrix} \tag{1}$$

$$Vt+1 \qquad \begin{bmatrix} F\cdot t+1 \\ Ch\cdot t+1 \end{bmatrix} \tag{2}$$

The matrix is such that Vt + 1 = Mx Vt, and therefore becomes:

$$\begin{bmatrix} F\cdot t+1 \\ Ch\cdot t+1 \end{bmatrix} = \begin{bmatrix} a & b \\ 1-a & 1-b \end{bmatrix} \begin{bmatrix} F\cdot t \\ Ch\cdot t \end{bmatrix}$$

in which a is the proportion of non-chablis which is not transformed into chablis during the time interval t, t + 1 (substracted from chablis frequency = 1 - a) and b is the proportion of chablis returning to non-chablis state, evaluated from the mean maximum chablis age. Applied to the data derived from Fig. 6.49, and calculated with different rates of change as a pessimistic and a moderate hypothesis, this yields for Mx:

	Non-chablis	Chablis	
N-Ch	96	4	Pessimistic
Ch	4	96	

	Non-chablis	Chablis	
N-Ch	96	1.3	Moderate
Ch	4	98.7	

Now, if $Vt = \begin{bmatrix} 75\% \\ 25\% \end{bmatrix}$, as in Los Tuxtlas and other forests, then after t + 1 years:

$$V_1 \ t+1 = \begin{bmatrix} 73\% \\ 27\% \end{bmatrix} \text{ (pessimistic), or } V_2 \ t+1 = \begin{bmatrix} 75.03 \\ 24.97 \end{bmatrix} \text{ (moderate),}$$

and after t + 2 years:

$$V_1 \ t+2 = \begin{bmatrix} 71.2\% \\ 28.8\% \end{bmatrix} \text{ (pessimistic), or } V_2 \ t+2 = \begin{bmatrix} 75.05 \\ 24.95 \end{bmatrix} \text{ (moderate).}$$

Fig. 6.49. Complete forest mosaic analysis of 5 ha in Los Tuxtlas forest reserve, Veracruz, Mexico. Trees, eco-units and mosaic have been used in a three-level model. **A** Topographical relief map of the 5 ha, equidistance of curves 1 m altitude, showing many toposequences. The mosaic is a complex of mosaics. **B** Eco-units determined and diagnosed by tree architecture along parallel scanning lines at a distance of 20 m. For calculations of mosaic dynamics see text. (Torquebiau 1981)

After 40 years, and with an uncertainty of about 0.5%, the pessimistic hypothesis would predict half of the forest to be covered by chablis, and the moderate hypothesis forecasts the proportion of chablis to remain a constant 25%. Total turnover time of each hectare of this forest in the last case can be calculated as some 97 years, which is of the order of "natural turnovers" between 80 and 150 years found all over the tropics (cf. Vooren 1985, 1986; Hartshorn 1980; Lieberman et al. 1985b; Oldeman, Hartshorn, both in Tomlinson and Zimmermann 1978; Brandani et al. 1988). The pessimistic hypothesis would indicate an elimination stage, the moderate one would indicate ecuilibrium dynamics.

This calculation has been given ample space here, because it provides a bridge between two-level, component/mosaic models or silvicultural models calculating the turnover for **trees**, and the silvatic mosaic models proposed in these pages. Moreover, it can be made after rather simple and fast observations. If a more complicated model is allowed in a research programme, the Torquebiau calculus could be linked up with the Brünig models (1987) allowing to evaluate the probable periodicity and character of shifting events (drought, fire, storms, squalls, tornadoes, lightning, biological agents, human impacts). This might finally provide a solid basis for silvicultural systems and forest management plans that take into account a balanced view of natural shifting events which are probabilistic, and eco-unit development which is mostly deterministic.

A quite different calculation has been published by Brandani et al. (1988). In order to check predictions on differences in tree species composition in Costa Rican rainforest as depending on gap characteristics, predictions on colonisation dynamics by different tree species and predictions on tree species richness and density as a function of gap dynamics, they used a complete census of seedlings and saplings in 51 gaps and processed the data according to each prediction to be tested. With an essentially statistical and descriptive approach, they validated the model of gaps with crown/bole/root zones proposed by Orians (1982) and which corresponds to the chablis model by Oldeman (1974b, 1978, 1983c), except for terminology.

The approach, which is probabilistic all-out, makes the calculations by Brandani et al. completely different from the essentially deterministic model by Torquebiau (1981), although both use the same three hierarchical levels corresponding grosso modo to component organisms (trees), eco-units and silvatic mosaics (cf. Oldeman 1983c). The predictions by Brandani et al. (1988), like the 19 predictions by Hartshorn on the same subject 10 years earlier (in Tomlinson and Zimmermann 1978), are rather hypotheses, whereas the results from Torquebiau matrices over tens of years in the future are predictions in the sense of forecasts.

The weak point remains the definition of the borders of gaps or eco-units (cf. Sect. 5.1.2; Barkman 1984, 1988b; Bongers and Popma 1988). The "vertical hole" by Brokaw (in Leigh et al. 1982) which was proven useless by Bongers and Popma's research but was not amended by these authors, has not been really improved by Brandani et al. (also see Florence 1981). As long as the spatial and temporal dimensions of the living system to be studied have been so imperfectly determined, which probably is inherent to the system itself, problems will keep cropping up in gap, patch, eco-unit and mosaic studies.

Oldeman (1989) has suggested a way out of this dilemma, by concentrating on the zero-events (Sect. 5.2.5.3; Sect. 6.2.8.8) and using an arbitrary eco-unit definition, only serving to establish a diagnostic map of the state of the mosaic. Zero-events have an effect that spreads beyond the limits of the new eco-unit they create (cf. Sect. 6.4.6). There is some evidence of very slow, wave-like fronts in light intensity, soil properties or air humidity spreading in the forest around the canopy hole made by a zero-event. These effects interfere with each other if zero-events occur rather close in time, perhaps within an interval of maximally 10 to 20 years. They cause, for instance, indicator species to appear in associations where they should not grow according to classical vegetation science (Koop 1981a; Barkman 1988b).

The hypothesis of interacting zero-events has been named by Oldeman (1989) the **ecological interference hypothesis.** It would, if true, change our notions about the "floristic identity" of eco-units (forest patches). In defining a new, young eco-unit (gap) as a selective biological gateway rather than as a biological straightjacket there is a perspective to integrate views that may now seem opposed, e. g. Brandani et al. (1988), the present book, Bongers and Popma (1988) and Torquebiau (1981). It would be the antithesis to the application of island biogeography in gap-phase models and other forest mosaic theories (cf. Zakaria-Ismail 1988; Barkman 1984; Brussaard 1984).

6.3.7 Graphical Succession Models

Two graphical representations of silvatic mosaic dynamics are given, the authors of which took into account the species composition of these mosaics. Figure 6.50A (Vredenbregt and Van de Winckel 1982) does not differentiate between eco-units and mosaics in all cases and mainly follows the Central-European concepts (Figs. 6.41A; 6.42) in considering tree mortality, not local forest mortality. Mosaic patterns are admitted by these authors only in their case C, when larger zero-events, not single tree mortality determine the dynamics. Comparing Fig. 6.50A with Fig. 6.6, it is clear that a bridge towards eco-unit composition in the architectural sense is present in the work of Vredenbregt and Van de Winckel (1982). The evident links with the Central European approach connect this model with phytosociology (cf. e. g. Figs. 6.50A and 6.43).

An elementary, classical classification of tree temperament is used by these authors (e. g. Kalela 1949; Fig. 6.17), i. e. light-requiring, semi-tolerant and tolerant trees. As to "succession", nearly every eco-unit may succeed to every other eco-unit in these mosaics (cf. Oldeman 1978, 1979, 1983c, and Fig. 6.19).

Figure 6.50B is from Six Dijkstra (1981; also see Six Dijkstra et al. 1985). He reconstructed the development of forests built mainly by the tree *Dacrydium cupressinum* (Podocarpaceae, Gymnosperms, "rimu") on acid-bog soils of the glacial terraces in South Westland (New Zealand). In the shifting event regime, windthrow over large surfaces is important, depending on the exposition of the site. On flat sites near the sea, the regeneration pattern covers areas of 1 to 20 ha, on slopes further away from the sea it is reduced to 0.2 to 2 ha. Along this gradient, soils change too and become better drained.

A

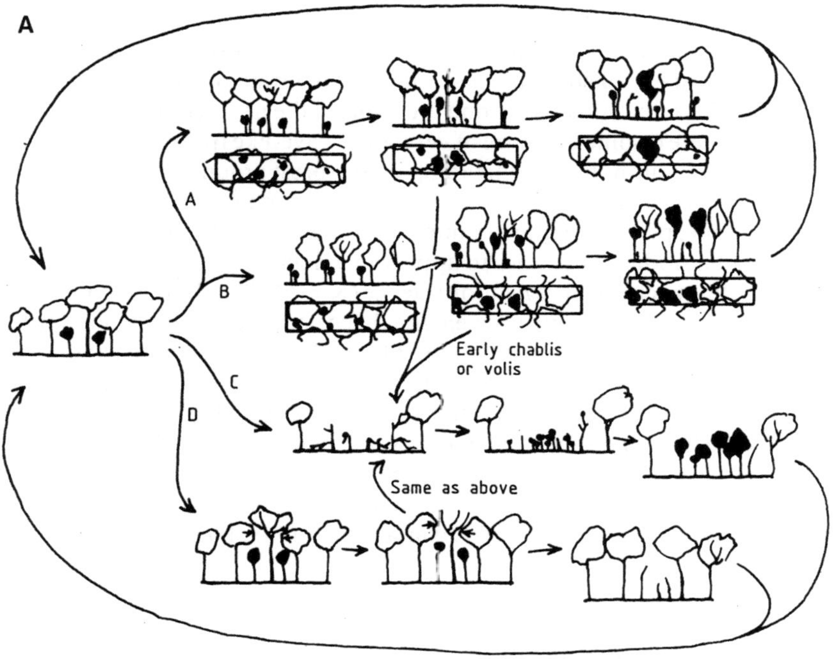

Fig. 6.50. Two architectural succession models with links toward species composition and other two-level tree mosaic approaches. **A** Succession and tree temperament. *A* standing mortality amongst tolerant trees; *B* same amongst semi-tolerant trees; *C* chablis (uprooting) or volis (trunk breaks), the authors say "clear mosaic pattern"; *D* standing mortality, sometimes also chablis or volis, but eco-units eliminated by occupation by neighbouring ones. (Vredenbregt and Van de Winckel 1982); terms somewhat adapted. **B** Probable successional series and eco-unit development (cf. Fig. 6.19) on acid-bog soils of the glacial terraces in South Westland (New Zealand). Silver pine = *Dacrydium colensoi*; rimu = *D. cupressinum* (both Podocarpaceae); Kamahi = *Weinmannia racemosa*; Quintinia = *Q. acutifolia* (both broadleaved). Note characterisation of phases and stages by indicator species, which may be supplemented by species matrices (e.g. Fig. 6.45). (Six Dijkstra 1981)

The graphics by Six Dijkstra should be compared with the "successional series" in Fig. 6.1. The arrows (Fig. 6.50B) show the kind of eco-unit that may succeed to another if a zero-event due to natural tree mortality or to accidents (star) occurs. Each site is characterised by a different weight per arrow. Some are rare in one site and frequent in another. Note that both eco-unit development phases and successional stages are characterised by species composition, the indicator species being mainly trees or shrubs, except in the "pakihi" stage when there is no forest as yet. The species lists given by Six Dijkstra (1980), also including monocotyledons, ferns and mosses, link architecture to floristics, being listed per development phase.

Other studies containing such successional diagrams based upon architectural analysis in Europe include, among others, Koop (1981a), Wilderink and Zakee (1985), Van Heusden (1980), and for tropical post-shifting agriculture mosaics in Papua New Guinea, Louman (1986).

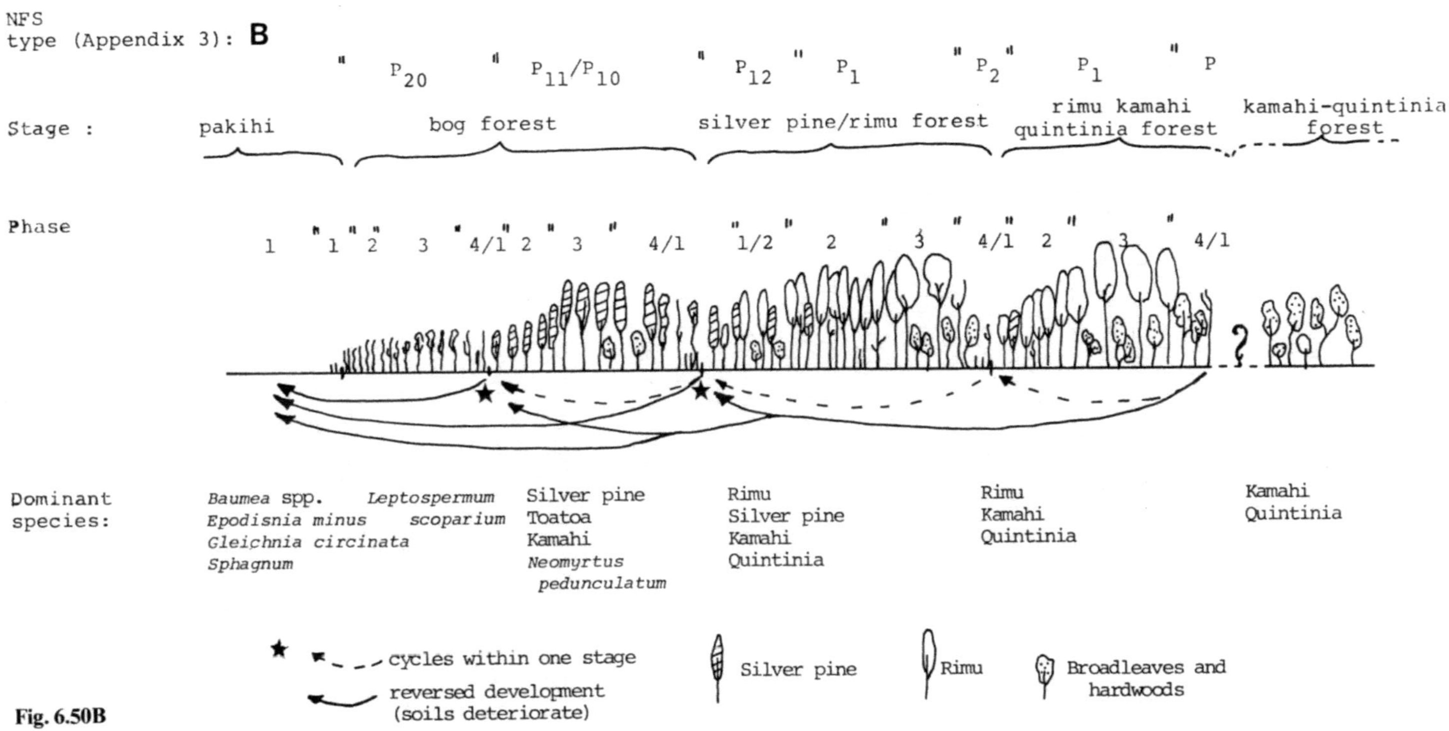

NFS
type (Appendix 3): B
" P20 " P11/P10 " P12 " P1 " P2 " P1 " P
Stage :
pakihi
bog forest
silver pine/rimu forest
rimu kamahi quintinia forest
kamahi-quintinia forest
Phase
1 " 1 " 2 " 3 " 4/1 " 2 " 3 " 4/1 " 1/2 " 2 " 3 " 4/1 " 2 " 3 " 4/1
Dominant species:
Baumea spp.
Epodisnia minus
Gleichnia circinata
Sphagnum
Leptospermum scoparium
Silver pine
Toatoa
Kamahi
Neomyrtus pedunculatum
Rimu
Silver pine
Kamahi
Quintinia
Rimu
Kamahi
Quintinia
Kamahi
Quintinia
cycles within one stage
reversed development (soils deteriorate)
Silver pine
Rimu
Broadleaves and hardwoods

Fig. 6.50B

6.3.8 Species Populations, Selection Pressure, Speciation

The systems hierarchy used in silvological analysis (Table 2.1) may contribute to the understanding of speciation. This is the origin of species (Abercrombie et al. 1977). Although the present book is on silvology and not on genetics, still the cross-link has to be presented. The need for this is shown by Grene (1987, p. 509): "There are the genealogical and the ecological hierarchies. The one has to do with reproduction, via the transfer of information, and the other with interaction, or the economics of transactions in, with and of organisms. Not all hierarchy theorists make this distinction but it is here, I want to suggest, that hierarchy theory holds most promise for research programs in evolutionary theory".

On the other hand, forest management for conservation or for production of commodities and services requires, more than any other form of land use, the availability of a rich spectrum of biological species populations. Krugman (1984, p. 71) states: "The modern forester must employ management strategies that maintain a reliable and highly varied genetic reservoir for future use and possible genetic improvement. In addition, means must also be provided for insuring and perpetuating selected large and small populations for future mass seed production to meet unforseen contingencies".

In other words, there is a need to understand the origin of species so as to avoid the extinction of species. This is linked to the different kinds of lethal or favourable selection pressure brought to bear upon useful and useless species populations alike, in both natural and managed forest ecosystems.

6.3.8.1 A Hierarchical Model of Selection Pressure

Genetics start at a submicroscopic systems level of organic molecules called genes. In the hierarchy this is far below the organ level, this being the smallest "black box" subsystem considered in this book (Chap. 3; Table 2.1). Genes are units of material of inheritance, an old name being germplasm. They can be considered as subsystems of a system one level higher. The latter consists of cell parts, which certainly include chromosomes and perhaps other cytosomes too (extranuclear inheritance). The system itself is called genome and its build-up in a particular organism or population is a genotype. The genome is transmitted from a cell to its descendant cells during cell division.

Transmission is implemented very precisely by replication of the genome, gene by gene. If the genome is altered due to imperfect replication, this is mutation. Mutations are transmitted in their turn. Their transmission may have either favourable or unfavourable consequences for the system they direct (cell, organ, organism). Geneticists mainly studied these consequences at higher hierarchical levels, particularly whole organisms. Their development, their behaviour in the environment and their reproduction are subjects of genetical interest.

Dawkins (1976), in his "selfish gene theory", considers genes as living systems in their own right, within their own intracellular environment. "An organism is a machine for gene survival". This viewpoint, although extreme, translates a general preoccupation of geneticists with genome transmission. This transmission may happen sexually, through fertilised eggs (zygotes), with the highest risk of mutation. A lower risk of mutation occurs in asexual genome transmission, a general term being apomixis. This includes cloning and in vitro propagation. Mutation risks increase with diminishing propagule size (Roux pers. commun.), but clones made from sizeable cuttings with not too young meristems may be taken to have the same genome. Agamospermy is the production of seed by apomictic means, which is rare in woody plants (cf. Heijbroek 1984, his Bibliography).

Apomictic mutations are special and include, for instance, "bud mutation" (= meristematic mutation) and architectural mutation (cf. Hallé in Tomlinson and Zimmermann 1978). This mutation may conceivably occur at levels higher than the genetic level. But it is then subjected to "selective interaction from below". On the one hand, if it kills the genome, this kind of mutation eliminates itself. On the other hand, if it leaves the genome totally unaffected it can only be inherited via extranuclear plasmatic transmission. A meristem of course is a favourable instrument for such transmissions. But intermediate cases, in which both cytoplasm and nucleus play a part, cannot be excluded.

The two predominant concepts in evolution studies are from Darwin's "Origin of Species" (see Leaky 1979). The first concept is the Linnean species. The second one is its origin, as explained by mutation and selection. These concepts lead to emphasis on transmission and replicators, and hence to a central role for population biology (e.g. Wilson 1974; Flanagan 1978; Namkoong 1981, especially for "quantitative genetics in forestry"). Paraphrasing Dawkins one might say: "An ecosystem is a machine for species population survival".

Here, one central and recurrent theme of the present book is encountered again. It is the balance between structural, deterministic and random, chance processes (cf. Sects. 2.4.3, 6.1.2). In genetics, the two French Nobel Prize winners Monod and Jacob considered this issue in their books, respectively "The Logic of Life" on structures and levels (Jacob 1970) and "Hazard and Necessity" on their balance with chance events (Monod 1970). One input of chance in genome transmission is considered as genetic drift (e.g. Cavalli-Sforzo in Wilson 1974). This is survival of genes that are stronger (dominant) than their (recessive) alternatives, whereas their effect is worse for species survival than that of the recessives. Genetic drift may cause bad medical surprises in human populations and bad yield surprises in crop populations.

The above outline is brief and incomplete, but it provides a reminder of some notions needed to tackle the question of natural selection and the silvological hierarchy. A hierarchy of systems has no upper or lower ends, but let us consider for the sake of simplicity a two-level hierarchy with living systems and subsystems. The following rule then applies to all such hierarchies, however defined.

"In a systems hierarchy with two levels, any change in a subsystem affects the chances of survival both of that subsystem and of the system, in one of three ways, i.e. favourably, unfavourably or neutrally".

Now let us extend the hierarchy to an indeterminate number of system levels, which is the only correct way, since levels can be freely defined by the researcher (Sect. 2.4.4). It then follows from the first rule, that:

"The effects of change at any system level are operational until they reach the system level from whence they are not transmitted further, because they are unfavourable or neutral to the survival of the system at that level."

This general rule can be applied to the silvological hierarchy. The genetic level is taken as the level of departure, because it has been proven that it is a privileged level for stocking the vital information to build well-adapted organisms, without errors that are immediately lethal.

"The effects of genetic mutation survive until they reach the system level at which they are either neutral or are eliminated."

This solves the polarity of classification versus ecological hierarchies. For hierarchies of classification, Grene (1987, p. 506) states: "The arrangement of their elements constrains, and thus controls, the behaviour of those very elements as long as the system so constituted continues to exist". She states that such systems are unidirectional, whereas ecological systems have dual control. In our view, dual control means dynamism. The rules of change formulated above refer to this very dynamism with dual, feedback control. They clarify the option of Chapter 2, in which Table 2.1. shows how hierarchies of classification and ecological ones can be harmonised. The main condition for this is that a system be first of all defined in real space and real time. Its architecture should be known.

Table 6.2 shows how selection pressure works in this hierarchical context. The hierarchy of Table 2.1 has been presented upside down from the genome on, so as to create a genetic vision. Figure 5.74B illustrates some of the interactions referred to, e.g. between plant cells and cell-sized insects, which cannot be interpreted in a logical way as direct interactions from organism to organism. They concern direct relations between a subsystem of a large organism, and a small organism.

Just following Table 6.2, let us suppose that a mutation occurs in a tree at the first level encountered, the genome. This is not the only conceivable change (cf. chimeras). The imaginary genetic mutation would, let us suppose, affect the leaf **cells** as to chlorophyll content. This cell type is then subjected to selection pressure from the other cells in its tissue. If tissular interactions were badly disturbed, this is where the mutation would be eliminated (tissue level). If the tissue survives with its modified photosynthetic characteristics, then it comes under selection pressure from other leaf tissues and is eliminated, tolerated, or favoured at the organ level. The changed leaf is then under selection pressure from other organs with which it is functionally linked, especially buds and axes. The organ complex either survives or not.

If so, it is selected by the branched complex in which it belongs, and which may or may not survive in a modified form. At each level, special kinds of selection pressure are added until the whole organism is selected as fit to survive or not. From here on, the mutant tree is internally consistent, but it has become a different eco-unit component. The eco-unit then effects selection pressure and either eliminates it or allocates more resources to this tree variant. This may make the eco-unit more or less "powerful" in the silvatic mosaic. The analysis of leguminous tree architecture by Oldeman (1989a, in press) yields a concrete example of this chain of processes.

But these internal, physiological selection pressures from level to level are not the only ones. At each level there also are external selection pressures at work (cf. Fig. 5.74B). In Table 6.2 they are indicated as horizontal arrows. At each level,

Table 6.2. Inverted hierarchy of Table 2.1, with consequences of mutations at successive levels of organisation. Left side: organisms large. To the right (cf. Fig. 5.74B) shifted hierarchies for ever smaller organisms, unicellular structure the smallest considered. Arrows vertical: internal selection in auto-evolution of an organism. Arrows horizontal: reciprocal selection in coevolution between organisms of different sizes, the larger one "being partly or entirely the biotope" of the smaller one. sb = selection pressure by

Replicators	Vertical arrow (▲ Genome … ▼ level)	*If mutation occurs then:* <——> = coevolution (*Interactors*)
Genome		
Cell	▲ sb: cell ▼	<——> Organism (unicellular)
Tissue	▲ sb: tissue comp. ▼	<——> Organism (unitissular)
Organ	▲ sb: other tissues ▼	<——> Organism (small, organismic, like small insect)
Organ complex	▲ sb: other organs ▼	<——> Organism (larger, e.g. herbivores)
Branched complex	▲ sb: other branches ▼	<——> Organism (larger, e.g. grazer)
Reiterated complex	▲ sb: other reiterated crowns ▼	<——> Organism (larger, e.g. competitor)
Eco-unit	▲ sb: other eco-unit components (hor. arrows) ▼	
Silvatic mosaic	▲ sb: neighbouring eco-units ▼	
Site category (abiotic resources)	▲ sb: energy and nutrient budget ▼	

interaction with smaller or larger organisms of other species is modified after a mutation. This may be lethal, neutral or stimulating for either of both parties. For instance, a mutant leaf of a host tree may kill off consumer insects, or the insects may kill the tree.

According to Grene (1987, p. 509), selection pressure and speciation may concern genes, individuals, demes, species or monophyletic taxa. This fits in with the possibility at each level n to reach the next higher level (n + 1) of organisation either by building larger (n + 1)-organisms from n-components, or to build (n + 1)-populations from n-organisms. The distinction is not always clear, as in termites and ants (cf. Franks 1989; Marais 1933), or in multispecies combinations like mycorrhizae. The founding hypothesis (e.g. Den Boer 1977; Turin and Den Boer 1988) seems to open perspectives to consider not only ants, termites or bees, but insect populations in general as living systems being much more than the sum of their parts, one of their attributes being a kind of reproduction ("founding of a population").

Table 6.2 shows a selection hierarchy bearing vertically on **replicators** and horizontally on **interactors.** We might distinguish four basic forms of selection pressure.

- intra-hereditary selection pressure mediated physiologically affecting genotypes and/or populotypes;
- inter-hereditary selection pressure mediated by interactor species, and which may lead to co-evolution in a narrow sense;
- supra-hereditary selection pressure mediated by the ecosystem, i.e. a combination of biotic and abiotic selectors (co-evolution lato sensu); and
- extra-hereditary selection pressure mediated by abiotic factors.

These forms of selection pressure are of course not mutually exclusive; they present a frame of reference. The term hereditary is chosen intentionally because it may include extranuclear mechanisms of transmission of information from one generation to the next, and therefore is broader than genetic.

In a less abstract way, the following selection mechanisms can be found in Table 6.2.

- Two-level feedback between system and subsystem. Flexibility of the system determines whether it may incorporate a mutant subsystem with or without changing its overall performance. If it does not change, mutation is neutral and its effects stop at the level of that system.
- Multi-level feedback linking many system levels. If functional mutations are carried over to higher levels, they may be accommodated at every step by system's flexibility. In higher-order systems this probably often changes overall system performance.
- Interspecific feedback between an organism and an "outside" organism. It at least changes the functioning of a subsystem in one or both organisms, if it survives. Two-level or multi-level feedback results. Reactions from mutually enhanced performance (e.g. mutualistic symbiosis) to diminished performance (e.g. non-lethal parasitism) may be thus explained.
- Community feedback between an organism and the web of ecosystem interactions (eco-unit, mosaic). This may eliminate, tolerate or stimulate the newly evolved organism or complex of symbiotic organisms.
- Abiotic feedback between an organism, its ecosystem and the site. It either allows or does not that an eco-unit or a silvatic mosaic with newly evolved components functions within the locally available energy and nutrient budget.

The latter kind of selection pressure is most often considered, although interspecific feedback has become a close second during the last decades (see Sects. 6.2.3; 6.2.4). The traditional emphasis on abiotic factors is due to the historical dimension of speciation studies. The species extinctions and migrations, as postulated in evolution theory to explain species distributions during and after geological and climatic upheavals like glaciations, exemplify this. For a South American overview of these matters see Van der Hammen (1986). More in general, the tropical evolutionary history has been described by Flenley (1979).

If no mutations of the genome, but site change occurred, the effects are the inverse. All local living systems would be subjected to abiotic selection pressure. The largest ones, the silvatic mosaics, would enter into an elimination phase (Fig. 6.6). Organisms first would change their behaviour, e.g. reiteration in trees, and then would become extinct when their flexibility is insufficient for adequate behaviour (e.g. Fig. 3.33). This order of events means that communities lose vitality first. Devitalisation of communities is triggered by degeneration of interactions among compartments. It may therefore well start with mortality of unicellular or other small organisms. Trees go last, if site decline is not due to some natural or man-made form of deforestation.

Now which are the organisms that in such cases survive over the longest periods? No genotype can have been selected earlier for such site changes, but there are the more flexible ones. An example was mentioned earlier for pioneer species, with **Budowski's rule** (Sect. 5.5.11; Table 5.7), and for the increased flexibility reached by decreasing life cycles, from **K**-strategy to **r**-strategy (Fig. 5.74). A broad genotype spectrum and an accelerated mutation rate would explain the genetic strategy of such hardy species, apt to live in inhospitable conditions. This fits in with the tendencies in vegetation architecture and dynamics as depending on environmental hospitality (Fig. 6.1), containing higher shares of pioneers and herbs as conditions become harder. Still, there are possibilities to enhance opportunistic behaviour in large trees (Fig. 3.1). This is exemplified by the reiteration pattern of European, post-glacial oak species (*Quercus robur, Q. petraea*), or by the architectural dynamics of Leguminous trees (Oldeman 1989a in press).

Only a hierarchical scheme of selection pressure and co-evolution like Table 6.2 can yield a coherent and explanatory picture of so many silvological phenomena which at first sight have no obvious common base. One of the essential features of this table is that it visualises a "bottom-up" approach, the many small changes originating below being selected again and again as to their effects at higher levels of organisation. This agrees with the emphasis placed by Dawkins (1988) upon the non-stochastic nature of speciation: mutations are stochastic, but selection pressure is not, because it tends to rapidly accumulate small viable changes by weeding out all non-viable ones. It is a dice play in which only sixes are accepted. The "top-down" approach, exemplified by Lamarck or Lysenko, with its hypothesis of changed behaviour at a higher level of the hierarchy (e.g. organ, organism) becoming hereditary, here proves to be extremely improbable in a comprehensible way, although it is not false or heretical in a dogmatic way. It goes against probability to suppose that a succesful change at a higher level (e.g. in a leaf) would induce all subsystems at the tissue level to support that change, which in their turn would induce all cells to do the same. A dice-throw of six at the top has only an infinitesimal chance to be imposed upon or to spread to the myriads of subsystems (genes) at the bottom.

6.3.8.2 Species Conservation and Human Health Aspects

It has to be borne in mind that political options for narrow gene pools supporting huge nations have existed in the past. The subsistence of the Roman Empire was based mainly on some ten useful species. The olive tree yielded fats and accessorily, sometimes, wood. Wheat yielded carbohydrates and minerals and some proteins in bread. Grapes contained mainly minerals and vitamins, available in wine. Plane trees, resistant to lead poisoning, lined the city streets and oak coppice provided fuelwood. With a few species of cattle, the bulk package was there. The rest of the plant and animal species were important only for refinements, if at all. And, just like many industrial nations today, if some other natural product was necessary these societies mined it "in nature", not caring whether the species yielding it would survive. The Empire knew agriculture, not forestry.

Today, fortunately, most decision-makers shudder at the thought of such dependance on a vulnerable and narrow gene pool. In many countries, sometimes late and sometimes timely, there is now the will to conserve as well as can be done the natural species of plants and animals in the world. Forests being the terrestrial ecosystems with the most numerous species and the most diverse species composition, a special line of forest management concerns species conservation. This is part of nature management, but it also plays an important part in multiple-use forest management, in particular selection silviculture. An example of the considerations involved are the papers by Stocker and by Tracey in Shepherd and Richter (1985) on managing Australian forests. It is emphasised that immediate usefulness is only one of the valid reasons to conserve plants and animals (cf. Jacobs 1988).

If a genotype is to be conserved (cf. Wolf 1987; Myers 1979) or given room to speciate further, the conditions to keep it alive are to be maintained through ecosystem management for preservation. If such a genotype is to ensure the production of a commodity, e.g. wood, fruits or bark, risks are minimised if an optimal fit of the species in all successive hierarchical levels is ascertained. These aims are best served by **in situ** (on-the-spot) conservation (cf. Harris 1984; Jacobs 1988; FAO et al. 1989) and to a lesser extent in the modified conditions of artificial seed banks and botanical or zoological gardens, for menaced species. Figure 6.51A shows available options, after a study on Amazonian ecological problems and possibilities by IUCN (1983a).

Another example is China. Its organisation for forest species conservation has been outlined by Krugman et al. (1983). For wildlife management, particularly in tropical forests of the southern hemisphere, Kikkawa (1985) edited a set of extremely interesting papers. Among them, Nix and Gillison's "gradsect" method for animal habitat classification is to be noted because of its links with architectural silvatic mosaic analysis. The sheer size of the species banks to be conserved in the tropics is not only indicated by the thick and numerous volumes of tropical floras, but also by books showing species-linked plant-animal interfaces. This last kind of document is spectacularly represented by the 483-page book by Van Roosmalen (1985), listing, describing and illustrating the fruits of the Guyanan flora. This information was needed to understand the ecology of local monkeys (e.g. Fig. 5.10).

It is essential that forest species be conserved in situ, i. e. in the ecosystem where they evolved, with their full mutation potential and under natural forest selection pressure. This "natural pressure" is the pressure resulting from all relevant arrows in Table 6.2, which is so complex that it cannot be imitated in artificial surroundings. This makes extinction of species as difficult to explain as speciation, a point which is often forgotten. Until now, the approaches of species diversity by species counts and diversity indices, building up diversity, as it were, species population by species population (e.g. Begon et al. 1986), have not yielded much more than "questions that delude answers", the subtitle of a brochure by Lugo (1988), with a mainly North American bibliography on the question.

In Situ Genetic Conservation—Available Options
(Modified from FAO, 1981)

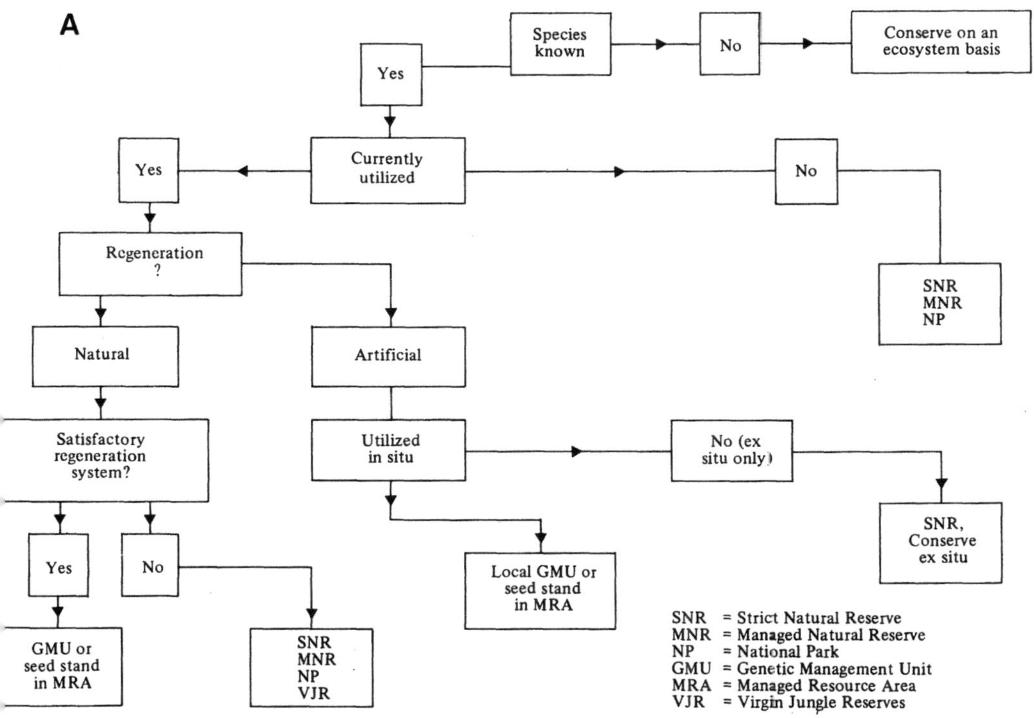

Fig. 6.51A (Fig. 6.51B see page 518)

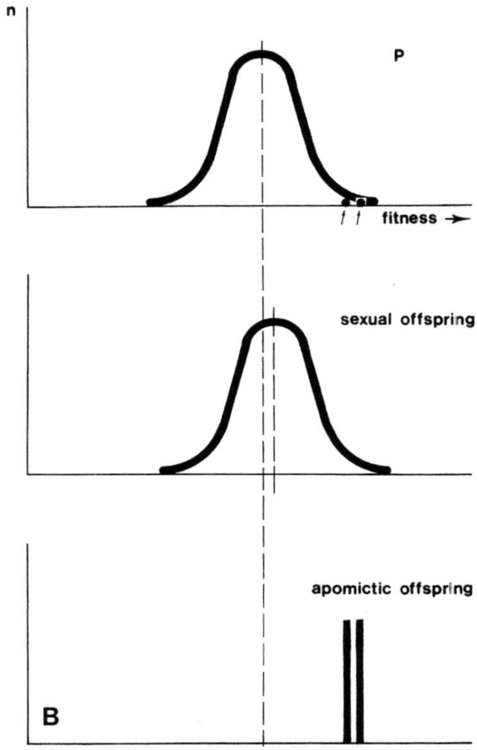

Fig. 6.51. Species, genotypes, forest mosaics, conservation and selection. **A** Options for species conservation as provided by IUCN (1983a) in a study on Amazonian rainforest. Applicable world-wide. Note implications of all options for the natural or artificial silvatic mosaics required. **B** Genotypes of crop trees vary around a natural optimum fitness, in which extreme types may be most useful (*above*). By selection of seed trees, a shift towards this crop optimum is sexually but partly achieved (*middle*) and by apomictic (vegetative) propagation the shift is stable but genetic amplitude is lost. (Heijbroek 1984)

One more reason to conserve species in their original surroundings is to understand their behaviour in the web of interactions between organisms. A much-neglected field proving how vital such understanding is for us all, is human health and epidemiology (e.g. Mouchet Chap. 17 in Fournier and Sasson 1983). This is often approached from the angle of agriculture. In particular the tropical deforestation processes with the aim of forest conversion to agricultural land use are so regarded. Bradley and Narayan (in FAO 1987, p. 36) say it as follows: "Extension of land use brings different vector hazards, chiefly resulting from man's intrusion into new ecosystems with disturbance of parasite life cycles maintained as zoonoses in the undisturbed environment". From the same authors, or from Mouchet (1983 as above), a short list of extremely serious tropical diseases may be summarised, in which new varieties or totally new syndromes have appeared in the last 15 years.

Such a list would include yellow fever, arboviroses (= arthropode-borne viroses) other than yellow fever, malaria, filarioses and leischmanioses. For the latter disease, Alexandre et al. (1987) studied the links between forest architecture and vector biology of the sandflies transmitting the pathogenic *Leishmania* species. They concluded that the degradation of the high rainforest studied in French Guyana contributed both to the increased multiplication of sandflies and to an intensified interface between these phlebotomes and humans. More in general, many disease cycles are situated in the forest canopy and are characterised by interactions among higher animals, vectors and pathogens. These cycles have only negligible interfaces with the humans at ground level. Deforestation brings them down, in which case human beings may take the place of animal species in the cycle.

Forest management would have to take into account such mechanisms, where selection pressure has driven dangerous organisms to extreme environments like the forest canopy, has left only small numbers of such organisms surviving with low probabilities of population explosions, and has reduced to low efficiency the contacts between such organisms and ground hosts like humans. In the present world of the end of the 20th century, such low-probability barriers of selection at different levels (Table 6.2) are broken down by deforestation, so creating risks, in the following manner.

- **Step 1.** Niches (cycles) of pathogens are disturbed, either enhancing or diminishing their multiplication and transmission potential (cf. Sornmani in FAO 1987, on malaria cycles) and at the same time increasing their interface with *Homo sapiens.* Increased exposure is more or less risky, according to the change in pathogen cycle efficiency.
- **Step 2.** In local human societies there is often a breakdown of traditional forms of social hygiene, or in the case of immigrant societies there was never time to develop it. Once exposure to a virulent pathogen cycle has taken place, no human cultural and sacred barriers of low transmission probability (e.g. the Old Testament) exist any more to stop endemics or epidemics. Traditional medicine often broke down, too.
- **Step 3.** Once a disease has become part of the local human ecosystem, the present world-wide systems of communication and travel dramatically heighten the probability of it spreading to other regions of the planet.

Several points deserve attention as essentials in this question. First, too much research in medicine and ecology is focussing on high-probability channels. The explanation of many new epidemics on the contrary should be sought in organisms literally "breaking out" of sometimes very low-probability channels. An example is the much-debated potential transmission of the HIV-virus by insects. It should receive much attention. The answer determines whether AIDS is a deforestation disease or not.

The second point is that the above steps really happen because they are "invisible" side-effects of other activities, the latter being rewarding for certain people. This is often linked to forest exploitation without forest management, be it for wood or for conversion to other land uses with products of high immediate profit, like cattle or minerals. The result is that health problems in people, as in crops, are dealt with by cheap blanket measures. Fortunately, all-out chemical warfare against certain pests and vectors is now frequently being abandoned (cf. Sect. 5.6.9;

FAO 1987). However, the clearing of all forest vegetation, to eliminate potential sources of pathogens menacing rural populations, is a primitive thought that is sometimes suggested and sometimes still partly implemented!

The third point is that most current measures are taken too late, once the disease has broken out. Artificial measures, like insecticides, most often appear to act as selection pressures that are too narrow to contain the flood of mutations leading to speciation phenomena in vector insects. The arguments of Sections 5.6.8 and 5.6.9 are here taken up again in the genetic context.

The fourth point is that many pathogens are so short-lived that they are barely dependent on climate. It should be remembered that Amsterdam only became malaria-free in the 1870's and Rome in the 1930's. Tourists are often reminded today to have a medical check-up after holidays in the tropics, and in reverse, it is known what European influenza did to the tropical Amerindians. Step 3 makes it necessary to tackle health problems world-wide, so that everyone in the world should be interested in correct forest management all over the world. Its absence would threaten us all.

The silviculturist can use silvological knowledge to act in the context of step 1, by carefully designing silvicultural systems in which natural barriers against human and crop diseases are maintained or reinforced. This task, which may seem so complex as to be almost impossible, can be streamlined by viewing the desired effect as a form of selection pressure. Checking the possibilities by using the arrows in Table 6.2 as a checklist may lead to measures with a multiple-level selection impact. The cases in which this seems to have occurred by accident (cf. FAO 1987, diverse authors) are worth studying in this context. So are D.H. Janzen's "thoughts from the tropics", a series of articles published in "Oikos" from 1985 on.

6.3.8.3 Speciation, Provenances, Tree Breeding

If manipulation of selection pressure may validly be considered as a means to escape health hazards, the opposite measure, i.e. summoning up certain desirable properties in organisms through artificial selection, is at the very base of agriculture. Tree crops have been selected in this way for a long time, the olive tree maybe for over four millennia. In the same species, one genotype produces more wood than the other. Usually, they come from different regions and therefore are indicated as "provenances". Provenance trials hence constituted one main line in forestry research since over a century. The use of clones (Sect. 6.3.8.1) is an important tool in perpetuating desired properties in tree populations (Heijbroek 1984, in Oswald 1985). This is brought to an extreme by in vitro tissue culture as a means to obtain apomictic tree populations (e.g. Evers 1984, for Douglas fir, *Pseudotsuga menziesii*).

Figure 6.51B shows how individuals in one species are thought to display "fitness traits". Few individuals are fit to survive in extreme situations, many in their optimal biotope. By adequate seed harvesting techniques, populations can be selected that show a shift towards extreme and desired properties, e.g. exceptional drought resistance or wood production. This shift, by seed selection, rests on the transfer of genetic information by sexual means, with its mutation risks and therefore is not spectacular (Fig. 6.51B, middle). Much more spectacular results are obtained by apomictic offspring of selected trees (Fig. 6.51B, below). The figure clearly shows both the advantages and the risks of such techniques in terms of fitness.

Indeed, for what circumstances do "fitness traits" make a tree fit? Clones or other selected populations have superior properties as crops, but may have inferior properties as eco-unit components. As an example, the provenances of *Pinus nigra* used for hasty reafforestation of large surfaces in the south of the Netherlands had been taken as they came in unidentified material from commercial nurseries. No careful selection had taken place. They now prove to be very vulnerable to a combination of pollution impacts and diseases. Their all-over fitness did not correspond to that required by the eco-units in which they grow, and which they built themselves on the poor sandy sites where they were planted. Fast wood production represented too narrow a selection criterion if all-over fitness was desired.

Pinus nigra also exemplifies plantations made with other exotic species, as is often the case in reafforestation (cf. Oldeman 1985). Common genera are *Acacia*, *Eucalyptus* and *Pinus*. They are planted all over the world, each species often far beyond its area of origin. This again raises the question of their fitness as components of other forest mosaics, on other sites. There is no fundamental difference between this question and the evolutionary questions raised as to species migrations over land bridges or after geological and climatic cataclysms, when new populations are hypothesised to have become established by one or a few "founders" (e.g. Mayr 1963 ex Dobzhansky 1970). In principle, it does not matter whether such founder populations become established by natural or anthropic dispersal. For human-dispersed *Pseudotsuga menziesii* (Douglas fir), having travelled from the Western USA to Europe by human vectors for economic reasons, Kuiper's data (1988) are relevant. The question of differences in climate and seasonality in such cases has been addressed by Mure (1983).

In the two cases, i.e. the introduction of both exotic and selected genotypes as tree crops, the careful design of silvicultural systems providing them with an optimal life-support system is absolutely necessary. In the euphoria of having found new super-clones, this aspect nonetheless is too often neglected. One of the first symptoms of such new components being foreign to their eco-units is the virulence and diversity of pests and diseases, each being either extremely low or extremely high. In the first case the exotic behaves during a first period as an aggressive weed, in the second case it costs much to protect it until it reaches a harvestable age.

Examples like *P. nigra* and *P. menziesii* show that species choice, genetic selection and tree breeding have the same basis. This is the harmonisation between genotype and environment. Natural species populations covering large areas include many genotypes. *Pinus contorta* spans 25° latitude in Western North America, from near Mexico to Canada. Critchfield (1957) analysed its geographical variation, using its remarkable diversity of habitats (silvatic mosaics, sites) and morphological variability, which made it apt for population biological analysis. In Europe, Mayer et al. (1982) reported upon a particularly large provenance trial, in which 112 provenances of white fir (*Abies alba*) were planted out in areas outside their region of origin; and Krugman et al. (1983) listed nearly 100 tree species in the context of Chinese selection and breeding programmes.

Table 6.2 shows that there are no simple selection pressures. However, the most complex and radical forms of pressure are linked to major geological and climatic events. After such immense stresses, it takes centuries before tree floras produce new and adapted species (cf. Sect. 6.3.8.2). The North-Western European tree flora is still very poor after the glaciations. There is now wide-spread concern about impending, thorough changes of climate by anthropogenic influences on the content of CO_2 and pollutants and the damage to the ozone layer in the atmosphere (e.g. DMG-DPG 1987; Lovelock 1987; UNEP 1987; Mohnen 1988; Beardsley 1988; Rowland 1989).

Hoffman (1984) outlined the choices imposed upon foresters if pessimistic forecasts became true. The consequences would have to be dealt with over the next century and three of Hoffman's issues are silvicultural. They require not only the design of silvicultural systems in which the reduction of risks is paramount, but also the accelerated breeding of trees and other forest components that can build such ecosystems. These organisms have to possess "the right" genotypes. It is outside the scope of the present book to go into further detail. However, it is clear that even such gigantic problems are not fundamentally different from the seemingly smaller ones mentioned above.

6.4 The Life Cycle of Silvatic Mosaics: Production Models

6.4.1 Relation Diagram for a Silvatic Mosaic

In Section 3.5.4, physiological production in trees was approached by using a sapstream diagram (Figs. 3.30, 3.35). In Chapter 4, some suggestions were made as to adapting sapstream production models to other forest components, such as herbs, shrubs or lianes. In Section 5.6.2, ecological production was considered through a relation diagram at eco-unit level (Figs. 5.75, 5.76). This diagram was compared with the sapstream diagram in Table 5.8, listing the compartments ensuring the crucial functions in each system. These functions, which have to be ensured in any type of living system, whether it is an organism or not, are: delimitation (skin, border layers), transfer of information, translocation of mass and energy, adaptation to changing environments, and reproduction. Reproduction involves the transfer of building information from an old system to start growing a new, "identical" one. In Table 6.2, these relation diagrams for successive hierarchical levels in living systems were used to define a stepped model of selection pressure for use in genetic theory.

The final relation diagram in this book concerns the silvatic mosaic level (Fig. 6.52). Because silvatic mosaics are built by eco-units, eco-unit composition determines the nature of a mosaic. Therefore, the relations amongst eco-units are explanatory for silvatic mosaic production. All relations considered in Fig. 6.52

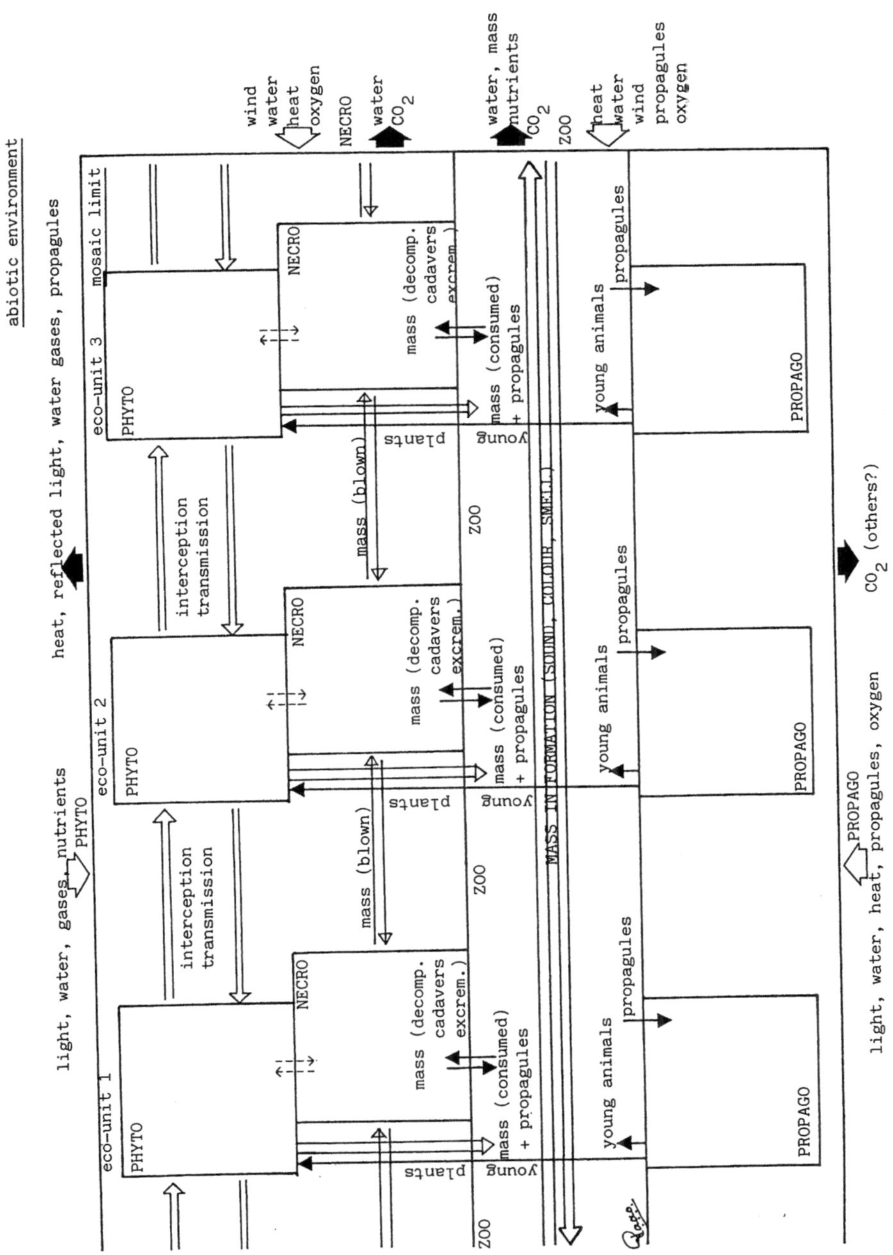

Fig. 6.52. General relation diagram for silvatic mosaic production models. Note that eco-units have been kept uniform here, but may be differentiated into categories (e.g. pioneers, long-lived, short-lived, etc.). Animals are the main vectors of biomass and information throughout mosaics. Compare with Figs. 3.30 and 5.75. No computer-simulated models have been built along these lines as yet, but methods of, e.g. Shugart, Brünig or Hari could be used with slight modifications

have thus been strictly limited to this kind of interaction. Physiological or internal eco-unit relationships are disregarded. The diagram is much simplified and can only be used as a first reconnaissance of this hierarchical system level in explanatory terms. The postulated "abiotic environment", for instance, is linked to the area within a site class that is available for a mosaic. Homogeneous areas of sufficient size are rare. Correction for a site mosaic will have to be incorporated in later versions. Moreover, the eco-units are represented as identical systems. Correction for different size, mass, architecture, phase and species composition per eco-unit can be implicit by attaching numerical values to arrows. But it might also be achieved by distinguishing eco-unit categories.

In other words, if clear distinctions could be made between different organs and tissues in Fig. 3.30, and between different classes of eco-unit compartments in Fig. 5.75 (green plants, rootless plants etc.), the same is not true for Fig. 6.52. Existing knowledge is insufficient to account for small-sized versus large-sized, short-lived versus long-lived or species-rich versus species-poor eco-units, and to classify them into compartments with different functions inside the silvatic mosaic. The description of silvatic mosaic development in terms of eco-unit composition in successive stages (e.g. Figs. 6.6, 6.12, 6.13) and the outline of some functions (e.g. Figs. 6.8, 6.15, or 6.21 to 6.25) is as far as we can go for the moment.

Therefore, the relation diagram in Fig. 6.52 has been kept open in several respects. First, it exemplifies the kind of relations considered amongst three eco-units, but the number of eco-units can be, and will have to be extended as far as necessary to include all eco-units building a particular mosaic. This has been indicated by the continuation of arrows at the left and right side, towards any other eco-unit to be considered. Second, when eco-unit numbers become large and eco-unit character is heterogeneous, the establishment of eco-unit classes will become necessary. This can be accommodated in Fig. 6.52 by replacing eco-units 1-2-3 by eco-unit classes 1-2-3. Such classes will have to be defined preferably in accordance with existing models (e.g. Bormann and Likens 1979; Shugart 1984). The visible and measurable criteria of phytomass, necromass, zoomass and propagule bank composition make this possible.

The types of interaction considered are taken from existing, if sometimes rare, studies or after traditionally considered phenomena. They are the following ones:

- Inputs and outputs between the mosaic and its abiotic environment are the classical ones in production models, except for "propagules". These are the seeds, spores, eggs or larvae that are carried from mosaic to mosaic by air, water, animals, or human vectors (exotic species).
- Interception and transmission are currently considered in silvicultural practice as "border effects". Shadow cast (light transmission), "rain shadows" at forest borders (interception) or windbreak effects (interception and transmission of mechanical energy) are among the interactions considered. In short, this is "the modification of the microclimate by the surrounding forest", so often mentioned in silviculture (cf. Burschel and Huss 1987). It is treated here systematically as a set of interactions.

- Mass (blown), as an interaction between the eco-units in the necrosphere compartment, depends largely on abiotic inputs. For instance, if much heat, little water and much wind come in, the blown (litter) mass going from one eco-unit to the next will increase.
- Young plants, young animals indicate germination and hatching processes, which are generally considered in regeneration studies with a bias towards tree germination. As a matter of fact, it is the operationalisation of part of the information content of the propagule bank, i.e. one process in the feedback web of all interactions considered at the silvatic mosaic level. This makes it possible to come to grips with the notion of "shifting event" in a quantifiable way.
- Mass (decomposition, cadavers, excrements) indicates the interaction between necromass and zoomass compartments, with a powerful matter translocation effect because animals move.
- Mass + information (sounds, colours, smells) is the whole process of animal mobility according to patterns regulated by signals such as bird song, reaction to fruit colour (Sects. 6.2.4.5, 6.2.4.6) or impellors (Sect. 5.6.8). This is considered to be the essential channel for mass translocation and information transfer within a mosaic and probably amongst different mosaics. It gives animals their clear niche (as a functional multidimensional vector) in ecosystems research.
- Propagules indicates a process of translocation of small masses. Although these may be important in the case of fruit crops, their effects remain small in proportion to those of the genetical information load that travels with them.

Table 6.3 shows that silvatic mosaics are living systems that differ profoundly from eco-units. The main points of difference are the horizontal direction of processes, the preponderant role of animals in process regulation and information transfer, and the large part played by random, abiotic processes. In themselves, abiotic phenomena can be explained as being regulated by a physical system of distribution of matter and energy (geology, pedogenesis, world climate), but these are not directly linked to biotic organisation and in a certain sense confront it. Table 6.2 (lower end) shows this as the ultimate "selection pressure", originating from site factors. A comparison between Tables 5.8, 6.2 and 6.3 shows that selection pressure is exerted by abiotic factors on all living systems, both organisms and biotic communities. However, this pressure is "filtered" by the communities before exerting its influence on organisms. This is the reason why evolution (Sect. 6.3.8) could not be treated earlier in this book. It was necessary to survey all living systems first, before arriving at Table 6.2.

6.4.2 Mosaic Production and Average Tree Production

The literature on production models for what are silvatic mosaics in terms of Table 2.1 can be understood best by considering Fig. 6.52. Commonly, the whole system is considered to be characterised by values averaged over the mosaic. Instead of writing "eco-unit 1-2-3" one might refer to "representative surfaces 1-2-3" and then select 3 ha or other random plots for measurement by statistical proce-

Table 6.3. Comparison between relation diagrams for a silvatic mosaic (Fig. 6.52) and an eco-unit (Fig. 5.75)

Property	Eco-unit	Silvatic mosaic
Inputs/ outputs	Through propagules, animals, phyllosphere; less so through rhizosphere/ necrosphere	Through phytomass, abiotic propagule transport, animal migration, water cycles, nutrients in soil
Primary production	Phyllosphere/rhizosphere	Phytomass, interacting eco-units in stufung
Transport system	Necromass/animals	Animals/abiotic
Direction large mass translocation	Top-down, in litter transfer and through necromass	Lateral, in animal movement between eco-units and abiotic processes (wind, water)
Nature of mass translocation	Transfer chain, dry and liquid, through interrelating but often spatially isolated organisms	Transfer chain, consumption/deposition, mostly liquid, mainly through plant/animal interface and abiotic
Regulation of mass translocation	Balanced feedbacks and impellors (chemical)	Animal population balance; sounds, colours and chemical (impellor) signals; abiotic (more or less random)
Reproduction system	Propagule bank (selective accumulation)	Propagule bank (information distributed according to eco-unit distribution and random abiotic factors)
Information in reproduction	Carried by propagule combinations	Carried by combinations of and interactions between eco-unit propagule banks (incl. selective transport)
Regulation of reproduction	Balances between propagule formation, longevity, consumption and germination	Shifting events (abiotic, more or less random) and interception/transmission between eco-units (stufung)
Dispersal to	Own eco-units and surrounding mosaic	Own mosaic and other sites
Defence	Feedback balances and population shifts	Passive shift towards other eco-unit composition; animal population explosions and extinctions

dures. The behaviour of "the forest" on each hectare is then explained by tree behaviour, and eco-unit formation is treated as a tree species mixture. Mielikainen (1980), for instance, when analysing structure and development of mixed Finnish pine-birch stands, states (p. 72): "It is difficult to prevent the formation of mixed stands". He then proceeds by accurately describing, but not in these terms, the process of early fragmentation of large *Pinus sylvestris* eco-units by the installation of fast-growing *Betula pendula* eco-units. The author chose sample plots containing around 300 trees each, displaying different proportions between the two species. Forest behaviour finally is explained by standing wood increment data, gathered for each species in three classes of plots: where it dominates, where the other dominates and where both balance each other.

This approach is wide-spread. One example is the study on tree growth, dynamics and productivity in a mature Malaysian mangrove forest by Putz and Chan (1986). Their sample plots, which preexisted for many decades in these managed mangroves, measured 0.16 ha each. Instead of wood increment, net aboveground primary productivity was calculated (biomass plus litter). Some organisms other than trees were mentioned (termites and fungi), but their influence was not calculated and they were considered as pests and diseases in *Rhizophora* and *Bruguiera* trees (Rhizophoraceae). Putz and Chan (1986, p. 228) refer to something close to mosaic dynamics when stating "... yet it is clear that the natural dynamics of mangrove forests in the area do not include anything akin to the clearfelling system now in use". But like the Finnish analysis which also shows implicit knowledge of fragmentation processes, the Malaysian study rather uses average values that eliminate mosaic dynamics from the final image, centred on the trees.

That such images are indeed the aim of many studies of this kind is due to the fact that both foresters and ecologists have concentrated for more than a century on methods derived from growth and yield studies (e.g. Burschel and Huss 1987, their Chap. 5.2). Growth refers to wood increment and yield is the economically harvestable wood volume per time period. Fries (1981) is completely clear in his paper advocating experimental studies on silvicultural systems in mixed indigenous tropical forests to be made with statistical methods, as in Europe. He gives examples of the measuring and calculation programmes involved, and writes (Fries 1981, p. 1): "Because of the great variation in the tropical rain forests, the mixture of many species and the large trees, studies in these forests tend to be very complicated, space- and time-consuming and costly." His final remarks include (p. 8):"... that experiments in mixed tropical forests, because of the great variation in these stands, will be very expensive, which certainly will reduce the total number of silvicultural alternatives to be experimentally tested.".

All this is undoubtedly true for mixed forest models based on statistical analysis and aiming to obtain growth-and-yield data for certain crop-tree components. But this is one of the basic reasons why alternatives for these methods have to be reconnoitered. It cannot be accepted that the cost of methods determines the amount of information that foresters have to work with. In that case, methods should be changed instead of reducing the options. Systems research, based upon clear and detailed hierarchies, and lending itself to numerical expression and computer-aided simulation, is such an alternative. Care should be taken, however, to use as many criteria as possible in common with classical growth-and-yield studies, because these are available in many cases, over large surfaces and long periods. For instance, Hladik (1984) rightly considers the value of basal area measurements as a source of information on tropical forest dynamics.

Most existing studies are clearly derived from growth and yield approaches. They are based on mean values for "forest ecosystems", obtained from measurements of "representative plots". Such averages are useful. For instance, they allow the evaluation of the general effects of deforestation on biomass and potential production, or the consequences of the consistent application of some forest management system, or conversion towards agriculture. Such conclusions are valid for a certain land mapping unit in land evaluation, or a site class on a forest management map (cf. Van Goor in Van Beusekom et al. 1987). However, this kind of information cannot be used directly for silvicultural design aiming at tailor-made systems for local forests.

6.4.3 Methods of Compartmentalisation

Advances have been made for the last three decades in the compartmentalisation of production studies, which until then mainly concerned wood biomass in tree trunks (e.g. Paterson 1956). For instance, Ogawa et al. (1965b) published a set of basic regressions for biomass estimation from stem volume, branch volume, and from correlations between leaf and stem weight, summed leaf area per tree and dry weight of leaves, leaf area and stem weight and many others in a Thaï rainforest (Khao Chong, also see Ogawa et al. 1965a). Phytomass was calculated in stem, branch, leaf and root compartments for five forest types in Thailand. Kira et al. (1967) applied the earlier study to estimate dry matter production in Khao Chong forests, comparing it with data from Japanese, Canadian, Australian and Campuchean forests (also see Kira and Shidei 1967). They split production in gross increment, losses by death and respiration, and net production, linking this to leaf area index. The Müller and Jörgen Nielsen model (1965, Fig. 5.79) for the Côte d'Ivoire and Denmark dates from the same period and is formulated in analogous terms.

In 1969, Yoda and Kira continued by publishing litter and organic soil matter data for Khao mountain, along an altitudinal gradient. Looking at Fig. 6.52, this is clearly an eco-unit property that was averaged over a whole mosaic. Around 1970, many publications in this line appeared, out of which Odum and Pigeon (1970) have been cited already (Fig. 5.77) and Reichle (1970) may be mentioned for Europe. They started to differentiate primary production for important plant species in the "forest ecosystem" considered. This approach is still being developed. Klinge et al. (in Golley and Medina 1975) separated Amazonian palms and dicotyledons in their biomass and structure research in "a central Amazonian rainforest". Yamada (1976) linked litter fall in the Mount Pangrango forest at ca. 1500 m altitude (West Java, Indonesia) to taxonomical plant groups, organs (branch, leaf), compartments ("mosses") and animals, and described it according to the phenology of the vegetation. Kira (in Tomlinson and Zimmermann 1978; in Huxley 1983) synthesised the models further. In 1986, Murphy and Lugo gave a detailed account of structure and biomass, linked to floristics, in the Puerto Rican subtropical, dry Guanica forest. One of their amazing conclusions is that in this dry forest about half of the 89.9 tons/ha living biomass occurred below the ground. It would be extremely interesting to study the eco-unit composition there, but the given values are averages. The same bias is found in the recent study on lianes by Putz and Chai (1987), who established correlations between tree properties or sizes and presence of certain liane types in Sarawak, Malaysia. Average production could be easily hooked up to these data if the adequate measurements were made, and this would prolong the list of mean production data for forested areas.

The forest succession models from the gap-model theory by Shugart (1984) and associated researchers have been mentioned in Section 5.6.6 (Fig. 5.83). They represent a first step in theoretical forest production modelling, going from the randomly chosen "representative surface" towards equalling such surfaces with naturally occurring regeneration surfaces. This approach, which brings production ecology closer to the eco-unit/silvatic mosaic concepts, has been treated in Section 5.6.6 mainly at gap level.

The present chapter on mosaics will consider the inclusion in these models of forest components other than trees. In Shugart and Seagle (1985, enormous Bibliography), animals are considered. Shugart et al. (1980) analysed the installation of plant components including pioneer shrubs in gaps. Both studies present examples from Australia, where a variant of tropical rainforest was analysed. "Gaps" and the resulting "patches" are considered as biological islands. Interactions like those shown in Fig. 6.52 are not taken into consideration except indirectly, perhaps, when assessing the initiation phase of a gap. Such gaps have properties determined in our view by eco-unit interaction (seed bank, litter).

The KIAMBRAM model that was developed by Shugart et al. (1980), based on Botkin et al. (1972 in Shugart et al. 1980), could not avoid including interactions between "islands", i.e. proper to the mosaic (Fig. 6.52). Some 150 species of woody plants had to be considered, a task as complex as phytosociological analysis of tropical forests mentioned earlier (Sects. 5.5.2, 6.3.4). In the KIAMBRAM model, a routine called SWITCH was introduced that takes into account some of the interactions determined in our view by silvatic mosaics:

SWITCH (1) - species are dispersed by birds;
SWITCH (2) - species are dispersed by wind;
SWITCH (3) - species are dispersed by gravity (seeds just fall);
SWITCH (4) - species require mineral soil;
SWITCH (5) - species require a raised site (N.B. this may be a site, not a mosaic attribute, if no fallen trunks are meant);
SWITCH (6) - stranglers (cf. Fig. 6.22).

The criteria per species used in the model are maximal attainable diameter and height, from botanical literature, two parameters linking height and stem diameter (cf. Sect. 3.7) according to a formula unconnected with the state of the woody plant, attainable age as known from different sources and calculations, a growth parameter linking diameter to other properties, shade-tolerance (see Sects. 5.4.2.1, 5.5.11), seed phenology, seed longevity, and the SWITCH functions.

Figure 6.53 shows an output of the model. It gives the basal area and plant numbers per hectare, exemplified by several species in the four categories with different temperament that are used in these models (Sect. 5.5.11), i.e. pioneers, early secondary species, late secondary species and mature phase species. The term phase is used by Shugart et al. in the sense of "stage" as defined in this chapter. Basal area being a parameter for biomass, this figure should be compared with Fig. 5.84 (Bormann and Likens 1979) and its discussion. The wavy curve for basal area against time indeed points to the mosaic character of the forest, admitted from the outset in the Shugart model, but also apparent without previous theory from the Bormann and Likens simulations (Sect. 5.6.6). The image of Fig. 6.53A was validated by observation in the field and is realistic.

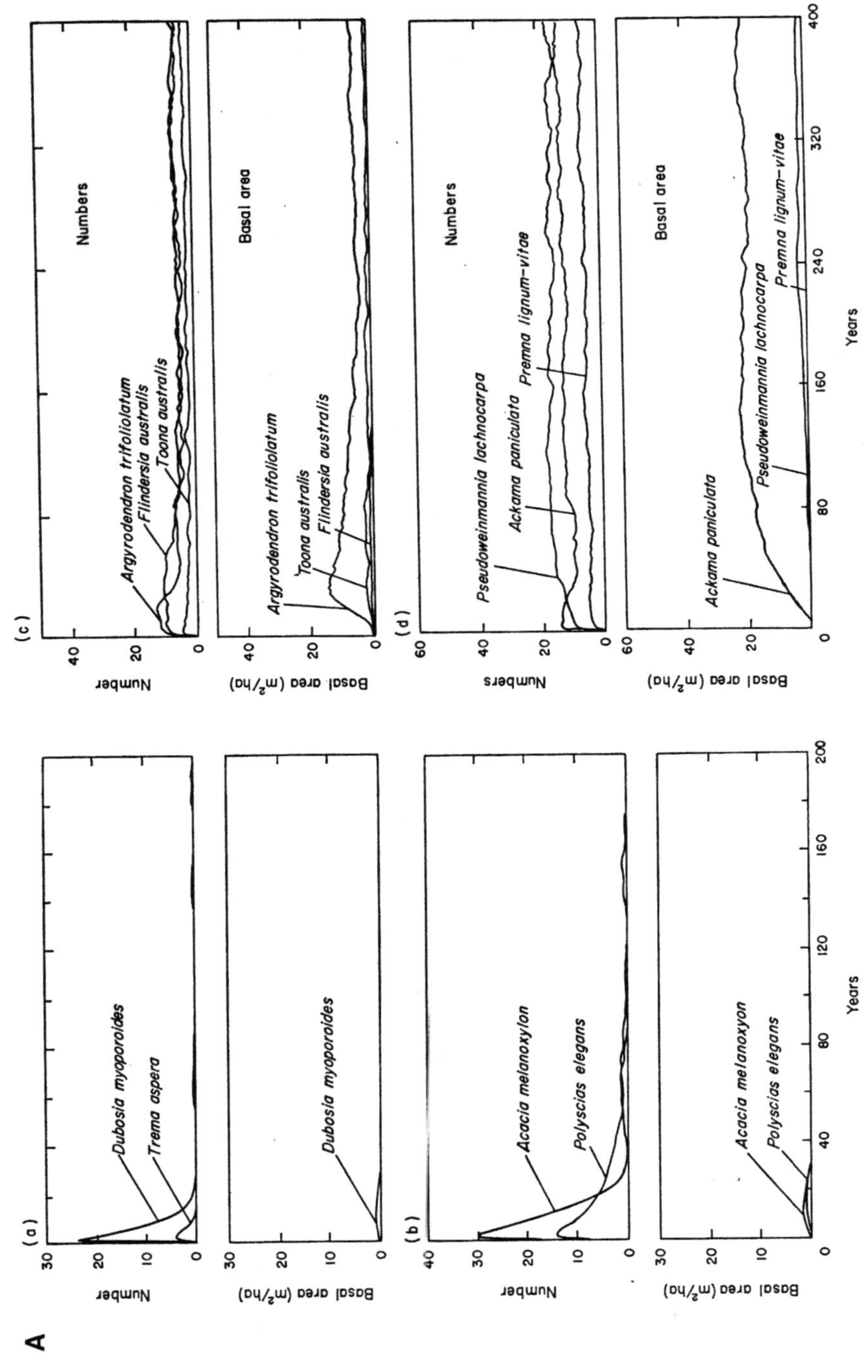
A
(a)
Number
Dubosia myoporoides
Trema aspera
Basal area (m²/ha)
Dubosia myoporoides
(b)
Number
Acacia melanoxylon
Polyscias elegans
Basal area (m²/ha)
Acacia melanoxyon
Polyscias elegans
Years
(c)
Number
Numbers
Argyrodendron trifoliolatum
Flindersia australis
Toona australis
Basal area (m²/ha)
Basal area
Argyrodendron trifoliolatum
Toona australis
Flindersia australis
(d)
Numbers
Numbers
Pseudoweinmannia lachnocarpa
Ackama paniculata
Premna lignum-vitae
Basal area (m²/ha)
Basal area
Ackama paniculata
Pseudoweinmannia lachnocarpa
Premna lignum-vitae
Years

Fig. 6.53A

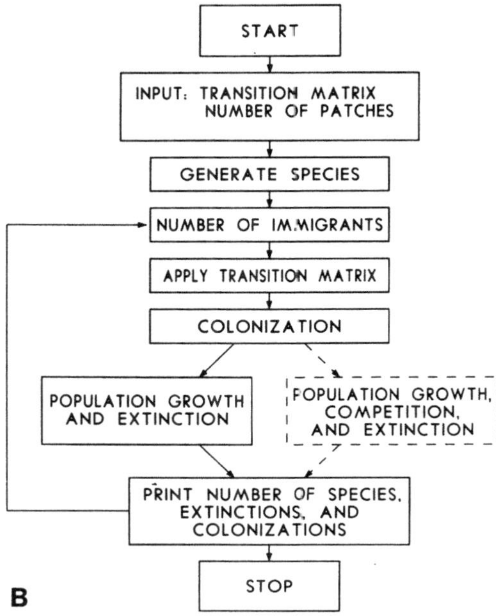

Fig. 6.53. Mosaic-like production from literature. **A** KIAMBRAM model, output for species performance in gaps of developing Australian rain forest, according to number of individuals and basal area, both per ha. Four categories of temperament: **a** pioneer species, **b** early secondary species, **c** late secondary species, **d** mature phase species, as defined by Shugart et al. (1980), to whom this figure is due. **B** Animal population responses to changing habitats in a landscape mosaic, flow chart for computer simulation. Note that animals are migrating, colonizing or becoming extinct in response to given landscape dynamics, and that no animal interactions between eco-units, as in Fig. 6.52, are considered, because "gaps" are seen as biological islands. (Sugart and Seagle 1985)

Figure 6.53B shows the operational flow chart for computer simulation of animal life in a forest mosaic composed by changing "biological islands", i.e. patches, some of which are more attractive to an animal species than others (Shugart and Seagle 1985). Many data have been given above, proving that an animal species prefers certain eco-units to others. The computer programme again yields a statistical model predicting the response of animal populations to shifts in available habitats, resulting from changing "patch" composition with succession. The mathematical basis is a Markov model (see Shugart and Seagle 1985), not unlike that used by Horn (1975) for his succession model of tree species in a New England forest.

With the simulation model of Fig. 6.53B, numbers of animals migrating to preferred habitats when these appear in the mosaic, colonising these habitats, and becoming extinct or emigrating when the habitats disappear again with forest development, can be predicted from year to year over every wanted period. Except for the statistical and population-biological concepts used, another scientific heritage is to be mentioned that colours this model. Like most previous studies in animal ecology, many examples of which have been cited above, this study also is based upon the "shopping list" of an animal species in terms of nutritional, reproductive and defensive requirements. The Shugart and Seagle model, like the graphical one by Van Vuure (1985), is improved in that it considers the geometry and development of the "shopping and living area" for each animal species.

SWITCH (1) of the KIAMBRAM model, mentioned above, is the only interaction considered which really corresponds to a silvatic mosaic process (Fig. 6.52). This and other silvatic mosaic interactions all have in common that animals and mosaics are not independent of each other, but that vegetation mosaics and their animal compartment develop together, by close feedback relationships. This cannot be simulated as yet by existing computer-models, although an important step towards an "ecological interference model" (Sect. 6.3.6; Oldeman 1989b) has been made in the SILVI-STAR model by Koop (1989).

Finally, the model by Shugart and West (1979) has to be mentioned. It concerns different spatial scales in mosaic development, i.e. different eco-unit sizes in the terms of the present text, and the growth of trees in different-sized gaps. This is highly relevant in silviculture in order to assess the consequences of harvesting methods upon forest regeneration, from the extreme of clearcutting to that of selection cutting (see also Shugart et al. 1980). Table 6.4 shows the outputs of such a model for mixed temperate-region deciduous forests.

Table 6.4. Growth and installation of trees in gaps of different sizes, tested by the model of Shugart and West (1979), from whom this table is taken

Stand size (ha)	Test 1: Can an individual tree grow to its maximum size?	Test 2: Does gap-phase replacement occur?
0.01	No	Yes
0.02	No	Yes
0.04	Yes	Yes
0.08	Yes	Yes
0.20	Yes	Rarely
0.40	Yes	No

6.4.4 *Transfer of Functions in Silvatic Mosaics*

In Section 3.8.3, transfers or functional shifts in and among the levels of the organs and the organisms were discussed. In Section 5.6.9 such transfers were regarded in and among the levels of organisms and eco-units. They were closely related to the allocation of resources to particular compartments.

In the present paragraph, transfers of functions in and between the levels of the eco-unit and the silvatic mosaic will be briefly discussed. They have many links with the subject of the preceding paragraph, compartmentalisation, and allow a functional outlook on fragmentation and fusion of eco-units.

Silvatic mosaics are the largest living systems that can be conceived, larger systems being abiotic. Therefore, if a life function cannot be ensured by the mosaic, it has to be transferred to one or more eco-units, because it cannot be performed abiotically. On the contrary, an eco-unit can transfer functions to both sides, i.e. to organisms or other compartments, and to the mosaic it belongs in. This situation makes silvatic mosaics quite unstable, which is expressed by the spelling of ecuilibrium (Sects. 6.2.3, 6.2.4). Transfer of function among eco-units and between eco-units and mosaics takes place in a shuttle-like way.

One of the most important transfers is linked to the **phytomass in the canopy** and its **function in regulating the microclimate**. In large eco-units, extended monocrop timber estates being an exaggerated example, this function is ensured by the upper structural ensemble, containing crowns of the present (Fig. 5.4). In even-aged tree crops, it is not even necessary that the crowns be of the present, because they form one dense leaf layer after canopy closure.

Now, if fragmentation occurs in such large eco-units, microclimate regulation is transferred to a stufed mosaic canopy, having quite other properties than an eco-unit canopy (Fig. 6.9). For certain calculation purposes, this will be accounted for by indicator values for average "canopy roughness" (cf. Brünig in Furtado 1980; Brünig et al. 1982). But again, such averages mask the fact that the organisation of the whole canopy, as a border layer, has become different. Lateral interception and transmission of light, water and mechanical energy (wind) among eco-units (Fig. 6.52) determine the microclimate patterns in a silvatic mosaic, rather than the well-modulated simple vertical gradients in single eco-units described by authors like Kira (in Tomlinson and Zimmermann 1978) or Oldeman (1974a, 1978, 1979).

Once the eco-units of such a mosaic enter into fusion, the mosaic starts to display an eco-unit-like microclimate regulation again. The shuttling transfer of this function, between layered and stufed systems at different system levels, has a profound influence on organisms smaller than trees. Quite different niche-patterns for herbs, climbers, shrubs and animals exist, for instance, in forests of which the internal climate is regulated in the former or in the latter way. Because tree seedlings in many respects behave like herbs during the first years (cf. Oldeman 1974a), this type of transfer of ecological functions is interesting for natural regeneration silviculture.

The smaller the eco-units the more complete is the transfer of microclimate regulation to the mosaic. One-tree eco-units have no proper microclimate left. It is wholly determined by interactions with neighbouring eco-units. If questions asked about the forest mainly concern the trees, the mosaic may then be analysed as a mosaic of large trees and tree-sized holes (Table 6.4). But as soon as problems to be solved include other life forms, such mixed-forest models lack the precision needed to formulate answers that are more than generalities. Otherwise, skipping the eco-unit level leads to very complicated methods, as was shown by the myriads of data gathered by hundreds of volunteers in order to construct the crown structure versus bird diversity matrix by Komdeur and Vestjens (1982).

Transfer of functions between eco-unit and mosaic level also is one of the main keys to understand forest stability, which is linked to a system's adaptation to the environment. In this case, the eco-unit reaches stability by maximising resistance and the silvatic mosaic by maximising resilience (for definition of these terms, e.g. see Begon et al. 1986). This is shown in the biostatic phase of the eco-unit (Fig. 5.4), in comparison with the ecuilibrium stage of the mosaic (Fig. 6.6). If species populations are considered to be both builders of ecosystem architecture and carriers of functions, the guilds sustaining one or the other form of stability are quite different (cf. also Table 6.3).

From what we understood about selection pressure and speciation (Sect. 6.3.8) it can be proven, however the system hierarchy be defined, that oscillating selection pressure at the site and whole-forest levels makes for irreversible developments in speciation. A population is forever determined in its properties by selective mortality among its potential ancestors, no matter whether this be by natural or anthropogenic selection. Because the carriers of function and builders of ecosystems are ceaselessly speciating, ecosystem stability can only be understood as a remarkably successful **attempt** at stability, in other words, a very slow, but nonetheless irreversible development.

If eco-unit resistance, or biostasis, breaks down, fragmentation occurs. The function is then transferred to the mosaic level, where it is ensured by resilience. The species-poor nature of biostatic eco-units (Sect. 5.2.4) contrasts with the species-rich character of ecuilibrium mosaics (Sect. 6.2.4). Now Begon et al. (1986, p. 762), say: "A community which is stable only within a narrow range of environmental conditions, or for only a very limited range of species characteristics, is said to be dynamically fragile". They contrast this to dynamically robust communities.

The transfer of stability functions in forest mosaics hence causes shuttling between fragile biostatic eco-units, fragmented robust silvatic mosaics, and fragile mosaics with merged eco-units (fusion). Forest management for conservation or products can assess the risks involved for sustainability by precisely analysing whence, in which circumstances, and to where a particular function is transferred. Large monocrop timber stands are relatively risky and fragile, flexible silviculture involving some form of selection cutting is less risky and leads to more robust forests.

Any discussion on transfer of functions in and among systems of such size, longevity and complexity touches upon large, abstract and necessarily somewhat unfocussed concepts and issues. In the present section, only some remarks are devoted to this subject that is largely situated beyond the spatial and temporal limits of a human life. It therefore can only be reconstructed to some extent, not observed. The next sections contain case studies and examples that may be seen in the same context.

6.4.5 Biogeochemical and Hydrological Cycles; Biomass

One of the most elegant and most elegantly visualised studies of biological production and biogeochemical cycling in a silvatic mosaic, called "forest ecosystem" by the authors, has been carried out by Duvigneaud and Denaeyer-De Smet (in Reichle 1970). Their model of the mixed broadleaved forest at Virelles, Belgium, with *Quercus robur, Carpinus betulus* and *Fagus sylvatica*, includes many component species, among which are some animals (earthworms), as well as soil compartments. These are specified as to their biomass, production and biogeochemical turnover. Pathways through the whole system are discussed in detail.

Figure 6.54 gives two of several drawings localising biomass production (Fig. 6.54B) and nutrients (Fig. 6.54A) in the forest and within the organ compartments of each tree. The authors do not say whether the drawings represent measured transects. Probably they are schematic reconstructions because no metric scale is indicated. If these drawings show the real configuration of this forest, it would be a mosaic with fused, small eco-units (e.g. Figs. 5.56–62).

Biogeochemical data are expensive but extremely important for silvology and forest management. Some information in Duvigneaud and Denaeyer-De Smet indicates, for instance, that trunks contain about 60% of the aboveground biomass in these forests, and Fig. 6.54A details their share in each nutrient. Impacts of chablis or wood harvest on the mosaic can so be precisely assessed in terms of primary production.

Study of earthworms and their role in soil turnover, with their 200 kg of biomass per hectare in Virelles forest, has not been chosen arbitrarily. One of the most powerful interactions between forest mosaics and their abiotic environment (Fig. 6.52) at the zoomass level is soil turnover by burrowing animals. For instance, Alvarado et al. (1981) showed the building of new soil profiles by leaf-cutting ants (*Atta* sp.) in Costa Rica (Fig. 6.55A). These authors even propose a new code for soil horizons that owe their existence to insects. Indeed, the change in soil profile is spectacular. In other tropical countries, such effects can be observed due to other insect species, e.g. termites (see Collins in Furtado 1980). In temperate regions, forest soils are rearranged in a less thorough way by earthworms and small insects, like Collemboles (e.g. Brown 1978). Again, Alvarado et al. (1981) give their data as means per hectare, indicating percentages of affected soil surface.

Figure 6.55B shows that root system architecture probably is different in various silvatic mosaics of various ages. It summarises the vertical distribution of fine-root surface, in three classes of root thickness, in forests near Turrialba, Costa Rica, at forest ages of 1, 8 and 70 years respectively (Berish 1982). Root surface is important, because it determines the interface between roots and soil environment (cf. Fig. 6.52). Such interactions may be boosted by other forest components, e.g. bacteria or mycorrhizal fungi. Moreover, part of the surface does not interact with the abiotic environment but with the organic matter belonging to the eco-unit.

With these reserves, the figures by Berish (1982, Fig. 6.55B) give an indication of average belowground eco-unit development within a silvatic mosaic. This author calculated the Root Area Index (RAI), which is the relative root surface. RAI is 5 if the root surface below 1 m^2 of soil equals 5 m^2. This concept is analogous to that of the Leaf Area Index. LAI is one of the key concepts in ecological production models, as a numerical parameter for photosynthesis. Berish's model therefore fits into the family of such production models. The interactions between soil and forest ecosystems in general are an important part of site analysis, with a view to assessing potential and real productivity of different silvatic mosaics.

Van IJssel and Sombroek, and Van Goor (both in Van Beusekom et al. 1987) gave overviews of site variability and its application in land use evaluation in the humid tropics. For temperate regions, much has been written on this essential topic (for bibliography, e.g. see Fanta 1985, 1986). Tropical research planning concerning site factors such as soil biology (in our terms mainly bound to eco-units) and soil

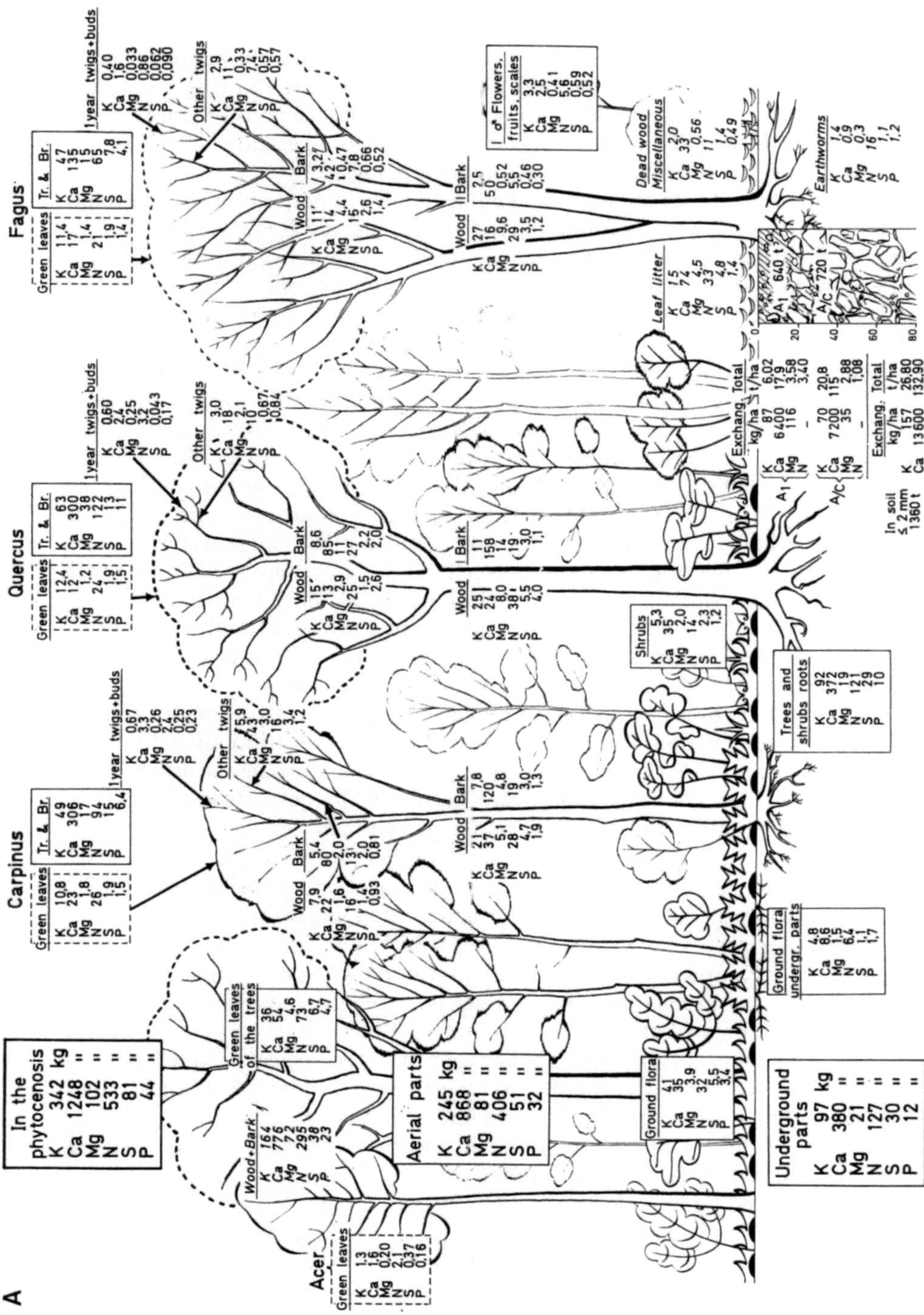
A
In the phytocenosis
K 342 kg
Ca 1248 "
Mg 102 "
N 533 "
S 81 "
P 44 "
Aerial parts
K 245 kg
Ca 868 "
Mg 81 "
N 406 "
S 51 "
P 32 "
Underground parts
K 97 kg
Ca 380 "
Mg 21 "
N 127 "
S 30 "
P 12 "
Acer
Green leaves K 1,3 Ca 1,6 Mg 0,20 N 2,1 S 0,37 P 0,16
Wood+Bark K 164 Ca 776 Mg 72 N 295 S 38 P 23
Green leaves of the trees K 36 Ca 54 Mg 4,6 N 73 S 6,7 P 4,7
Ground flora K 41 Ca 35 Mg 3,9 N 32 S 5,5 P 3,4
Ground flora undergr. parts K 4,8 Ca 8,6 Mg 1,5 N 6,4 S 1,1 P 1,7
Carpinus
Green leaves K 10,8 Ca 23 Mg 1,8 N 26 S 1,9 P 1,5
Tr. & Br. K 49 Ca 306 Mg 17 N 94 S 15 P 6,4
1year twigs+buds K 0,67 Ca 3,3 Mg 0,26 N 2,4 S 0,25 P 0,23
Other twigs K 5,9 Ca 4,3 Mg 3,0 N 16 S 3,4 P 1,2
Wood K 7,9 Ca 22 Mg 1,6 N 16 S 1,4 P 0,93
Bark 5,4 80 2,0 13 2,0 0,81
Wood K 21 Ca 37 Mg 5,1 N 28 S 4,7 P 1,9
Bark 7,8 120 4,8 19 3,0 1,3
Quercus
Green leaves K 12,4 Ca 12 Mg 1,2 N 24 S 1,9 P 1,5
Tr. & Br. K 63 Ca 300 Mg 38 N 122 S 13 P 11
1year twigs+buds K 0,60 Ca 2,4 Mg 0,25 N 3,2 S 0,043 P 0,17
Other twigs K 3,0 Ca 18 Mg 2,1 N 10 S 0,67 P 0,84
Wood K 15 Ca 13 Mg 2,9 N 25 S 1,5 P 2,6
Bark 8,6 85 11 27 2,2 2,0
Wood K 25 Ca 24 Mg 8,0 N 38 S 5,5 P 4,0
Bark 11 158 14 19 3,0 1,1
Shrubs K 5,3 Ca 35 Mg 2,0 N 14 S 2,3 P 1,2
Trees and shrubs roots K 92 Ca 372 Mg 19 N 121 S 29 P 10
Fagus
Green leaves K 11,4 Ca 17 Mg 1,4 N 21 S 1,9 P 1,4
Tr. & Br. K 47 Ca 135 Mg 15 N 65 S 7,8 P 4,1
1year twigs+buds K 0,40 Ca 1,6 Mg 0,033 N 0,86 S 0,062 P 0,090
Other twigs K 2,9 Ca 11 Mg 0,33 N 7,4 S 0,57 P 0,57
Wood K 11 Ca 14 Mg 4,4 N 15 S 2,6 P 1,4
Bark 3,2 42 0,47 7,8 0,66 0,52
Wood K 27 Ca 16 Mg 9,6 N 29 S 3,5 P 1,2
Bark 2,5 50 0,52 5,5 0,46 0,30
♂ Flowers, fruits, scales K 3,3 Ca 2,5 Mg 0,41 N 5,6 S 0,59 P 0,52
Dead wood Miscellaneous K 2,0 Ca 33 Mg 0,56 N 11 S 1,4 P 0,49
Leaf litter K 1,5 Ca 74 Mg 4,5 N 33 S 4,8 P 1,4
Earthworms K 1,4 Ca 0,9 Mg 0,3 N 16 S 1,1 P 1,2
Exchang. kg/ha Total t/ha
A1 K 87 6,02 Ca 6400 17,9 Mg 116 3,58 N – 3,40
A/C K 70 20,8 Ca 7200 115 Mg 35 2,88 N – 1,08
In soil ≤ 2 mm 1360 t
Exchang. kg/ha Total t/ha
K 157 26,80 Ca 13600 132,90 Mg 151 6,46 4,48
A1 640 t
A/C 720 t
0 20 40 60 80

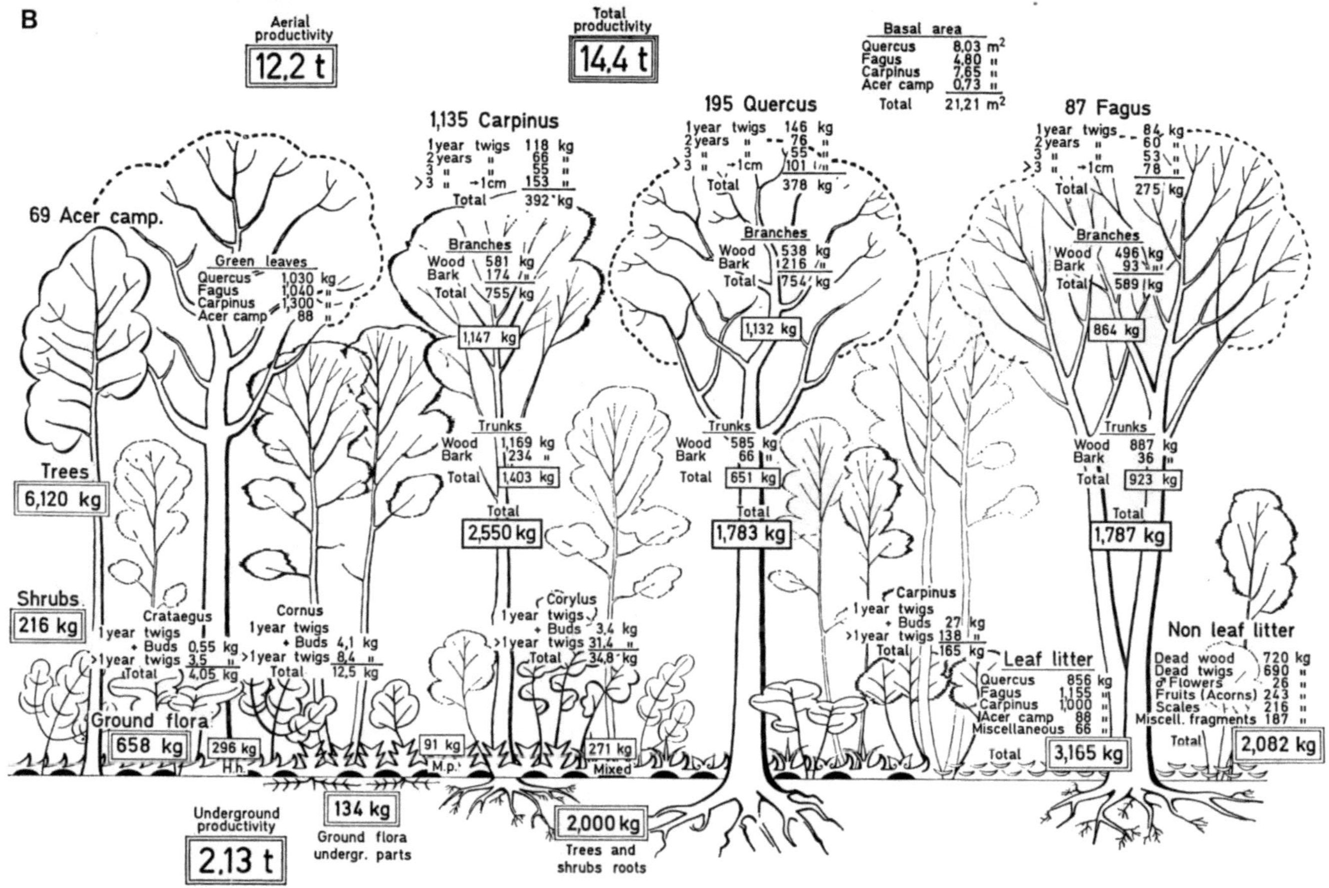

Fig. 6.54. Nutrient content per compartment (**A**) and productivity of biomass in tons (**B**) for a mixed hardwood forest at Virelles, Belgium. (After Duvigneaud and Denaeyer-De Smet in Reichle 1970). These two images form part of a series also including standing biomass, other phytogeochemical data and a cycling diagram. If the architecture is well sketched, this is a silvatic mosaic composed by small, fused eco-units. Data, however, are given as amounts and flows per hectare, and hence are averages for this mosaic

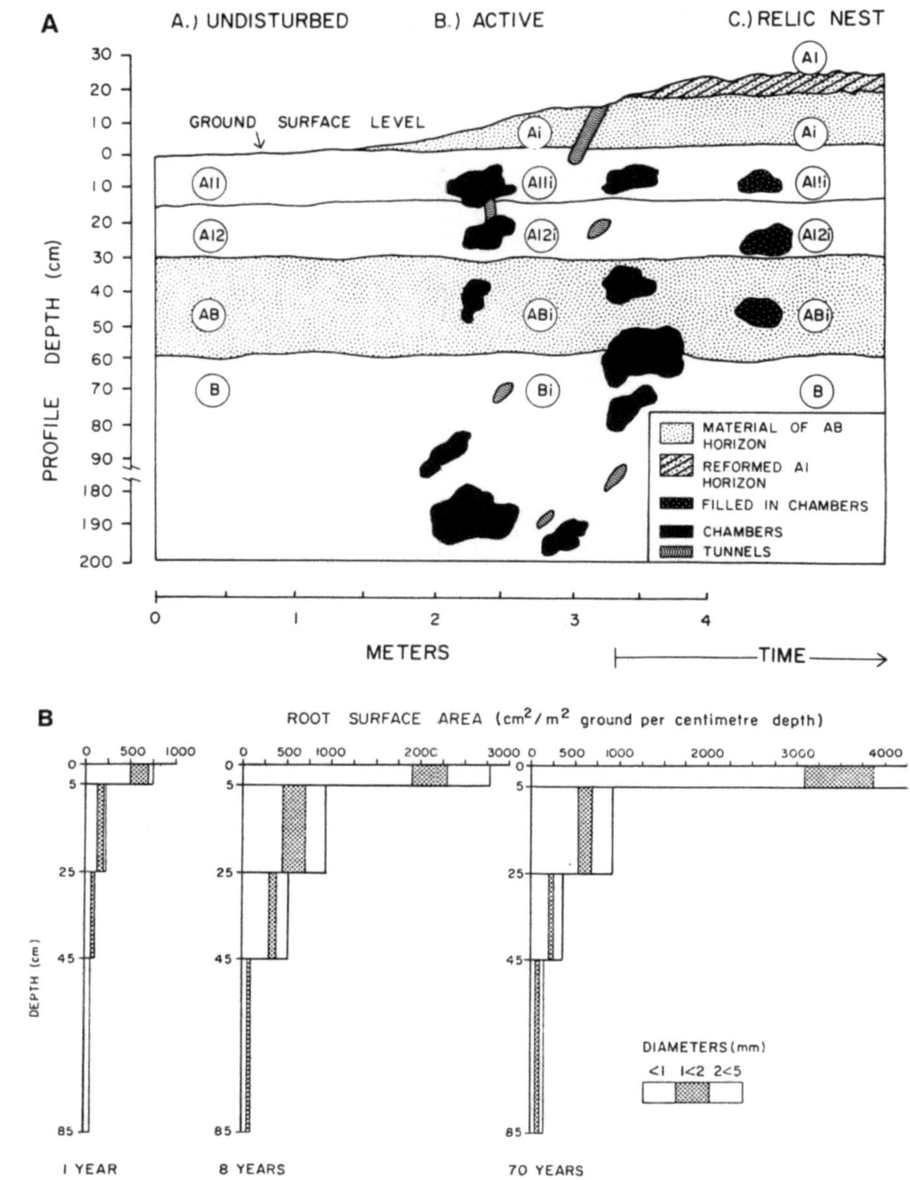

Fig. 6.55. Animals and roots in silvatic mosaics, examples mapped (**A**) or as averages per hectare (**B**). **A** Turnover of soil masses by leaf-carrying ants (*Atta* sp.) in Costa Rica have such far-reaching consequences for the soil profile that the authors suggest special names for insect-formed horizons, marked with (*i*). (After Alvarado et al. 1981). **B** Root surface area calculated for a developing pioneer mosaic (our terms: precuilibrium) in Costa Rica, near Turrialba. Leaf Area Index (LAI), often used in ecological production models, is here complemented by Root Area Index. Such indices eliminate forest area surface and therefore are useful only to calculate average production per ha. (Berish 1982)

fertility (silvatic mosaics; cf. Sect. 6.4.4) was summarised in 1986 by Swift. This booklet gives an overview of contemporary insights and model principles.

Production models often use "nutrient balance sheets", again expressed in mean values per hectare, and most often limited to certain crop plants. Bruijnzeel and Wiersum (1985) presented such a nutrient balance for *Agathis dammara* plantations under various management systems in Central Java, Indonesia. They compared nutrient inputs from the atmosphere and geological sources with losses through various forms of wood harvesting, agricultural intercropping and deep seepage. As on many volcanic soils, in Java phosphorus is the rarest and hence the most critical element to handle in management systems. These data may be compared with the assessment by Bormann and Likens (1979) of some harvesting systems and their nutrient consequences in temperate, North American forests. Van Andel and Nelissen (1979) published a nutrient balance for herbaceous forest components of clearings in coniferous woods on poor soils. Their conclusions largely explain treeless stages in mosaic development (cf. Fig. 6.43), because herbs may monopolise nutrients by stocking them in their organisms.

Silvatic mosaics may be converted in many natural or artificial ways. This most often is indicated in literature as "disturbance" (cf. Sect. 5.1.3). The recovery of Amazonian rainforests after man-made disturbance such as shifting cultivation, wood harvesting and different forms of silviculture has been treated in terms of mass and nutrients by Jordan (1987). Figure 6.56A sketches shifting agriculture dynamics, with the concomitant change in biomass compartments given as accurately as possible. Figure 6.56B and C shows the change in two important elements, calcium and phosphorus. These figures are to be compared with Duvigneaud and Denaeyer-De Smet (Fig. 6.54), showing the distribution of nutrient elements in a European forest. Both cases indicate that nutrient models are very complex. Different nutrients do not show the same behaviour and limiting amounts of one nutrient may be great handicaps for forest survival.

However, the interaction between forest mosaics and nutrients means that the biotic community indeed influences the nutrient content of a site, one of the axioms in succession theory mentioned above. Van Goor (1985) examined this case for Central and Western Europe. He concludes that the impact of fast-growing conifers on European soils has been deleterious since the end of the 19th century, but that this was due mainly to inadequate terrain and soil preparation techniques (see his Bibliography). Again, figures refer to hectares, not to mosaics and their eco-unit composition, because conifer plantations rarely if ever have been considered from a mosaic viewpoint, but always from the agricultural viewpoint centring on yield per hectare. Van Goor (l.c.) states that afforestation of soils impoverished by excessive use, such as heathlands, peat soils and abandoned agricultural land, has on the contrary enriched the nutrient contents. This would agree with Fig. 6.56, considering recolonisation of agricultural lands by natural secondary forests.

A related issue is forest fertilisation, which has received renewed attention because of atmospheric pollution, particularly by nitrogen. Unbalanced nitrogen budgets can block, hinder or exhaust the ecosystem flows of other nutrients, particularly phosphorus. Hüttl (1986) reviewed European fertiliser experiments in forests. He evaluates forest fertilisation as a very useful tool to increase growth (of trees), particularly on low fertility soils, and formulates a prudently positive opinion on compensation fertilisation in cases of nutrient cycling disturbed by pollution. However, no forest components are mentioned other than trees. In view of the interactions shown in Figs. 5.75 and 6.52, this seems to be a questionable simplification.

The rules-of-thumb in Fig. 5.88B concern the effects of chemical substances either as impellors or as poisons, according to their chemical nature and dosage. These rules may well apply to chemical fertilisers (Lugo 1978), when assessing their effects on all forest components, including non-trees. Selective elimination of species belonging to the edaphon is certain to occur in fertilised forests.

One of the elements which is deposited in large quantities in European forests with rainwater is nitrogen. This was thought to have a beneficial fertiliser-like effect, until overdoses were reached. Authors such as Poels (1987) for Suriname, or Hendry et al. (1984) for Costa Rica, analysed inputs by substances dissolved in probably unpolluted rain. In Costa Rica, precipitation contained little nitrogen, but in Suriname this and other nutrients appeared in high quantities. Poels (1987) calculated the forests to have been in a continuous process of enrichment for several centuries, if the rains had always carried the same amounts. The authors mentioned did complete nutrient analyses of the rainwater. This and other work has shown that most nutrients and minerals needed for forest productivity on poor soils and/or under heavy losses by leaching enter the system in the rains. Again,

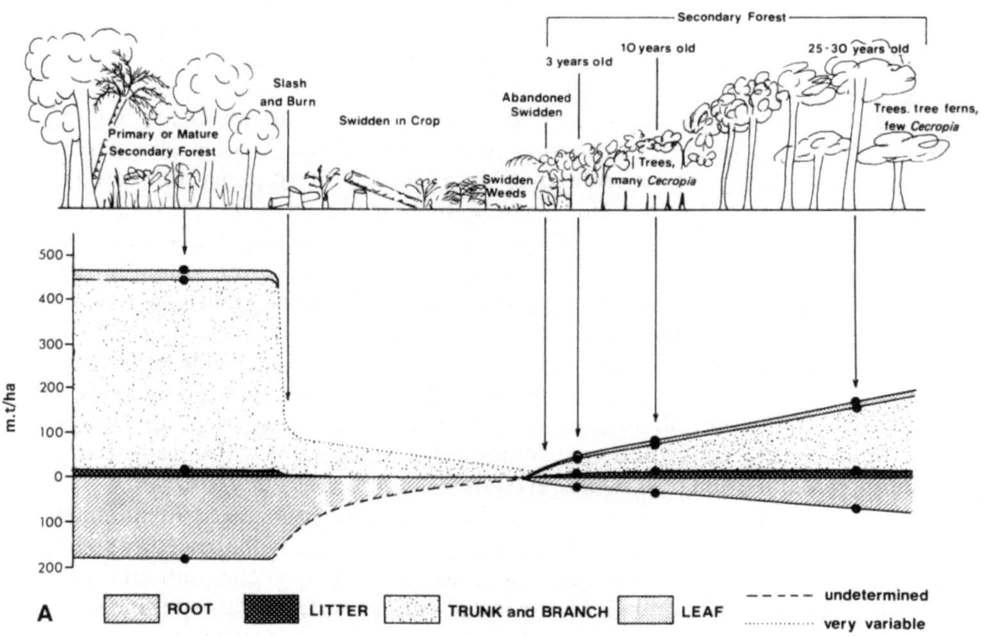

Fig. 6.56A

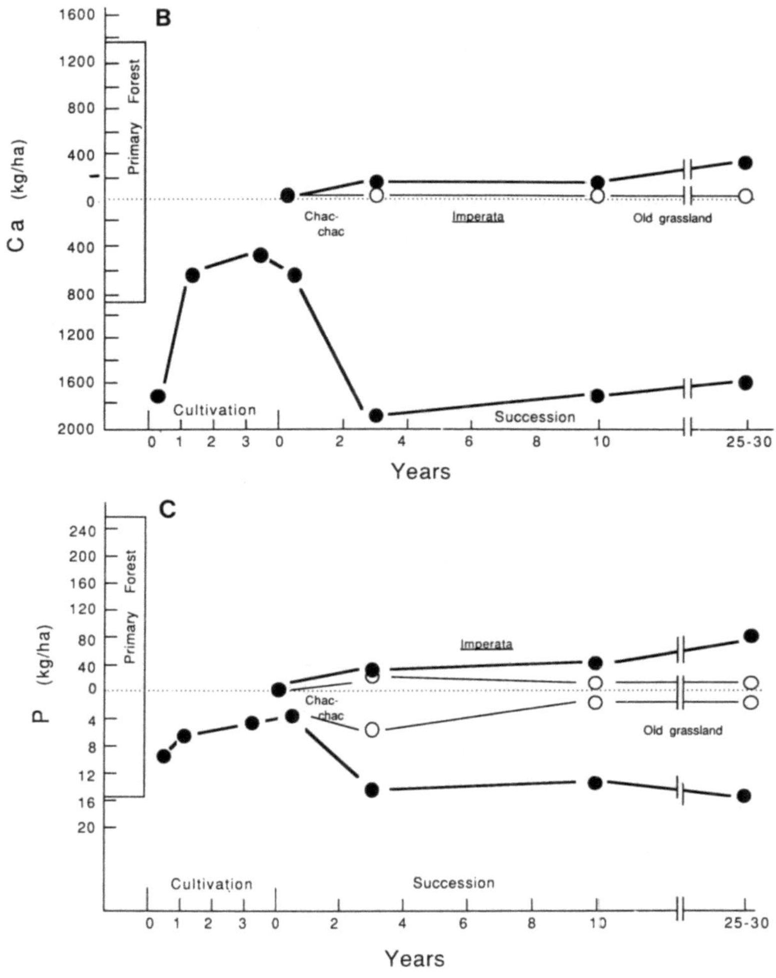

Fig. 6.56. Production models for silvatic mosaics in case of shifting cultivation where land is limited (Amazonia). Values indicated as averages per hectare, sketch in (**A**) not architectural. **A** Changes in biomass stocks during the process of forest clearing, agriculture and forest comeback. Approximate values for primary forest, derived from allometric extrapolation. *m.t.* metric tons (1000 kg). **B** Calcium stocks during same process. *Chac-chac* chacras shifting field; *Imperata* grassland, degraded; *Succession* early precuilibrium (or early aggrading large eco-unit). **C** Phosphorus stocks during same process, note different behaviour of this element, much linked to non-tree components (animals, fungi). (Scott in Jordan 1987). Compare with Fig. 6.54

such data are given as averages over large surfaces; but sometimes they are punctual for some particular location.

Indeed, the carrier of all nutrients and minerals is water. Sapstream effects, such as the daily hydrostatic variations in stem thickness in trees (e.g. Leikola 1969) bear witness to the direct effects of water availability. Figure 6.52 shows the importance of water for silvatic mosaic functioning. The average availability and movements of water in and through such a mosaic, and their seasonal variations to a large extent determine the composition and production of a mosaic. They lead to quite different mosaic architectures as illustrated earlier (e.g. Figs. 5.44 and 5.55). Again, availability may also result in excess. This was shown by Pereira and Kozlowski (1977) in their study on responses to flooding amongst woody angiosperms. Genera examined were *Eucalyptus, Populus, Salix, Ulmus, Quercus* and *Fraxinus.* They showed various stress symptoms when the forests they build were flooded. First came stomatal closure, inhibition of root growth and changes in stem and root morphology. In the end there were either adjustments such as the formation of adventitious roots, or final dieback if flooding lasted too long. Stress symptoms were not due to plant water stress, however, but transpiration was significantly influenced.

A mosaic that transpires less also produces less. The balance between water inputs and outputs in the phytomass compartment (Fig. 6.52) largely determines how silvatic mosaics produce and are composed. Such balances have been analysed by Bruijnzeel (1986) and compared for many tropical forests. A thorough case study of these was published earlier by the same author (Bruijnzeel 1982).

One spectacular example of the influence of forest transpiration on the water balance of the whole site is in the Amsterdamse bos (Amsterdam forest; Balk 1979). This forest covers 895 ha. Its establishment was begun in 1936 and the last tree was planted in 1970. The forest was planted in a polder, i.e. an ancient lake pumped dry, with a soil below sea level at a depth varying between -1.95 m and -3.55 m (Balk 1979 p. 34). Now, the Amsterdamse bos is a precuilibrium forest biotically very rich already with, for instance, 241 bird species, 80 of which are breeding in the forest, 700 beetle species, 450 species of herbs, 300 species of fungi and 150 tree species. This richness also indicates a neo-natural mosaic (cf. Fig. 6.31). Unpublished measurements by the hydrological team of Prof. Dr. W.H. van der Molen (pers. commun. 1987) have shown that the groundwater level under the forest was 3.5 m lower than in the rest of the polder, i.e. it was between 5 and 7 m below sea level. This value should be considered against the fact that the general groundwater level in the polder has been established by a set of very powerful pumps, functioning regularly, day and night, to maintain it.

Transpiration also depends on the wind regime (e.g. Dolman 1986) and is therefore influenced by the "canopy roughness" of large forest areas. This factor is taken into account systematically by Brünig (in Furtado 1980; see Figs. 5.38, 5.80). Canopy roughness is a silvatic mosaic property closely linked to the stufung pattern of each particular mosaic. If small eco-units prevail, "roughness" is applied to mosaics, but if large eco-units do so, the concept is applied to each eco-unit, at least in most simulation models. Fragmentation and fusion are key concepts here (Sect. 6.4.4). The reader should find out what is meant in each publication, before using data and conclusions.

6.4.6 *Light and Forest Photology*

Stufung and layering, which are respectively mosaic and eco-unit properties, are mixed up in many zoological studies (Sect. 6.2.3.1). This is also true in most light interception models. An example is the analysis of the "leaf canopy" in nine tropical agricultural and "successional" ecosystems, given by Ewel et al. (1982, Fig. 6.57A). Amongst their examples there are agroforestry systems ("wooded garden", "cacao/cordia/plantain/paspalum", "shaded coffee"), agricultural systems, and early precuilibrium stages of the local forest mosaic, natural ("succession") and artificially imitated ("mimic"). Leaf masses have been expressed according to the Leaf Area Index method (LAI; see Sect. 6.4.5). In Fig. 6.57, LAI is given per height level, and because the size of the measured plots has been eliminated, this means that an "average mosaic" has been constructed. There are two hierarchical levels: the mosaic is the system, the leaf mass is a subsystem. One of the most interesting features of this model is that Ewel et al. (1982) combine these averages with root biomass profiles (Fig. 6.57B). This permits conclusions on general effects of leaf and root damage in terms of biomass production per hectare.

A mosaic analysis of light interception has been produced by Jansen and Krommendijk (1983), in fragmenting large eco-units of *Pseudotsuga menziesii* plantations (Douglas fir) on sandy soils in the Netherlands. Figure 6.58A shows, for instance, the zones around a canopy gap where light is transmitted, crossing the neighbouring eco-units, in the form of sunflecks, from 09.00 h to 15.00 h each day during the months of June and July. The authors linked these calculated amounts of light, derived from measurements with a modified horizontoscope and general data, to germination conditions for Douglas fir. Figure 6.58A shows that in some cases the innovation phase of a small eco-unit may extend **below** a neighbouring older eco-unit. This kind of interactions (cf. Sect. 6.3.6, "ecological interference") is indicated in Fig. 6.52 as "interception/transmission" between phytomass compartments of eco-units (cf. also Sect. 6.4.4). It is an authentic interaction at the mosaic level, linked to no single eco-unit but to their reciprocal influences. In the tropical rainforest, a less precise analysis has been done in Gabon by Florence (1981). He measured interactions between innovation phases and surrounding eco-units with the light meter of his camera and mapped isoluxes, i.e. lines on which light intensity is a certain percentage of that outside the forest (cf. Fig. 5.2D).

Roussel (1972) coined a special term for the science of light in forests, i.e. forest photology. His book summarises much earlier temperate region research in this field and many original data. His subsystems point to a "mixed-forest" approach explaining the "whole forest" (the silvatic mosaic) in terms of component species and their interaction (competition). Because stands, i.e. managed eco-units, are used as objects, results are often relevant for eco-units in general. One of the aspects treated by Roussel (1972) is the growth direction and crown development of trees under asymmetrical illumination (cf. also Loehle 1986). This is an important factor in interception/transmission interactions (Fig. 6.52), sometimes leading to "allogenic" eco-unit closure by neighbouring crowns. Small eco-units may be elimi-

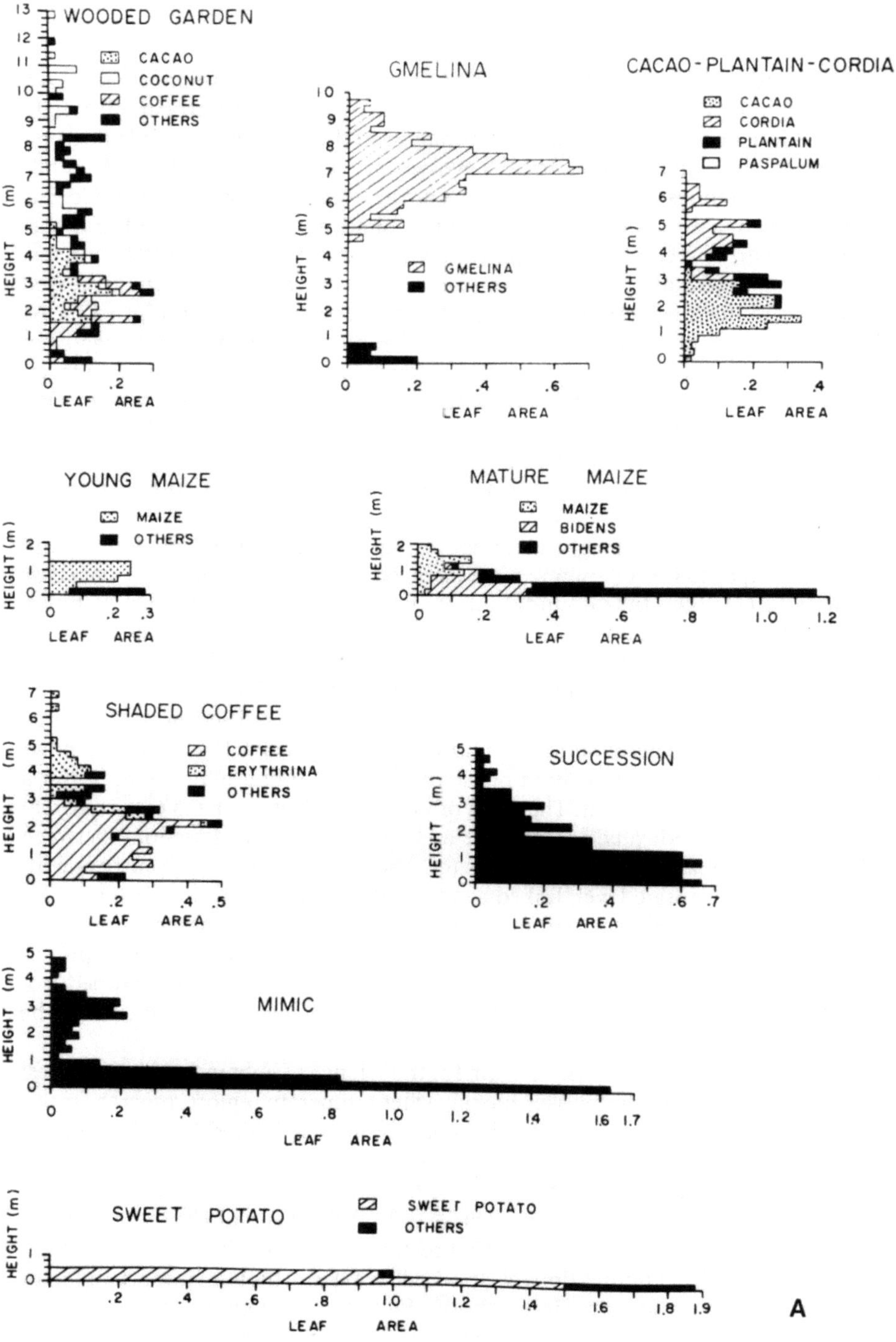
WOODED GARDEN
CACAO
COCONUT
COFFEE
OTHERS
HEIGHT (m)
LEAF AREA
GMELINA
GMELINA
OTHERS
CACAO-PLANTAIN-CORDIA
CACAO
CORDIA
PLANTAIN
PASPALUM
YOUNG MAIZE
MAIZE
OTHERS
MATURE MAIZE
MAIZE
BIDENS
OTHERS
SHADED COFFEE
COFFEE
ERYTHRINA
OTHERS
SUCCESSION
MIMIC
SWEET POTATO
SWEET POTATO
OTHERS
A

Fig. 6.57A

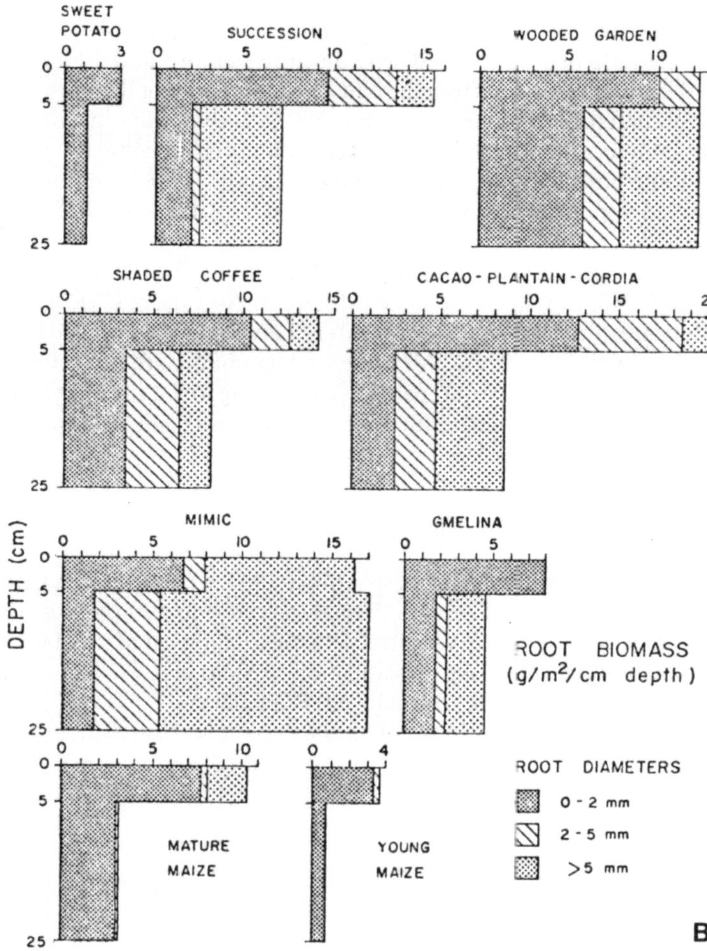

Fig. 6.57. Vertical distribution of leaf area (**A**) and root mass over different root classes (**B**) in nine tropical agricultural and successional mosaics in Costa Rica. Information either as index (**A**) or per surface area (m^2, **B**), i.e. not linked to mosaic architecture. Such data are useful to create average production models for such vegetation mosaics. (Ewel et al. 1982)

nated in this way (cf. Fig. 6.27C). Another curious fact in this field was experimentally proven by Roussel (1983). Illumination has a direct effect on the reiteration of shoots on tree trunks. This does not happen when trunks are obscured. The hypothesis is that light signals, reaching the cambium through very small bark fissures, stimulate the formation of adventitious meristems.

Light has two influences other than photosynthesis. The spectral composition of light in the lower reaches of forests is codetermined by the interception/transmission relations between eco-units. It determines many germination processes through photoreactive pigments in seeds (e.g. see Vazquez-Yanes in Gomez-Pompa and Vazquez-Yanes 1976; cf. also Sect. 6.2.4.5). Moreover, some parts of the spectrum are photosynthetically more active than others, so that the most active wavelengths of light are used up in the upper canopy and the less active arrive in the undergrowth. Finally, from the equator to the south or the north there is an increasingly changing day-length over the year, with short days in winter and long ones in the summer. Daylength for many green plants is the signal for growth and differentiation at the beginning of the growing season. These combined effects have been studied for instance by Van Miegroet (1970) for several European hardwood species.

Other light interception studies can be found for Ecuador in Grubb and Whitmore (1966, 1967), for the Côte d'Ivoire in Alexandre (1982a, b) and for the Mediterranean evergreen oak forests in Eckardt et al. (1975). The reader is referred to their long bibliographies. All these sources concern the differentiation of light interception per hectare into patterns explaining forest diversity, most often in terms of species, but also in structural terms. None has been used so far in complete mosaic production models.

6.4.7 *Site Quality and Productivity*

The question is very old in forestry. Since 1804, when Cotta (ex Cajander 1930) proposed a classification of sites from 0 (sterility) to 100 (highest forest production possible), foresters have wrestled with the integration of site data, species data and production data into one classification in yield classes. The classical continental European expression is "Bonität" (cf. Burschel and Huss 1987, their pp. 45, 259). It is derived from the Latin word "bonus" (= good), and therefore literally means

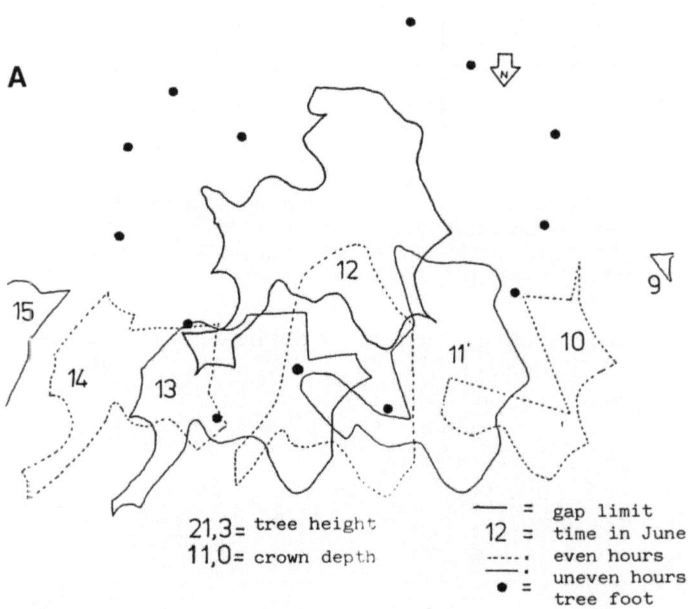

Fig. 6.58A

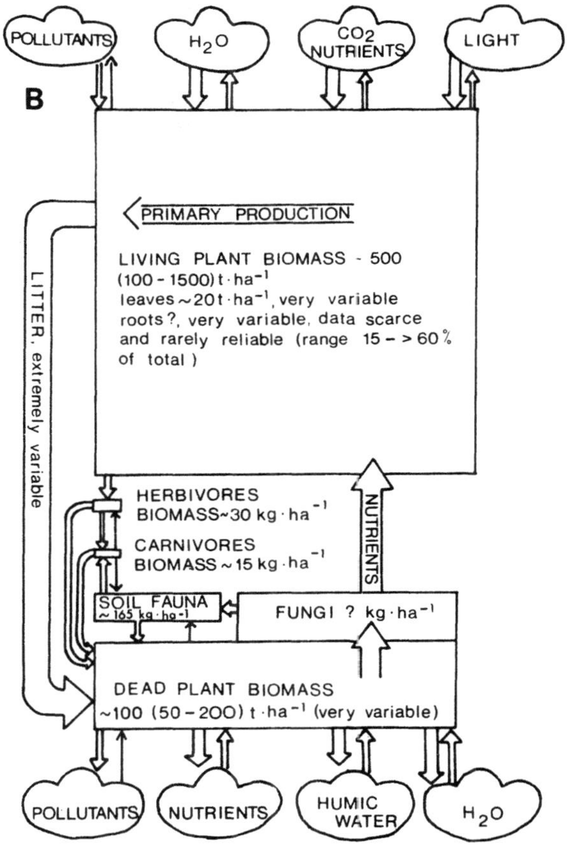

Fig. 6.58. Examples of a precise mosaic model for light (**A**) and a general, aggregated production model for mature tropical lowland rainforest. **A** Contribution of sun-flecks in fragmenting Douglas fir plantation (innovation/late aggradation) on the Veluwe sands of the Netherlands, to eco-unit interaction, interception/transmission, see Fig. 6.52. (Jansen and Krommendijk 1983). **B** General mature rainforest model, on which larger variations probably correspond to interactions within each eco-unit, smaller ones to interactions between eco-units within a mosaic (e.g. herbivores, carnivores). (Brünig 1982a)

either "site goodness" or "forest goodness". Cajander (1930 and earlier) created a forest type classification in which each forest type combines site quality and forest quality, using the non-tree species as indicators. His classification is mentioned here, at the end of the present book, because it is one answer to many of the dilemmas encountered throughout the whole discussion of systems analysis. Cajander's approach has been followed up by more than one author, and using criteria of more than one phytosociological school, a more recent example being Zonneveld (in Van Goor 1983). It all amounts in our terms to the ordination, in classes or along gradients (Kessell 1979), of forest mosaics, taking into account components, eco-units, mosaics and sites.

Forest analysis is advancing slowly. The only important advantage that the present book may offer as compared with Cajander's excellent analysis is the hierarchical architectural approach. This solves the problems met by this author when classifying forests according to physiognomy and development phase or stage. A first step in this direction had of course been taken by the succession ecologists and the Central Europeans (cf. Figs. 6.41–43). However, this has to be pursued to its logical end by the distinction of an eco-unit level and a silvatic mosaic level and the very different living systems encountered at each of these levels. Shugart's models are in so far different that a grid-like approach replaces the mapping of natural borders. He chooses grid dimensions that come close to average eco-unit sizes, but so may block the easiest way to mosaic analysis.

6.5 Site Resource Distribution and Silvatic Mosaics

The many examples shown earlier indicate the incredible number of variations upon the theme of "forest ecosystem" that can be encountered on the land surface of this planet. As stated above, silvatic mosaic development is the meeting place between biotic growth processes and abiotic distribution processes. Authors since Cotta (1804 ex Cajander 1930) in forestry and Von Humboldt in biology (around the year 1800, see Botting 1973) analysed the spatial and temporal distribution of climate and soil in order to link their patterns to vegetation patterns. This is the inverse of the present book, in which site factors were seen as more or less variable inputs encountered by the living systems, each at their own level and more or less filtered by higher levels.

Because so many authors have worked from site to forest, it is proper to end the present chapter by a rapid outline of their methods. These result in forest site maps, a kind of document that is abundantly available in some countries and not so abundant in others. They are much used, and their usefulness is beyond question. Site maps are indispensable both for silvologists and foresters.

Climatological maps and soil maps rest on physical and chemical analyses, the mathematical background of which is largely probabilistic. In view of the success of these methods in chemo-physical research, the temptation exists to treat variation in forests within one site as a probabilistic pattern, too. This is correct as far as forest architecture is a response to probabilistic shifting events, but it neglects the structure of living and growing systems which codetermines silvatic mosaic development. It is not vain to repeat this once more, because the balance between probabilistic and structural determinants of mosaic architecture and functioning is too often overlooked (cf. Sects. 5.6.6, 6.4.2). Usually, authors make a choice and see it either one way or the other. Their readers have to be aware of this and take it into account when applying such data and conclusions. An example is the image of canopy roughness (probabilistic) as compared with stufung (structural), mentioned in Sections 6.4.4. and 6.5.5.

6.5.1 Biosphere Models

The image, as seen from large to small, starts with the **biosphere**, a term used by the French Jesuit philosopher Teilhard de Chardin. The biosphere is the thin layer of life covering the surface of our planet. The concept was nearly unanimously adopted in the whole world. One of the largest international research networks, organised by Unesco, Man and the Biosphere (MAB), was named after it. As a philosophical issue, it should be mentioned that Teilhard also distinguished the **noosphere**. This is the space on earth dominated by the (human) spirit, νους (noos) in ancient Greek. When thinking about forests, the noosphere is a central issue. If humans are not forest components, but exert an outside influence on forests, the most fanatical nature conservationists are right and forests are in need of protection against Man. If, on the other hand, human beings are the only forest components capable of designing the right dosage of their interaction with the other forest components, forest management obtains a quite different outlook. In the present book, the second viewpoint is upheld.

In Flanagan (1970), the biosphere is analysed by its cycles of energy, water, carbon (the main building block of organisms), oxygen, nitrogen and minerals. Although more recent research data are now available, this book is still one of the best introductions to these questions. Figure 6.58B shows such a master model for very large regions. It is taken from Brünig (1982a), who considered the role of tropical rainforests as source and sink of carbon. As a matter of fact, the most highly variable flows on this figure can be understood as eco-unit properties (Table 6.3), e.g. litter fall, soil fauna or fungi. The animals have been placed in their own compartments, variation not being marked as extreme, which fits with their place in silvatic mosaic interactions, one level higher than the other components mentioned.

In his conclusions, Brünig (1982a, p. 67) pinpoints two areas of incertitude, where more research is needed. The first covers the "nature, size and dynamics of the interdependency and interaction between forest ecosystem structure, functioning, stability and survivability, site and man." This is the subject of the present book. The second field mentioned by Brünig as urgently needing research concerns "information on the area and condition of the forested and deforested lands in the tropics". Although more data are available now (cf. Wiersum 1984; Bol et al. 1984; Van Beusekom et al. 1987; diverse recent inventories by FAO), this less fundamental but more descriptive field of research remains important worldwide, not only for the tropics.

Because the main building block of organic systems is the carbon atom, the carbon cycle on our planet has received much recent and renewed attention. This line of research has been triggered by the increasing amounts of CO_2 in the atmosphere (cf. Sect. 6.3.8.3; DMG-DPG, 1987; Hoffman 1984). These are largely due to the industrial and other urban activities by humans. Goudriaan (1987) gave an overview of the biosphere as a driving force in the global carbon cycle, to the bibliography of which the reader is referred. He treated the explanations of biosphere interactions through the carbon cycle as driving forces for evolutionary phenomena in the geological past (e.g. glaciations) and warned against the huge, world-wide consequences for the biosphere if present trends were not stopped or reversed. An opposite view involving autoregulation of these carbon-cycles is the nucleus of the Gaia hypothesis (Lovelock 1987). The methods for monitoring CO_2-exchange in forests have been summarised by Eckardt (1984).

Autoregulation here is used to indicate the ultimate consequence of the postulate of succession, biotic communities changing and regulating their environment (cf. Sect. 6.3.1). The community is Life on Earth. The environment is the planet itself under the impact of radiation and other forms of energy and matter from the solar system. The Gaia hypothesis says (Lovelock 1987, p. 152) "that the physical and chemical condition of the surface of the Earth, of the atmosphere, and of the oceans has been and is actively made fit and comfortable by the presence of life itself. This is in contrast to the conventional wisdom which held that life adapted to the planetary conditions as it and they evolved their separate ways." (cf. Sect. 6.3.8, particularly Table 6.2). The proposal by Lieth and Markert (1988) to establish Element Concentration Cadasters for Ecosystems (ECCE) in different biomes would be quite interesting in order to check the tenets of the Gaia hypothesis. It would also be an aid to monitoring pollution and other long-term environmental degradations of which humans nowadays live in fear (cf. Fig. 6.61). Pollution hazards indeed triggered Lieth and Markert's proposal.

The energy and water cycles, read from such meteorological data as temperature and precipitation, often completed by measured or calculated values for evapotranspiration, have been the basis for most bioclimatological maps and the methods of those by which they are established. They started with Von Humboldt (cf. Botting 1973) and were continued by well-known systems like those of Köppen or Thornthwaite. Now, they have arrived at widely accepted mapping principles, internationally divulged by organisations such as FAO or Unesco (see Bibliography in Van IJssel and Sombroek, in Van Beusekom 1987; or UNESCO 1979). A system widely used in tropical America is the Holdridge triangle (cf. Holdridge et al. 1971). Its author visualised the average properties of forest mosaics in each of his bioclimatic life zones (cf. Fig. 5.65). The refined mapping techniques and methods of the Institut de la Carte Internationale du Tapis Végétal in Toulouse should not be forgotten (cf. Laumonier 1980; Pascal 1983). It was at Toulouse that Gaussen invented the climagrammes now so widely used, for instance by Walter (1968).

Before the enormous accumulation of data from the 1960's on, Paterson (1956) wrote a pioneer book classifying the world's land surface as to its potential productivity in terms of m^3 of wood per hectare per year. His book was much decried at the time for lack of proof and for insufficient soil data. Still, the same principles he used were applied later, when data streamed in and computer facilities became available to process them. This resulted in a series of world ecosystem productivity maps (Fig. 6.59), mostly as a result of the International Biological Programme and following international research like the MAB-Unesco network. Much of this work has been summarised by the National Academy of Sciences (1975).

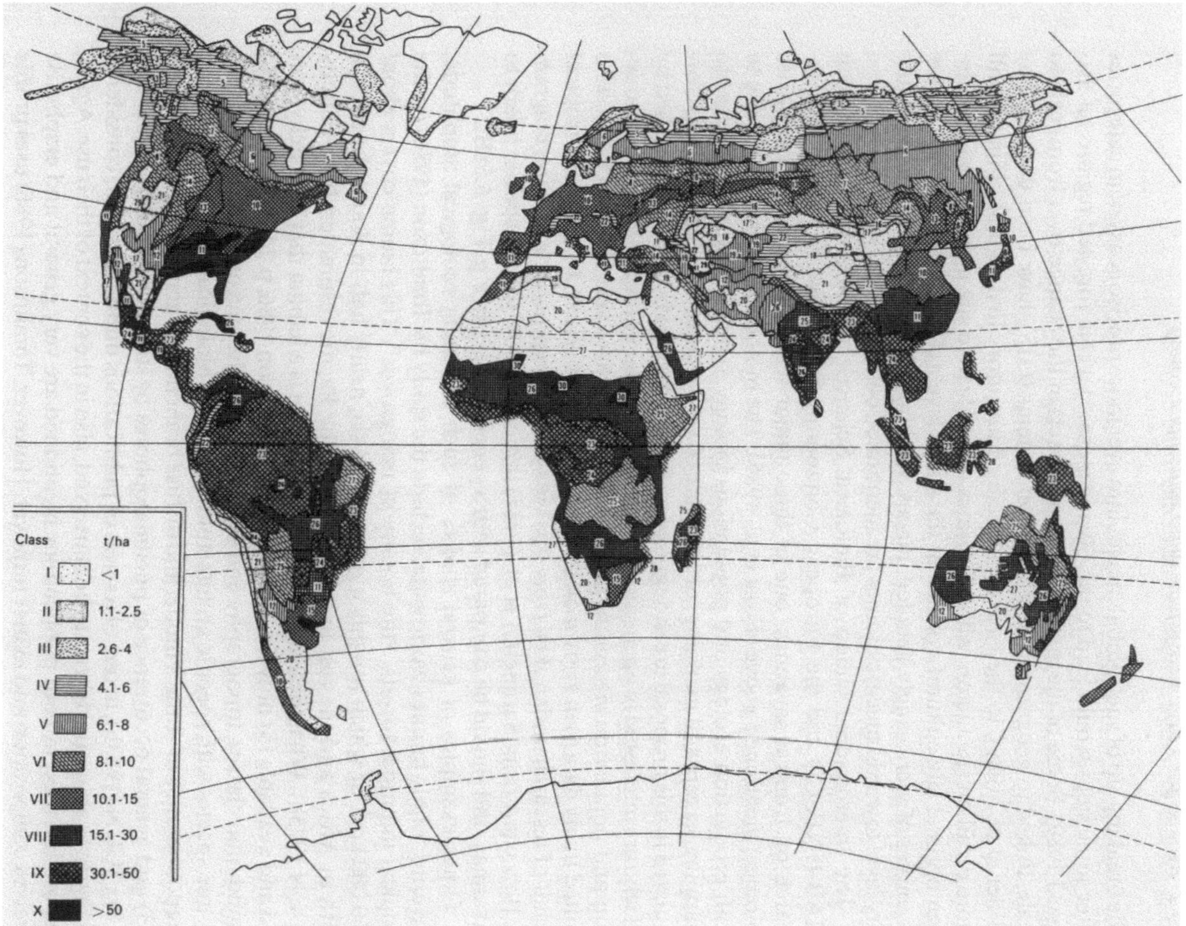

Fig. 6.59. Potential production as distributed over the world according to "sites". Primary production on the land surface of the continents in t/ha dry matter. (Rodin et al. in Nat. Acad. of Sciences 1975)

6.5.2 Models for Large Ecosystems and Landscape Ecology

The breaking-up of such global images into ever more precise ecosystem categories after some steps would lead to the silvatic mosaic level in forested regions (cf. Richard 1989). Some of the methodological aspects of the change-over from scale to scale, linked to geographical information systems (GIS), have been discussed by Miller et al. (1989), in the case of a rare bird inventory in Tanzania. Finally, still more detail on site factors within mosaics would lead to site maps allowing to design optimal silvicultural systems at the stand level (cf. Fanta 1985, 1986). Some examples of increasingly detailed images, working "downwards" from the planetary level towards specific biomes are mentioned below.

The International Union of Biological Sciences, in close cooperation with MAB-UNESCO and the European Community formulated a proposal for research on tropical savannas, one of these major regional ecosystem categories (biomes) mentioned above (Frost et al. 1985). The methodology, which is close to both production ecology and populations biology, omits nearly all references to mapping, except soil mapping, or structure of vegetations. The 20 hypotheses formulated in this proposal are thought-provoking and might be profitably studied by all who are interested in open savanna forests. A similar approach from the production and population viewpoint is apparent in Bourlière and Hadley (1970), who united many data then known on tropical savannas. Figure 6.60A is a neo-tropical result of research undertaken from this viewpoint. It represents a relation diagram or flow chart from Sarmiento et al. (1985) indicating how woody species adapt to savannas, and should be compared with a general diagram (e.g. Fig. 6.58B).

Other examples of general studies of such sweeping ecological amplitude, covering whole biomes worldwide, include the study by Tomlinson (1986) on the tropical intertidal forests, "the botany of mangroves". This biome covers about 160,000 km^2, of which roughly a quarter in Asia, nearly half in South America, a fifth in Africa and the rest in Australia and the Pacific (estimated from data by IUCN 1983b). What these forests have in common is a habitat that is "patchy, varied and available for only limited periods" (Tomlinson 1986, p.24), which is due to inundation classes, topography and salinity limits. They accommodate some tens of tree species with specific architectural and other biological characteristics, e.g. preponderance of Attim's model and diffuse branching in general (cf. Figs. 3.2 and 3.3) and general occurrence of pneumatophores or stilt roots (cf. Figs. 3.15D and 3.16). IUCN (1983b), in one of a series of publications during that year, considered mangroves as natural forests to be conserved, also in the context of their use. Again, the models used in both publications mentioned are very general and emphasise species composition and matter turnover. However, Tomlinson (1986) uses profile diagrams to establish an architectural context. Lescure (1980) did this in a more restricted and detailed study of the mangroves of Guadeloupe (Antilles).

Another biome of enormous size is the tropical rainforest. From Schnell (1950) or Richards (1952, 1970) to Jacobs (1988) and in a large number of handbooks and other publications, this biome including numerous forest types has been the subject of great interest. Its area has probably been some 18 million km^2 and today has shrunk to less than half that surface (e.g. Van Beusekom et al. 1987). Rainforests

Adaptation of woody species in savannas

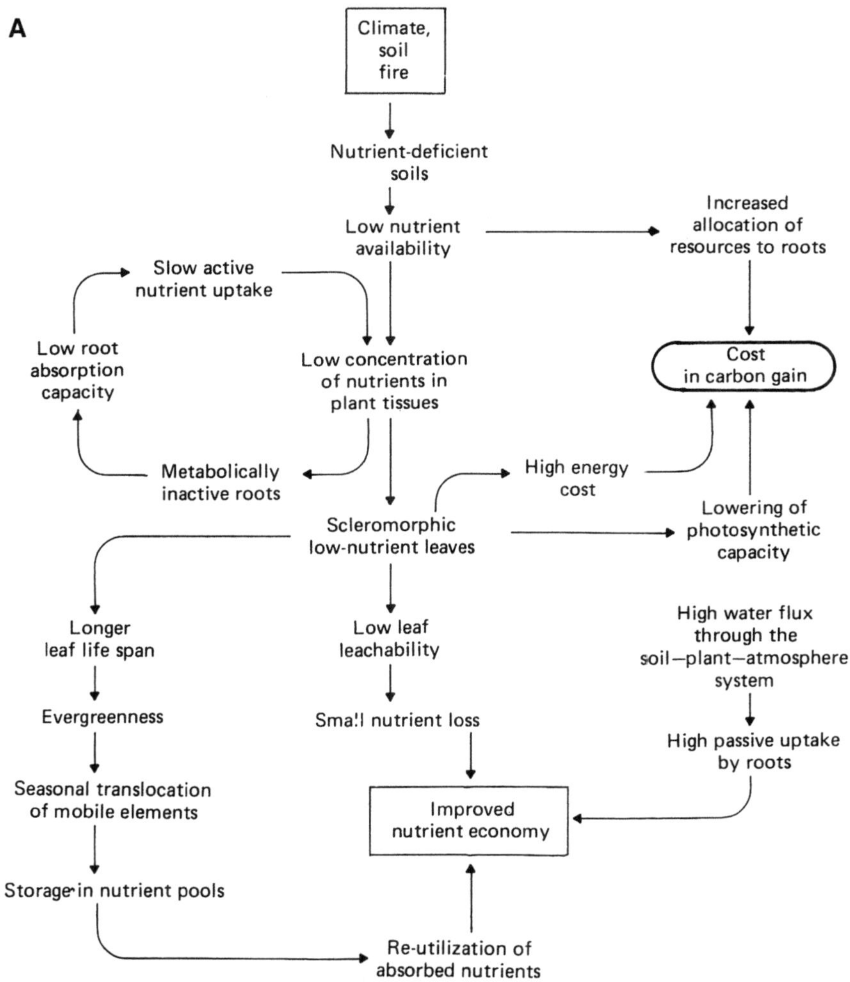

Fig. 6.60. Examples of models detailing certain site types as variations upon general production models. **A** Processes influencing nutrient economy of neotropical seasonal savanna trees and their "cost" in carbon cycle terms. (After Sarmiento et al. 1985). **B** Shifting event regime that may lead to conversion of forest into savanna. Caco, Gran Sabana, Rio Caroni basin, Venezuela, schematic transects with measured tree heights, (Fölster in INTERCIENCIA 1986), data relevant for silvatic mosaic analysis on precise sites. See also Fig. 6.20, from the same region (Fig. 6.60B see page 554)

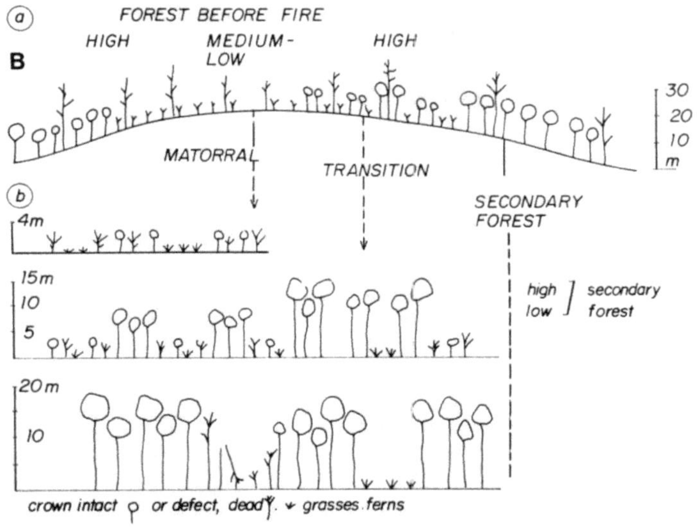

Fig. 6.60B

have little in common, except a temperature that even at night never becomes lower than 16° C, at most one month per year when evapotranspiration exceeds precipitation, and low daylength variation over the year. The first downscaling within this biome is most often done per tropical continent, e.g. by Whitmore (1975) in his "Tropical Rain Forests of the Far East". For Africa, no such recent book exists, except for West Africa (Martin 1989).

A view of the whole Amazon basin and its forests, exclusively in terms of average ecological production and its underlying climatical and soil resources is presented by Salati and Vose (1984). They treat Amazonia as a compartment of the whole South American continent east of the Andes and assess general trends in climate and land-use that could occur if deforestation of Amazonia reached large proportions. One of the recurrent themes in such Amazonian studies (see Bibliography in Salati and Vose 1984) is the recycling of rainwater over such large forest areas: the forest is claimed to recycle about 50% of the local precipitations through its evapo- transpiration. Although there are serious indications that such feedback relationships occur between large forests and climate, no generally accepted models exist as yet that can predict climatic deforestation effects accurately.

This is treated more thoroughly by the authors in IUCN (1983a). There, even a rough division is made in forest types being variants of either freshwater inundation forests (varzea, igapo; cf. Fig. 6.35) or forests being never flooded (terra firme forests). General traits of ecological production and species composition are given. As was said earlier (Sect. 5.6.5), such broad numerical calculated data are apt to guide land-use policies, and this was the aim of the IUCN-paper (1983a). On the other hand, the explanation of these continent-wide patterns has been sought in terms of palaeoclimatic refuges (cf. De Granville 1978; Van der Hammen 1986) and, more recently, palaeodynamics of the rivers, regulated by sub-Andean tectonics (Kalliola et al. 1987, 1988; Rasanen et al. 1987; Salo 1987). These last authors base their hypotheses upon a study in which the ecosystem size was scaled down again to include "only" some regions in Peruvian Amazonia. The two explanations do not seem to be mutually exclusive (Oldeman 1989).

Another breakdown into more detailed systems of a large ecosystem is presented in the case of the Venezuelian drainage basin of Rio Caroni in a special issue of the scientific magazine Interciencia (Interciencia 1986). This analysis not only describes the local sites and their vegetational cover, but indeed arrives at vegetational dynamics (Fig. 6.60B). Fölster (in Interciencia 1986), who studied the behaviour of the silvatic mosaics under the influence of local shifting events, fires in particular, did so because there is a tendency towards more or less natural conversion of forest into savanna and of savanna into semi-desert. An example of such forests can be seen in Fig. 6.20.

The link with eco-unit architecture and composition then is near; Kalliola et al. (1988) scale down to autoecology of plants. This stage is also exemplified by a study on the forest vegetation of Gabon by Reitsma (1988). In his book, only the choice of four detailed inventory plots in Gabon, which will be monitored, bridged the hierarchical distance between abiotic systems and living systems. The plots were several hundreds of kilometers distant from each other, so that they should be considered as all-but stochastic samples in four quite different sites.

A quite different example is the classification of forest habitat types of Central Idaho (USA) by Steele et al. (1981). They state (p. 99): "This classification attempts to provide a natural stratification of forest lands in terms of vegetative development. It is designed to reflect the combined forces of the environment upon a given site and discounts the temporary alterations of disturbance." And they write in their summary: "In addition to site classification, descriptions of mature forest communities are provided with tables to portray the ecological distribution of all species." This is again the breakdown of a site category into smaller ecosystems according to a model without spatial and temporal ecosystem definition and in terms of component species. It also is the first step for linking it to production models. In the publication by Steele et al., the analysis of forest architecture is replaced by a series of excellent photographs for each forest type described. It should be remembered, however, that photographs represent raw data and are not analytical documents, so that architectural analysis of these forests cannot be replaced by such pictures (cf. Oldeman 1981b).

Another broad approach is given by Brünig et al. (in Brünig 1982b). It concerns a general renewable resources model for Western Europe and its forests and evaluates the impacts of present trends if these are allowed to continue (Fig. 6.61). The figure shows that it was meant to assess the effects of exponentially increasing pollution. This would lead to large oscillations in ecosystem functioning around the year 2010 and collapse of the ecosystems around 2020. Such models are expressed again as averages per surface unit, so that ecosystem architecture is not included. Populations are not included either, except the human population. The authors specify that the tendencies expressed in this model rest on aggregates of information from lower-level models in their hierarchy. The latter differs somewhat from

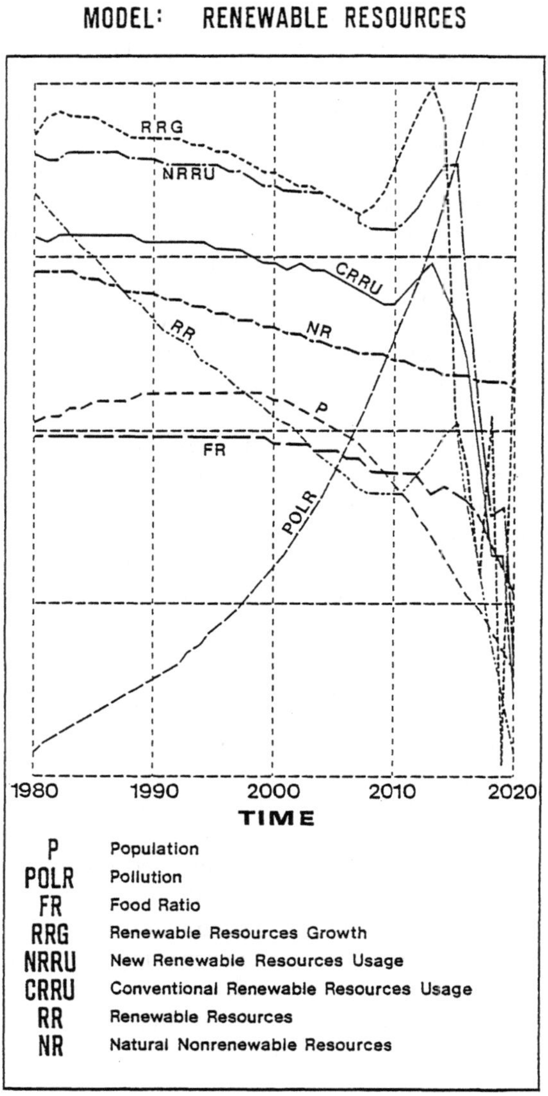

Fig. 6.61. Example of a very general model forecasting ecological developments in Western Europe and its forests from 1980 to 2020. Assumptions made for investments, exploitation of resources, renewal and use of raw material. Such assumptions rest on aggregates of outputs from smaller-scale models, at levels of a hierarchy that is different from Table 2.1 and which explains stands (no distinction eco-unit/mosaic) in terms of their tree components (cf. Fig. 6.58B). Exponential increase of pollution trends leads to, or is a symptom of, highly oscillating ecosystem behaviour from ca. 2010 on, and collapse around 2020. Such general models are both warnings, because they rest on realistic scenarios, and indicative options to combat ecological degradation. Much work still has to be done at lower hierarchical levels to check all assumptions and to allow accurate calculated predictions or design of silvicultural systems at management level. (Brünig et al. in Brünig 1982b)

the one in Table 2.1 but includes averaged architectural parameters, e.g. canopy roughness. In its present form, this model (Fig. 6.61) is useful as a warning and allows distinguishing broad options for improving the ecological situation, including forests. However, much research at lower levels is still needed to confirm the simulated trends.

6.5.3 Some Remarks on Altitudinal and Latitudinal Gradients

Some final examples concern altitudinal zonation of forest vegetation in mountainous regions. It may be illustrated by the work of Ohsawa (1984) for Mt. Fuji, Japan (Ohsawa et al. 1985), for Mount Kerinci in Sumatera, Indonesia (Ohsawa et al. 1986) and Ohsawa (1987) in the Himalayas of Eastern Nepal. For Mt. Fuji, Mt. Kerinci and the Bhutan Himalaya, overall production models can probably be established with the site data provided, which show a gradient with altitude above sea level. In all cases, the gradual disappearance of certain forest components and their replacement by others is clear. Such a sequence occurs on Mount Fuji for *Abies veitchii* (lowest), through *Betula ermanii* to a combination of *Alnus maximowiczii* and *Salix reinii* (highest altitude below tree line). In both Mt. Fuji and Mt. Kerinci studies, distribution maps for tree species may be found which could be easily transformed into eco-unit maps (cf. also Ohsawa 1981). This would make such studies bridges between the site-towards-forest line of research and the forest-towards-site line.

In the Andes, this zonation was studied already around the beginning of the 19th century by Von Humboldt (see Botting 1973). In Ecuador, these studies were continued and detailed, mainly in a descriptive sense concerning species composition (e.g. Acosta-Solis 1976). Monasterio (1980), in her thorough study on the high Venezuelian Andes, united site/population analyses of both the páramos (also cf. Cleef 1981; Fig. 5.45B) and the Venezuelian *Polylepis* (Chrysobalanaceae) forests.

A gradient study, which retraces processes over centuries has been published by Payette et al. (1985) for mosaic mortality patterns in old-growth *Picea mariana* forests at the subarctic tree line in Northern Quebec, Canada. They link these patterns to temporal variations in available ecological resources, i.e. site change. Ødum (1979) treated the tree and forest lines in the Northern Atlantic (Greenland, Faroe Islands) in an experimental way. He considered the introduction of exotic tree species with other temperaments in order to bridge the actual and potential tree lines which, according to him, are separated because of migration problems met by trees in such biotopes separated by the sea. In our terms, he would try to create neonatural forest mosaics (cf. Sect. 6.2.8.4). For Europe, descriptive data mainly in terms of species composition, can be found in the paragraphs on boreal and altitudinal tree lines in Mayer (1986).

6.6 Preliminary Conclusion on Silvatic Mosaics

It is abundantly clear that the mosaic level is the most difficult forest ecosystem level. It has been most often neglected in existing studies because it has been confused with "stands". Still, even if the lacunae in knowledge are considerable, it was shown that silvatic mosaics are built and function along rules that are quite different from those determining eco-unit life. This is important in silvology, but it also has considerable practical importance. Production by mosaics is more or is less than the mere sum of the production by their subsystems, stands or eco-units, because interaction amongst eco-units redistributes productivity. Animals play a crucial role in mosaics. They may be disregarded to a certain extent in stand management, but they can disturb the whole pattern and functioning of a silvatic mosaic if extinctions or demographic explosions occur.

The indiscriminate use of statistical analysis in silvatic mosaic studies has been shown to lead to the establishment of averages where the distribution of ordinated values would be more relevant. Such averages mask important regulation mechanisms and eco-unit interactions, and may even lead to dead-end models which can not be used as subsystems within higher hierarchical levels. Silvicultural systems designed according to such dead-end models incur high risks by factors that have been overlooked and are hidden behind averages. "Normal" distributions and processes in stands sometimes are mathematical artefacts which do not correspond to normal behaviour of forests. Silvatic mosaic ecology therefore remains a field to be urgently further researched. However, even now the available data should be considered in silvicultural design and in silvological studies.

7 Some Important Silvological Rules

7.1 What Are Rules?

The systematic description and exemplification of the silvological hierarchy in the preceding chapters shows systems behaviour which may be compared in similar system states at distinct hierarchical levels. The abstract formulation of this comparable behaviour leads to some more general statements on silvological systems.

Do such statements represent natural laws? It is too early to affirm this. Natural laws are believed to pervade nature from the smallest to the largest imaginable systems. Moreover, many think that natural laws should be expressed as accountant's models for natural systems, the basic tenet being that natural phenomena follow mathematical formulae and therefore can be calculated.

The rules outlined below have not been checked throughout any "universal" hierarchy, from the quantum level to that of galaxies. The outcome of these rules in terms of bookkeeping, concerning inputs, outputs and processed amounts of matter, energy and information, in sandwich hierarchies with three silvological levels, is quite imperfectly known, to say it optimistically. However, some quantitative statements with low precision can be made when considering relation diagrams (cf. Figs. 3.30, 5.75, 6.52). Such statements take the form of saying that "more", "less" or "equal amounts of" matter, energy or information are transmitted along a certain arrow in certain circumstances. This can be perfectly well expressed quantitatively, in binary terms, although exactitude is low (cf. Chap. 2).

However, a low level of exactitude may well characterize all that is ever certain in silvological hierarchies. They are prime examples of "the multiple, the temporal and the complex" to use the words of Prigogine and Stengers (1985). As islands of organisation in a sea of randomness, trees, eco-units and silvatic mosaics will never be the epitome of classical scientific certitude and precision. As emphasised in Section 6.3.8.1, the upward stream of unpredictable change from lower levels, particularly that of genetics and mutation, profoundly influences the community levels above, although these eliminate much by selection.

In this context the silvological rules, discovered during the studies by so many people, that have lead to the present book, are presented with sincere modesty. However, as far as they go they are true and are useful both to silvologists and foresters. For the former, they make it possible to formulate sharp hypotheses to be tested in order to deepen our insight into forest ecosystems. For the latter, even as they stand now, they constitute a useful checklist for the potential effects of silvicultural measures. The combination of silvologist and silviculturist, in a team or in one person, allows innovative methods in silviculture to be designed.

7.2 The Rules of Scale

The principles of biological organisation in living systems do not change with scale, but the mechanisms making these principles work change drastically from one hierarchical level to the next.

First, let us remember the development states of the living systems at the levels considered in the present book (Table 7.1).

The first rule of scale therefore is: at every silvological level, systems go through analogous development steps between their birth and their death.

It is always necessary to pinpoint the development state of a system studied or to be managed, because systems behaviour is not the same in each successive state. States may be arranged in a sequence of steps or in a gradient of continuous change.

Moreover, whatever the level and whatever the development state, living systems always have five attributes.

The second rule of scale therefore is: a living system displays a border layer, a system for interior mass and energy translocation, an interior transfer of information, an adjustment to changing exterior circumstances, and an information bank allowing reproduction of the system; but the implementation of each attribute is very different per hierarchical level.

For the three system levels mentioned (Table 7.1), the mechanisms implementing these five attributes have been compared in Tables 5.8 and 6.3. For research and management purposes alike, it is indispensable to pinpoint the system's limits first, by localising the border layer, and then to check the other five attributes and the mechanisms that make them operational.

Finally, interactions in forest ecosystems are not merely taking place among organisms of comparable size, but also among living systems of widely diverging sizes, such as trees and bacteria (Sect. 6.3.8; Fig. 5.74B).

This may be formulated as a more precise assessment. If, within a pair of interacting living systems, one system has n times the size or mass of the other, interaction will occur most probably between the smallest living system and that subsystem of the largest one which is closest to n^{-1} of the size or mass of that largest system. This generalises some intuitive pre-scientific notions, e.g. the concept of disease (interaction cell/bacterium, or fungus/tissue) or that of pest (interaction animal/organ-or-tissue).

With these rules of scale, silvological living systems and their interactions can be described and defined adequately to formulate and test hypotheses. They maximise clarity but at the same time maximise scientific liberty because no particular hierarchy is included in these rules (cf. Sects. 2.3, 2.4; also Berger and Berry 1988). Finally, they favour a balance between holism and reductionism, because they do not emphasise any particular direction, either from the trees to the forest or from the forest to the trees (cf. Hofstädter 1979).

Table 7.1. Analogous development at subsequent system levels (cf. Table 2.1). Note that phasic development follows comparable lines but that implementation is very different from level to level, as indicated by different vocabulary. At each level living systems are found with their five attributes (see Sect. 7.2), but only at one level are these systems organisms. At the other levels they are not even super-organisms or suborganisms, they are just other kinds of living systems. This point is apt to cause confusion if not very clearly stated

System	Developmental sequence				
Organ	Initium	Juvenile	Fully grown	Over-ripe	Dead
Organism					
- animal	Embryo	Juvenile	Adult	Decrepit	Dead
- tree	Seedling	Potential	Of the present	Of the past	Dead
Eco-unit	Innovation	Aggradation	Biostasis	Degradation	Dead
Silvatic mosaic	Installation	Precuilibrium	Ecuilibrium	Elimination	Collapse

7.3 The Rules of Fragmentation and Fusion

The larger a system in size or mass, the less stable is the architecture that can be maintained by a given kind of biological organisation. Such systems then tend to fragment into several smaller systems with a similar organisation type. Small living systems with a similar kind of organisation may, on the contrary, fuse and grow together as if they formed one system, at least temporarily.

It is to be noted that fragmentation does not mean the individualisation of subsystems. The smaller systems conserve a kind of organisation similar to that of the former large system, so that the second rule of scale excludes their having been subsystems. The same is true with fusion, which does not form a supersystem for the same reasons. In other words, the fragments of a living system in interaction form a system one level higher, e.g. a mosaic or a clonal population, but the small systems in fusion form a larger system at their own level.

The notion of fragmentation exists since long in zoology, where it is associated with regeneration of animal fragments into complete organisms, e.g. worms (cf. Storer and Usinger 1955). The same authors may be consulted for the phenomenon of colony-building by animals (cf. Coelentherates, like corals and polyps) which comes close to the notion of fusion mentioned here. One key to understanding the difference between a subsystem and either a fragment or a participant in fusion is its identity, which is indicated above by its similar or dissimilar organisation type. The latter may be strict, and in this case is often hereditary, or it may be flexible, and in this case is often triggered by environmental forces.

An example of fragmentation and fusion versus large, unfragmented systems is the treatment of trees by Hallé (1986) in terms of two basic arborescent strategies, i.e. gigantism and repetition. The hierarchical levels concerned are the primary levels of organs and organisms, and the auxiliary levels between these two, which are respectively the organ complex (module, unit of morphogenesis), the branched complex (tree model) and the reiterated complex (cf. Table 2.1; Figs. 3.1, 5.8). A tree may be built either by manifold repetition of fragments (fragmentation/fusion) or by increasing the size of one similarly organised system. Independent survival of "tree fragments" is one way to herbaceous life strategies (e.g. Hallé and Oldeman 1970; Hallé et al. 1978).

For reasons exposed in the Sections on genetics (Sect. 6.3.8) fragmentation is a prerequisite for renewal. This may be the renewal of the same system according to strict instructions of genetic origin. In this case, reproduction through small juvenile fragments (the seed!) is its expression. Renewal may also concern the organisation of the system. The small fragments may become organised otherwise than the original, large system. This differentiation may be reversible or not. The smaller the fragment, the more specific organisation is lost. This either leads to the death of the fragment (e.g. allogenic closure of a miniature eco-unit), or to a surprising lack of constancy in organisation (e.g., miniature cuttings or a few cells in vitro, Roux pers. commun.).

These notions are not clear enough as yet to formulate more precise rules of fragmentation and fusion. The overall formulation given is a convenient base, however, to generate testable silvological hypotheses, as long as tests are consciously devised in feasible ways, closely linked to phenomena at the hierarchical level concerned. For silvicultural design, the present formulation is immediately useful because as part of a checklist it points to fragmentation or fusion effects. For instance, silvicultural systems may have built-in fragmentation risks for stands, by storms, pests or genetical variety among trees. The other way round, a system meant for conservation management with enhanced diversity may risk ending up with a monotonous forest in which many, diverse small eco-units are fused into one simple system.

7.4 The Rules of Transfer of Functions

All biological functions within a living system are carried out either by subsystems or at levels that ensure a minimal cost-benefit ratio in terms of matter and energy.

A minimum cost versus a maximum benefit yields a minimum cost-benefit ratio. It depends on the reliability of the resource regime in the environment of a living system, whether or not one and the same subsystem is the optimal carrier of a certain function in perpetuity. Green leaves are not the optimal systems to provide plants with carbohydrates in cold winters or droughts. This function therefore shuttles between these organs and certain tissues, e.g. rays, which can ensure them temporarily. In dry regions, the leaf function may be completely transferred to the stems (Fig. 3.46).

Examples of transfer of functions at higher hierarchical levels have been given in Sections 5.6.9 and 6.4.4. They include transfer of functions from organisms to combinations of organisms (symbiosis), i.e. from organism to eco-unit, and transfer of functions among eco-units and silvatic mosaics.

Givnish (1986), in the book he edited on the economy of plant form and function, discussed the strong and weak points of ecological cost-benefit analysis. His book contains indispensable information in this field. However, it does not display any explicit and well-defined concept of systems analysis. It therefore remains open to criticism, e.g. concerning "atomistic" interpretation when zooming in at phenomena which would be seen as important in systems analysis, but which here are seemingly situated at too detailed levels. The basic aspect of ecological cost-benefit analysis remains a "trait", not a subsystem of an organism. Hence the notion of transfer of functions is lacking in Givnish's book.

In transfer of functions, as in fragmentation and fusion, the nature of the information involved is very important. It may be very strict to very flexible. Because so little is known in this respect, and because ecological cost-benefit analysis is still too little developed, it is not possible as yet to formulate more detailed rules of functional transfer more precisely.

There is enough evidence, however, to continue research in this line. In silviculture the formulation chosen above is enough to start checking such transfers, in particular when they relate to productivity and go from crop plants to "useless" organisms. This concerns the risk of pests and diseases (cf. Sect. 5.6.8; 5.6.9). Silvicultural design for nature conservation generally should avoid the tranfer of functions to some few but frequent species. This is a risk for forest ecosystems in general. Damage to a guild containing one or two species only (cf. Sects. 5.5.8) is a risk to the whole ecosystem. Species richness in each guild, and also the ensurance of a function at more than one level are therefore both an ecological richness and an economical way to prevent risks.

7.5 The Rules of Irreversibility and Process Oscillation

Patterns and processes in living systems change irreversibly by the immigration or emigration, the appearance or extinction of subsystems, or by the triggering of self-reinforcing destructive or constructive cycles. Because of these processes, all production values in living systems oscillate around average values against time.

These rules were implicit in the elimination and collapse stages of silvatic mosaics (Fig. 6.6D, E), the concept of stepped selection pressure (Table 6.2), the relation diagram leading to the upsetting and collapse of ecosystems (Fig. 6.61), and the considerations on the life cycle of trees in terms of production and architecture (Sect. 3.6). It is particularly the phenomena that lead to modulated production curves (Sect. 3.6.2), in which there is superposition of daily, seasonal, yearly and lifetime oscillations, that illustrate the more general rules.

In a sense, today's society is confronted with these ecological rules. Like every ecosystem, human communities live on top of their wastes and other residues. This is also one of the basic tenets of classical succession theories (cf. Sect. 6.1.2), in which ecosystems are considered to reach a final equilibrium state, the climax, when residues are recycled under the prevailing climatic circumstances of that region. Unfortunately, the climatic and soil conditions change too fast for the climax ever to be reached. The environmental cards are reshuffled too early every time.

Again, there are more or less reliable periodicities in the impacts of abiotic factors upon silvatic mosaics, their eco-units and the organisms that build them. With every favourable shift in circumstances, production goes up. If this goes on long enough, the production reaches the limits inherent in the system. With the next drop in hospitality of the abiotic environment, there are many species, residues and other traces of the former, wealthy period. Extinction is forever, but so is speciation, at all hierarchical levels. The former state of the system will never come back, be it only for lack of the same components.

Reliable periodical impacts will select rhythmic behaviour in plants or populations. Unreliable periodical impacts select flexibility. The closer the system is to the abiotic environment, the less reliable become many periodicities. The difference between an equilibrium and an ecuilibrium is that the latter shifts from similar state to similar state, without ever completely having the same identity in the sense of a temporal and spatial constancy of its architecture, its subsystems, its processes.

As formulated above, the rules of irreversibility say that no change in a living system is reversible, be it constructive or destructive. No system ever returns completely to its state of yesterday. Cyclic change, postulating the return of a living system to exactly the same state after a certain timespan, at best is a close approximation if and when during that time no major genetic or environmental change occurs. In that case change is limited to details. As a matter of fact, the return of a system to an earlier state, like Lamarckism (cf. Sect. 6.3.8), is not impossible but extremely improbable, going against all probabilities.

Some examples of irreversible change are the following.

- **Forest components** may either change by speciation, driven genetically, or by changed selection pressures, driven ecologically;
- **eco-units** may either show a change in component species composition, driven by speciation/selection within that unit, or have such change driven from the outside by changed selection pressure, e.g. because of changes in silvatic mosaic, site or management practice;
- **silvatic mosaics** may display changes in eco-unit composition, driven either endogeneously by factors like a modified propagule bank, or exogeneously by natural or artificial site and landscape changes.

The irreversibility of silvological developments should change the theoretical outlook on the balance between randomness and determinism in forest ecosystems. Basic viewpoints are given by Dawkins (1988), Prigogine and Stengers (1985) or Hofstädter (1979).

For silvicultural systems design, the rules of irreversibility emphasise the great importance of checking that all ecosystem changes induced by management improve and do not destroy stands, mosaics and landscapes. Subsystem choice, e.g. choice of tree species and options in forest health management, is significant in this respect. And land use, including urbanisation, may easily exert destructive selection pressure on desired systems, be they crops, other species, stands, national parks or forest districts with their typical mosaic structures.

7.6 Use and Significance of These Rules

Earlier, Oldeman (1974a) sought and found a general principle behind phenomena such as stem suckering, root suckering, stump suckering, other epicormic shoots, branch and trunk regeneration, reproduction by cuttings or by layering, and so on.

The principle that emerged was that of architectural reiteration (Sect. 3.2.4). We now can treat all these phenomena as variations upon the same theme, which is the inherited architecture of a woody plant. The first surmise that something analogous was possible for forest ecosystems has been visualised and expressed by Oldeman in Van Beusekom et al. (1987).

The present chapter lists our progress to date, with the four mentioned categories of silvological rules. It deserves to be mentioned that they have been found only after the elimination of all artificial, management-born notions from our models. Natural limits have taken the place of grids, natural species the place of crops, weeds or pests, natural distribution of matter, biomass and energy took the place of desired distribution with a bias to economical production. Not only has this detour outside human society allowed the formulation of a first version of truly silvological rules, but it has also lead to a better understanding of silvicultural systems, for the very reason that they were left out of focus in the beginning.

With the present book, starting from the end, it becomes possible to check many a silvological study or silvicultural plan. First, their hierarchical levels should be pinpointed, and the systems at each level that have to be understood or managed. Then the relevant rules are looked up and their cross-references to terms, paragraphs, tables or figures. The pertinent parts of the text and bibliography finally permit following up the line of enquiry as far as is necessary to apply the information gathered to the study or project.

It is certain that lacunae in relevant knowledge will often be encountered. This cannot be avoided. It is the nature of science that any endeavour to stock all knowledge and understanding of the universe on paper, in computer memories and in human brains, is vain. The only computer large enough to contain the information of the universe is the universe itself; and there would still always be lots of noise. Science may at best arrive at proven reconstructions of such parts of the universe as seem most relevant to us. It is hoped that this book may contribute to this task in the field of silvology and for silviculturists needing this information.

References

Abercrombie M, Hickman CJ, Johnson ML (1977, 6th edn., 6th repr.) A dictionary of biology. Penguin Books, Harmondsworth (UK), p 311

Acosta-Solis M (1976) Sinopsis de los estudios fitogeográficos en el Ecuador. Bol Inst Panam Geogr Hist, Sec Ecuador 11/12, 5: 5-21

Adams R, Adams M, Willens A, Willens A (1978) Dry lands: man and plants. Architectural Press, London, p 152

Akker A, Groeneveld W (1984) Reconnaissance of Amazonian rainforest architecture and insect habitats. AUW-Silviculture, Wageningen 84-19: p 95

Alder D, Bruenig EF, Heuveldop J, Smith J (1979) Struktur und Funktionen im Regenwald des internationalen Amazon-Ökosystemprojektes: vorläufige Mitteilung über Klassifikation der Bestände, Variation der Bestandesstruktur und Niederschlagsmerkmale. Amazoniana 6, 4: 423-444

Alexandre DY (1977) Régénération naturelle d'un arbre charactéristique de la forêt équatoriale de Côte d'Ivoire: Turraeanthus africana Pellegr. Oecol Plant 12, 3: 241-262

Alexandre DY (1978a) Le rôle disséminateur des éléphants en forêt de Taï, Côte-d'Ivoire. Rev Ecol, Terre Vie 32: 47-71

Alexandre DY (1978b) Observations sur l'écologie de *Trema guineensis* en basse Côte d'Ivoire. Cah ORSTOM, Sér Biol 13, 3: 261-266

Alexandre DY (1980) Caractère saisonnier de la fructification dans une forêt hygrophile de Côte d'Ivoire. Rev Ecol, Terre Vie 34: 335-359

Alexandre DY (1982a) Pénétration de la lumière au niveau du sous-bois d'une forêt dense tropicale. Ann Sci For 39, 4: 419-438

Alexandre DY (1982b) Etude de l'éclairement du sous-bois d'une forêt dense humide sempervirente (Tai, Côte-d'Ivoire). Acta Oecol Oecol Gener 3, 4: 407-447

Alexandre DY (1984) Strata in tropical rain-forest at Tai (Ivory Coast). In: Sutton et al. (eds) vol II, Tropical Rain-Forest. The Leeds Symp, pp 15-24

Alexandre DY (1986) Croissance et démographie des semis naturels en forêt de Tai (Côte d'Ivoire). Mém Mus Nat, Sér A Zool 132: 193-200

Alexandre DY, Guillaumet J-L, Kahn F, Namur C de (1978) Conclusion: caractéristiques des premiers stades de reconstitution. Cah ORSTOM, Sér Biol 13, 3: 267-270

Alexandre DY, Dedet J-P, Esterre P (1987) La leishmanniose en Guyane française. 7. Caractéristiques structurales de quelques sites de contamination humaine en forêt. Cah ORSTOM, Sér Ent Méd Parasitol 25, 2: 101-109

Alexandrian D (1981) La souche de bruyère arborescente pour la fabrication de pipes. Rev For Française 3, 2: 127-133

Alfen EHG van (1981) Factoren die van invloed zijn op de H/d verhouding van de beuk. AUW-Silviculture, Wageningen, p 37

Altmann H (1979) Giftpflanzen - Gifttiere. BLV, München

Alvarado A, Berish CW, Peralta F (1981) Leaf-cutter ant *(Atta cephalotes)* influence on the morphology of Andepts in Costa Rica. Soil Sci Soc Am J 45, 4: 790-794

Amorini E, Gambi G (1976) Un esempio di rinnovazione spontanea del pino nero sull'appennino Tosco-Marchigiano. Ann Inst Speriment Selvicoltura Arezzo 7: 1-25

Andel J van, Nelissen HJM (1979) Nutritional status of soil and plant species in several clearings in coniferous woods compared to that in two related habitats. Vegetatio 39, 2: 115-121

Andersson C, Prentice IC, Sjøgren E, Verwijst T (eds) (1988) Research in progress at the Institute of Ecological Botany, Uppsala University 1985-1988. Medd Växtbiol Inst Uppsala 1988, 1: 117

Andrewartha HG, Birch LC (1984) The ecological web. Univ Press, Chicago, p 506
Armson KA (1977) Forest soils: properties and processes. Univ Press, Toronto, p 390
Ashton PS (1964) Ecological studies in the mixed dipterocarp forests of Brunei State. Clarendon, Oxford p 75 + app
Asimov I (1972, 2nd impr.) Life and energy. Avon Books, New York (Discus Books 380 00299), p 378
Assmann E (1953) Die Standraumfrage und die Methodik von Mischbestandsuntersuchungen. Allg Forst Jagdzeit 125
Aubréville A (1938) La forêt coloniale; les forêts de l'Afrique occidentale française. Ann Acad Sci Colon Paris 9: 1-245
Aubréville A (1961) Etude écologique des principales formations végétales du Brésil. CTFT, Nogent-sur-Marne
Aubréville A (1965) Principes d'une systématique des formations végétales tropicales. Adansonia, NS 5, 2: 153-197
Ayensu ES (ed) (1980) Jungles. J Cape, London, p 200
Backman G (1942) Das Wachstum der Bäume. Wilhelm Roux Archiv für Entwicklungsmechanik der Organismen; Organ für die gesamte kausale Morphologie 141: 455-499
Backman G (1943) Wachstum und organische Zeit. Johann Ambrosius Barth, Leipzig, p 195
Bakker H de, Schelling J (1966) Systeem van bodemclassificatie voor Nederland (with Engl. summary, key). PUDOC, Wageningen, p 217
Bakker K, Cappenberg ThE, Croin Michielsen N, Freijsen AHJ, Nienhuis PH, Woldendorp JW, Zijlstra JJ (eds) (1985) Inleiding tot de oecologie. Bohn, Scheltema and Holkema, Utrecht/Antwerpen, p 565
Balk JT (1979) Een kruiwagen vol bomen: verleden en heden van het Amsterdamse bos. Stadsdrukkerij, Amsterdam, p 160 (in Dutch)
Baren B van (1982) Forest insects, diseases, fire. AUW-Silviculture, Field Work Rep, Wageningen, p 31
Baren B van, Hilgen P (1983) De structuur en dynamiek in La Tillaie, een ongestoord beukenbos in het bosgebied van Fontainebleau. AUW-Silviculture, Wageningen, p 70 (in Dutch)
Barkman JJ (1968) Das synsystematische Problem der Mikrogesellschaften innerhalb der Biozönosen (with Engl. summ.). In: Tuexen R (ed) Ber Int Symp Vegetationskunde, Stolzenau/Weser 1964: 21-53
Barkman JJ (1970) Enige nieuwe aspecten inzake het probleem van synusiae en microgezelschappen (with German summ.). AUW, Misc Papers, Wageningen 5: 85-116
Barkman JJ (1973) Synusial approaches to classification. In: Whittaker R (ed) Handbook of Vegetation Science. Junk, The Hague V: 437-491
Barkman JJ (1979) The investigation of vegetation texture and structure. In: Werger MJA (ed) The Study of Vegetation. Junk, The Hague, pp 125-160
Barkman JJ (1984) Biologische minimumarealen en de eilandtheorie. Vakbl Biol 64, 9: 162-167
Barkman JJ (1985) Typologie en classificatie van bossen op basis van de vegetatiestructuur. In: Koop H (ed) Verslag Themadag werkgroep Bosecosystemen Utrecht 1985. Rijks Inst Natuurbeheer, Leersum, pp 25-40
Barkman JJ (1988 a) Some reflections on plant architecture and its ecological implications. In: Werger MJA et al. (eds) Plant form and vegetation structure. SPB Acad Publ, The Hague, pp 1-7
Barkman JJ (1988 b) Introduction. In: Barkman JJ, Sykora KV (eds) Dependent Plant Communities. SPB Acad Publ, The Hague
Barkman JJ, Stoutjesdijk Ph (1987) Microklimaat, vegetatie en fauna. PUDOC, Wageningen, p 223
Barreto LS (1987) Um novo método para elaboracão de Tabelas de Producão: aplicacão ao pinhal. Serv Nacion, Lisbon, Parques Reservas Conserv Natureza, p 52
Bartelink H, Dijk G van, Hiele R van (1987) Lady Park Wood: analysis of spontaneous development in a mixed broadleaved woodland on limestone. AUW-Silviculture For Ecol, Wageningen 87-25: p 116 + app
Baudière A, (1974, publ. later) Contribution à l'étude structurale des forêts des Pyrénées orientales: hêtraies et chênaies acidiphiles. Coll Phytosociol 3: 17-44
Baudière A, Jalut G, Turmel A (1986) La hêtraie du versant nord du Mont Coronat (Pyrénées orientales): évolution historique et contemporaine. In: Soc Bot France, Groupement Sci ISARD, Proc Colloq Int Bot Pyrénéenne, pp 49-79

Bauer F (ed) (1985) Die Sache mit dem Wald. BLV, München p 376
Baumgartner A (1967) Ecological significance of the vertical energy distribution in plant stands. Nat Resourc Res 5: 367-374
Beardsley T (1988) Getting warmer? This has (so far) been the warmest decade in 127 years. Sci Am 259, 1: 8
Beaudou AG, Blic P de, Chatelin Y, Collinet J, Filleron JC, Guillaumet J-L, Kahn F, Zueli KB, Richard JF (1978) Recherche d'un language transdisciplinaire pour l'étude du milieu naturel. Trav Doc ORSTOM, Paris 91: p 143
Bech NJ (1983) La durée du cycle sylvigénétique en Forêt de Tai, Côte d'Ivoire. AUW-silviculture, Wageningen, p 84
Becking JH (1961) Studies on nitrogen-fixing bacteria of the genus *Beijerinckia* I. Geographical and ecological distribution in soils. Plant Soil 14, 1: 49-81
Beekman F (1981 a) Structural and dynamic aspects of the occurrence and development of lianes in the tropical rain forest. AUW-Silviculture, Wageningen p 45
Beekman F (1981 b) La dynamique d'une forét alluviale et le rôle des lianes. AUW-Silviculture, Wageningen, p 24
Begon M, Harper JL, Townsend CR (1986) Ecology: individuals, populations and communities. Blackwell, Oxford, p 876
Bellamy D (1978, repr. 1979) Botanic man. Hamlyn, London, p 208
Berg JJ van den, Saris FJA (1985) Het gemeentelijk landschapsplan, basisstudie. Gemeente Grathem (NL)
Berger JO, Berry DA (1988) Statistical analysis and the illusion of objectivity. Am Sci 76, 2: 159-165
Berish CW (1982) Root biomass and surface area in three successional tropical forests. Can J For Res 12: 699-704
Berryman AA (1986) Forest insects: principles and practice of population management. Plenum, New York
Beurel F, Doumenge C (1982) Recherche de relations entre anatomie et accrochage chez les lianes ligneuses tropicales. DEA-Rep, USTL, Montpellier, p 74
Beusekom CF van, Goor CP van, Schmidt P (1987) Wise utilization of tropical rain forest lands. Proc MAB-UNESCO Symp 1984. Ede (NL), Tropenbos Sci Ser 1: p 154; Summary Tropenbos Inf Ser 4: p 14
Bidwell RGS (1974) Plant Physiology. Mac Millan, New York, 24 + p 643
Blancaneaux Ph (1973) Notes de pédologie guyanaise: les djougoung-pété du bassin versant expérimental de la Crique Grégoire (Sinnamary-Guyane Française). Cah ORSTOM, Sér Pédol 11, 1: 29-42
Blanton CM, Ewel JJ (1985) Leaf-cutting ant herbivory in successional and agricultural tropical ecosystems. Ecology 66, 3: 861-869
Bleijendaal HPO (1979) Lijnbeplanting, een verjongingsmethode voor tropische bossen. AUW-Silviculture, Wageningen, p 94 (in Dutch)
Bodegom AJ van (1981) Gevolgen van selectieve exploitatie in het mesofytisch regenwoud te Sarwa, Suriname: de eerste 14 jaar. AUW-Silviculture, Wageningen, p 68 + app (in Dutch)
Boehm W (1979) Methods of studying root systems. Springer, Berlin Heidelberg New York, Ecol Stud 33: p 180
Boer PJ den (1973) Das Überleben von Populationen und Arten und die Bedeutung von Umweltheterogenität. (Engl. summ.) Verh Dtsch Zool Ges, Ann Rep 66: 125-136
Boer PJ den (1977) Dispersal power and survival. Carabids in a cultivated countryside. AUW, Wageningen, Misc Papers 14: p 190
Boerboom JHA (1974) Succession studies in the humid tropical lowlands of Surinam. Proc 1st Int Congr Ecol, Junk, The Hague, pp 343-347
Bol MMGR, Maaren A van, Oldeman RAA, Wassink JTj (1984) Mensen en bossen. Stichting Maatschappij en Onderneming, SMO-Inf, The Hague 84/4: p 83 (in Dutch)
Bongers F, Popma J (1988) Trees and gaps in a Mexican tropical rain forest. Dr thesis, Rijksuniv, Utrecht, ISBN 90-9002344-5, p 185
Bongers T (1985) Nematoden in Nederlandse bossen. AUW-Nematology, Wageningen, LH-Mededelingen 175: p 50

Bonsen K (1984) Architectuur en ontwikkeling van de gewone esdoorn (*Acer pseudo-platanus* L.). AUW-Silviculture, Wageningen, p 33
Bormann FH, Likens GE (1979) Pattern and process in a forested ecosystem. Springer, Berlin Heidelberg New York, p 253
Bosch AL (1984) A new root observation method: the perforated soil system. Oecol Plant 5: 61-74
Bosch AL (1985) Luchtverontreiniging door ammoniak en de rol van mycorrhiza in een grovedennenbos. Levende Natuur 86, 2: 42-45
Bosman JA (1979) Hoogte, stamdiameter en kroonafmetingen van bomen. AUW-Silviculture, Wageningen, p 86
Bosveld ATC, Botterweg RRP (1985) Onderzoek naar samenstelling, structuur en dynamiek van het regenbos te Yangambi. AUW-Silviculture, Wageningen, p 130
Botting D (1973) Humboldt and the cosmos. Sphere Books, London p 295 (facsimile illustr)
Bouchon J, Faille A, Lemee G, Robin A-M, Schmitt A (1973), Cartes et notice des sols, du peuplement forestier et des groupements végétaux de la réserve biologique de la Tillaie en forêt de Fontainebleau. Univ Paris XI, Orsay
Bourlière F, Hadley M (1970) The ecology of tropical savannas. Ann Rev Ecol Syst 1: 125-152
Boyce SG (1978) Theory for new directions in forest management. USDA, For Serv Res Pap, Asheville, North Carolina, SE For Exp Station, 193: p 19
Brandani A, Hartshorn GS, Orians GH (1988) Internal heterogeneity of gaps and species richness in Costa Rican tropical wet forest. J Trop Ecol 4: 99-119
Bratawinata AA (1986) Bestandesgliederung eines Bergregenwaldes in Ostkalimantan/Indonesien nach floristischen und strukturellen Merkmalen. Dr. thesis, Georg-August Univ, Göttingen 118 + p XXXI
Braun HJ (1963) Die Organisation des Stammes von Bäumen und Sträuchern. Wiss Verlags Ges, Stuttgart 12 + p 162
Braun HJ (1970) Funktionelle Histologie der sekundären Sprossachse. I Das Holz. Bornträger, Berlin 11 + p 190
Braun HJ (1974) Zum Wuchsverhalten von Pappeln bei optimalen und reduziertem Wasserangebot. Allg Forst Jagdztg 145: 112-113
Braun HJ (1979) Das Leben der Bäume. Univ Freiburg, p 166
Braun HJ (1982) Lehrbuch der Forstbotanik. Fischer, Stuttgart, p 257
Braun-Blanquet J (1934, 3rd ed) Pflanzensoziologie. Springer, Berlin Heidelberg, p 865
Brauns A (1976) Taschenbuch der Waldinsekten. Fischer, Stuttgart, vol 1 (Systematik Ökologie), p 443; vol 2 (Oekol Freiland-Differential-Diagnose; illustr), pp 444-817
Breemen N van, Visser WFJ, Pape Th (eds) (1988) Biogeochemistry of an oak-woodland ecosystem in the Netherlands affected by acid atmospheric deposition. PUDOC, Agric Res Rep, Wageningen, 930: p 197 + microfiche
Brosset A (1974) La nidification des oiseaux en forêt gabonaise: architecture, situation des nids et prédation. Terre Vie 28: 579-620
Brown AL (1978) Ecology of soil organisms. Heineman, London, p 116
Brown S, Lugo AE, Silander S, Liegel L (1983) Research history and opportunities in the Luquillo experimental forest. USDA For Serv, South For Exp Station, Gen Tech Rep, Orleans SO-44: p 128
Bruchwald A (1986) Simulation growth model MDI-1 for Scots pine. Ann Warsaw Agric Univ-SGGW-AR, For Wood Technol 34: 47-52
Bruenig EF (1974) Ecological studies of the Kerangas forests of Sarawak and Brunei. Bureau Sarawak For Dept, Kuching, Borneo p 300
Bruenig EF (ed) (1978, 2nd edn) Transactions of the international MAB-IUFRO workshop on tropical rainforest ecosystems research. Chair World For, Spec Rep, Reinbek, Hamburg 1: VI + p 364
Bruenig EF (1982 a) Some problems connected with the assessment of the role of humid tropical evergreen forests as source and sink of carbon. Mitt Geol-Palaeont Inst Hamburg (SCOPE/UNEP Sonderbd) 52: 59-68
Bruenig EF (ed) (1982 b, 2nd edn) Transactions of the 3rd International MAB-IUFRO workshop on ecosystem research; IUFRO-Congr Kyoto, Jpn, Sept 1981. Bonn, German Nat MAB-Comm, Mimeogr, p 161

Bruenig EF (1983) Chemical warfare and the dynamics and management of tropical forest ecosystems. Ho Chi Min City, Vietnam, Inter Symp Long-Term ecol and human Consequences of Chemic. Warfare in Viet-Nam. Mimeogr Paper, p 16

Bruenig EF (1985) Lowland-montane ecological relationships and interdependencies between natural forest ecosystems. Chengdu (China), Int Symp Ecology of Develop of trop and subtrop mountain areas. Symp Paper, Mimeogr, p 30

Bruenig EF (1987) The forest ecosystem: tropical and boreal. Ambio 16, 2/3: 68–79

Bruenig EF, Schneider TW (1975) Forstgut Auermühle: Geschichte und Wirtschaftsgrundsätze. Reinbek, Weltforstwirtschaftsinst, p 68 + phot

Bruenig EF, Alder D, Smith JP (1978) The international Amazon MAB rainforest ecosystem pilot project at San Carlos de Rio Negro: vegetation classification and structure. 2nd Int MAB-IUFRO Workshop on Trop Rainfor Ecosys Res, Jakarta, Oct 1978, p 35

Bruenig EF, Grossman WD, Kreysa J, Sander N, Schneider TW (1982) System modelling as tool in ecosystem research. Poster Paper, BES Symp Leeds, april 1982; Handout, Mimeogr, p 18

Bruenig EF, Bossel H, Elpel K-P, Grossmann W-D, Schneider TW, Wang Zhu-Hao, Yu Zuo-Yue (1986) Ecologic-socioeconomic system analysis and simulation: a guide for application of system analysis to the conservation, utilization and development of tropical and subtropical land resources in China. Inst World For Ecol, Reinbek, Hamburg, ISBN 3-926301-01-7, p 388

Bruggink W (1986) *Acacia nilotica* and sheep grazing in Kenya: an integrated system. 2nd Int Legume Conf, St Louis, Poster Paper 13, Abstr, pp 37–38

Bruijnzeel LA (1982) Hydrological and biogeochemical aspects of man-made forests in South-Central Java, Indonesia. Dr. thesis, Free Univ, Amsterdam; Alblasserdam, Offsetdrukkerij Kanters BV, p 250

Bruijnzeel LA (1986) Environmental impacts of (de)forestation in the humid tropics: a watershed perspective. Wallaceana 46: 3–13

Bruijnzeel LA, Wiersum KF (1985) A nutrient balance sheet for *Agathis dammara* Warb. plantation forests under various management conditions in Central Java, Indonesia. For Ecol Management 10: 195–208

Brussaard L (1984) Recente ontwikkelingen in de eiland-biogeografie. Vakbl Biol 64, 7: 121–125

Budowski G (1965) Forest species in successional process. Turrialba 15, 1: 40–42

Buecking W (ed) (1985) Der Bannwald "Waldmoor-Torfstich". Bd 3. Forstliche Versuchs- und Forschungsanstalt, Freiburg, p 278

Buis J (ed) (1979) Is er dan bos in Tunesie? Verslag Bosbouwexcursie 1978. AUW-For Tech/For Management/Silviculture, Wageningen, p 80 (in Dutch)

Burg R van der, Ouden J den, Weersink H (1986) Amère-Bière et le Canard Vert. AUW-Silviculture For Ecol, Wageningen 86-03: p 15 + transects (in Dutch)

Burschel P, Huss J (1987) Grundriß des Waldbaus. Parey, Hamburg, p 352

Caballé G (1977) Multiplication végétative en forêt dense du Gabon de la liane *Entada scelerata* (Mimosoideae). Adansonia NS 17, 2: 215–220

Caballé G (1980) Caractéristiques de croissance et multiplication végétative en forêt dense du Gabon de la "liane à eau", *Tetracera alnifolia* Willd. (Dilleniaceae). Adansonia NS 2; 19, 4: 467–475

Caballé G (1986) Sur la biologie des lianes ligneuses en forêt gabonaise. Dr. Thesis, USTL Montpellier, p 341

Cairns J, Dickson KL, Herricks EE (eds) (1977) Recovery and restoration of damaged ecosystems. Univ Press of Virginia, Charlottesville (USA), p 531

Cajander AK (1930) Wesen und Bedeutung der Waldtypen (also in English). Silva Fenn 15: p 66

Carson R (1962) Silent spring. Houghton, London

Carter VC, Dale T (1955, rev. edn 1974) Topsoil and civilization. Univ Press, Oklahoma, p 292

Castro-Dos Santos A (1981) L'appareil végétatif des Monocotylédones. Ph. D. thesis, USTL, Montpellier, p 249

Champagnat P (1949) La taille des arbres fruitiers. Maison Rustique, Paris, p 217

Chan HT, Appanah S (1980) Reproductive biology of some Malaysian Dipterocarps, I, Flowering biology. Malaysian For 43, 2: 132-143

Chappuis C (1971) Un exemple de l'influence du milieu sur les émissions vocales des oiseaux: l'évolution des chants en forêt équatoriale. Rev Ecol, Terre Vie 118: 183-202

Chappuis C (1974-1985) Illustrations sonores de problèmes bioacoustiques posés par les oiseaux de la zone éthiopienne. Alauda 42: 197-222; 467-500; 43: 427-474; 46: 327-355; 47: 195-212; 49: 35-58; 53: 115-136

Charles-Dominique P, Atramentowicz M, Charles-Dominique M, Gerard H, Hladik A, Hladik CM, Prévost MF (1981) Les mammifères frugivores arboricoles nocturnes d'une forêt guyanaise: interrelations plantes-animaux. Rev Ecol, Terre Vie 35: 341-435

Chauvin R (1967) L'univers des insectes. Arnoldo Mondadori, Milan

Chen A (1987) Unravelling another Mayan mystery. Discover Magazine 8, 6: 40-49

Chéret V (1987) La sapinière du Luchonnais (Pyrénées Hautes-Garonnaises): étude phytoécologique, recherches sur le phénomène du dépérissement forestier. Dr. thesis, Univ Paul Sabatier, Toulouse, p 287

Cheviron D (1985) Essai d'introduction de méthodes numériques dans l'étude de la structure de groupements forestiers. Ph. D. thesis, Univ Claude Bernard I, Lyon, p 132 (address: 43, Boul du 11 Novembre 1918, 69622 Villeurbanne Cedex - France)

Christensen JB (1987) Catastrofale gebeurtenissen zoals stormschade en hoe daar bij de beheersplannen rekening mee gehouden kan worden (with Engl. summ.). Ned Bosb Tijdschrift 59, 3: 58-62

Cleef AM (1981) The vegetation of the paramos of the Colombian Cordillera Oriental. Dr. thesis, Rijksuniv, Utrecht, p 321

Clements FE (1936) Nature and structure of the climax. J Ecol 24: 252-284

Cochet P (1977) Etude et culture de la forêt. Ecole Nationale du GREF, Nancy, p 235

Coo H de (1981) Ontstaan en ontwikkeling van natuurlijke wilgenbossen in de Lepelaarsplassen. AUW-Silviculture/Rijksdienst IJsselmeerpolders, Lelystad/Wageningen, p 69

Cooper CF (1974, 2nd pr.) The ecology of fire. Freeman, Readings Sci Am, San Francisco, pp 294-302

Corner EJH (1964) The life of plants. Weidenfeld and Nicolson, London, p 315

Corner EJH (1966) The natural history of palms. Weidenfeld and Nicolson, London, p 393

Cosijn R, Oldeman RAA, Stortenbeker CW (eds) (1983) Dood hout in het bos (with Engl summaries). Spec Vol (Themanummer) Ned Bosbouwtijdschrift 55, 2/3: 41-146

Cousens J (1974) An introduction to woodland ecology. Oliver and Boyd, Edinburgh, p 151

Craker LE, Simon JE (eds) (1986) Herbs, spices and medicinal plants: recent advances in botany, horticulture and pharmacology. Oryx, Phoenix, Arizona, p 359

Cramers JA (1766) Anleitung zum Forst-Wesen. Waisenhaus-Buchhandlung, Braunschweig

Creber GT, Challoner WG (1984a), Climatic indication from growth rings in fossil woods. In: Brenchley P (ed) Fossils and climate. Wiley, London, chapter 4

Creber GT, Challoner WG (1984b) Influences of environmental factors on the wood structure of living and fossil trees. Bot Rev 50, 4: 357-448

Cribb AB, Cribb JW (1981) Useful wild plants in Australia. Collins, Sidney, p 269

Critchfield WB (1957) Geographic variation in *Pinus contorta*. Harvard Univ, Cambridge (Mass), Moors Cabot 3: p 118

Cuatrecasas J (1968) Páramo vegetation and its life forms. In: Proc UNESCO México Symp 1966. Dummlers Colloq Geogr, Bonn 9: 163-186

Dansereau P, Lems K (1957) The grading of dispersal types in plant communities and their ecological significance. Contr Inst Bot Univ Montréal, Montréal p 71

D'Arcy Thompson W (1917) On growth and form. Univ Press, Cambridge

Dauget J-M (1985) La réaction aux traumatismes: comparaison entre les arbres et les coraux. Rev Ecol, Terre Vie 40: 113-117

Davis DR, Clayton DH, Janzen DH, Brooke AP (1986) Neotropical Tineidae, II: Biological notes and descriptions of two new months phoretic on spiny pocket mice in Costa Rica (Lepidoptera: Tineoidea). Proc Entomol Soc Wash 88, 1: 98-109

Dawkins R (1976) The selfish gene. Univ Press, Oxford, p 224

Dawkins R (1988, 2nd impr) The blind watchmaker. Penguin Books, London, p 332

Debach P (1974) Biological control by natural enemies. Cambridge Univ Press, London, p 323
Debano LF, Rice RM, Conrad LE (1979) Soil heating in chaparral fires: effects on soil properties, plant nutrients, erosion and runoff. USDA For Serv, Pacific Southwest For Range Exp Station, Research Paper PSW-145, Berkeley, p 21
Derkman G (1979) De betekenis van enige groeiplaatsfactoren voor de groei van beukenbeplantingen in verband met droogteschade in 1976. RIBL "De Dorschkamp", Wageningen, Rapp 201: p 72 + 30 fig
Derkman G, Koop H (1977) Structuur en verjonging van een oerbos Puszcza Białowieza (Oost-Polen). AUW-Nat Management, Wageningen 70-71: p 36 (in Dutch)
Descoings B-M (1976) Approche des formations herbeuses tropicales par la structure de la végétation. Dr. thesis, USTL, Montpellier, p 221
Devineau J-L (1975) Etude quantitative des forêts-galeries de Lamto (moyenne Côte d'Ivoire). Ph. D. thesis, Univ Paris VI, Paris, p 190
Devineau J-L (1976) Principales caractéristiques physionomiques et floristiques des formations forestières de Lamto (Moyenne Côte d'Ivoire). Ann Univ Abidjan, Sér E Ecol 9: 274-303
Devineau J-L (1982) Etudes pondérales des litières d'arbres dans deux types de forêts tropophiles en Côte d'Ivoire. Ann Univ Abidjan, Série E Ecol 15: 27-62
Dickinson CH, Pugh GJF (eds) (1974) Biology of plant litter decomposition. Vol 1. Academic Press, London, 241 + p 46
Dijk J van (1987), The ecological significance of architectural and crown characteristics of juvenile trees in a Venezuelan cloud forest. AUW-Silviculture For Ecol, Wageningen 87-16: p 19 + ill
Djailany VP (1987) Formations végétales naturelles de Sumatra (Indonésie): structures et caractéristiques radiométriques. Dr. thesis, Univ Paul Sabatier, Toulouse, p 296 + annex
DMG-DPG (1987) Warnung vor drohenden weltweiten Klimaänderungen durch den Menschen. Dtsch Meteorol Ges/Dtsch Phys Ges, Bad Honnef, p 16
Dobzhansky T (1970) Genetics of the evolutionary process. Columbia Univ Press, New York, p 505
Dodson CD, Stermitz FR, Castro CO, Janzen DH (1987) Neolignans from fruits of *Ocotea veraguensis*. Phytochemistry 26, 7: 2037-2040
Dolman AJ (1986) Estimates of roughness length and zero plane displacement for a foliated and non-foliated oak canopy. Agric For Meteorol 36: 241-248
Dolman AJ (1987) Summer and winter rainfall interception in an oak forest. Predictions with an analytical and a numerical simulation model. J Hydrol
Donselaar J van (1970) Floristic and ecological data on the lianes of the Brokopondo District, Surinam. Acta Bot Neerl 19, 2: 287-296
Dort K van, Olsthoorn A, Raad H (1979) Ondergroei, structuur en microgradienten in eiken en beukenbos in het Speulderbos. AUW-Silviculture, AUW-Vegetation Sci, AUW-Agrometeorol, Wageningen, p 109 + app
Dubost G (1968) Les niches écologiques des forêts tropicales sud-américaines et africaines, sources de convergences remarquables entre Rongeurs et Artiodactyles. Rev Ecol, Terre Vie 22: 3-28
Duvigneaud P, Denaeyer-Desmet S (1970) Biological cycling of minerals in temperate deciduous forests. In: Reichle DE (ed) Analysis of temperate forest ecosystems. Springer, Berlin Heidelberg New York, p 304
Dyanat-Nejad H (1971) Nombre et tropismes des racines néoformées expérimentalement dans l'appareil souterrain du Cacaoyer (*Theobroma cacao* L.). Bull Soc Bot France, Mém 1971: 109-123
Eckardt FE (1984) Monitoring of CO_2-exchange in forests. In: Aagren GI (ed) State and change of forest ecosystems. Swed Univ Agric Sci, Dept Ecol Environ Res, Rep 13: 137-147
Eckardt FE, Heim G, Methy M, Sauvezon R (1975) Interception de l'énergie rayonnante, échanges gazeux et croissance dans une forêt méditerranéenne à feuillage persistant *(Quercetum ilicis)*. Photosynthetica 9, 3: 145-156
Edelin C (1977) Images de l'architecture des Conifères. Thèse Spécialité, Univ Montpellier, p 254
Edelin C (1984) L'architecture monopodiale: l'exemple de quelques arbres d'Asie tropicale. Thèse d'Etat, USTL, Montpellier, p 258

Ek RC (1989) Spatial distribution of epiphyte vegetation in Saül, French Guiana. Rijks Univ, Utrecht, Bioinformatic Group, p 38
Ellenberg H (ed) (1971) Integrated experimental ecology: methods and results of ecosystem research in the German Solling project. Ecol Stud 2, Springer, Berlin Heidelberg New York, p 214
Ellenberg H (ed) (1973) Ökosystemforschung. Springer, Berlin Heidelberg New York, p 280
Ellenberg H, Galat-Luong A, Maydell H-J von, Muehlenberg M, Panzer KF, Schmidt-Lorenz R, Sumser M, Szolnoki TW (1988) Pirang: ecological investigations in a forest island in the Gambia. Warnke, Hamburg, p 324
Emanuel KA (1988) Towards a general theory of hurricanes. Am Sci 76, 4: 370-379
Engbeck JH (1976, 2nd pr) The enduring giants. Univ Ext, Univ California, Berkeley, p 119
Ernst WHO, Andel J van, Drent RH, Tinbergen JM, Bakker K (1985) Autoecologie. In: Bakker et al. (eds) Inleiding tot de oecologie. Bohn, Scheltema and Holkema, Utrecht, pp 69-126
Eusebio MA (1977) Occurrence of heartrot in virgin forests and decay in logged-over areas. Sylvatrop Philipp For Res J 2, 3: 195-208
Evers PW (1984) Growth and morphogenesis of shoot initials of Douglas fir, *Pseudotsuga menziesii* (Mirb.) Franco, in vitro. D. Sc. thesis, AUW, Wageningen, chapters also as Meded RIBL "De Dorschkamp"
Ewel J et al. (1982) Leaf area, light transmission, roots and leaf damage in nine tropical plant communities. Agro-Ecosyst 7: 305-326
Faber PJ (1975) Stabiliteit van bos ten opzichte van wind: een theoretisch gezichtspunkt. Ned Bosbouwtijdschrift 47: 179-187
Faber PJ (1983) Concurrentie en groei van bomen binnen een opstand. Uitv Verslag RIBL "De Dorschkamp", Wageningen, 18, 1
Faber PJ (1985) Groei en plantafstand van "Rap" populier in een Nelderproef. Ned Bosbouwtijdschrift 57, 5/6: 157-166
Faille A, Lemee G, Pontailler JY (1984) Dynamique des clairières d'une forêt inexploitée (réserves biologiques de la forêt de Fontainebleau). I - Origine et état actuel des ouvertures. Acta Oecol, Oecol Gener 5, 1: 35-51; II - Fermeture des clairières actuelles. Acta Oecol, Oecol Gener 5, 2: 181-199
Falinski JB (1977) Białowieza primeval forest (text in English, French, German). Phytocoenosis, Biuletyn Fitosocjologiszny, Warszawa 6, 1/2: 133-148
Fanta J (1982) Natuurlijke verjonging van het bos op droge zandgronden. RIBL "De Dorschkamp", Wageningen, Rapp 301, p 236
Fanta J (1985) Groeiplaats: onderzoek, classificatie en betekenis voor de bosbouw (Engl summ.). Ned Bosbouwtijdschrift 57, 10/11: 333-347
Fanta J (ed) (1986) Forest dynamics research in Western and Central Europe. Proc IUFRO-workshop, 17-20 sept 1985, Wageningen. PUDOC, Wageningen
FAO (1987) Effects of agricultural development on vector-borne diseases. FAO, Rome AGL/MISC/12/87: p 144
FAO, IUCN, UNESCO (1989) Plant genetic resources, their conservation *in situ* for human use. FAO, Rome, p 38
Farrar JL (1961) Longitudinal variation in the thickness of the annual ring. For Chron 37: 323-331
Findley R (1981) Mount St. Helen's aftermath: the mountain that was and will be. Nat Geogr 160, 6: 713-733
Flanagan D (ed) (1970) The biosphere. A Scientific American Book. Freeman, San Francisco, p 134
Flanagan D (ed) (1978) Evolution. A Scientific American Book. Freeman, San Francisco, p 135
Flenley J (1979) The equatorial rain forest: a geological history. Butterworth, London, p 162
Florence J (1981) Chablis et silvigénèse dans une forêt dense humide sempervirente du Gabon. Thèse de Spécialité, Univ Strasbourg (France), p 261
Foerster M (1970) Einige Beobachtungen zur Ausbildung des Wurzelsystems tropischer Waldbäume. Allg Forst Jagdztg 141: 185-188
Foresta H de (1981) Premier temps de la régénération naturelle après exploitation papetière un forêt tropicale humide: ARBOCEL, Guyane Française. Thèse 3me Cycle, USTL, Montpellier, p 114 + app

Foresta H de, Charles-Dominique P, Erard Ch, Prevost MF (1984) Zoochorie et premiers stades de la régénération naturelle après coupe en forêt Guyanaise. Rev Ecol, Terre Vie 39: 369-400
Fournier F, Sasson A (eds) (1983) Ecosystèmes forestiers tropicaux d'Afrique. ORSTOM/UNESCO, Paris, p 473
Fraiture A de (1985) Recherche sur les techniques de production et l'organisation de quelques pépinières de secteur au Burkina Faso. AUW-Silviculture, Wageningen 85-15: p 168
Franklin JF (1981) Wilderness for baseline ecosystem studies. Proc XVII IUFRO World Congr, Ibaraki, Div 1: 37-48
Franks NR (1989) Army ants: a collective intelligence. Am Sci 77, 2: 139-145
Freysen AHJ, Woldendorp JW (eds) (1978) Structure and functioning of plant populations. Kon Ned Akad Wetensch, Verh Afd Natuurk Ser 2, 70. North Holland, Amsterdam, p 323
Fries J (1981) Growth and yield studies in mixed indigenous tropical forests. Some thoughts on the means and ends. Proc XVII IUFRO World Congr, Ibaraki, Div 4: 58-70
Fritts HC (1976) Tree rings and climate. Academic Press, London, 12 + p 567
Frost P, Medina E, Menaut J-C, Solbrig O, Swift M, Walker B (1985) Responses of savannas to stress and disturbance: a proposal for a collaborative programme of research. Biol Int, Spec Issue 10: p 82
Furtado JI (ed) (1980) Tropical ecology and development. Int Soc Trop Ecol, Kuala Lumpur, p 692
Garwood NC, Janos DP, Brokaw N (1979) Earthquake-caused landslides: a major disturbance to tropical forests. Science 205: 997-998
Gasc J-P (1981) Quelques nouvelles données sur la répartition et l'écologie des Sauriens en Guyane Française. (bibliography!). Rev Ecol, Terre Vie 35: 273-325
Gatsuk LE, Smirnova OV, Vorontzova LI, Zaugolnova LB, Zhukova LA (1980) Age states of plants of various growth forms: a review. J Ecol 68: 675-696
Gäumann E (1935) Über den Stoffhaushalt der Buche *(Fagus sylvatica)*. Ber Dtsch Bot Ges 53: 366-377
Geerling C (1982) Guide de terrain des ligneux Saháliens et Soudano-Guinéens. Wageningen, Meded Landbouwuniv Wageningen 82-3: p 340
Gelder T de (1981) Bossen in het Gelders rivierengebied, in het bijzonder de bosontwikkeling in de landgoedbossen. AUW Silviculture/RIBL "De Dorschkamp", Wageningen, p 71 + maps
Genard M, Lescourret F (1984) Les petits vertébrés de la pineraie a crochets, fonction descriptive et rôle dans la régénération forestière: l'exemple du massif d'Osséja (Pyrénées orientales). Inst Nat Agron, Paris-Grignon, p 270 + ann
Genard M, Lescourret F, Durrieu G (1986) Mycophagie chez le sanglier et dissémination des spores de champignons hypogés. Gaussenia Toulouse 2: 17-23
Gibbs-Smith C (1978) The inventions of Leonardo da Vinci. Phaidon, Oxford, p 110
Giessen RC van der, Koutstaal H, Kwadijk F (1984) Dehesa del Moncayo: historia, estructura y desarrollo de un hayedo en España. (Dutch text available). AUW-Silviculture For Ecol, Wageningen p 5 + transects
Gill AM, Tomlinson PB (1977) Studies on the growth of the Red Mangrove (*Rhizophora mangle* L.), 4. The adult root system. Biotropica 9, 3: 145-155
Gillon Y (1984) Laboratoire d'Ecologie Tropicale ECOTROP, Rapport général d'activité 1982-1984. CNRS, Paris, p 145
Givnish TJ (ed) (1986) On the economy of plant form and function. Cambridge Univ Press, London, p 717
Goddard C, Kalotas A (eds) (1988, 2nd edn) PUNU, yankunytjatjara plant use. Angus and Robertson, North Ryde (NSW Australia), p 166
Goderie R (1984) Het Beukenbouwwerk (with English chapter summaries). AUW-Silviculture, Wageningen 84-06: p 52
Godron M (1971) Essai sur une approche probabiliste de l'écologie des végétaux. Thesis D.Sc., USTL, Montpellier, CNRS A02820, p 247
Godron M, Daget P, Long G, Sauvage C, Emberger L, Le Floc'h E, Poissonet J, Wacquant JP (1968) Code pour le relevé méthodique de la végétation et du milieu. CNRS, Paris, p 291
Golley FB, Medina E (eds) (1975) Tropical ecological systems. Trends in terrestrial and aquatic research. Springer, Berlin Heidelberg New York

Gomez-Pompa A (1985 a), Los recursos bióticos de México. Ed Alhambra Mexicana, México DF, p 122

Gomez-Pompa A (1985 b) On Mayan silviculture. INIREB, México

Gomez-Pompa A, Amo RS del (eds) (1985) Investigaciones sobre la regeneración de selvas altas en Veracruz, México. Vol II. Ed Alhambra Mexicana, México DF, p 421

Gomez-Pompa A, Vazquez-Yanes C (1976), Regeneración de selvas. Comp Ed Continental, México, p 676

Goor CP van (ed) (1983) Ecologie en gebruik van bossen. PUDOC, Wageningen, p 180

Goor CP van (1985) The impact of tree species on soil productivity. Neth J Agric Sci 33: 133-140

Goor CP van, Lynden KR van, Meiden HA van der (1974, 2nd edn) Bomen voor nieuwe bossen. Arnhem, Kon Ned Heide Maatschappij, p 119

Goudriaan J (1987) The biosphere as a driving force in the global carbon cycle. Neth J Agric Sci 35: 177-187

Goulding M (1980) The fishes and the forest. California Univ Press, Berkeley, p 280

Graaf NR de (1986) A silvicultural system for natural regeneration of tropical rain forest in Suriname. PUDOC, Wageningen, p 250

Granville J-J de (1976) Notes guyanaises: quelques forêts sur le Grand Inini. Cah ORSTOM, Sér Biol 11, 1: 23-34

Granville J-J de (1978) Recherches sur la flore et la végétation guyanaises. Thèse ès Sciences, Univ Montpellier, p 272

Granville J-J de (1979, 1984) Monocotyledons and Pteridophytes indicators of environmental constraints in the tropical vegetation. Paper presented to 5th Symp Tropical Biology, Kuala Lumpur (ms 4 p) publ 1984, Candollea 39: 265-269

Grenand P (1980) Introduction à l'étude de l'univers wayãpi: ethnoécologie des Indiens du Haut-Oyapock (Guyane Française). Langues et Civilisations à tradition orale, 40, SELAF, Paris

Grene M (1987) Hierarchies in biology. Am Sci 75: 504-510

Groenendijk H, Oldeman RAA, Veer C (1985) Basic assumptions in present-day university curricula in forestry. IUFRO S6.09/AUW-Silviculture, Wageningen

Grubb PJ, Whitmore TC (1966) A comparison of montane and lowland rain forest in Ecuador. II. The climate and its effects on the distribution and physiognomy of forests. J Ecol 54: 303-333

Grubb PJ, Whitmore TC (1967) A comparison of montane and lowland rain forest in Ecuador. III. The light reaching the ground vegetation. J Ecol 55: 33-57

Grubb PJ, Lloyd TD, Pennington TD, Whitmore TC (1963) A comparison of montane and lowland forest in Ecuador, I. Forest structure, physiognomy and floristics. J Ecol 51: 567-601

Guevara Sada S (1986) Plant species availability and regeneration in Mexican tropical rain forest. Uppsala, Acta Univ Upsaliensis, Compreh Summ of Uppsala Diss from the Fac of Sci 48

Guillaumet J-L, Kahn F (1979) Description des végétations forestières tropicales. Approche morphologique et structurale. Candollea 34, 1: 109-131

Guillaumet J-L, Couturier G, Dosso H (1984) Recherche et aménagement en milieu forestier tropical humide: le projet Tai de Côte-d'Ivoire. UNESCO, Paris, Notes Tech MAB 15: p 245

Hadley M (ed) (1988) Rain forest regeneration and management. IUBS/IVIC/UNEP/ UNESCO, Paris, Biol Int, Spec Issue 18: p 68

Hagg H, Pel M (1983) Successie na ontbossing in het tropisch regenwoud van Suriname: bijlage. AUW-Vegetation Sci, Wageningen, 77 ill

Hallé F (1986) Deux stratégies pour l'arborescene: gigantisme et répétition. Naturalia Monspeliensia, Spec Vol, (ISSN 0373-515X), pp 159-170

Hallé F, Martin R (1968) Etude de la croissance rythmique chez l'Hévéa (*Hevea brasiliensis* Müll. Arg. Euphorbiacées-Crotonoidées). Adansonia (NS) 8, 4: 475-503

Hallé F, Oldeman RAA (1970) Essai sur l'architecture et al dynamique de croissance des arbres tropicaux. Masson, Paris, p 178, Engl transl 1975, Penerbit Univ Malaya, Kuala Lumpur, p 156

Hallé F, Oldeman RAA, Tomlinson PB (1978) Tropical trees and forests: an architectural analysis. Springer, Berlin Heidelberg New York, p 441

Hammen T van der (1986) The paleoecology of tropical South America. WOTRO, The Hague, ISBN 90-70623-04-8, 1982: 35-91

Hari P, Mäkelä A, Sievanen R (1981) Systems concepts in theoretical plant ecology. Univ Technol, Helsinki, Systems Theory Lab, Rep B60: p 32
Hari P, Kaipiainen L, Korpilahti E, Mäkelä A, Nilson T, Oker-Blom P, Ross J, Salminen R (1985) Structure, radiation and photosynthetic production in Coniferous stands. Univ Helsinki, Dept Silviculture, Research Notes 54: p 233
Harper JL (1977) Population biology of plants. Academic Press, London, p 892
Harris LD (1984) The fragmented forest: island biogeography theory and the preservation of biotic diversity. Univ Press, Chicago, p 211
Hartshorn GS (1980) Neotropical forest dynamics. Biotropica 12 (Spec Vol): 23-30
Hasenkamp JG (1976) Naturgemäßer Waldbau: eine zukunftsträchtige Art der forstlichen Wirtschaftsführung. Natur Museum, Frankfurt 106, 3: 74-80
Haxaire C (1976) Etude d'un abattis chez les indiens Wayapi, Guyane Française. USTL, Montpellier, DEA, p 58
Helgerson OT (1988) Historic fire year for Oregon and California. FIR-Rep Medford (Oregon), Spec Fire Issue 9, 4
Hendry CD, Berish CW, Edgerton ES (1984) Precipitation chemistry at Turrialba, Costa Rica. Water Resour Res 20, 11: 1677-1684
Hermy M (1984a) The creation and the analysis of data matrices in vegetation science. Bull Soc R Bot Belg 117: 63-72
Hermy M (1984b) A numerical approach to the phytosociology of riverine woods to the South of Bruges (Flanders, Belgium). Coll Photosociol 9: 227-258
Herrera PRA, Menendez L, Rodriguez ME (eds) (1986) Ecología de los bosques tropicales de la Sierra del Rosario, Cuba. Proyecto MAB 1. La Havana (Cuba), Summaries. Univ Havana, Lab Bot, p 59
Hett JM (1971) A dynamic analysis of age in sugar maple seedlings. Ecol 52: 1071-1074
Heusden W van (1980) Toekomstige ontwikkeling der naaldbossen in het Noordhollands Duinreservaat. AUW-Silviculture, Wageningen, p 35 + app; also see Beheersnota '80-'85 Noord Hollands Duinreservaat (1980)
Heijbroek HM (1984) Clones in forestry and in nature. Arboricultural J 8: 275-286
Hijweege WL, Mes G (1986) Un plan de futuras actividades para el bosque escuela en consideración a la situación actual en Iturbide, NL Linares (Nuevo León, México). UANL, Fac Silvicultura Manejo de Recursos, AUW-Silviculture, Wageningen, p 140 + app
Hilgen PR, Brenk JJW van den (1984) Bovengrondse interactie: groeiruimte en architectuur in gelijkjarige meningen van eik en beuk (Engl summary). RIBL "De Dorschkamp", Wageningen, Rapp 373; AUW-Silviculture, Rapp 84-11, p 130
Hladik A (1978) Phenology of leaf production in Gabon: distribution and composition of food for folivores. In: Montgomery GG (ed) The ecology of arboreal folivores. Smithsonian Inst, Washington DC, pp 51-71
Hladik A (1984) The information on tropical forest dynamics as given by basal area measurements. In: Aagren GI (ed) State and change of forest ecosystems. Swed Univ Agric Sci, Dept Ecol Environ Res, Rep 13: 149-155
Hladik A, Hladik CM (1980), Utilisation d'un ballon captif pour l'étude du couvert végétal en forêt dense humide. Adansonia (NS) 19, 3: 325-336
Hladik A, Bahuchet S, Ducatillion C, Hladik CM (1984) Les plantes à tubercules de la forêt dense d'Afrique Centrale. Rev Ecol, Terre Vie 39: 249-290
Hoekman DH (1985) Radar backscattering of forest stands. Int Remote Sensing 6, 2: 325-343
Hoff M (1978) Les collines calcaires sous-Vosgiennes. Bull Soc Nat Mulhouse 770, 2: 17-33
Hoffman JS (1984) Carbon dioxide and future forests. For, March 1984: 164-167
Hofmann J-P (1981) Recherche de critères de sélection pour la résistance à la casse au vent chez *Hevea brasiliensis*. Thèse ès Sciences, Univ Paris-Sud, Centre d'Orsay, p 224
Hofstädter DR (1979) Gödel, Escher, Bach: an eternal golden braid. Basic Books, New York
Holdridge LR, Grenke WC, Hatheway WH, Liang T, Tosi J (1971) Forest environments in tropical life zones. Pergamon, Oxford, p 747
Holthuijzen AMA, Boerboom JHA (1982) The *Cecropia* seedbank in the Surinam lowland rain forest. Biotropica 14, 1: 62-68

Holthuijzen AMA, Sharik TL (1985) The avian seed dispersal system of eastern red cedar *(Juniperus virginiana).* Can J Bot 63, 9: 1508-1515

Hommel PWFM (1987) Landscape ecology of Ujung Kulon (West Java, Indonesia). Publ by the autor, PO Box 98, 6700 AB Wageningen, p 206 + app

Hondelmann W (1975) Safeguarding the plant genetic resources of the developing countries. Inst Sci Coop, Tübingen, Plant Research Develop 1: 5-11

Horn HS (1971) The adaptive geometry of trees. Univ Press, Princeton

Horn HS (1975) Forest succession. Sci Am 232, 5: 90-98

Hotta M (ed) (1984) Forest ecology and flora of G Gadut, West Sumatra. Kyoto Univ, Yoshida Coll, Biol Lab, p 220 + transects

Houten JG Ten, Gysels H (eds) (1984) Encyclopedie van het milieu. Oosthoek Uitg Mij BV, Utrecht, p 576

Houtzagers G (1954) Houtteelt der gematigde luchtstreek. Zwolle, Tjeenk Willink, 1: p 576

Houtzagers G (1956) Houtteelt der gematigde luchtstreek. Zwolle, Tjeenk Willink, 2: p 438

Hozumi K (1975) Studies on the frequency distribution of the weight of individual trees in a forest stand, V. Jpn J Ecol 25, 3: 123-131

Hozumi K, Shinozaki K (1974) Studies on the frequency distribution of the weight of individual trees in a forest stand, IV. Jpn J Ecol 24, 3: 207-212

Huettl RF (1986) Forest fertilisation: results from Germany, France and the Nordic countries. Paper read before Fertiliser Soc Lond (Greenhill House, 90-93 Cowcross Str, London EC1M6BH), p 40

Huisman W (1987) Vegetatiestructuur, microklimaat en geluidsoverdracht. Vakbl Biol 67, 20: 389-392

Huxley PA (ed) (1983) Plant research and agroforestry. ICRAF, Nairobi, p 617

INTERCIENCIA (1986) Venezuela, la cuenca del Río Caroní (with Engl & Port summ, articles 3 languages). INTERCIENCIA, Rev Ciencia Tecnol Am, Caracas, Spec Issue 11, 6: 256-366

IUCN (1983 a) Ecological structures and problems of Amazonia. Int Union Conserv Nature and Nat Res Gland (CH), Comm Ecol Papers 5: p 79

IUCN (1983 b) Global status of mangrove ecosystems. Comm Ecol Papers 3; Environmentalist 3, Suppl 3: 1-88

Ives WGH, Wong HR (1988) Tree and shrub insects of the prairie provinces. North For Cent, Edmonton (Alberta, Can), ISBN 0-662-15770-2, p 327

Jacob F (1970) La logique du vivant. Gallimard, Paris, p 354

Jacobs M (1976) The study of lianas. Flora Malesiana Bull 29: 2610-2618

Jacobs M (1988) The tropical rain forest, a first encounter. Springer, Berlin Heidelberg New York Tokyo, p 295

James NDG (1982, 3rd edn) The forester's companion. Blackwell, Oxford, p 381

Janse ARP, Boer ThA de (1987) Met het oog op de aarde. Het Spectrum, Wormer (NL), p 144

Jansen H, Krommendijk G (1983) Natuurlijke verjonging onder Douglas op holt- en haarpodzolen. AUW-Silviculture, Wageningen, p 124

Janzen DH (1967) Synchronization of sexual reproduction of trees within the dry season in Central America. Evolution (USA) 21, 3: 620-637

Janzen DH (1970 a) Herbivores and the number of tree species in tropical forests. Am Nat 104, 940: 501-528

Janzen DH (1970 b) *Jacquinia pungens,* a heliophile from the understorey of tropical deciduous forest. Biotropica 2, 2: 112-119

Janzen DH (1974) Tropical blackwater rivers, animals and mast-fruiting by the Dipterocarpaceae. Biotropica 6: 69-103

Janzen DH (1978) In: Tomlinson and Zimmerman (eds), pp 83-128

Janzen DH (1979) How to be a fig. Ann Rev Ecol Syst 10: 13-51

Janzen DH (1982) Seeds in tapir dung in Santa Rosa National Park, Costa Rica. Brenesia 19/20: 129-135

Janzen DH (1983) Physiological ecology of fruits and their seeds. Springer, Berlin Heidelberg New York Tokyo, Encycl Pl Phys, NS, 12 C: 625-655

Janzen DH (1985) *Spondias mombin* is culturally deprived in megafauna-free forest. J Trop Ecol 1: 131-155

Janzen DH (1986a) Guanacaste National Park: tropical ecological and cultural restoration. Proj Rep Serv Parques Nacionales Costa Rica, San José, p 39 + figs
Janzen DH (1986b) Natural history of a wind-pollinated Central American Legume tree *(Ateleia herbert-smithii)*. 2nd Int Legume Conf, St Louis, Abstr, pp 14-15
Janzen DH (1986c) Mice, big mammals, and seeds: it matters who defecates what where. In: Estrada A, Fleming TH (eds) Frugivores and seed dispersal. Junk, Dordrecht, Chapter 20
Janzen DH (1986d) Chihuahuan desert Nopaleras: defaunated big mammal vegetation. Ann Rev Syst 17:595-636
Janzen DH, Ataroff M, Farinas M, Reyes S, Rincon N, Soler A, Soriano P, Vera M (1976) Changes in the arthropod community along an elevational transect in the Venezuelan Andes. Biotropica 8, 3: 193-203
Janzen DH, Ryan CA, Liener IE, Pearce G (1986) Potentially defensive proteins in mature seeds of 59 species of tropical Leguminosae. J Chem Ecol 12, 6: 1469-1480
Jeannoda-Robinson V (1977) Contribution à l'étude de l'architecture des herbes. Ph.D. thesis, USTL, Montpellier, p 76 + 36 ill
Jenik J (1979) Pictorial encyclopedia of forests. Hamlyn, London, p 495
Jensen RV (1987) Classical chaos. Am Sci 75, 2: 168-181
Johansson D (1974) Ecology of vascular epiphytes in West-African rain forest. Acta Phytogeogr Suecica 59: 69
Johns AD (1988) Effects of "selective" timber extraction on rain forest structure and composition and some consequences for frugivores and folivores. Biotropica 20, 1: 31-37
Jong M de (1988) Evolutionary approaches to insect communication systems. Dr. thesis, Univ Leiden (NL), p 165
Jonge Mulock However P de (1984) The threshold of successful attack for *Ips typographus* and associated micro-organisms on *Picea abies*. AUW-Silviculture, Wageningen 84-23: p 46
Jonkers WBJ (1987) Vegetation structure, logging damage and silviculture in a tropical rain forest in Suriname. PUDOC, Wageningen
Jordan CF (1987) Amazonian rain forests: ecosystem disturbance and recovery. Springer, Berlin Heidelberg New York Tokyo, p 133
Jovanovič B, Jovanovič R, Zupančič M (eds) (1986) Natural potential vegetation of Yugoslavia (commentary to the map 1:1 000 000). Publ for 18th IUFRO World Congr Sci Counc Vegetation Map of Yugoslavia, Ljubljana, p 122 + map
Kahn F (1977) Analyse structurale des systèmes racinaires des plantes ligneuses de la forêt tropicale dense humide. Candollea 32, 2: 321-358
Kahn F (1978) Architecture et dynamique spatiale racinaires chez les plantes ligneuses des zones forestières tropicales humides. MAB/ORSTOM, Abidjan (CdI), Rapp Ronéo 7: p 17
Kahn F (1982) La reconstitution de la forêt tropicale humide: Sud-Ouest de la Côte d'Ivoire. Mém ORSTOM, Paris 97: p 150
Kahn F (1983) Architecture comparée de forêts tropicales humides et dynamique de la rhizosphère. Thèse d'Etat, USTL, Montpellier, p 426
Kahn F (1986a) Les palmiers des forêts tropicales humides du Bas Tocantins (Amazonie Brésilienne). Rev Ecol, Terre Vie 41: 3-13
Kahn F (1986b) Life forms of Amazonian palms in relation to forest structure and dynamics. Biotropica 18, 3: 214-218
Kahn F, Castro A de (1985) The palm community in a forest of Central Amazonia, Brazil. Biotropica 17, 3: 210-216
Kahn F, Mejia K, Castro A de (1988) Species richness and density of palms in terra firme forests of Amazonia. Biotropica 20, 4: 266-269
Kaibiyainen LK, Khair P, Sazonova A, Mäkelä A (1986) Balance of the water transport system in the scotch pine. III. Area of conducting xylem and needle mass. Lesovedenie (Sov For Sci) 1: 31-37
Kalela EK (1949) Ecological character of tree species and its relation to silviculture. Acta For Fenn 57, 1: 5-35
Kalliola R, Salo J, Makinen Y (1987) Regeneración natural de selvas en la Amazonia peruana 1: dinámica fluvial y sucesión ribereña. Mem Mus Hist Nat "Javier Prado", Lima 18: p 102 + app

Kalliola R, Makinen Y, Salo J (1988) Regeneración natural de selvas en la Amazonia peruana 2: autecología de algunas especies sucesionales. Mem Mus Hist Nat "Javier Prado", Lima 19: p 54

Kamminga P (1982) Forest Department Western Australia, September 79-May 80 (with bibliography). AUW-Silviculture, Field Work Rep, Wageningen, p 13

Kamminga P (1983) Brandbare stof in Nederlandse opstanden en brandaanpassingen bij bomen (with bibliography). AUW-Silviculture, Wageningen, p 33

Kaplan Y (1984) The ecosystem of the Yahudia nature reserve with emphasis on dynamics of germination and development of *Quercus ithaburensis* Decne. Dr. thesis, AUW-Nat Management, Wageningen, p 339 + map

Kaufmann J (1972) Insect travellers. Windmill, Kingswood (UK), p 126

Kervel W, Eyndhoven L van, Gisbergen P van (1986) Structuur en ontwikkeling van natuurlijke douglasbossen in Washington en Oregon. (VS). AUW-Silviculture, Wageningen 86-28: p 94

Kessell SR (1979) Gradient modeling: resource and fire management. Springer, Berlin Heidelberg New York, p 432

Ketelaars JJMH, Budelman A (1976) Anthropogene vegetaties in het Brokopondo-district, Suriname. AUW-Plant Taxonomy, Wageningen, p 100

Kikkawa J (ed) (1985) Wildlife management in the forests and forestry-controlled lands in the tropics and the Southern hemisphere. Proc IUFRO S1.08. St Lucia, Queensland, Univ of Queensland, ISBN 0867761520, p 77

Kilbertus G, Mourey A, Schwartz R, Prevost MF (1980) Activités biologiques dans les sols tropicaux (Guyane Française). I. Influence du déboisement sur la microflore tellurique, étude préliminaire. Bull Acad Soc Lorraine Sci 19: 117-130

Kira T, Shidei T (1967) Primary production and turnover of organic matter in different forest ecosystems of the Western Pacific. Jpn J Ecol 17, 2: 70-87

Kira T, Ogawa H, Yoda K, Ogino K (1967) Comparative ecological studies on three main types of forest vegetation in Thailand. IV. Dry matter production, with special reference to the Khao Chong rain forest. Nature Life SE Asia 5: 149-174

Kitching RL, Jones RE (eds) (1981) The ecology of pests: some Australian case histories. CSIRO, Melbourne, p 254

Klinge H (1973 a) Root mass estimation in lowland tropical rain forests of Central Amazonia, Brazil. I. Fine root masses of a pale yellow latosol and a giant humus podzol. Trop Ecol 14, 1: 29-38

Klinge H (1973 b) Root mass estimation in lowland tropical rain forests of Central Amazonia, Brazil. II. "Coarse root mass" of trees and palms in different height classes. An Acad Brasil Ciênc 45, 3/4: 595-609

Koestler JN (1955) Der Bestockungsaufbau in der waldbaulichen Bestandesdiagnose. Allg Forstz 10, 1: 1-8

Koestler JN, Brueckner E, Bibelriether H (1986) Die Wurzeln der Waldbäume. Parey, Hamburg, p 284

Kohyama T (1983) Seedling stage two subalpine *Abies* species in distinction from sapling stage: a matter-economic analysis. Bot Mag Tokyo 96: 49-65

Kohyama T (1987) Stand dynamics in a primary warm-temperate rain forest analyzed by the diffusion equation. Bot Mag Tokyo 100: 305-317

Kojima S (1979) Biogeoclimatic zones of Hokkaido Island, Japan. J Coll Liberal Arts, Toyama Univ Jpn (Nat Sci) 12: 97-141

Kojima S (1981) Biogeoclimatic ecosystem classification and its practical use in forestry. J Coll Liberal Arts, Toyama Univ Jpn (Nat Sci) 14, 1: 41-75

Kojima S (1983) Phytogeocoenoses of the montane *Abies sachalinensis* biogeoclimatic zone in Kitami District, Hokkaido, Japan. I. Analysis of Coniferous forest phytogeocoenoses. J Coll Liberal Arts, Toyama Univ (Nat Sci) 16, 2: 173-281

Kojima S, Peters R (1984) Analysis of stand structure of a *Betula ermanii* forest in Nakagawa experiment forest of Hokkaido University, Hokkaido, Japan. J Coll Liberal Arts, Toyama Univ 17, 1: 33-44

Komdeur J, Vestjens JPM (1982) De relatie tussen bosstructuur en broedvogelbevolking in Nederlandse naaldbossen. AUW-Silviculture, AUW-Nat Management, Wageningen, p 232

Koning J de (1983) La forêt du Banco. Meded Landbouwhogeschool, Wageningen 83-1: p 921

Koop H (1981 a) Vegetatiestructuur en dynamiek van twee natuurlijke bossen: het Neuenburger en Hasbrucher Urwald. (Engl summ), PUDOC, Wageningen, p 112 (in Dutch)
Koop H (1981 b) De schaal van spontane ontwikkelingen in het bos. Ned Bosbouwtijdschrift 53, 3:82-90
Koop H (1983) Ecological monitoring of natural and semi-natural forests. Ann Rep RIN, Leersum, Projekt 160, (p 5, in proofs)
Koop H (1985) Omvormingsbeheer naar natuurlijk bos: een paradox? Ned Bosbouwtijdschrift 58, 1/2:2-11
Koop H (1987) Het RIN Bosecologisch informatiesysteem (The RIN forest ecology information system, with Engl summ and legends). Rijksinst Natuurbeheer, Leersum, RIN-Rapp 87/4: p 47
Koop H (1989) Forest dynamics: SILVI-STAR a comprehensive, monitoring system. Springer, Berlin Heidelberg New York Tokyo
Koptur S (1984 a) Outcrossing and pollinator limitation of fruit set: breeding systems of neotropical *Inga* trees (Fabaceae: Mimosoideae). Evolution 38, 5: 1130-1143
Koptur S (1984 b) Experimental evidence for defence of *Inga* (Mimosoideae) saplings by ants. Ecology 65, 6: 1787-1793
Koptur S (1985) Alternative defenses against herbivores in *Inga* (Fabaceae: Mimosoideae) over an elevational gradient. Ecology 66, 5: 1639-1650
Kormondy EJ (1976, 2nd edn) Concepts of ecology. Englewood Cliffs, New Jersey, Prentice-Hall, p 238
Korpel S (1980) Development and structure of natural spruce forest in Slovakia in relation to anti-avalanche protective function (in Czech. with English summary). Acta Fac For Zvolen Czech 22:9-39
Korpel S, Paule L (1976) Die Eibenvorkommen in der Umgebung von Harmanec, Slowakei. Arch Naturschutz Landschaftsforsch, Berlin 16, 2: 123-139
Kort I de, Baas P (1986) Diktegroei en houtstructuur van vitale en niet-vitale douglas in de Peel. Ned Bosbouwtijdschrift 58, 3: 52-57
Kortlandt A (1982) Vegetation research and the "Bulldozer Herbivores". Poster Paper, Trop Rain For Symp, Leeds, handout (publ in Proc of the Symp 1984)
Kozlowski TT (ed) (1981) Water deficits and plant growth, vol VI, Woody plant communities. Academic Press, New York, p 582
Kraft G (1884) Beiträge zur Lehre von den Durchforstungen, Schlagstellungen und Lichtungstrieben. Klindworth, Hannover, p 147
Krochmal A, Kologiski R (1974) Understory plants in a mature Loblolly Pine plantation in North Carolina. SE For Exp Station, USDA For Serv Res Note, Asheville, NC (USA), SE-208, p 8
Krochmal A, Krochmal C (1984, 2nd edn) A field guide to medicinal plants. Times Books, New York, ISBN 0-8129-6336-9, p 274
Krugman SL (1984) Policies, strategies and means for genetic conservation in forestry. In: Yeatman CW, Kafton D, Wilkes G (eds) Plant genetic resources. Am Ass Adv Sci, Selected Symp, Washington DC 87: 71-78
Krugman SL, Ching KK, Dinus RJ, Fins L, Kellison RC, Winieski J (1983) Forest genetics and tree improvement in the People's Republic of China. Soc Am For, USA ISBN 0-939970-19-8, p 84
Kuenstle E (1971) Der Jahresgang des CO_2-Gaswechsels von einjährigen Douglasientriebe in einem 20-jährigen Bestand. Allg Forst Jagdztg 142: 105-108
Kuiper LC (1984) Verslag van een zes maanden durende studiereis naar de Westkust van de Verenigde Staten in 1984. AUW-Silviculture, Wageningen D84-02: p 10
Kuiper LC (1986) Die Struktur eines unbehandelten Douglasienbestandes in Baden-Württemberg. AUW-Silviculture, Wageningen D86-04: p 9 + ill + app
Kuiper LC (1988) The structure of natural Douglas-fir forests in Western Washington and Western Oregon. Agric Univ Wageningen Papers 88-5: p 47
Kuiper LC, Schooten JP van (1985) Dynamische aspecten van windworp (Wind and trees; Engl summ) Ned Bosbouwtijdschrift 57, 4: 130-133
Kuiters L (1987) Phenolic acids and plant growth in forest ecosystems. Free Univ Press, Amsterdam, p 150

Lamont B, Ralph C, Christensen P, Pate JS (1986) Ectomycorrhizas in sclerophyllous Australian Legumes and their dispersal by mycophagous Marsupials. 2nd Int Legume Conf, St Louis, Poster Paper 47, Abstr, p 51
Lanier L (1986) Précis de sylviculture. Ecole Nationale du GREF, Nancy, p 468
Lappi J, Smolander H (1984) Integration of the hyperbolic radiation-response function of photosynthesis. Photosynthetica 18, 3: 402-410
Larcher W (1976, 2nd edn) Ökologie der Pflanzen. Ulmer, Stuttgart, UTB 232, p 320
Larsen JA (1980) The boreal ecosystem. Academic Press, New York, p 500
Laumonier Y (1980) Contribution à l'étude écologique et structurale des forêts de Sumatra. Thesis, Univ Paul Sabatier, Toulouse, 2374: p 137
Lavelle P (1984) The soil system in the humid tropics. Biol Int, IUBS, Paris 9: 2-17
Laws RM, Parker ISC, Johnstone RCB (1975) Elephants and their habitats. Clarendon, Oxford, p 367
Leakey RE (1979) The Illustrated Origin of Species by Charles Darwin. Rainbird, London, p 240
Leathart S (1977) Trees of the world (Dutch translation "Alle bomen van de Wereld"). Uitg Septuaginta, Hoofddorp, p 224
Leeuwen CG van, Westhoff V (1961) De nivellering van flora en fauna. Natura 58, 10: 132-140 (in Dutch)
Leeuwenberg AJM (ed) (1987) Medicinal and poisonous plants of the tropics. PUDOC, Wageningen, p 152
Leffef F (1984) Bosinsecten en bosbrand. AUW-Silviculture, Wageningen 84-08: p 34
Leibundgut H (1958) Empfehlungen für die Baumklassenbildung und Methodik bei Versuchen über die Wirkung von Waldpflegemaßnahmen. London, Proc XII IUFRO World Congr (1956) Oxford, Vol 2
Leibundgut H (1959) Über Zweck und Methodik der Struktur- und Zuwachsanalyse von Urwäldern. Schweiz Z Forstwes 110, 3: 110-124
Leibundgut H (1970 2nd edn) Der Wald, eine Lebensgemeinschaft. Huber, Frauenfeld, p 200
Leibundgut H (1978) Über die Dynamik Europäischer Urwälder. Allg Forstz 33, 24: 686-690
Leibundgut H (1982) Europäische Urwälder der Bergstufe. Haupt, Bern, p 308
Leigh EG, Rand AS, Windsor DM (eds) (1982) The ecology of a tropical forest: seasonal rythms and long-term changes. Univ Press, Oxford, p 468
Leikola M (1969) The effect of some climatic factors on the daily hydrostatic variations in stem thickness of Scots pine. Ann Bot Fenn 6: 173-181
Leikola M (1971) Metsikkosadannan maaraeraassahoidetussa mannikossa. (Throughfall in a managed Scots Pine stand in Southern Finland; Engl summ and legends). Silva Fenn 5, 2: 129-174
Leikola M (1976) Nakokohtia metsikkoekologisten ymparisto-tekijoiden mittaamisesta. (Measuring environmental factors in a forest ecosystem; Engl summ and legends) Silva Fenn 10, 2: 148-156
Lemee G (ed) (1975) Recherches sur l'écosystème de la forêt subéquatoriale de Basse Côte-d'Ivoire. Rev Ecol, Terre Vie 29: 169-264
Leroy-Deval J (1973) Les anastomoses racinaires chez l'okoumé *(Aucoumea klaineana)*. CR Acad Sci Paris, Sér D 276: 3425-3428
Lescure J-P (1978a) An architectural study of the vegetation's regeneration in French Guiana. Vegetatio 37, 1: 53-60
Lescure J-P (1978b) La mangrove guyanaise: architecture des jeunes stades et vie avienne. Cah ORSTOM, Sér Biol 12, 4: 361-376 (ms. 1977)
Lescure J-P (1980) Aperçu architectural de la Mangrove guadeloupéenne. Acta Oecol, Oecol Gener 1, 3: 249-265
Lescure J-P (1986) La reconstitution du couvert végétal après agriculture sur brûlis chez les Wayap̃i du Haut Oyapock (Guyane française). Ph. D. thesis, Univ Paris VI, p 142
Lette H (1981) Natural regeneration on a deforested area in Bajo Calima, Colombia. Coop AUW/INDERENA/PNUD/FAO/CONIF, AUW-Silviculture, Wageningen, p 73
Levitt J (1980) Responses of plants to environmental stresses. Vol II. Water, radiation, salt and other stresses. Academic Press, New York
Lieberman M, Lieberman D, Hartshorn GS, Peralta R (1985a) Small-scale altidudinal variation in lowland wet tropical forest vegetation. J Ecol 73: 505-516

Lieberman D, Lieberman M, Peralta R, Hartshorn GS (1985 b) Mortality patterns and stand turnover rates in a wet tropical forest in Costa Rica. J Ecol 73: 915-924
Lieth H, Markert B (1988) The establishment of element concentration cadasters for ecosystems (ECCE) in the different vegetation zones of the Earth. Biol Int 16: p 7-11
Loehle C (1986) Phototropism of whole trees: effects of habitat and growth form. Am Midland Nat 116, 1: 190-196
Loftis DL (1985) Preharvest herbicide treatment improves regeneration in Southern Appalachian hardwoods. S J Appl For 9, 3: 177-180
Louman BT (1986) Structural and floristic composition of a *Lithocarpus* fallow in Papua New Guinea. AUW-Silviculture, Wageningen, p 31 + app
Lovelock JE (1987, 2nd edn) Gaia, a new look at life on earth. Univ Press, Oxford Paperback, p 157
Lugo A (1978) Stress and ecosystems. In: Thorp JH, Gibbons JW (eds) Energy and environmental stress in aquatic systems. DOE Symp Ser (CONF-771114). Nat Inf Serv, Springfield, Va, pp 62-101
Lugo A (1988) Diversity of tropical species: questions that delude answers. Biol Int, Spec Issue 19: p 37
Lyford WH, Wilson BF (1964) Development of the root system of *Acer rubrum* L. Harvard For Paper 10: p 17
Maas MVM, Gubbels MEMN (1986 a) Root development in *Robinia pseudoacacia:* observations in a perforon root box system. 2nd Int Legume Conf, St Louis, Poster Paper 52, Abstr pp 53-54
Maas MVM, Gubbels MEMN (1986 b) Silvicultural diagrams for *Robinia pseudoacacia.* 2nd Int Legume Conf, St Louis, Poster Paper 53, Abstr, p 54
MacArthur RH, Wilson EO (1967) The theory of island biogeography. Univ Press, Princeton
Maheswaran J, Gunatilleke IAUN (1988) Litter decomposition in a lowland rain forest and a deforested area in Sri Lanka. Biotropica 20, 2: 90-99
Mäkelä A (1988) Models of pine stand development: an eco-physiological systems analysis. Univ Helsinki, Dept Silviculture Res Notes, Helsinki 62: p 150
Mamiya Y (1981) Population increase of the pine wood nematode, *Bursaphelenchus lignicolus,* in relation to development of pine wilt. Proc XVII IUFRO World Congr, Ibaraki, Div 2: 275-285
Mangenot G (1955) Etude sur les forêts de plaines et plateaux de la Côte-d'Ivoire. IFAN, Etudes Eburnéennes, Dakar 4: 5-61
Mangenot G (1973) Données élémentaires sur l'Angiospermie. Ann Univ d'Abidjan, Sér E, VI, Fasc 1: p 233
Mansoer S (1986) Nabootsingen van boomkronen en toepassing daarvan bij simulatie van fisheye foto's. AUW-Silviculture, AUW-Computer Sci, Wageningen 86-11, p 53
Marais EN (1933) Die siel van die mier. Van Schaick BPK, Pretoria
Mariaux A (1975) Essai de dendroclimatologie en climat sahélien sur *Acacia raddiana.* Bois For Trop 163: 27-53
Marks GC, Kozlowski TT (eds) (1973) Ectomycorrhizae, their ecology and physiology. Academic Press, New York, p 444
Martens MJM (ed) (1986) Proceedings Workshop on sound propagation in forested areas and shelterbelts. Fac Sci, Nijmegen (NL)
Martin C (1989) Die Regenwälder Westafrikas. Birkhäuser, Basel, p 235
Marx DH, Artman JD (1979) *Pisolithus tinctorius* ectomycorrhizae improve survival and growth of pine seedlings on acid coal spoils in Kentucky and Virginia. Reclam Rev 2: 23-31
Matthews SW (1976) What's happening to our climate. Nat Geogr 150: 576-615
Maury-Lechon G (1982) Régénération forestière en Guyane française: recrû sur 25 ha de coupe papetière en forêt dense humide (Arbocel). Bois For Trop 197: 3-21
Mayer H (1980, 2nd edn) Waldbau auf soziologisch-ökologischer Grundlage. Fischer, Stuttgart, p 483
Mayer H (1981) Die 10 ökologischen Wald-Wild-Gebote. Univ Bodenkultur, Waldbau-Inst, Vienna, p 11
Mayer H (1983) Waldbauliche Probleme in Gebirgswäldern des Maghreb (Nordafrika). Centralbl Gesamte Forstwes 100, 1: 1-16

Mayer H (1986) Europäische Wälder. Fischer, Stuttgart, p 385

Mayer H, Tichy K (1979) Das Eichen-Naturschutzgebiet Johannser Kogel im Lainzer Tiergarten, Wienerwald. Centralbl Gesamte Forstwes 96, 4: 193-226

Mayer H, Neumann M, Schrempf W (1979) Der Urwald Rothwald in den Niederösterreichischen Kalkalpen. Jahrb Ver Schutz Bergwelt 44: 79-117

Mayer H, Freimoser F, Kral F (1982) Ergebnisse des internationalen Tannenherkunftsversuches Wien 1967-1978, Morphologie und Wachstumsverhalten der Provenienzen. Centralbl Gesamte Forstwes 99, 3: 169-191

McClure FA (1966) The Bamboos: a fresh perspective. Harvard Univ Press, Cambridge (Mass) (out of print)

McEvoy TJ, Sharik TL, Smith DWM (1980) Vegetative structure of an Appalachian oak forest in Southwestern Virginia. Am Midland Nat 103, 1: 96-105

McGee CE, Hooper RM (1970) Regeneration after clearcutting in the southern Appalachians. SE For Exp Station, Asheville, NC USDA For Serv Res Paper SE-70: p 12

McMahon TA (1975) The mechanical design of trees. Sci Am 233: 92-102

McMahon TA, Kronauer RE (1976) Tree structures: deducing the principles of mechanical design. Div Engineer Appl Phys, Harvard Univ, Cambridge (USA)

McPherson MP (ed) (1985) US strategy on the conservation of biological diversity. Agency Int Develop, Washington DC 20523, p 54

Meerjarenplan Bosbouw (1984) Beleidsvoornemen, Tweede Kamer der Staten-Generaal vergaderjaar 1984-1985. 18630, nrs 1-2. State Printer (Staatsdrukkerij), The Hague, p 246

Meggers BJ, Ayensu ES, Duckworth WP (eds) (1973) Tropical forest ecosystems in Africa and South America: a comparative review. Smithsonian Inst, Washington, p 350

Meidner H, Sheriff DW (1976) Water and Plants. Blackie, Glasgow, 10 + p 148

Mendoza VB (1978) Adaptability of six tree species to cogonal areas: additional information on and possible role of phenols and sugars. Sylvatrop Philipp. For Res J 3, 1: 1-7

Meulen F van der (1979) Plant sociology of the Western Transvaal bushveld, South Africa. Nijmegen Cath Univ, NL, publ in Liechtenstein, Vaduz, Gantner Verlag, Diss Bot 49: p 192 + phot

Meulen F van der, Wanders EAJ (1985) Dynamics and management of some coastal dune woodlands near The Hague, The Netherlands. Vegetatio 62: 457-465

Meyenfeldt CFWM von, Noordam D, Savenije HJF, Scheltens EB, Torren K van der, Visser PA, Voogd W de (1978) Herstel van de verwoeste binnenlandse bosgebieden in Zuid-Vietnam. Wageningen, Rapp LH (Projectgroep Vietnam), Centr Magaz, 2 vol

Miegroet M van (1970) On the photoreactivity of some hardwood species studied by their leaf characteristics. Sylva Gandavensis 23: 1-27

Miegroet M van (1976) Van bomen en bossen, 2 vol. Wet Uitg E Story-Scientia, Gent, p 1166

Mielikaeinen K (1980) Structure and development of mixed pine and birch stands. (in Finnish, Engl summ, legends). Comm Inst For Fenn 99, 3: 1-82

Mikola P (ed 1980) Tropical mycorrhiza research. Clarendon Press, Oxford, p 270

Miller RI, Stuart SN, Howell KM (1989) A methodology for analyzing rare species distribution patterns utilizing GIS technology: The rare birds of Tanzania (in press)

Miquel S (1985) Plantules et premiers stades de croissance des espèces forestières du Gabon: potentialité d'utilisation en agroforesterie. Thèse 3me cycle, Univ Pierre et Marie Curie, Paris, p 158 + app

Mitchell AW (1982) Reaching the rain forest roof: a handbook on techniques of access and study in the canopy. Leeds, Phil and Litt Soc, Nairobi, UNEP, p 36

Mitja D, Hladik A (1989) Aspects de la reconstitution de la végétation dans deux jachères en zone forestière africaine humide (Makokou, Gabon). Acta Oecol, Oecol Gener 10, 1: 75-94

Mitscherlich G (1970) Wald, Wachstum und Umwelt. I. Form und Wachstum von Baum und Bestand. Sauerländer, Frankfurt 8 + p 142 (2nd impr 1978)

Mitscherlich G (1971) Wald, Wachstum und Umwelt. II. Waldklima und Wasserhaushalt. Sauerländer, Frankfurt, 9 + p 365

Mitscherlich G (1975) Wald, Wachstum und Umwelt. III. Boden, Luft und Produktion. Sauerländer, Frankfurt, 8 + p 352

Moeller CM, Mueller D, Nielsen J (1954) Graphic presentation of dry matter production of european beech. Det Forstl Forsøgsv 21: 327-335
Mohnen VA (1988) The challenge of acid rain. Sci Am 259, 2: 14-22
Mohren GMJ (1987) Simulation of forest growth, applied to douglas fir stands in the Netherlands. Dr thesis, AUW, Wageningen, p 183
Mohren GMJ, Burg J van den, Burger FW (1986) Phosphorus deficiency induced by nitrogen input in Douglas fir in the Netherlands. Plant Soil 95: 191-200
Monasterio M (ed) (1980) Estudios ecológicos en los páramos andinos. Ed Univ Los Andes, Mérida (Venezuela), p 312
Money DC (1965; 3rd edn 1978) Climate, soils and vegetation. Univ Tutorial Press, Slough (UK), p 286
Monod J (1970) Le hasard et la nécessité. Ed du Seuil, Paris, p 197
Moorehead A (1971, 2nd printing), Darwin and the Beagle. Penguin Books, London, p 280
Morgan RPC (1979) Soil erosion. Longman, London, p 113
Mori SA et al. (1987) The Lecythidaceae of a lowland neotropical forest: La Fumée Mountain, French Guiana. Mem New York Bot Garden 44: p 190
Morris R, Morris D (1965) Men and snakes. Hutchinson, London (French transl 1965, Stock, Paris, p 224)
Morselt AFW, Smits WTM, Limonard T (1986) Histochemical demonstration of heavy metal tolerance in ectomycorrhizal fungi. Plant Soil 96: 417-420
Mueller D, Joergen Nielsen A (1965) Production brute, pertes par respiration et production nette dans la forêt ombrophile tropicale. Forstl Forsøgsv Danmark 29: 60-160
Mulder JPM (1983) A simulation of rainfall interception in a pine forest. Dr. thesis, Rijksuniv, Groningen, p 109
Mulder AJ, Waart CJM de (1984) Preliminary architectural study of Coastal Redwood *(Sequoia sempervirens)*. AUW-Silviculture, Wageningen, p 21
Mulder AJ, Waart CJM de (1985) Uitkap in een boerenbos in Tirol: een architectuurstudie. AUW-Silviculture, Wageningen 85-17: p 80
Mure V (1983) Comportement phénologique d'arbres plantés hors de leurs aires d'origine et notamment changés d'hémisphère. USTL, Montpellier, DEA-Rep, Ecol Trop, p 58 + tables
Murphy DH (1973) Animals in the forest ecosystem. In: Chuang SH (ed) Animal life and nature in Singapore. Univ Press, Singapore, pp 53-73
Murphy PG, Lugo AE (1986) Structure and biomass of a subtropical dry forest in Puerto Rico. Biotropica 18, 2: 89-96
Muséum National (ed) (1986) Vertébrés et forêts tropicales humides d'Afrique et d'Amérique. Proc Entretiens du Muséum, 1982. Mém Mus Nat Hist Nat (NS), Sér A Zool 132: p 327
Mutoji-A-Kazadi (1977) Notes de sylvigénèse pour la Guyane: transect et photographies aériennes. USTL, Montpellier, DEA-Rep, Bot Trop, p 35 + phot
Myers N (1979) The sinking ark. Pergamon Press, Oxford, p 307
Namkoong G (1981) Introduction to quantitative genetics in forestry. Castle House, Tunbridge Wells, p 342
National Academy of Sciences (1975) Productivity of world ecosytems. NAS, Washington DC, p 165
National Academy of Sciences (1979) Tropical legumes: resources for the future. NAS, Washington DC, p 331
National Academy of Sciences (1980) Firewood Crops, Shrub and Tree species for energy production. Vol 1. NAS, Washington DC, p 237
National Academy of Sciences (1983) Firewood Crops, Shrub and Tree species for energy production. Vol 2. NAS, Washington DC, p 92
Ndiaye P (1977) Préliminaire à l'étude architecturale des végétations herbacées. USTL, Montpellier, DEA Rep, p 22
Newbery DM, Alexander IJ, Thomas DW, Gartlan JS (1988) Ectomycorrhizal rain-forest legumes and soil phosphorus in Korup National Park, Cameroon. New Phytol 109: 433-450
Nieuwolt S (1977) Tropical climatology: an introduction to the climates of the low latitudes. Wiley, New York, p 207
Niklas KJ (1986) Computer-simulated plant evolution. Sci Am 254, 3: 78-86

Nooteboom HP (ed) (1987) Report of the 1982-1983 Bukit Raya expedition. Rijksherbarium, Leiden (NL), p 93
Nuesink JA (1980) Essen in de polder. RIJP-Rapp Lelystad, 1980-36 Abb: p 35 + app (in Dutch)
Odum EP (1975, 2nd edn) Ecology. Holt Rinehart and Winston, London, p 244
Odum HT, Pigeon RF (eds) (1970) A tropical rain forest. 3 vol Off Inf Serv US Atomic Energy Comm, Oak Ridge (USA)
Ødum S (1979) Actual and potential treeline in the North Atlantic region, especially in Greenland and in Faroes. Holarctic Ecol 2: 222-227
Offermans DMJ (1985) Field and aerial observations on *Piptadeniastrum africanum* (Hook f) Brenan, with special reference to ageing. AUW-Silviculture, Wageningen 86-22: p 83
Offermans DMJ (1986) Crown shyness: a parameter for ageing in *Piptadeniastrum africanum*. Neth J Agric Sci 34: 493-497
Ogawa H, Yoda K, Kira T, Ogino K, Shidei T, Ratanawongse D, Apasutaya C (1965a) Comparative ecological study of three main types of forest vegetation in Thailand. I. Structure and floristic composition. Nature Life SE Asia 4: 13-48
Ogawa H, Yoda K, Ogino K, Kira K (1965b) Comparative ecological studies on three main types of forest vegetation in Thailand. II. Plant biomass. Nature Life SE Asia 4: 49-80 + table
Ohkubo T, Kaji M, Hamaya T (1988) Structure of primary Japanese beech (*Fagus japonica* Maxim.) forests in the Chichibu Mountains, Central Japan, with special reference to regeneration processes. Ecol Res 3: 101-116
Ohsawa M (1981) A basic unit in forest community dynamics: a case study in the subalpine forest of Japan. Proc Biotrop Symp Forest Regeneration of SE Asia, Bogor, 43-62
Ohsawa M (1984) Differentiation of vegetation zones and species strategies in the subalpine region of Mt Fuji. Vegetatio 57: 15-52
Ohsawa M (ed) (1987) Life zone ecology of the Bhutan Himalaya. Chiba Univ, Chiba (Jpn), Lab Ecol, p 313
Ohsawa M, Nainggolan PHJ, Tanaka N, Anwar C (1985) Altitudinal zonation of forest vegetation on Mount Kerinci, Sumatra: with comparisons to zonation in the temperate region of east Asia. J Trop Ecol 1: 193-216
Ohsawa M, Shakya PR, Numata M (1986) Distribution and succession of West Himalayan forest types in the Eastern part of the Nepal Himalaya. Mountain Res Develop 6, 2: 143-157
Oldeman RAA (1969) Etude biologique des pinotières de la Guyane française. Cah ORSTOM, Sér Biol 10: 3-18
Oldeman RAA (1971) Un *Eschweilera* (Lecythidaceae) à pneumatophores en Guyane française. Cah ORSTOM, Sér Biol 15, 21-27
Oldeman RAA (1972) L'architecture de la végétation ripicole forestière des fleuves et criques guyanais. Adansonia (NS) 12, 2: 253-265
Oldeman RAA (1974a, 2nd edn) L'architecture de la forêt guyanaise. Mém ORSTOM, Paris 73: p 204
Oldeman RAA (1974b) Ecotopes des arbres et gradients écologiques verticaux en forêt guyanaise. Terre Vie 28, 4: 487-520
Oldeman RAA (1978) Architecture and energy exchange of dicotyledonous trees in the forest. In: Tomlinson PB, Zimmerman MH (eds) Tropical Trees as living systems, Chapter 23. Cambridge Univ Press, London, New York
Oldeman RAA (1979) Quelques aspects quantifiables de l'arborigénèse et de la sylvigénèse. Oecol Plant 14, 3: 1-24
Oldeman RAA (1980) Grondslagen van de bosteelt. AUW-Silviculture, Wageningen, p 202 (out of print)
Oldeman RAA (1981a) Schaal, grootschaligheid en kleinschaligheid in de bosbouw (On large-scale and small-scale forestry, with English summary and legends). Ned Bosbouwtijdschrift 53, 3: 71-81
Oldeman RAA (1981b) Forests, humans and computers. Vol paper XVII IUFRO World Congr, Kyoto S6.09: p 14

Oldeman RAA (ed) (1982) Tropical hardwood utilization: practice and prospects. Nijhoff/Junk, The Hague, p 584

Oldeman RAA (1983a) On rural silvicultural systems. 3 poster papers, Int Symp "Let there be Forest" (Wiersum 1984)

Oldeman RAA (1983b) The design of ecologically sound agroforests. In: Huxley P (ed) Plant research and agroforestry, Chapter 14. ICRAF, Nairobi

Oldeman RAA (1983c) Tropical rain forest, architecture, silvigenesis and diversity. In: Sutton SL et al (eds) Tropical rain forest: ecology and management. Blackwell, Oxford, p 498

Oldeman RAA (1985) Choix d'espèces pour une forêt saine et stable. In: Oswald M (ed) Impacts de l'Homme sur la forêt (Symp Int IUFRO, Strasbourg 1984). Colloq INRA, Paris 30: 185-201

Oldeman RAA (1986) On the limits of forest ecosystems: do they have skins? Proc 18th IUFRO World Congr, Ljubljana, Div 6: 312-324

Oldeman RAA (1989) Dynamics in tropical rain forests. In: Holm-Nielsen LB, Nielsen IC, Balslev H (eds) Tropical forests: botanical dynamics, speciation and diversity: 3-21. Academic Press, London

Oldeman RAA (1989a, in press) Biological implications of leguminous tree architecture. Proc 2nd Int Conf Biol Leguminosae, St Louis (June 1986)

Oldeman RAA (1989b) Forest ecosystems and their components: an introduction. In: Oldeman RAA, Schmidt P, Arnolds EJM (eds) AUW-Papers, Wageningen, PUDOC

Oldeman RAA, Dijk J van (1988) Tree characteristics, sylvigenesis and architectural diversity. In: Hadley M (ed) Rain forest regeration and management. Biol Int, Spec Vol 18: 18-22

Oldeman RAA, Dijk J van (in press) Diagnosis of the temperament of tropical rain forest trees. In: Gomez-Pompa A (ed) title unknown (Guri Workshop), UNESCO, Paris

Oldeman RAA, Fundter JM (1986) On Diptero-Vochysian tree strategies. Naturalia Monspeliensia, Spec Issue, Colloq Int Arbre, pp 191-208

Oldeman RAA, Meer P van der (1990) Diagnosis of rain forests. (Provisional title) (in preparation)

Oldeman RAA, Westra JJ (1980) De betekenis van hout in de oecologie van boom en bos. Ned Bosbouwtijdschrift 52, 2: 39-46

Oldeman RAA, Westra JJ, Tenge OR (1983) Bosontwikkelingen, natuurwaarde en transectanalyse (Engl. summ. + fig. legends). Ned Bosbouwtijdschrift 55, 6: 242-257 (in Dutch)

Oldeman RAA, Gremmen J, Goudzwaard L (1985) Naar teeltdiagrammen voor groveden (Towards silvicultural diagrams for Scots pine; Engl. summ.). Ned Bosbouwtijdschrift 57, 3: 65-74

Oliver CD (1978) The development of Northern Red Oak in mixed stands in Central New England. Yale Univ, School For Environ Stud, New Haven (US), Bull 91: p 63

Oliver CD (1980) Even-aged development of mixed-species stands. J For 78, 4: 201-203

Oliver CD, Stephens EP (1977) Reconstruction of a mixed-species forest in Central New England. Ecology 583: 562-572

Oosterhuis L (1979) Nature conservation: coppice woods. AUW-Nat Conserv, Wageningen, p 41

Oosterhuis L, Oldeman RAA, Sharik TL (1982) Architectural approach to analysis of North American temperate deciduous forests. Can J For Res 12, 4: 835-847

Opdam P, Dorp D van, Braak CJF ter (1984) The effect of isolation on the number of woodland birds in small woods in the Netherlands. J Biogeogr 11: 473-478

Orians GH (1982) The influence of tree-falls in tropical forests on tree species richness. Trop Ecol 23, 2: 255-279

Orozco-Segovia A, Vazquez-Yanes C (1982) Plants and fruit bat interactions in a tropical rain forest area, southeastern México. Brenesia 19/20: 137-149

Orozco-Segovia A, Vazquez-Yanes C, Coates-Estrada R, Perez-Nasser N (1987) Ecophysiological characteristics of the seed of the tropical forest pioneer *Urera caracasana* (Urticaceae). Tree Physiol 3: 375-386

Oswald M (ed) (1985) Impacts de l'homme sur la forêt. (texts in Fr., Engl. & Germ.). Proc IUFRO Symp Strasbourg (1984), Ed Colloq INRA, Paris, ISBN 2-85340-661-X, 30: p 512

Ovington JD (1965) Organic production, turnover and mineral cycling in woodlands. Biol Rev Cambridge Phil Soc 40

Pas JB van der (1981) A statistical appraisal of *Armillaria* root rot in New Zealand plantations of *Pinus radiata*. NZ J For Sci 11, 1: 23-36
Pascal J-P (1983) Les forêts denses humides sempervirentes de basse et moyenne altitudes du Sud de l'Inde. Thèse d'Etat, Univ Paul Sabatier, Toulouse, p 553 + app
Paterson SS (1956) The forest area of the world and its potential productivity. Diss., Göteborg (Sweden), p 216 + (now historical!) forest map
Payette S, Filion L, Gauthier L, Boutin Y (1985) Secular climate change in old-growth tree-line vegetation of northern Quebec. Nature (Lond) 315, 6015: 135-138
Pearson DL, Derr JA (1986) Seasonal patterns of lowland forest floor arthropod abundance in Southeastern Perú. Biotropica 18, 3: 244-256
Pel M (1983) Successie na ontbossing in het tropisch regenwoud van Suriname. AUW-Vegetation Sci, Wageningen, p 47 + app
Pelleboer J (1983) Weerboek; extreme weersituaties van de laatste twee eeuwen in Nederland. Publiboek, Hilversum, p 152 (in Dutch)
Pereira JS, Kozlowski TT (1977) Variations among woody Angiosperms in response to flooding. Physiol Plant 41: 184-192
Pesson P (ed) (1974) Ecologie forestière. Gauthier-Villars, Paris, p 382
Peterken GF, Jones EW (1987) Forty years of change in Lady Park Wood. I. The old growth stands. J Ecol 75: 477-512
Peters R (1985) Japanese beech forest. AUW-Silviculture, Wageningen, 85-16, p 81
Philippe M (1988) Test d'un diffusiomètre haute-définition en tant que moyen d'investigation de la forêt guyanaise. Mém DEA Biol Vég Trop, Univ Paris VI, p 61
Philippona J, Kalkhoven J, Opdam P (1983) De betekenis van vegetatiekenmerken voor bosvogelgemeenschappen. Het Vogeljaar (Spec Issue "birds and forest structure") 31, 2: 74-88
Philipson WR, Ward BM, Butterfield BG (1971) The vascular cambium, its development and activity. Chapman and Hall, London, p 182
Pianka ER (1974) Evolutionary ecology. Harper and Row, New York, p 356
Pickett STA, Thompson JN (1978) Patch dynamics and the design of nature reserves. Biol Conserv 13: 27-37
Pickett STA, White PS (eds) (1985) The ecology of natural disturbance and patch dynamics. Academic Press, Orlando, p 472
Pijl LJ van der (1969) Principles of dispersal in higher plants. Springer, Berlin Heidelberg New York, p 154
Poels RLH (1987) Soils, water and nutrients in a forest ecosystem in Suriname. PUDOC, Wageningen, p 253
Prescott-Allen R, Prescott-Allen C (1983) Genes from the wild. Earthscan Paperback, Publ by IIED, London, p 101
Prévost MF (1981) Mise en évidence de graines d'espèces pionnières dans le sol de la forêt primaire en Guyane. Turrialba 31, 2: 121-127
Prévost MF (1983) Les fruits et les graines des espèces végétales pionnieres de Guyane Française. Rev Ecol, Terre Vie 38: 121-145
Prigogine I, Stengers I (1985, transl. from French 1979) Order out of chaos. Fontana Paperbacks, Flamingo Books, London, p 349
Pritchett WL (1979) Properties and management of forest soils. Wiley, New York, p 55
Pruša E (1985) Die böhmischen und mährischen Urwälder. Academia Praha, Prague, Vegetace CSSR A15, p 578
Pruša E (1987) Follow-up of the dynamics of natural forests from Pruša 1985 over the period 1975-1985. (in Czechoslovakian without transl but illustr). Vytiskl Lesprojekt, Prague, p 32 + 43
PUDOC (1982, 2nd edn) Bosbescherming. PUDOC, Wageningen, p 385
Putz FE, Chai P (1987) Ecological studies of lianas in Lambir National Park, Sarawak, Malaysia. J Ecol 75: 523-531
Putz FE, Chan HT (1986) Tree growth, dynamics and productivity in a mature mangrove forest in Malaysia. For Ecol Management 17: 211-230
Raad H (1981) Bor-forest: groeiplaats, soorten en fysionomie van een savanne in Zuid-Soedan. AUW-Silviculture, Wageningen, p 52 + app

Rabbinge R, Mohren GMJ (1984) Potentiële en actuele groei van een bos. Ned Bosbouwtijdschrift 56, 4: 83-91 (in Dutch)

Raffe JK van, Roelvink M (1986) Douglas-plantverbandonderzoek. AUW-Silviculture, Wageningen 86-14: p 56; also see English summary in Fanta (1986)

Rasanen ME, Salo JS, Kalliola RJ (1987) Fluvial perturbance in the Western Amazon Basin: regulation by long-term sub-Andean tectonics. Science 238: 1398-1401

Reader J, Croze H (1977) Pyramids of life. Lippincott, Philadelphia, p 222

Rees KCJ van, Comerford NB (1986) Vertical root distribution and strontium uptake of a slash pine stand on a Florida spodosol. Soil Sci Soc Am J 50, 4: 1042-1046

Reffye Ph de (1979) Modélisation de l'architecture des arbres par des processus stochastiques. Thèse d'Etat, Univ Paris-Sud, Centre d'Orsay, Paris, p 194

Reffye Ph de, Edelin C, Jaeger M, Cabart C (1986) Simulation de l'architecture des arbres. Naturalia Monspeliensia, Spec Issue, Colloq Int Arbre, pp 223-240

Reffye Ph de, Edelin C, Jaeger M (1989) La modélisation de la croissance des plantes. Recherche 207: 158-168

Reichle DE (ed) (1970) Analysis of temperate forest ecosystems. Springer, Berlin Heidelberg New York, p 304

Reijneveld A (1981) Een stage in Zuid-India. AUW-Silviculture, Wageningen, Mimeogr, p 46

Reitsma JM (1988) Végétation forestière du Gabon/Forest vegetation of Gabon (bilingual). Tropenbos Foundation, Tech Ser, Ede (NL) 1: p 142

Reuver E, Zee F van der (1984) De boommarter *(Martes martes)*: een literatuuroverzicht. AUW-Silviculture, Arnhem, RIN/Wageningen 86-04b, p 67

Richard JF (1989) Le paysage: un nouveau langage pour l'étude des milieux tropicaux. Init Doc Techn 72: p 210 + app. ORSTOM Paris

Richards BN (1974) Introduction to the soil ecosystem. Longman, London, p 266

Richards PW (1952) The tropical rain forest. Univ Press, Cambridge, p 450

Richards PW (1970) The life of the jungle. McGraw-Hill, New York, p 232

Ricklefs RE (1976) The economy of nature. Concord, Chiron, New York, p 455

Riéra B (1983) Chablis et cicatrisation en forêt guyanaise. Thesis, Univ Paul Sabatier, Toulouse, p 167

Rijksen HD (1978) A field study on Sumatran orang utans: ecology, behaviour and conservation. Meded Landbouwuniv, Wageningen 78, 2: p 421

Rollet B (1974) L'architecture des forêts denses humides sempervirentes de plaine. CTFT, Nogent/s/Marne (France), p 298

Rollinson TJD, Evans J (1987) The yield of sweet chestnut coppice. Her Majesty's Stat Off, London, For Comm Bull 64: p 20

Roloff A (1985) Morphologie der Kronenentwicklung von *Fagus sylvatica* L. (Rotbuche) unter besonderer Berücksichtigung möglicherweise neuartiger Veränderungen. Ph. D. Thesis, Georg-August Univ, Göttingen, p 178

Roosmalen MGM van (1980) Habitat preferences, diet, feeding strategy and social organization of the Black Spider Monkey (*Ateles paniscus paniscus* Linnaeus 1758) in Surinam. D. Sc. thesis, Agric Univ, Wageningen; RIN-Rapp, Leersum 80/13: p 175

Roosmalen MGM van (1985) Fruits of the Guianan flora. RUU-Inst Syst Bot, Utrecht, AUW-Silviculture, Wageningen, ISBN 9-900987-4, p 483

Root R (1967) The niche exploitation pattern of the blue-grey gnatcatcher. Ecol Monogr 37: 317-350

Rosenthal GA, Janzen DH (1985) Ammonia utilization by the Bruchid beetle, *Caryedes brasiliensis* (Bruchidae). J Chem Ecol 11, 4: 539-544

Ross J (1981) The radiation regime and architecture of plant stands. Junk, The Hague, p 391

Roussel L (1972) Photologie forestière. Masson, Paris, p 144

Roussel L (1983) Formation éventuelle de certains gourmands d'origine cambiale. Acad Agric France, Proc Verb Séance 27 avril 1983, pp 572-574

Rowland FS (1989) Chlorofluorocarbons and the depletion of stratospheric ozone. Am Sci 77, 1: 36-45

Russell AE, Ewel JJ (1985) Leaching from a tropical andept during big storms: a comparison of three methods. Soil Sci 139, 2: 181-189

Sabatier D (1983) Fructification et dissémination en forêt Guyanaise: l'exemple de quelques espèces ligneuses. Thèse 3me Cycle, USTL, Montpellier, Vol 1: 238 + XX p, Vol II: 380 ill

Salati E, Vose PB (1984) Amazon basin: a system in equilibrium. Science 225, 4658: 129-138

Salo J (1987) Pleistocene forest refuges in the Amazon: evaluation of the biostratigraphical, litostratigraphical and geomorphological data. Ann Zool Fenn 24: 203-211

Sarmiento G, Goldstein G, Meinzer F (1985) Adaptive strategies of woody species in neotropical savannas. Biol Rev 60: 315-355

Sattler R (ed) (1982) Axioms and principles of plant construction. Nijhoff/Junk, The Hague, p 186

Scarrone F (1969) Recherches sur les rythmes de croissance du Manguier et de quelques végétaux ligneux malagasy. D. Sc. thesis, Univ Clermont-Ferrand

Schemske DW, Brokaw N (1981) Treefalls and the distribution of understory birds in a tropical forest. Ecology 62, 4: 938-945

Schermbeek A. J. van (1898) preface to his Dutch translation of Gayer: Het bosch (de leer der bestanden) Uitg Wees, Breda, p 668 (in Dutch)

Schilz H, Bryer J, Corcoran T (1982) Five spruce budworm profiles for Maine - a visual presentation. Univ Maine, For Engineer Misc Ser, Orono, p 19

Schimper AFW (1903) Plant geography upon a physiological basis. (Engl. transl. by WR Fischer). Clarendon, Oxford

Schmidt P (1977) Wachstum und Wasserverbrauch von Bäumen: methodische Experimente und Untersuchungen an Pappelklonen. Dr. thesis, Albert-Ludwigs Univ, Freiburg, p 217

Schmidt P (1980) Bomen over groei van bomen. In: Oldeman RAA (1980) Grondslagen van de Bosteelt: Chapter IV. AUW-Silviculture, Wageningen, p 202 (out of print)

Schmidt P (in prep) Phytomass and nutrient cycling in managed tropical rain forest in Suriname. (Provisional title) AUW, Wageningen, Paramaribo UvS, Ser "Ecology and Management of Tropical Rainforest in Surinam". PUDOC

Schnell R (1950) La forêt dense. Lechevalier, Coll Manuels ouest-africains, Paris 1: p 330

Schreuel I, Stegeman GT (1986) Growth of *Acacia auriculiformis* in Yogyakarta Province, Indonesia. 2nd Int Legume Conf, St Louis, Poster Paper 77, Abstr, p 64

Schroeder TO (1977) Floristiek en opbouw van ondergroei in tropisch regenwoud. AUW-Silviculture, Wageningen, p 50

Schubart HOR, Walker I (1987) The dynamics of the Amazonian terra-firme forest. Experientia, Spec Issue with Multiauthor Rev 43, 3: 221-289

Schulz JP (1960) Ecological studies on rain forest in Northern Surinam. Verh Kon Ned Akad Wetensch (Afd Natuurk) Amsterdam, Reeks 2, 53, 1: p 267

Schulze ED (1970) Der CO_2-Gaswechsel der Buche *(Fagus sylvatica)* in Abhängigkeit von den Klimafaktoren im Freiland. Flora 159: 177-232

Schuurman JJ, Goedewagen MAJ (1965) Methods for the examination of root systems and roots. PUDOC, Wageningen, p 86

Sharifi MR, Rundel PW (1986) Seasonal productivity and phenology of a nitrogen-fixing and a non-nitrogen-fixing legume coexisting in the Sonoran desert. 2nd Int Legume Conf, St Louis, Poster Paper 79, Abstr, p 65

Shepherd KR, Richter HV (eds) (1985) Managing the tropical forest. Austr Nat Univ, Canberra, p 341

Shimazu M, Karagiri (1981) Pathogens of the pine sawyer, *Monochamus alternatus* Hope, and possible utilization of them in a control program. Proc XVII IUFRO World Congr, Ibaraki, Div 2: 275-280

Shinozaki K, Joda K, Hozumi K, Kira T (1964) A quantitative analysis of plant form - the pipe model theory. I Jpn J Ecol 14, 3: 97-105; II, Jpn J Ecol 14, 4: 133-139

Shugart HH (1984) A theory of forest dynamics: the ecological implications of forest succession models. Springer, Berlin Heidelberg New York Tokyo, p 278

Shugart HH, Seagle SW (1985) Modeling forest landscapes and the role of disturbances in ecosystems and communities. In: The Ecology of natural disturbance and patch dynamics. Academic Press, New York, pp 353-455

Shugart HH, West DC (1977) Development of an Appalachian deciduous forest succession model and its application to assessment of the impact of the chestnut blight. J Environ Management 5: 161-179

Shugart HH, West DC (1979) Size and pattern of simulated forest stands. For Sci 25, 1: 120-122
Shugart HH, Hopkins MS, Burgess IP, Mortlock AT (1980) The development of a succession model for subtropical rain forest and its application to assess the effects of timber harvest at Wiangaree State Forest, New South Wales. Environ Management 11: 243-265
Si-Jiu Liu, Tillberg E (1984) Sensitivity to phytohormones determined by outer-inner polarity of higher plants: an overall model for phytohormone action. Plant Cell Environ 7: 75-80
Simon JJ (1986) Disappearing species, deforestation and data. New Sci May 1986: 60-63
Six Dijkstra HG (1981) Architecture and age structure of terrace rimu forest at Saltwater, South Westland, New Zealand and the effect of selective logging by the parallel hauling method. M. Sc. thesis, Univ Canterbury (NZ), p 155
Six Dijkstra HG, Mead DJ, James IL (1985) Forest architecture in terrace rimu forest of Saltwater Forest, South Westland, and its implications for management. NZ J For Sci 15, 1: 3-22
Smathers GA (1978) Balsam woolly aphid. Asheville, NC, Research/Resources Inform Bull Blue Ridge Parkway 1: p 2
Smith DC, Douglas AE (1987) The biology of symbiosis. Arnold, London, p 302
Smits WThM (1982) Dipterocarpaceae: vegetative propagation and mycorrhiza. AUW-Silviculture, Wageningen, p 75
Smits WThM (1984) Gezonde mycorrhiza, gezonde bossen. Ned Bosbouwtijdschrift 56, 4: 92-96
Smits WThM, Oldeman RAA, Limonard T (1987) Mycorrhizae and Dipterocarpaceae in East Kalimantan rain forest. WOTRO-Yearb, The Hague 1986: 67-78
Smolander H (1984) Measurement of fluctuating irradiance in field studies of photosynthesis. Acta For Fenn 187: p 56
Smolander H, Lappi J (1985) Integration of a nonlinear function in a changing environment: estimating photosynthesis using mean and variance of radiation. Agric For Meteor 34: 83-91
Sommer WH, Tuexen R (1974) Tatsachen und Probleme der Grenzen in der Vegetation. Cramer, Lehre
Sonderen L van (1986) Studying root-systems with perforated soil systems. AUW-Silviculture, Wageningen 86-13: p 25
Spedding CRW, Walsingham JM, Hoxey AM (1981) Biological efficiency in agriculture. Academic Press, London, p 383
Sprent JI, Sutherland JM, Faria SM de (1986) Structure and function of root nodules from woody Legumes. 2nd Int Legume Conf, St Louis, Invited Paper, Abstr, p 23
Sprugel DG (1976) Dynamic structure of wave-regenerated *Abies balsamea* forests in the North-Eastern United States. J Ecol 64, 3: 889-911
Steele R, Pfister RD, Ryker RA, Kittams JA (1981) Forest habitat types of Central Idaho. USDA For Serv, Intermountain For Range Exp Stat, Gen Techn Rep, Ogden (Utah, USA) INT-114: p 138 + maps, phot
Steenis CGGJ van (1956) Basic principles of rain forest sociology; and: Rejuvenation as a factor for judging the status of vegetation types: biological nomad theory. In: UNESCO, Actes Coll Kandy, pp 159-165; 212-218 (see UNESCO 1958)
Sterringa JT (1971) Een onderzoek naar de vegetatiestructuur van recent geexploiteerd drooglandbos in het Mapanegebied, Suriname. AUW-Silviculture, Wageningen, p 50 + transects
Stevens PS (1974) Patterns in nature. Little Brown, Boston, p 240
Stocki J (1981) Preliminary study on insect life-spaces in Oostereng forest. AUW-Silviculture, Wageningen, p 45
Storer TI, Usinger RL (1955) Elements of zoology. McGraw-Hill, New York, p 552
Suselo TB (1983) Autoécologie et développement de quatre espèces de bois commerciaux de la forêt de Sumatera, Indonésie. Ph. D. Thesis, USTL, Montpellier, p 101
Susmel L (1979) Modello strutturale di abetine e peccete primarie (in Ital., Engl. summ.). Atti Inst Ecol Selvicoltura, Univ Degli Studi, Padova 1, 1: p 48
Sutton SL, Whitmore TC, Chadwick AC (eds) (1983) Tropical rain forest: ecology and management. Blackwell, Oxford, p 498
Swart B (1987) Bijdrage aan de totstandkoming van teeltdiagrammen voor de grove den (*Pinus sylvestris* L.) in Nederland. AUW-Silviculture For Ecol, Wageningen, 87-07

Swift MJ (ed) (1986) Tropical soil biology and fertility (TSBF) interregional planning workshop. IUBS/MAB-UNESCO. Biol Int Spec Issue 13: p 68

Szujecki A (ed) (1983) The process of forest soil macrofauna formation after afforestation of farmland. Agric Univ Press, Warsaw, p 196

Szujecki A (1987) Ecology of forest insects. Junk, Dordrecht, p 342

Tamis WLM (1986) Nematoden in een ammonium-depositiegradient in een grovedennenbos. Hydrobiol Adviesburo Klink bv Rapp Meded, Wageningen 23: p 14 + app

Temple A (1975) Ericaceae: étude architecturale de quelques espèces. USTL, Montpellier, DEA-Rep, Biol Végét, p 95

Terborgh J, Stern M (1987) The surreptitious life of the saddlebacked Tamarin. Am Sci 75, 3: 260-269

Temorshuizen AJ, Schaffers AP (1987) Occurrence of carpophores of ectomycorrhizal fungi in selected stands of *Pinus sylvestris* in the Netherlands in relation to stand vitality and air pollution. Plant Soil 104: 209-217

Tol G van (1979) Natuurlijke verjonging van beuk op de Veluwe. (with Engl. summary and legends). Ned Bosbouwtijdschrift 51, 4: 106-112

Tomlinson PB (1986) The botany of mangroves. Cambridge Univ Press, London, p 413

Tomlinson PB, Zimmermann MH (eds) (1978) Tropical trees as living systems. Univ Press, Cambridge, p 675

Torquebiau EF (1979) The reiteration of the architectural model: a demographic approach. USTL, Montpellier, DEA-Rep, p 51

Torquebiau EF (1981) Analyse architecturale de la forêt de Los Tuxtlas (Veracruz), Mexique. Thèse 3me Cycle, Univ Montpellier, p 185

Torquebiau EF (1984) Man-made Dipterocarp forest in Sumatra. Agrofor Syst 2: 103-127

Torquebiau EF (1985) The tropical rain forest mosaic pattern with special reference to Indonesia. Biotrop, Bogor, p 68

Torquebiau EF (1986) Mosaic patterns in dipterocarp rain forest in Indonesia, and their implications for practical forestry. J Trop Ecol 2 (proofs without pp)

Trochain J-L (1980) Ecologie végétale de la zone intertropicale non-désertique. Univ Paul Sabatier, Toulouse, p 468

Tromp J, Ovaa JC (1969) The effect of nitrogen application on the seasonal variations in the amino acid composition of the xylem sap of apple. Z Pflanzenphysiol 60: 232-241

Turenne J-F (1977) Modes d'humification et différenciation podzolique dans deux toposéquences guyanaises. Mém ORSTOM, Paris 84: p 173

Turin H, Boer JP den (1988) Changes in the distribution of Carabid beetles in the Netherlands since 1880. II. Biol Conserv 44: 179-200

Tweel PA van den (1979) Methoden voor wortelonderzoek, getoetst aan *Castanea sativa* zaailingen. AUW-Silviculture, Wageningen, p 36

Tweel PA van den (1980) De plantgatenwoeler en de beworteling van de douglas. unpublished manuscript

Tweel PA van den, Boer WMJ den (1986) De vitaliteit van de Nederlandse bossen: situatie, oorzaken en onderzoek. Ned Bosbouwtijdschrift 58, 4: 82-90

Ulrich B, Pankrath J (eds) (1983) Effects of accumulation of air pollutants in forest ecosystems. Reidel, Dordrecht, p 389

UNEP (1987) The changing atmosphere: UNEP Environment Brief No 1. Wallaceana 48/49: 9-16

UNESCO (1958) Study of tropical vegetation: proceedings of the Kandy symposium. UNESCO, Paris, p 226

UNESCO (1974) Le sahel: bases écologiques de l'aménagement. (available in English) UNESCO, Paris, MAB Tech Notes, p 99

UNESCO (1979) Map of the world distribution of arid regions. (available in French). UNESCO, Paris, MAB Tech Notes 7: p 54

UNESCO/UNEP/FAO (1978) Tropical forest ecosystems. A state-of-knowledge report. UNESCO, Paris, p 683

Valkeman G (1985) Contribution to the study of a savanna protected from fire in Central Ivory Coast. The structural development of the woody vegetation during the first 21 years of protection. AUW-Silviculture, Wageningen 85-21: p 50

Vandeursen J, Wisse J (1985) De invloed van fluviatiele dynamiek op de groei en structuur van een natuurlijk Fraxino-Ulmetum in de Elzas. RIN Leersum, AUW-Silviculture, Wageningen 85-22: p 82

Vazquez-Yanes C, Orozco-Segovia A (1986) Dispersal of seeds by animals: effect of light controlled dormancy in *Cecropia obtusifolia*. In: Estrada A, Fleming TH (eds) Frugivores and seed dispersal. Junk, The Hague, chapter 6

Vazquez-Yanes C, Orozco-Segovia A (1987a) Fisiología ecológica de semillas en la Estación de Biología Tropical "Los Tuxtlas", Veracruz, México. Rev Biol Trop 35 (Suppl 1): 85-96

Vazquez-Yanes C, Orozco-Segovia A (1987b) Light gap detection by the photoblastic seeds of *Cecropia obtusifolia* and *Piper auritum*, two tropical rain forest trees. Biol Plant 29, 3: 234-236

Verhulst PF (1844) Recherches mathématiques sur la loi d'accroissement de la population. Mém Acad R Bruxelles 18: 1-58

Verkerke W (1986a) The fig life cycle. Handout and poster paper, Symp Trop Biol, Rijksuniv Utrecht

Verkerke W (1986b) Anatomy of *Ficus ottoniifolia* (Moraceae) syconia and its role in the fig-wasp symbiosis. Proc K Ned Acad Wetensch, Amsterdam

Vervelde GJ (ed) (1975) Produktiviteit in biologische systemen. PUDOC, Wageningen, p 312

Vliet I van (1979) In: Landgoed Verwolde: de relatie landbouw-natuur in een cultuurlandschap met veel houtwallen. Part II: appendix. AUW, Wageningen, Project Group For, Scale Soc (Projectgroep Bos, Schaal en Samenleving)

Vogel EF de (1979) Seedlings of Dicotyledons. PUDOC, Wageningen, p 203

Vooren AP (1979) Essai sur la voûte forestière et sa régénération: analyse structurale et numérique d'une toposéquence en forêt de Tai, Côte d'Ivoire. ORSTOM-Adiopodoumé, Abidjan, AUW-Silviculture, Wageningen

Vooren AP (1985) Patterns in tree and branch-fall in a West African rain forest. AUW-Silviculture, Wageningen 85-05: p 34

Vooren AP (1986) Nature and origin of tree and branch fall in the Tai Forest (Ivory Coast). Neth J Agric Sci 34: 112-115

Vooren AP (1987) Development versus forest conservation: avoiding a conflict in the Tai-region (Ivory Coast). In: Van Beusekom et al (eds) (1987) Report 85-05. AUW-Silviculture

Vooren AP (in prep) Nature-analog forestry: a silvicultural basis. Tropenbos, Sci Ser

Vooren AP, Offermans DMJ (1985) An ultra-light aircraft for low-cost, large-scale stereoscopic aerial photographs. Biotropica 17: 84-88

Vredenbregt JJ, Winckel RJM van de (1982) Bos, biologische rijkdom, aanplant en beheer: voorstellen omtrent aanleg en verder beheer van bos op ecologische grondslag, op jonge klei en zware zavelgronden (Z Flevoland). Lelystad, Min Verkeer Waterstaat, RIJP 1982-50, vol I, p 166; vol II, ill

Vuure T van (1985) Zoogdieren, bossen en wederzijdse invloeden. PUDOC, Wageningen, p 135

Walls J (1980) Land, man and sand: desertification and its solution. MacMillan, New York, p 336

Walter H (1968) Die Vegetation der Erde, II. gemäßigten u. arktischen Zonen. Fischer, Stuttgart, p 1001

Walter H (1973, 3rd edn) Die Vegetation der Erde. I. tropischen u. subtropischen Zonen. Fischer, Stuttgart p 743

Walter J-MN (1974) Arbres et forêts alluviales du Rhin. Bull Soc Hist Nat Colmar 55: 37-88

Walter J-MN (1982) Architectural profiles of flood-forests in Alsace. In Dierschke H (ed) Struktur und Dynamik von Wäldern. Cramer, Vaduz, pp 187-234

Watt AS (1947) Pattern and process in the plant community. J Ecol 13: 1-22

Webb LJ (1958) Cyclones as an ecological factor in tropical lowland rainforest, North Queensland. Austr J Bot 6, 3: 220-228

Webb LJ (1959) A physiognomic classification of Australian rain forests. J Ecol 47: 551-570

Webb LJ (1977) The dynamics of tropical rainforests. For Log, Austr Nat Univ Canberra 10: 11-18

Webb LJ, Tracey JG (1981) Australian rainforests: pattern and change. In: Keast A (ed) Ecological biogeography of Australia. Junk, The Hague, Chapter 22

Webb LJ, Tracey JG, Haydock KP (1967) A factor toxic to seedlings of the same species associated with living roots of the non-gregarious subtropical rain forest three *Grevillea robusta*. J Appl Ecol 4: 13-25

Webb LJ, Tracey JG, Williams WT (1972) Regeneration and pattern in the subtropical rain forest. J Ecol 60: 675-695

Webb LJ, Tracey JG, Williams WT (1976) The value of structural features in tropical forest typology. Austr J Ecol 1: 3-28

Weeda EJ, Westra R, Westra Ch, Westra T (1985, 1987, 1988) Nederlandse oecologische flora: wilde planten en hun relaties. Inst Natuurbeschermingseducatie, Amsterdam, vol 3

Weidelt H-J (1984) Die Auswirkung waldbäuerlicher Pflegemaßnahmen auf die Entwicklung exploitierter Dipterocarpaceen-Wälder. Habilitationsschrift, Georg-August-Univ, Göttingen, p 166

Weinstock JA (1983) Rattan: ecological balance in a Borneo rainforest swidden. Econ Bot 37, 1: 58-68

Werger MJA, Westhoff V (1985) Systeemoecologie, structureel. In: Bakker et al. (ed) Inleiding tot de oecologie. Bohn, Scheltema and Holkema, Utrecht, pp 283-322

West BJ, Goldberger AL (1987) Physiology in fractal dimensions. Am Sci 75, 4: 354-365

Westhoff V, Held AJ den (1969) Plantengemeenschappen in Nederland. Thieme, Zutphen, p 324

Wheeler Q, Blackwell M (eds) (1984) Fungus-insect relationships: perspectives in ecology and evaluation. Columbia Univ Press, New York, p 514

Wheelwright NT, Orians GH (1982) Seed dispersal by animals: contrasts with pollen dispersal, problems of terminology, and constraints on coevolution. Am Nat 119, 3: 402-413

Whitehead D (1978) The estimation of foliage area from sapwood basal area in Scots pine. Forestry 51, 137-149

Whitmore TC (1974) Change with time and the role of cyclones in tropical rain forest on Kolombangara, Solomon Islands. Commonw For Inst, Oxford, Inst Paper 46: p 78

Whitmore TC (1975, 2nd ed 1985) Tropical rain forests of the Far East. Clarendon, Oxford, p 282

Whitmore TC (1982) On pattern and process in forests. In: Newman EI (ed) The plant community as a working mechanism. Blackwell, Oxford, pp 45-59

Whittaker RH (1970) Communities and ecosystems. Macmillan, London

Wiegers J (1982) Untersuchungen zum Verhalten von *Betula pubescens* Ehrh. in Mooren der Niederlande, I. In: Dierschke H (ed) Struktur und Dynamik von Wäldern. Cramer, Vaduz, pp 275-297

Wiegers J (1985) Succession in fen woodland ecosystems in the Dutch haf district. Cramer, Vaduz, p 152 + tables

Wiersum KF (ed) (1984) Strategies and designs for afforestation, reforestation and tree planting. PUDOC, Wageningen, p 432

Wijngaarden W van (1985) Elephants-trees-grass-grazers. ITC, Enschede (NL) 4: p 159, with maps courtesy Kenya Soil Survey, PO Box 14733, Nairobi

Wilderink EC, Zakee RA (1985) Aspecten van de vegetatie op de Wageningse berg en bergrand, 2 vol. AUW-For Management/Vegetation Sci/Silviculture For Ecol, Wageningen 85-13: p 118 + figs

Wilson EO (ed) (1974) Ecology, evolution, and population biology. Readings from Scientific American. Freeman, San Francisco, p 319

Winckel R van de (1980) Het Wyhlerwald, architectuur van een Rijnoeverbos. AUW-Silviculture, Wageningen

Winckel R van de (1986) Energiewallen: handleiding voor de teelt van brandhout. Uitg De Twaalf Ambachten, Boxtel (NL), (De Bleken 2, 5282 HB Boxtel), p 71

Wingfield JC, Ball GF, Dufty Jr, AM, Hegner RE, Ramenofsky M (1987) Testosterone and aggression in birds. Am Sci 75, 6: 602-608

Wit CT de et al (1978) Simulation of assimilation, respiration and transpiration of crops. PUDOC, Wageningen, p 141

Woell H (1988) Struktur und Wachstum von kommerziell genutzten Dipterocarpaceenmischwäldern ... in den Philippinen. Diss. Dr., Hamburg Univ, p 264

Wolf EC (1987) On the brink of extinction: conserving the diversity of life. Worldwatch Inst, Worldwatch Paper, Washington DC 78: p 54

Yamada I (1976) Forest ecological studies of the montane forest of Mt Pangrango, West Java. III. Litter fall of the tropical montane forest near Cibodas. SE Asian Stud 14, 2: 194-229

Yoda K, Kira T (1969) Comparative ecological studies on three main types of forest vegetation in Thailand. V. Accumulation and turnover of soil organic matter with notes on the altitudinal soil sequence on Khao (Mt.) Luang, peninsular Thailand. Nature Life SE Asia 6: 83-112

Young A (1976, 2nd edn. 1980) Tropical soils and soil survey. Univ Press, Cambridge, p 468

Zadelhoff FJ van (1976) Kringlopen van bio-elementen en organische stof in tropische regenwouden. AUW-Silviculture, Wageningen (tropical) p 89 + tables

Zakaria-Ismail M (1988) The theory of island biogeography: its application in conservation. Wallaceana 51: 8-10

Zandt G van der, Sterck F (1986) Growth dynamics of *Acia albida*. 2nd Int Legume Conf St Louis, Poster Paper 89, Abstr, p 88

Zimmermann MH (ed.) (1964) The formation of wood in forest trees. Academic Press, New York, p 652

Zimmermann MH (1983) Xylem structure and the ascent of sap. Springer, Berlin Heidelberg New York Tokyo, p 143

Zimmermann MH, Brown CL (1971, 3rd edn 1977) Trees, structure and function. Springer, Berlin Heidelberg New York, p 336

Zukrigl K, Eckhart G, Nather J (1963) Standortskundliche und waldbauliche Untersuchungen in Urwaldresten der niederösterreichischen Kalkalpen. Mitt Forstl Bundes-Versuchsanst Mariabrunn, Schönbrunn, 62: p 245 + maps + tables

Index

This index has been carefully made into an encyclopedial entry into the subject matter of this book. Page numbers in **bold face** indicate definitions, figures and sometimes information judged to be essential otherwise. Page numbers in low cast point to other pick-up points in the text. Page numbers between parentheses ( ) show pages where the item is not mentioned by name, but by a synonym or by some statement referring to it. The additional use of cross-references is recommended.

B

C

F

Q

R

T

MIX
Papier aus verantwortungsvollen Quellen
Paper from responsible sources
FSC® C105338

If you have any concerns about our products,
you can contact us on
ProductSafety@springernature.com

In case Publisher is established outside the EU,
the EU authorized representative is:
Springer Nature Customer Service Center GmbH
Europaplatz 3, 69115 Heidelberg, Germany

Printed by Libri Plureos GmbH
in Hamburg, Germany